Switched Reluctance Motor Drives

Switched Reluctance Motor Drives

Fundamentals to Applications

Edited by

Berker Bilgin

James Weisheng Jiang

Ali Emadi

CRC Press

Taylor & Francis Group

Boca Raton London New York

CRC Press is an imprint of the
Taylor & Francis Group, an **informa** business

CRC Press
Taylor & Francis Group
6000 Broken Sound Parkway NW, Suite 300
Boca Raton, FL 33487-2742

First issued in paperback 2022

© 2019 by Taylor & Francis Group, LLC
CRC Press is an imprint of Taylor & Francis Group, an Informa business

No claim to original U.S. Government works

ISBN-13: 978-1-138-30459-8 (hbk)
ISBN-13: 978-1-03-233875-0 (pbk)
DOI: 10.1201/9780203729991

Library of Congress Cataloging-in-Publication Data

Names: Bilgin, Berker, editor. | Jiang, James Weisheng, editor. | Emadi, Ali, editor.
Title: Switched reluctance motor drives : fundamentals to applications / [editors] Berker Bilgin, James Weisheng Jiang, and Ali Emadi.
Description: First edition. | Boca Raton, FL : CRC Press/Taylor & Francis Group, 2018. | Includes bibliographical references and index.
Identifiers: LCCN 2018029898 | ISBN 9781138304598 (hardback : acid-free paper) | ISBN 9780203729991 (ebook)
Subjects: LCSH: Reluctance motors.
Classification: LCC TK2785 .S84 2018 | DDC 621.46--dc23
LC record available at https://lccn.loc.gov/2018029898

Visit the Taylor & Francis Web site at
http://www.taylorandfrancis.com

and the CRC Press Web site at
http://www.crcpress.com

Co-authors of the Switched Reluctance Motor Drives: Fundamentals to Applications, with our traction SRM drive at McMaster Automotive Resource Center (from left to right)

First row: James Weisheng Jiang and Jianbin Liang

Second row: Yinye Yang, Alan Dorneles Callegaro, Ali Emadi, Berker Bilgin, Elizabeth Rowan, Brock Howey, Jianning (Joanna) Lin, and Haoding Li

Not pictured: Jianning Dong and Jin Ye

Photo credit: Ron Scheffler – Hamilton, Ontario, Canada

Contents

Preface

Electric motors impact various aspects of our lives. We use electric motors in numerous applications, such as air conditioners, refrigerators, washing machines, fans, vacuum cleaners, pools, beverage vending machines, and so on. Electric motors are the workhorse in industrial applications as well, and we use them heavily in machining tools, cranes, pumps, and compressors. With the emergence of more efficient electrified vehicles, electric motors are also being used in transportation systems at an increasingly rapid rate.

Electric motors are the largest consumer of electrical energy; therefore, they play a critical role in the growing market for electrification. Due to their simple construction, switched reluctance motors (SRMs) are exceptionally attractive for the industry to respond to the increasing demand for high-efficiency, high-performance, and low-cost electric motors with a more secure supply chain.

This book is organized to provide a detailed discussion of the multidisciplinary aspects of SRMs and help engineers design SRM drives for various applications. This book begins with an overview of the electric motor market in Chapter 1. The use of electric motors in industrial, residential, commercial, and transportation sectors are discussed, and the current status and future trends in the industry are presented. Chapter 1 also highlights the advantages and challenges of SRM technology compared to the other motor types.

Chapter 2 explains the operational principles of SRMs through fundamentals of electromagnetics and energy conversion. The selection of the number of stator and rotor poles of an SRM is critical to satisfying the performance requirements of an application. Therefore, Chapter 3 quantifies how the pole configuration of an SRM is calculated. Chapter 4 introduces the modeling of an SRM, which is essential both for the motor design and control. Chapter 5 presents the generating mode of operation in SRMs.

In Chapter 6, the considerations when selecting the materials for a switched reluctance motor are explained. Chapter 7 presents the design considerations and describes how defining the dimensions of a motor is related to the performance. Chapter 8 explains the mechanical aspects and construction of a switched reluctance motor.

Chapter 9 presents the control of SRM and discusses how to define the control parameters to improve the performance and reduce torque ripple. In Chapter 10, power electronic converters to drive SRMs are introduced. Position sensorless control is discussed in Chapter 11, which is important for the robust and low-cost operation of motor drives.

Acoustic noise and vibration is the most well-known issue in SRMs. However, it can be reduced significantly by analyzing the noise source and by optimizing the motor geometry and current control. In Chapter 12, fundamentals of vibrations and acoustic noise are introduced first. Then, in Chapter 13, noise and vibration in SRMs are discussed in detail.

Chapter 14 introduces the thermal management aspects of SRMs. Chapter 15 presents axial flux SRMs, which are attractive for selected applications including propulsion and energy generation. Finally, Chapter 16 introduces two SRM designs, which are proposed to replace the permanent magnet machines in a residential HVAC application and a hybrid-electric propulsion system. The design of a high-power converter for switched reluctance motor drives is also presented in Chapter 16.

Throughout this book, we aimed to cover major technical aspects of switched reluctance motor drives and provide the tools that the engineers would need to design SRMs. Therefore, the chapters include various practical examples, scripts, and detailed illustrations. We hope that this book will be an easy-to-follow reference for the practicing engineers, academicians, and students to understand, design, and analyze switched reluctance motor drives.

We would like to acknowledge the efforts and assistance of the staff of Taylor & Francis/CRC Press, especially Ms. Nora Konopka, Ms. Kyra Lindholm, and Ms. Vanessa Garrett, and the help of the staff of Lumina Datamatics, especially Ms. Angela Graven and Mr. Edward Curtis. We would also like to thank our colleagues at Cogent Power Inc. and Canada Excellence Research Chair in Hybrid Powertrain Program at McMaster University for their feedback on the chapters.

Berker Bilgin, James Weisheng Jiang, and Ali Emadi
May 2018

MATLAB® is a registered trademark of The MathWorks, Inc. For product information, please contact:

The MathWorks, Inc.
3 Apple Hill Drive
Natick, MA 01760-2098 USA
Tel: 508 647 7000
Fax: 508-647-7001
E-mail: info@mathworks.com
Web: www.mathworks.com

JMAG is a registered trademark of JSOL Corporation. For product information please contact:

JMAG Division JSOL Corporation
Harumi Center Building, 2-5-24,
Harumi Chujo-ku, Tokyo, Japan
Tel: +81 (3) 5859-6007
Fax: +81 (3) 5859-6030
E-mail: info@jmag-international.com
Web: www.jmag-international.com

ACTRAN is a trademark of Free Field Technologies. For product information, please contact:

MSC Software Belgium SA, Free Field Technologies Division
Axis Park Louvain-la-Neuve
9 rue Emile Francqui
B-1435 Mont-Saint-Guibert
Belgium
Tel: +32 10 45 12 26
E-mail: contact@fft.be
Web: www.fft.be

Editors

Berker Bilgin (IEEE S′09-M′11-SM′16) received his PhD degree in electrical engineering from Illinois Institute of Technology in Chicago, Illinois, USA. He also has an MBA degree from DeGroote School of Business, McMaster University in Hamilton, Ontario.

He is the research program manager in Canada Excellence Research Chair in the Hybrid Powertrain Program at McMaster Institute for Automotive Research and Technology, McMaster University, Hamilton, ON, Canada. He is managing many multidisciplinary projects on the design of electric machines, power electronics, electric motor drives, and electrified powertrains.

Dr. Bilgin is the co-founder of Enedym, Inc., which is a spin-off company of McMaster University. Enedym specializes in electric machines, electric motor drives, advanced controls and software, and virtual engineering.

Dr. Bilgin was the general chair of the 2016 IEEE Transportation Electrification Conference and Expo. He is also an associate editor for *IEEE Transactions on Transportation Electrification*.

James Weisheng Jiang received his bachelor's degree in vehicle engineering from the College of Automotive Engineering, Jilin University, China in 2009. He worked as a research assistant at the Clean Energy Automotive Engineering Research Center, Tongji University, China from 2009 to 2011. He got his PhD degree from McMaster University in 2016. He is currently a principal research engineer at McMaster Automotive Resource Centre (MARC). He has designed and implemented a 60 kW 24/16 switched reluctance motor for traction purpose in HEV. He has also been involved in designs and implementations of traction motors with interior permanent and ferrite magnets. He has also been working on NVH analysis for switched reluctance motors and permanent magnet synchronous motors.

Ali Emadi (IEEE S′98-M′00-SM′03-F′13) received Bachelor of Science and Master of Science degrees in electrical engineering with highest distinction from Sharif University of Technology, Tehran, Iran, in 1995 and 1997, respectively, and a PhD degree in electrical engineering from Texas A&M University, College Station, TX, USA, in 2000. He is the Canada Excellence Research Chair in Hybrid Powertrain and a professor in the Departments of Electrical and Computer Engineering and Mechanical Engineering at McMaster University in Hamilton, Ontario, Canada. Before joining McMaster University, Dr. Emadi was the Harris Perlstein Endowed Chair Professor of Engineering and director of the Electric Power and Power Electronics Center and Grainger Laboratories at Illinois Institute of Technology in Chicago, Illinois, USA, where he established research and teaching facilities as well as courses in power electronics, motor drives, and vehicular power systems. He was the founder, chairman, and president of Hybrid Electric Vehicle Technologies, Inc. (HEVT)—a university spin-off company of Illinois Tech. He is the Founder, President, and CEO of Enedym Inc.—a McMaster University spin-off company. Dr. Emadi has been the recipient of numerous awards and recognitions. He was the advisor for the Formula Hybrid Teams at Illinois Tech and McMaster University, which won the GM Best Engineered Hybrid System Award at the 2010, 2013, and 2015 competitions. He is the principal author/coauthor of over 450 journal and conference papers as well as several books including *Vehicular Electric Power Systems* (2003), *Energy Efficient Electric Motors* (2004), *Uninterruptible Power*

Supplies and Active Filters (2004), *Modern Electric, Hybrid Electric, and Fuel Cell Vehicles* (2nd ed, 2009), and *Integrated Power Electronic Converters and Digital Control* (2009). He is also the editor of the *Handbook of Automotive Power Electronics and Motor Drives* (2005) and *Advanced Electric Drive Vehicles* (2014). Dr. Emadi was the Inaugural General Chair of the 2012 IEEE Transportation Electrification Conference and Expo (ITEC) and has chaired several IEEE and SAE conferences in the areas of vehicle power and propulsion. He is the founding Editor-in-Chief of the *IEEE Transactions on Transportation Electrification*.

Contributors

Berker Bilgin
McMaster Institute for Automotive
    Research and Technology (MacAUTO)
McMaster University
Hamilton, Ontario, Canada

Alan Dorneles Callegaro
McMaster Institute for Automotive
    Research and Technology (MacAUTO)
McMaster University
Hamilton, Ontario, Canada

Jianning Dong
Department of Electrical Sustainable
    Energy, TU Delft
Delft, Netherlands

Ali Emadi
McMaster Institute for Automotive
    Research and Technology (MacAUTO)
McMaster University
Hamilton, Ontario, Canada

Brock Howey
McMaster Institute for Automotive
    Research and Technology (MacAUTO)
McMaster University
Hamilton, Ontario, Canada

James Weisheng Jiang
McMaster Institute for Automotive
    Research and Technology (MacAUTO)
McMaster University
Hamilton, Ontario, Canada

Haoding Li
McMaster Institute for Automotive
    Research and Technology (MacAUTO)
McMaster University
Hamilton, Ontario, Canada

Jianbin Liang
McMaster Institute for Automotive
    Research and Technology (MacAUTO)
McMaster University
Hamilton, Ontario, Canada

Jianing (Joanna) Lin
McMaster Institute for Automotive
    Research and Technology (MacAUTO)
McMaster University
Hamilton, Ontario, Canada

Elizabeth Rowan
McMaster Institute for Automotive
    Research and Technology (MacAUTO)
McMaster University
Hamilton, Ontario, Canada

Yinye Yang
Magna Powertrain
Concord, Ontario, Canada

Jin Ye
San Francisco State University
San Francisco, California

List of Symbols

a	m/s^2	Acceleration
a	m	Radius of a shell (Chapter 13)
A	m^2	Area of the conducting loop (Chapter 4)
A	m^2	Area
A	m	Amplitude of the wave
A_c	m^2	Cross section are through which the flux flows
a_e	m/s^2	Acceleration of a charge
A_s	m^2	Slot area
A_s	m^2	Area of the sound radiation surface (Chapter 13)
ax	—	Axial order (Chapters 12 and 13)
B	T (Tesla)	Magnetic flux density
B_r	T	Magnetic flux density in radial direction
B_t	T	Magnetic flux density in tangential direction
c	m/s	Speed of sound (Chapter 13)
c_B	m/s	Speed of bending sound wave (Chapter 13)
C_e	s	Electric time constant
$circ$	—	Circumferential order (Chapters 12 and 13)
c_p	J/(kg·K)	Specific heat
C_r	—	Rotor geometric center (Chapter 7)
C_r	—	Circumference of the rotor (Chapter 7)
C_{sb}	—	Shaft-bearing geometric center, also referred as rotation center (Chapter 7)
C_{sc}	—	Stator-case geometric center (Chapter 7)
D	—	Diode
D	m or mm	Amplitude for vibration at the steady state (Chapter 12)
D	mm	Bore diameter (Chapter 7)
d_c	mm	Diameter for bare copper (Chapter 7)
D_c	mm	Mean diameter of the stator yoke (Chapter 13)
D_r	mm	Rotor outer diameter (Chapter 7)
D_s	mm	Stator outer diameter (Chapter 7)
D_{sh}	mm	Rotor shaft diameter (Chapter 7)
d_{sum}	m or mm	Sum of surface waves (Chapter 12)
d_w	mm	Diameter for wire with insulation (Chapter 7)
e	m	Mass eccentricity
E	Pa, MPa or GPa	Modulus of elasticity or Young's modulus
E'	Pa, MPa or GPa	Real part of Young's modulus
E''	Pa, MPa or GPa	Imaginary part of Young's modulus
E_t	Pa, MPa or GPa	Equivalent elasticity modulus (Chapter 13)

f	Hz	Frequency
F	N	Force
F	N	External force (Chapter 12)
F_0	N	Amplitude for external force (Chapter 12)
F_A	—	Sum for all individuals' fitness values (Chapter 9)
f_{elec}	Hz	Electrical frequency (Chapter 13)
ff_{copper}	—	Bare copper slot fill factor
F_i	—	Fitness value for a particular individual (Chapter 9)
f_m	Hz	Mechanical frequency (Chapter 7)
f_{mech}	Hz	Mechanical frequency
f_n	Hz	Natural frequency (Chapters 12 and 13)
F_r	N	Radial force wave over one mechanical cycle (Chapter 13)
F_r	N	Amplitude of a harmonic for radial force (Chapter 13)
f_{samp}	Hz	Sampling frequency (Chapter 4)
F_u	N	Unbalanced force resulting from unbalanced mass
F_V	N	Electromagnetic force acting on charges on volume V (Chapter 2)
h	W/(m²K)	Heat transfer coefficient
H	A/m	Magnetic field strength
H_c	A/m	Magnetic field intensity in the core (Chapter 2)
h_f	mm	Mean thickness of the frame (Chapter 13)
H_g	A/m	Magnetic field intensity in the airgap (Chapter 2)
h_{ov}	mm	One-sided axial overhang length of the winding ends (Chapter 13)
h_r	mm	Rotor pole height (Chapter 7)
h_s	mm	Stator pole height (Chapter 7)
h_s	mm	Tooth height (Chapter 13)
i	A	Electric current
I	A	Electric current
I	kg·m²	Moment of inertia
I_{amp}	A	Amplitude of phase current (Chapter 4)
i_d	A	d-axis current (Chapter 4)
i_{dq}	A	Current in the dq frame (Chapter 4)
i_k	A	Measured current
i_{k_low}	A	Lower current references of the kth phase
i_{k_up}	A	Upper current references of the kth phase
I_{lower}	A	Lower boundary of a hysteresis band for a reference current
i_{LUT}	A	Current lookup table (Chapter 4)
I_{max}	A	Maximum phase current
I_{min}	A	Minimum phase current

I_p	A	RMS value of the phase current
i_{ph}	A	Phase current
I_{phase}	A	Phase current array (Chapter 4)
i_q	A	q-axis current (Chapter 4)
i_{rated}	A	Rated current of the machine (Chapter 4)
I_{ref}	A	Reference current
I_{RMS_coil}	A	RMS value of the current flowing through the coil
I_{upper}	A	Upper boundary of a hysteresis band for a reference current
$i_{\alpha\beta}$	A	Current in the $\alpha\beta$ frame (Chapter 4)
J	A/m²	Current density in conductor
k	—	A lumped parameter term that takes into account the mechanical construction of the motor (Chapter 4)
k	N/m	Stiffness
k	W/(m·K)	Thermal conductivity
k	N·s/m	Spring coefficient (Chapter 12)
$K_{(circ)}$	N/m	Lumped stiffness of the circumferential mode circ (Chapter 13)
k_0	—	Acoustic waveform number (Chapter 13)
K_e	—	Eddy current loss coefficient (Chapter 14)
K_{fb}	—	Frictional loss coefficient (Chapter 14)
K_h	—	Hysteresis loss coefficient (Chapter 14)
k_{ph}	—	Array for phase shift factors
k_r	—	Radial component of acoustic wave number (Chapter 13)
k_{rot}	—	−1 for counter clockwise rotation and 1 for clockwise rotation
k_z	—	Axial component of acoustic wave number (Chapter 13)
l	m	Length of a shell (Chapter 13)
l	m	Length
L	H	Inductance
L_a	H	Self-inductance at aligned position
L_A	dB	Function of factor for A-weighting correction
l_c	m	Length of a close-loop flux path (Chapter 2)
L_{end}	m	Height of end turn (Chapter 7)
L_f	mm	Frame length (Chapter 13)
l_g	m	Length of the flux path in the air gap
$L_{inc\_k,k}$	H	Incremental self-inductance of the kth phase
$L_{k,k}$	H	Self-inductances of the kth phase
L_m	H	Self-inductance at midway position
L_{max}	H	Max. Inductance

L_{min}	H	Min. Inductance
l_r	m	Rotor's geometric center axis (Chapter 8)
L_R	m	Stack length of the stator core (Chapter 7)
l_{rpa}	m	Arc length of the rotor pole (Chapter 7)
L_{rpa}	m	Total of the arc lengths of the rotor poles (Chapter 7)
L_S	m	Stack length of the rotor core (Chapter 7)
l_{sb}	—	Geometric axis for the shaft-bearing assembly
l_{sc}	—	Geometric center axis for stator-case assembly (Chapter 8)
l_{slot}	m	Height of the slot
L_{total}	m	Total axial length of the active volume (Chapter 7)
L_u	H	Self-inductance at unaligned position (Chapter 7)
L_{un}	m	Total arc length available between the rotor poles (Chapter 7)
l_w	m	Length of wire (Chapter 2)
m	kg	Mass
m	—	Number of motor phases
M	kg	Lumped mass for the cylindrical shell
m_e	g	Mass of an electron
MF	—	Magnification factor (Chapter 12)
M_r	kg	Rotor mass
n	rpm	Revolutions per minute
N	—	Number of turns (Chapter 4)
n_e	—	Number of charges per cubic meter (Chapter 2)
N_{lam}	—	Number of lamination sheets for either the stator or the rotor (Chapter 7)
n_m	rpm	Motor speed
N_p	—	Number of poles (Chapter 7)
$N_{parallel}$	—	Number of parallel paths in a motor winding (Chapter 7)
N_{Ph}	—	Number of phases (Chapter 7)
n_{pulses}	—	Number of encoder pulses at a time step (Chapter 4)
N_{pulses}	—	Number of pulses (Chapter 4)
N_r	—	Number of rotor poles
$Nr\#1_{mech}$	degree	Mechanical angle for Nr#1 (Chapter 3)
N_{RPM}	rpm	Mechanical rotation speed (Chapter 4)
N_s	—	Number of stator poles (Chapter 7)
$Ns\#1_{elect}$	degree	Electrical angle for Ns#1 (Chapter 3)
$Nr\#1_{mech}$	degree	Electrical angle for Nr#1 (Chapter 3)
n_{samp}	—	Number of simulation time steps at each current sampling period (Chapter 4)

Ns_{elect}	degree	Electrical angle of a stator pole (Chapter 4)
N_{series}	—	Number of coils connected in series per parallel path in a winding (Chapter 7)
n_{slot}	—	Slot number per pole
N_{slot}	—	Number of stator slots
Ns_{mech}	degree	Mechanical angle of a stator pole (Chapter 4)
N_{str}	—	Number of strands (Chapter 7)
N_{total}	—	Size of the population (Chapter 9)
N_{turn}	—	Number of turns (Chapter 7)
p	Pa	Pressure
P_{core}	W, kW	Core loss
P_{cu}	W, kW	Copper loss
P_{eddy}	W, kW	Eddy current loss
P_{excess}	W, kW	Excess loss
P_{fw}	W, kW	Windage and friction losses (Chapter 14)
$P_{hystersis}$	W, kW	Hysteresis loss
P_{in}	W, kW	Motor input power
P_{mech}	W, kW	Mechanical power
P_{out}	W, kW	Motor output power
pp	—	Number of pole pairs
p_r	Pa, GPa	Analytical expression of a harmonic content for pressure (Chapter 13)
P_r	W, kW	Rated power of motor
P_r	Pa, GPa	Amplitude of a harmonic content for radial pressure (Chapter 13)
q	—	Temporal order (Chapters 12 and 13)
q	C	Charge
Q	dB	Peak power of the damped frequency (Chapter 13)
q_{enc}	C	Total charge within a surface (Chapter 2)
r	—	Radial order (Chapter 12)
r	m	Radius
R	Ω	Electric resistance
R	Ω	Phase resistance
R	H^{-1}	Reluctance (Chapter 12)
R_{coil}	Ω	Resistance of the coil
R_{dc}	Ω	DC resistance
R_{eq}	Ω	Equivalent reluctance (Chapter 2)
R_f	mm	Mean radius of the frame (Chapter 13)
$Ripple_{Normalized}$	—	Ratio between the net torque ripple and the average torque
$Ripple_{Percentage}$	—	Normalized torque ripple multiplied by 100%
R_{ph}	Ω	Phase resistance (Chapter 4)

$R_{proximity}$	Ω	Proximity resistance
R_{skin}	Ω	Skin resistance
R_t	mm	Mean radius of the teeth-coil region (Chapter 13)
S	—	Total number of torque pulsations (Chapter 3)
S	—	Circuit switch
S_f	—	Stacking factor (Chapter 7)
T	N·m	Torque
T	s	Oscillating period
T	°C or K	Temperature
T_{amb}	°C or K	Ambient temperature (Chapter 7)
T_{ave}	Nm	Average torque (Chapter 7)
T_{ave_r}	Nm	Required torque (Chapter 9)
T_{co}	Nm	Co-energy torque (Chapter 4)
T_d	Nm	Damping torque
T_e	Nm	Electromagnetic torque
T_{e_ref}	Nm	Total torque reference
$T_{e\_ref(k)}$	Nm	Reference torque for kth phase
t_{k_off}	—	Time instant when the kth phase switching states are OFF
t_{k_on}	—	Time instant when the kth phase switching states are ON
t_{lam}	mm	Thickness for lamination sheets (Chapter 7)
T_{limit}	°C or K	Maximum permitted temperature (Chapter 7)
T_{LUT}	Nm	Torque lookup table (Chapter 4)
T_{max}	Nm	Maximum torque
T_{min}	Nm	Minimum torque
T_p	Nm	Peak torque
$T_{peak-to-peak}$	Nm	Magnitude of the torque pulsation
T_{phase}	Nm	Phase torque array (Chapter 4)
T_{pr}	degree	Pole pitch for the rotor
T_{ps}	degree	Pole pitch for the stator
T_{ripple}	—	Periodic component of the instantaneous torque waveform (Chapter 7)
T_s	s	Sampling time (Chapter 4)
T_{samp}	s	Sampling period (Chapter 4)
u	—	Temporal order (Chapters 12 and 13)
u_e	—	Temporal order in electrical position (Chapter 13)
u_n	V	Neutral point voltage (Chapter 10)
u_w	V	Phase voltage (Chapter 10)
v	—	Spatial order (Chapters 12 and 13)
v	m/s	Vibration velocity on the surface (Chapter 13)
v	m/s	Velocity of a moving charge (Chapter 2)
V	V	Voltage

V_a	m³	Volume of the air gap (Chapter 2)
V_{active}	m³	Volume of active material (Chapter 7)
V_c	m³	Volume of the core (Chapter 2)
v_d	m/s	Drift velocity (Chapter 2)
V_{dc}	V	DC voltage
V_{DC}	V	DC voltage
v_k	V	Measured voltage of kth phase
v_k	V	Measured voltage
v_{ph}	V	Phase voltage
V_t	m³	Volume of the teeth-coil region (Chapter 13)
W_c	J or kJ	Co-energy (Chapter 2)
W_f	J or kJ	Magnetic energy
w_{slot}	mm	Width of slot opening
x	m or mm	Displacement
X	m or mm	Amplitude of a vibrating wave (Chapter 12)
y_r	mm	Rotor back iron thickness (Chapter 7)
y_s	mm	Stator back iron thickness (Chapter 7)

List of Greeks

α	rad/s^2	Angular acceleration
α	1/°C or 1/K	Thermal coefficient (Chapter 14)
α	degree	Air-gap spatial position (Chapter 13)
β_r	degree or radian	Rotor pole arc angle (Chapter 7)
β_s	degree or radian	Stator pole arc angle (Chapter 7)
γ	degree	Stator circumferential position (Chapter 7)
ΔT_{rms}	Nm	Net RMS value of torque ripple
ΔT_{RMS}	Nm	RMS value of net torque ripple
ε	mm	Required element size (Chapter 13)
ε	mm/mm or μm/mm	Strain (Chapter 13)
ε	V	Motional EMF (Chapter 9)
ε	V	Voltage induced across the armature (Chapter 4)
ϵ_0	F/m	Permittivity of free space
ζ	—	Damping ratio (Chapter 12 and 13)
ζ_a	—	Acoustic damping ratio (Chapter 13)
ζ_s	—	Structural damping ratio (Chapter 13)
η	%	Efficiency
θ	degree	Angle between normal vector of area and direction of B field (Chapter 4)
θ	degree	Rotor position
θ_a	degree	Aligned position
$\theta_{aligned}$	degree	Aligned position
θ_c	degree	Conduction angle
θ_{off}	degree	Turn-off angle
θ_{OFF}	degree	Turn-off angle
θ_{on}	degree	Turn-on angle
θ_{ON}	degree	Turn-on angle
θ_{ov}	degree	Overlapping angle
θ_p	degree	Pole pitch angle
θ_r	degree	Mechanical position of the center axis of the rotor poles (Chapter 3)
θ_{rotor}	degree	Machine position
θ_s	degree	Mechanical position of the center axis of the stator poles (Chapter 3)
θ_{un}	degree	Unaligned position
$\theta_{unaligned}$	degree	Unaligned position
λ	m	Sound wavelength
λ_a	Wb	Phase flux-linkage at unaligned position
λ_A	Wb	Flux linkage due to armature current
λ_{ax}	m	Wavelength of axial mode, ax

λ_B	m	Wavelength of bending wave (Chapter 13)
λ_{base}	m	Fundamental wavelength for the first (base) circumferential or spatial order
λ_d	Wb	d-axis flux linkage (Chapter 4)
λ_{dq}	Wb	Flux in the dq frame (Chapter 4)
λ_{fall}	Wb	Decreasing flux linkage for the outgoing phase
λ_k	Wb	Estimated self-flux linkage
λ_{LUT}	Wb	Flux linkage lookup table
λ_{ph}	Wb	Phase flux linkage
λ_q	Wb	q-axis flux linkage (Chapter 4)
λ_{rise}	Wb	Rising flux linkage for the incoming phase
λ_u	Wb	Phase flux-linkage at aligned position
λ_v	m	Wavelength of spatial order, v (Chapter 12)
$\lambda_{\alpha\beta}$	Wb	Flux in the $\alpha\beta$ frame (Chapter 4)
μ	N/A^2	Permeability
μ_0	N/A^2	Permeability of free space (Chapter 2)
μ_{eff}	N/A^2	Effective permeability (Chapter 2)
μ_r	N/A^2	Relative permeability (Chapter 2)
ξ_{base}	cycles per meter	Spatial frequency for the base circumferential or spatial order (Chapter 12)
Π	W, kW	Sound power (Chapter 13)
Π_{ref}	W	Reference of sound power, 10^{-12} W (Chapter 13)
ρ	kg/m^3	Mass density
ρ	C/m^3	Charge density (Chapter 2)
ρ	$\Omega\cdot$m	Electric resistivity
ρ_0	Ω	Resistivity at the initial temperature (Chapter 14)
ρ_s	kg/m^2	Mass per unit surface area (Chapter 13)
σ	N/m^2 or Pa	Stress (Chapter 13)
σ	—	Radiation ratio (Chapter 13)
τ	s	Time between the collusion of free electrons with the atoms (Chapter 2)
τ	N/m^2	Shear stress
τ	Nm	Torque produced by the DC motor (Chapter 4)
τ_e	s	Electrical time constant
τ_r	degree or radian	Rotor taper angle (Chapter 7)
τ_s	degree or radian	Stator taper angle
φ	degree or radian	Phase angle (Chapters 12 and 13)
Φ	Wb	Magnetic flux
φ_{ind}	Wb	Induced flux (Chapter 2)
ϕ_{phase}	degree	Angle of phase current (Chapter 4)
ϕ_s	Wb	Flux passing through the armature coil due to the external field (Chapter 4)
ω	rad/s	Annular rotating speed
ω	rad/s	Oscillating angular frequency (Chapters 12 and 13)

$\Omega_{(circ)}$	—	Root of the characteristic equation of motion of the cylindrical shells (Chapter 13)
ω_d	rad/s	Damped circular frequency (Chapter 12)
ω_{elec}	rad/s	Electrical speed
ω_f	rad/s	Forcing angular frequency (Chapter 13)
ω_m	rad/s	Mechanical angular speed (Chapter 15)
ω_n	rad/s	Natural angular frequency

List of Abbreviations

AC	Alternative current
AFM	Axial flux machine
AFSRM	Axial flux SRM
AM	Amplitude modulation
ANN	Artificial neural network
ARCFL	Absolute value of the rate of change of flux linkage
ATF	Automatic transmission fluid
AWG	American wire gauge
BCC	Body centered cubic
BDC	Bottom dead center
BDRM	Brushless double-rotor machine
BEV	Battery electric vehicle
BLDC	Brushless DC motor
BOF	Basic oxygen furnace
CAC	Central air conditioner
CAD	Computer-aided design
CCW	Counter clockwise
CDF	Cumulative distribution function
CFD	Computational fluid dynamics
CGO	Conventional grain-oriented steel
CNC	Computer numerical control
CSMO	Current sliding mode observer
CTE	Coefficient of thermal expansion
CUAC	Commercial unitary air conditioners
CW	Clockwise
DC	Direct current
DDR	Direct drive rotary
DIFC	Direct instantaneous force control
DSP	Digital signal processor
EAF	Electric arc furnace
EMF	Electromotive force
EMI	Electromagnetic interference
EPR	Ethylene-propylene
ETP	Electrolytic tough pitch
FE	Finite element
FEA	Finite element analysis
FFT	Fast Fourier transform
FNM	Flow network modeling

FRF	Frequency response function
FSMO	Flux-linkage sliding mode observer
GA	Genetic algorithm
GMO	Gaussian mutation operator
GOES	Grain-oriented electrical steel
HEV	Hybrid electric vehicle
HGO	High-permeability grain oriented
HP	Horsepower
ICE	Internal combustion engine
ICM	Intermediate crossover method
ID	Inner diameter
IEC	The international electrotechnical commissions
IGBT	Insulated gate bipolar transistor
IM	Induction motor
IPM	Interior permanent magnet
IPMSM	Interior permanent magnet synchronous machine
ISO	International Organization for Standardization
LB	Lower boundary
LCM	Linear crossover method
LED	Light-emitting diode
LHM	Linear hybrid motor
LPM	Lumped parameter model
LPTN	Lumped parameter thermal network
MF	Magnification factor
MMF	Magnetomotive force
MOSFET	Metal–oxide–semiconductor field-effect transistor
MW	Magnet wire
Nd-Fe-B	Neodymium-iron-boron
NEMA	National electrical manufacturers association
NOES	Non-oriented electrical steel
NPC	Neutral Point Diode Clamped, or Neutral Point Clamp
NTC	Negative temperature coefficient
NVH	Noise, vibration, and harshness
OD	Outer diameter
OEM	Original equipment manufacturer
PA	Polyamide
PBT	Polybutylene terephthalate
PC	Polycarbonate
PCB	Printed circuit board
PD	Partial discharges
PD	Specific power
PDIV	Partial discharge inception voltage
PE	Polyethylene

PEEK	Polyetheretherketone
PHEV	Plug-in hybrid electric vehicle
PI	Proportional-Integral (controller)
PI	Polarization index
PID	Proportional–integral–derivative (controller)
PLL	Phase locked loop
PM	Permanent magnet
PM	Phase modulation
PMMA	Polymethylmetachrylate
PMSM	Permanent magnet synchronous motor
PMW	Pulse width modulation
PP	Polypropylene
PPS	Polyphenylene sulfide
PS	Polystyrene
PSU	Polysulfone
PTC	Positive temperature coefficient
PTFE	Polytetrafluoroethylene
PVC	Polyvinyl chloride
PWM	Pulse width modulation
RAC	Room air conditioner
RBF	Radial basis function
RFM	Radial flux machines
RFSRM	Radial flux SRM
RMO	Random mutation operator
ROW	Rest of the world
RTD	Temperature detector
SED	Sound energy density
SIL	Sound intensity level
SMC	Soft magnetic composites
SMO	Sliding mode observer
SP	Power density
SPL	Sound pressure level
SRG	Switched reluctance generator
SRM	Switched reluctance machine
SWL	Sound power level
TEFC	Totally enclosed, fan-cooled
TENV	Totally Enclosed, Non-ventilated
TEXP	Totally Enclosed, Explosion-proof
THS	Toyota hybrid system
TRFS	Theoretical maximum ripple-free speed
TSF	Torque sharing function
TSFF	Temperature sensitive ferrofluids
UB	Upper boundary

UDDS	Urban dynamometer driving schedule
UMP	Unbalanced magnetic pull
USGS	United States geological survey
VA	Volt-ampere
VPI	Vacuum pressure impregnation

1

Electric Motor Industry and Switched Reluctance Machines

Berker Bilgin and Ali Emadi

CONTENTS

1.1 Introduction

Electric motors are electromechanical devices that convert electrical energy into mechanical energy. During this conversion process, some of the electrical energy is lost and dissipated as heat. Relative to alternatives, electric machines are efficient, cost effective and can scale from very large applications to tiny applications; generally, not feasible for other mechanical energy sources. These factors have driven a massive market demand for electric machines, making electric motors a major consumer of electrical energy.

Utilizing higher efficiency, lower cost, and variable speed electric motors will increase the level of electrification, improve overall system efficiency, reduce operational costs, and electricity consumption, while reducing emissions. The switched reluctance machine (SRM) is a promising electric machine architecture that will definitely play a significant role in future market expansion. One of the main advantages of SRM is its low-cost and simple construction, which can provide reliable operation in a harsh environment, like transportation. But, there are also challenges to solve in SRM including high torque ripples and acoustic noise.

Throughout this book, we will offer a microscopic view of all technical aspects of SRM. Before that, we would like to provide a macroscopic view of the use of electric motors in different sectors to explore the role of SRM in the electric motor industry. We will start by looking at the energy consumption of electric motors. Then, we will analyze electric motor usage in different sectors. After providing the big picture of the electric motor industry, we will look at the potential and challenges in utilizing SRM technology in motor drive applications.

1.2 Overview of Energy Consumption of Electric Motors

As the largest economy in the world, the United States is a major player in electricity generation, consumption, and use of electric motors in the industrial, commercial, residential, and transportation sectors. According to statistics from the U.S. Energy Information Administration, total U.S. electricity consumption reached 9,627 million kilowatt-hours (kWh) per day by November 2016 [1]. As shown in Figure 1.1, the majority of the electrical energy has been produced from coal in the last decade. The share of coal-fired energy generation in total electricity production continues to decline to limit the growth of energy-related CO_2 emissions. As a result, natural gas has become the predominant source of power generation. In 2016, 34.21% of electricity was generated from natural gas. Electricity generation from renewable sources increased from 6.26% in 2013 to 8.30% in

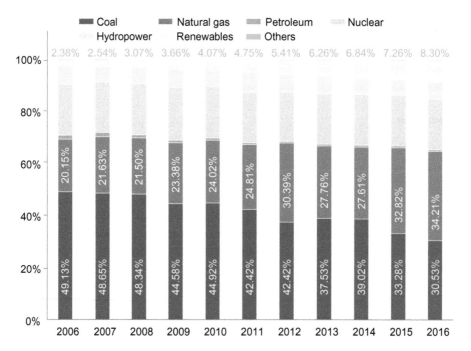

FIGURE 1.1
Energy resources for electricity generation in U.S. (From U.S. Energy Information Administration, Short-term energy outlook. December 6, 2016, Available: http://www.eia.gov/.)

2016. The portion of electricity from renewables is expected to grow faster in the next few decades due to declining costs of renewable energy sources, penetration of new energy storage capabilities, and stricter targets in reducing energy-related CO_2 emissions.

Figure 1.2 shows the percentage of energy use in different sectors [2–4]. Including the electricity-related losses, the share of the industry sector in the total electrical energy

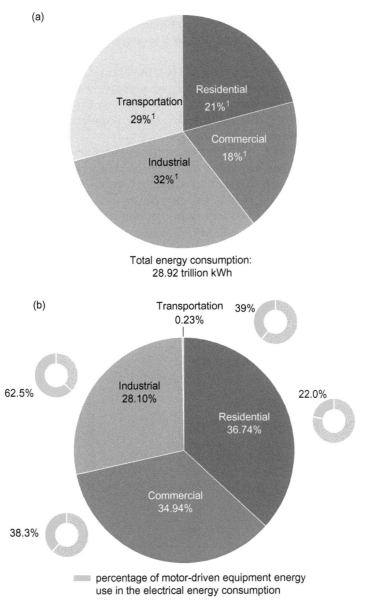

FIGURE 1.2
U.S. energy consumption by sectors in 2015: (a) percentage of total energy consumption, (b) percentage of total electrical energy consumption. (From U.S. Energy Information Administration, Annual energy outlook 2016 with projections to 2040, August 2016, Available: http://www/eia/gov/; Waide, P. and Brunner, C.U., Energy-efficiency policy opportunities for electric motor-driven systems, International Energy Agency Working Paper, 2011; U.S. Department of Energy, Advanced manufacturing office, premium efficiency motor selection and application guide – A handbook for industry. February, 2014, Available: http://www.energy.gov/.)

consumption is 28.10%, and 62.5% of the electrical energy consumed by the industrial sector was used to power electric motor driven applications, mostly for industrial handling and processing. The highest percentage of the total electrical energy consumption was in the residential sector at 36.74%. However, only 22% of the electrical energy in the residential sector was used to power electric motors. In residential applications, electric motors are mostly employed in refrigerators, freezers, and heating, ventilation, and air conditioning (HVAC) systems. The second largest use of electric motors is in the commercial sector; 38.3% of the electrical energy consumed by the commercial sector is used for electric motors, mostly for HVAC systems.

It should be noted that the industrial sector has the highest electrical energy usage for electric motor driven equipment due to the use of a relatively large number of medium- and large-size electric motors. However, as we will see in the next section, small size motors make up the biggest portion of the U.S. motor population, and they are mostly used in residential and commercial applications.

You must have already noticed that the portion of electrical energy consumption in the transportation sector is small even though it has the second highest total energy consumption in the United States. The reason is that the vast majority of the vehicles in the U.S. are powered solely by internal combustion engines (ICEs) and require fossil fuels as their energy source [5]. However, the portion of electrical energy is expected to increase in the transportation sector with the penetration of more efficient electrified vehicles, including hybrid, plug-in hybrid, and battery electric vehicles. Electric motors in the transportation system today are mostly used in electric railways and low-power vehicular applications (e.g. windshield wipers, seat adjustment, power windows, and sunroof), but electric motors used to power vehicles are perhaps the fastest growing motor market in the transportation industry.

Electric motors are the biggest consumer of electrical energy in the United States. This is also true globally. Figure 1.3 shows the global electricity demand by end-use. Around 46% of the global electricity demand is from electric motors. Among the sectors, the highest

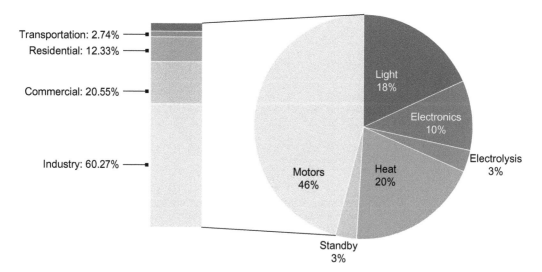

FIGURE 1.3
Global electrical energy demand by end use and electrical energy demand of motor applications by sector. (From U.S. Department of Energy, Advanced manufacturing office, premium efficiency motor selection and application guide – A handbook for industry. February, 2014, Available: http://www.energy.gov/.)

portion of electrical energy demand for motor applications is from the industrial sector (60.27%), then the commercial sector follows at 20.55%, and residential sector at 12.33%.

1.3 The U.S. Motor Population

According to a 2011 report from International Energy Agency, the U.S. electric motor stock was approximately 2,230 million [3]. Table 1.1 summarizes the U.S. motor population in 2011. The table shows that around 90% of these motors were less than 750 watts (W) and they consumed 9% of the total electrical energy for all motors. Small size motors are mostly used in mass-produced appliances in residential and commercial sectors (e.g. refrigerator compressors, extractor fans). Most of the small size motors are either single-phase induction or shunt wound direct current (DC) motors with a maximum operating voltage of 240 volts (V). They operate 1,500 hours per year with a load factor of 40%, on average. They have an average lifetime of 6.7 years and their nominal efficiency is around 40% [3].

Medium size motors are categorized from 0.75 kilowatts (kW) to 375 kW. These are mostly general purpose motors used in pumps, fans, compressors, conveyors, and industrial handling and processing. Medium size motors account for about 10% of all motor population in the United States, but they consume 68% of motor-related electrical energy. This portion increases to 75% on the global scale. The majority of the medium size motors are three-phase induction motors with 2–8 poles. On average, they operate 3,000 hours per year with a load factor of 60%, on average. The nominal efficiency is around 88%. They have an average lifetime of 7.7 years [3].

Large size electric motors with more than 375 kW output power account for only 0.03% of the electric motor population (around 0.6 million). However, they consume about 23% of all motor-related electrical energy. These motors are manufactured in small quantities (around 40,000 per year), and they are custom-designed for industrial and infrastructural applications. The operating voltage can be between 1 kilovolts (kV) and 20 kV. Due to their large size, they can achieve a nominal efficiency around 90%. On average, they operate 4,500 hours per year with a load factor of 70%. They have an average lifetime of 15 years [3]. It should be mentioned that in the industry, horsepower (HP) is often used to express the output power of an electric motor. The unit of HP will be utilized in many places in this chapter and 1 HP is 0.7457 kW.

TABLE 1.1

Summary of the U.S. Motor Population in 2011

	Small-Size	Medium-Size	Large-Size
Output power	<750 W	750 W–375 kW	>375 kW
Population in 2011	2,000 million (90%)	230 million (10%)	0.6 million (0.03%)
Consumption of electrical energy for all motors	9%	68%	23%
Average operating hours	1,500	3,000	4,500
Average load factor	40%	60%	70%
Life time (years)	6.7	7.7	15
Nominal efficiency	40%	88%	90%

Source: Waide, P. and Brunner, C.U., Energy-efficiency policy opportunities for electric motor-driven systems, International Energy Agency Working Paper, 2011.

1.4 Electric Motor Usage in the Industrial Sector

As shown in Figure 1.2, the industrial sector in the United States consumes 32% of the total energy and 28.10% of the total electrical energy. Within the industrial sector, motor drive applications consume 62.5% of the electric power. This makes the U.S. industry sector the largest consumer of electricity used by electric motors.

Figure 1.4 shows the motor applications and percentage of motor system energy used in the industrial sector [6]. In the United States, material processing (e.g. mills, grinders, lathes) and material handling applications (e.g. belts, conveyors, elevators, cranes) employ 34% of the general purpose industrial motors. Pump motors are mostly used for circulating water or other process fluids. In the industry sector, compressor motors are found in HVAC and pneumatic power tools. Fan motors are used for ventilation and exhaust systems. Motors in refrigeration systems are mainly used in food industry, and in paper and metal processing [6].

Figure 1.5 shows the motor system energy use by different manufacturing sectors in the industry [7]. Chemical products industry is the largest process user of electrical energy for motor applications. Paper industry follows it.

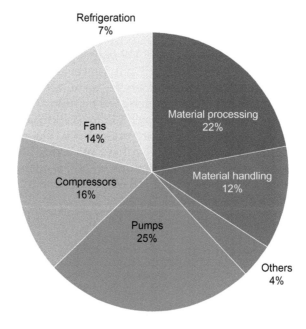

FIGURE 1.4
Share of motor applications in the industrial sector. (From Lowe, M. et al., U.S. Adoption of high-efficiency motors and drives: Lessons learned. February 25, 2010, Available: http://www.cggc.duke.edu/.)

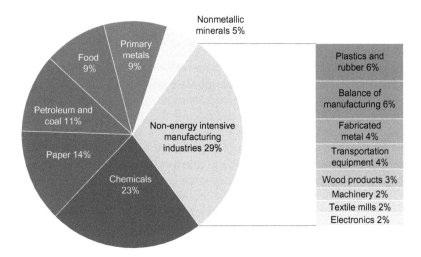

FIGURE 1.5
Motor system energy use by different manufacturing sectors. (From U.S. Energy Information Administration, Electricity use by machine drives varies significantly by manufacturing industry. October 8, 2013, Available: http://www.eia.gov/.)

1.5 Electric Motor Usage in the Residential Sector

According to a 2013 report from U.S. Department of Energy Building Technologies Office, HVAC applications account for 63% of the electric motor energy consumption in the residential sector [8]. As shown in Figure 1.6, residential HVAC applications include central air conditioners (CAC), heat pumps, furnace fans, room air conditioners (RAC), and dehumidifiers. Refrigerators and freezers are the second largest consumers for motor-related energy use. Electric motors in the range of $\frac{1}{10}-\frac{1}{2}$ HP account for the biggest portion of the motor population in the residential sector.

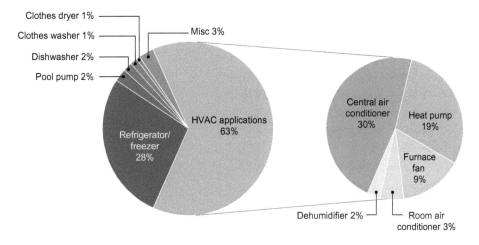

FIGURE 1.6
Motor system energy consumption by end-use in the residential sector. (From U.S. Department of Energy, Building technologies office, energy savings potential and opportunities for high-efficiency electric motors in residential and commercial equipment. December 2013, Available: https://energy.gov/.)

In residential HVAC systems, central air conditioners and heat pumps typically contain a compressor, an outdoor fan, and an indoor blower motor. An air conditioner and a heat pump operate based on similar principles, but unlike the air conditioner, a heat pump can reverse the direction of the air flow and transfer heat from outside to increase indoor temperatures [9]. 2–5 HP capacitor start single-phase induction motors are widely used for residential HVAC compressors. Permanent magnet motors and variable speed capability is being implemented in HVAC applications to improve the compressor efficiency. However, since the initial purchase cost of the motor drive system is an important parameter in residential applications, use of permanent magnet motors might be an issue in the long run. Low-noise operation is desirable to increase the comfort in residential HVAC compressors. Fan motors commonly use permanent split capacitor single-phase induction motors. These motors suffer from low efficiency. Today, these motors have been upgraded with brushless DC (BLDC) motors to increase the system's efficiency. Condenser fan motors are in the range of $\frac{1}{4}$–$\frac{1}{2}$ HP, while blower motors are in the range of $\frac{1}{3}$–1 HP.

In window or through-wall room air conditioners, 90% of the electrical energy is used to power the compressor motor, and the fans consume 10% of the total energy. Compressor motors in room air conditioners are rated $\frac{1}{2}$–2 HP and fan motors are rated $\frac{1}{8}$–$\frac{1}{3}$ HP. Condenser and blower are typically driven by a double-shafted single-phase induction motor. Consumers are sensitive to the higher upfront purchase cost of room air conditioners. This is a major challenge in increasing the use of permanent magnet motors in this application, especially for the condenser and blower fans.

Refrigerators and freezers (R/F) account for 28% of the electric motor energy consumption in residential applications. Figure 1.7 shows the motor units used in different applications

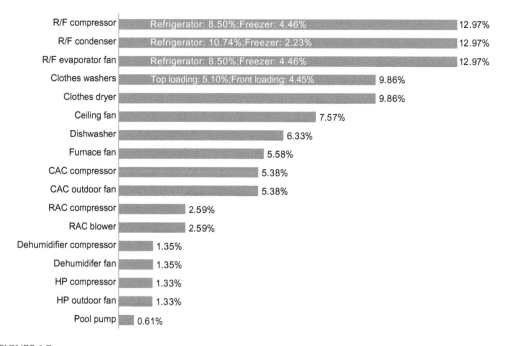

FIGURE 1.7
Motor driven system units in the residential sector. (From U.S. Department of Energy, Building technologies office, energy savings potential and opportunities for high-efficiency electric motors in residential and commercial equipment. December 2013, Available: https://energy.gov/.)

in the residential sector. It can be observed that R/F compressor, condenser, and evaporator fan motors account for 39% of the total motor population in the residential sector. Among them, compressor motors consume 90% of the R/F energy use. Residential R/F compressors are typically $\frac{1}{8}-\frac{1}{3}$ HP single-phase induction motors. These motors mostly operate in on/off cycles, which cause significant power losses. Variable speed operation could improve the compressor efficiency, but R/F applications are highly sensitive to component purchasing costs. Condenser and evaporator fan motors are rated less than $\frac{1}{100}$ HP.

Residential clothes washers and dryers account for 2% of the motor-related energy use; however, as shown in Figure 1.7, they account for about 20% of the motor population in residential applications. Clothes washers use a motor attached to the drum to provide the power during the washing cycle (low speed-high torque) and spin cycle (high speed-low torque). These motors range from $\frac{1}{4}$ to 1 HP, and they spend 80% of their operation in the washing cycle. Top loading washer motors account for 5.1% of the total residential motor population while front loading motor population account for 4.45%. Most top-loading washers are sold at lower prices; therefore, this application is highly cost sensitive. Low-efficiency capacitor-start induction motors dominate this market. In some high-end products, BLDC motors are used to improve the system efficiency. Capacitor start single-phase induction motors rated around $\frac{1}{10}$ HP are widely used in clothes dryers. However, dryer motors represent 5% of the total dryer energy consumption; therefore, the improved overall efficiency with the utilization of a permanent magnet motor drive system might not justify the higher cost. For other residential applications, typical motor ratings are listed in Table 1.2 [10].

TABLE 1.2

Typical Motor Ratings for Other Residential Application

Application	Typical Power (HP)
Vacuum cleaner	0.5–2
Dehumidifier	$\frac{1}{4}$
Attic fan	$\frac{1}{3}$
Window fan	$\frac{1}{20}-\frac{1}{5}$
Water well pump	$\frac{1}{2}-3$
Mixer, food processor, Blender	$\frac{1}{50}-\frac{1}{10}$
Electric can opener	$\frac{1}{50}$
Trash compactor	$\frac{1}{4}$
Garbage disposal	$\frac{1}{4}$
Garage door opener	$\frac{1}{3}-\frac{1}{2}$
Large power tools	1–5
Hand held power tools	$\frac{1}{10}-1$
Electric lawn/garden tools	$\frac{1}{10}-2$
PC fan	$<\frac{1}{100}$
Dishwasher pump	$\frac{1}{2}$

Source: U.S. Department of Energy, Opportunities for energy savings in the residential and commercial sectors with high-efficiency electric motors. December 1, 1999, Available: https://energy.gov/.

1.6 Electric Motor Usage in the Commercial Sector

In 2015, 28% of the total commercial sector energy was used for space heating, space cooling, water heating, and ventilation [2]. As shown in Figure 1.8, HVAC applications in the commercial sector (commercial air conditioners, air distribution, chiller compressors, circulation pumps/water distribution) account for 74% of total motor-related energy consumption.

Commercial air conditioning systems can be classified into two broad categories: decentralized air conditioning systems and centralized air conditioning systems. Decentralized air conditioning systems are employed in single and relatively smaller spaces. Centralized air conditioning systems use mainly chilled water as the cooling medium. They are found in large commercial buildings and require extensive ductwork and pipes for air distribution [11].

In decentralized commercial air conditioning, the primary technologies are Packaged Terminal Air Conditioners (PTAC) and Commercial Unitary Air Conditioners (CUAC). PTAC is used in hotels, some apartments, and office buildings. They are designed to go through the wall. These units operate similar to the room air conditioners in residential applications, but they usually have a higher cooling capacity. Unlike room air conditioners with double-shafted motors, they have separate motors for the condenser and evaporators. The compressor motor of PTAC is rated at $\frac{1}{2}$–3 HP, while the blower motors are rated at $\frac{1}{10}$–$\frac{1}{4}$ HP. BLDC motors are being used more often in these types of units. However, the initial purchasing cost is still an important parameter that might limit the higher penetration of permanent magnet-based motors in PTACs.

CUACs contain all the mechanical elements of the HVAC systems, including the compressor, indoor blower, and outdoor blower. The rooftop units are suitable for single, flat

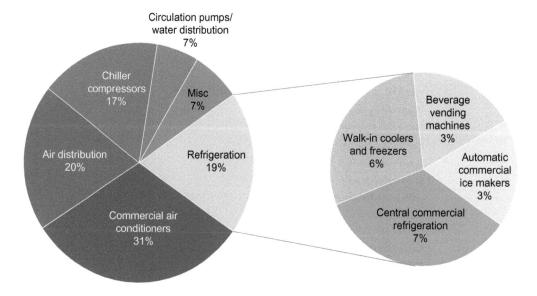

FIGURE 1.8
Motor system energy use by end-use in the commercial sector. (From U.S. Department of Energy, Building technologies office, energy savings potential and opportunities for high-efficiency electric motors in residential and commercial equipment. December 2013, Available: https://energy.gov/.)

buildings with large floor areas, such as motels, department stores, and movie theaters. Permanent split capacitor type single-phase induction motors are typically used for outdoor fans [8]. They range between $\frac{1}{4}$–1 HP. Three-phase induction machines rated at 1.7–7.4 HP are used for indoor blowers. Depending on the capacity, one or multiple compressors can be utilized in CUAC with individual three-phase induction motors running at single-speed. For small size units, the compressor motor output power ranges between 2–5 HP, for medium size units 5–20 HP, and for large size units 20–100 HP.

Centralized air conditioning is mainly employed in large commercial buildings and chilled water HVAC system is the primary solution. The chiller and auxiliary systems are the biggest consumers of commercial sector electrical power. You can refer to [12] for more information on the function of the chiller, air handling unit, and the water tower in a centralized air conditioning system. Here, we will focus on the electric motors used in chilled-water air conditioning systems.

Chillers usually run under partial-load conditions that might have a significant effect on the efficiency of the system. For this reason, many chillers employ multiple compressors and run them in on/off cycles to match the partial-load conditions. Large buildings often use multiple chillers, air-handling units, and circulation pumps. Ratings of the motors and circulation pumps depend on the type of the chiller. The compressor motor for the reciprocating chiller is rated between $7\frac{1}{2}$ and 150 HP, for a screw type chiller 40–750 HP, and centrifugal chiller 50–1,000 HP. Large chillers can have motors greater than 1,000 HP [8].

In the cooling water loop, a centrifugal pump circulates the cooling water between the chiller condenser and the cooling tower. In a reciprocating chiller, circulation pump motors are rated at 5 HP, in screw-type chiller 15 HP, and in centrifugal chiller 20 HP. In the cooling tower, motor driven fans assist the cooling process. Cooling tower fan/blower motors are rated at 5–25 HP. Central air handling units distribute the cooling air to individual spaces inside the commercial building. Depending on the size of the unit, they include pump motors rated at 5–25 HP. Due to the high output power requirements, chilled water HVAC systems employ three-phase induction motors [8].

Refrigeration applications use 19% of the commercial sector motor-related energy. As shown in Figure 1.9, electric motors in self-contained refrigeration, central commercial

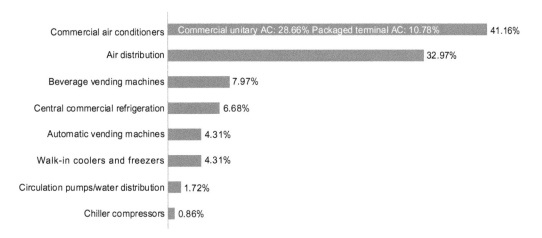

FIGURE 1.9
Motor driven system units in the commercial sector. (From U.S. Department of Energy, Building technologies office, energy savings potential and opportunities for high-efficiency electric motors in residential and commercial equipment. December 2013, Available: https://energy.gov/.)

refrigeration, beverage vending machines, automatic vending machines, and walk-in coolers account for 23.27% of the motor population in the commercial sector. Self-contained commercial refrigeration equipment is employed for storing and displaying refrigerated and frozen food in supermarkets and foodservice applications. In these applications, motors drive compressors, evaporator, and condenser fans.

In self-contained refrigeration equipment, the compressor motor is rated less than 1 HP. Compressor motors in the reach-in refrigerators are rated at $\frac{1}{3}$–$\frac{1}{2}$ HP, in reach-in freezers $\frac{1}{2}$–1 HP, and in beverage merchandisers $\frac{1}{3}$–$\frac{3}{4}$ HP. Depending on the application, single-phase capacitor starting induction motors and three-phase induction motors are used for these types of refrigeration units. The efficiency of these compressor motors is usually around 70% [10]. Condenser fan motors are typically $\frac{1}{2}$ HP and evaporator fan motors are less than $\frac{1}{50}$ HP. The number of BLDC motors is increasing as the condenser and evaporator fan motors in commercial refrigeration applications.

Central refrigeration systems in supermarkets and grocery stores can have compressor racks employing some parallel-connected compressors. A typical store can have 10–20 compressors ranging between 3 and 15 HP. These are mostly single-speed, three-phase induction motors. Due to the tight operating margins in supermarkets, a potential investment in new motor technologies depends heavily on the lifetime cost of the motor, which includes the upfront purchasing cost plus the cost of energy usage over the lifetime of the system [10]. Electric motors in beverage vending machine compressors are rated around $\frac{1}{3}$ HP, and the fan motors are rated at $\frac{1}{50}$–$\frac{1}{15}$ HP. Single-phase induction motors are commonly used in these applications.

1.7 Electric Motor Usage in the Transportation Sector

1.7.1 Vehicle Propulsion

The transportation sector accounts for 23% of the global energy-related greenhouse gas emission [13]. As shown in Figure 1.2, the transportation sector is the second largest energy user in the United States, but the portion of electrical energy is low. This is because the vast majority of the vehicles in the U.S. transportation system are powered solely by internal combustion engines and require fossil fuels as the energy source. In 2015, 96% of the transportation sector energy was supplied from petroleum products, among which 63% was motor gasoline (the remaining portion was mostly jet fuel and fuel oil). As shown in Figure 1.10, light-duty vehicles consumed around 57% of the total transportation-related energy in 2015 [2].

As shown in Figure 1.11, each year has seen an increase in the number of personal cars and commercial vehicles in use. As of 2015, there were around 1.28 billion vehicles in use globally [14]. The number of registered vehicles in the United States increased from 193 million in 1990 to 264 million in 2015 [15]. U.S. auto sales were 17 million in 2016 [16].

Today, the majority of the U.S. vehicle stock is conventional gasoline and diesel powered internal combustion engine vehicles. With the new regulations to reduce energy-related greenhouse gas emissions and the decrease in the cost of electrified powertrains, the percentage of hybrid, plug-in hybrid, and battery electric vehicles is expected to increase significantly.

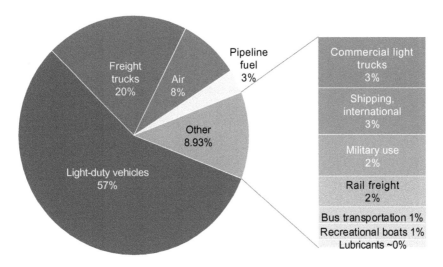

FIGURE 1.10
Energy demand by mode in U.S. transportation sector in 2015. (From U.S. Energy Information Administration, Annual energy outlook 2016 with projections to 2040, August 2016, Available: http://www/eia/gov/.)

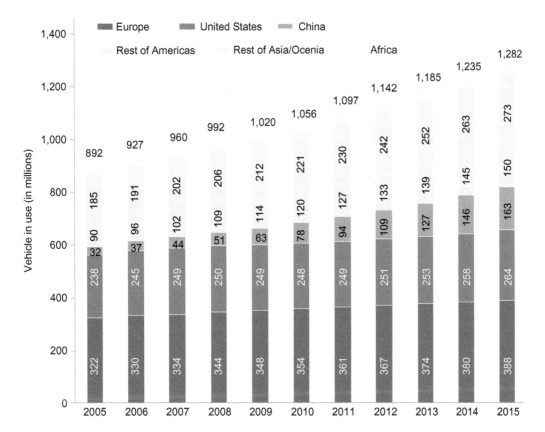

FIGURE 1.11
Number vehicles in use worldwide (in millions). (From OICA, Motorization rate 2015 – Worldwide, Available: http://www.oica.net/.)

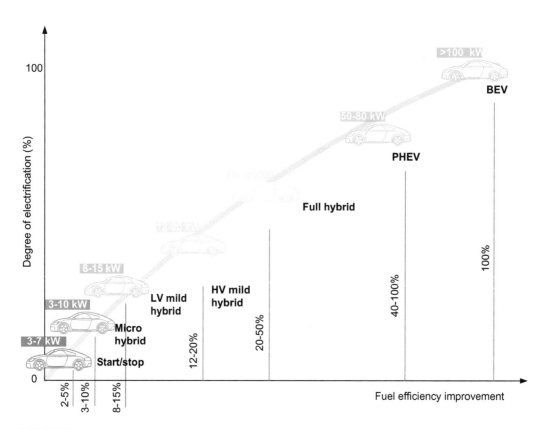

FIGURE 1.12
Degree of electrification and electric traction motor power. (From Bilgin, B. et al., *IEEE Trans. Transport. Electrific.*, 1, 4–17, 2015.)

The degree of electrification for a vehicle defines the ratio of electrical power available to the total power. As shown in Figure 1.12, as the degree of electrification increases, more fuel efficiency improvement is achieved and the electrical power requirement increases. Today, the majority of the hybrid, plug-in hybrid, and battery electric vehicles employ an interior permanent magnet synchronous machine due to its high torque density and high efficiency, especially at low and medium speed ranges. The Toyota Prius, the top selling hybrid-electric vehicle as of 2016, has a 60 kW V-shaped interior permanent magnet traction motor with a maximum torque of 207 Newton meter (Nm) and a top speed of 13,500 rpm. The 2011 version of the Nissan Leaf electric automobile has 80 kW traction motor with delta-shaped rotor magnets and it delivers a maximum torque of 280 Nm and a maximum speed of 10,390 rpm. Another electric vehicle, the Chevrolet Spark, also employs an interior permanent magnet traction motor with double-barrier rotor geometry with bar windings and delivers a maximum power of 105 kW, a maximum torque of 540 Nm, and a maximum speed of 4,500 rpm [5]. Tesla is using 310 kW, 600 Nm, three-phase induction machine with copper rotor bars in its Model S high-performance electric vehicle [17].

As shown in Figure 1.12, the start/stop function provides 2%–5% fuel efficiency improvement, and, today, this is mostly achieved by using the starter motor. The conventional starter motors are brushed DC motors. Mild hybrids have an auto start/stop function and regenerative braking capability. They also provide some use of electric power for propulsion.

48 V systems typically provide around 8%–15% fuel efficiency improvement. The majority of the vehicles manufactured after 2017 will have start/stop capability [18]. Today, permanent magnet and induction motors are employed in mild hybrid electric vehicles.

Electric motors are also utilized for many other uses in automotive applications, which are not limited to hybrid and electric vehicles. Almost all vehicles include one or more electric power steering motor, dual clutch transmission motor, seat adjustment motor, engine cooling fan motor, electronic stability/anti-lock braking system motor, window lift motor, door lock motor, hatch/tailgate motor, and sunroof motor. These are low-voltage, low-power motors and they are designed for specific purposes. These motors are among the small-size motors, which make up 90% of the motor population in the United States as shown in Table 1.1. Therefore, it is important to notice the effect of car sales in the electric motor industry, especially in the mature motor markets, such as in the United States and Western Europe, where the growth in electric motor sales is driven by an increase in vehicle sales.

1.7.2 e-Bike Motors

With the trend towards improving energy efficiency and reducing energy-related greenhouse gas emissions, electric bikes (e-bikes) are becoming a viable mode of transportation. The total global market for e-bike sales was around 31.7 million units in 2014 [19]. China is the global leader in the adoption and sales of e-bikes. Around 170 million consumers use e-bikes every day in China. E-bike sales in China were 28.8 million, which corresponds to 91% of the total global market. The second largest e-bike market is Western Europe with 1.2 million e-bikes sold in 2014. Japan and the United States follow it by 440,000 and 185,000 units, respectively. According to the same research study, the e-bike market is expected to reach 40.3 million units in 2023. China is projected to represent 85% of the total market. Sales in North America, Western Europe, and Latin America will also grow significantly. E-bike sales in Western Europe are expected to reach 3.3 million in 2023.

As of 2016, electric motor power for e-bikes is regulated, so that an e-bike can still be categorized as a bicycle and used on roads and bike paths without requiring additional licensing and registration. The electric drive system is usually activated through pedaling, but in some classes, it is also possible to activate the electrical power with the throttle. In Europe, motor power is limited to a top speed of 25 kilometers per hour (kph). Continuous motor power is 250 W, and the peak motor power is 500 W. In the United States, peak motor power can go up to 750 W with a top speed of 32 kph [20,21].

E-bikes today are mostly designed with either mid-drive or hub motors (see Figure 1.13). As of 2016, mid-drive motors hold the majority of the market. Hub motors replace the regular wheel hubs and connect the tire, rim, and spokes to the axle. The majority of the hub motors are mounted on the rear wheel since the rear wheel usually has stronger support.

There are two types of hub motors. Geared hub motors include a planetary gear set. The gear set enables higher motor speed leading to a reduction in motor dimensions. To eliminate the resistance during coasting, geared hub motors might include a freewheel mechanism, which limits the regenerative braking capability. In 2016, the majority of the hub motors used in e-bikes are geared hub motors. Surface permanent magnet brushless DC motors with concentrated windings are heavily used in geared hub motors.

The second type of hub motors is the direct drive or gearless hub motors. Since the planetary gear set is eliminated, these motors provide higher torque at low speed. The maximum speed of gearless hub motors is around 500 rpm. The planetary gear set is the primary source of noise; therefore, gearless hub motors can deliver quieter operation. On the other hand, they are usually heavier, and since they are located on the wheel, they

(a)   (b)

FIGURE 1.13
e-bike motors: (a) mid-drive unit and (b) hub motor.

position the weight towards the end of the bike. This may reduce the balance. Furthermore, since the motor weight is built into the wheel, this increases the unsprung mass. It means that the weight of the gearless hub motor cannot be sprung as a part of the frame [22]. Surface permanent magnet brushless DC motors with concentrated windings are the preferred choice for gearless hub motors. Figure 1.14 shows the picture of a commercial gearless hub motor from Crystallite. It is an exterior rotor machine, and the magnets are located on the inner surface of the rotor.

FIGURE 1.14
Gearless rear hub motor (Crystallite HS3548).

FIGURE 1.15
Mid-drive motor of SHIMANO STEPS™ drive system.

The mid-drive unit is located at the bottom bracket, where the pedals are attached to the frame. Hub motors drive the wheel itself whereas mid-drive motors drive the chain forward. This way, they take advantage of the mechanical drive system. When the rider changes the gear, the motor benefits from this, and it can run at relatively higher speed. This makes mid-drive motors suitable for mountain biking and hill climbing.

Mid-drive motors also come with a much higher gear ratio. Planetary gear sets in hub motors usually have a gear ratio around 5–11. With multi-level gear design, mid-drive motors can achieve a higher gear ratio. Since the unit is attached to the bike frame, the unsprung mass issue as in the gearless hub motors does not exist in mid-drive units. However, mid-drive units require a torque sensor and speed sensor to control the output power. This increases the cost of the motor and, hence, the e-bike.

Figure 1.15 shows the electric motor of SHIMANO STEPS™ mid-drive unit. It has three-level gear system to achieve a high gear ratio. It has concentrated windings. This motor has spoke-type magnets with tangential magnetization. This magnet configuration enables high flux density in the air gap and maintains constant torque over a large speed range. But the extended speed capability of the motor is limited. The motor is designed to provide constant torque-assist up to the maximum pedaling speed. Spoke-type magnet rotor configuration is usually more expensive to manufacture as compared to surface permanent magnet motors.

1.8 Overview of Electric Motor Industry

The global electric motor market was valued at \$122.5 billion in 2017, and it is estimated to reach \$164 billion in 2022. More than 60% of the demand and shipment will be in three regions: United States, Western Europe and China. China's share in the global demand is projected to increase to 40.91% by 2022 from 33.13% in 2012 [23]. This is due to the increasing manufacturing and rising middle class in China.

Electric motor demand and shipment in the Asia/Pacific region (China, Japan, and other Asia/Pacific) is expected to increase. In 2002, the Asia/Pacific region had 39.86% of the global electric motor demand and 43.97% of global electric motor shipment. In 2022, this region is expected to have 63.90% of the global demand and 69.21% of the global shipment [23].

Leading electric motor suppliers in the North and South America are ABB, WEG, Nidec, Regal Beloit, and Siemens. The suppliers are developing higher-efficiency motor drives. One of the main targets in the industry is to achieve competitive prices. Increasing supply chain cost has also been a concern [24]. Due to its low-cost and simple construction, SRMs will definitely play a significant role in the price competition and market expansion of high-efficiency motor drives.

1.9 Electric Motor Types

Induction and permanent magnet machines dominate in the motor drive systems used in industrial, residential, commercial, and transportation applications. Figure 1.16 shows the typical structure of these motors and compares them with that of a switched reluctance motor. Here, the most significant advantage of SRM becomes apparent. All of these motors have a stator, winding, and rotor. But SRM has one item less than permanent magnet and induction machines. As shown in Figure 1.16a, permanent magnet machine has magnets inserted in the rotor to provide the rotor excitation. As shown in Figure 1.16b, induction machine has rotor bars to achieve magnetic induction and create the forces on the rotor.

Switched reluctance machine does not have permanent magnets or conductors on the rotor. Also, its winding is much simpler than the permanent magnet and induction machines. These features bring many advantages to the SRM architecture (e.g. low manufacturing cost, robust operation at high speed and harsh environment) and also many challenges (e.g. nonlinear characteristics, torque ripples, and acoustic noise and vibration). We will investigate these issues in more detail throughout this book. Now, we will have a brief look at the other motor types, and address the advantages and challenges they have.

1.9.1 Permanent Magnet Machines

As shown in Figure 1.16a, rotor magnets in permanent magnet (PM) machines provide an independent source of magnetic flux. In interior permanent magnets machines, the magnets are embedded in the rotor. By applying a careful design for the location and arrangement of the magnets, the output torque at higher speeds can be improved by utilizing the reluctance torque. In residential and commercial applications, surface permanent magnet machines are widely used, where the magnets are located on the surface of the rotor. These motors usually come with concentrated windings, similar to an SRM [17].

The independent rotor excitation in permanent magnet machines can provide high torque density and better efficiency especially at low and medium speed range [5]. This is one of the main reasons why permanent magnet machines are preferred in applications with high-efficiency requirements. As we will see in the next chapter, there is a

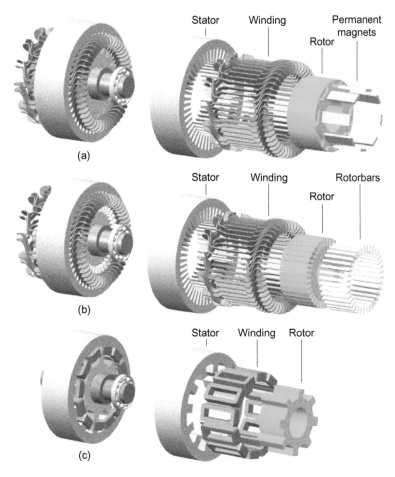

FIGURE 1.16
Typical structure of electric machine types: (a) interior permanent magnet synchronous machine, (b) induction machine, and (c) switched reluctance machine. (From Bilgin, B. et al., *IEEE Trans. Transport. Electrific.*, 1, 4–17, 2015.)

relationship between the strength of the magnetic field and forces acting on the rotor. If a strong magnetic field is maintained in the air gap, higher forces can be generated, resulting in higher torque density. Depending on the material, rotor magnets can provide a strong magnetic field in the air gap without utilizing coils on the rotor. Therefore, copper losses could be reduced leading to higher efficiency.

In permanent magnet machines, high torque density and high efficiency can be achieved by using magnets that can provide high flux density. As the speed increases, the back-electromotive force of a PM machine is likely to exceed the terminal voltage. If the back-electromotive force is not reduced, current cannot be injected into the motor and torque drops. To extend the speed range of the motor, field weakening is applied and the phase angle between the current and voltage is adjusted so that the stator flux effectively opposes the magnet flux. Hence, the excitation torque from permanent magnets is reduced and the additional reluctance torque component from the rotor saliency helps to extend the speed range [17]. Field weakening reduces the efficiency at high speeds. In addition, the permanent magnet should have enough coercivity, so that it is not demagnetized during the field weakening.

High-energy rare-earth magnets (e.g. Neodymium Iron Boron and Samarium Cobalt) are used in permanent magnet machines to achieve high torque density, high efficiency and high resistance to demagnetization. The main disadvantage of permanent magnet machines is the sensitivity of rare-earth magnets to temperature. When the temperature of Neodymium Iron Boron (NdFeB) magnet increases to 160°C, the output torque can drop by up to 46% [25]. Samarium Cobalt (SmCo) rare-earth magnet has the highest resistance to demagnetization. It can handle continuous temperatures above 250°C and it is used in aerospace and military applications.

However, permanent magnets have high cost, and they take up a significant portion of the total cost of a permanent magnet machine, even though they represent a small portion of the total weight of the motor. Figure 1.17 shows the distribution of mass and cost in an 80 kW interior permanent magnet motor designed for a traction application [26]. It can be observed that rotor magnets take up 3% of the mass, but 53% of the cost, for this specific application. The price of SmCo is even higher than NdFeB magnet.

Rare-earth permanent magnets are made from rare-earth elements. The most commonly used rare-earth elements in permanent magnets are neodymium and dysprosium. Dysprosium is used to improve the resistance of the magnet towards demagnetization at elevated temperatures. In a typical NdFeB magnet for a traction motor, neodymium takes up around 31% of the total mass, whereas dysprosium takes up around 8.7% [17].

There are significant supply chain issues with rare-earth elements, which can cause prices to spike from time to time. These events could adversely affect the cost of permanent magnet motors as the demand for high-efficiency motors increase in industrial, residential, commercial, and transportation applications and electricity generation. For this reason, neodymium and dysprosium are defined as critical materials by the U.S. Department of Energy [27].

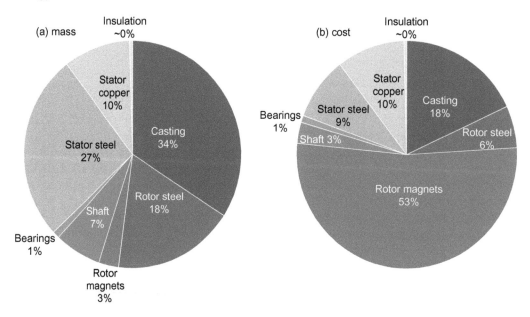

FIGURE 1.17

Distribution of (a) mass and (b) cost in a permanent magnet machine for a traction application. (From Miller, J.M., Electric Motor R&D, 2013 U.S. DOE Hydrogen and Fuel Cells Program and Vehicle Technologies Program Annual Merit Review and Peer Evaluation Meeting, Oak Ridge National Laboratory. May 15, 2013, Available: http://energy.gov/.)

Depending on the performance requirements and operating temperature, magnets other than NdFeB and SmCo are used in electric motor applications. Ferrite magnets have low energy product and they are usually used in low-cost applications, which do not require high torque density. Aluminum nickel cobalt (AlNiCo) alloy magnets can operate at high temperatures, but they have low coercivity. The biggest application for AlNiCo magnets is watt-hour meters [28].

1.9.2 Rare-Earth Materials

The global market for permanent magnets is estimated to reach $15 billion by 2018. The largest demand is in Asia-Pacific Region. The demand in this region is likely to keep growing due to the increasing manufacturing and end-user applications in China. The United States is an important manufacturer of high-performance magnets for military and other strategic applications, but it still imports 60% of its permanent magnet consumption [28]. China produces 76% of the permanent magnets in the world.

Rare-earth magnets have been dominating the global permanent magnet market and they are heavily utilized in electric motor applications. Besides magnets, rare-earth materials are used in many critical applications such as in photovoltaic films, vehicle batteries, and lighting. Neodymium and dysprosium are the primary rare-earth materials that are most commonly employed in high-energy permanent magnets.

The rare-earth magnet supply chain can be divided into five stages: (i) mining, milling and concentration of the ore, (ii) separation into individual rare-earth oxides, (iii) rare-earth metal production, (iv) alloy or powder production, and (v) magnet manufacturing [27]. In the downstream portion of the supply chain, production is highly concentrated amongst a few companies that dominate the market. China dominates the rare-earth production, even though they hold only about half of the known world reserves [29]. According to United States Geological Survey (USGS), the world rare-earth mine production was 125,000 metric tons in 2014. As shown in Figure 1.18, China holds 84% of the world production. This value

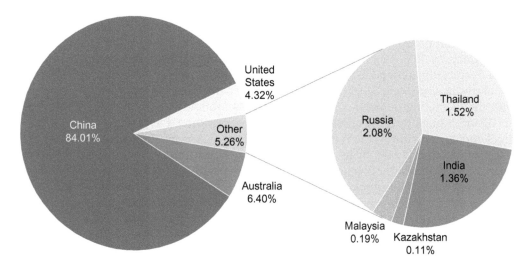

FIGURE 1.18

Estimated world mine production of rare-earth materials by country (total output: 125,000 metric tons of rare-earth oxide equivalent). (From United States Geological Survey (USGS), 2014 mineral yearbook, rare earths [advance release], December 2016, Available: http://www.usgs.gov/.)

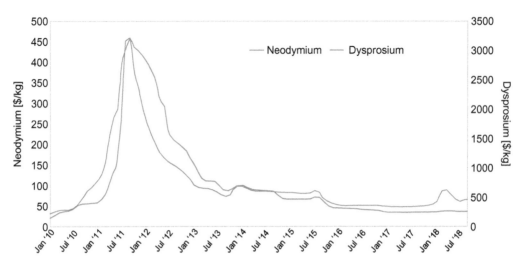

FIGURE 1.19
Price of neodymium and dysprosium between 2011 and 2016. (From Metal-Pages.com, 2018.)

is based on China's Ministry of Land and Resources production quotas. Around 20% of China's rare earth oxide was produced illegally and the total production was well beyond the 2015 quota. In the United States, the mining and processing of rare earth materials take place in Mountain Pass Mine in California [30].

As it was shown in Figure 1.17, rare-earth magnets are the largest cost component of an electric motor. This is related to the high price of rare-earth materials. Figure 1.19 illustrates the change in price in neodymium and dysprosium between 2011 and 2016. Rare-earth prices seem lowering from the peak in 2013. But it should be noted that in December 2001, the price of neodymium was less than $20 per kg, and the price of dysprosium was less than $150 per kg [27]. Being the largest producer of rare-earth materials, China has significant control over the price and availability of these materials in the global market. China imposes export quotas on rare-earth products. For example in 2010, China reduced the export quota for rare-earth products by 40%. As a result of this, the price of neodymium jumped to $115 and the price of dysprosium jumped to $400 per kg [27]. As of September 2016, the price of neodymium and dysprosium were $49.89 per kg and $257.50 per kg, respectively.

With the increasing demand for high-efficiency motors in industrial, residential, commercial, and transportation sectors, and in energy generation, the price volatility of rare-earth elements and supply chain issues are significant concerns. As explained earlier, the United States imports 60% of its permanent magnet consumption. China holds 76% of permanent magnet production and 84% of rare-earth material production. Therefore, the United States might end up being highly dependent on the external resources if it relies heavily on electric machines using rare-earth permanent magnets. This is why rare-earth elements are regarded as critical materials by U.S. Department of Energy [27].

There are also production and environmental issues for rare-earth materials. Rare-earth elements are not rare as their name implies. The earth's crust contains sufficient rare-earth elements. However, they are difficult to mine and it is usually hard to find them in high enough concentration so that the extraction process is economically viable. As depicted in Figure 1.20, production of rare-earth materials is a small portion of the total metal production. However, mining and refining of rare-earth materials require significant amount of

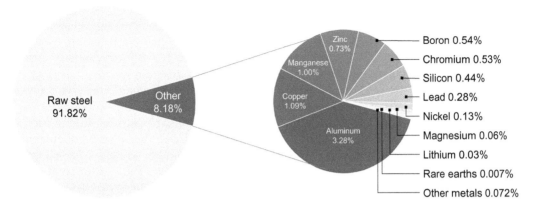

FIGURE 1.20

Breakdown of global metal production in 2015 - total metal production: 1.75 billion metric tones. (From United States Geological Survey (USGS), Historical statistics for mineral and material commodities in the United States, June 2017, Available: http://www.usgs.gov/.)

capital and expertise. Depending on the location and the production capacity, the extraction and processing of rare-earth ore might need capital between $100 million to $1 billion [27].

The processing of rare-earth elements into high-purity rare-earth oxides is a highly-specialized chemical process and it requires significant know-how in mineral processing. Therefore, producing rare-earth materials is not a process that can be addressed just by opening new mines. The same applies to the production of rare-earth permanent magnets. Producing high-quality permanent magnets require expertise and know-how as well. The Japanese organization, Hitachi, holds more than 600 patents on the production of high-quality neodymium-iron-boron magnets, which makes it difficult for other corporations to produce these magnets [27]. Mining and extraction of rare-earth materials can have adverse environmental effects, which also increase the cost of production. Besides the political reasons, environmental concern is also a factor for the quota China is applying on the rare-earth exports [31].

In summary, permanent magnet machines provide high torque density and high efficiency, especially at low speeds. However, the price volatility, supply chain issues and environmental concerns for the rare-earth materials, are significant concerns for the production of high-energy rare-earth magnets. In the long run, these problems might prevent the electric motor industry from responding to the increasing demand for high-efficiency, high-performance, and low-cost electric motors. From an economic perspective, boosting the supply and production of rare-earth metals might not solve the problem, because mining and producing rare-earths is a capital intensive and environmentally delicate process.

1.9.3 Induction Machines

Induction machines (IMs) are the most widely used motors in many different applications in the industrial, commercial and residential sectors. The majority of the electric motors in use today are IMs. Single-phase capacitor-start IMs are widely used in small size motors.

There are primarily two types of induction machines. In wound-rotor IMs, the rotor circuit is made up of three-phase windings. The rotor windings are then short circuited using slip rings. As shown in Figure 1.16b, in squirrel-cage IMs, the rotor bars are inserted by die casting, where melted aluminum is molded in the rotor slots. End rings then short-circuit the bars. Squirrel-cage IMs are commonly utilized in low- and medium-power applications.

In an IM, the air gap magnetic field and the rotor rotate at different speeds. This is why induction machines are referred as the asynchronous machine. When there is a difference between the speeds of the rotor and the air gap magnetic field, rotor conductors are exposed to a time-changing magnetic field. The time-changing magnetic field induces voltage across the rotor bars. Since the end-rings short circuit the rotor bars, current flows through the conductors and this generates the rotor magnetic field. The interaction between stator and rotor magnetic fields creates the torque. If the rotor rotates at the same speed as the stator magnetic field, the rotor bars will not be exposed to a time-changing field and torque will be zero. Then, the rotor speed is eventually balanced at a speed lower than the synchronous speed, where the motor torque equals to the load torque [17,25].

When compared to permanent magnet machines, IMs have lower cost construction, because they don't have permanent magnets. Die-casting of aluminum rotor bars is also a low-cost process. The self-starting capability is one of the main reasons why IMs are widely used in our industry. Both permanent magnet and switched reluctance machines require a power converter, control algorithm, and position feedback or estimation. Three-phase induction machines can start from the AC supply without a power converter, control algorithm, or position feedback. Single-phase induction machines can also start directly from a single phase AC supply. However, single-phase excitation generates two magnetic fields, rotating in different directions. To maintain self-starting capability, they need a means to create phase-lag between these two fields. This can be accomplished in many different ways, but capacitor start single-phase induction machine is the most popular type in the residential, commercial and industrial sectors.

Today, the majority of IMs are still used in on/off cycles to match the partial load requirements. Especially in commercial HVAC systems, there are sometimes three different induction machines in one unit, which run three different compressors at different cycles. This solution might look feasible from the cost point of view. There are well-established low-cost manufacturing techniques available for induction machines. However, from a systems perspective, this is not an efficient solution. With the increasing demand for higher efficiencies – both in motor and system level – and decreasing cost of adjustable speed drives, variable frequency operation becomes much more feasible.

As compared to permanent magnet machines, an IM-based motor drive system can still have a significant cost advantage due to the lack of more costly permanent magnets. However, it is harder to say the same when compared to a switched reluctance machine based motor drive system. Lower manufacturing cost can be achieved with a switched reluctance machine because it has much simpler windings and it doesn't require rotor conductors.

Compared to permanent magnet machines, induction machines typically have lower efficiency and power factor – especially single-phase induction machines typically run at low efficiencies (around 70%). Since there is no direct excitation on the rotor, the excitation current is drawn from the stator current. This leads to a low power factor. This is the reason why induction machines usually require a small air gap to minimize the reactive power requirement.

Due to induced currents on the rotor conductors, there are non-negligible rotor copper losses in the induction machine. Especially for high-torque operation and applications with high power density requirement, rotor copper losses present challenges in extracting the heat out of the motor. For induction machines with die-casted aluminum rotor bars, the rotor copper losses can be more significant, because the electrical conductivity of aluminum is 60% that of copper.

For high torque and high power density applications, induction machines with copper rotor bars can be utilized. The traction motor in Tesla Model S electric vehicle is an example [5]. In this motor, copper is die-casted in the rotor slots to achieve high conductivity, lower rotor resistance and, hence, higher efficiency.

Die casting is a metal casting process where molten metal is injected into the die at high pressure. It is a quick and reliable process and enables low-cost mass manufacturing of induction machines. However, copper die-casting is not a feasible solution for cost sensitive applications. The melting temperature of copper (1,080°C) is much higher than aluminum (660°C). In copper die-casting, the dies should be preheated, and during the die casting process, the temperature should be controlled. In addition, the density of copper is higher than aluminum and this requires specialized tooling and higher tonnage presses [32]. This makes copper die cast induction machines a lot more expensive than aluminum die cast induction machines.

In summary, induction machine has been the low-cost solution in the industrial, residential, and commercial sectors for various direct-drive single-speed applications. However, with the increasing demand for high system and motor efficiency, this advantage has been diminishing. Higher efficiencies can be maintained by using variable frequency drives, but this reduces the advantage from low-cost manufacturing and operation of self-starting induction motors. In addition, aluminum die cast induction machine suffers from high rotor losses. This also reduces the efficiency. Copper has much higher conductivity, but copper die-cast induction motors are a lot more expensive. Therefore, the increasing demand for higher efficiency and variable speed operation reduces the cost advantage of induction machines and makes switched reluctance machine based motor drive systems a more viable alternative.

1.10 Switched Reluctance Machines

As compared to the permanent magnet and induction motors, switched reluctance machines have a simple, low-cost and robust construction. As shown in Figure 1.16, SRM has one less component than other motor types. Permanent magnet machines have magnets on the rotor, and induction machines have conductors on the rotor. The SRM stator is made of a salient pole laminated core with concentrated windings, while the rotor also has a salient pole structure without windings or permanent magnets. This enables reliable operation at high speeds and high temperatures but also brings many challenges, such as high torque ripples, and acoustic noise and vibration.

One of the most significant advantages of a switched reluctance machines is the simple and low-cost construction. All mass-produced electric machines employ copper and steel. Manufacturing of windings and laminated cores are well-established industry practices. There are many companies with significant expertise in electrical steel production, punching or laser cutting laminations, lamination stacking, and winding automation. These are the practices that are needed to manufacture an SRM. As compared to permanent magnet machines, an SRM does not need magnet insertion and initial magnetization. As compared to induction machines, an SRM does not need die-casting (in squirrel-cage IM) or additional winding process (in wound-rotor IM) for the rotor conductors. Therefore, any motor manufacturer, which has the supply chain to manufacture permanent magnet or induction machines, can easily make switched reluctance machines with much lower cost.

An SRM has a simple rotor construction without coils or permanent magnets giving it two significant advantages of cost and supply chain security. As we discussed previously in this chapter, rare-earth magnets, which provide high torque density and high efficiency at low speeds in permanent magnet machines, suffer from price volatility, supply chain issues, and environmental concerns. This can be a problem in the long run as the demand for high-efficiency motors is increasing. Due to the lack of permanent magnets, the low-speed efficiency in switched reluctance machines might end up lower. However, SRMs can provide similar and even higher efficiencies at the medium and high-speed range. In addition, SRMs can deliver comparable or superior efficiency over the entire operating or drive cycle of the application.

As compared to other motor types, an SRM is more suitable to running at high-speed and high-temperature conditions. This advantage also comes from the lack of rotor excitation. At high speeds, the performance of the PM motors can be seriously limited by the rotor displacement and rotational stress. When the rotor is rotating at high speeds, centrifugal forces dominate and high stresses occur on the magnet slots and the bridges. The bridges are the sections of the rotor core which are located between the magnet slot and the air gap. They have to be carefully designed so that they can handle the centrifugal forces and saturate quickly not to cause any flux leakage. Switched reluctance machine has a simple and robust rotor core. It doesn't have slots or bridges. This makes an SRM a better candidate for high-speed operation.

As the speed increases, field weakening is applied in PM machines to extend the speed range. Field weakening reduces the efficiency of PM machines at high speeds. As we will see in the next few chapters, flux weakening naturally happens in SRMs at high speeds. Compared to permanent magnet and induction machines, an SRM has the largest constant-power speed range, which makes it suitable for high-speed operation.

When it comes to high-temperature operation, SRM has an advantage, as well. The magnetic properties of permanent magnets are highly dependent on the operating temperature. In permanent magnet machines, the maximum operating temperature should be defined to maintain stable operation of the magnet. As discussed previously in this chapter, the flux density and coercivity of NdFeB magnet reduces as the temperature increases. This affects the output torque and demagnetization. Permanent magnet machines using NdFeB magnets are usually designed so that the magnet temperature stays around 100°C during the continuous operation.

The SRM's rotor is made of laminated steel only. The magnetic properties of steels change drastically near the Curie temperature, which is 770°C for iron. In the case of non-oriented steel, the permeability of electrical steel does not alter much below 500°C. Electrical steels have surface insulation to reduce eddy current losses. The surface insulation chosen for an SRM should generally handle the annealing temperature, which is a heat treatment process to eliminate the stress on the laminations and helps to return the magnetic properties to the stress-free conditions [33]. Annealing temperature depends on many parameters such as the material, duration, and pressure. Surface insulation on electric steels can handle temperatures in the range of 200°C–400°C.

In SRMs, the magnet wires used in the stator windings determine the thermal rating of the machine. The insulation around the copper conductor enables the contact between the wires without causing any electrical short circuit. Magnet wire insulation is made of organic material, which softens at a lower temperature than copper or electrical steel.

However, thermal class of magnet wires can go up to 240°C, which is much higher than permanent magnets. 200°C is a common standard for magnet wires.

When compared to induction machines, SRM designs are still advantageous for the high-temperature operation. In induction machines, rotor conductors are conventionally manufactured by die-casting aluminum in the rotor slots. Aluminum has lower conductivity than copper; therefore, rotor copper losses can be a significant constraint in the high-temperature operation of induction machines.

As discussed previously, copper die-casted induction machines have lower rotor losses and they are more suitable to operate at high temperatures. But their manufacturing process is more challenging and expensive. When designing the thermal management system for copper die-casted induction machines, the higher coefficient of thermal expansion (CTE) of copper should also be taken into account. The difference between the CTE of copper and steel would apply fatigue stress at the copper-steel interface and cause cracks on the conducting bars when the motor goes through thermal cycling [33].

Another advantage of an SRM is its fault-tolerant operation capability. As it will be discussed in detail in the next few chapters, each phase of SRM can be considered electrically isolated from each other. This means that when one phase is excited with current, the magnetic flux that links with other phase coils is negligible and, hence, the mutual coupling can be ignored. Therefore, torque production of one phase is independent of the others. If there is a fault in one phase, the other phases can generate torque and keep the motor running with reduced performance.

The salient-pole construction of SRMs enable fault-tolerant operation due to electrically isolated phases. However, this is also a major source of high torque ripple. The torque in each phase of an SRM is dependent on the relative position between the stator and rotor poles, and the level of excitation current. During phase commutation, phase torques are added up together, and the overall torque profile ends up with a pulsated waveform. Torque ripple in an SRM is much higher than permanent magnet and induction machines.

Torque ripple should not be an obstacle in the adoption of SRMs in motor drive systems. Each particular application determines the need for torque ripple requirements. For example, conventional internal-combustion engines have significant torque ripples due to the firing of individual cylinders. In vehicular applications, large flywheels are used to manage the torque ripples. Furthermore, the brackets connecting the engine to the chassis are carefully designed to reduce the effect of torque ripple [34]. Torque ripples in SRM can be reduced significantly by modifying the rotor geometry and shaping the phase current. The torque ripple reduction techniques in SRM will be discussed in depth in different chapters of the book.

Acoustic noise and vibration is the most well-known issue in switched reluctance machines. Due to the salient pole construction of an SRM, when a phase is excited with current, the flux penetrates into the rotor, mostly in the radial direction, and generates large radial forces. These radial forces deform the stator core and the frame, which results in vibrations and acoustic noise.

Acoustic noise and vibration cannot be eliminated in SRMs, but it can be reduced significantly. Acoustic noise reduction in SRM requires a multidisciplinary approach. The stator and rotor geometry, pole combination, materials, frame, shaft, and current control should be optimized to match the acoustic noise specifications of the given application. System level analysis plays a significant role in the design and optimization of switched reluctance machines. It is very challenging to achieve low acoustic noise, low torque ripple, low temperature rise, and high efficiency in a wide torque and speed range, by satisfying low-cost and high power density constraints. This is also true for permanent magnet and induction machines. To better optimize a switched reluctance machine, the requirements at different torque and speed points should be identified. It is important to know at which operating points low acoustic noise is desired, and what the torque ripple and efficiency requirements are at those points. Then the geometry, materials, and the current control can be optimized to reduce the acoustic noise.

Single source of excitation in SRMs also causes some challenges. The stator current is responsible for both excitation and torque generation. For this reason, an SRM is usually designed with a smaller air gap, which requires tighter mechanical tolerances. Induction machine also has a single source of excitation, and it is manufactured with a small air gap as well. Permanent magnet machines can be designed with a larger air gap because the permanent magnets independently generate the rotor field. A design with tighter mechanical tolerances is not an issue to prevent utilization of switched reluctance machines. Our industry has strong expertise in machining and manufacturing. If there is a small increase in the cost of an SRM due to tighter tolerances, it can be justified by the material and energy savings.

Another challenge for SRM is that it has low power factor at light load operation. As it will be explained in detail in Chapter 2, at small currents, an SRM works in the linear region of the magnetization curve. When operating in the linear region, an SRM has a lower power factor and less than half of the total magnetic energy is converted into mechanical work. The rest is stored in the magnetic circuit and supplied back to the source at the end of the stroke or dissipated inside the motor. In practice, an SRM operates in the nonlinear region of the magnetization curve to achieve a higher power factor and better utilization of the converter.

The converter topology that is used in SRM drives is different from that of permanent magnet and induction machines. As it will be addressed in the next few chapters, the direction of the torque in an SRM is independent of the direction of the current. Therefore, unipolar phase current is required. Asymmetric bridge converters are widely applied in SRM drives. In permanent magnet and induction machines, three-phase full bridge converters are used.

Utilizing a different converter is not a big challenge for the adoption and mass manufacturing of SRM drives. The cost of permanent magnet and induction motor drives might be cheaper today because semiconductor manufacturers are already manufacturing three-phase modules where all the switches are in one package. However, the same can be done for SRM drives. In fact, there are already modules available where the switches and diodes for one phase are packaged in one module. This enables a more compact converter design for SRM. A high-power traction SRM converter will be presented in Chapter 16 utilizing customized modules.

It should also be noted that in the permanent magnet and induction motor drives, the switches on the same phase leg cannot be activated at the same time. If that happens, the DC link will be short-circuited. This is called a shoot-through fault. Since the phases in SRMs are electrically isolated from each other, this is not the case. In fact, in an asymmetric bridge converter, two switches are turned on at the same time. Therefore, SRM does not have a shoot-through fault condition.

A challenge that has been attributed to SRM is its nonlinear characteristics, which make the analysis and optimization harder. It is true that an SRM has nonlinear characteristics. But, considering the advancements in modeling and computation techniques, motor drives, and control algorithms, this is not a challenge any longer. In a surface permanent magnet machine, since the permeability of permanent magnets is very close to that of air, the effective air gap is large, and the magnetic circuit can be assumed linear. Therefore, analytical expressions can be used to model and optimize the surface permanent magnet motors.

Due to its nonlinear characteristics, numerical modeling techniques, such as finite element analysis are needed to optimize SRMs. This is not a problem anymore, thanks to the advanced software platforms (e.g. JMAG, ANSYS, Infolytica, COMSOL, Flux), which can analyze and characterize an SRM quickly. As it will be introduced in Chapter 4, combining analytical and finite element tools, it is possible to accurately model and optimize switched reluctance machines.

1.11 Switched Reluctance Motor Drive Applications

Today, switched reluctance machines have a small share in the electric motor population. SRMs are used in some industrial pumps, vacuum cleaners, and agricultural and mining vehicles. The small market share of SRM is historically due to high acoustic noise and vibration, torque ripples, challenges in design and optimization due to its nonlinear characteristics, and the cost of the power converter due to the use of unconventional power modules. As it was discussed in the previous section and will be addressed throughout the book, these challenges can now be resolved and an SRM can now be designed to match the requirements of a wide range of application.

Today, induction machines dominate in the electric motor driven applications. This is mainly because induction machines don't need an adjustable speed drive to start. However, many mechanical systems in the industrial, residential, and commercial sectors operate with variable loads in long operating hours. Significant efficiency gains can be accomplished by adapting the motor speed and torque to the load conditions [3]. In many applications, electric motors are oversized and they run continuously at partial load. This is a significant loss in capital, efficiency, and performance. Adjustable speed drives can eliminate the losses due to partial loads. They adjust the speed and torque of the motor to the load and eliminate the need for mechanical components such as gears, transmissions, and clutches. The motor size can also be optimized for the application requirements. Besides, adjustable speed drives can enable regenerative braking capability, which leads to higher system efficiency. The largest benefit of using adjustable speed drives comes with pumps, fans, escalators, cranes, and air-conditioning system. Motor drive systems running with on/off cycles, such as air compressors, conveyors, and refrigerators can benefit from adjustable speed drives.

Induction machines have been dominating the electric motor powered applications due to their low-cost manufacturing and self-starting capability. However, with the increasing demand for higher efficiencies, induction machines should be designed with smaller losses, and they need to operate with adjustable speed drives. These factors diminish the cost advantages of induction machines. Permanent magnet machines are being used more often these days in various applications. They have lower losses as compared to induction machines, and the system efficiency is improved with the use of adjustable speed drives. However, as explained previously in this chapter, the price volatility, supply chain issues and environmental concerns for the rare-earth materials, are significant issues for the permanent magnet machines as the demand increases.

Many motor drive applications are highly sensitive to initial purchasing cost. Industrial pumps and compressors, commercial HVAC systems, crane, and conveyors are suitable applications for SRM technology. Most of these applications are not sensitive to acoustic noise. SRM technology is also a good candidate for many of the residential applications such as refrigerators and freezers, top-loading residential clothes washers, room air conditioners, and swimming pool pumps.

Due to its wide constant power speed range, SRM is an excellent candidate for traction applications. High torque ripples, acoustic noise, and low torque density at low speeds have been the major obstacles in the adoption of SRMs in electrified drivetrain applications. However, increasing demand for electrified vehicles will put more pressure on developing reduced cost motor drive systems, which will open further opportunities for switched reluctance machines. Besides, especially for hybrid and plug-in hybrid electric vehicles, the manufacturers are moving towards higher speed motors to reduce the motor volume and improve the powertrain efficiency. This is an advantage for SRM since it provides higher efficiency at high speeds.

E-bikes represent one potential opportunity for SRM technology. As discussed earlier, the e-bike market is expected to grow drastically in the next few years. Today, permanent magnet machines dominate the e-bike motor drives. E-bikes are extremely cost sensitive, and manufacturers (mostly located in China) are putting significant effort into reducing the cost of motor drive systems. SRM could be a game changer in the e-bike market. Powertrain costs can be reduced significantly with SRM, offering a potential cost advantage to original equipment manufacturers (OEMs) to expand their market share.

Before we conclude this chapter, we would like to try to answer a major question: why SRM technology is not widely applied in motor drive applications? The answer is quite complicated, but we will try to explain it from a technology perspective and an industry perspective. If you look at the history of electric motor technologies, you would see that we're moving from complicated motor structures and simple drives towards simple motor structures and complicated drives; because the cost of both power electronics and digital control has been declining. In addition, government regulations have been pushing markets toward more efficient designs to reduce energy consumption and emissions. These forces place an emphasis on higher efficiency motors and better controllability.

For example, the brushed DC motor, the very first technology we learn when we start working on electric motors, has a complex structure. The armature circuit is on the rotor, and mechanical brushes rectify the armature current generated by the alternating field inside the motor. Hence, the brushed DC motor runs by applying DC voltage to the armature circuit. The motor drive systems have evolved to induction machines, brushless DC and, then, interior permanent magnet machines. Among these electric machine types, SRM has the simplest construction. But the controls and optimization are more complicated. As we will see throughout the book, with the advancements in modeling, analysis, optimization and control techniques, designing and implementing SRM based motor drives is no longer a complex task. The increasing demand for higher-efficiency and lower-cost motor drive system is already creating the need for switched reluctance motors.

From an industry perspective, the answer is more complicated. Besides the technical factors listed above, organizational, economic, and system level factors act as a barrier to the adoption of SRM drives. Switched reluctance motors are still at the stage of product development. This process has to go hand-in-hand with the system level analysis. Therefore, the interaction between the motor manufacturers, steel producers, OEMs, and designers has to be well established.

Today, electric motor wholesale sector prefers reducing types and number of electric motors to a minimum to reduce the capital cost of manufacturing and inventory [3].

In many cases, they offer general-purpose motors, which are mass manufactured and usually oversized for the application. In critical industrial and commercial applications, the customers keep multiple units in their inventory to prevent possible interruptions in their processes in case the electric motor fails. The large inventory of old, never-used motors is a factor causing delays and challenges in adopting new technologies.

Eighty percent of the electric motor sales are directly to wholesalers, distributors, and OEMs [3]. The wholesalers get the majority portion of the profits. Motor manufacturers are spending more effort and capital to reduce the cost of production of existing technologies rather than designing new technologies that could change the game. Therefore, there haven't been many efforts from the industry on developing low-cost SRM drives. But, it is changing due to the increasing demand for lower-cost and high-efficiency motor drive systems.

However, high volume applications, like washing machines and other consumer appliances often use specially designed motors to meet increasingly strict cost and efficiency targets. This process involves large volume contracts that the appliance manufacturers made directly with the motor manufacturer, bypassing the distributor to achieve lower costs. Automotive traction motors are also increasingly designed specifically to the automotive product requirements and purchased directly from the manufacturer or build in-house. As motor and drive design becomes easier and more cost effective, one can expect the cost and efficiency to continue to favour simpler motor and more complex drive custom designed to a given application. These are the areas where SRM technology is likely to become more prevalent in the future.

Last but not the least, SRM has not been well understood in our electric motor industry. With this highly technical book, we expect to provide an in-depth discussion of the multidisciplinary aspects of switched reluctance machines and help engineers design switched reluctance motor drives for various applications.

References

1. U.S. Energy Information Administration. Short-term energy outlook. December 6, 2016. [Online]. Available: http://www.eia.gov/. Accessed: December 14, 2016.
2. U.S. Energy Information Administration. Annual energy outlook 2016 with projections to 2040. August 2016. [Online]. Available: http://www/eia/gov/. Accessed: December 14, 2016.
3. P. Waide and C. U. Brunner, Energy-efficiency policy opportunities for electric motor-driven systems, International Energy Agency Working Paper, 2011.
4. U.S. Department of Energy, Advanced manufacturing office, premium efficiency motor selection and application guide – A handbook for industry. February, 2014. [Online]. Available: http://www.energy.gov/. Accessed: December 16, 2016.
5. B. Bilgin, P. Magne, P. Malysz, Y. Yang, V. Pantelic, M. Preindl, A. Korobkine, W. Jiang, M. Lawford, and A. Emadi, "Making the case for electrified transportation," *IEEE Transactions on Transportation Electrification*, vol. 1, no. 1, pp. 4–17, 2015.
6. M. Lowe, R. Golini, and G. Gereffi, U.S. Adoption of high-efficiency motors and drives: Lessons learned. February 25, 2010. [Online]. Available: http://www.cggc.duke.edu/. Accessed: December 17, 2016.
7. U.S. Energy Information Administration, Electricity use by machine drives varies significantly by manufacturing industry. October 8, 2013. [Online]. Available: http://www.eia.gov/. Accessed: December 17. 2016.

8. U.S. Department of Energy, Building technologies office, energy savings potential and opportunities for high-efficiency electric motors in residential and commercial equipment. December 2013. [Online]. Available: https://energy.gov/. Accessed: December 18, 2016.

9. U.S. Department of Energy, Heat pump systems. [Online]. Available: https://energy.gov/. Accessed: December 19, 2016.

10. U.S. Department of Energy, Opportunities for energy savings in the residential and commercial sectors with high-efficiency electric motors. December 1, 1999. [Online]. Available: https://energy.gov/. Accessed: December 20, 2016.

11. A. Bhatia, Centralized vs decentralized air conditioning systems, continuing education and development, Inc. [Online]. Available: http://seedenr.com/. Accessed: December 20, 2016.

12. YouTube Video, How a chiller tower and air handling unit work together. January 1, 2016. [Online]. Available: http://www.youtube.com/. Accessed: December 20, 2016.

13. International Energy Agency, Global EV outlook 2017 beyond one million electric cars. [Online]. Available: http://www.eia.org/. Accessed: March 28, 2018.

14. OICA, Motorization rate 2015 – Worldwide. [Online]. Available: http://www.oica.net/. Accessed: March 28, 2018.

15. U.S. Department of Transportation, Federal Highway Administration, Office of Highway Policy Information, Highway statistics 2016. [Online]. Available: https://www.fhwa.dot.gov/. Accessed: March 28, 2018.

16. Statistica, Light Vehicle retail sales in the United States from 1977 to 2017 (in 1,000 units). [Online]. Available: http://www.statistica.com/. Accessed: March 28, 2018.

17. B. Bilgin and A. Emadi, Electric motors in electrified transportation: A step toward achieving a sustainable and highly efficient transportation system, *IEEE Power Electronics Magazine*, vol. 1, no. 2, pp. 10–17, 2014.

18. Bosch sees future requiring multiple powertrain technologies; the larger the vehicle, the more the electrification. June 18, 2013. [Online]. Available: http://www.greencarcongress.com/.

19. R. Citron and J. Gartner, Executive summary: Electric bicycles, throttle-control and pedal-assist e-bicycles, batteries, and motors: Global market opportunities, barriers, technology issues, and demand forecasts, Navigant Research. 2014. [Online]. Available: http://www.navigantresearch.com/. Accessed: January 11, 2017.

20. What are electric bike classes and why do they matter? [Online]. Available: https://electricbikereview.com/. Accessed: January 12, 2017.

21. Electric bike laws in the United States. [Online]. Available: https://electricbikereview.com/. Accessed: January 12, 2017.

22. What's the difference between electric bike motors? [Online]. Available: https://electricbikereview.com/. Accessed: January 12, 2017.

23. Freedonia, World electric motors – Demand and sales forecasts, market share, market size, market leaders. [Online]. Available: http://www.freedoniagroup.com/. Accessed: January 19, 2017.

24. HIS Markit, CWIEME Berlin 2017 Electric Motor Market Update, Presented by: Andrew Orbinson, Senior Analyst IHIS Markit Industrial Automation. [Online]. Available: http://www.coilwindingexpo.com/. Accessed: March 29, 2018.

25. B. Bilgin and A. Sathyan, Fundamentals of electric machines, *Advanced Electric Drive Vehicles*, CRC Press, Boca Raton, FL, 2014.

26. J. M. Miller, Electric Motor R&D, 2013 U.S. DOE Hydrogen and Fuel Cells Program and Vehicle Technologies Program Annual Merit Review and Peer Evaluation Meeting, Oak Ridge National Laboratory. May 15, 2013. [Online]. Available: http://energy.gov/. Accessed: January 26, 2017.

27. U.S. Department of Energy, Critical Materials Strategy. December 2011. [Online]. Available: http://energy.gov/. Accessed: January 26, 2017.

28. T. Abraham and B. L. Gupta, Continued growth for permanent magnets, *Ceramic Industry*, vol. 164, no. 10, pp. 19–21, 2014.

29. C. Pathemore, *Elements of Security: Mitigating the Risks of U.S. Dependence on Critical Minerals, 2011*. Washington, DC: Center for a New American Security.

30. United States Geological Survey (USGS). 2014 mineral yearbook, rare earths [advance release], December 2016. [Online]. Available: http://www.usgs.gov/. Accessed: January 25, 2017.

31. Resnick Institute for Sustainable Energy Science. Critical materials for sustainable energy applications, August 2011. Pasadena, CA.

32. G. C. Mechler, Manufacturing and cost analysis for aluminum and copper die cast induction motors for GM's powertrain and R&D divisions, MS thesis, Department Materials Science and Engineering, Massachusetts Institute of Technology, Cambridge, MA, 2010.

33. Y. Yang, B. Bilgin, M. Kasprzak, S. Nalakath, H. Sadek, M. Preindl, J. Cotton, N. Schofield, and A. Emadi, Thermal management of electric machines, *IET Electrical Systems in Transportation*, doi:10.1049/iet-est.2015.0050. 2016.

34. N. J. Nagel, Fundamentals of electric motor control, in *Advanced Electric Drive Vehicles*, CRC Press, Boca Raton, FL, 2014.

2

Electromagnetic Principles of Switched Reluctance Machines

Berker Bilgin

CONTENTS

A Switched Reluctance Machine (SRM) has salient poles both on the rotor and stator. Figure 2.1 shows the cross section view of a typical SRM. The rotor is made of ferromagnetic material and does not include any kind of excitation source. The stator is also made of ferromagnetic material, similar to the rotor, and it has coils wound around each stator pole.

In a switched reluctance motor, electromagnetic forces create the motion. Similar to other types of electric machines, the operational principles of an SRM can be explained with the laws of electricity and magnetism. In this chapter, we will first explain how the stator pole generates the electromagnetic field and then how the forces are exerted on the rotor by quantifying the fundamental principles of electromagnetics. Then, we will investigate the electromechanical energy conversion principles.

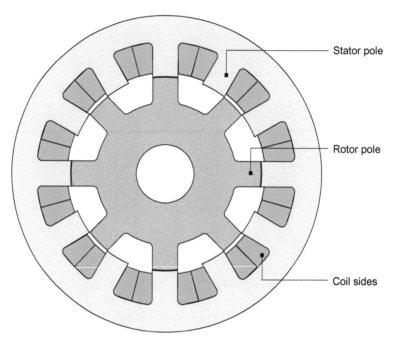

FIGURE 2.1
Cross section view of a switched reluctance machine.

2.1 Magnetic Fields

Consider the bar magnet shown in Figure 2.2. The north and south poles on the bar magnet generate a magnetic field, where the direction of the field is from the north pole to the south pole by convention.

The similar magnetic field configuration can be created with an electromagnet, where the magnetic field is produced by an electric current. Figure 2.3a shows a simple electromagnet, where the current carrying coil is wound around an iron core. Probably you already

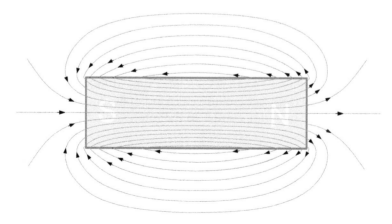

FIGURE 2.2
Magnetic field of a bar magnet.

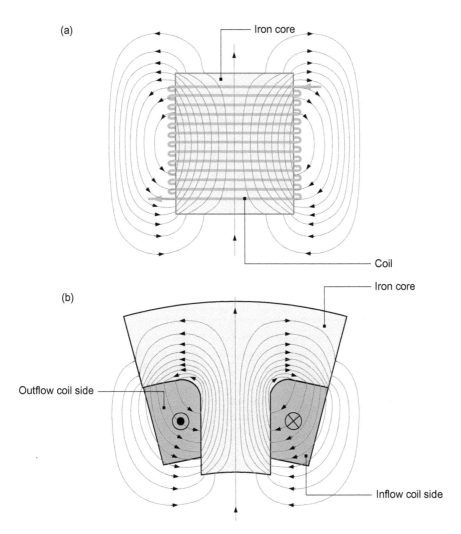

FIGURE 2.3
(a) Simple electromagnet and (b) stator pole of an SRM as an electromagnet.

observed that the electromagnet in Figure 2.3a is very similar to the stator pole of an SRM in Figure 2.1. As shown in Figure 2.3b, similar magnetic field lines can be observed when the coil around a stator pole of an SRM is excited. The magnetic field lines curls from the top of the iron core to the bottom. Similar to the bar magnet in Figure 2.2, the magnetic field lines are more concentrated within the core; hence, the magnetic field is stronger inside the iron core. In addition, the flux lines refract more when leaving the iron cores in Figure 2.3 as compared to the case with the bar magnet in Figure 2.2. The reason for these is related to the permeability of the iron core and it will explained in the next sections.

It can be observed from Figure 2.3b that, in order to generate the magnetic field in the same direction as the bar magnet, the current flows from the right side of the core and comes out the left side. The French physicist André-Marie Ampère quantified this phenomenon in 1826. Ampère's law states that a magnetic field is generated around a current-carrying conductor, and the direction of the magnetic field can be found by the right-hand rule. Right-hand rule is a fundamental theorem used in Calculus. It gives the position of the final vector when two vectors are exposed to vector multiplication. The three-dimensional representation of a vector is defined as:

$$\vec{A} = A_x\hat{x} + A_y\hat{y} + A_z\hat{z} \tag{2.1}$$

where $\hat{x}$, $\hat{y}$, and $\hat{z}$ are the unit vectors parallel to the x, y, and z axis, respectively. $A_x, A_y,$ and A_z are the magnitudes of the components of $\vec{A}$ in each axis. The vector product of two individual vectors is calculated as:

$$\vec{A} \times \vec{B} = \begin{vmatrix} \hat{x} & \hat{y} & \hat{z} \\ A_x & A_y & A_z \\ B_x & B_y & B_z \end{vmatrix} = \left(A_y B_z - A_z B_y\right)\hat{x} + \left(A_z B_x - A_x B_z\right)\hat{y} + \left(A_x B_y - A_y B_x\right)\hat{z}. \tag{2.2}$$

Figure 2.4 shows how right-hand rule is applied to find the direction of the magnetic field generated by a current carrying conductor. If the thumb of your right hand shows the direction of the current, then the remaining four fingers show the direction of the curl of the magnetic field. In this case, the current flows into the page and the magnetic field curls in the clockwise direction.

Now, we can identify how the magnetic field in the electromagnet in Figure 2.3 is generated. In Figure 2.3b, the current flows into the page in the coil on the right (inflow coil side). When the right-hand rule is applied, the magnetic field curls in a clockwise direction. The same current exits the coil side on the left (outflow coil side). The magnetic field then curls in a counterclockwise direction.

From Figure 2.4, it can be observed that as the distance from the coil increases, the thickness of the magnetic field lines reduces. This represents that the magnitude of

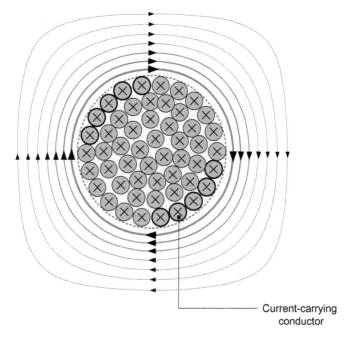

Current-carrying
conductor

FIGURE 2.4
Magnetic field generated by a current carrying conductor.

the magnetic field, $\vec{B}$ decreases as the distance from the wire increases. This has been quantified in Biot-Savart law:

$$\vec{B}(\vec{r}) = \frac{\mu_0}{4\pi} I \int \frac{d\vec{l}_w \times \vec{r}}{r^2} \tag{2.3}$$

where I is the magnitude of the steady current, $d\vec{l}_w$ is the element along the length of the wire, $\vec{r}$ is the displacement vector from the current source, and μ_0 is a constant and defined as the permeability of free space:

$$\mu_0 = 4\pi \times 10^{-7} \ \text{N/A}^2. \tag{2.4}$$

The unit of a magnetic field is equal to $[\text{N/Am}]$ and it is called a tesla. Biot-Savart law states that the magnetic field vector created by a steady current depends on the magnitude, direction, length, and proximity of the electric current. Steady current here refers to a continuous flow of electrons through a conductor without any change in motion or, in other words, without picking up new electrons into the motion.

Magnetic fields have zero divergence and non-zero curl, which are two important characteristics of vector fields. The mathematical expressions for the divergence and curl of the magnetic field are used in the derivation of forces in electromagnetic field and they will be explained in the following sections.

2.2 Curl of Magnetic Fields

It was clear from Figure 2.4 that the magnetic fields circulate upon themselves. The curl of magnetic field is explained by Ampère's law. As shown in Figure 2.5, if we define a surface enclosing the total current in Figure 2.4, Ampère's law states that the line integral around the surface, which the total current passes through, is a constant, and this constant is independent of the shape and radius of the surface.

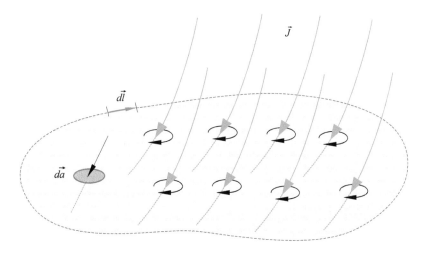

FIGURE 2.5
Geometric representation of Ampère's law.

Therefore, the curl of a magnetic field can be expressed as:

$$\nabla \times \vec{B} = \mu_0 \vec{J} \tag{2.5}$$

where $\vec{J}$ is the current density, which is the amount of charge passing through a unit area. Current density is in units of A/m^2 or C/m^2s. $\nabla \times \vec{B}$ is the curl operator and for a three-dimensional magnetic field vector, it is expressed as:

$$\nabla \times \vec{B} = \left(\frac{\partial B_z}{\partial y} - \frac{\partial B_y}{\partial z} \right) \hat{x} + \left(\frac{\partial B_x}{\partial z} - \frac{\partial B_z}{\partial x} \right) \hat{y} + \left(\frac{\partial B_y}{\partial x} - \frac{\partial B_x}{\partial y} \right) \hat{z}. \tag{2.6}$$

The curl of a vector field is defined as the measure of the tendency of the field to circulate about a point. All magnetic fields circulate upon themselves. However, this doesn't mean that the curl of the magnetic vector field is nonzero whenever it curves. The curl of the magnetic field vector is nonzero at the location where the electric current flows. Away from that location, the field vector still curves, but the curl at any point is zero. Since the magnitude of the electromagnetic field vector reduces with the distance from the source current, as quantified by the Biot-Savart law in (2.3), the reduction in amplitude exactly compensates for the curvature of the field lines. Therefore, if you put a small paddle wheel at a location away from the surface through which the current flows, the push from different sides of the paddle wheel will result in different directions and different magnitudes. They compensate each other, and the total net push will be zero [1].

Equation 2.5 expresses Ampère's law in differential form. It can also be defined in integral form using Stokes's theorem. In Figure 2.5, the current flows through the surface and, hence, the direction of all current density vectors is into the page. When the right-hand rule is applied to each current density vector, the microscopic magnetic fields would be all rotating clockwise. The surface normal vectors of these curls are also in the direction of the current density vector. Thus, the sum of their surface integrals cancels each other inside the surface. Then only the integration over the surface boundary is left. Therefore, Stokes's theorem states that

$$\int \left(\nabla \times \vec{B} \right) d\vec{a} = \oint \vec{B} \cdot d\vec{l} = \mu_0 \int \vec{J} \cdot d\vec{a} \tag{2.7}$$

since $\int \vec{J} d\vec{a}$ is the total current that passes through the surface, I_{enc}, the integral form of Ampère's law can be expressed as:

$$\oint \vec{B} \cdot d\vec{l} = \mu_0 I_{enc} \tag{2.8}$$

where $d\vec{l}$ is the integration component along the circumference of the amperian loop, which is the loop that cover the surface the current flows through, as shown in Figure 2.5.

Equation 2.8 states that an electric current flowing through a surface produces a circulating magnetic field around any path that bounds that surface (amperian loop). James Clerk Maxwell quantified that time-changing electric flux through the same surface also generates circulating magnetic field. Therefore, the curl of magnetic field can be expressed with the Ampère-Maxwell law as:

$$\oint \vec{B} \cdot d\vec{l} = \mu_0 \left(I_{enc} + \epsilon_0 \frac{d}{dt} \int_s \vec{E} \cdot d\vec{a} \right) \tag{2.9}$$

where ϵ_0 is the permittivity of free space and $\int_s \vec{E} \cdot d\vec{a}$ is the electric flux through the surface bounded by the closed path used for calculating the closed-loop line integral in $\oint \vec{B} \cdot d\vec{l}$. When analyzing the magnetic characteristics of electric machines, the effect of time-changing electric flux in (2.9) is usually negligible.

2.3 Divergence of Magnetic Fields

Magnetic field lines circulate around themselves and they turn back to the point where they started. Divergence is a measure of how much a vector spreads out (diverges) from the point where it started. Hence, the divergence of magnetic fields is always zero. This can be expressed by Gauss's law for magnetic fields in differential form as:

$$\nabla \cdot \vec{B} = 0. \tag{2.10}$$

The divergence of a vector is a scalar, and, for a three-dimensional magnetic field vector, it is defined as:

$$\nabla \cdot \vec{B} = \frac{\partial B_x}{\partial x} + \frac{\partial B_y}{\partial y} + \frac{\partial B_z}{\partial z}. \tag{2.11}$$

The divergence of magnetic field in (2.10) can be expressed in integral form as:

$$\oint_s \vec{B} \cdot d\vec{a} = 0, \tag{2.12}$$

which states that the magnetic flux of a magnetic field vector going through a closed surface is zero. Since the magnetic field lines circulate, the amount of magnetic flux going into the closed surface and out of the closed surface will be equal, but in opposite directions. Therefore the net magnetic flux through any closed surface will be zero. This is not the case for electric fields. Gauss's law for electric fields can be expressed in integral form as:

$$\oint_s \vec{E} \cdot d\vec{a} = \frac{q_{enc}}{\epsilon_0}. \tag{2.13}$$

Gauss's law for electric field states that the flux of an electric field going through any closed surface is proportional to the total charge within that surface. The differential form of

Gauss's law shows that the divergence of electric field is non-zero and it is proportional to the charge density ρ, which is expressed in coulombs per cubic meter:

$$\nabla \cdot \vec{E} = \frac{1}{\epsilon_0} \rho. \tag{2.14}$$

The difference between the divergence of magnetic and electric fields in (2.12) and (2.14) shows that the opposite electric charges may be isolated from each other and the electric field diverges from positive charge and converges to the negative charge. However, opposite magnetic poles always come in pairs and magnetic fields circulate around themselves. Therefore, magnetic fields have zero divergence and electric fields have non-zero divergence. Magnetic fields have non-zero curl and electric fields have zero curl:

$$\nabla \times \vec{E} = 0. \tag{2.15}$$

2.4 Magnetic Field in Ferromagnetic Materials

The stator pole of a switched reluctance machine looks very similar to the electromagnet as shown in Figure 2.3. So far, we have analyzed how the stator pole of an SRM generates the magnetic field. As mentioned before, the magnetic flux lines are denser and, hence, the magnetic field is stronger inside the core of the electromagnet.

The electromagnet was created by winding the current carrying conductor around an iron core. Iron is a ferromagnetic material. Due to their magnetic properties, when exposed to an external field, the magnetic field inside a ferromagnetic material is different when compared to the field outside of the material.

Magnetic fields always occur as dipoles. The same principle applies in the atomic scale as well. Electrons spinning around their axis create small currents, which create a magnetic dipole moment [2]. Atoms and molecules with magnetic dipole moments can be regarded as small magnets. When exposed to an external magnetic field, magnetic dipoles align themselves in the direction of the external magnetic field. The greater the magnetic dipole moment of a material is, the better the magnetic dipoles align themselves with the external magnetic field.

In ferromagnetic materials, such as iron, there are domains whereby the magnetic dipoles are 100% aligned. These domains are uniformly distributed in their neighborhood, but the net internal magnetic field is zero. When exposed to an external magnetic field, these domains align their polarity in union in the direction of the external field. Therefore, the magnetic field inside the ferromagnetic material becomes stronger than the external field.

The field inside the material is called magnetic flux density $\vec{B}$, and it is measured in units of tesla. The external field is called magnetic field intensity $\vec{H}$, and it is measured in units of A/m. When an external magnetic field is applied, the magnetic field inside the ferromagnetic material $\vec{B}$ will be different from the magnetic field outside of the material $\vec{H}$. The change is dependent on the characteristics of the material and it is defined as the permeability:

$$\vec{B} = \mu_r \mu_o \vec{H} \tag{2.16}$$

where μ_0 is the permeability of free space as defined in (2.4). μ_r is the relative permeability and it equals to 1 for air. For ferromagnetic materials, μ_r can be thousand-fold higher than that of air and it shows that field inside the ferromagnetic material will be much higher than the external field. Remember the difference in the refraction of the magnetic flux lines in Figures 2.2 and 2.3. The reason is related to the difference in the relative permeability of the bar magnet and the iron core. The relative permeability of a permanent magnet is around 1.05–1.1, which is very close to that of air. The relative permeability of iron can be thousand-fold higher than air. In Figure 2.2, since the relative permeability of the bar magnet and air are very close, the flux lines leave the magnet without much refraction. However, in Figure 2.3, when the flux lines leave the iron core, they see a significant change in relative permeability and this causes refraction of the flux lines.

Now we know the effect of ferromagnetic materials on the magnetic field and we discussed this on an electromagnet, which is similar to the stator pole of an SRM. The rotor pole of an SRM is also made of ferromagnetic material. Now you can imagine that if we bring the rotor pole in the vicinity of the stator pole, magnetic field configuration in Figure 2.3b will change.

When we bring the rotor pole in the vicinity of the stator pole, the ferromagnetic material of the rotor pole sees the external field from the stator pole (electromagnet). So, the domains in the rotor pole try to align themselves with the external field. Ferromagnetic material of the rotor is magnetized, and magnetic poles are generated. The strength of the magnetization is dependent on the permeability of the ferromagnetic material. The magnetic dipoles align themselves in the direction of the external field and the magnetic field inside the ferromagnetic material becomes stronger. The magnetic field configuration now changes as shown in Figure 2.6 and the magnetic field lines are sucked into the rotor pole.

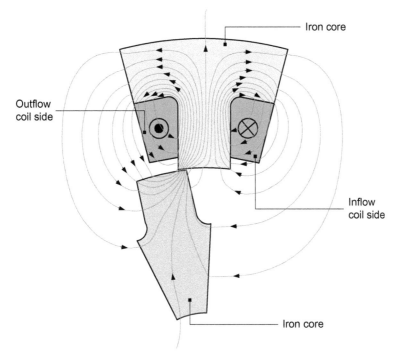

FIGURE 2.6
Magnetic field when there is ferromagnetic material in the vicinity of an electromagnet.

When the rotor pole (ferromagnetic material) approaches the vicinity of the stator pole (electromagnet), the change in the magnetic field configuration is due to the change in the reluctance of the magnetic path. This is due to the higher relative permeability of iron μ_r. Since the relative permeability of air is one, it shows higher reluctance to the magnetic flux. In an SRM, when the rotor pole gets closer to the stator pole, the length of the air gap between the stator and rotor poles decreases and the total path of the magnetic flux will include more and more iron. Therefore, the reluctance that the external field $\vec{H}$ sees decreases. This is where the name "reluctance machine" comes from. Now you can imagine that when the stator pole is excited with current, the rotor pole of an SRM will be attracted towards to the stator pole to reach to an alignment. At the aligned position, the air gap length between the stator and rotor poles are the smallest and, hence, the reluctance is smallest. Before or beyond the aligned position, the reluctance of the magnetic circuit is higher, and the rotor will tend to stay in this equilibrium point.

It should be noted that the increase in the external field does not create an indefinite increase in the magnetic flux density. There is a certain point where the increase in the external field doesn't increase the magnetic field inside the ferromagnetic material. This is the state where nearly all the domains are completely aligned inside the material and it is called saturation. Therefore, the relative permeability μ_r of a ferromagnetic material is not constant as the magnetic field intensity increases. As shown in Figure 2.7 with the magnetization curve of a typical ferromagnetic material, the relative permeability reduces and approaches that of air at higher level of magnetic field intensity. This results in a slower increase in the magnetic field inside the ferromagnetic material when the external field gets stronger. As it will be discussed later, the torque production capability of an SRM increases when working in the saturated region.

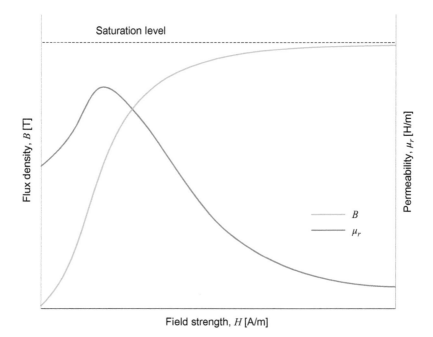

FIGURE 2.7
Typical magnetization (B-H) and permeability (μ_r-H) curves.

2.5 Lorentz Force Law

As shown in Figure 2.3, when current is applied to an electromagnet, north and south poles are generated, and the direction of the magnetic field can be found by the right-hand rule. As shown in Figure 2.6, when a ferromagnetic material approaches the vicinity of an electromagnet, the dipoles align themselves with the external field and north and south magnetic poles are generated on the ferromagnetic material. The magnetic poles on the stator pole and the rotor pole in Figure 2.6 generate forces. These forces are the main sources of torque (rotation) and radial forces (vibration and acoustic noise) in an SRM and they are derived using Lorentz Force law.

In 1892, the Dutch physicist Hendrik Antoon Lorentz quantified that the magnetic fields created by moving charges generate force:

$$\vec{F} = q\left[\vec{E} + \left(\vec{v} \times \vec{B}\right)\right] \tag{2.17}$$

where q is the charge moving with a velocity of $\vec{v}$. The first part of the equation is Coulomb's law and it shows the force exerted on the charge q in the electric field $\vec{E}$. The velocity vector is in the direction of the movement of the charge. Therefore, from the second part of (2.17), the magnetic component of the force will be perpendicular to the velocity vector due to the cross product. This shows that the magnetic component of the Lorentz Force cannot change the kinetic energy of a charge, but it can change its direction [3].

Equation 2.17 represents the force acting on a unit charge. If ρ is the charge per unit volume in units of C/m^3 and $q = \rho d\tau$, then the electromagnetic force acting on charges on volume V can be expressed as:

$$\vec{F}_V = \int_V \left(\vec{E} + \vec{v} \times \vec{B}\right)\rho \, d\tau = \int_V \left(\rho\vec{E} + \rho\vec{v} \times \vec{B}\right) d\tau \tag{2.18}$$

where

$$\underbrace{\rho}_{\frac{C}{m^3}} \underbrace{\vec{v}}_{\frac{m}{s}} = \underbrace{\vec{J}}_{\frac{C}{s}\frac{1}{m^2} = \frac{A}{m^2}} . \tag{2.19}$$

Therefore, $\vec{J}$ is the current density vector. Then,

$$\vec{F}_V = \int_V \left(\rho\vec{E} + \vec{J} \times \vec{B}\right) d\tau. \tag{2.20}$$

Hence, force per unit volume can be expressed as:

$$\vec{f} = \rho\vec{E} + \vec{J} \times \vec{B}. \tag{2.21}$$

From (2.21), it can be observed that the magnetic part of the Lorenz Force, which is the main part of the forces in electric machines, is the cross product of the current density vector $\vec{J}$ and the flux density vector $\vec{B}$. The highest amount of force is generated when these two fields are perpendicular to each other.

The north and south poles on the stator pole and the rotor pole in Figure 2.6 generate force and, considering the stator pole is mechanically fixed, the rotor pole is attracted towards the stator pole. If the rotor pole is connected on a rotating shaft, this force will create tangential and radial components, which are calculated by using the Maxwell stress tensor. The tangential component of the Maxwell stress tensor, T_t is used to calculate the torque and the radial component, T_r is used to calculate the acoustic noise and vibration:

$$T_r = \frac{1}{2\mu_0}\left(B_r^2 - B_t^2\right) \tag{2.22}$$

$$T_t = \frac{1}{\mu_0}\left(B_r B_t\right). \tag{2.23}$$

where B_r and B_t are the radial and tangential component of the magnetic flux density vector. Appendix 2.1 shows how the tangential and radial component of the Maxwell stress tensor is calculated using principles of vector calculus and Maxwell's equations. It can be observed that the calculation starts from the Lorentz Force law in (2.17), which constitutes the principle of force generation mechanism in an SRM.

2.6 Electromagnetic Induction and Faraday's Law

As quantified by Ampère's law, steady currents flowing through a conductor generate a steady magnetic field around it. The magnetic field lines circulate upon themselves and the direction of the magnetic field can be found by using the right-hand rule (see Figure 2.4). Biot-Savart law in (2.3) quantifies that the magnitude of this magnetic field reduces as the distance from the current carrying conductor increases.

The British physicist Michael Faraday questioned whether a steady magnetic field also generates steady currents and in 1831 he proved that it is not the steady magnetic field, but a time-changing magnetic field that generates current. When the time-changing magnetic flux passes through the surface enclosed by the conducting loop, an electric field and, hence, electromotive force is induced and a current flows around the wire loop. This is called electromagnetic induction. In 1833, the Russian physicist Heinrich Lenz found that, when time-changing magnetic flux flows through a surface, the current that flows in the conducting loop enclosing that surface moves in a certain direction to oppose the change in flux. Therefore, if the magnetic flux through the surface is decreasing, the induced current produces its own magnetic flux opposing the decrease. Figure 2.8 depicts the graphical illustration of Faraday and Lenz's law. If the magnetic flux (φ) coming out of the surface enclosed by the conducting loop is decreasing in time (imagine a bar magnet at the bottom of the conducting loop is moving away from the loop), the induced current flows in the counter clockwise direction. The current induces its own magnetic flux (φ_{ind}) and as per right-hand rule, the induced flux (φ_{ind}) will be out of the surface again to offset the decrease (your four fingers show the direction of the current flowing around the loop and your thumb shows the direction of the induced magnetic flux). If the magnetic flux (φ) coming out of the surface inside the wire loop was increasing with time (image the bar magnet at the bottom of the loop is now approaching towards the loop), the current would flow in the clockwise direction, so that the induced magnetic flux (φ_{ind}) was into the surface as per the

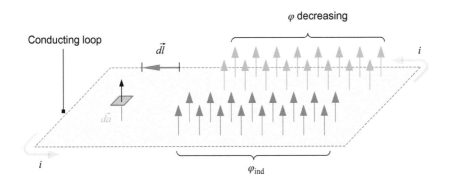

FIGURE 2.8
Graphical representation of Faraday and Lenz's law.

right-hand rule to offset the increase. It can be observed that the magnetic field is actually fighting against the change and behaves like it has its own inertia. This will be explained later when we discuss about inductance.

Equation 2.24 quantifies Faraday and Lenz's law in integral form:

$$\oint \vec{E} \cdot d\vec{l} = -\frac{d}{dt}\int_s \vec{B} \cdot d\vec{a}. \tag{2.24}$$

Faraday's law states that time-changing magnetic flux through a surface induces a circulating electric field around the closed path enclosing that surface. Please note the expression "circulating electric field" carefully. As it has been discussed before, charge-based electric fields originate from the positive charge and terminate on the negative charge. Divergence is a measure of how much a vector spreads out (diverges) from the point it started. Hence, charged-based electric fields have non-zero divergence and zero curl, as seen in Equations 2.14 and 2.15. However, electric fields generated by changing magnetic fields circulate back on themselves. Therefore, they have zero divergence and non-zero curl as expressed by Faraday's law in differential form:

$$\nabla \times \vec{E} = -\frac{\partial \vec{B}}{\partial t}. \tag{2.25}$$

The circulating electric field generated by the time-changing magnetic flux is capable of moving charges around the conducting loop that it circulates. Since the move of charges is the definition of electric current, the induced electric field acts like the electromotive force ε, which creates electric current:

$$\varepsilon = \oint \vec{E} \cdot d\vec{l}. \tag{2.26}$$

One important question to ask here is when time-changing flux through a surface induces circulating electric field, which way does the current generated by induced electric field flows? The answer is the minus sign on the right-hand side of Faraday's law in (2.24) and (2.25). The minus sign is Lenz's law and it states that the current induced by the time-changing magnetic flux flows in a certain direction to oppose the change. Therefore, in

Figure 2.8, if the magnetic flux (φ) flowing through the surface is out of the page and it is decreasing in time, the current i flows in counter clockwise direction. So when the right-hand rule is applied, the flux generated by the induced current (φ_{ind}) supports the direction of flux φ to offset the decrease.

In (2.24), since $\int_s \vec{B} \cdot d\vec{a}$ is the magnetic flux and $\oint \vec{E} \cdot d\vec{l}$ is the electromotive force as given in (2.26), Faraday's law can be expressed in a simpler way, which is heavily used in electrical engineering:

$$\varepsilon = -\frac{d\varphi}{dt}. \tag{2.27}$$

It should be noted that the minus sign tells us the direction of the circulation of the induced electric field and, hence, the direction of the current. The magnitude of the electromotive force is related to the rate of change magnetic flux.

2.7 Electric Current and Equivalent Circuit

The other important question to ask here is how the induced electric field moves the charges to generate the current. The induced electric field exerts force on the charge and moves it around the conducting loop in Figure 2.8. As quantified by French physicist Charles-Augustin de Coulomb in 1783, the electric field applies force on the charges and moves them through the conducting loop:

$$\vec{F}_e = q\vec{E}. \tag{2.28}$$

The induced electric field moves the charges (free electrons) around the wire loop to make the electric field zero. If the electromotive force is maintained by constantly applying changing magnetic flux through surface inside the conducting loop or wire loop, continuous current flow can be maintained. The charges accelerate with the force from the induced electric field:

$$\vec{a}_e = \frac{\vec{F}_e}{m_e} \tag{2.29}$$

where m_e is the mass of an electron. You probably already observed that (2.29) is Newton's second law of motion. Without an electric field applied, the free electrons in a conductor have thermal motion in all directions. Free electrons in copper have an average speed around $\langle \vec{v}_e \rangle = 10^6 \, m/s$ and, therefore, time between the collusion of free electrons with the atoms is around $\tau = 3 \times 10^{-14} \, s$ for copper at room temperature. When the force in (2.29) is applied, the free electrons pick up a speed between the collusion, which is called as the drift velocity:

$$\vec{v}_d = \vec{a}_e \tau = \frac{\vec{F}_e}{m_e} \tau = \frac{q\vec{E}}{m_e} \tau. \tag{2.30}$$

If the number of free electrons or charges per cubic meter is n_e, the electric current can be defined as the total charge flowing through the surface S with the drift velocity $\vec{v}_d$:

$$i = S v_d \, n_e \, q. \tag{2.31}$$

The unit of current is coulomb per second. Combining (2.30) and (2.31) the relationship between the electric field and current can be derived as:

$$i = \frac{S q^2 \tau \, n_e}{m_e} \vec{E}. \tag{2.32}$$

The electromotive force is the line integral of the electric field as given in (2.26). Hence, it can be expressed as:

$$E = \frac{\varepsilon}{\ell} \tag{2.33}$$

where ℓ is the length of the loop in which the induced current flows. Hence, (2.32) can be expressed as:

$$i = \frac{S q^2 \tau \, n_e}{m_e} \frac{\varepsilon}{\ell}. \tag{2.34}$$

Then the relationship between the induced electromotive force and current can be expressed as:

$$\frac{\varepsilon}{i} = \frac{m_e}{q^2 n_e \tau} \frac{\ell}{S}. \tag{2.35}$$

The first term on the right side of the equation in (2.35) is a constant and it is related to the characteristics of the material conducting the electric current. It is defined as the resistivity:

$$\rho = \frac{m_e}{q^2 n_e \tau}. \tag{2.36}$$

Therefore, (2.35) can be reorganized as

$$\frac{\varepsilon}{I} = \rho \frac{\ell}{S} \tag{2.37}$$

where the term on the right-hand side is the resistance of the conductor the electric current flows through. The relationship between the electromotive force and current is dependent on the resistivity of the conducting material, and the cross-section area and the length of the conductor. This relationship was quantified by the German physicist Georg Ohm in 1827 and it is called Ohm's law:

$$\frac{\epsilon}{I} = R \tag{2.38}$$

where R is the resistance of the wire loop in units of ohm $[\Omega]$. Ohm's law states a linear relationship between the electromotive force and the current. It should be noted here that the resistivity of a conductor in (2.36) is strongly dependent on the temperature. When the temperature increases, time between the collusion of free electrons τ reduces because the speed of the free electrons goes up. Therefore, when the temperature increases, the resistivity and, hence, the resistance increases.

If the wire loop in Figure 2.8 is connected to a voltage source, its equivalent circuit can be represented as in Figure 2.9. Here, R represents the resistance of the wire loop, $d\vec{a}$ is out of the page, and $d\vec{l}$ is in the direction of current as per right-hand rule.

To analyze the circuit in Figure 2.9, we need to calculate the line integral of electric field around the circuit [4]. The electric field inside the voltage source is from the positive terminal to the negative terminal. Since $d\vec{l}$ is in a counterclockwise direction, if we follow the direction of the current, the contribution of the voltage source to the line integral of the electric field will be $-v$. The electric field inside the resistive element is in the direction of the current; therefore $\vec{E} \cdot d\vec{l}$ is positive and equal to the voltage drop on the resistor, $+iR$. When Faraday's law in (2.24) is applied, the closed-loop line integral of the electric field in the circuit can be calculated as:

$$\oint \vec{E} \cdot d\vec{l} = -v + iR = -\frac{d}{dt} \int \vec{B} \cdot d\vec{a} = -\frac{d\varphi}{dt}. \tag{2.39}$$

Therefore, the voltage equation can be expressed as:

$$v = Ri + \frac{d\varphi}{dt}. \tag{2.40}$$

It should be noted that the surface in Figure 2.9 is a wire loop with a single turn. In the case of more number of turns, N, the surface can be considered similar to a spiral staircase, where the magnetic flux goes through multiple layers. Therefore, if the loop has N turns, the magnetic flux goes N times though the same surface, and the induced electromotive force will be N times larger. Hence, (2.40) can be written as:

$$v = Ri + N\frac{d\varphi}{dt} Ri + \frac{d(N\varphi)}{dt} = Ri + \frac{d\lambda}{dt} \tag{2.41}$$

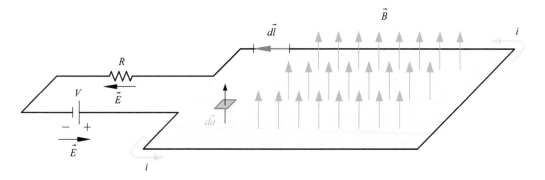

FIGURE 2.9
Equivalent circuit of a wire loop connected to a voltage source.

where λ is called the flux linkage and it defines the total magnetic flux linked with the circuit. Equation (2.41) is the equivalent model for a magnetic circuit. It also represents the model of a single-phase of an SRM. As you will see throughout this book, the flux linkage λ of an SRM is dependent on the current and the relative position between the stator and rotor poles. This makes the equivalent circuit model of an SRM inherently nonlinear.

2.8 Inductance

As shown in (2.41), when a voltage is applied to a magnetic circuit, the flux linkage and, hence, the current does not change immediately. The rate of change of current is dependent on the characteristics of the magnetic circuit called inductance.

As given in (2.41), the equivalent model of the magnetic circuit in Figure 2.10a can be expressed as:

$$v = Ri + N\frac{d\varphi}{dt}. \tag{2.42}$$

As stated by Ampère's law in (2.8), current applied to the current carrying conductor generates magnetic field. If the current carrying conductor is not a ferromagnetic material $\mu_r = 1$, Ampère's law can be expressed as in (2.43) to calculate the external field:

$$\oint \vec{B} \cdot d\vec{l} = \mu_0 I_{enc} \Rightarrow \oint \vec{H} \cdot d\vec{l} = I_{enc} \Rightarrow \oint \vec{H} \cdot d\vec{l} = Ni. \tag{2.43}$$

If we have N-turn conductor wound around a core made of a ferromagnetic material, equation (2.43) states that the current applied to the conductor will generate the external magnetic field and the magnetic flux will circulate inside the core over a closed-loop line with length l_c as shown in Figure 2.10:

$$Hl_c = Ni \Rightarrow H = \frac{Ni}{l_c}. \tag{2.44}$$

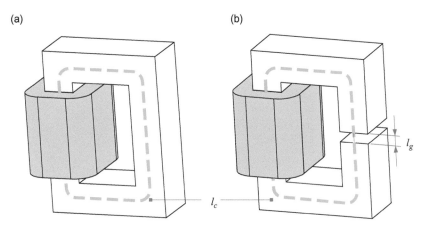

(a)   (b)

FIGURE 2.10
Simple magnetic circuit: (a) without air gap and (b) with air gap.

As given in (2.16), the relationship between the magnetic field intensity, H, and magnetic field density, B, for a magnetic material is defined as:

$$B = \mu_r \mu_0 H \tag{2.45}$$

where B is the flux per unit area. Therefore,

$$\varphi = B A_c \tag{2.46}$$

where A_c is the cross-section area through the flux flows. Combining (2.44), (2.45), and (2.46) results in:

$$\varphi = B A_c = \mu_r \mu_0 H A_c = \mu_r \mu_0 N \frac{A_c}{l_c} i. \tag{2.47}$$

Now we can apply (2.47) in the equivalent model in (2.42):

$$v = Ri + N \frac{d\varphi}{dt} = Ri + N \frac{d}{dt} \left(\mu_r \mu_0 N \frac{A_c}{l_c} i \right). \tag{2.48}$$

Assuming that the material related parameters (e.g. relative permeability) and geometry dependent parameters (e.g. number of turns, length of the flux path, cross section area) are constant, (2.48) can be expressed as:

$$v = Ri + \left(N^2 \mu_r \mu_0 \frac{A_c}{l_c} \right) \frac{di}{dt}. \tag{2.49}$$

The term in parenthesis is related to the dimensions and the material characteristics of the core, and the number of turns of the conductor. This term is called inductance:

$$L = N^2 \mu_r \mu_0 \frac{A_c}{l_c}. \tag{2.50}$$

Hence, (2.49) can be expressed as:

$$v = Ri + L \frac{di}{dt}. \tag{2.51}$$

The unit of inductance is in henries [H] and it means that if we apply 1V to the circuit, 1H of inductance will result in rate of change of current of 1 A/s. Therefore, inductance can be represented as the inertia of an electromagnetic system.

From (2.51), current can be expressed as:

$$v = Ri + L \frac{di}{dt} \Rightarrow \frac{di}{dt} = \frac{1}{L}(v - Ri) \Rightarrow i = \int \frac{1}{L}(v - Ri) dt. \tag{2.52}$$

As depicted in Figure 2.7, the increase in the magnetic field intensity does not create an indefinite increase in the magnetic flux density in the ferromagnetic materials. The relative permeability of ferromagnetic material is not constant, and it reduces as the current increases—for the same magnetic flux path l_c, the magnetic field intensity increases with current as per (2.44). When the magnetization characteristics of the material are included, the dynamic model becomes nonlinear.

2.9 Reluctance

As discussed previously, when the rotor pole of an SRM approaches the vicinity of the stator pole, the magnetic field configuration changes due to the higher relative permeability of the ferromagnetic material that makes the rotor pole. The change in the magnetic field configuration is actually the change in the reluctance of the magnetic circuit. This is where the name "reluctance machine" comes from. The reluctance of the magnetic path is defined as:

$$\mathcal{R} = \frac{l_c}{\mu_r \mu_0 A_c}. \tag{2.53}$$

If we use (2.53) in (2.50), inductance can be expressed as:

$$L = \frac{N^2}{\mathcal{R}}. \tag{2.54}$$

Since magnetic flux is expressed as (2.47), we can calculate

$$\mathcal{R}\varphi = \frac{l_c}{\mu_r \mu_0 A_c} \mu_r \mu_0 \frac{A_c}{l_c} Ni \Rightarrow \mathcal{R}\varphi = Ni. \tag{2.55}$$

Equation 2.55 suggests that the magnetic circuit in Figure 2.10 can be represented similarly to an electrical circuit where Ni is represented with a voltage source, φ is like the current, and $\mathcal{R}$ is the resistance as shown in Figure 2.11a. One big difference in this analogy is that a resistor in an electric circuit dissipates electrical energy. However, reluctance stores magnetic energy. This will be discussed further in the next section.

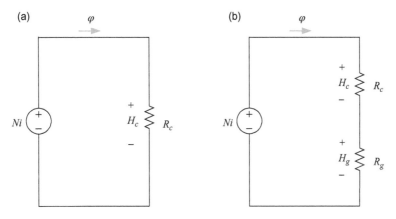

FIGURE 2.11
Electric circuit analogy for a magnetic system: (a) without air gap and (b) with air gap.

From (2.53), it can be observed that as the relative permeability decreases, the reluctance increases. Therefore, for a magnetic system, air will have much higher reluctance than the core made of ferromagnetic material.

Figure 2.10b shows a magnetic system with an air gap. Since the air gap and the ferromagnetic material have different permeability, Ampère's law in (2.8) applies as:

$$\oint \vec{H} \cdot d\vec{l} = Ni \Rightarrow \sum_{j=1}^{n} H_n l_n = Ni. \tag{2.56}$$

Considering the permeability of the ferromagnetic material is the same everywhere inside the core, the total flux path can be expressed with two components: the length of the flux path in the core l_c and the length of the flux path in the air gap l_g. Hence, (2.56) can be expressed as:

$$H_g l_g + H_c l_c = Ni \tag{2.57}$$

where H_g and H_c are the magnetic field intensity in the air gap and in the core, respectively. The electrical circuit analogy of the magnetic system in Figure 2.10b can now be represented by two reluctance components as shown in Figure 2.11b.

Using equation (2.16), the relationship between the magnetic flux density and magnetic field intensity can be incorporated in (2.57). Please note that the relative permeability of air is one:

$$\frac{B_g}{\mu_0} l_g + \frac{B_c}{\mu_r \mu_0} l_c = Ni. \tag{2.58}$$

Neglecting the leakage flux, magnetic flux generated by the applied current flows through the core and the air gap as shown in the equivalent circuit in Figure 2.11b. Neglecting the fringing flux, the cross-section area A_c passing through the flux is the same in the core and the air gap:

$$\frac{\varphi}{\mu_0 A_c} l_g + \frac{\varphi}{\mu_r \mu_0 A_c} l_c = Ni \Rightarrow \varphi \left(\frac{l_g}{\mu_0 A_c} + \frac{l_c}{\mu_r \mu_0 A_c} \right) = Ni. \tag{2.59}$$

Therefore, the equivalent reluctance of the magnetic system with air gap in Figure 2.11b can be defined as:

$$R_{eq} = \frac{l_g}{\mu_0 A_c} + \frac{l_c}{\mu_r \mu_0 A_c}. \tag{2.60}$$

We can further organize (2.60) to express it similarly to the fundamental equation of magnetic reluctance in (2.53):

$$R_{eq} = \frac{l_g}{\mu_0 A_c} + \frac{l_c}{\mu_0 A_c} \frac{1}{\mu_r} = \frac{l_c}{\mu_0 A_c} \left(\frac{l_g}{l_c} + \frac{1}{\mu_r} \right) = \frac{l_c}{\mu_0 A_c} \left(\frac{1}{l_c / l_g} + \frac{1}{\mu_r} \right). \tag{2.61}$$

The right-most term in (2.61) is expressed as the effective permeability [5]:

$$\mu_{\text{eff}} = \frac{1}{\left(\dfrac{1}{l_c / l_g} + \dfrac{1}{\mu_r} \right)}. \tag{2.62}$$

Hence the equivalent reluctance in (2.60) can be shown as:

$$R_{eq} = \frac{l_c}{\mu_{\text{eff}} \mu_0 A_c}. \tag{2.63}$$

By using the equivalent reluctance, the magnetic flux can be calculated as:

$$\varphi \frac{l_c}{\mu_{\text{eff}} \mu_0 A_c} = Ni \Rightarrow \varphi = \mu_{\text{eff}} \mu_0 \frac{Ni}{l_c} A_c. \tag{2.64}$$

Inserting (2.64) into (2.42) results in

$$v = Ri + N \frac{d\varphi}{dt} \Rightarrow v = Ri + N \frac{d}{dt} \left(\mu_{\text{eff}} \mu_0 \frac{Ni}{l_c} A_c \right) \Rightarrow v = Ri + \left(N^2 \mu_{\text{eff}} \mu_0 \frac{A_c}{l_c} \right) \frac{di}{dt}. \tag{2.65}$$

The term in the parenthesis in the right-most side of equation (2.65) is the inductance of the magnetic circuit, and it can be observed that this term is very similar to the definition of inductance in (2.50):

$$L = N^2 \mu_{\text{eff}} \mu_0 \frac{A_c}{l_c}. \tag{2.66}$$

For ferromagnetic materials, since $\mu_r \gg 1$, the effective permeability in (2.62) can be simplified as

$$\mu_{\text{eff}} \approx \frac{l_c}{l_g}. \tag{2.67}$$

In a practical circuit, the effective permeability will be smaller than the relative permeability of the ferromagnetic core. Therefore, the air gap increases the equivalent reluctance of the magnetic system and reduces the inductance. However, as we will see in the next section, most of the energy in a magnetic circuit is stored in the high-reluctance components in the magnetic system, which is an important operational parameter in a switched reluctance machine.

2.10 Magnetic Energy

As we analyzed so far in this chapter, electric current generates magnetic field, which exerts forces on the rotor of an SRM. Due to the salient pole configuration of SRMs, the motion of the rotor changes the effective air gap length in the magnetic circuit. This causes a variation in the flux linkage, induces electromotive force to limit the rate of change of current. Therefore, magnetic field acts as a bridge in the electromechanical energy conversion.

First, let's consider the magnetic system in Figure 2.11b. Since there are no movable components in this system, some of the input electrical energy will be dissipated in the resistance of the coils and the rest will be stored in the magnetic system. Using (2.41), we can calculate the energy balance for this system:

$$v = Ri + \frac{d\lambda}{dt} \Rightarrow vi = Ri^2 + i\frac{d\lambda}{dt} \Rightarrow \left(vi - Ri^2\right)dt = id\lambda. \tag{2.68}$$

Flux linkage is in units of Weber (Wb) or volt-seconds (Vs). The right side of (2.68) is in volt-ampere-seconds, which corresponds to energy in Joules. From (2.68), it can be observed that the input electrical energy after the power loss in the resistance of the coil $(vi - Ri^2)dt$ is stored in the magnetic field $(id\lambda)$. Considering a ferromagnetic material with constant permeability, the flux linkage increases linearly with current as shown in Figure 2.12. The energy stored in the magnetic field is calculated by considering constant current for infinitesimal change in the flux linkage. This corresponds to the upper half of the line in Figure 2.12. Therefore, magnetic energy can be calculated as:

$$W_f = \int_{\lambda_1}^{\lambda_2} id\lambda. \tag{2.69}$$

For the linear case, since $\lambda = Li$, (2.69) can be reorganized as:

$$W_f = \int_{\lambda_1}^{\lambda_2} id\lambda = \int_0^i Lidi = \frac{1}{2}Li^2. \tag{2.70}$$

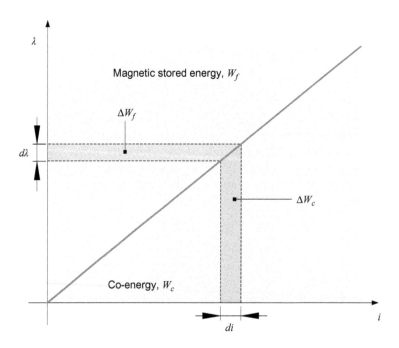

FIGURE 2.12
Flux linkage vs current for a linear magnetic system.

As calculated in (2.66), for the magnetic system in Figure 2.10b, the inductance can be expressed as:

$$L = N^2 \mu_{eff} \mu_0 \frac{A_c}{l_c}$$

since $\lambda = Li$ and $\lambda = N\varphi$, then

$$\varphi = \frac{Li}{N}. \tag{2.71}$$

Therefore,

$$\varphi = \mu_{eff} \mu_0 \frac{Ni}{l_c} A_c. \tag{2.72}$$

From (2.46), since $\varphi = BA_c$, then

$$BA_c = \mu_{eff} \mu_0 \frac{Ni}{l_c} A_c \Rightarrow i = \frac{Bl_c}{\mu_{eff} \mu_0 N}. \tag{2.73}$$

If we insert (2.66) and (2.73) into (2.70), magnetic energy in a gapped core can be calculated as:

$$W_f = \frac{1}{2} Li^2 = \frac{1}{2} N^2 \mu_{eff} \mu_0 \frac{A_c}{l_c} \frac{B^2 l_c^2}{\mu_{eff}^2 \mu_0^2 N^2} = \frac{1}{2} \frac{B^2 A_c l_c}{\mu_{eff} \mu_0}. \tag{2.74}$$

From (2.62), effective permeability of a magnetic system with air gap was expressed as:

$$\mu_{eff} = \frac{1}{\left(\dfrac{1}{l_c / l_g} + \dfrac{1}{\mu_r} \right)} = \frac{1}{\dfrac{l_g}{l_c} + \dfrac{1}{\mu_r}}. \tag{2.75}$$

Therefore, (2.74) can be reorganized as given in [5]:

$$W_f = \frac{1}{2} \frac{B^2 A_c l_c}{\dfrac{\mu_0}{\dfrac{l_g}{l_c} + \dfrac{1}{\mu_r}}} = \frac{1}{2} \frac{B^2 A_c l_c \left(l_c + \mu_r l_g \right)}{\mu_r \mu_0 l_c} = \frac{B^2 A_c l_c}{2 \mu_r \mu_0} + \frac{B^2 A_c l_g}{\mu_0}. \tag{2.76}$$

If the volume of the core is expressed as $V_c = A_c l_c$ and the volume of the air gap is expressed as $V_g = A_c l_g$, then (2.76) can be simplified as:

$$W_f = \frac{B^2 V_c}{2 \mu_r \mu_0} + \frac{B^2 V_g}{2 \mu_0}. \tag{2.77}$$

We can reorganize (2.77) as:

$$W_f = \frac{B^2 V_g}{2\mu_0} \frac{V_c}{\mu_r V_g} + \frac{B^2 V_g}{2\mu_0} = \frac{B^2 V_g}{2\mu_0}\left(1 + \frac{V_c}{V_g}\frac{1}{\mu_r}\right). \tag{2.78}$$

Since $V_c = A_c l_c$ and $V_g = A_c l_g$, (2.78) can be simplified as:

$$W_f = \frac{B^2 V_g}{2\mu_0}\left(1 + \frac{l_c}{l_g}\frac{1}{\mu_r}\right). \tag{2.79}$$

In (2.67), for $\mu_r \gg 1$, effective permeability was expressed as $\mu_{eff} \approx l_c / l_g$. If we insert (2.67) into (2.79), then magnetic energy can be calculated as [5]:

$$W_f = \frac{B^2 V_g}{2\mu_0}\left(1 + \frac{\mu_{eff}}{\mu_r}\right). \tag{2.80}$$

Equation (2.80) has important conclusions for better understanding the torque production principles in an SRM. It says that in an inductor with an air gap, most of the energy is stored in the air gap itself, which is a high-reluctance magnetic path as compared to the core. As it will be discussed in the next section, in a switched reluctance machine, the energy stored in the air gap is converted to mechanical energy. The permeability of ferromagnetic materials reduces as the level of saturation increases. This means that when the core saturates, its permeability reduces, the reluctance of the magnetic path increases, and the magnetic system can store more energy, which is then converted to mechanical energy. Therefore, the torque production capability of an SRM improves as the saturation level of the core increases.

2.11 Co-energy and Electromechanical Energy Conversion

For the linear case in Figure 2.12, the upper and lower halves of the curve have the same area. Therefore, instead of using $(id\lambda)$, the energy can also be calculated by considering constant flux linkage for infinitesimal change in current (λdi). This corresponds to the lower half of the line in Figure 2.12 and is called co-energy.

The co-energy principal is heavily used in systems storing energy. For example, in a mechanical system, energy, dW is defined as the work done when force, F acting on an object moves it a certain distance, dx in the direction of the force: $dW = Fdx$. Similarly, fixing the displacement and changing the force can calculate the same result. However, from the physical perspective, if there is no displacement, there is no work or energy. But, in terms of units $dW' = xdF$ is also expressed in units of joules and it is referred as co-energy [6]. Co-energy is a non-physical quantity used in theoretical analysis of energy storing systems. It is heavily used in the analysis of non-linear systems, such as electric machines and electromechanical devices.

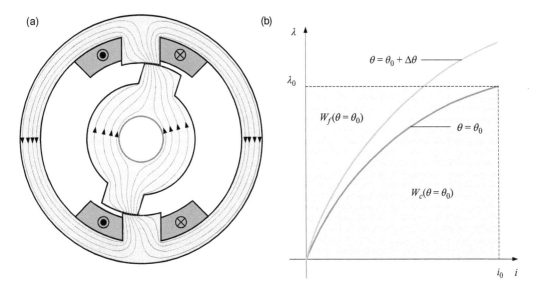

FIGURE 2.13
(a) Single-phase SRM and (b) nonlinear flux linkage vs current characteristics.

In the case of magnetic system with air gap in Figure 2.10b, all the input energy was converted to magnetic energy. Now consider the single-phase SRM shown in Figure 2.13a. If the copper losses in the windings, the core losses in the iron core, and the mechanical losses in the rotor are neglected, the incremental input energy must be equal to the incremental magnetic energy and the incremental mechanical energy:

$$eidt = dW_f + T_e d\theta \tag{2.81}$$

where T_e is the electromagnetic torque and θ is the rotor position. From Faraday's law, the magnitude of the induced voltage is expressed as $e = d\lambda / dt$. Then (2.81) results in

$$id\lambda = dW_f + T_e d\theta \Rightarrow dW_f = id\lambda - T_e d\theta. \tag{2.82}$$

Figure 2.13b shows the nonlinear flux-linkage current characteristics for the single-phase SRM. Unlike the case in Figure 2.12, flux linkage does not increase linearly with current. This is due to the nonlinear magnetization characteristics of ferromagnetic materials shown in Figure 2.7. If the rotor position is fixed, $d\theta = 0$, the magnetic energy in (2.81) will be $id\lambda$ as calculated in (2.69). Therefore, for a fixed rotor position, a single phase of an SRM behaves like a gapped inductor. The energy to excite the SRM from zero to (i_1, λ_1) will be the upper half of the magnetization curve at θ_0.

As mentioned previously, flux linkage versus current characteristics in an SRM is dependent on the relative position of the rotor and stator poles. Therefore, magnetic energy is a function of the partial derivatives of λ and θ:

$$dW_f(\lambda, \theta) = \frac{\partial W_f}{\partial \lambda} d\lambda \bigg|_{\theta=const} + \frac{\partial W_f}{\partial \theta} d\theta \bigg|_{\lambda=const}. \tag{2.83}$$

Combining (2.82) and (2.83) results in

$$dW_f = id\lambda - T_e d\theta = \frac{\partial W_f}{\partial \lambda} d\lambda \bigg|_{\theta=const} + \frac{\partial W_f}{\partial \theta} d\theta \bigg|_{\lambda=const}, \qquad (2.84)$$

hence,

$$i = \frac{\partial W_f}{\partial \lambda} \bigg|_{\theta=const} \text{ and } T_e = -\frac{\partial W_f}{\partial \theta} \bigg|_{\lambda=const}. \qquad (2.85)$$

From (2.85), it can be observed that electromagnetic torque can be calculated from the rate of change of magnetic energy with position by keeping the flux linkage constant. Such calculations require complicated mathematical calculations since they require inversion of the relationship between flux linkage and current [6]. This is why the co-energy method is highly used in analytical and numerical calculations of electromagnetic torque. Co-energy is defined as the lower half of the flux linkage versus current characteristics in Figure 2.13b and it can be calculated by subtracting the magnetic energy from the total energy:

$$W_c = i\lambda - W_f(\lambda, \theta). \qquad (2.86)$$

If we take the derivative of (2.86) and insert incremental magnetic energy expression in (2.82), we can calculate the incremental co-energy:

$$dW_c = d(\lambda i) - dW_f(\lambda, \theta) = id\lambda + \lambda di - id\lambda + T_e d\theta \Rightarrow dW_c = \lambda di + T_e d\theta. \qquad (2.87)$$

Co-energy is the lower half of the flux linkage versus current characteristics and, hence, it is a function of the partial derivatives of i and θ:

$$dW_c(i,\theta) = \frac{\partial W_c}{\partial i} di \bigg|_{\theta=const} + \frac{\partial W_c}{\partial \theta} d\theta \bigg|_{i=const}. \qquad (2.88)$$

Combining (2.87) and (2.88) results in

$$dW_c = \lambda di + T_e d\theta = \frac{\partial W_c}{\partial i} di \bigg|_{\theta=const} + \frac{\partial W_c}{\partial \theta} d\theta \bigg|_{i=const}, \qquad (2.89)$$

hence,

$$\lambda = \frac{\partial W_c}{\partial i} \bigg|_{\theta=const} \text{ and } T_e = \frac{\partial W_c}{\partial \theta} \bigg|_{i=const}. \qquad (2.90)$$

Using co-energy, the electromagnetic torque can be calculated by keeping the current constant over small perturbations of position. This method is often referred to as Virtual Work Method.

When the linear case in Figure 2.12 and the nonlinear case in Figure 2.13b are compared, it can be observed that the co-energy in an SRM increases with the level of saturation. This further validates that the torque production capability of an SRM improves when the magnetic core operates in the saturated region. Please note that magnetic energy is stored in the magnetic field and, when the current is switched off, it turns back to the supply in an ideal case. When operating in the linear region, magnetic energy and co-energy are the same. Therefore, in practical terms, when an SRM operates in the linear region, it means that the two units of energy have to be supplied in order to make one unit of mechanical work. This results in poor use of power semiconductors and would increase the volt-ampere requirements for the converter.

2.12 Losses

In the previous section, when describing the energy balance equation we neglected the copper losses in the windings, core losses in the iron core, and mechanical losses in the rotor. In that case, the input of electrical energy equals to the magnetic energy and the mechanical energy, as shown in (2.81). During the electromechanical energy conversion process, some of the input power is lost and dissipated as heat inside the machine. When the losses are considered, the energy balance equation ends up as in Figure 2.14.

When the total losses are subtracted from the input power, the mechanical output power and, hence, the efficiency of the systems can be calculated as:

$$\eta = \frac{P_{in} - P_{loss}}{P_{in}} \times 100\% \tag{2.91}$$

where P_{in} is the electrical input power and P_{loss} is the total power losses.

2.12.1 Copper Losses

As discussed in Section 2.7, the induced electric field moves the charges (free electrons) around the conducting loop. When the electric field applies forces to charges, the free electrons pick up a speed between the collusion. This is called the drift velocity and it

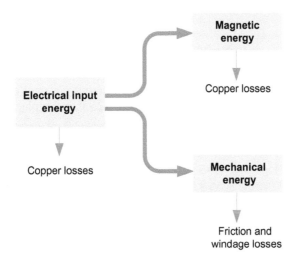

FIGURE 2.14
Energy balance considering the losses.

was quantified in (2.30). Power loss occurs during the collusion of these electrons and it is referred as the copper or joule losses. In switched reluctance machines, copper losses are due to the electrical resistivity of the stator winding and it is usually the most significant energy loss in the machine.

As quantified in (2.38), the relationship between the voltage and current is called the resistance. Therefore, the copper losses can be quantified as:

$$P_{cu} = VI = (IR)I = I^2 R. \tag{2.92}$$

As quantified in (2.37), resistance is inversely proportional to the surface area, S and it is directly proportional to the resistivity, ρ and the length, ℓ of the conductor:

$$P_{cu} = I^2 R = I^2 \rho \frac{\ell}{S}. \tag{2.93}$$

It should be noted that higher temperature results in more collusion between the free electrons. Therefore, the time between the collusion of free electrons, τ in (2.36) decreases. Thus, the resistivity and the copper losses increase with temperature. The relationship between the resistivity and temperature is defined as:

$$\rho = \rho_0 \left[1 + \alpha (T - T_0) \right] \tag{2.94}$$

where ρ_0 is the resistivity at the initial temperature T_0, α is the temperature coefficient, and T is the final temperature.

2.12.2 AC Copper Losses

Equation 2.93 shows that the copper loss is a function of the current through the conductor, and the resistivity, length, and cross-section area of the conductor. Here, it was assumed that the current is homogenously distributed within the cross-section area of the conductor. This assumption does not hold when the frequency of the current increases. For a round conductor with a radius of r_0, the cross-sectional area is defined as πr_0^2, and this can be calculated as:

$$S = 2\pi \int_0^{r_0} r \, dr = 2\pi \left. \frac{r^2}{2} \right|_0^{r_0} = \pi r_0^2. \tag{2.95}$$

Current density in the conductor is defined as the current per unit cross-section area of the conductors in units of A / m^2. Inserting (2.95) into (2.93) and replacing the current with current density, $J(r)$ results in:

$$P_{cu} = I^2 \rho \frac{\ell}{S} = \left[J(r)S \right]^2 \rho \frac{\ell}{S} = \rho \ell J^2(r) S = 2\pi \rho \ell \int_0^{r_0} J^2(r) r \, dr. \tag{2.96}$$

For homogenous current density distribution, (2.96) will result in (2.93). In case of alternating currents, the current density inside the conductor might not be homogenous and it becomes a function of frequency. When it comes to a single conductor, this phenomenon is referred as the skin effect. For multiple conductors, the current density distribution in each conductor is affected by the neighboring conductors and it is called the proximity effect.

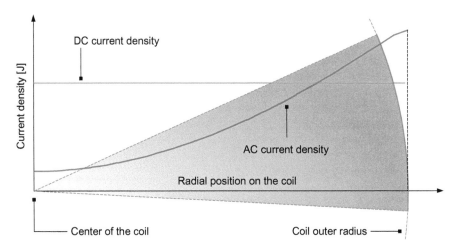

FIGURE 2.15
Current density distribution inside the coil at 1 kHz.

In high-frequency operation, skin and proximity effects cause nonuniform current density distribution, leading to additional AC copper losses.

The skin effect is the tendency of the current to flow on the surface of the conductor. This results in a nonuniform current density distribution. As the frequency of the current increases, the depth of the highest current density from the outer surface of the conductor decreases. This is called the skin depth. It is a function of the excitation frequency, and the electrical and magnetic properties of the conducting medium:

$$\delta = \frac{1}{\sqrt{\pi \sigma \mu f}} \tag{2.97}$$

where f is the excitation frequency, σ is the conductivity, and μ is the permeability of the conducting medium. Figure 2.15 shows the current density distribution inside the coil at 1kHz. It can be observed that at higher frequencies, the current density is smaller at the center of the conductor and higher on the surface. As the frequency increases, skin depth decreases and leads to a lower effective cross section area of the conductor. Therefore, the effective resistance of the conductor increases leading to higher copper losses [3]. The copper losses due to the effect of higher excitation frequencies are referred as the AC copper losses and they are calculated by estimating the AC resistance of the winding. The ratio of AC to DC resistance increases, as the skin depth gets smaller.

The effect of skin depth can be more pronounced when the skin depth is smaller than the wire radius. In order to reduce the resistive losses at high-frequency applications, the windings are made with multiple strands. To the reduce the AC copper losses caused by the proximity effect, Litz wires are frequently used in high-frequency applications, which can provide lower AC resistance due their twisted geometry [7].

2.12.3 Eddy Current Losses

AC copper losses in the conductor occur due to Faraday's law. AC current in the conductor creates a time-changing magnetic field. As quantified in (2.24), time-changing magnetic field induces a time-changing electric field. The induced electric field moves the electrons in the conductor and generates current. These currents are called eddy currents and they flow in a direction to oppose the change in the magnetic field created by the AC excitation current. In case of skin effect, the electric field induced by the eddy currents generates voltage that is stronger at the center of the conductor. Therefore, the electrons are pushed to flow on the surface of the conductor. The higher the excitation frequency, the higher the change in magnetic field, the higher the voltage induced inside the conductor, and the further the excitation current is pushed toward the outer surface of the conductor.

Eddy currents also generate losses in the magnetic core. The ferromagnetic core materials have finite electrical conductivity, which is much smaller than the conductivity of copper. When the motor is in operation, the change in magnetic flux through the magnetic core induces eddy currents. Due to the finite electrical conductivity of core materials, these eddy currents generate losses on the core, which is referred as the eddy current losses.

Eddy current losses occur in both stator and rotor cores of a switched reluctance machine. The practical way to reduce eddy current losses is to increase the resistance on the conductive path of eddy currents. For this purpose, the stator and rotor cores of switched reluctance machines are laminated as shown in Figure 2.16b, where each lamination is electrically isolated from each other by means of surface coating. The finite element simulations for this figure have been conducted in JMAG. Therefore, the effective resistivity of the conducting path increases leading to higher total resistance. For the given induced electromotive force (EMF) created by the time-changing magnetic field, the eddy currents will be lower due to the higher resistance of the conducting loop. This results in lower eddy current losses in the core. Similar to the skin effect in the conductor, eddy current losses are a function of the frequency and magnetic field strength. The thickness of the laminations is also related to the skin depth of the core.

2.12.4 Hysteresis Losses

So far, we have analyzed copper losses and eddy current losses, which are categorized as joule losses in engineering applications. In a macroscopic sense, these losses are based on a current flowing through a conducting loop and generating heat due to the resistance

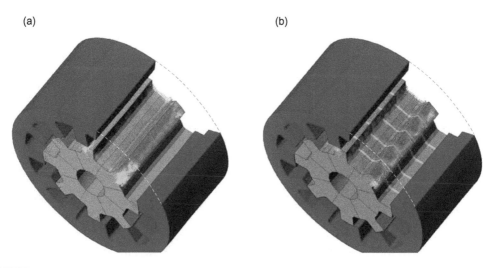

(a)   (b)

FIGURE 2.16
Eddy currents in the rotor core of an SRM: (a) without laminations and (b) with laminations.

in the loop. Hysteresis loss is related to the energy to change the alignment of dipoles in a ferromagnetic material.

As discussed in Section 2.4, ferromagnetic materials have domains whereby the magnetic dipoles are 100% aligned. The domains are uniformly distributed in their neighborhood, but the net internal field is zero. When exposed to an external magnetic field, these domains align their polarity in the direction of the external field. Therefore, the magnetic field inside the ferromagnetic material becomes stronger than the external field. It was also discussed that the increase in the external field does not create an indefinite increase in the magnetic flux density. There is certain point where nearly all the domains are completely aligned, and it is called saturation.

Figure 2.17 shows the change in the magnetic characteristics of a ferromagnetic material when it is exposed to alternating external field. When the material is first exposed to an external field, the magnetic flux density inside the material is zero. This corresponds to P1 in Figure 2.17. When the external field increases in the positive direction, the field inside the material increases linearly up to P2. Beyond this point, the saturation level in the material increases, and the magnetic flux density does not change linearly as the external field increases. Thus, the permeability of the ferromagnetic material reduces, and it approaches that of air when it fully saturates at P3.

At this point, if the external field is reduced, the magnetization characteristics do not follow the initial magnetization curve. Instead, we end up at P4 when the external field is zero. This means that we have now magnetic field inside the material, since some of the domains keep themselves in alignment. This is called residual magnetism.

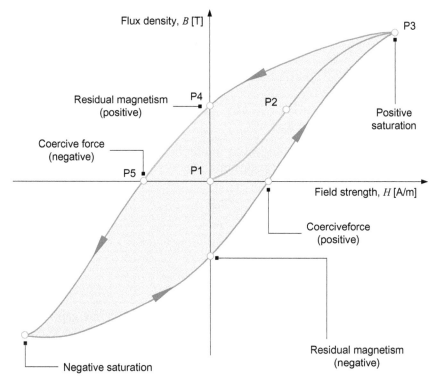

FIGURE 2.17
Typical hysteresis loop of a ferromagnetic material.

If we keep reducing the external field, we end up at P5 in Figure 2.17. Here, the external field is in the opposite direction, but there is no magnetic field inside the ferromagnetic material. At this point, the vectorial sum of the external field and the magnetic field in the material due to the residual magnetism is zero.

If we keep changing the direction and magnitude of the external field, the magnetic field inside the ferromagnetic material will cover the entire hysteresis loop in Figure 2.17 and the material will be exposed to periodic magnetization. In this process, the area under the hysteresis loop represents the energy which is required to change the orientation of the domains and expand in the form of heat. This energy is called the hysteresis loss and it is dependent on the strength of the magnetic field and the frequency.

2.12.5 Mechanical Losses

In electric machines, the electromagnetic forces in the air gap apply torque on the rotor to rotate it at a certain speed. Some of the power in the air gap is lost mechanically on the rotor. Mechanical losses are mostly dominated by mechanical friction and windage losses.

Mechanical friction losses are due to the frictional resistance between two surfaces when they are in motion. The friction losses are dependent on the speed, but they are independent on the load. The most significant friction losses in the rotor are the bearing losses.

Windage in a rotating machine is the resistance of air against the motion of the rotor. Windage power loss increases exponentially with the rotational speed or the radius of the rotor. Especially in high-speed operation, windage losses can be significant in SRMs due to the salient pole structure. In order to reduce the windage losses, the air gap between rotor poles can be filled with epoxy resin to create a smooth rotor surface. Smooth rotor surface has much lower friction, and this results in lower windage power loss and windage noise [7].

Appendix A: Derivation of Radial and Tangential Forces from Lorentz Force Law

In a switched reluctance machine, tangential and radial force densities [N/m²] are heavily used in the analysis of torque and acoustic noise. They are defined as:

$$F_t = \frac{B_r B_t}{\mu_0} \tag{A.2.1}$$

$$F_r = \frac{\left(B_r^2 - B_t^2\right)}{2\mu_0} \tag{A.2.2}$$

where B_r and B_t are the radial and tangential components of the air gap magnetic flux density, and μ_0 is the permeability of free space. These expressions are Maxwell stress tensors [2] and it will be shown here that they are calculated using Lorentz Force law.

As given in (2.17), Lorentz force law state that

$$\vec{F} = q\left(\vec{E} + \vec{v} \times \vec{B}\right) \tag{A.2.3}$$

where $\vec{F}$ is the electromagnetic force vector, q is the charge of the electron in the electric $\vec{E}$ and magnetic $\vec{B}$ fields; and $\vec{v}$ is the velocity vector of the charge. Then, the electromagnetic force on the charges in volume V can be expressed as:

$$q = \rho d\tau \tag{A.2.4}$$

where ρ is the charge per unit volume and its unit is C / m^3. Then,

$$\vec{F}_V = \int_V \left(\vec{E} + \vec{v} \times \vec{B} \right) \rho d\tau = \int_V \left(\rho \vec{E} + \rho \vec{v} \times \vec{B} \right) d\tau \tag{A.2.5}$$

where

$$\underset{\underset{m^3}{C}}{\rho} \underset{\underset{s}{m}}{\vec{v}} = \underset{\underset{s}{C} \underset{m^2}{1} = \underset{m^2}{A}}{\vec{J}} \tag{A.2.6}$$

So, $\vec{J}$ is the current density vector. Then,

$$\vec{F}_V = \int_V \left(\rho \vec{E} + \vec{J} \times \vec{B} \right) d\tau \tag{A.2.7}$$

Therefore, force per unit volume can be expressed as

$$\vec{f} = \rho \vec{E} + \vec{J} \times \vec{B}. \tag{A.2.8}$$

We would like to express (A.2.8) using only the field vectors $\vec{E}$ and $\vec{B}$. From Gauss's law for electric fields in (2.14):

$$\nabla \cdot \vec{E} = \frac{\rho}{\varepsilon_0} \Rightarrow \rho = \varepsilon_0 \left(\nabla \cdot \vec{E} \right) \tag{A.2.9}$$

where ε_0 is the permittivity of free space. $\nabla \cdot \vec{E}$ is the divergence of the electric field vector. From Ampère-Maxwell law in integral form in (2.9), the differential form can be expressed as:

$$\nabla \times \vec{B} = \mu_0 \vec{J} + \varepsilon_0 \mu_0 \frac{\partial \vec{E}}{\partial t} \Rightarrow J = \frac{1}{\mu_0} \left(\nabla \times \vec{B} \right) - \varepsilon_0 \frac{\partial \vec{E}}{\partial t}. \tag{A.2.10}$$

Inserting (A.2.9) and (A.2.10) into (A.2.8) results in:

$$f = \rho \vec{E} + \vec{J} \times \vec{B} = \varepsilon_0 \left(\nabla \cdot \vec{E} \right) \vec{E} + \left(\frac{1}{\mu_0} \left(\nabla \times \vec{B} \right) - \varepsilon_0 \frac{\partial \vec{E}}{\partial t} \right) \times \vec{B}. \tag{A.2.11}$$

From Faraday's law in (2.25):

$$\nabla \times \vec{E} = -\frac{\partial \vec{B}}{\partial t} \Rightarrow \frac{\partial \vec{B}}{\partial t} = -\nabla \times \vec{E}. \tag{A.2.12}$$

From vector calculus

$$\frac{\partial}{\partial t}\left(\vec{E}\times\vec{B}\right)=\left(\frac{\partial\vec{E}}{\partial t}\times\vec{B}\right)+\left(\vec{E}\times\frac{\partial\vec{B}}{\partial t}\right)\Rightarrow\left(\frac{\partial\vec{E}}{\partial t}\times\vec{B}\right)=\frac{\partial}{\partial t}\left(\vec{E}\times\vec{B}\right)-\left(\vec{E}\times\frac{\partial\vec{B}}{\partial t}\right). \qquad \text{(A.2.13)}$$

Applying (A.2.12) to the right-most term in (A.2.13) results in

$$\left(\frac{\partial\vec{E}}{\partial t}\times\vec{B}\right)=\frac{\partial}{\partial t}\left(\vec{E}\times\vec{B}\right)+\vec{E}\times\left(\nabla\times\vec{E}\right). \qquad \text{(A.2.14)}$$

(A.2.11) can be expanded as

$$\vec{f}=\varepsilon_0\left(\nabla\cdot\vec{E}\right)\vec{E}+\frac{1}{\mu_0}\left(\nabla\times\vec{B}\right)\times\vec{B}-\varepsilon_0\left(\frac{\partial\vec{E}}{\partial t}\times\vec{B}\right). \qquad \text{(A.2.15)}$$

Inserting (A.2.14) into (A.2.15) results in

$$\vec{f}=\varepsilon_0\left(\nabla\cdot\vec{E}\right)\vec{E}+\left[\frac{1}{\mu_0}\left(\nabla\times\vec{B}\right)\times\vec{B}-\varepsilon_0\left(\frac{\partial}{\partial t}\left(\vec{E}\times\vec{B}\right)+\vec{E}\times\left(\nabla\times\vec{E}\right)\right)\right]. \qquad \text{(A.2.16)}$$

Expanding ε_0 in (A.2.16) results in

$$\vec{f}=\varepsilon_0\left(\nabla\cdot\vec{E}\right)\vec{E}+\frac{1}{\mu_0}\left(\nabla\times\vec{B}\right)\times\vec{B}-\varepsilon_0\frac{\partial}{\partial t}\left(\vec{E}\times\vec{B}\right)-\varepsilon_0\vec{E}\times\left(\nabla\times\vec{E}\right), \qquad \text{(A.2.17)}$$

since $\vec{A}\times\vec{B}=-\left(\vec{B}\times\vec{A}\right)$

$$\vec{f}=\varepsilon_0\left(\left(\nabla\cdot\vec{E}\right)\vec{E}-\vec{E}\times\left(\nabla\times\vec{E}\right)\right)-\frac{1}{\mu_0}\left(\vec{B}\times\left(\nabla\times\vec{B}\right)\right)-\varepsilon_0\frac{\partial}{\partial t}\left(\vec{E}\times\vec{B}\right). \qquad \text{(A.2.18)}$$

In (A.2.18), we would like to express the terms $\vec{E}\times(\nabla\times\vec{E})$ and $\vec{B}\times(\nabla\times\vec{B})$ in a different way to create symmetry between electric and magnetic field components. For this purpose, we will use the product rule for gradients:

$$\nabla\left(\vec{A}\cdot\vec{B}\right)=\vec{A}\times\left(\nabla\times\vec{B}\right)+\vec{B}\times\left(\nabla\times\vec{A}\right)+\left(\vec{A}\cdot\nabla\right)\vec{B}+\left(\vec{B}\cdot\nabla\right)\vec{A}. \qquad \text{(A.2.19)}$$

The proof of this identity is given in Appendix B. In $(\vec{A}\cdot\nabla)\vec{B}$, the term in parenthesis is the dot product of $\vec{A}$ and the nabla operator, which is defined as:

$$\nabla=\hat{x}\frac{\partial}{\partial x}+\hat{y}\frac{\partial}{\partial y}+\hat{z}\frac{\partial}{\partial z}. \qquad \text{(A.2.20)}$$

The result of the dot product is a scalar, which is then multiplied by each component of the vector $\vec{B}$. If $\vec{B}=\vec{A}$, then

$$\vec{A}\cdot\vec{B}=\vec{A}\cdot\vec{A}=\left(A_x\hat{x}+A_y\hat{y}+A_z\hat{z}\right)\cdot\left(A_x\hat{x}+A_y\hat{y}+A_z\hat{z}\right). \qquad \text{(A.2.21)}$$

Since $\hat{x}\cdot\hat{x}=\hat{y}\cdot\hat{y}=\hat{z}\cdot\hat{z}=1$ and $\hat{x}\cdot\hat{y}=\hat{x}\cdot\hat{z}=\hat{y}\cdot\hat{z}=0$

$$\vec{A}\cdot\vec{A}=A_x^2+A_y^2+A_z^2=A^2. \tag{A.2.22}$$

If we apply $\vec{B}=\vec{A}$ to (A.2.19), we get

$$\nabla\left(A^2\right)=\vec{A}\times\left(\nabla\times\vec{A}\right)+\vec{A}\times\left(\nabla\times\vec{A}\right)+\left(\vec{A}\cdot\nabla\right)\vec{A}+\left(\vec{A}\cdot\nabla\right)\vec{A}. \tag{A.2.23}$$

This can be further simplified as:

$$\nabla\left(A^2\right)=2\left(\vec{A}\cdot\nabla\right)\vec{A}+2\vec{A}\times\left(\nabla\times\vec{A}\right). \tag{A.2.24}$$

For electric and magnetic field vectors, (A.2.24) can be expressed as:

$$\nabla\left(E^2\right)=2\left(\vec{E}\cdot\nabla\right)\vec{E}+2\vec{E}\times\left(\nabla\times\vec{E}\right)\Rightarrow\vec{E}\times\left(\nabla\times\vec{E}\right)=\frac{1}{2}\nabla\left(E^2\right)-\left(\vec{E}\cdot\nabla\right)\vec{E} \tag{A.2.25}$$

$$\nabla\left(B^2\right)=2\left(\vec{B}\cdot\nabla\right)\vec{B}+2\vec{B}\times\left(\nabla\times\vec{B}\right)\Rightarrow\vec{B}\times\left(\nabla\times\vec{B}\right)=\frac{1}{2}\nabla\left(B^2\right)-\left(\vec{B}\cdot\nabla\right)\vec{B}. \tag{A.2.26}$$

In (A.2.18), force per unit volume was calculated as:

$$\vec{f}=\varepsilon_0\left(\left(\nabla\cdot\vec{E}\right)\vec{E}-\vec{E}\times\left(\nabla\times\vec{E}\right)\right)-\frac{1}{\mu_0}\left(\vec{B}\times\left(\nabla\times\vec{B}\right)\right)-\varepsilon_0\frac{\partial}{\partial t}\left(\vec{E}\times\vec{B}\right)$$

Now, we can insert (A.2.25) and (A.2.26) in (A.2.18):

$$\vec{f}=\varepsilon_0\left[\left(\nabla\cdot\vec{E}\right)\vec{E}-\frac{1}{2}\nabla\left(E^2\right)+\left(\vec{E}\cdot\nabla\right)\vec{E}\right]+\frac{1}{\mu_0}\left[\quad-\frac{1}{2}\nabla\left(B^2\right)+\left(\vec{B}\cdot\nabla\right)\vec{B}\right]-\varepsilon_0\frac{\partial}{\partial t}\left(\vec{E}\times\vec{B}\right). \tag{A.2.27}$$

You probably noticed that the expression for the electric and magnetic fields on the right side of (A.2.27) become symmetric if $(\nabla\cdot\vec{B})\vec{B}$ is added in the second term (there is space left for this expression). According to Gauss's law for magnetic fields, since magnetic fields come as dipoles and in practice magnetic monopoles do not exists, then $\nabla\cdot\vec{B}=0$ as shown in (2.10). Therefore, inserting this term in (A.2.27) will have no effect on the equation. Hence,

$$\vec{f}=\varepsilon_0\left[\left(\nabla\cdot\vec{E}\right)\vec{E}+\left(\vec{E}\cdot\nabla\right)\vec{E}-\frac{1}{2}\nabla\left(E^2\right)\right]$$

$$+\frac{1}{\mu_0}\left[\left(\nabla\cdot\vec{B}\right)\vec{B}+\left(\vec{B}\cdot\nabla\right)\vec{B}-\frac{1}{2}\nabla\left(B^2\right)\right]-\varepsilon_0\frac{\partial}{\partial t}\left(\vec{E}\times\vec{B}\right). \tag{A.2.28}$$

Equation A.2.28 represents the force per unit volume and it contains all aspects of electro-magnetics and momentum. It is a complicated expression, but it can be simplified using the Maxwell stress tensor.

An object or a volume can be represented by three surfaces in the xy, xz, and yz planes. The vector representing the area of a surface is always perpendicular to the surface or the plane. The force acting on this surface can be either parallel to the area vector or perpendicular to it. So, to fully characterize all possible combinations of forces acting on all possible surfaces in three-dimensional space, we need nine components, each with two-unit vectors. A tensor, therefore, represents the direction of the force vector and the direction of the area vector. The forces acting on a three-dimensional object can be expressed with rank 2 tensor:

$$\vec{\vec{T}}_{ij} = \begin{bmatrix} T_{xx} & T_{xy} & T_{xz} \\ T_{yx} & T_{yy} & T_{yz} \\ T_{zx} & T_{zy} & T_{zz} \end{bmatrix}. \tag{A.2.29}$$

By using the stress tensor, all forces acting in all directions on a solid object can be represented. Here, T_{xx} refers to the x directed force on a surface whose area vector points on the x direction. T_{yx} refers to the x directed force on a surface whose area vector points on the y direction. The diagonal terms are called pressure and non-diagonal terms are called shears. In the diagonal terms, since the force applied on the surface is in the same direction of the area vector, the object contracts in that direction. This is why they are called pressure components.

In order to simplify the force per unit volume expression in (A.2.28), the Maxwell stress tensor is used:

$$T_{ij} = \varepsilon_0 \left(E_i E_j - \frac{1}{2} \delta_{ij} E^2 \right) + \frac{1}{\mu_0} \left(B_i B_j - \frac{1}{2} \delta_{ij} B^2 \right). \tag{A.2.30}$$

where δ_{ij} is Kronecker delta and it equals to 1 if the indices are the same and zero otherwise. Now we will investigate how to incorporate the Maxwell stress tensor into the force expression. First, we will explore how to multiply a vector with a tensor. We take a vector $\vec{a} = a_x \hat{x} + a_y \hat{y} + a_z \hat{z}$. The multiplication of a tensor with a vector results in a vector, and each component of the resulting vector can be calculated as:

$$\underbrace{\left(\vec{a} \cdot \vec{\vec{T}} \right)_j}_{\substack{\text{the result is} \\ \text{not a scalar, it is a vector}}} = \underbrace{\sum_{i=x,y,z} a_i T_{ij}}_{\substack{\text{gives the } j^{th} \text{ component} \\ \text{of the resulting vector.} \\ \text{Since } j=x,y,z \text{ then this} \\ \text{calculation has to be done} \\ \text{three times to get the entire} \\ \text{vector}}} . \tag{A.2.31}$$

Now, we will calculate the divergence of the Maxwell stress tensor. Later, you will see that there is a correlation between the divergence of the Maxwell stress tensor and the force per unit volume. For a vector $\vec{v} = v_x \hat{x} + v_y \hat{y} + v_z \hat{z}$, divergence of $\vec{v}$ is:

$$\nabla \cdot \vec{v} = \left(\frac{\partial}{\partial x} \hat{x} + \frac{\partial}{\partial y} \hat{y} + \frac{\partial}{\partial z} \hat{z} \right) \cdot \left(v_x \hat{x} + v_y \hat{y} + v_z \hat{z} \right) = \frac{\partial v_x}{\partial x} + \frac{\partial v_y}{\partial y} + \frac{\partial v_z}{\partial z}. \tag{A.2.32}$$

From (A.2.31), divergence of a tensor can be expressed as:

$$\left(\nabla \cdot \vec{\vec{T}} \right)_j = \sum_{i=x,y,z} \nabla_i T_{ij} \tag{A.2.33}$$

where,

$$\nabla_{i=x} = \frac{\partial}{\partial x} \qquad \nabla_{i=y} = \frac{\partial}{\partial y} \qquad \nabla_{i=z} = \frac{\partial}{\partial z}$$

If we insert (A.2.30) into (A.2.31), it results in

$$\left(\nabla \cdot \ddot{T}\right)_j = \sum_{i=x,y,z} \nabla_i \left[\varepsilon_0 \left(E_i E_j - \frac{1}{2}\delta_{ij}E^2 \right) + \frac{1}{\mu_0}\left(B_i B_j - \frac{1}{2}\delta_{ij}B^2 \right) \right]. \tag{A.2.34}$$

We can distribute ∇_i to each term.

$$\begin{aligned}
\left(\nabla \cdot \ddot{T}\right)_j = \sum_{i=x,y,z} \bigg[&\varepsilon_0 \left(\left(\nabla_i E_i\right)E_j + \left(E_i\nabla_i\right)E_j - \frac{1}{2}\delta_{ij}\nabla_i E^2 \right) \\
&+ \frac{1}{\mu_0}\left(\left(\nabla_i B_i\right)B_j + \left(B_i\nabla_i\right)B_j - \frac{1}{2}\delta_{ij}\nabla_i B^2 \right) \bigg]
\end{aligned} \tag{A.2.35}$$

Please note that for the second terms in the electric field component in (A.2.35), instead of writing $(E_i\nabla_i)E_j$, we can write $E_i(\nabla_i E_j)$ by simply changing the location of the brackets. This also applies to the magnetic field term. The reason is to make the expression similar to what is given in (A.2.28).

Equation A.2.35 represents only one component of the resulting vector. For $i = x, y, z, \nabla_i E_i$ and $E_i\nabla_i$ can be expressed as:

$$\sum_{i=x,y,z} \nabla_i E_i = \nabla_x E_x + \nabla_y E_y + \nabla_z E_z = \nabla \cdot \vec{E} \tag{A.2.36}$$

$$\sum_{i=x,y,z} E_i\nabla_i = E_x\nabla_x + E_y\nabla_y + E_z\nabla_z = \vec{E} \cdot \nabla. \tag{A.2.37}$$

The same applies to the corresponding magnetic field terms:

$$\sum_{i=x,y,z} \nabla_i B_i = \nabla \cdot \vec{B} \text{ and } \sum_{i=x,y,z} B_i\nabla_i = \vec{B} \cdot \nabla. \tag{A.2.38}$$

Since $\delta_{xx} = \delta_{yy} = \delta_{zz} = 1$ and $\delta_{xy} = \delta_{xz} = \delta_{yz} = 0$, this term can have a non-zero value only when $i = j$.
Therefore,

$$\frac{1}{2}\delta_{ij}\nabla_i E^2 = \frac{1}{2}\nabla_j E^2 \text{ and } \frac{1}{2}\delta_{ij}\nabla_i B^2 = \frac{1}{2}\nabla_j B^2. \tag{A.2.39}$$

Inserting (A.2.36), (A.2.37), (A.2.38), and (A.2.39) into (A.2.35) results in

$$\left(\nabla \cdot \ddot{T}\right)_j = \varepsilon_0 \left[\left(\nabla \cdot \vec{E}\right)E_j + \left(\vec{E}\cdot\nabla\right)E_j - \frac{1}{2}\nabla_j E^2\right] + \frac{1}{\mu_0}\left[\left(\nabla \cdot \vec{B}\right)B_j + \left(\vec{B}\cdot\nabla\right)B_j - \frac{1}{2}\nabla_j B^2\right]. \qquad \text{(A.2.40)}$$

In (A.2.40), when $j = x$, $E_j = E_x \hat{x}$, then the same applies for the other components. Therefore, for the resulting $(\nabla \cdot \ddot{T})$ vector with x, y, z components, E_j term can be expressed as $\vec{E}$. The same approach can be applied on the $\frac{1}{2}\nabla_j E^2$ term. For $j = x$, this term will be $\frac{1}{2}\frac{\partial}{\partial x}E^2$. Hence, for the resulting $(\nabla \cdot \ddot{T})$ vector, the term $\frac{1}{2}\nabla_j E^2$ can be expressed as $\frac{1}{2}\nabla E^2$. We can apply the same on the terms with B_j.

Equation (A.2.40) was for each component of the vector $(\nabla \cdot \ddot{T})$, which now can be written as:

$$\left(\nabla \cdot \ddot{T}\right) = \varepsilon_0 \left[\left(\nabla \cdot \vec{E}\right)\vec{E} + \left(\vec{E}\cdot\nabla\right)\vec{E} - \frac{1}{2}\nabla E^2\right] + \frac{1}{\mu_0}\left[\left(\nabla \cdot \vec{B}\right)\vec{B} + \left(\vec{B}\cdot\nabla\right)\vec{B} - \frac{1}{2}\nabla B^2\right]. \qquad \text{(A.2.41)}$$

In (A.2.28), we had the expression for the force per unit volume:

$$\vec{f} = \underbrace{\varepsilon_0 \left[\left(\nabla \cdot \vec{E}\right)\vec{E} - \frac{1}{2}\nabla\left(E^2\right) + \left(\vec{E}\cdot\nabla\right)\vec{E}\right] + \frac{1}{\mu_0}\left[\left(\nabla \cdot \vec{B}\right)\vec{B} - \frac{1}{2}\nabla\left(B^2\right) + \left(\vec{B}\cdot\nabla\right)\vec{B}\right]}_{\nabla \cdot \ddot{T}} - \varepsilon_0 \frac{\partial}{\partial t}\left(\vec{E}\times\vec{B}\right)$$

Therefore,

$$\vec{f} = \nabla \cdot \ddot{T} - \varepsilon_0 \frac{\partial}{\partial t}\left(\vec{E}\times\vec{B}\right). \qquad \text{(A.2.42)}$$

The first term in (A.2.42) is the divergence of the Maxwell stress tensor. The second term comes from the work-energy theorem of electrodynamics. The energy flux per unit area (energy per unit time per unit area) transported in electromagnetic fields is called the Poynting vector, which is named after the English physicist John Henry Poynting after his discovery in 1884:

$$\vec{S} = \frac{1}{\mu_0}\left(\vec{E}\times\vec{B}\right). \qquad \text{(A.2.43)}$$

Applying (A.2.43) in (A.2.42) results in:

$$\vec{f} = \nabla \cdot \ddot{T} - \varepsilon_0 \mu_0 \frac{\partial}{\partial t}\vec{S}. \qquad \text{(A.2.44)}$$

For electric machine applications, force per unit volume can be expressed with the divergence of the Maxwell stress tensor and the second term drops out. Then the total force on the charges in volume V can be expressed as:

$$\vec{F} = \int_V \left(\nabla \cdot \ddot{T}\right) d\tau. \qquad \text{(A.2.45)}$$

According to the fundamental theorem of divergence, for a given vector $\vec{A}$:

$$\int_v \left(\nabla \cdot \vec{A} \right) d\tau = \oint_s \vec{A} \cdot d\vec{a}. \tag{A.2.46}$$

If we apply (A.2.46) in (A.2.45), total force [N] can be calculated as

$$\vec{F} = \oint_s \vec{T} \, d\vec{a}. \tag{A.2.47}$$

In electric machines, the force acting on the rotor is mainly the magnetic force. Therefore, the terms related to an electric field can be also dropped in the Maxwell stress tensor expression. Hence, (A.2.30) can be expressed as:

$$T_{ij} = \frac{1}{\mu_0} \left(B_i B_j - \frac{1}{2} \delta_{ij} B^2 \right) = \left(\frac{B_i}{\mu_0} B_j - \frac{1}{2} \delta_{ij} B \frac{B}{\mu_0} \right) = \left(H_i B_j - \frac{1}{2} \delta_{ij} B H \right) \tag{A.2.48}$$

again, δ_{ij} is the Kronecker delta and it is defined as

$$\delta_{ij} = \begin{bmatrix} \delta_{xx} & \delta_{xy} & \delta_{xz} \\ \delta_{yz} & \delta_{yy} & \delta_{yz} \\ \delta_{zx} & \delta_{zy} & \delta_{zz} \end{bmatrix} = \begin{bmatrix} 1 & 0 & 0 \\ 0 & 1 & 0 \\ 0 & 0 & 1 \end{bmatrix}. \tag{A.2.49}$$

As defined previously, T_{ij} is the force per unit area in the ith direction acting on the surface in the jth direction. If we take Cartesian coordinates as our reference system, then the forces acting on the surface in the direction of x-axis can be calculated as:

$$T_{xx} = \frac{1}{\mu_0} \left(B_x B_x - \frac{1}{2} B^2 \right) = \left(H_x B_x - \frac{1}{2} B H \right) \tag{A.2.50}$$

$$T_{xy} = \frac{1}{\mu_0} \left(B_x B_y \right) = \left(H_x B_y \right) \tag{A.2.51}$$

$$T_{xz} = \frac{1}{\mu_0} \left(B_x B_z \right) = \left(H_x B_z \right). \tag{A.2.52}$$

In this case, (A.2.50) is the pressure component, and (A.2.51) and (A.2.52) are the shear components. The forces acting on the surface in the direction of y-axis can be calculated as:

$$T_{yx} = \frac{1}{\mu_0} \left(B_y B_x \right) = \left(H_y B_x \right) \tag{A.2.53}$$

$$T_{yy} = \frac{1}{\mu_0} \left(B_y B_y - \frac{1}{2} B^2 \right) = \left(H_y B_y - \frac{1}{2} B H \right) \tag{A.2.54}$$

$$T_{yz} = \frac{1}{\mu_0}\left(B_y B_z\right) = \left(H_y B_z\right). \tag{A.2.55}$$

In this case, (A.2.54) is the pressure component, and (A.2.53) and (A.2.55) are shear components. Finally, the forces acting on the surface in the direction of x-axis can be calculated as:

$$T_{zx} = \frac{1}{\mu_0}\left(B_z B_x\right) = \left(H_z B_x\right) \tag{A.2.56}$$

$$T_{zy} = \frac{1}{\mu_0}\left(B_z B_y\right) = \left(H_z B_y\right) \tag{A.2.57}$$

$$T_{zz} = \frac{1}{\mu_0}\left(B_z B_z - \frac{1}{2}B^2\right) = \left(H_z B_z - \frac{1}{2}BH\right) \tag{A.2.58}$$

where (A.2.58) is the pressure component, and (A.2.56) and (A.2.57) are shear components. The Maxwell stress tensor, when Cartesian coordinates are taken as the reference frame, can now be expressed in the matrix form as:

$$\ddot{T}_{xyz} = \begin{bmatrix} \frac{1}{\mu_0}\left(B_x B_x - \frac{1}{2}B^2\right) & \frac{1}{\mu_0}\left(B_x B_y\right) & \frac{1}{\mu_0}\left(B_x B_z\right) \\[2mm] \frac{1}{\mu_0}\left(B_y B_x\right) & \frac{1}{\mu_0}\left(B_y B_y - \frac{1}{2}B^2\right) & \frac{1}{\mu_0}\left(B_y B_z\right) \\[2mm] \frac{1}{\mu_0}\left(B_z B_x\right) & \frac{1}{\mu_0}\left(B_z B_y\right) & \frac{1}{\mu_0}\left(B_z B_z - \frac{1}{2}B^2\right) \end{bmatrix} \tag{A.2.59}$$

$$\ddot{T}_{xyz} = \begin{bmatrix} \left(H_x B_x - \frac{1}{2}BH\right) & \left(H_x B_y\right) & \left(H_x B_z\right) \\[2mm] \left(H_y B_x\right) & \left(H_y B_y - \frac{1}{2}BH\right) & \left(H_y B_z\right) \\[2mm] \left(H_z B_x\right) & \left(H_z B_y\right) & \left(H_z B_z - \frac{1}{2}BH\right) \end{bmatrix} \tag{A.2.60}$$

where

$$B = \sqrt{B_x^2 + B_y^2 + B_z^2} \tag{A.2.61}$$

$$H = \sqrt{H_x^2 + H_y^2 + H_z^2}. \tag{A.2.62}$$

In two-dimensional analysis, the z-component of the Maxwell stress tensor can be eliminated. Hence, (A.2.59) can be expressed as:

$$\ddot{T}_{xy} = \begin{bmatrix} \dfrac{1}{\mu_0}\left(B_x B_x - \dfrac{1}{2}B^2\right) & \dfrac{1}{\mu_0}\left(B_x B_y\right) \\[3mm] \dfrac{1}{\mu_0}\left(B_y B_x\right) & \dfrac{1}{\mu_0}\left(B_y B_y - \dfrac{1}{2}B^2\right) \end{bmatrix} \tag{A.2.63}$$

Since $B^2 = B_x^2 + B_y^2 + B_z^2$, and $B_z = 0$ in two-dimensional analysis, (A.2.63) can be simplified as:

$$\ddot{T}_{xy} = \begin{bmatrix} \dfrac{1}{\mu_0}\left(B_x{}^2 - \dfrac{1}{2}\left(B_x{}^2 + B_y{}^2\right)\right) & \dfrac{1}{\mu_0}\left(B_x B_y\right) \\[3mm] \dfrac{1}{\mu_0}\left(B_y B_x\right) & \dfrac{1}{\mu_0}\left(B_y{}^2 - \dfrac{1}{2}\left(B_x{}^2 + B_y{}^2\right)\right) \end{bmatrix}$$

$$= \begin{bmatrix} \dfrac{1}{2\mu_0}\left(B_x{}^2 - B_y{}^2\right) & \dfrac{1}{\mu_0}\left(B_x B_y\right) \\[3mm] \dfrac{1}{\mu_0}\left(B_y B_x\right) & \dfrac{1}{2\mu_0}\left(B_y{}^2 - B_x{}^2\right) \end{bmatrix} \tag{A.2.64}$$

In rotating electric machine applications, tangential and radial components of the electromagnetic forces generate the torque, and the noise and vibration. If the radial and tangential coordinates are taken as the reference frame, the Maxwell stress tensors can be expressed as:

$$T_r = \frac{1}{2\mu_0}\left(B_r^2 - B_t^2\right) \tag{A.2.65}$$

$$T_t = \frac{1}{\mu_0}\left(B_r B_t\right). \tag{A.2.66}$$

These are the force densities in the radial and tangential directions in two-dimensional analysis. These are the same expressions given in (A.2.1) and (A.2.2) and their calculation starts with the Lorentz Force law as shown by (A.2.3). The Maxwell stress tensor in the radial and tangential coordinates can be expressed in the matrix form as:

$$\ddot{T}_{rt} = \begin{bmatrix} \dfrac{1}{2\mu_0}\left(B_r{}^2 - B_t{}^2\right) & \dfrac{1}{\mu_0}\left(B_r B_t\right) \\[3mm] \dfrac{1}{\mu_0}\left(B_t B_r\right) & \dfrac{1}{2\mu_0}\left(B_t{}^2 - B_r{}^2\right) \end{bmatrix}. \tag{A.2.67}$$

In (A.2.65) and (A.2.66), radial and tangential components of the magnetic flux density vector have to be used. The tensor transformation between the Cartesian coordinates in (A.2.64), and radial and tangential coordinates in (A.2.67) is

$$\ddot{T}_{rt} = P\,\ddot{T}_{xy}\,P^{-1} \tag{A.2.68}$$

where

$$P = \begin{bmatrix} cos(\theta) & sin(\theta) \\ -sin(\theta) & cos(\theta) \end{bmatrix} \qquad (A.2.69)$$

$$P^{-1} = \begin{bmatrix} cos(\theta) & -sin(\theta) \\ sin(\theta) & cos(\theta) \end{bmatrix}. \qquad (A.2.70)$$

Here, θ represents the position where the flux density components have been measured. Equation A.2.67 shows the tensors; hence, the force per unit area in radial and tangential directions. Using (A.2.45), forces can be calculated:

$$\vec{F}_r = \frac{1}{2\mu_0} \oint_s \left(B_r^2 - B_t^2 \right) d\vec{s} \qquad (A.2.71)$$

$$\vec{F}_t = \frac{1}{\mu_0} \oint_s \left(B_r B_t \right) d\vec{s}. \qquad (A.2.72)$$

In two-dimensional analysis, planar symmetry is considered, where the magnetic flux density is on the xy-plane and the current density has only z-axis component. In this case, the surface integral can be reduced to a line integral:

$$\vec{F}_r = \frac{1}{2\mu_0} L_{st} \int \left(B_r^2 - B_t^2 \right) d\vec{l} \qquad (A.2.73)$$

$$\vec{F}_t = \frac{1}{\mu_0} L_{st} \int \left(B_r B_t \right) d\vec{l} \qquad (A.2.74)$$

where L_{st} is the axial length of the motor. If an arc is defined as the integration line covering an entire pole pitch and if the arc is divided into n segments, the total force acting over the arc can be expressed as

$$F_r = \frac{1}{2\mu_0} L_{st} \sum_{i=1}^{n} \left[\left(B_{r_i}^2 - B_{t_i}^2 \right) \frac{D}{2} (\theta_i - \theta_{i-1}) \right] \qquad (A.2.75)$$

$$F_t = \frac{1}{\mu_0} L_{st} \sum_{i=1}^{n} \left[\left(B_{r_i} B_{t_i} \right) \frac{D}{2} (\theta_i - \theta_{i-1}) \right] \qquad (A.2.76)$$

where D is the diameter of the arc defined for line integration, n is the number of segments on the arc, and $(\theta_i - \theta_{i-1})$ is the angle in radians that each segment covers on the arc. The equations can be further simplified as

$$F_r = \frac{1}{2\mu_0} \frac{D}{2} L_{st} \sum_{i=1}^{n} \left[\left(B_{r_i}^2 - B_{t_i}^2 \right) (\theta_i - \theta_{i-1}) \right] \qquad (A.2.77)$$

$$F_t = \frac{1}{\mu_0} L_{st} \frac{D}{2} \sum_{i=1}^{n} \left[\left(B_{n} B_{t_i} \right) \left(\theta_i - \theta_{i-1} \right) \right].$$
(A.2.78)

Hence, torque can be calculated using the tangential force that was calculated for one pole:

$$T = \frac{D}{2} 2p F_t$$
(A.2.79)

where $2p$ corresponds to number of poles, and if the tangential forces are calculated for one pole. Therefore, the torque equation can be expressed as:

$$T = \frac{L_{st}}{\mu_0} \frac{D^2}{4} 2p \sum_{i=1}^{n} \left[\left(B_{n} B_{t_i} \right) \left(\theta_i - \theta_{i-1} \right) \right].$$
(A.2.80)

Appendix B: Proof for the Product Rule for Gradients

From (A.2.19), the product rule for gradients is expressed as:

$$\nabla \left(\vec{A} \cdot \vec{B} \right) = \vec{A} \times \left(\nabla \times \vec{B} \right) + \vec{B} \times \left(\nabla \times \vec{A} \right) + \left(\vec{A} \cdot \nabla \right) \vec{B} + \left(\vec{B} \cdot \nabla \right) \vec{A}.$$
(A.2.81)

The vectors $\vec{A}$ and $\vec{B}$, and the nabla operator are defined as:

$$\vec{A} = A_x \hat{x} + A_y \hat{y} + A_z \hat{z}$$
(A.2.82)

$$\vec{B} = B_x \hat{x} + B_y \hat{y} + B_z \hat{z}$$
(A.2.83)

$$\nabla = \hat{x} \frac{\partial}{\partial x} + \hat{y} \frac{\partial}{\partial y} + \hat{z} \frac{\partial}{\partial z}$$
(A.2.84)

where $\hat{x}, \hat{y}$, and $\hat{z}$ are the unit vectors. For the nabla operator, the unit vectors are written before the differentiator to emphasize that it is multiplication, not differentiation. In (A.2.81), for the term $\nabla(\vec{A} \cdot \vec{B})$, the dot product of two vectors is a scalar:

$$\vec{A} \cdot \vec{B} = A_x B_x + A_y B_y + A_z B_z.$$
(A.2.85)

Multiplying the nabla operator with a scalar is defined as the gradient:

$$\nabla \left(\vec{A} \cdot \vec{B} \right) = \hat{x} \frac{\partial}{\partial x} \left(A_x B_x + A_y B_y + A_z B_z \right) + \hat{y} \frac{\partial}{\partial y} \left(A_x B_x + A_y B_y + A_z B_z \right)$$
$$+ \hat{z} \frac{\partial}{\partial z} \left(A_x B_x + A_y B_y + A_z B_z \right).$$
(A.2.86)

For the term $\vec{A} \times (\nabla \times \vec{B})$ in (A.2.81), $\nabla \times \vec{B}$ is the curl of vector $\vec{B}$. It is calculated as given in (2.6), and the result is also a vector.

$$\nabla \times \vec{B} = \hat{x}\left(\frac{\partial}{\partial y}B_z - \frac{\partial}{\partial z}B_y\right) + \hat{y}\left(\frac{\partial}{\partial z}B_x - \frac{\partial}{\partial x}B_z\right) + \hat{z}\left(\frac{\partial}{\partial x}B_y - \frac{\partial}{\partial y}B_x\right). \qquad (A.2.87)$$

Hence, $\vec{A} \times (\nabla \times \vec{B})$ is the cross product of two vectors and can be calculated as given in (2.2):

$$\vec{A} \times \left(\nabla \times \vec{B}\right) = \hat{x}\left[A_y\left(\frac{\partial}{\partial x}B_y - \frac{\partial}{\partial y}B_x\right) - A_z\left(\frac{\partial}{\partial z}B_x - \frac{\partial}{\partial x}B_z\right)\right]$$

$$+ \hat{y}\left[A_z\left(\frac{\partial}{\partial y}B_z - \frac{\partial}{\partial z}B_y\right) - A_x\left(\frac{\partial}{\partial x}B_y - \frac{\partial}{\partial y}B_x\right)\right]$$

$$+ \hat{z}\left[A_x\left(\frac{\partial}{\partial z}B_x - \frac{\partial}{\partial x}B_z\right) - A_y\left(\frac{\partial}{\partial y}B_z - \frac{\partial}{\partial z}B_y\right)\right]. \qquad (A.2.88)$$

Each term in (A.2.88) can be further expanded:

$$\vec{A} \times \left(\nabla \times \vec{B}\right) = \hat{x}\left(A_y\frac{\partial}{\partial x}B_y - A_y\frac{\partial}{\partial y}B_x - A_z\frac{\partial}{\partial z}B_x + A_z\frac{\partial}{\partial x}B_z\right)$$

$$+ \hat{y}\left(A_z\frac{\partial}{\partial y}B_z - A_z\frac{\partial}{\partial z}B_y - A_x\frac{\partial}{\partial x}B_y + A_x\frac{\partial}{\partial y}B_x\right)$$

$$+ \hat{z}\left(A_x\frac{\partial}{\partial z}B_x - A_x\frac{\partial}{\partial x}B_z - A_y\frac{\partial}{\partial y}B_z + A_y\frac{\partial}{\partial z}B_y\right). \qquad (A.2.89)$$

By symmetry, the term $\vec{B} \times (\nabla \times \vec{A})$ in (A.2.81) can calculated using (A.2.90):

$$\vec{B} \times \left(\nabla \times \vec{A}\right) = \hat{x}\left(B_y\frac{\partial}{\partial x}A_y - B_y\frac{\partial}{\partial y}A_x - B_z\frac{\partial}{\partial z}A_x + B_z\frac{\partial}{\partial x}A_z\right)$$

$$+ \hat{y}\left(B_z\frac{\partial}{\partial y}A_z - B_z\frac{\partial}{\partial z}A_y - B_x\frac{\partial}{\partial x}A_y + B_x\frac{\partial}{\partial y}A_x\right)$$

$$+ \hat{z}\left(B_x\frac{\partial}{\partial z}A_x - B_x\frac{\partial}{\partial x}A_z - B_y\frac{\partial}{\partial y}A_z + B_y\frac{\partial}{\partial z}A_y\right). \qquad (A.2.90)$$

For the term $(\vec{A} \cdot \nabla)\vec{B}$ in (A.2.81), $(\vec{A} \cdot \nabla)$ can be considered as the dot product of two vectors. Hence, the result is a scalar:

$$\vec{A} \cdot \nabla = A_x \frac{\partial}{\partial x} + A_y \frac{\partial}{\partial y} + A_z \frac{\partial}{\partial z}. \tag{A.2.91}$$

Then, this value is multiplied by each term of vector $\vec{B}$ and the final results will also be a vector:

$$\left(\vec{A} \cdot \nabla\right)\vec{B} = \hat{x}\left(A_x \frac{\partial}{\partial x} + A_y \frac{\partial}{\partial y} + A_z \frac{\partial}{\partial z} \right)B_x + \hat{y}\left(A_x \frac{\partial}{\partial x} + A_y \frac{\partial}{\partial y} + A_z \frac{\partial}{\partial z} \right)B_y$$

$$+ \hat{z}\left(A_x \frac{\partial}{\partial x} + A_y \frac{\partial}{\partial y} + A_z \frac{\partial}{\partial z} \right)B_z. \tag{A.2.92}$$

It can be reorganized as:

$$\left(\vec{A} \cdot \nabla\right)\vec{B} = \hat{x}\left(A_x \frac{\partial}{\partial x}B_x + A_y \frac{\partial}{\partial y}B_x + A_z \frac{\partial}{\partial z}B_x \right) + \hat{y}\left(A_x \frac{\partial}{\partial x}B_y + A_y \frac{\partial}{\partial y}B_y + A_z \frac{\partial}{\partial z}B_y \right)$$

$$+ \hat{z}\left(A_x \frac{\partial}{\partial x}B_z + A_y \frac{\partial}{\partial y}B_z + A_z \frac{\partial}{\partial z}B_z \right). \tag{A.2.93}$$

By symmetry, the term $(\vec{B} \cdot \nabla)\vec{A}$ in (A.2.81) can be calculated using (A.2.93):

$$\left(\vec{B} \cdot \nabla\right)\vec{A} = \hat{x}\left(B_x \frac{\partial}{\partial x}A_x + B_y \frac{\partial}{\partial y}A_x + B_z \frac{\partial}{\partial z}A_x \right) + \hat{y}\left(B_x \frac{\partial}{\partial x}A_y + B_y \frac{\partial}{\partial y}A_y + B_z \frac{\partial}{\partial z}A_y \right)$$

$$+ \hat{z}\left(B_x \frac{\partial}{\partial x}A_z + B_y \frac{\partial}{\partial y}A_z + B_z \frac{\partial}{\partial z}A_z \right). \tag{A.2.94}$$

In order to calculate $\vec{A} \times \left(\nabla \times \vec{B}\right) + \vec{B} \times \left(\nabla \times \vec{A}\right) + \left(\vec{A} \cdot \nabla\right)\vec{B} + \left(\vec{B} \cdot \nabla\right)\vec{A}$, we add up (A.2.89), (A.2.90), (A.2.93), and (A.2.94):

$$\bar{A} \times (\nabla \times \bar{B}) + \bar{B} \times (\nabla \times \bar{A}) + (\bar{A} \cdot \nabla)\bar{B} + (\bar{B} \cdot \nabla)\bar{A} = \hat{x}$$

$$\left[\left(A_y \frac{\partial}{\partial x} B_y - A_y \frac{\partial}{\partial y} B_x - A_z \frac{\partial}{\partial z} B_x + A_z \frac{\partial}{\partial x} B_z \right) + \left(B_y \frac{\partial}{\partial x} A_y - B_y \frac{\partial}{\partial y} A_x - B_z \frac{\partial}{\partial z} A_x + B_z \frac{\partial}{\partial x} A_z \right) \right.$$
$$\left. + \left(A_x \frac{\partial}{\partial x} B_x + A_y \frac{\partial}{\partial y} B_x + A_z \frac{\partial}{\partial z} B_x \right) + \left(B_x \frac{\partial}{\partial x} A_x + B_y \frac{\partial}{\partial y} A_x + B_z \frac{\partial}{\partial z} A_x \right) \right]$$

$$+ \hat{y} \left[\left(A_z \frac{\partial}{\partial y} B_z - A_z \frac{\partial}{\partial z} B_y - A_x \frac{\partial}{\partial x} B_y + A_x \frac{\partial}{\partial y} B_x \right) + \left(B_z \frac{\partial}{\partial y} A_z - B_z \frac{\partial}{\partial z} A_y - B_x \frac{\partial}{\partial x} A_y + B_x \frac{\partial}{\partial y} A_x \right) \right.$$
$$\left. + \left(A_x \frac{\partial}{\partial x} B_y + A_y \frac{\partial}{\partial y} B_y + A_z \frac{\partial}{\partial z} B_y \right) + \left(B_x \frac{\partial}{\partial x} A_y + B_y \frac{\partial}{\partial y} A_y + B_z \frac{\partial}{\partial z} A_y \right) \right]$$

$$+ \hat{z} \left[\left(A_x \frac{\partial}{\partial z} B_x - A_x \frac{\partial}{\partial x} B_z - A_y \frac{\partial}{\partial y} B_z + A_y \frac{\partial}{\partial z} B_y \right) + \left(B_x \frac{\partial}{\partial z} A_x - B_x \frac{\partial}{\partial x} A_z - B_y \frac{\partial}{\partial y} A_z + B_y \frac{\partial}{\partial z} A_y \right) \right.$$
$$\left. + \left(A_x \frac{\partial}{\partial x} B_z + A_y \frac{\partial}{\partial y} B_z + A_z \frac{\partial}{\partial z} B_z \right) + \left(B_x \frac{\partial}{\partial x} A_z + B_y \frac{\partial}{\partial y} A_z + B_z \frac{\partial}{\partial z} A_z \right) \right].$$

$$(A.2.95)$$

The terms in the boxes in (A.2.95) cancel each other. After reorganizing the terms, the resulting equation will be:

$$\vec{A}\times\left(\nabla\times\vec{B}\right)+\vec{B}\times\left(\nabla\times\vec{A}\right)+\left(\vec{A}\cdot\nabla\right)\vec{B}+\left(\vec{B}\cdot\nabla\right)\vec{A}=\hat{x}\left[\left(A_x\frac{\partial}{\partial x}B_x+B_x\frac{\partial}{\partial x}A_x\right)+\left(A_y\frac{\partial}{\partial x}B_y+B_y\frac{\partial}{\partial x}A_y\right)+\left(A_z\frac{\partial}{\partial x}B_z+B_z\frac{\partial}{\partial x}A_z\right)\right]$$

$$+\hat{y}\left[\left(A_x\frac{\partial}{\partial y}B_x+B_x\frac{\partial}{\partial y}A_x\right)+\left(A_y\frac{\partial}{\partial y}B_y+B_y\frac{\partial}{\partial y}A_y\right)+\left(A_z\frac{\partial}{\partial y}B_z+B_z\frac{\partial}{\partial y}A_z\right)\right]$$

$$+\hat{z}\left[\left(A_x\frac{\partial}{\partial z}B_x+B_x\frac{\partial}{\partial z}A_x\right)+\left(A_y\frac{\partial}{\partial z}B_y+B_y\frac{\partial}{\partial z}A_y\right)+\left(A_z\frac{\partial}{\partial z}B_z+B_z\frac{\partial}{\partial z}A_z\right)\right].$$

$$(A.2.96)$$

According to the product rule for differentiation:

$$\frac{\partial}{\partial x}\left[f(x)g(x)\right]=f(x)\frac{\partial}{\partial x}g(x)+g(x)\frac{\partial}{\partial x}f(x). \qquad (A.2.97)$$

Notice that each term in (A.2.96) is similar to the right side of (A.2.97). Therefore, (A.2.96) can be reorganized as:

$$\vec{A}\times\left(\nabla\times\vec{B}\right)+\vec{B}\times\left(\nabla\times\vec{A}\right)+\left(\vec{A}\cdot\nabla\right)\vec{B}+\left(\vec{B}\cdot\nabla\right)\vec{A}=\hat{x}\left(\frac{\partial}{\partial x}A_xB_x+\frac{\partial}{\partial x}A_yB_y+\frac{\partial}{\partial x}A_zB_z\right)$$

$$+\hat{y}\left(\frac{\partial}{\partial y}A_xB_x+\frac{\partial}{\partial y}A_yB_y+\frac{\partial}{\partial y}A_zB_z\right)$$

$$+\hat{z}\left(\frac{\partial}{\partial z}A_xB_x+\frac{\partial}{\partial z}A_yB_y+\frac{\partial}{\partial z}A_zB_z\right). \qquad (A.2.98)$$

Equation (A.2.98) can be reorganized as:

$$\vec{A}\times\left(\nabla\times\vec{B}\right)+\vec{B}\times\left(\nabla\times\vec{A}\right)+\left(\vec{A}\cdot\nabla\right)\vec{B}+\left(\vec{B}\cdot\nabla\right)\vec{A}=\hat{x}\frac{\partial}{\partial x}\left(A_xB_x+A_yB_y+A_zB_z\right)$$

$$+\hat{y}\frac{\partial}{\partial y}\left(A_xB_x+A_yB_y+A_zB_z\right)$$

$$+\hat{z}\frac{\partial}{\partial z}\left(A_xB_x+A_yB_y+A_zB_z\right). \qquad (A.2.99)$$

The above expression equals to (A.2.86) for $\nabla\left(\vec{A}\cdot\vec{B}\right)$ and shows that

$$\nabla\left(\vec{A}\cdot\vec{B}\right)=\vec{A}\times\left(\nabla\times\vec{B}\right)+\vec{B}\times\left(\nabla\times\vec{A}\right)+\left(\vec{A}\cdot\nabla\right)\vec{B}+\left(\vec{B}\cdot\nabla\right)\vec{A}. \qquad (A.2.100)$$

Questions and Problems

1. Figure Q.2.1 shows a rough block diagram showing the principal equations of a magnetic system. Please fill out the spaces and complete the diagram. Please write down the equations linking each block.

2. As discussed in Section 2.7, without an electric field applied, the free electrons in a conductor have thermal motion in all directions. Free electrons in copper have an average speed around $\vec{v}_e = 10^6$ m/s and, therefore, time between the collusion of free electrons with the atoms is around $= 3 \times 10^{-14}$ s for copper at room temperature. When the force generated by the electric field is applied, the free electrons pick up a speed between the collusion. This is called the drift velocity and it was quantified in (2.30):

$$\vec{v}_d = \vec{a}_e \tau = \frac{\vec{F}_e}{m_e} \tau = \frac{q\vec{E}}{m_e} \tau.$$

If the charge of a free electron is $q = 1.6 \times 10^{-19}$ C and the mass of an electron is $m_e = 10^{-30}$ kg, if we apply 10 V potential difference over 10 meters of copper conductor, what would be the drift velocity? What would you conclude if you compare the drift velocity with the $\vec{v}_e$?

3. The magnetic circuit shown in Figure Q.2.2 is made of a linear magnetic material with a relative permeability of 1000. The dimensional parameters of the core are as follows: h = 35 mm, w = 25 mm, L_{st} = 10 mm, and $d_1 = d_2 = 5$ mm. The length of

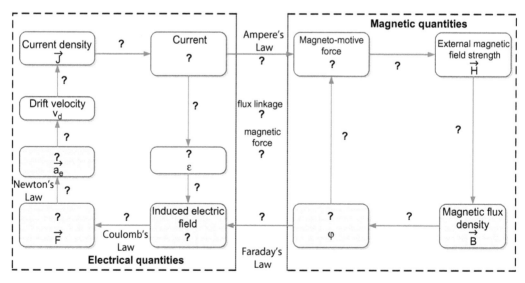

FIGURE Q.2.1
Block diagram for the principle equations of a magnetic system.

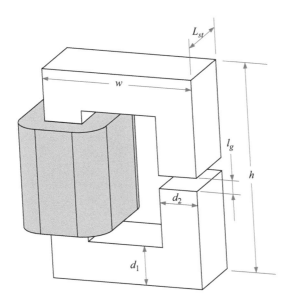

FIGURE Q.2.2
Dimensional parameters of a magnetic system with air gap.

the air gap l_g is 0.5 mm. The coil has 10 turns and 20 A is applied to the coil. Please calculate the flux linkage of the coil? What is the magnetic field intensity in the core and the air gap? What is the magnetic flux density?

4. Table Q.2.1 shows the finite element simulation results for a three-phase 12/8 SRM in one-half electrical cycle, when one of the phases is energized with a constant current of 20 A when the rotor was rotating at 1000 rpm. The phase resistance is 0.2117 Ω.

In (2.41), the voltage applied to a phase of an SRM was defined as:

$$v = Ri + \frac{d\lambda}{dt}$$

In Table Q.2.1, the flux linkage and time vectors are given as $[\lambda]_{t\times1}$ and $[time]_{t\times1}$, respectively. Therefore, from (2.41), the terminal voltage can be calculated as:

$$[v]_{t-1\times1} = Ri + \frac{\lambda(t) - \lambda(t-1)}{time(t) - time(t-1)}.$$

Write a MATLAB® algorithm, to calculate the terminal voltage from the flux linkage given in Table Q.2.1. Compare the calculated terminal voltage and the voltage data given in Table Q.2.1. Then write another algorithm to calculate the terminal voltage using flux linkage and rotor position (mechanical angle in radians).

TABLE Q.2.1

Finite Element Simulation Results for a 12/8 SRM in One-Half Electrical Cycle at 20A Constant Current Excitation at 1000 rpm

Time [s]	Mechanical Angle [deg]	Electrical Position [deg]	Torque [Nm]	Flux Linkage [Wb]	Voltage [V]
0	0	0	2.87E-06	2.09E-02	4.253907
0.000125	0.75	6	0.099056	0.020929	4.976992
0.00025	1.5	12	0.199265	0.021123	6.037038
0.000375	2.25	18	0.315334	0.021453	7.16778
0.0005	3	24	0.449664	0.021947	8.526255
0.000625	3.75	30	0.625527	0.022633	10.15457
0.00075	4.5	36	0.857403	0.023582	12.3981
0.000875	5.25	42	1.192798	0.024881	15.48
0.001	6	48	1.773084	0.026763	20.73045
0.001125	6.75	54	2.952918	0.029689	30.53119
0.00125	7.5	60	5.19303	0.034303	44.51888
0.001375	8.25	66	6.256288	0.040131	52.30015
0.0015	9	72	6.452102	0.046365	55.24284
0.001625	9.75	78	6.536435	0.052795	55.75648
0.00175	10.5	84	6.559905	0.059299	56.38191
0.001875	11.25	90	6.601224	0.065891	56.98873
0.002	12	96	6.610802	0.072521	57.24058
0.002125	12.75	102	6.613843	0.07915	57.17736
0.00225	13.5	108	6.588581	0.085807	57.53025
0.002375	14.25	114	6.547559	0.092453	57.42973
0.0025	15	120	6.481668	0.099095	57.29031
0.002625	15.75	126	6.449959	0.105726	57.09963
0.00275	16.5	132	6.354507	0.112277	56.38438
0.002875	17.25	138	6.262703	0.118673	54.92126
0.003	18	144	6.032075	0.124704	51.67962
0.003125	18.75	150	5.677352	0.130241	47.45344
0.00325	19.5	156	5.12714	0.135283	43.52018
0.003375	20.25	162	4.377247	0.139535	35.96137
0.0035	21	168	3.504661	0.142279	20.79484
0.003625	21.75	174	2.561391	0.143427	12.19929
0.00375	22.5	180	-5.08E-06	1.44E-01	6.825071

5. In the previous question, we calculated voltage from the flux linkage and time data. It is possible to calculate the flux linkage from voltage. From (2.41), flux linkage can be expressed as

$$v = Ri + \omega \frac{d\lambda}{d\theta} \Rightarrow \frac{d\lambda}{d\theta} = \frac{1}{\omega}(v - Ri) \Rightarrow d\lambda = \frac{1}{\omega}(v - Ri)d\theta \Rightarrow \lambda = \lambda_0 + \int \frac{1}{\omega}(v - Ri)d\theta$$

In Table Q.2.1, voltage and rotor position vectors are given as $[v]_{t \times 1}$ and $[\theta]_{t \times 1}$, respectively. The flux linkage for θ_2 is calculated as:

$$\lambda(\theta_2) = \lambda(\theta_1) + \frac{1}{\omega}\big(v(\theta_2) - Ri\big)(\theta_2 - \theta_1).$$

Flux linkage at θ_3 is calculated as:

$$\lambda(\theta_3) = \lambda(\theta_2) + \frac{1}{\omega}\big(v(\theta_3) - Ri\big)(\theta_3 - \theta_2)$$

$$= \lambda(\theta_1) + \frac{1}{\omega}\big(v(\theta_2) - Ri\big)(\theta_2 - \theta_1) + \frac{1}{\omega}\big(v(\theta_3) - Ri\big)(\theta_3 - \theta_2).$$

Flux linkage at θ_4 is calculated as:

$$\lambda(\theta_4) = \lambda(\theta_3) + \frac{1}{\omega}\big(v(\theta_4) - Ri\big)(\theta_4 - \theta_3) = \lambda(\theta_1) + \frac{1}{\omega}\big(v(\theta_2) - Ri\big)(\theta_2 - \theta_1)$$

$$+ \frac{1}{\omega}\big(v(\theta_3) - Ri\big)(\theta_3 - \theta_2)$$

$$+ \frac{1}{\omega}\big(v(\theta_4) - Ri\big)(\theta_4 - \theta_3).$$

Therefore, the calculation of the flux linkage vector can be expressed as:

$$[\lambda]_{t-1\times1} = \lambda(\theta_{t-1}) + \sum_{t=2}^{j} \frac{1}{\omega}\big(v(\theta_t) - Ri\big)(\theta_t - \theta_{t-1}).$$

Here only $\lambda(\theta_1)$ is not known, which can be calculated either analytically using linear approximation at unaligned position or using this value directly from Table Q.2.1. Now please write a MATLAB algorithm to calculate the flux linkage from voltage data in Table Q.2.1. Then compare the calculated flux linkage with the data in the table.

6. It was quantified in (2.90) that the electromagnetic torque can be calculated from the rate of change of co-energy with the rotor position. Flux linkage is the rate of change of co-energy with current.

$$\lambda = \frac{\partial W_c}{\partial i}\bigg|_{\theta=const} \quad \text{and} \quad T_e = \frac{\partial W_c}{\partial \theta}\bigg|_{i=const}$$

Tables Q.2.2 and Q.2.3 show the torque and flux linkage profiles of the same SRM for various excitation currents and rotor positions (mechanical angles). If (2.90) holds and the losses in the magnetic circuit are negligible, co-energy W_c can be calculated

TABLE Q.2.2

Flux Linkage of a 12/8 SRM at Different Rotor Positions and Excitation Currents

Rotor position [deg]	Current [A]									
	2	4	6	8	10	12	14	16	18	20
0	0.0021	0.0042	0.0062	0.0083	0.0104	0.0125	0.0146	0.0167	0.0188	0.0209
0.75	0.0021	0.0042	0.0063	0.0083	0.0104	0.0125	0.0146	0.0167	0.0188	0.0209
1.5	0.0021	0.0042	0.0063	0.0084	0.0105	0.0127	0.0148	0.0169	0.0190	0.0211
2.25	0.0021	0.0043	0.0064	0.0086	0.0107	0.0129	0.0150	0.0172	0.0193	0.0215
3	0.0022	0.0044	0.0066	0.0088	0.0110	0.0131	0.0153	0.0175	0.0197	0.0219
3.75	0.0022	0.0045	0.0068	0.0090	0.0113	0.0136	0.0158	0.0181	0.0204	0.0226
4.5	0.0023	0.0047	0.0070	0.0094	0.0118	0.0141	0.0165	0.0189	0.0212	0.0236
5.25	0.0025	0.0049	0.0074	0.0099	0.0124	0.0149	0.0174	0.0199	0.0224	0.0249
6	0.0027	0.0053	0.0080	0.0107	0.0133	0.0160	0.0187	0.0214	0.0241	0.0268
6.75	0.0029	0.0059	0.0089	0.0118	0.0148	0.0178	0.0208	0.0237	0.0267	0.0297
7.5	0.0035	0.0070	0.0105	0.0140	0.0175	0.0210	0.0245	0.0278	0.0311	0.0343
8.25	0.0043	0.0087	0.0132	0.0176	0.0220	0.0261	0.0299	0.0334	0.0368	0.0401
9	0.0053	0.0106	0.0160	0.0214	0.0268	0.0318	0.0360	0.0396	0.0430	0.0464
9.75	0.0062	0.0125	0.0188	0.0252	0.0315	0.0374	0.0422	0.0460	0.0495	0.0528
10.5	0.0071	0.0144	0.0217	0.0290	0.0363	0.0431	0.0486	0.0526	0.0560	0.0593
11.25	0.0080	0.0162	0.0245	0.0327	0.0410	0.0487	0.0550	0.0591	0.0626	0.0659
12	0.0089	0.0181	0.0273	0.0366	0.0457	0.0544	0.0614	0.0658	0.0693	0.0725
12.75	0.0098	0.0200	0.0301	0.0403	0.0504	0.0600	0.0677	0.0723	0.0759	0.0791
13.5	0.0108	0.0218	0.0330	0.0441	0.0552	0.0656	0.0740	0.0790	0.0826	0.0858
14.25	0.0117	0.0237	0.0357	0.0478	0.0598	0.0712	0.0802	0.0857	0.0892	0.0925
15	0.0126	0.0255	0.0386	0.0516	0.0645	0.0767	0.0865	0.0923	0.0959	0.0991
15.75	0.0134	0.0273	0.0413	0.0553	0.0691	0.0822	0.0926	0.0989	0.1025	0.1057
16.5	0.0143	0.0292	0.0441	0.0590	0.0738	0.0877	0.0987	0.1055	0.1091	0.1123
17.25	0.0152	0.0310	0.0468	0.0626	0.0783	0.0931	0.1047	0.1118	0.1155	0.1187
18	0.0161	0.0328	0.0496	0.0663	0.0829	0.0984	0.1105	0.1181	0.1217	0.1247
18.75	0.0170	0.0346	0.0522	0.0698	0.0873	0.1036	0.1159	0.1237	0.1274	0.1302
19.5	0.0178	0.0363	0.0549	0.0734	0.0917	0.1086	0.1207	0.1281	0.1324	0.1353
20.25	0.0186	0.0380	0.0574	0.0768	0.0959	0.1132	0.1245	0.1314	0.1365	0.1395
21	0.0194	0.0397	0.0599	0.0801	0.0999	0.1172	0.1274	0.1339	0.1388	0.1423
21.75	0.0201	0.0411	0.0621	0.0829	0.1033	0.1201	0.1294	0.1357	0.1403	0.1434
22.5	0.0206	0.0419	0.0634	0.0846	0.1050	0.1220	0.1310	0.1370	0.1410	0.1440

TABLE Q.2.3

Electromagnetic Torque of a 12/8 SRM at Different Rotor Positions and Excitation Currents

Rotor position [deg]	Current [A]									
	2	4	6	8	10	12	14	16	18	20
0	0.0000	0.0000	0.0000	0.0000	0.0000	0.0000	0.0000	0.0000	0.0000	0.0000
0.75	0.0010	0.0039	0.0088	0.0157	0.0246	0.0355	0.0484	0.0633	0.0802	0.0991
1.5	0.0020	0.0079	0.0178	0.0317	0.0496	0.0715	0.0974	0.1273	0.1613	0.1993
2.25	0.0031	0.0125	0.0281	0.0501	0.0784	0.1131	0.1541	0.2015	0.2552	0.3153
3	0.0044	0.0178	0.0401	0.0714	0.1118	0.1613	0.2197	0.2873	0.3639	0.4497
3.75	0.0062	0.0247	0.0558	0.0994	0.1555	0.2243	0.3057	0.3997	0.5063	0.6255
4.5	0.0084	0.0339	0.0764	0.1362	0.2131	0.3074	0.4189	0.5478	0.6939	0.8574
5.25	0.0117	0.0471	0.1062	0.1893	0.2964	0.4276	0.5828	0.7620	0.9654	1.1928
6	0.0174	0.0699	0.1578	0.2813	0.4405	0.6354	0.8662	1.1327	1.4350	1.7731
6.75	0.0289	0.1163	0.2628	0.4686	0.7341	1.0591	1.4439	1.8882	2.3918	2.9529
7.5	0.0575	0.2318	0.5243	0.9357	1.4662	2.1140	2.8528	3.6023	4.3859	5.1930
8.25	0.0726	0.2938	0.6657	1.1891	1.8620	2.6442	3.5083	4.4013	5.3227	6.2563
9	0.0719	0.2918	0.6622	1.1838	1.8548	2.6472	3.5512	4.5045	5.4677	6.4521
9.75	0.0719	0.2924	0.6645	1.1884	1.8622	2.6571	3.5713	4.5459	5.5404	6.5364
10.5	0.0715	0.2917	0.6634	1.1869	1.8600	2.6552	3.5717	4.5589	5.5578	6.5599
11.25	0.0718	0.2935	0.6681	1.1955	1.8736	2.6731	3.5893	4.5808	5.5877	6.6012
12	0.0715	0.2930	0.6675	1.1948	1.8726	2.6728	3.5997	4.5883	5.5913	6.6108
12.75	0.0716	0.2942	0.6705	1.2001	1.8809	2.6817	3.6016	4.5714	5.5942	6.6138
13.5	0.0710	0.2922	0.6663	1.1927	1.8693	2.6669	3.5821	4.5691	5.5717	6.5886
14.25	0.0706	0.2910	0.6635	1.1879	1.8615	2.6535	3.5510	4.5412	5.5405	6.5476
15	0.0694	0.2865	0.6535	1.1698	1.8323	2.6144	3.5071	4.4844	5.4781	6.4817
15.75	0.0688	0.2843	0.6486	1.1610	1.8189	2.5926	3.4728	4.4508	5.4514	6.4500
16.5	0.0678	0.2808	0.6407	1.1469	1.7962	2.5613	3.4305	4.3907	5.3759	6.3545
17.25	0.0676	0.2803	0.6396	1.1448	1.7922	2.5510	3.4107	4.3413	5.3071	6.2627
18	0.0669	0.2777	0.6339	1.1344	1.7749	2.5242	3.3620	4.2804	5.1395	6.0321
18.75	0.0666	0.2766	0.6315	1.1297	1.7660	2.5051	3.2985	4.0553	4.8658	5.6774
19.5	0.0653	0.2715	0.6197	1.1082	1.7307	2.4441	3.1544	3.7610	4.4067	5.1271
20.25	0.0636	0.2644	0.6035	1.0789	1.6823	2.3540	2.9413	3.4078	3.8462	4.3772
21	0.0591	0.2462	0.5619	1.0040	1.5621	2.1544	2.6012	2.9468	3.2501	3.5047
21.75	0.0481	0.2003	0.4572	0.8163	1.2664	1.7124	2.0065	2.2315	2.4177	2.5614
22.5	0.0000	0.0000	0.0000	0.0000	0.0000	0.0000	0.0000	0.0000	0.0000	0.0000

```
%% Calculate torque from the flux linkage
for i=2:length(current_array)
    for j=2:length(mech_angle)
        if j==2
            % co_energy0=trapz(current_array(1,i),flux_linkage(1,1:i)); OR
            co_energy0=0;
            for k=2:i
                temp=((flux_linkage(1,k)+...
                    flux_linkage(1,k-1))/2)*...
                    (current_array(k)-current_array(k-1));
                co_energy0=co_energy0+temp;
            end
        end
        % co_energy=trapz(current_array(1:i),flux_linkage(j,1:i)); OR
        co_energy=0;
        for k=2:i
            temp=((flux_linkage(j,k)+...
                flux_linkage(j,k-1))/2)*...
                (current_array(k)-current_array(k-1));
            co_energy=co_energy+temp;
        end
        torque_from_flux_linkage(j,i)=...
            (co_energy-co_energy0)/...
            ((mech_angle(j)-mech_angle(j-1))*(pi/180));
        co_energy0=co_energy;
    end
```

FIGURE Q.2.3
Sample algorithm to calculate the electromagnetic torque from flux linkage.

from the flux linkage by calculating $\int \lambda \, di$. Then, torque can be calculated from co-energy by calculating $dW_c / d\theta$.

Figure Q.2.3 shows a sample MATLAB algorithm to calculate the torque from the flux linkage. Run the algorithm and plot the electromagnetic torque and the torque calculated from flux linkage together. Do they match?

7. The generalized torque expression in an SRM was quantified in (2.90) as:

$$T_e = \left. \frac{\partial W_c}{\partial \theta} \right|_{i=const}.$$

It was also quantified in (2.70) that the magnetic stored energy can be calculated as:

$$W_f = \frac{1}{2} L i^2.$$

As shown in Figure 2.12, for the linear case the magnetic stored energy and co-energy are the same. Using this information, please calculate the electromagnetic torque in an SRM for the linear case when the core is not saturated.

References

1. D. Fleisch, *A Student's Guide to Maxwell's Equations*, Cambridge, UK: Cambridge University Press, 2013.
2. D. J. Griffiths, *Introduction to Electrodynamics*, Upper Saddle Drive, NJ: Prentice Hall, 1999.
3. B. Bilgin and A. Sathyan, "Fundamentals of electric machines," in *Advanced Electric Drive Vehicles*, Boca Raton, FL: CRC Press, 2014.
4. W. Lewin, "Faraday's Law – Most Physics College Books have it WRONG!" *MIT Open Courseware, Lecture Notes for Physics – 8.02*, Retrieved: October, 2016. (Online). Available: http://web.mit.edu/8.02/www/Spring02/lectures/lecsup4-1.pdf.
5. W. G. Hurley and W. H. Wofle, *Transformers and Inductors for Power Electronics Theory, Design and Applications*, Chichester, UK: John Wiley & Sons, 2013.
6. S. Zurek, Co-energy, Encyclopedia Magnetica. Accessed: October, 2016. (Online). Available: http://www.encyclopedia-magnetica.com/doku.php/coenergy.
7. W. Tong, *Mechanical Design of Electric Motors*, Boca Raton, FL: CRC Press, 2014.

3

Derivation of Pole Configuration in Switched Reluctance Machines

Berker Bilgin

CONTENTS

3.1 Introduction

A Switched Reluctance Machine (SRM) is identified by the number of stator poles and number of rotor poles. As shown in Figure 3.1, an SRM with 6 stator poles and 4 rotor poles is called a 6/4 SRM, whereas an SRM with 8 stator poles and 6 rotor poles is called an 8/6 SRM. The selection of number of stator poles and number of rotor poles has a significant effect on the operational and performance parameters of the machine, including the number of phases and number of torque pulsations [1]. Many other switched reluctance motor topologies are also possible including double rotor, double stator, multi-teeth, and toroidal winding SRMs [2–11].

An SRM has salient poles both on the stator and the rotor. When an electrical phase is energized with constant current, the electromagnetic torque is dependent on the relative position between the stator and rotor poles.

Depending on the number of stator and rotor poles, each stator pole has a certain electrical position. If the coils around the stator poles, which have the same electrical angle at a certain rotor position, are excited with the same current, these poles generate the same torque and, hence, those coils make up a phase. Therefore, an electrical phase in an SRM is defined as the connection of coils together whose stator poles have the same electrical angle. Since SRM has concentrated windings, the number of coils is equal to the number of stator poles. Depending on the number of stator poles, the coils belonging to one phase can be connected in series, parallel, or in a series/parallel combination. If the stator poles sharing the same electrical position are placed evenly around the stator circumference, then a balanced torque generation can be maintained in each phase. The difference between the electrical angles of stator poles belonging to different phases (phase shift) enables continuous torque production.

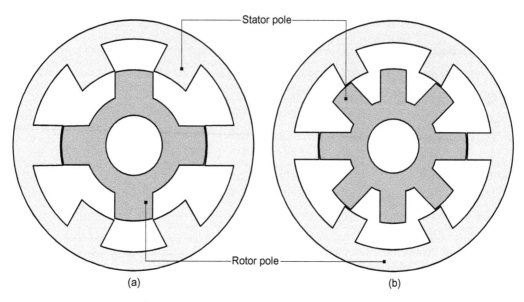

FIGURE 3.1
Cross section view of switched reluctance machine: (a) 6/4 interior-rotor SRM and (b) 8/6 exterior-rotor SRM.

In an SRM, an electrical cycle is defined when a rotor pole moves from the position where a certain stator pole is in the middle of two consecutive rotor poles (unaligned position) to the next similar position. In Figure 3.2a, the stator pole Ns#1 is at the unaligned position and it is in the middle of consecutive rotor poles Nr#1 and Nr#2. One electrical cycle will be completed when Ns#1 is in the middle of Nr#4 and Nr#1, and the rotor rotates in counterclockwise direction. In this case, 360° of an electrical cycle is completed.

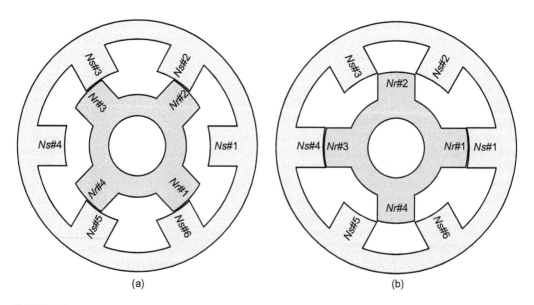

FIGURE 3.2
Cross section view of a 6/4 SRM: (a) unaligned position and (b) aligned position.

Figure 3.2b shows the position where *Ns*#1 is aligned with *Nr*#1. The same aligned position can be achieved when *Ns*#1 is aligned with *Nr*#2, *Nr*#3, and *Nr*#4. These correspond to different mechanical positions. The aligned position can also be achieved for the other stator poles, *Ns*#2 and *Ns*#3 with each of the rotor poles. Therefore, for a certain stator pole, the electrical cycle repeats itself by the number of rotor poles in one revolution of the rotor (360° mechanical). In an SRM, if the stator and rotor pole configuration and the number of phases are defined properly, the stator poles belonging to the same phase will share the same electrical cycle and, hence, the same torque pulses in one mechanical revolution. For the stator poles belonging to other phases, these cycles will be phase shifted. Therefore, the total number of torque pulsations or number of strokes in one mechanical revolution is defined as:

$$S = mN_r \tag{3.1}$$

where m is the number of phases and N_r is the number of rotor poles. Depending on the number of rotor and stator poles, each stator pole has a certain electrical position at a given mechanical position of the rotor. This can be either aligned (180° electrical), unaligned (0° electrical), somewhere before the alignment (<180° electrical), or somewhere after the alignment (>180° electrical). The calculation of the number of rotor poles for the given number of stator poles and phases is based on the electrical angle of each stator pole in each phase.

3.2 Electrical Position of Stator Poles

The calculation of electrical angles will be explained on a three-phase 6/8 SRM shown in Figure 3.3. In all the examples presented in this chapter, it is assumed that the first stator pole is aligned with the first rotor pole at the initial position for drawing purposes. The angles inside the stator and rotor poles are mechanical angles representing the center axis location of the poles.

The center of axis for each of the stator and rotor poles stands at a certain mechanical angle. Considering the symmetric and even distribution of stator and rotor poles, this mechanical angle, called the pole pitch, is defined as:

$$T_{pr} = \frac{360}{N_r}, \theta_r[t] = T_{pr}(t-1), t = 1, 2, \ldots, N_r \tag{3.2}$$

$$T_{ps} = \frac{360}{N_s}, \theta_s[t] = T_{ps}(t-1), t = 1, 2, \ldots, N_s \tag{3.3}$$

where T_{pr} and T_{ps} are the pole pitches for the rotor and stator, and θ_r and θ_s are the mechanical positions of the center axis of the rotor and stator poles, respectively. Considering a certain direction of rotation—in this case it is counterclockwise as shown in Figure 3.3—the rotor poles are at certain electrical angles for the given rotor position. Electrical angles lower than 180° represent that when the coil of that stator pole is energized, the electromagnetic torque is in the same direction as the rotation. For a clockwise rotation, the electrical angles will be reversed.

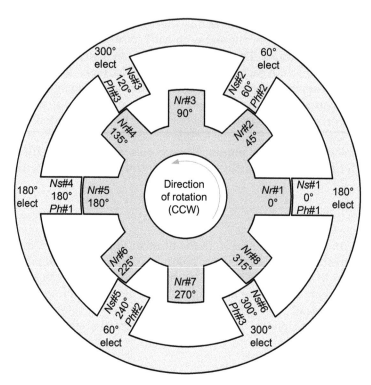

FIGURE 3.3
Mechanical and electrical angles in three-phase 6/8 SRM.

It can be observed in Figure 3.3 that Nr#1 is aligned with Ns#1, which corresponds to 180° electrical. This also holds for pairs Nr#5 and Ns#4. The difference in the mechanical angle between these stator and rotor poles is zero. For a certain stator pole, its relative mechanical position with the rotor pole repeats itself by the number of rotor poles. This means that, in one mechanical revolution, the electrical angle varies number of rotor poles times faster than the mechanical angle. In Figure 3.3 for Ns#1, when the rotor rotates by the rotor pole pitch—45° mechanical in counterclockwise direction—Nr#8 will be aligned with Ns#1, which still is 180° electrical for this stator pole. Therefore, the electrical angle for Ns#1 can be calculated as:

$$Ns\#1_{elect} = \left(Nr\#1_{mech} - Ns\#1_{mech}\right)N_r + 180° \tag{3.4}$$

where $Ns\#1_{elect}$ is the electrical angle for $Ns\#1$, and $Ns\#1_{mech}$ and $Nr\#1_{mech}$ are the mechanical angles for $Ns\#1$ and $Nr\#1$, respectively. The constant 180° in (3.4) comes from the fact that $Nr\#1$ is aligned with $Ns\#1$ at the initial position. Mechanical position has 360° cycle in one revolution. Electrical angle also has 360° cycle. Therefore, (3.4) can be reorganized and generalized for counterclockwise rotation as:

$$Ns_{elect\_CCW} = mod\left(\left(Nr_{mech} - Ns_{mech}\right)N_r + 180°, 360\right). \tag{3.5}$$

For the given stator pole, (3.5) results in the same value for all the rotor poles at a given position. For example, the center axis of $Ns\#2$ is at 60° mechanical in 6/8 SRM as shown in Figure 3.3. The center axis of $Nr\#3$ and $Nr\#6$ are at 90° and 225° mechanical. Applying (3.5) on these poles will result in 60° electrical. This means that, the electrical angle of the given stator pole remains the same when the rotor rotates a mechanical angle corresponding to one electrical cycle. This holds for pole pairs around the stator circumference, if the SRM configuration is symmetric and the pole pitch between the stator and rotor poles are constants. For clockwise rotation, the electrical angles will be reversed and calculated as:

$$Ns_{elect\_CW} = mod\left(\left(Ns_{mech} - Nr_{mech}\right)N_r + 180°, 360\right). \tag{3.6}$$

For the given rotor position for the 6/8 SRM in Figure 3.3, if $Ns\#2$ is energized, the closest rotor pole $Nr\#2$ will move in a counterclockwise direction. Hence, in clockwise rotation, the electrical angle of $Ns\#2$ should be larger than 180°. Applying (3.6) for $Nr\#3$ and $Nr\#6$, the electrical angle of $Ns\#2$ in clockwise rotation will be 300° electrical. Therefore, the electrical angle of stator poles can be generalized as

$$Ns_{elect} = mod\left[k_{rot}\left(Ns_{mech} - Nr_{mech}\right)N_r + 180°, 360\right] \tag{3.7}$$

where k_{rot} is –1 for counterclockwise rotation and 1 for clockwise rotation. In the 6/8 SRM in Figure 3.3, since the stator pole pairs $Ns\#1$-$Ns\#4$, $Ns\#2$-$Ns\#5$, and $Ns\#3$-$Ns\#6$ always have the same electrical position but differ from other pole pairs, when the same current is applied at the same instant, these poles create the same torque on the opposite corners of the stator bore. This is accomplished by connecting the coils of these stator pole pairs in the same electrical circuit, and this creates the phases. Therefore, 6/8 SRM in Figure 3.3 is a three-phase machine. Figure 3.4 shows another example by comparing a four-phase 8/6 SRM and a four-phase 8/14 SRM proposed in [12]. By using (3.5), the electrical angle (in counterclockwise rotation) for each stator pole can be calculated at the given rotor position. Please note that the electrical angles of each stator pole for these two configurations are the same with different number of rotor poles.

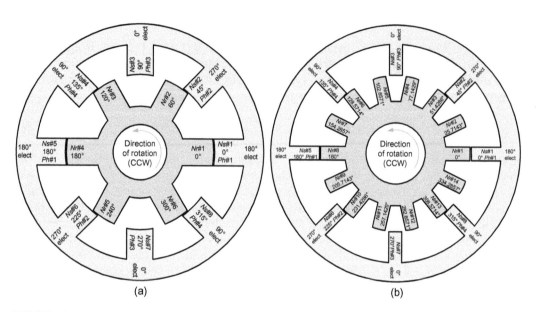

FIGURE 3.4
Mechanical and electrical angles in four-phase SRMs: (a) 8/6 and (b) 8/14.

3.3 Derivation of Pole Configuration

By comparing the four-phase 8/6 and 8/14 SRMs in Figure 3.4, it can be observed that different numbers of stator and rotor poles can satisfy the balanced operation in an SRM. When selecting the pole configuration and number of phases in an SRM, certain conditions must be satisfied to maintain a balanced operation. In an SRM, symmetry is maintained when electrical angles of stator poles in one phase are the same. The coils of these stator poles are connected together in an electrical circuit to make up a phase and when the phase is excited with current, these stator poles generate the same torque around the air gap.

Figure 3.5 shows the mechanical and electrical angles in a 12/22 SRM. If the 12/22 SRM is designed as a three-phase machine, it will have 4 stator poles per phase. Therefore, the coils of Ns#2, Ns#5, Ns#8, and Ns#11 will be connected together to create a phase to maintain a balanced distribution of stator poles around the air gap. At the given rotor position and considering counterclockwise rotation, Ns#2 and Ns#8 are at 240° electrical, while Ns#5 and Ns#11 are at 60° electrical. When energized with the same current at the same time, Ns#5 and Ns#11 will generate positive torque, but Ns#2 and Ns#8 will generate negative torque. This will cause imbalanced operation, and the same will apply to the other two phases.

When the 12/22 SRM is designed as a six-phase machine, it will have 2 stator poles per phase. In this case, the coils of Ns#2 and Ns#8 will be connected to Ph#2, and Ns#5 and Ns#11 will be connected to Ph#5. Therefore, at the given rotor position and considering counter clockwise rotation, Ph#2 will be at 240° electrical, and Ph#5 will be at 60° electrical. Since these phases have different electrical circuits, it will be possible to generate balanced torque when 12/22 SRM is designed as a six-phase machine.

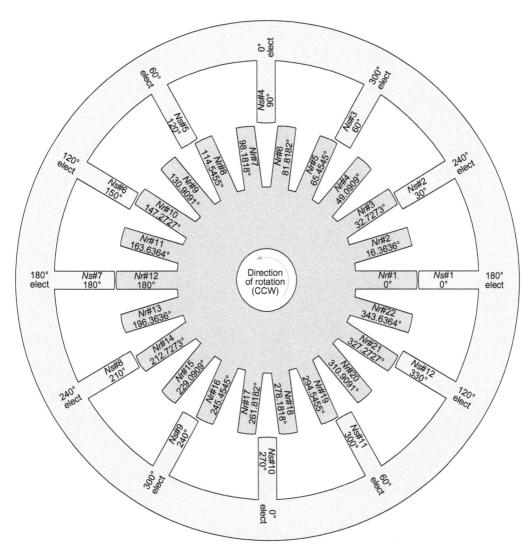

FIGURE 3.5
Mechanical and electrical angles in 12/22 SRM.

A switched reluctance machine will have a locked rotor condition when all phases are at either an unaligned position (0° electrical) or aligned position (180° electrical). This is the typical case when the ratio between the number of stator poles and number of rotor poles is an integer number, such as in a three-phase 6/6 SRM, three-phase 6/12 SRM, or four-phase 8/16 SRM. There are configurations that don't follow this rule but still have the locked rotor condition. For example, as shown in Figure 3.6, in a four-phase 8/12 SRM, there are 2 stator poles per phase. The ratio between the number of stator and rotor poles is not an integer. Using (3.5), it can be calculated that all the stator poles are either at 180° or 0° electrical, which are both equilibrium conditions. The configurations with locked rotor conditions will be explained later in this section when the expression for the number of rotor poles is derived.

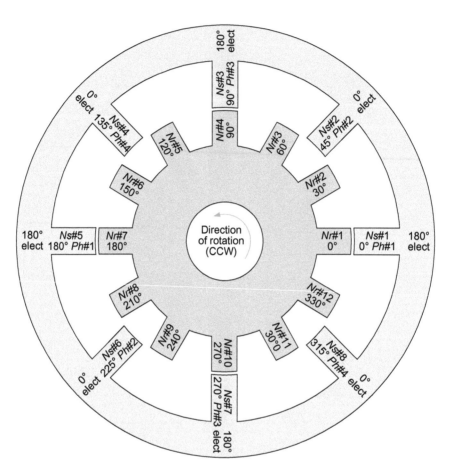

FIGURE 3.6
Mechanical and electrical angles in four-phase 8/12 SRM.

In some certain cases, even though the SRM configuration is symmetric and balanced, different phases might have the same electrical angle. These phases are connected to different electrical circuits, but they can be excited with the same current. This will be similar to applying parallel branches in one phase. This might be necessary depending on the power level of the application. As it can be seen in Figure 3.7, when designed as a three-phase machine, a 12/8 SRM has 4 stator poles per phase. In this case Ns#1, Ns#4, Ns#7, and Ns#10 have the same electrical angle and they belong to the same phase. When analyzed as a six-phase machine, Ns#1 and Ns#7 will be connected to Ph#1, and Ns#4 and Ns#10 will be connected to Ph#4; but these phases will still have the same electrical angle. Therefore, a six-phase 12/8 SRM operates similar to the three-phase 12/8 SRM with parallel braches. Unless it is required by the application, this might be poor utilization of the drive.

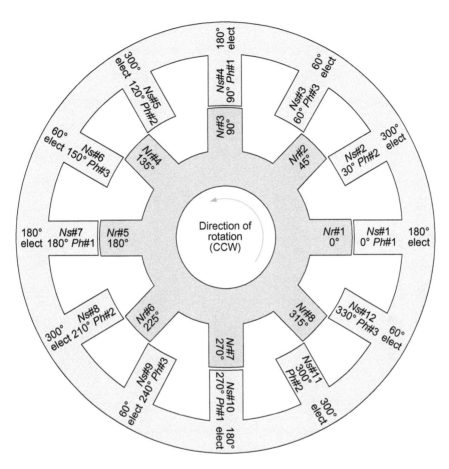

FIGURE 3.7
Mechanical and electrical angles in three-phase 12/8 SRM.

Table 3.1 shows some of the SRM configurations that satisfy the balanced operation described above. It shows the number of phases for the given number of stator poles and number of rotor poles. The empty cells show that the given number of stator and rotor poles does not provide a working configuration. The maximum number of stator and rotor poles is 25 in the list to limit the size of the table.

If the configurations in Table 3.1 are analyzed carefully, a certain pattern can be observed. The number of rotor poles is related to the number of stator poles per phase, which can be expressed as the configuration index:

$$k = \frac{N_r}{N_s / m} \Rightarrow k = \frac{mN_r}{N_s} \tag{3.8}$$

TABLE 3.1

SRM Configurations with Number of Stator Poles (Ns), Number of Rotor Poles (Nr) and Number of Phases

										Ns										
Nr	6	7	8	9	10	11	12	13	14	15	16	17	18	19	20	21	22	23	24	25
2	3		4		5		6		7		8		9		10		11		12	
3				3			4			5			6			7			8	
4	3				5		3		7		4		9		5		11		6	
5										3					4					5
6			4	3	5				7	5	8		3		10	7	11		4	
7																3				
8	3				5		3		7				9		5		11		3	
9							4			5						7			8	
10	3		4				6		7	3	8		9				11		12	5
11																				
12				3	5				7	5	4		3		5	7	11			
13																				
14	3		4		5		6				8		9		10	3	11		12	
15				3			4						6		4	7			8	5
16	3				5		3		7				9		5		11		3	
17																				
18			4		5				7	5	8				10	7	11		4	
19																				
20	3						3		7	3	4		9				11		6	5
21				3			4			5			6						8	
22	3		4		5		6		7		8		9		10				12	
23																				
24				3	5				7	5			3		5	7	11			
25										3					4					

where N_s / m is the number of stator poles per phase and mN_r is the number of strokes. For example, for three-phase 6/10 SRM the configuration index is 5 and for four-phase 8/14 SRM the configuration index is seven. Configuration index is an integer number and it can be defined by the designer to calculate the number of rotor poles for the given number of stator poles and number of phases. However, it can be observed from Table 3.1 that not every rotor pole number within the range of k provides a working configuration. If the configuration index is an integer multiple of any of the prime factors of number of phases, SRM configuration will not provide the symmetric and balanced conditions. As a result, the number of rotor poles can be calculated as [13]

$$Nr = \frac{Ns}{m} k \prod_i ceil\left(\frac{mod(k,i)}{i}\right) \tag{3.9}$$

where the parameter i represents the prime factors of number of phases. Configuration index, k varies from one to any arbitrary integer. Number of stator poles per phase is a positive integer. The result of the *ceil* function is 1 for the working configurations and zero for non-working configurations. As an example, for 12 stator poles and 6 phases, the prime factor of 6 is $i = [2\ 3]$. Then, (3.9) can be expressed as

$$Nr = \frac{Ns}{m} k \prod_{i=2}^{i=3} ceil\left(\frac{mod(k,i)}{i}\right)$$

for $k = 11$, *ceil* function for $i = 2$ and $i = 3$ will result in 1. Therefore, the expression for the number of rotor poles can be calculated as

$$Nr = \frac{Ns}{m} k = \frac{12}{6} 11 = 22.$$

Equation 3.9 also explains why a four-phase 8/12 SRM in Figure 3.6 has a locked rotor condition even though the ratio between the number of stator and rotor poles is not an integer. Configuration index k for this case is 6. For 4 phases, $i = [2\ 2]$, which divides the configuration index by an integer. Therefore, the *ceil* function is zero for four-phase 8/12 SRM. If the configuration index, k is not an integer, the stator poles belonging to the same phase will have different electrical angles, generally producing torque in different directions. In this case, the SRM configuration may not maintain a balanced operation.

Figure 3.8 shows a three-phase 6/14 SRM calculated using (3.9). When compared to the conventional three-phase 6/8 SRM given in Figure 3.3, it can be observed that the stator poles of these two configurations have the same electrical angles, even though the mechanical angles of the rotor poles are different.

As another example, the three-phase 18/24 SRM in Figure 3.9 has the inverse configuration compared to the four-phase 24/18 SRM in Figure 3.10. In both scenarios, the SRMs have 6 stator poles per phase and provide 72 strokes in one mechanical revolution. However, the four-phase 24/18 SRM has more coils and power converter legs than the three-phase 18/24 SRM.

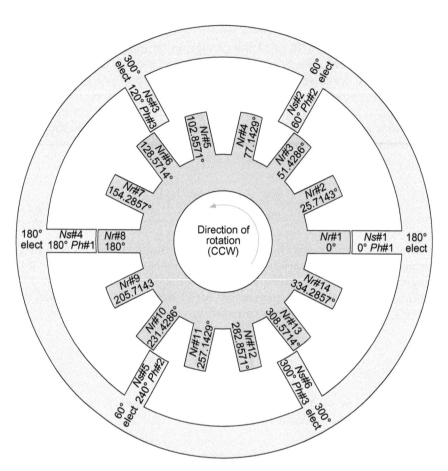

FIGURE 3.8
Mechanical and electrical angles in three-phase 6/14 SRM.

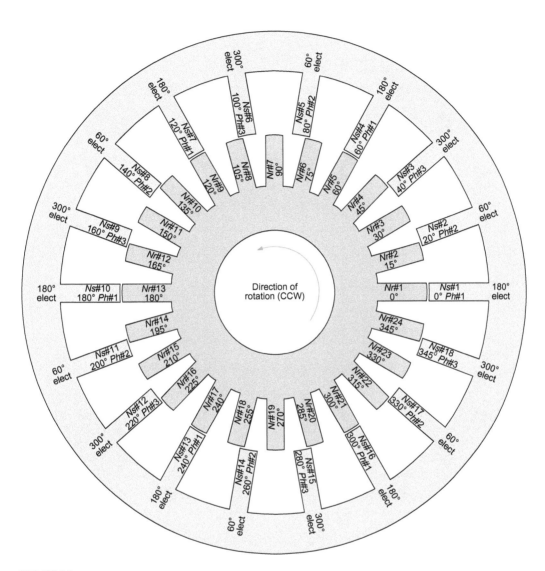

FIGURE 3.9

Mechanical and electrical angles in three-phase 18/24 SRM.

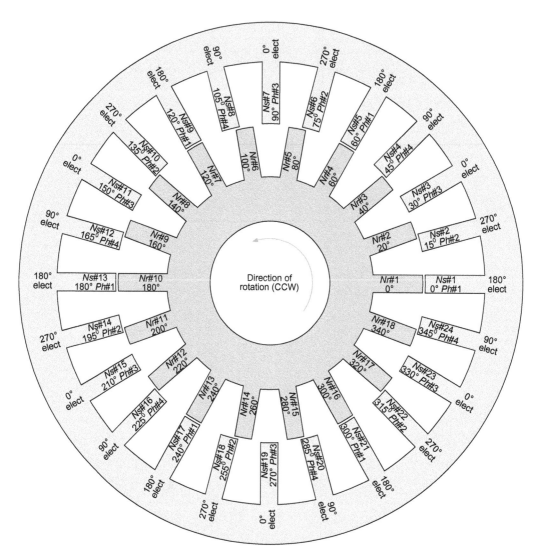

FIGURE 3.10
Mechanical and electrical angles in four-phase 24/18 SRM.

Using (3.9), various stator and rotor pole combinations can be calculated. Figures 3.11 through 3.14 show different five-phase, 2 stator poles per phase SRM configurations: 10/18, 10/16, 10/14, and 10/12 SRMs. The mechanical positions of the rotor poles and their relative positions with the stator poles are different. However, the same electrical angles exist in all of these configurations but at different phases. Therefore, the phase firing sequences of these machines might be different. Please note that the pole arc angles are kept the same in these configurations for drawing purposes. As it is the case in all SRMs, pole arc angles and the other dimensions should be adjusted to satisfy the performance requirements [14].

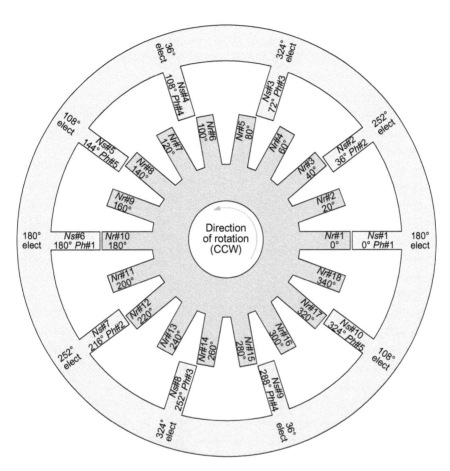

FIGURE 3.11
Mechanical and electrical angles in five-phase 10/18 SRM.

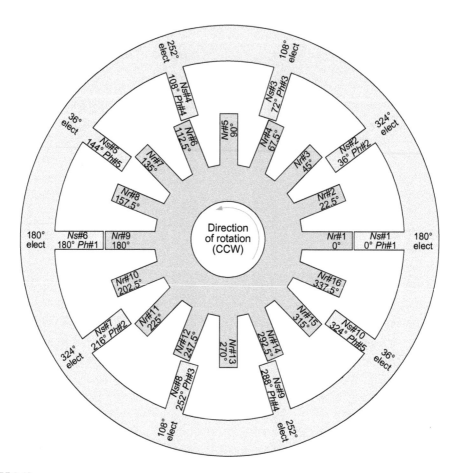

FIGURE 3.12
Mechanical and electrical angles in five-phase 10/16 SRM.

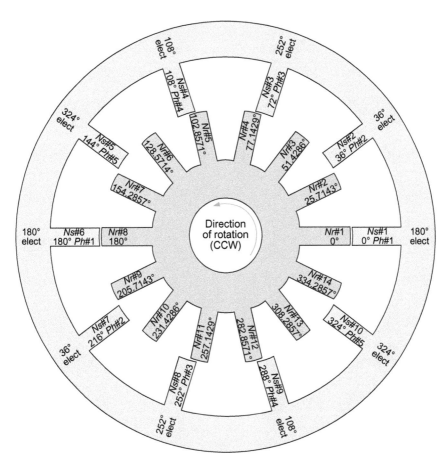

FIGURE 3.13
Mechanical and electrical angles in five-phase 10/14 SRM.

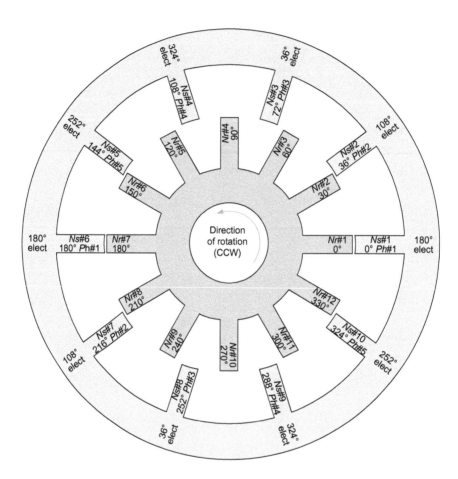

FIGURE 3.14
Mechanical and electrical angles in five-phase 10/12 SRM.

3.4 SRMs with Odd Number of Stator Poles per Phase

Most of the conventional SRM configurations have an even number of stator poles per phase. The SRM configurations calculated using (3.9) and listed in Table 3.1 show that SRMs with an odd number of stator poles per phase also provide symmetric configuration. Figure 3.15 shows the mechanical and electrical angles in the three-phase 9/12 SRM. This configuration has 3 stator poles per phase. It can be observed that 9/12 SRM has similar electrical angles as the three-phase 6/8 SRM given in Figure 3.3 Ns#1, Ns#4, and Ns#7 are all at 180° electrical. They are symmetrically distributed around the air gap and, when the coils of these stator poles are connected to the same electrical circuit, they generate the same torque. This also applies to the other two phases.

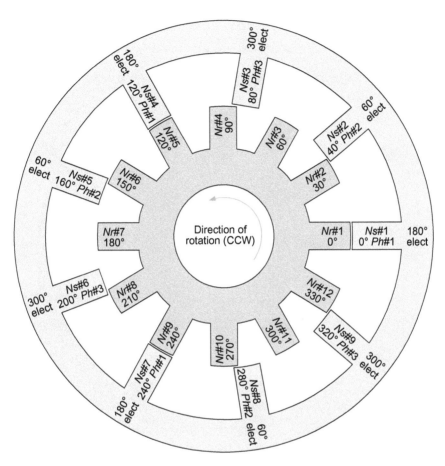

FIGURE 3.15
Mechanical and electrical angles in three-phase 9/12 SRM.

The general use of SRMs with even number of stator poles per phase is due to the fact that magnetic poles can be generated with non-coupled coil configuration. Therefore, mutual coupling between phases becomes negligible and the flux linkage characteristics for a single phase can be used to model the other phases. Figure 3.16 shows the typical flux distribution in a three-phase 12/8 SRM, which has 4 stator poles per phase. When the coils in one of the phases are energized, the total number of flux paths is equal to the number of stator poles per phase. The same situation can be observed in Figures 3.17 and 3.18 for three-phase 6/10 and four-phase 24/18 SRMs. In the 6/10 SRM, there are 2 stator poles per phase and, hence, there are two flux patterns around the stator back iron. In the four-phase 24/18 SRM, there are 6 stator poles per phase and, hence, there are 6 flux patterns. In this coil configuration, the flux patterns are between the stator poles belonging to the same phase; therefore, it can be named as non-coupled coil configuration.

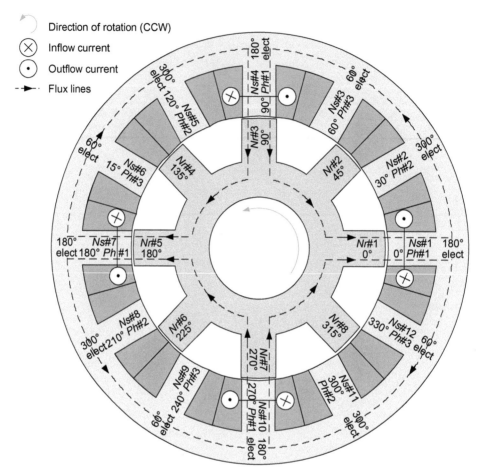

FIGURE 3.16
Typical flux distribution, and mechanical and electrical angles in three-phase 12/8 SRM.

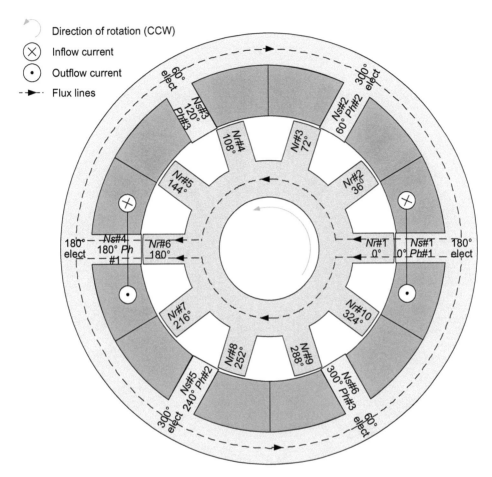

FIGURE 3.17
Typical flux distribution and mechanical and electrical angles in three-phase 6/10 SRM.

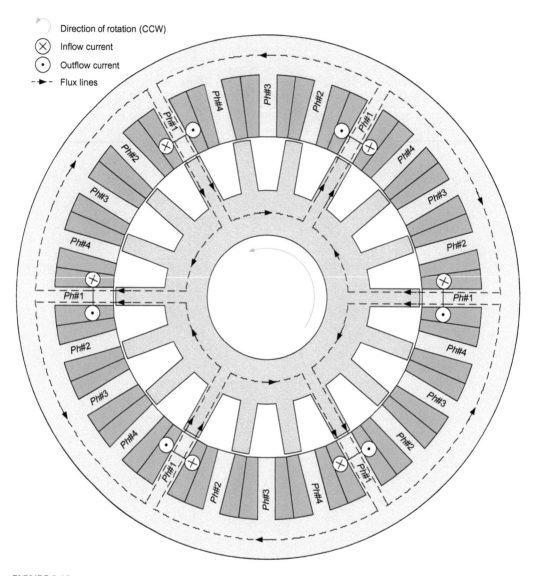

FIGURE 3.18

Typical flux distribution and mechanical and electrical angles in four-phase 24/18 SRM (see **Figure 3.10** for mechanical and electrical angles).

In SRMs with an odd number of stator poles per phase, a similar flux pattern is desired to facilitate a balanced operation. Figure 3.19 shows the cross-section view of a three-phase 9/12 SRM with the desired flux pattern. In this case, the coils are connected in a non-coupled configuration [15]. When the right-hand corkscrew rule is applied to the coils, the direction of the flux of Ns#1 will be towards the air gap, whereas the direction of the flux of Ns#2 and Ns#3 will be in the opposite direction.

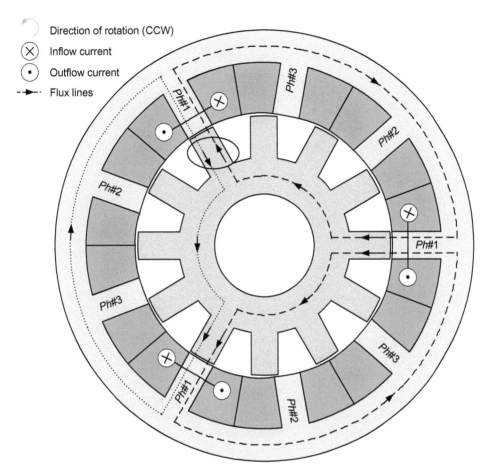

FIGURE 3.19
Desired flux distribution for a three-phase 9/12 SRM.

However, if the total number of flux patterns is equal to three to maintain balanced operation with non-coupled configuration, one of the stator poles should have opposing directions (see the flux directions for Ns#2 inside the circle in Figure 3.19). This cannot happen, because the coil around Ns#2 has a certain direction and, hence, the flux pattern with dotted lines cannot be generated. As a result, only two magnetic poles would be generated in a three-phase 9/12 SRM with non-coupled coil configuration, as shown in Figure 3.20.

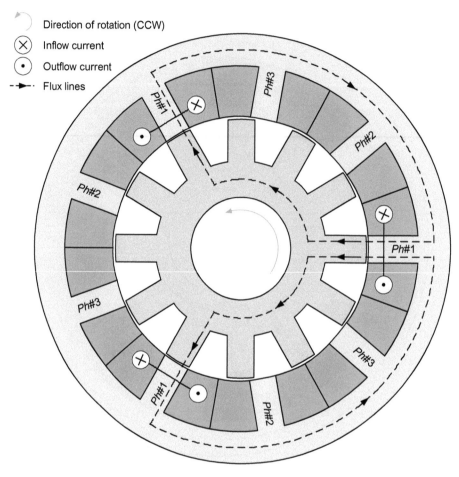

FIGURE 3.20
Flux patterns in a three-phase 9/12 SRM with non-coupled coil connections.

In SRMs with an odd number of stator poles per phase, balanced magnetic poles can be generated by applying mutually coupled coil configuration as shown in Figure 3.21. In this case, the coils in one phase are all in opposite directions. With this coil configuration, the flux patterns are distributed symmetrically around the stator circumference.

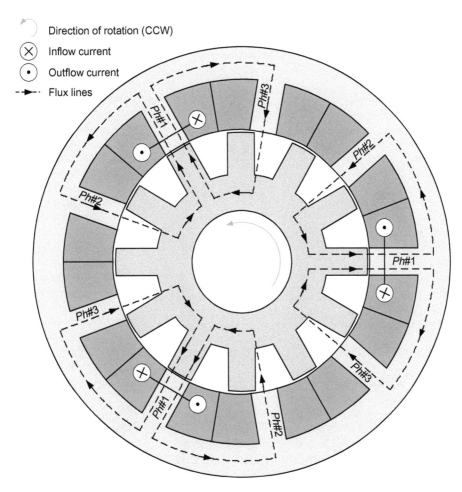

FIGURE 3.21
Flux patterns in three-phase 9/12 SRM with mutually coupled coil connections.

Figures 3.22 and 3.23 show the mechanical and electrical angles and the flux patterns in a three-phase 15/20 SRM, which has 5 stator poles per phase and a configuration index of 4. Figures 3.24 and 3.25 show another example with an odd number of stator poles per phase. In this case, four-phase 20/15 SRM still has 5 stator poles per phase and it has an inverse configuration as compared to a three-phase 15/20 SRM.

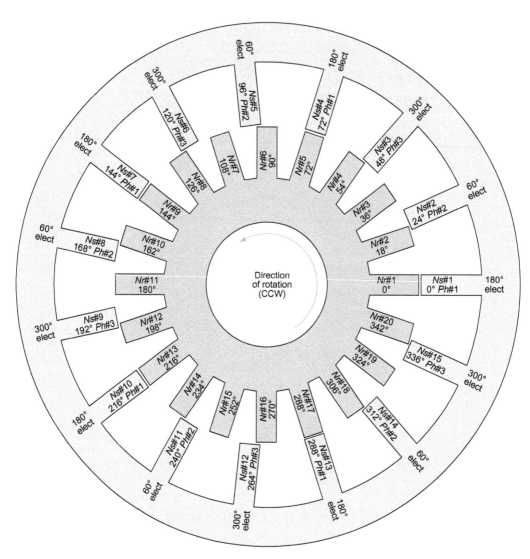

FIGURE 3.22
Mechanical and electrical angles in a three-phase 15/20 SRM.

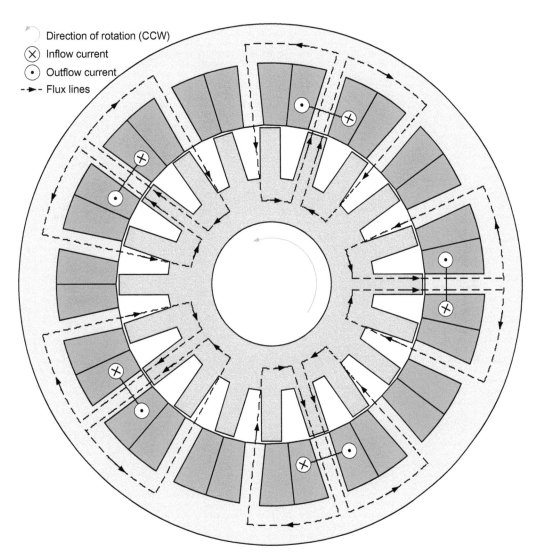

FIGURE 3.23
lux patterns in a three-phase 15/20 SRM with mutually coupled coil configuration.

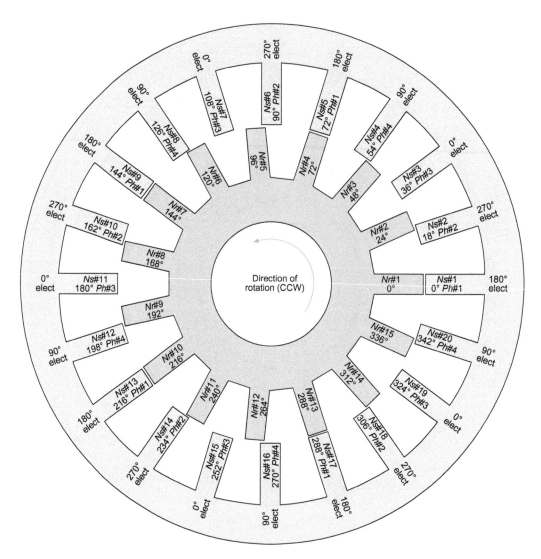

FIGURE 3.24
Mechanical and electrical angles in four-phase 20/15 SRM.

With a mutually coupled coil configuration, the flux generated when one phase coil is energized now links to the coils of other phases. Therefore, unlike conventional SRMs with an even number of stator poles per phase and non-coupled coil configuration, mutual inductance cannot be neglected.

FIGURE 3.25
Flux patterns in four-phase 20/15 SRM with mutually coupled coil configuration.

3.5 Summary

SRMs are identified by the number of stator poles and number of rotor poles. In this chapter, a generalized expression for the relationship between the number of stator poles, number of rotor poles, and number of phases has been presented. The relative position of the rotor poles towards the stator poles in each phase defines the symmetric and balanced operation of SRMs. By quantifying this relationship, the working SRM configurations have been derived. Many different stator and rotor pole combinations are available with different number of phases. The derived expression works either when the number of stator poles or the number of rotor poles is greater than the other.

In general, SRM configurations have an even number of stator poles per phase. In this case, the required number of poles can be generated using non-coupled coil configuration. The derived expression shows that SRMs with an odd number of stator poles per phase also have the same electrical angles as the conventional ones and they provide symmetric configuration using mutually coupled coil configuration.

It should be noted that, the selection of the number of rotor poles, number of stator poles, and number of phases in an SRM depends on many factors including the desired torque speed and torque ripple characteristics, mechanical tolerances, converter and controller requirements, switching frequency, precision of rotor position measurement, and so on.

Questions and Problems

1. What does an electrical phase mean? How is it defined in a switched reluctance machine?

2. Calculate the mechanical positions of stator and rotor poles for 6/14, 8/10, 16/20, and 20/16 SRM configurations. At the position where the first rotor pole is aligned with the first stator pole and rotor rotates in counter clockwise direction, what are the electrical angles of the stator poles? What would be the number of phases in each configuration?

3. Calculate the mechanical positions of stator and rotor poles for 9/15, 12/15, and 15/12 SRM configurations. At the position where the first rotor pole is aligned with the first stator pole and rotor rotates in counter clockwise direction, what are the electrical angles of the stator poles? Would these configurations work with the same winding configuration as the ones in question 1?

4. Calculate the mechanical positions of stator and rotor poles for 6/12, 8/20, 8/7, and 12/10 SRM configurations. At the position where the first rotor pole is aligned with the first stator pole and rotor rotates in counter clockwise direction, what are the electrical angles of the stator poles? Do you see an issue with these configurations?

5. Consider the SRM configuration with 18 stator poles and 24 rotor poles? What is the number of phases that can be used in this configuration? Can this machine run with different number of phases?

6. Is there an SRM configuration that can run as three- and four-phases? If not, why?

7. Why is it necessary to use mutually coupled windings configuration in SRM with an odd number of stator poles per phase? What is the challenge in modeling these types of SRMs?

8. What are the parameters that should be considered in selecting the number of stator poles, number of rotor poles and number of phases? Would the selection of the SRM configuration be related to the application? Please justify the reason of your answer.

References

1. B. Bilgin, P. Magne, P. Malysz, Y. Yang, V. Pantelic, A. Korobkine, W. Jiang, M. Lawford, and A. Emadi, "Making the case for electrified transportation," *IEEE Transactions on Transportation Electrification*, vol. 1, no. 1, pp. 4–17, 2015.

2. Y. Yang and A. Emadi, *Double-Rotor Switched Reluctance Machine*, McMaster Tech ID 13-022, Patent Application No. US 14/061,812 and CDN 2,830,944, October 24, 2013.

3. Y. Yang, N. Schofield, and A. Emadi, *Torque Ripple and Radial Force Reduction in Double-Rotor Switched Reluctance Machines*, McMaster Tech ID 15-030, US 9,621,014, April 11, 2017.

4. T. Guo, N. Schofield, and A. Emadi, *Double Rotor Switched Reluctance Machine with Segmented Rotors*, McMaster Tech ID 15-067, Patent Application No. US 14/918,800 and CDN 2,911,274, October 21, 2015.

5. Y. Oner, B. Bilgin, and A. Emadi, *Double-Stator Single-Winding Switched Reluctance Motor*, McMaster Tech ID 18-051, Discovery Disclosure, McMaster University, December 27, 2017.

6. Y. Oner, B. Bilgin, and A. Emadi, *Multi-Teeth Switched Reluctance Motor with Short-Flux Path*, McMaster Tech ID 18-028, Discovery Disclosure, McMaster University, September 10, 2017.

7. Y. Oner, B. Bilgin, and A. Emadi, *Multi-Teeth Switched Reluctance Motor*, McMaster Tech ID 18-020, Discovery Disclosure, McMaster University, August 14, 2017.

8. P. Suntharalingam and A. Emadi, *Switched Reluctance Machine with Toroidal Winding*, McMaster Tech ID 15-052, Provisional Patent Application No. US 62/143,282, April 6, 2015; International Application No. PCT/CA2016/050396, April 6, 2016.

9. Y. Oner, B. Bilgin, and A. Emadi, *Switched Reluctance Machine with Short-Flux Path*, McMaster Tech ID 18-023, Discovery Disclosure, McMaster University, August 16, 2017.

10. B. Bilgin, R. Yang, N. Schofield, and A. Emadi, *Alternating-Current Driven, Salient-Teeth Reluctance Motor with Concentrated Windings*, McMaster Tech ID 17-019, Provisional Patent Application No. US 62/477,611, March 28, 2017.

11. B. Bilgin and A. Emadi, *Switched Reluctance Machine with Rotor Excitation Using Permanent Magnets*, McMaster Tech ID 13-034, Patent Application No. US 14/103,041 and CDN 2,836,309, December 11, 2013.

12. P. C. Desai and A. Emadi, "Switched Reluctance Machine," U.S. Patent 7,230,360 B2, June 12, 2007.

13. B. Bilgin and A. Emadi, *Switched Reluctance Machine with Even Pole-Phase Index*, McMaster Tech ID 15-056, Provisional Patent Application No. US 62/161,905, May 15, 2015; International Application No. PCT/CA2016/050548, May 13, 2016.

14. B. Bilgin, A. Emadi, and M. Krishnamurthy, "Design considerations for switched reluctance machines with a higher number of rotor poles," *IEEE Transactions on Industrial Electronics*, vol. 59, no. 10, pp. 3745–3756, 2012.

15. B. Bilgin and A. Emadi, *Switched Reluctance Machine with Odd Pole-Phase Index*, McMaster Tech ID 15-057, Provisional Patent Application No. US 62/161,907, May 15, 2015; International Application No. PCT/CA2016/050551, May 13, 2016.

4

Operational Principles and Modeling of Switched Reluctance Machines

Brock Howey and Haoding Li

CONTENTS

4.1 Introduction

In Chapter 2, we explored the electromagnetic principles and showed that switched reluctance machine (SRM) operation can be explained using Maxwell's equations and Lorentz Force law. We derived the voltage equation, which represents the equivalent circuit of a single phase of an SRM. Additionally, we explored the concept of magnetic co-energy, which led to the derivation of the electromagnetic torque in SRM, and we showed that torque production capability of an SRM increases under magnetic saturation. Therefore, magnetic saturation must be considered when analyzing an SRM.

Due to the effects of magnetic saturation, as well as the non-linear dependence of the flux linkage to rotor position, the machine characteristics must be considered in the modeling of SRMs. For this reason, in this chapter, the equivalent circuit model for the SRM is revisited in the context of modeling. From this exercise, torque production mechanism is re-analyzed, which will emphasize the ideas presented in Chapter 2 from a different perspective. At this point, the reader will be introduced to phase excitation principles. After developing an in-depth understanding of basic operation principles, we will introduce modeling techniques for SRMs.

To give context to the development of the SRM equivalent circuit, it is important to first understand the operating principles and power conversion mechanism of a generic separately excited direct current (DC) motor. This will be the beginning of our discussion.

4.2 Equivalent Circuit of SRM

4.2.1 Review of the Equivalent Circuit of a Separately Excited DC Motor

In its simplest form, the DC motor is a conductive loop, called an armature, placed within a stationary magnetic field, where the current flow in the armature is perpendicular to the direction of the magnetic field. The construction of such a simple DC motor is shown in Figure 4.1, where the coil *abcd* represents the armature.

The voltage equation for the circuit shown in Figure 4.1 has been derived in Chapter 2:

$$V = iR + \frac{d\lambda}{dt} \tag{4.1}$$

where V is the voltage across the armature coil, R is the armature resistance, i denotes the armature current, and λ is the flux linkage of the coil. The armature coil in Figure 4.1 is in a magnetic field created by the north and south magnetic poles. When current flows through the armature coil and the magnetic field created by the stator magnets (B field) flows through the surface *abcd*, the armature coil will be subject to Lorentz Force law, thus causing the armature loop to rotate. The flux linkage due the B field can be derived by applying Maxwell's equations. For a simple loop with N turns, the expression for the flux linkage can be calculated as:

$$\lambda = NABcos(\theta) \tag{4.2}$$

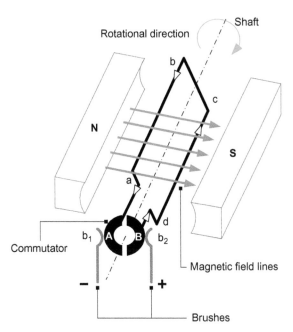

FIGURE 4.1
Diagram illustrating the operational principles of a simple DC motor.

where A is the area of the conducting loop, and θ is the angle between the normal vector of the area A and the direction of the B field. Notably, as the loop rotates, the angle θ will vary, creating a change in the flux linkage passing through the loop. Due to this varying flux linkage, voltage will be induced across the armature which can be calculated using Faraday's law:

$$\varepsilon = -N\,A\,B\,\omega\sin(\theta) \tag{4.3}$$

This is the voltage induced across a simple armature coil with N turns in a constant magnetic field B, where the loop has a fixed area A. In practical DC motors, multiple loops are distributed circumferentially within the armature slots, along with the help of the commutators. Taking this into account, the induced electromotive force (EMF) can be expressed in a general form as:

$$\varepsilon = k\,\phi_s\,\omega \tag{4.4}$$

where k is a lumped parameter term that takes into account the mechanical construction of the motor, ϕ_s is the flux passing through the armature coil due to the external field, and ω is the rotational speed of the armature.

The armature circuit stores energy within itself due to the inductance of the circuit. In a DC machine, armature inductance L can be assumed as constant. This is because the multiple armature coils symmetrically distributed in the armature slots create a uniform air gap. Therefore, the flux linkage due to armature current can be expressed as:

$$\lambda_A = L\,i \tag{4.5}$$

The energy stored in the armature inductance can be represented as a voltage drop in the voltage equation using Faraday's law:

$$V_L = L\frac{di}{dt}$$

(4.6)

By combining (4.4) and (4.6) into (4.1), the voltage equation of the DC motor can be derived as in (4.7). Please note that the minus sign in the induced voltage expression in (4.3) tells us the direction of the circulation of the induced electric field and, hence, the direction of the current. The derivation of the voltage equation from the induced electric field was shown in (2.39), so we will not discuss it again here. Figure 4.2 shows the equivalent circuit of the DC motor.

$$V = iR + V_L\frac{di}{dt} + \varepsilon = iR + L\frac{di}{dt} + k\,\phi_s\,\omega$$

(4.7)

From (4.7), it can be observed that the rate of change of flux linkage in (4.1) is separated into two components. The first term, V_L is due to the stored energy and, hence, the inductance of the armature circuit. The other term, ε, is the speed-dependent EMF, which is due to the interaction between armature current and the external field. Next, the power distribution in the DC motor can be derived by multiplying both sides of (4.7) by the armature current:

$$i\,V = i^2 R + i\,L\frac{di}{dt} + k\,\phi_s\,\omega\,i$$

(4.8)

Applying the product rule,

$$iL\frac{di}{dt} = \frac{d}{dt}\left(\frac{1}{2}L\,i^2\right) - \frac{1}{2}i^2\frac{dL}{dt}$$

(4.9)

where, due to the constant inductance,

$$\frac{1}{2}i^2\frac{dL}{dt} = 0$$

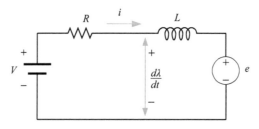

FIGURE 4.2
Equivalent circuit of the armature of a DC motor.

Thus, (4.8) can be rewritten as

$$i\,V = i^2 R + \frac{d}{dt}\left(\frac{1}{2}L\,i^2\right) + k\,\phi_s\omega\,i \tag{4.10}$$

It can be seen from (4.10) that the electrical energy supplied to the armature circuit (the term on the left) is dissipated as heat with the copper losses i^2R (the first term on the right) and stored in the armature magnetic circuit (the middle term on the right). The right most term in (4.10) is the air gap power $k\phi_s\,\omega i$. If the rotor losses are neglected, then this term is converted into mechanical power. Hence, by considering the mechanical power relationship in (4.11), the torque equation of the DC motor can be derived.

$$P_{mech} = \tau\,\omega \tag{4.11}$$

By combining the air gap power expression in (4.10) with (4.11), the torque produced by the DC motor can be expressed as:

$$\tau = k\,\phi_s i \tag{4.12}$$

From (4.12), it can be seen that the torque production in DC motors is expressed based on the interaction between magnetic field and the armature current. Unlike the DC motor, SRM has only a single source of excitation. As we will see in the next section, even though the SRM and DC motor have similar equivalent circuits, the torque equation will differ due to the single source of excitation.

4.2.2 Equivalent Circuit of SRM

As the name implies, the separately excited DC motor has two sources of excitation, with independent excitation on the stator. SRMs have a single source of excitation on the stator. Flux linkage is produced only when the phase coils are excited with current. Equation (4.1) can be applied both to the SRM and the separately excited DC motor. In the DC motor, the armature inductance is constant. This is where the derivation of the equivalent circuit model of SRM diverges from that of the DC motor. Due to the salient structure of SRM, the air gap between rotor and stator poles changes based on the relative position of the rotor and the stator. For this reason, the phase inductance in SRM is dependent on rotor position. Therefore, the flux linkage expression for the DC motor in Equation 4.5 appears differently in an SRM:

$$\lambda = L(\theta)i \tag{4.13}$$

If we insert (4.13) into (4.1), we get

$$V = iR + L(\theta)\frac{di}{dt} + i\frac{dL(\theta)}{dt} = iR + L(\theta)\frac{di}{dt} + i\frac{dL(\theta)}{d\theta}\frac{d\theta}{dt} \tag{4.14}$$

By noting that the rotational speed is defined as

$$\omega = \frac{d\theta}{dt}$$

Equation 4.14 can be simplified as

$$V = iR + L(\theta)\frac{di}{dt} + i\frac{dL(\theta)}{d\theta}\omega \tag{4.15}$$

It should be noted that, in the flux linkage expression in (4.15), phase inductance is only dependent on rotor position not the current. When the magnetic core of an SRM saturates, the permeability and, hence, the inductance decreases. Therefore, (4.13) only applies to the case where the machine is not magnetically saturated.

When we compare (4.15) to (4.7), we see that the last term on the right-hand side of (4.15) is analogous to the EMF given in (4.4), as it is also speed dependent. Similar to the DC machine, this term represents the air gap power which will be converted into mechanical power.

$$\varepsilon = i\frac{dL(\theta)}{d\theta}\omega \tag{4.16}$$

In the case of SRM, this induced voltage is created by the variation of the inductance over time due to the saliency of the pole structure. Please note that the EMF of the DC motor in Equation 4.4 is not dependent on armature current, but the speed and flux. The flux is created by a separate excitation source. Therefore, when a prime mover rotates the shaft of the DC motor, no armature current is required to produce the induced EMF, which can be measured on the terminals of the armature coil. However, for an SRM, (4.16) would be zero if there was no phase current. For this reason, the speed-dependent term on the right-hand side of (4.15) cannot be measured directly. From (4.15), the equivalent circuit of a single phase of an SRM can be derived as in Figure 4.3.

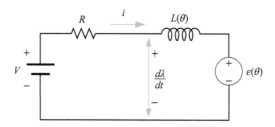

FIGURE 4.3
Equivalent circuit of one phase of SRM (without saturation effects).

4.2.3 Power and Torque in SRM

The power conversion process in SRM can be derived by multiplying both sides of (4.15) by the phase current

$$iV = i^2R + iL(\theta)\frac{di}{dt} + i^2\frac{dL(\theta)}{d\theta}\omega \tag{4.17}$$

Under the assumption that the machine is working in the magnetic linear region, the second term on the right-hand side of (4.17) can be expanded further using the product rule for differentiation. We have done a similar calculation for the DC motor in Equation 4.9.

$$iL(\theta)\frac{di}{dt} = \frac{d}{dt}\left(\frac{1}{2}L(\theta)i^2\right) - \frac{1}{2}i^2\frac{dL(\theta)}{d\theta}\frac{d\theta}{dt} \tag{4.18}$$

However, unlike the DC motor, the third term in (4.18) is non-zero due to the time varying inductance. This term can be simplified again by including the rotational speed:

$$iL(\theta)\frac{di}{dt} = \frac{d}{dt}\left(\frac{1}{2}L(\theta)i^2\right) - \frac{1}{2}i^2\frac{dL(\theta)}{d\theta}\omega \tag{4.19}$$

By substituting (4.19) into (4.17), the power equation of SRM is calculated as

$$iV = i^2R + \frac{d}{dt}\left(\frac{1}{2}L(\theta)i^2\right) + \frac{1}{2}i^2\frac{dL(\theta)}{d\theta}\omega \tag{4.20}$$

Notably, (4.20) is very similar to the power equation of the DC motor in Equation 4.10. It can be seen that the input electrical power on the left-hand side is converted respectively to electrical, magnetic, and mechanical power on the right-hand side. On the right-hand side, the electrical power is dissipated as heat in terms of copper losses. The magnetic power is stored as magnetic field energy in the magnetic core. The remaining power is air gap power, which is converted into mechanical power, and translates into torque production. Torque can be extracted from (4.20) using the mechanical power expression in (4.11). Hence, the torque expression in the linear region is

$$\tau = \frac{1}{2}i^2\frac{dL(\theta)}{d\theta} \tag{4.21}$$

From (4.21), it can be seen that the direction of torque is independent of the polarity of the current, due to the proportionality of the torque to the square of the current. When (4.12) and (4.21) are compared, some differences in the torque production of the DC machine and SRM can be observed even though these motors have similar equivalent circuits. The rate of change of inductance caused by the saliency in the geometry is the primary torque production mechanism in an SRM, as opposed to the interaction between the armature current and the flux generated by the stator excitation in the DC motor. This is because the excitation from the field winding or permanent magnets in a DC motor is not present in SRM.

As stated above, (4.21) is the torque expression when SRM operates in the linear region of the magnetization curve. If you compare the derivation of power expressions in the DC motor and SRM, the effect of the linear-region-operation on the torque production capability of SRM can be observed. When applying the product rule in the DC motor, we came up with the expression in (4.9)

$$iL\frac{di}{dt} = \frac{d}{dt}\left(\frac{1}{2}L\,i^2\right) - \frac{1}{2}i^2\frac{dL}{dt}$$

where the right-most term was zero since the inductance did not change with time. When inserted into (4.8), there were no changes in the air gap power expression, which is converted to mechanical power. When applying the product rule for an SRM, we came up with the expression in (4.19):

$$iL(\theta)\frac{di}{dt} = \frac{d}{dt}\left(\frac{1}{2}L(\theta)i^2\right) - \frac{1}{2}i^2\frac{dL(\theta)}{d\theta}\omega$$

Here, the right-most term is no longer zero, because the inductance changes with position. When compared to the air gap power expression in (4.17), now the right-most term in (4.19) is half of the air gap power with a negative sign. Therefore, half of the air gap power is converted into mechanical energy and the rest stored in the magnetic circuit as it was shown in the flux linkage versus current characteristics in Figure 2.12. As discussed in Chapter 2, the general torque equation for SRM is derived using magnetic co-energy, which is applicable for both magnetically linear and saturated regions:

$$\tau = \left.\frac{\partial W_c}{\partial\theta}\right|_{i=const} \tag{4.22}$$

When SRM operates in the saturated region, torque production capability improves; less energy is stored in the magnetic circuit and more air gap power is converted into mechanical power. This phenomenon will become clearer when we investigate the torque and voltage profiles of an SRM in the next section at different rotor positions and excitation currents.

4.3 Operational Characteristics of Switched Reluctance Machines

SRM has a salient structure, which is an essential part of its torque production mechanism. Because of this, SRM characteristics such as flux linkage and torque are functions of rotor position. At the same time, these characteristics are dependent on the phase current.

SRM characteristics can be analyzed using Finite Element Analysis (FEA) software to account for nonlinearities, and then these characteristics can be used to understand the operation of the machine under dynamic conditions. As discussed before, when the coils of a conventional SRM phase are energized with current, the majority of the generated flux links to the coils of the same phase. Therefore, the mutual flux, which is the flux that would link to the coils of other phases, is negligible in conventional switched reluctance machines. Due to the negligible mutual flux linkage, once the characteristics of one phase of an SRM are determined, they can be used for other phases as well. To obtain these characteristics, constant current is applied to one phase of the machine for one electrical cycle. The results show how SRM characteristics change with position. Repeating the same analysis with different currents shows how the characteristics change as the magnetic core of SRM saturates.

In Figure 4.4, the static profile for phase flux linkage of a 12/8 SRM is shown under 10 A of constant current excitation, along with the relative position of the rotor and stator poles. Here, we use the term static to express that the phase current is constant in the entire electrical cycle. As we will see later in this chapter, phase current needs to be regulated under the dynamic conditions to generate continuous torque.

When a stator pole is positioned between two rotor poles, the relative distance between the rotor and stator poles is the largest. This position is called the unaligned position. It can be seen from Figure 4.4 that at the unaligned position, the flux linkage is at the minimum due to the large air gap. In contrast, when the rotor and stator poles are in alignment, the resulting air gap between the stator and rotor poles is minimized. Here, the flux linkage is at its maximum.

The flux linkage is also dependent on the excitation current. As higher level of current is applied, the incremental increase in the flux linkage will decrease due to the higher magnetic flux density and, hence, the saturation level of the core. As shown in Figure 4.5, the increase in the flux linkage from 10 to 15 A is much smaller than the increase from 5 to 10 A.

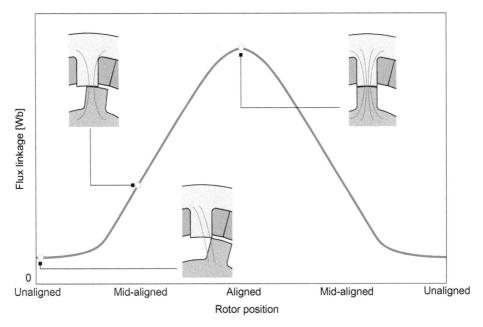

FIGURE 4.4
Phase flux linkage profile under 10 A constant current excitation for one electrical cycle.

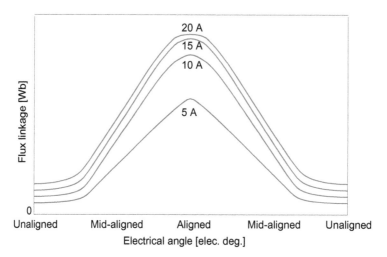

FIGURE 4.5
Phase flux linkage at various excitation currents for one electrical cycle.

Figure 4.6 shows the flux linkage characteristics in Figure 4.5 in a different way. In this case, the flux linkage waveforms are plotted as a function of current at different rotor positions. Notably, at the aligned position, the flux linkage does not increase linearly with current. At this position, the air gap between the rotor and the stator poles is the smallest and, hence, the effective permeability in (2.67) is the largest. This is why the flux linkage-current characteristics at the aligned position in Figure 4.6 look similar to the magnetization curve of electrical steel (see Figure 2.7 in Chapter 2). In contrast, at the unaligned position, magnetic saturation is not noticeable even at high currents, due to the large air gap.

The electromagnetic torque can be modeled in a similar manner with constant current excitation at different rotor positions. When the flux linkage profile in Figure 4.5 and the torque profile in Figure 4.7 are compared, it can be observed that the torque is positive

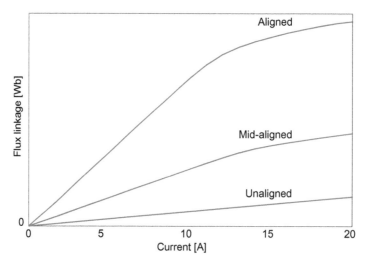

FIGURE 4.6
Phase flux linkage at various rotor positions for different current levels.

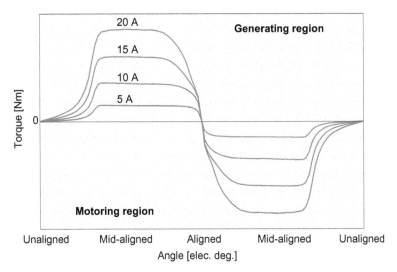

FIGURE 4.7
Rotor torque profile at various excitation currents for one electrical cycle.

when the flux linkage increases with rotor position and the torque is negative when the flux linkage decreases with rotor position. This matches with the torque expression in (4.21). Please also note that the torque characteristics in Figure 4.7 have been calculated when the phase is energized with constant positive current. The current is the same when torque is negative. This also matches with the torque expression in (4.21) and it shows that the torque is independent of the direction of the current.

From the torque profile in Figure 4.7, you can observe that before the rotor pole reaches aligned position, any excitation current applied to the phase would draw the rotor in the direction of rotation, thus creating positive torque. Once the rotor pole is past the aligned position, excitation current would draw the rotor pole against the direction of rotation. In this case torque will be negative. Therefore, when using SRM in motoring mode, the phases must be activated in the positive torque production region. When applying the current at the positive slope of the flux linkage profile, the electromagnetic torque will be in the direction of rotation. If SRM is used in generating mode, the phase should be excited when the slope of the flux linkage profile is negative. In this case, negative torque will drive power from the prime mover and the SRM will generate power to the DC link. The generating mode of SRM operation will be investigated in detail in Chapter 5.

Figure 4.8 shows the terminal voltage profile at various excitation currents for one electrical cycle. It can be seen that the static terminal voltage is dependent on rotor position and current, similar to the flux linkage and torque. The static voltage waveform is the voltage that would be observed at the terminals of the phase winding if the rotor rotates for one electrical cycle when constant current flows through the phase winding. The static voltage waveform is not a directly measurable characteristic, but it is useful to analyze the rate of change of current under dynamic control. We will explore this more in Chapter 9. If we measure the voltage across the phase winding during the operation of SRM, we would not be able to observe the waveform in Figure 4.6, but we would observe the modulated DC link voltage.

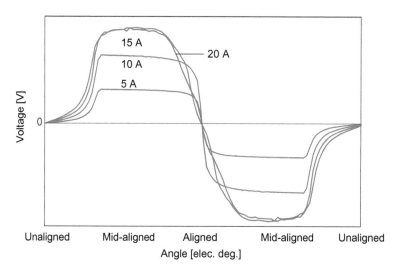

FIGURE 4.8
Terminal voltage at various excitation currents for one electrical cycle.

Since the terminal voltage is dependent on the rate of change of flux linkage as given in (4.1), its shape resembles that of the torque, but there is one significant difference. In Figure 4.7, torque at 20 A is higher than the torque at 15 A. However, in Figure 4.8, voltage at 20 A is almost the same as the voltage at 15 A. This is the effect of saturation. As we discussed in Chapter 2, co-energy in SRM increases with the level of saturation, which leads to improved torque production capability without the need for higher voltage. This results in better utilization of the power semiconductor devices in the converter and also reduces the converter volt-ampere requirements.

4.4 Phase Excitation Principles of Switched Reluctance Machines

In the motoring mode of operation, continuous torque production in SRM can be realized if phases are sequentially excited as the rotor poles move towards the stator poles. This corresponds to the positive slope of the flux linkage profile as the phases are excited between unaligned and aligned positions. It is important that the phase is demagnetized before the rotor passes the aligned position to prevent negative torque generation. In SRM control, the understanding of timing for phase excitation is critical to optimize the performance of the machine at different operating points.

4.4.1 Phase Excitation Sequence

With respect to a given phase, there are two important rotor positions to identify the electrical cycle: unaligned and aligned position. Figure 4.9 shows the electrical angles with respect to phase A in a 12/8 SRM at different rotor positions. In Figure 4.9a, Ph#1 is at the unaligned position, which is designated as 0° electrical by convention. Therefore, 180° electrical is designated as the aligned position, which can be seen in Figure 4.9c.

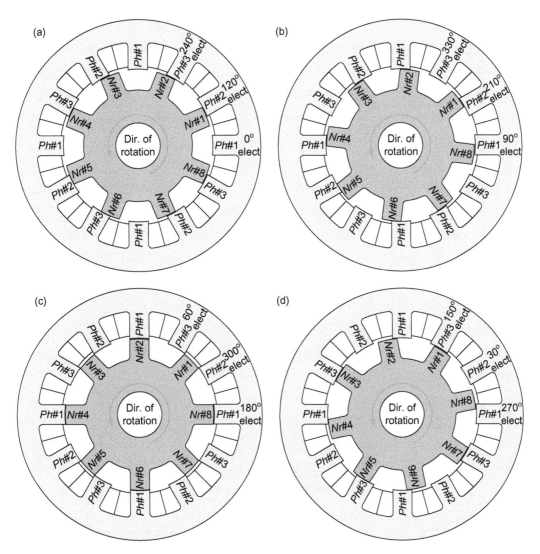

FIGURE 4.9
Electrical angles at different rotor positions with respect to Ph#1: (a) 0° electrical, (b) 90° electrical, (c) 180° electrical, and (d) 270° electrical.

From the position in Figure 4.9d, if the rotor keeps rotating in the same direction by 11.25° mechanical, it completes another 90° electrical (90° *elec.* = 11.25° *mech* × N_r, where N_r is the number of rotor poles and it equals to 8 in this case), and Ph#1 will be again at the unaligned position, but the stator poles of Ph#1 will be between different rotor poles as compared to the unaligned position in Figure 4.9a. In this case, one electrical cycle would be completed.

In motoring mode, each phase would be excited sequentially in the positive slope of the flux linkage profile. It can be observed from Figure 4.9 that the excitation sequence is dependent on the direction of rotation. For the 12/8 SRM, the phase excitation order is Ph#1-Ph#3-Ph#2 for counterclockwise rotation. Figure 4.10 shows the phase excitation sequence and the ideal current profiles for the 12/8 SRM.

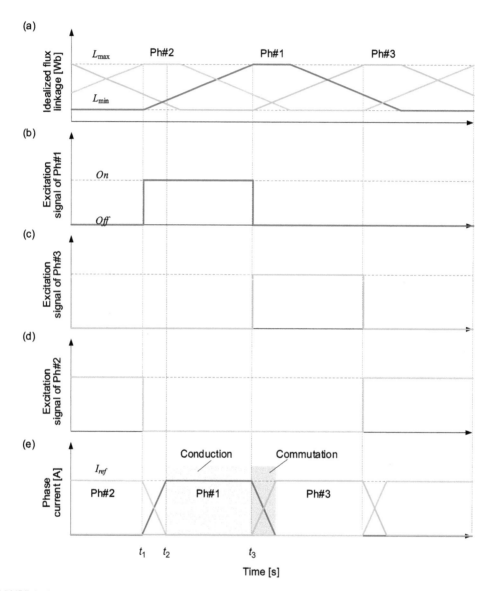

FIGURE 4.10
Idealized phase excitation sequence for a three phase SRM: (a) idealized flux linkage, (b) excitation signal Ph#1, (c) excitation signal Ph#3, (d) excitation signal Ph#2, and (e) phase current.

For Ph#1, the excitation start at time t_1, which is referred as the turn-on angle θ_{on}. Then, at time t_3, the phase excitation for Ph#1 is turned off, the phase starts demagnetizing and the phase current begins to decay. The angle when the demagnetization starts is referred as the turn-off angle θ_{off}. Since the phase excitation should occur during the rising edge of the flux linkage profile, ideally turn-on angle will be after 0° electrical and the turn-off angle will be before 180° electrical. However, the turn-on angle can be advanced to help build up the current before the induced voltage starts increasing, which is particularly useful at high speeds. The selection of conduction angles is subject to the control of the motor and is investigated in Chapter 9.

For continuous torque production in the three-phase 12/8 SRM, each phase should conduct at least for 120° electrical, such that when one phase is demagnetized, excitation in another phase begins. This case is shown in Figure 4.10. At the instant t_1, Ph#2 is turned off and Ph#1 is turned on. However, current does not instantaneously rise or decay, due to the inductance of the phase. It can be seen that after t_1, current only reaches the reference value I_{ref} at time t_2, at which point the outgoing phase Ph#2, is completely demagnetized. In between t_1 and t_2, both phases are conducting simultaneously. This phenomenon is referred to as phase commutation and it is a major source of torque ripple. The ideal flux patterns during the single-phase conduction and phase commutation are shown in Figure 4.11.

Figure 4.11a represents time t_2 in Figure 4.10 when Ph#1 is conducting. The resulting magnetic flux lines travel through the excited stator poles. In Figure 4.11b, Ph#1 is at the aligned position and Ph#3 is excited. This represents t_3 in Figure 4.10. At this instant, demagnetization of Ph#1 has just started and two phases conduct at the same time. It can be observed that the magnetic flux takes a shorter path compared to Figure 4.11a when only one phase was conducting.

Figure 4.12 shows the three-phase excitation sequence over one electrical cycle of Ph#1. The magnetic flux pattern constantly changes in the dynamic operation and rotates around the air gap. This excitation sequence is what creates continuous torque production in SRMs. From Figure 4.12a, c, and e, it can be observed that when single phase conducts, the magnetic flux takes a longer path and the flux only links to the coils that are energized with current. During the commutation between phases, the flux patterns are considerably shorter, since two phases conduct at the same time. This is mutual coupling, because flux generated by one phase links to the coils of another one. In conventional switched reluctance machines, the mutual coupling occurs very briefly in each cycle and the effect can typically be neglected when modeling an SRM.

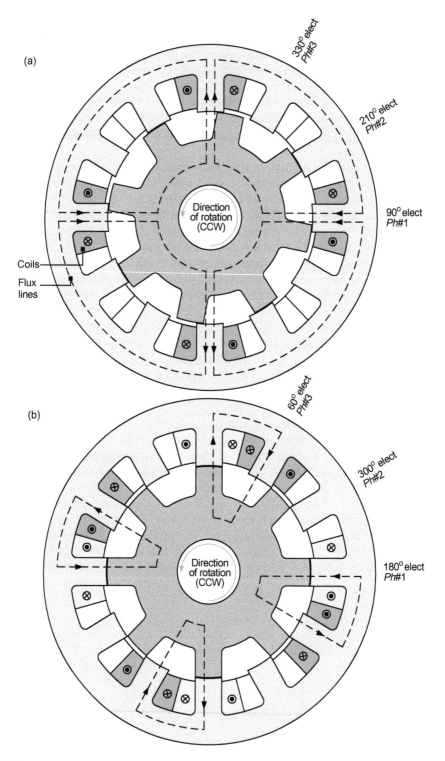

FIGURE 4.11
Ideal flux patterns: (a) long flux paths during single-phase conduction and (b) short flux paths during phase commutation.

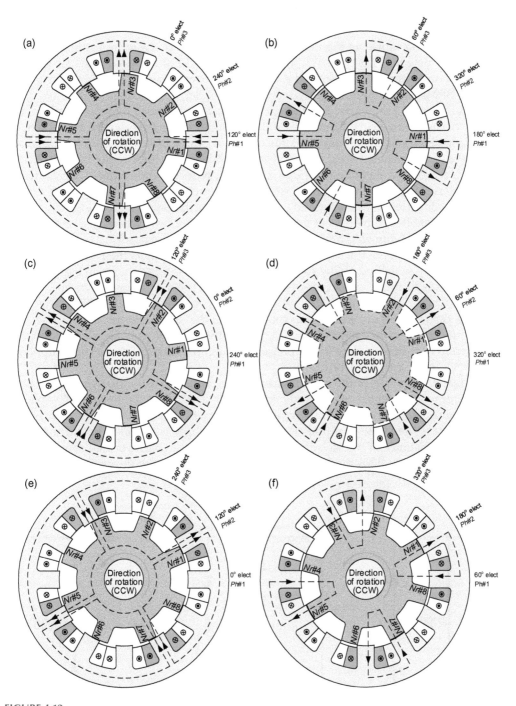

FIGURE 4.12
Three phase excitation sequence over one electrical cycle of Ph#1: (a) Ph#1 excitation, (b) Commutation between Ph#1 and Ph#3, (c) Ph#3 excitation, (d) Commutation between Ph#3 and Ph#2, (e) Ph#2 excitation, and (f) Commutation between Ph#2 and Ph#1.

4.4.2 Asymmetric Bridge Converter

The phase excitation signals in Figure 4.10 switch on and off as the rotor position changes and, hence, the magnetic flux patterns shown in Figure 4.12 rotate around the air gap for continuous torque production. We need a power converter to switch the phase excitation signals. For SRMs, an asymmetric bridge converter is widely used to control the current in each phase. Figure 4.13 shows the circuit diagram of an asymmetric bridge converter for one phase. It consists of two switching devices and two diodes located diagonally. This is where the term "asymmetric" comes from. As defined in (4.21), the torque in an SRM is a function of the square of the current; hence, torque is independent of the direction of the phase current. For this reason, the asymmetric bridge converter produces unidirectional current flow. Since each phase in an SRM is electrically isolated from each other, we need to use an asymmetric bridge converter for each phase. We will investigate the asymmetric bridge converter and other types of SRM converters in detail in Chapter 10. For now, we need to understand the operation of an asymmetric bridge converter for modeling purposes.

During phase excitation, the asymmetric bridge converter applies the DC link voltage, V_{DC}, to the coils by switching on the switching devices (S_1 and S_2) as shown in Figure 4.13a. This corresponds to time t_1 with regards to Ph#1, in Figure 4.10. Then, at the end of the phase excitation, the switching devices are switched off as shown in Figure 4.13b. This corresponds to time t_3 in Figure 4.10. At this mode, due to the stored energy in the phase, the diodes (D_1 and D_2) become forward biased and negative DC link voltage is applied to the phase. Current continues to flow in the phase winding, through the diodes and supplied back to the DC link until it reaches zero.

A third operating mode of the asymmetric bridge converter is the freewheeling mode shown in Figure 4.13c. In this mode, only one pair of switch and diode is on. The phase coil is not connected to the supply; therefore, no energy flows back to the DC link. Since the voltage across the phase terminals is zero and the phase is shorted, the current only decays due to resistance of the coil, and thus does so at a much slower rate compared to the case in Figure 4.13b. This mode is generally used to control the phase current during the conduction period.

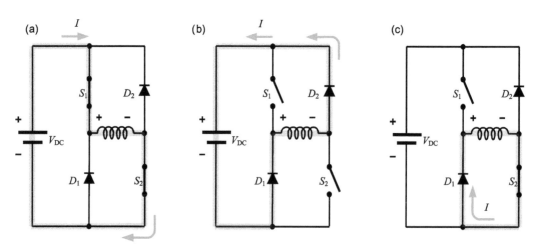

FIGURE 4.13
Operational modes of asymmetric bridge converter for one phase: (a) both switching devices are on, V_{DC} is applied to the phase, (b) both switching devices are off, $-V_{DC}$ is applied to the phase, and (c) Freewheeling mode, voltage across the phase is zero.

TABLE 4.1

Summary of the Operating Modes of the Asymmetric Bridge
Converter

Mode	Device Conducting	Voltage Applied to Phase
1	Switches (S_1, S_2)	$+V_{DC}$
2	Diodes (D_1, D_2)	$-V_{DC}$
3	One diode, one switch $(D_1, S_2$ or $D_2, S_1)$	0

At time t_1 in Figure 4.10, S_1 and S_2 are turned on as shown in Figure 4.13a. The phase current rises. When the phase current reaches I_{ref}, we need to control the current; otherwise, it would keep increasing if the induced voltage is lower than the DC link voltage. In general, the third operating mode (Figure 4.13c) is used for this purpose. Using the mode in Figure 4.13c, the phase current decays and when it goes below a certain threshold, mode in Figure 4.13a is applied to increase the current again. These two modes are repeated within the conduction interval to keep the phase current close to I_{ref}. This method of current regulation is termed hysteresis control. We will present the model of a hysteresis controller later in this chapter. At time t_3, the phase must be demagnetized by turning both switches off and the mode in Figure 4.13b is applied. Current will then decay at a faster rate since $-V_{DC}$ is applied across the phase terminals. Table 4.1 summarizes the three modes of the asymmetric bridge converter.

4.5 Modeling of Switched Reluctance Machines

So far we looked into the phase excitation principles and static phase characteristics, which are crucial to understand the operating principles and nonlinear nature of an SRM. Now we will develop the models to analyze the dynamic behavior of an SRM. The torque production capability and power factor of the SRM significantly improve under magnetic saturation. This is an essential concept, and must be considered when modelling these machines. Unlike permanent magnet and induction machines, SRMs are not excited with sinusoidal current. Due to the switching behavior of the magnetic field, phase transformation techniques cannot be directly applied to SRMs. Therefore, SRM requires different modeling techniques to analyze its dynamic performance.

4.5.1 Static Machine Characterization and Look-up Tables

The heart of the SRM dynamic model is the voltage equation in (4.1). It represents the relationship between the phase voltage v_{ph}, phase current i_{ph}, and phase flux linkage λ_{ph}. It can be rewritten to express these phase quantities:

$$v_{ph} = R_{ph}i_{ph} + \frac{d\lambda_{ph}\left(i_{ph}, \theta_{elec}\right)}{dt} \tag{4.23}$$

If we rearrange (4.23), calculate the phase current under dynamic conditions, and we have access to the static characteristics of the motor (Figures 4.5 and 4.7), we can calculate the torque produced by each phase by looking up for the torque value from Figure 4.7. Recall that the flux linkage profile in Figure 4.5 and the torque profile in Figure 4.7 both change with current and position, and that the change with current is nonlinear when operating under magnetic saturation (at higher phase currents). Therefore, we need to know the torque and the flux linkage at various constant current values and rotor positions and create Look-up Tables (LUTs) so that we can solve for the dynamic current and torque.

Figure 4.14 shows the procedure for the characterization of an SRM. It is recommended to use electromagnetic FEA tools for the characterization of the SRM. It is possible to estimate the static characteristics of the motor using analytical methods, but it can be difficult to do accurately considering the nonlinearities in SRMs. Analytical estimation typically requires the use of several assumptions which cannot always be satisfied. Today, FEA tools can mesh reliably, run quickly, provide more accurate results than analytical methods, and most software can be scripted to generate different geometries quickly.

As shown in Figure 4.14, flux linkage and torque look-up tables are generated for N number of different currents and for different electrical angles θ_{elec}. Phase current array I_{phase} includes the current only for one phase and the other phase currents are zero. Since the mutual flux linkage between different phases is negligible in SRMs, the look-up tables for one phase can be used to model all phases.

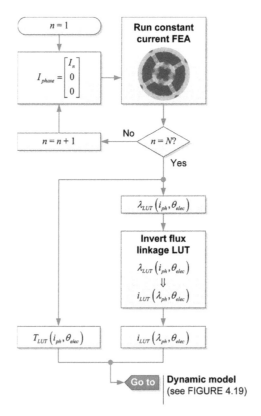

FIGURE 4.14
SRM characterization procedure to obtain the look-up tables.

4.5.2 Mathematical Model (Integrator Method)

If we solve (4.23) for the flux linkage, we get:

$$\lambda_{ph}(t) = \int \left[v_{ph}(t) - R_{ph}i_{ph}(t) \right] dt \tag{4.24}$$

On the right-hand side of (4.24), i_{ph} is still dependent of the flux linkage and the electrical angle. Therefore, (4.24) can be reorganized as:

$$\lambda_{ph}(t) = \int \left[v_{ph}(t) - R_{ph}i_{LUT}\left(\lambda_{ph}(t), \theta_{elec}(t)\right) \right] dt \tag{4.25}$$

This is the reason why the flux linkage LUT was inverted in Figure 4.14. With i_{LUT}, we have the relationship between the flux linkage and current at different electrical angles. Using (4.25), we can calculate the flux linkage at a certain time instant and look up the current using the past values of flux linkage. Hence, in discrete form (4.25) can be expressed as:

$$\lambda_{ph}(k) = \lambda_{ph}(k-1) + \left[v_{ph}(k) - R_{ph}i_{LUT}\left(\lambda_{ph}(k-1), \theta_{elec}(k-1)\right) \right] T_s \tag{4.26}$$

where T_s is the sampling time used to solve the discrete-time integration and k represents the discrete step number. In (4.26), i_{LUT} represents the current needed to achieve a certain flux linkage at a given position. As shown in Figure 4.15, (4.26) can be implemented in a simulation model to solve for the flux linkage and ultimately the current.

Please note the "reset integrator" block in Figure 4.15. This function is used to model the diode effect. When the switches of the asymmetric bridge converter are both turned

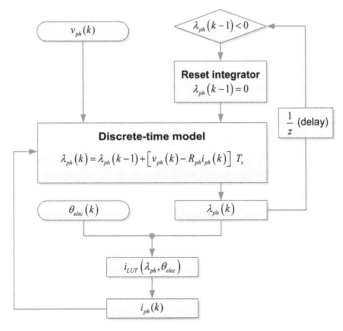

FIGURE 4.15
Solving the voltage equation using integration.

off, the diodes are forward biased due to the stored energy. The direction of the source current changes, but the direction of the phase current does not. The current now flows back to the supply and negative DC link voltage is applied to the phase. When the stored energy is dissipated completely, the phase current will reach to zero and the diodes turn off. So, the current and, hence, the flux linkage cannot go below zero. The "reset integrator" block in Figure 4.15 assures that this condition is met. Since λ_{ph} is calculated by integration, it resets the output of the integrator so the flux linkage and, hence, the current stays at zero.

4.5.3 Mathematical Model (Derivative Method)

If the machine is characterized as in Figure 4.14, then we have direct access to the flux linkage look-up table $\lambda_{LUT}(i_{ph}, \theta_{elec})$. This can also be used to calculate current from (4.23) using the derivative of the flux linkage:

$$i_{ph} = \frac{1}{R_{ph}} \left[v_{ph} - \frac{d\lambda_{LUT}\left(i_{ph}, \theta_{elec}\right)}{dt} \right] \tag{4.27}$$

In the continuous- and discrete-time domains, (4.27) can be expressed as in (4.28) and (4.29), respectively.

$$i_{ph}(t) = \frac{1}{R_{ph}} \left[v_{ph}(t) - \frac{d\lambda_{LUT}\left(i_{ph}(t), \theta_{elec}(t)\right)}{dt} \right] \tag{4.28}$$

$$i_{ph}(k) = \frac{1}{R_{ph}} \left[v_{ph}(k) - \frac{\lambda_{LUT}\left(i_{ph}(k), \theta_{elec}(k)\right) - \lambda_{LUT}\left(i_{ph}(k-1), \theta_{elec}(k-1)\right)}{T_s} \right] \tag{4.29}$$

This method does not require the inversion of the flux linkage look-up table. However, since we are using a derivative with this method, a small noise or discrepancy in the calculations would be amplified in the next time step, which might cause significant errors in the results. In the derivative case, the LUT calculates the flux linkage produced from the current at a given position. As shown in Figure 4.16, (4.29) can be implemented in a simulation model to solve for the current and the flux linkage. The unit delay block is used to prevent a possible algebraic loop in the model.

In the rest of this book, the integration-based method will be used when modeling switched reluctance machines. In order to use this method, LUT inversion must be performed, as discussed in the next section.

4.5.4 $\lambda - i$ LUT Inversion

Since mutual coupling is typically negligible in SRM, it can be assumed that the flux linkage of a given phase is only dependent on the current in that phase and the rotor position. Figure 4.17 shows a sample λ_{LUT} and an inverted i_{LUT} after the transformation:

$$\lambda_{ph}\left(i_{ph}, \theta_{elec}\right) \rightarrow i_{ph}\left(\lambda_{ph}, \theta_{elec}\right)$$

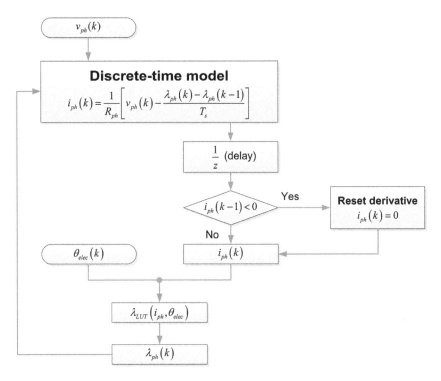

FIGURE 4.16
Solving the voltage equation using derivation.

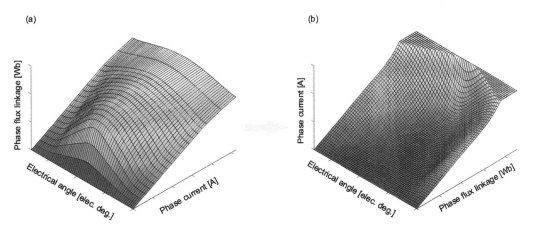

FIGURE 4.17
Inversion of flux linkage look-up table: (a) before conversion and (b) after conversion.

Sample MATLAB® code for the LUT inversion algorithm is shown in Figure 4.18. The flux linkage look-up table is created for the phase currents, currentArray of size [N _ current,1] and electrical angles, angleArray of size [N _ angles,1]. Therefore, the size of the flux linkage matrix is [N _ angles, N _ current]. We would like to convert the flux linkage matrix into a current matrix of size [N _ angles, N _ fluxlink]. Usually, N _ angles is greater than N _ current when characterizing SRMs. Since the current matrix will be calculated using interpolation, it would be practical to generate a square matrix for the current LUT by selecting N _ fluxlink same as N _ angles. It can be observed from Figure 4.17 that the mesh in the inverted look-up table is much denser due to this adjustment.

First, we generate a flux linkage array of size [N _ fluxlink,1], equally spaced from the minimum flux linkage to the maximum flux linkage of the original flux linkage matrix. In the sample algorithm in Figure 4.18, we call this array targetArray, because the algorithm attempts to find the current for each target value of this array.

Next, at each rotor position, we need to find the current which corresponds to each flux linkage value in targetArray, (named TARGET in Figure 4.18). For this purpose, we use interpolation (interp1) with three inputs. The first two inputs, fluxlink _ row and currentArray, are used to generate a look-up array for the interpolation. Using this table, we interpolate to find the current for the target flux linkage TARGET.

Since targetArray was generated from the entire flux linkage matrix, fluxlink _ row at an individual position might not include some of those values. This might cause out of range interpolation errors, which can be avoided by using a saturation function to limit TARGET to the maximum flux linkage for the given position. This effectively prevents the look-up table from extrapolating and generating i_{LUT} values that are greater than the maximum characterized current. The effect of the saturation function can be observed from plateau region the i_{LUT} map in Figure 4.17b.

```matlab
function LUT_currentMatrix = invertLUT(angleArray, currentArray, fluxlinkMatrix)

    % size(angleArray)     => [N_angles,  1]
    % size(currentArray)   => [N_current, 1]
    % size(fluxlinkMatrix) => [N_angles,  N_current]

    N_angles   = length(angleArray);
    N_fluxlink = N_angles;                          % to achieve square matrix

    LUT_currentMatrix = zeros(N_angles, N_fluxlink); % pre-allocate
    targetArray = linspace(min(fluxlinkMatrix(:)),max(fluxlinkMatrix(:)),N_fluxlink);
    current_max = max(currentArray);                % maximum current

    for i_angle = 1:N_angles

        fluxlink_row = fluxlinkMatrix(i_angle,:);
        % all flux linkages for given angle, from minimum to maximum

        for i_fluxlink = 1:N_fluxlink

            TARGET = targetArray(i_fluxlink);
            % the flux linkage value we want to determine the current for

            if TARGET > max(fluxlink_row)
                LUT_currentMatrix(i_angle, i_fluxlink) = current_max; % saturation function
            else
                LUT_currentMatrix(i_angle, i_fluxlink) = ...
                interp1(fluxlink_row, currentArray, TARGET);          % lookup the current
            end
        end
    end

end
```

FIGURE 4.18
Sample MATLAB look-up table inversion algorithm.

4.5.5 Complete Dynamic Model

A sample SRM dynamic model is shown in Figure 4.19 in discrete-time domain. Even though it is possible to have both continuous and discrete representation of the model, it is usually beneficial to use a discrete model with a fixed time step, as it is closer to digital controller implementation.

Since the LUTs represent the nonlinear characteristics of a switched reluctance machine, the voltage equation in Figure 4.15 can be used to calculate the dynamic current waveform. Once the current waveform is calculated, it can be applied to the torque LUT to calculate the dynamic torque generated by each phase.

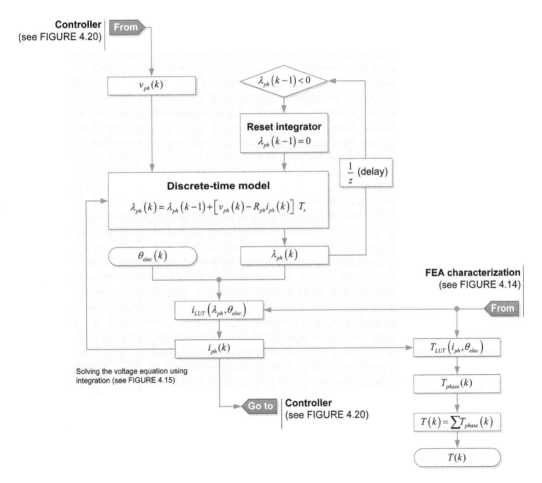

FIGURE 4.19
Dynamic model of an SRM.

4.5.5.1 Mechanical Angle Model

The dynamic model in Figure 4.19 analyzes the motor performance at a constant speed and calculates the current waveform based on the control parameters (e.g. turn-on and turn-off angles). At a constant speed, the mechanical position of the rotor increases linearly and repeats every 360°. Therefore, the mechanical position can be calculated through integration of the speed:

$$\theta_{mech}(k) = mod\left(N_{RPM}\left(\frac{360°}{60}\right)T_s\, k + \theta_{mech0}, \quad 360°\right) \qquad (4.30)$$

where N_{RPM} is the mechanical rotation speed in revolutions-per-minute (RPM), T_s is the discrete time step, k is the step number, and θ_{mech0} is the initial rotor position. In (4.30), N_{RPM} is first converted to rotations per second by dividing it by 60. Since one rotation is 360°, $N_{RPM}(360°/60)$ converts to degrees per second.

4.5.5.2 Electrical Angle Model

From Figure 4.19, it is clear that the LUTs are dependent on the instantaneous electrical angles (θ_{elec}) of each phase and a full electrical cycle (360°) represents one stroke. Since there are N_r number of rotor poles, the electrical cycle repeats N_r times for one phase. Hence, the relationship between the mechanical position and the electrical angle can be expressed as:

$$\theta_{elec} = N_r \theta_{mech} \qquad (4.31)$$

where θ_{mech} is calculated in (4.30). Since an electrical cycle also repeats every 360°, (4.31) can be reorganized as:

$$\theta_{elec} = mod\left(N_r \theta_{mech}, 360°\right) \qquad (4.32)$$

If we assume that zero degrees mechanical is when the first rotor pole is aligned with the first stator pole, and the first stator pole belongs to the first phase (Ph#1), the electrical angle of Ph#1 will be 180° electrical when θ_{mech} is zero. The electrical phase shift between the other phases is $360/m$, where m is the number of phases. Hence, the relationship between the electrical angle of each phase and the mechanical angle can be expressed as:

$$\theta_{elec} = mod\left(\left(N_r \theta_{mech} + 180°\right) + \frac{360}{m} k_{ph}, 360°\right). \qquad (4.33)$$

where k_{ph} is an array of size m. k_{ph} applies the phase shift angle $360/m$ to each phase. Let's take the 12/8 SRM in Figure 4.9c as an example to analyze how to calculate k_{ph}. Let's assume that this was the initial position of the rotor and, hence, θ_{mech} is zero. Ph#1 is at the aligned position and since this was our assumption when deriving (4.33), k_{ph} for Ph#1 will be zero. This will always hold when θ_{mech} is zero. In Figure 4.9c, Ph#3 is at 60° electrical and Ph#2 is at 300° electrical. Therefore, k_{ph} for Ph#3 should be 2 and k_{ph} for Ph#2 should be 1. Then, for counterclockwise rotation, $k_{ph} = \begin{bmatrix} 0 & 1 & 2 \end{bmatrix}$.

k_{ph} for Ph#1 will always be zero based on our assumptions. k_{ph} for other phases is dependent on the pole configuration and direction of rotation. If you check the 5-phase SRMs shown in Figures 3.11 through 3.14, it can be observed that phases have different electrical angles. Therefore, the same k_{ph} cannot be used for these configurations.

In order to calculate the values of k_{ph}, we need to go back to our discussions in Chapter 3. The electrical angle of each stator pole was calculated using (3.7):

$$Ns_{elect} = mod\left[k_{rot}\left(Ns_{mech} - Nr_{mech}\right)N_r + 180°, 360\right]$$

where Ns_{mech} is the mechanical angle of a stator pole and Nr_{mech} is the mechanical angle of a rotor pole. k_{rot} is -1 for counterclockwise rotation and 1 for clockwise rotation. Please recall from Chapter 3 that, for a symmetric SRM configuration, (3.7) results in the same value for all Nr_{mech} values at the given rotor position. Therefore, if we calculate the electrical angles of the first stator poles of each phase towards the first rotor pole and sort them, we can calculate k_{ph} for each phase.

Let's take the five-phase 10/18 SRM in Figure 3.11 as an example. Ns#2 is at 36° mechanical and Nr#1 is at 0° mechanical. Using (3.7), the electrical angle of Ns#2 for counterclockwise rotation can be calculated as 252° electrical. The electrical angles of the other stator poles and, hence, phases can be calculated in the same way. At the given rotor position in Figure 3.11, Ph#1 is at 180° electrical, Ph#2 is at 252°, Ph#3 at 324°, Ph#4 is at 36°, and Ph#5 is at 108° electrical. If we sort the electrical angles after the electrical angle of the first phase, which is always 180° due to our assumption of the initial rotor position, we end up $\left[Ph\#1\ Ph\#2\ Ph\#3\ Ph\#4\ Ph\#5\right] = \left[180°\ 252°\ 324°\ 36°\ 108°\right]$. Hence, for the five-phase 10/18 SRM $k_{ph} = \left[0\ 1\ 2\ 3\ 4\right]$ for counterclockwise rotation.

Now we investigate the five-phase 10/14 SRM in Figure 3.13. Using (3.17), the electrical angles can be calculated as Ph#1 at 180°, Ph#2 at 36°, Ph#3 at 252°, Ph#4 at 108°, and Ph#5 at 324° electrical. If we sort the electrical angles after the electrical angle of the first phase, we end up with $\left[Ph\#1\ Ph\#2\ Ph\#3\ Ph\#4\ Ph\#5\right] = \left[180°\ 252°\ 324°\ 36°\ 108°\right]$. Hence, for the five-phase 10/14 SRM, $k_{ph} = \left[0\ 3\ 1\ 4\ 2\right]$ for counterclockwise rotation.

4.5.5.3 Controller Modeling

The model in Figure 4.19 includes most of the details of an SRM drive system, except the relationship between the phase current i_{ph} and phase voltage v_{ph}. This relationship is dependent on how the current is controlled, and it is determined by the converter and control parameters. Figure 4.20 shows the outline of how the current control is established. The phase current is compared to a commanded current waveform i_{cmd}. The current controller regulates the phase voltage with the help of the converter to reduce the error. The plant model in Figure 4.20 is SRM dynamic model in Figure 4.19.

As shown in Figure 4.10, i_{cmd} is usually created based on the reference current i_{ref}, turn-on angle θ_{ON}, and turn-off angle θ_{OFF}. These parameters are dependent on the operating conditions of the motor. The control of switched reluctance machines and how the control parameters are determined will be discussed in detail in Chapter 9.

4.5.5.4 Hysteresis Control Modeling

Hysteresis control is widely used in switched reluctance motor drives to track the current reference. Figure 4.21 shows an example for the hysteresis current control. In order to generate the current profile in Figure 4.21a, the voltage profile in Figure 4.21b is applied to the phase by using the operating modes of asymmetric bridge converter in Figure 4.13.

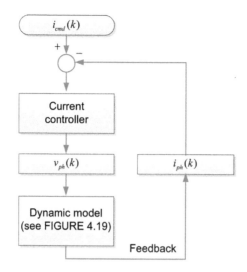

FIGURE 4.20
Overview of the current control.

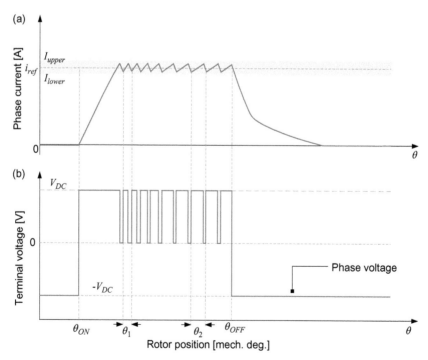

FIGURE 4.21
Hysteresis control with soft switching: (a) phase current and (b) phase voltage.

At the turn-on angle θ_{ON}, both switches in the asymmetric bridge converter are turned on as shown in Figure 4.13a and the phase current rises. Once the phase current reaches the reference value i_{ref}, it is maintained within the hysteresis band. The hysteresis band is usually defined as a percentage of i_{ref}.

At the turn-off angle θ_{OFF}, both switches are turned off and, as shown in Figure 4.13b, the diodes conduct due to the stored energy in the phase. The polarity of the supply current changes and negative DC link voltage is applied to the phase. The phase is demagnetized and the phase current decays.

Between the conduction angles, the phase current is regulated within the upper limit I_{upper} and lower limit I_{lower} of the hysteresis band, which are defined as:

$$I_{upper} = i_{ref} \times (1 + \beta) \tag{4.34}$$

$$I_{lower} = i_{ref} \times (1 - \beta) \tag{4.35}$$

where β is the hysteresis band and it is between 0 and 1. Between θ_{ON} and θ_{OFF}, the phase current is controlled using the asymmetric bridge converter modes in Figure 4.13a and c. In the freewheeling mode (Figure 4.13c), only one switch is turned off and the phase winding is short circuited. The stored energy is dissipated in the resistance of the phase winding and the phase current decays at a smaller rate compared to the mode in Figure 4.13b. This is referred as soft or unipolar switching. It is possible to control the phase current in SRM using soft (unipolar) switching or hard (bipolar) switching, as discussed in Chapter 10. With soft switching, at time step k, hysteresis control can be modeled as:

$$v_{ph}(k) = \begin{cases} 0 & | \quad \text{excitation signal} \leq 0 \quad \cap \quad i_{ph} \leq 0 \\ -V_{DC} & | \quad \cdots \quad \cap \quad i_{ph} > 0 \\ 0 & | \quad \text{excitation signal} > 0 \quad \cap \quad i_{ph} \geq I_{upper} \\ +V_{DC} & | \quad \cdots \quad \cap \quad i_{ph} < I_{lower} \\ v_{ph}(k-1) & | \quad \cdots \quad \cap \quad I_{upper} > i_{ph} \geq I_{lower} \\ 0 & | \quad \cdots \quad \cap \quad \text{otherwise} \end{cases} \tag{4.36}$$

For hard switching, only two operational modes in Figure 4.13 are used and the hysteresis control can be modeled as:

$$v_{ph}(k) = \begin{cases} 0 & | \quad \text{excitation signal} \leq 0 \quad \cap \quad i_{ph} \leq 0 \\ -V_{DC} & | \quad \cdots \quad \cap \quad i_{ph} > 0 \\ -V_{DC} & | \quad \text{excitation signal} > 0 \quad \cap \quad i_{ph} \geq I_{upper} \\ +V_{DC} & | \quad \cdots \quad \cap \quad i_{ph} < I_{lower} \\ v_{ph}(k-1) & | \quad \cdots \quad \cap \quad I_{upper} > i_{ph} \geq I_{lower} \\ 0 & | \quad \cdots \quad \cap \quad \text{otherwise} \end{cases} \tag{4.37}$$

where the "|" symbol means "given" and the "⌒" symbol represents logical "and." It is important to note from equations (4.36) and (4.37) that the hysteresis control is only operational when the excitation signal is larger than zero. Otherwise, the phase should be off, and the applied voltage should be zero after the current is dissipated. The excitation signal has been explained in Figure 4.10.

4.5.6 Additional Modeling

The block diagrams shown in Figures 4.19 and 4.20 are the framework of the SRM drive model and they include the major aspects of the dynamic model. However, the model must be further expanded to be used in a practical context.

In Figure 4.19, the mechanical angle is directly converted to electrical angles. The observed mechanical angle, $\hat{\theta}_{mech}$ is usually retrieved by a position sensor (transducer) or position estimation, which can introduce errors in the mechanical and electrical angle calculations. In addition, in Figure 4.20, we assumed that the phase current feedback is directly accessible but in discrete-time implementation, the analog current signal is converted into digital data through analog-to-digital converter (ADC), which introduces sampling error.

Finally, the current transducer and the inverter might also introduce nonlinearities in the system. In SRM design and performance analysis, nonlinearities due to the position feedback and current sampling might be more significant. Figure 4.22 shows the block

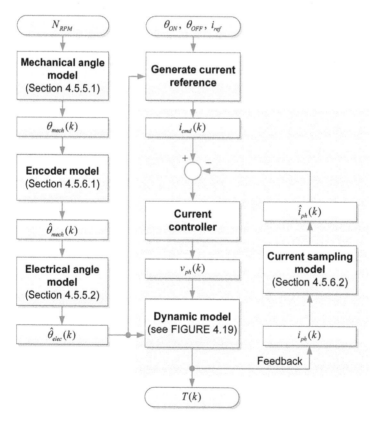

FIGURE 4.22
Block diagram of the dynamic model of SRM including the models for position feedback and current sampling.

diagram of the dynamic model of a switched reluctance motor drive when modeling of the position feedback and the current sampling is taken into account.

4.5.6.1 Encoder Modeling

The inaccuracy of the position feedback from the position sensor (often an encoder) depends on the resolution of the encoder and this inaccuracy may need to be accounted for in the dynamic model of an SRM. Since the current waveform and the control performance are dependent on the rotor position, modeling the inaccuracy of the encoder might be important especially at high speed operation. There are different types of encoders and different methods of modeling them. Here we consider an incremental rotary encoder model.

An incremental rotary encoder samples the rotor position as shown in Figure 4.23. Typically, rotary incremental encoders provide a certain number of pulses per rotation. By counting these pulses, the position relative to the initial position can be estimated. The number of pulses per rotation N_{pulses} represents the accuracy of the sensor, and depends on what type and brand of encoder is used. This information should be specified in the encoder datasheet.

For the given number of pulses for rotation N_{pulses}, the mechanical angle between encoder pulses can be calculated as:

$$\hat{\theta}_{pulse} = \frac{360°}{N_{pulses}} \tag{4.38}$$

For the given speed in rotations per minute N_{RPM}, the rotor makes $N_{RPM}/60$ rotations per second. Hence, the number of pulses per second can be calculated as:

$$n_{pulses/sec} = \frac{N_{RPM}}{60} N_{pulses} \tag{4.39}$$

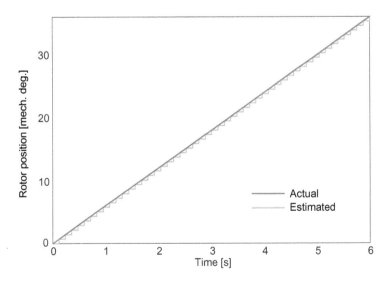

FIGURE 4.23
Encoder position signal @ 1 RPM (N_pulses = 500).

Then, the total number of encoder pulses at time step k can be calculated as:

$$n_{pulses} = floor\left(n_{pulses/sec}\, k\, T_s\right) \tag{4.40}$$

where T_s is the sampling time. Equation (4.40) acts like an integrator. The *floor* function is to make sure that the output of (4.40) is an integer to represent the pulses. The change in position at the kth time step can be calculated as:

$$\hat{\theta}_{mech}\left(k\right) = \hat{\theta}_{pulse}\cdot n_{pulses}\left(k\right) + \theta_{mech}\left(k=1\right) \tag{4.41}$$

where $\theta_{mech}(k=1)$ is the initial position (θ_{mech0}).

4.5.6.2 Modeling of Current Sampling

Current sensors are analog devices (i.e. hall effect sensors), but the output voltage of the sensor is sampled digitally. If the control parameters are analyzed without taking the current sampling into account, dynamic torque waveform could be miscalculated. The effect of current sampling can be modelled in a simulation environment, by triggering a subsystem at a given sampling frequency or by using generic zero-order-hold (ZOH) model.

If the current sampling frequency is f_{samp}, then the sampling period can be calculated as

$$T_{samp} = \frac{1}{f_{samp}} \tag{4.42}$$

The number of simulation time steps at each current sampling period is

$$n_{samp} = round\left(\frac{T_{samp}}{T_s}\right) \tag{4.43}$$

where T_s is the sampling period of the dynamic model. Therefore, when the simulation time step k is an integer multiple of n_{samp} or when a counter in the dynamic model resets itself at every n_{samp} steps, the phase current at that time instant $i_{Ph}(k)$ should be fed into the controller.

4.6 Introduction to Modeling of Mutually-Coupled Switched Reluctance Machines

As discussed in Chapter 3, there are certain SRM configurations that utilize mutual coupling between phases. In this case, the assumption that the phases are isolated does not hold and mutual-flux linkage cannot be neglected. In mutually-coupled SRMs (MCSRM), torque is produced by the changing mutual-flux linkage rather than the changing self-flux linkage as in a conventional SRM. This provides several benefits, but one major benefit is that conventional three phase inverter can be used for MCSRMs. Besides, MCSRMs can be modeled and controlled similar to alternating current (AC) machines. In this section, modeling of a three-phase MCSRM will be introduced, but the basic principles can be expanded to any higher number of phases. Considering that AC motor control principles have already been discussed quite extensively in other resources, only the basics of the modeling as it applies to SRMs will be covered in this section.

4.6.1 Static Characterization and Look-up Tables

Since the MCSRM is still a reluctance machine, operation under saturated condition is still needed to improve the torque density and power factor. Hence, similar to conventional SRMs, FEA is used to characterize the machine to develop the LUTs, which describe the relationship between current and flux linkage.

MCSRMs can be controlled with conventional full-bridge inverter and sinusoidal current waveforms. Therefore, unlike conventional SRMs, phase transformation methods can be applied to model MCSRMs. Phase quantities can be transformed to a two-phase rotating reference frame (Clarke's or $\alpha - \beta$ transformation) and then to a two-phase stationary reference frame (Park's or $d - q$ transformation). Using the direct d and quadrature q axes, significantly simplifies the control problem, and these transformation methods are commonly used for induction and permanent magnet machines. Maximum torque is generated when the resultant current vector between the two phases is at a specific angle, which is dependent on the flux linkage characteristics. Excitation can be advanced by phase shifting the current waveforms, which can be accomplished by shifting the dq angle, away from the quadrature axis. The characterization requires that arrays of d-axis and q-axis currents be defined:

$$i_{dlist} = \left[-i_{rated} \ldots i_{rated} \right] \tag{4.44}$$

$$i_{qlist} = \left[-i_{rated} \ldots i_{rated} \right] \tag{4.45}$$

where i_{rated} is the rated current of the machine. Then, the amplitude and angle of the phase current can be calculated as:

$$I_{amp} = \sqrt{i_{dlist}^2 + i_{qlist}^2} \tag{4.46}$$

$$\phi_{phase} = \tan^{-1}\left(\frac{i_{qlist}}{i_{dlist}} \right) + \theta_{elec0} \tag{4.47}$$

where θ_{elec0} represents the electrical angles of each phase when Ph#1 is at the unaligned position (0° electrical). For a three-phase machine, depending on the pole configuration, this means that there are two possibilities, depending on k_{ph} which is defined similarly to Section 4.5.5.2:

$$\theta_{elec0} = \begin{bmatrix} 0° \\ +120° \\ -120° \end{bmatrix} \quad or \quad \theta_{elec0} = \begin{bmatrix} 0° \\ -120° \\ +120° \end{bmatrix} \tag{4.48}$$

The phase current waveform in MCSRM can be expressed as:

$$I_{phase} = I_{amp} \sin\left(\omega_{elec} t + \phi_{phase} \right) \tag{4.49}$$

where ω_{elec} is the electrical speed in rad/s. One of the main differences between the conventional SRM and MCSRM is the mechanical period for one electrical cycle. In a MCSRM, the mechanical period for one electrical cycle is doubled. This is because MCSRM current waveform conducts for the full electrical cycle due to the mutual coupling and sinusoidal currents. In contrast, for a conventional SRM the phase current should reach to zero at the end of the first half of the electrical cycle to avoid generating negative torque. Figure 4.24 shows ideal current waveforms for conventional and mutually-coupled 12/8 SRMs. It can be seen that, one electrical cycle in the conventional SRM corresponds to 45° mechanical where as it corresponds to 90° mechanical in the MCSRM. Hence the relationship between the mechanical and electrical angles in MCSRMs can be expressed as:

$$\theta_{elec} = \frac{N_r}{2}\theta_{mech} \tag{4.50}$$

Figure 4.25 shows the block diagram for the characterization of a MCSRM. The procedure is similar to that of a conventional SRM, but the flux linkage and torque are expressed in dq-axis values instead of phase values.

4.6.2 $\lambda - i$ Look-up Table Inversion

The flux linkage LUT inversion can be accomplished in a similar way as for the conventional SRM. The major difference is that for the MCSRM, three-dimensional interpolation is required instead of two-dimensional interpolation, to achieve the transformation:

$$\lambda_{dq}\left(i_d, i_q, \theta_{elec}\right) \rightarrow i_{dq}\left(\lambda_d, \lambda_q, \theta_{elec}\right)$$

The result of this transformation, shown in Figure 4.26, is i_d and i_q, which can be combined to form a complex number to simplify the model:

$$i_{dq} = i_d + ji_q \tag{4.51}$$

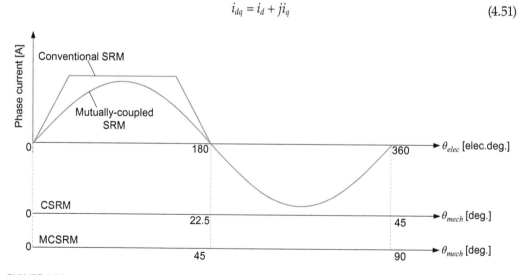

FIGURE 4.24

Mechanical period for one electrical cycle in conventional and mutually-coupled 12/8 SRMs with idealized current waveforms.

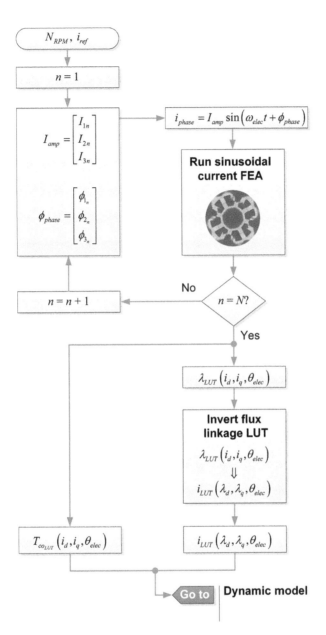

FIGURE 4.25
MCSRM characterization procedure in *dq* frame.

In order to maintain high accuracy, the flux linkage LUTs might be created for many electrical angles, which might increase the time and computational effort required for dynamic analysis. In order to reduce the complexity of the LUTs, harmonic decomposition can be applied, and the LUTs can be created in the frequency domain rather than the temporal domain to reduce the number of discrete FEA simulations. If symmetry is also utilized, then the number of discrete FEA simulations can be significantly decreased; drastically reducing characterization time [1,2].

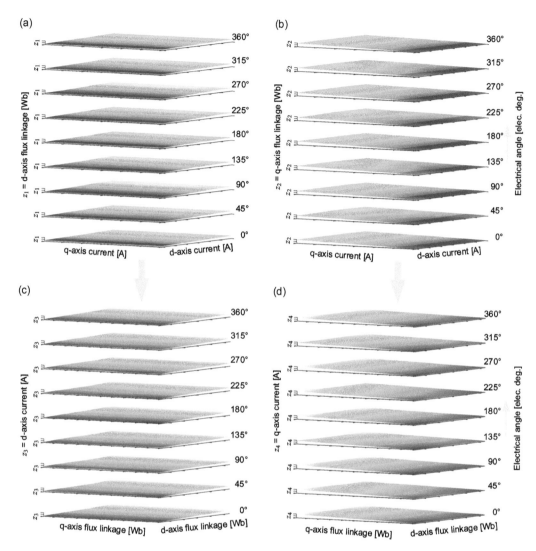

FIGURE 4.26
Inversion of d-q flux linkage LUTs: (a) before conversion, z_1, (b) before conversion, z_2, (c) after conversion, z_3, and (d) after conversion, z_4.

4.6.3 MCSRM Dynamic Model

The MCSRM dynamic model is quite similar to that of the conventional SRM. The differences are how current is controlled, how the electrical angle is calculated, and that all currents are in the dq-reference frame rather than the phase quantities. The integration of the voltage equation is also similar, with the exception that the voltage equation is in the $\alpha\beta$ reference frame:

$$\lambda_{\alpha\beta}(k) = \lambda_{\alpha\beta}(k-1) + \left[v_{\alpha\beta}(k) - R_{ph}i_{\alpha\beta}(k-1)\right]T_s \qquad (4.52)$$

Figure 4.27 shows the dynamic model for the MCSRM in the *dq*-reference frame. Reference frame conversions are well documented in literature, but they will be shown here for clarity:

Clarke's transformation ($abc \rightarrow \alpha\beta$):

$$x_{\alpha\beta} = \frac{2}{3}\begin{bmatrix} 1 & -\dfrac{1}{2} & -\dfrac{1}{2} \\ 0 & \dfrac{\sqrt{3}}{2} & -\dfrac{\sqrt{3}}{2} \end{bmatrix} x_{abc} \tag{4.53}$$

Inverse Clarke's transformation ($\alpha\beta \rightarrow abc$):

$$x_{abc} = \begin{bmatrix} 1 & 0 \\ -\dfrac{1}{2} & \dfrac{\sqrt{3}}{2} \\ -\dfrac{1}{2} & -\dfrac{\sqrt{3}}{2} \end{bmatrix} x_{\alpha\beta} \tag{4.54}$$

Park's transformation ($\alpha\beta \rightarrow dq$):

$$x_{dq} = \begin{bmatrix} \cos(\theta_{elec}) & \sin(\theta_{elec}) \\ -\sin(\theta_{elec}) & \cos(\theta_{elec}) \end{bmatrix} x_{\alpha\beta} \tag{4.55}$$

Using Euler's identity in (4.56), (4.55) results into an alternative form as in (4.57).

$$e^{j\theta} = \cos(\theta) + j\sin(\theta) \tag{4.56}$$

$$x_{dq} = x_{\alpha\beta} e^{-j\theta_{elec}} \tag{4.57}$$

Inverse Park's transformation ($dq \rightarrow \alpha\beta$):

$$x_{\alpha\beta} = \begin{bmatrix} \cos(\theta_{elec}) & -\sin(\theta_{elec}) \\ \sin(\theta_{elec}) & \cos(\theta_{elec}) \end{bmatrix} x_{dq} \tag{4.58}$$

Using the Euler's identity in (4.56), (4.58) can be expressed as:

$$x_{\alpha\beta} = x_{dq} e^{j\theta_{elec}} \tag{4.59}$$

Since the MCSRM produces torque from the mutual coupling between phases, T_{mc} and also reluctance torque due to co-energy, T_{co}, both of these torque components must be summed to find the total torque as shown in Figure 4.27. The torque produced by the mutual coupling of the phases T_{mc} can be calculated from the flux linkage and current:

$$T_{mc} = \frac{3}{2} pp \, im\left(\lambda_{dq} i_{dq}\right) \tag{4.60}$$

where λ_{dq} is the phase flux linkage and i_{dq} is the phase current in the *dq* reference frame. In Equation 4.61, *pp* is the number of pole pairs and it is expressed as:

$$pp = \frac{N_r}{2} \tag{4.61}$$

The co-energy torque T_{co} is determined through look-up tables just like the case in conventional SRM. When both are combined, the overall torque equation can be expressed as

$$T_{mc} = \frac{3}{2} pp\, im\left(\lambda_{dq} i_{dq}\right) + T_{co} \tag{4.62}$$

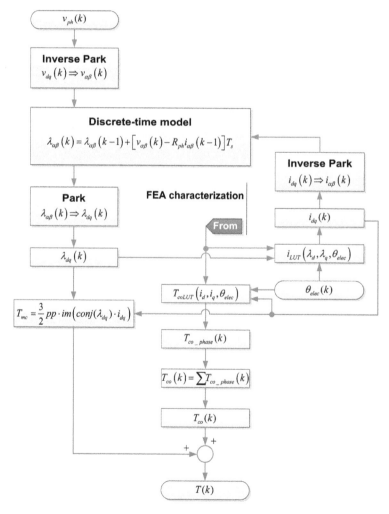

FIGURE 4.27
MCSRM dynamic model.

Questions and Problems

In this chapter, we will have one large problem with many steps. Working on each step, you will be able to develop a dynamic model of a 12/8 switched reluctance motor drive system. Tables Q.4.1 through Q.4.3 show the static torque, flux linkage, and terminal voltage characteristics of the 12/8 SRM, under different excitation currents and rotor positions.

TABLE Q.4.1

Torque Characteristics of a 12/8 SRM

Electrical Angle [°]	0 A	2 A	4 A	6 A	8 A	10 A	12 A	14 A	16 A	18 A	20 A
0	0	0.00	0.00	0.00	0.00	0.00	0.00	0.00	0.00	0.00	0.00
3	0	0.00	0.00	0.00	0.01	0.01	0.02	0.02	0.03	0.04	0.05
6	0	0.00	0.00	0.01	0.02	0.02	0.04	0.05	0.06	0.08	0.10
9	0	0.00	0.01	0.01	0.02	0.04	0.05	0.07	0.10	0.12	0.15
12	0	0.00	0.01	0.02	0.03	0.05	0.07	0.10	0.13	0.16	0.20
15	0	0.00	0.01	0.02	0.04	0.06	0.09	0.12	0.16	0.21	0.25
18	0	0.00	0.01	0.03	0.05	0.08	0.11	0.15	0.20	0.26	0.32
21	0	0.00	0.01	0.03	0.06	0.09	0.14	0.19	0.24	0.31	0.38
24	0	0.00	0.02	0.04	0.07	0.11	0.16	0.22	0.29	0.36	0.45
27	0	0.01	0.02	0.05	0.08	0.13	0.19	0.26	0.34	0.43	0.53
30	0	0.01	0.02	0.06	0.10	0.16	0.22	0.31	0.40	0.51	0.63
33	0	0.01	0.03	0.07	0.12	0.18	0.26	0.36	0.47	0.59	0.73
36	0	0.01	0.03	0.08	0.14	0.21	0.31	0.42	0.55	0.69	0.86
39	0	0.01	0.04	0.09	0.16	0.25	0.36	0.49	0.64	0.81	1.00
42	0	0.01	0.05	0.11	0.19	0.30	0.43	0.58	0.76	0.97	1.19
45	0	0.01	0.06	0.13	0.23	0.36	0.52	0.71	0.92	1.17	1.44
48	0	0.02	0.07	0.16	0.28	0.44	0.64	0.87	1.13	1.44	1.77
51	0	0.02	0.09	0.20	0.36	0.56	0.81	1.10	1.44	1.83	2.26
54	0	0.03	0.12	0.26	0.47	0.73	1.06	1.44	1.89	2.39	2.95
57	0	0.04	0.16	0.36	0.65	1.01	1.46	2.00	2.61	3.29	3.97
60	0	0.06	0.23	0.52	0.94	1.47	2.11	2.85	3.60	4.39	5.19
63	0	0.07	0.28	0.64	1.15	1.80	2.55	3.35	4.20	5.07	5.95
66	0	0.07	0.29	0.67	1.19	1.86	2.64	3.51	4.40	5.32	6.26
69	0	0.07	0.29	0.66	1.19	1.86	2.65	3.54	4.48	5.42	6.36
72	0	0.07	0.29	0.66	1.18	1.85	2.65	3.55	4.50	5.47	6.45
75	0	0.07	0.29	0.66	1.18	1.86	2.65	3.56	4.52	5.50	6.50
78	0	0.07	0.29	0.66	1.19	1.86	2.66	3.57	4.55	5.54	6.54
81	0	0.07	0.29	0.66	1.19	1.86	2.65	3.57	4.55	5.55	6.55
84	0	0.07	0.29	0.66	1.19	1.86	2.66	3.57	4.56	5.56	6.56
87	0	0.07	0.29	0.67	1.19	1.87	2.66	3.58	4.57	5.58	6.59
90	0	0.07	0.29	0.67	1.20	1.87	2.67	3.59	4.58	5.59	6.60
93	0	0.07	0.29	0.67	1.19	1.87	2.67	3.60	4.59	5.60	6.61
96	0	0.07	0.29	0.67	1.19	1.87	2.67	3.60	4.59	5.59	6.61
99	0	0.07	0.29	0.67	1.20	1.88	2.68	3.60	4.59	5.59	6.61

(Continued)

TABLE Q.4.1 (*Continued*)

Torque Characteristics of a 12/8 SRM

Electrical Angle [°]	0 A	2 A	4 A	6 A	8 A	10 A	12 A	14 A	16 A	18 A	20 A
102	0	0.07	0.29	0.67	1.20	1.88	2.68	3.60	4.57	5.59	6.61
105	0	0.07	0.29	0.67	1.20	1.87	2.67	3.59	4.58	5.59	6.61
108	0	0.07	0.29	0.67	1.19	1.87	2.67	3.58	4.57	5.57	6.59
111	0	0.07	0.29	0.66	1.19	1.87	2.66	3.57	4.55	5.56	6.57
114	0	0.07	0.29	0.66	1.19	1.86	2.65	3.55	4.54	5.54	6.55
117	0	0.07	0.29	0.66	1.18	1.85	2.63	3.54	4.51	5.51	6.52
120	0	0.07	0.29	0.65	1.17	1.83	2.61	3.51	4.48	5.48	6.48
123	0	0.07	0.29	0.65	1.16	1.82	2.60	3.49	4.47	5.47	6.47
126	0	0.07	0.28	0.65	1.16	1.82	2.59	3.47	4.45	5.45	6.45
129	0	0.07	0.28	0.64	1.15	1.81	2.57	3.45	4.42	5.41	6.40
132	0	0.07	0.28	0.64	1.15	1.80	2.56	3.43	4.39	5.38	6.35
135	0	0.07	0.28	0.64	1.14	1.79	2.55	3.42	4.37	5.35	6.32
138	0	0.07	0.28	0.64	1.14	1.79	2.55	3.41	4.34	5.31	6.26
141	0	0.07	0.28	0.64	1.14	1.78	2.54	3.39	4.30	5.23	6.16
144	0	0.07	0.28	0.63	1.13	1.77	2.52	3.36	4.28	5.14	6.03
147	0	0.07	0.28	0.63	1.13	1.77	2.51	3.33	4.16	5.02	5.88
150	0	0.07	0.28	0.63	1.13	1.77	2.51	3.30	4.06	4.87	5.68
153	0	0.07	0.27	0.63	1.12	1.75	2.48	3.23	3.92	4.65	5.42
156	0	0.07	0.27	0.62	1.11	1.73	2.44	3.15	3.76	4.41	5.13
159	0	0.06	0.27	0.61	1.10	1.71	2.41	3.06	3.59	4.13	4.78
162	0	0.06	0.26	0.60	1.08	1.68	2.35	2.94	3.41	3.85	4.38
165	0	0.06	0.26	0.59	1.05	1.63	2.27	2.79	3.19	3.56	3.93
168	0	0.06	0.25	0.56	1.00	1.56	2.15	2.60	2.95	3.25	3.50
171	0	0.06	0.23	0.52	0.93	1.45	1.98	2.36	2.64	2.89	3.08
174	0	0.05	0.20	0.46	0.82	1.27	1.71	2.01	2.23	2.42	2.56
177	0	0.03	0.14	0.32	0.57	0.88	1.19	1.38	1.53	1.65	1.74
180	0	0.00	0.00	0.00	0.00	0.00	0.00	0.00	0.00	0.00	0.00
183	0	−0.03	−0.14	−0.32	−0.57	−0.88	−1.19	−1.38	−1.53	−1.65	−1.74
186	0	−0.05	−0.20	−0.46	−0.82	−1.27	−1.71	−2.01	−2.23	−2.42	−2.56
189	0	−0.06	−0.23	−0.52	−0.93	−1.45	−1.98	−2.36	−2.64	−2.89	−3.08
192	0	−0.06	−0.25	−0.56	−1.00	−1.56	−2.15	−2.60	−2.95	−3.25	−3.50
195	0	−0.06	−0.26	−0.59	−1.05	−1.63	−2.27	−2.79	−3.19	−3.56	−3.93
198	0	−0.06	−0.26	−0.60	−1.08	−1.68	−2.35	−2.94	−3.41	−3.85	−4.38
201	0	−0.06	−0.27	−0.61	−1.10	−1.71	−2.41	−3.06	−3.59	−4.12	−4.78
204	0	−0.07	−0.27	−0.62	−1.11	−1.73	−2.44	−3.15	−3.76	−4.40	−5.13
207	0	−0.07	−0.27	−0.63	−1.12	−1.75	−2.48	−3.23	−3.92	−4.65	−5.42
210	0	−0.07	−0.28	−0.63	−1.13	−1.77	−2.50	−3.30	−4.06	−4.87	−5.68
213	0	−0.07	−0.28	−0.63	−1.13	−1.77	−2.52	−3.33	−4.20	−5.02	−5.88
216	0	−0.07	−0.28	−0.63	−1.13	−1.77	−2.52	−3.36	−4.29	−5.14	−6.03
219	0	−0.07	−0.28	−0.64	−1.14	−1.78	−2.54	−3.39	−4.30	−5.23	−6.16
222	0	−0.07	−0.28	−0.64	−1.14	−1.79	−2.55	−3.41	−4.35	−5.31	−6.26
225	0	−0.07	−0.28	−0.64	−1.14	−1.79	−2.55	−3.42	−4.37	−5.35	−6.32
228	0	−0.07	−0.28	−0.64	−1.15	−1.80	−2.56	−3.43	−4.38	−5.37	−6.36

(*Continued*)

TABLE Q.4.1 (*Continued*)

Torque Characteristics of a 12/8 SRM

Electrical Angle [°]	0 A	2 A	4 A	6 A	8 A	10 A	12 A	14 A	16 A	18 A	20 A
231	0	−0.07	−0.28	−0.64	−1.15	−1.80	−2.57	−3.45	−4.42	−5.42	−6.41
234	0	−0.07	−0.28	−0.65	−1.16	−1.82	−2.59	−3.47	−4.44	−5.45	−6.45
237	0	−0.07	−0.29	−0.65	−1.16	−1.82	−2.60	−3.49	−4.47	−5.47	−6.47
240	0	−0.07	−0.29	−0.65	−1.17	−1.83	−2.62	−3.51	−4.48	−5.48	−6.48
243	0	−0.07	−0.29	−0.66	−1.18	−1.85	−2.63	−3.54	−4.51	−5.52	−6.52
246	0	−0.07	−0.29	−0.66	−1.19	−1.86	−2.65	−3.56	−4.54	−5.54	−6.55
249	0	−0.07	−0.29	−0.66	−1.19	−1.87	−2.66	−3.57	−4.56	−5.55	−6.57
252	0	−0.07	−0.29	−0.67	−1.19	−1.87	−2.67	−3.58	−4.57	−5.57	−6.59
255	0	−0.07	−0.29	−0.67	−1.20	−1.87	−2.68	−3.59	−4.58	−5.59	−6.60
258	0	−0.07	−0.29	−0.67	−1.20	−1.88	−2.68	−3.60	−4.59	−5.59	−6.61
261	0	−0.07	−0.29	−0.67	−1.20	−1.88	−2.68	−3.60	−4.59	−5.59	−6.61
264	0	−0.07	−0.29	−0.67	−1.19	−1.87	−2.67	−3.60	−4.59	−5.59	−6.61
267	0	−0.07	−0.29	−0.67	−1.19	−1.87	−2.67	−3.60	−4.59	−5.60	−6.61
270	0	−0.07	−0.29	−0.67	−1.20	−1.87	−2.67	−3.59	−4.58	−5.59	−6.60
273	0	−0.07	−0.29	−0.67	−1.19	−1.87	−2.66	−3.58	−4.57	−5.58	−6.59
276	0	−0.07	−0.29	−0.66	−1.19	−1.86	−2.66	−3.57	−4.56	−5.56	−6.56
279	0	−0.07	−0.29	−0.66	−1.19	−1.86	−2.65	−3.57	−4.55	−5.55	−6.55
282	0	−0.07	−0.29	−0.66	−1.19	−1.86	−2.66	−3.57	−4.55	−5.54	−6.54
285	0	−0.07	−0.29	−0.66	−1.18	−1.86	−2.65	−3.56	−4.52	−5.50	−6.50
288	0	−0.07	−0.29	−0.66	−1.18	−1.85	−2.65	−3.55	−4.50	−5.47	−6.45
291	0	−0.07	−0.29	−0.66	−1.19	−1.86	−2.65	−3.54	−4.48	−5.42	−6.36
294	0	−0.07	−0.29	−0.67	−1.19	−1.86	−2.64	−3.51	−4.40	−5.32	−6.26
297	0	−0.07	−0.28	−0.64	−1.15	−1.80	−2.55	−3.35	−4.20	−5.07	−5.95
300	0	−0.06	−0.23	−0.52	−0.94	−1.47	−2.11	−2.85	−3.60	−4.39	−5.19
303	0	−0.04	−0.16	−0.36	−0.65	−1.01	−1.46	−2.00	−2.61	−3.29	−3.97
306	0	−0.03	−0.12	−0.26	−0.47	−0.73	−1.06	−1.44	−1.89	−2.39	−2.95
309	0	−0.02	−0.09	−0.20	−0.36	−0.56	−0.81	−1.10	−1.44	−1.83	−2.26
312	0	−0.02	−0.07	−0.16	−0.28	−0.44	−0.64	−0.87	−1.13	−1.44	−1.77
315	0	−0.01	−0.06	−0.13	−0.23	−0.36	−0.52	−0.71	−0.92	−1.17	−1.44
318	0	−0.01	−0.05	−0.11	−0.19	−0.30	−0.43	−0.58	−0.76	−0.97	−1.19
321	0	−0.01	−0.04	−0.09	−0.16	−0.25	−0.36	−0.49	−0.64	−0.81	−1.00
324	0	−0.01	−0.03	−0.08	−0.14	−0.21	−0.31	−0.42	−0.55	−0.69	−0.86
327	0	−0.01	−0.03	−0.07	−0.12	−0.18	−0.26	−0.36	−0.47	−0.59	−0.73
330	0	−0.01	−0.02	−0.06	−0.10	−0.16	−0.22	−0.31	−0.40	−0.51	−0.63
333	0	−0.01	−0.02	−0.05	−0.08	−0.13	−0.19	−0.26	−0.34	−0.43	−0.53
336	0	0.00	−0.02	−0.04	−0.07	−0.11	−0.16	−0.22	−0.29	−0.36	−0.45
339	0	0.00	−0.01	−0.03	−0.06	−0.09	−0.14	−0.19	−0.24	−0.31	−0.38
342	0	0.00	−0.01	−0.03	−0.05	−0.08	−0.11	−0.15	−0.20	−0.26	−0.32
345	0	0.00	−0.01	−0.02	−0.04	−0.06	−0.09	−0.12	−0.16	−0.21	−0.25
348	0	0.00	−0.01	−0.02	−0.03	−0.05	−0.07	−0.10	−0.13	−0.16	−0.20
351	0	0.00	−0.01	−0.01	−0.02	−0.04	−0.05	−0.07	−0.10	−0.12	−0.15
354	0	0.00	0.00	−0.01	−0.02	−0.02	−0.04	−0.05	−0.06	−0.08	−0.10
357	0	0.00	0.00	0.00	−0.01	−0.01	−0.02	−0.02	−0.03	−0.04	−0.05

TABLE Q.4.2

Flux Linkage Characteristics of a 12/8 SRM

Electrical Angle [°]	0 A	2 A	4 A	6 A	8 A	10 A	12 A	14 A	16 A	18 A	20A
0	0	0.002	0.004	0.006	0.008	0.010	0.013	0.015	0.017	0.019	0.021
3	0	0.002	0.004	0.006	0.008	0.010	0.013	0.015	0.017	0.019	0.021
6	0	0.002	0.004	0.006	0.008	0.010	0.013	0.015	0.017	0.019	0.021
9	0	0.002	0.004	0.006	0.008	0.010	0.013	0.015	0.017	0.019	0.021
12	0	0.002	0.004	0.006	0.008	0.011	0.013	0.015	0.017	0.019	0.021
15	0	0.002	0.004	0.006	0.008	0.011	0.013	0.015	0.017	0.019	0.021
18	0	0.002	0.004	0.006	0.009	0.011	0.013	0.015	0.017	0.019	0.021
21	0	0.002	0.004	0.006	0.009	0.011	0.013	0.015	0.017	0.020	0.022
24	0	0.002	0.004	0.007	0.009	0.011	0.013	0.015	0.018	0.020	0.022
27	0	0.002	0.004	0.007	0.009	0.011	0.013	0.016	0.018	0.020	0.022
30	0	0.002	0.005	0.007	0.009	0.011	0.014	0.016	0.018	0.020	0.023
33	0	0.002	0.005	0.007	0.009	0.012	0.014	0.016	0.018	0.021	0.023
36	0	0.002	0.005	0.007	0.009	0.012	0.014	0.016	0.019	0.021	0.024
39	0	0.002	0.005	0.007	0.010	0.012	0.014	0.017	0.019	0.022	0.024
42	0	0.002	0.005	0.007	0.010	0.012	0.015	0.017	0.020	0.022	0.025
45	0	0.003	0.005	0.008	0.010	0.013	0.015	0.018	0.021	0.023	0.026
48	0	0.003	0.005	0.008	0.011	0.013	0.016	0.019	0.021	0.024	0.027
51	0	0.003	0.006	0.008	0.011	0.014	0.017	0.020	0.022	0.025	0.028
54	0	0.003	0.006	0.009	0.012	0.015	0.018	0.021	0.024	0.027	0.030
57	0	0.003	0.006	0.010	0.013	0.016	0.019	0.022	0.026	0.029	0.032
60	0	0.003	0.007	0.010	0.014	0.017	0.021	0.024	0.028	0.031	0.034
63	0	0.004	0.008	0.012	0.016	0.020	0.023	0.027	0.031	0.034	0.037
66	0	0.004	0.009	0.013	0.018	0.022	0.026	0.030	0.033	0.037	0.040
69	0	0.005	0.010	0.015	0.019	0.024	0.029	0.033	0.037	0.040	0.043
72	0	0.005	0.011	0.016	0.021	0.027	0.032	0.036	0.040	0.043	0.046
75	0	0.006	0.012	0.017	0.023	0.029	0.035	0.039	0.043	0.046	0.050
78	0	0.006	0.012	0.019	0.025	0.031	0.037	0.042	0.046	0.050	0.053
81	0	0.007	0.013	0.020	0.027	0.034	0.040	0.045	0.049	0.053	0.056
84	0	0.007	0.014	0.022	0.029	0.036	0.043	0.049	0.053	0.056	0.059
87	0	0.008	0.015	0.023	0.031	0.039	0.046	0.052	0.056	0.059	0.063
90	0	0.008	0.016	0.024	0.033	0.041	0.049	0.055	0.059	0.063	0.066
93	0	0.008	0.017	0.026	0.035	0.043	0.052	0.058	0.062	0.066	0.069
96	0	0.009	0.018	0.027	0.037	0.046	0.054	0.061	0.066	0.069	0.073
99	0	0.009	0.019	0.029	0.038	0.048	0.057	0.065	0.069	0.073	0.076
102	0	0.010	0.020	0.030	0.040	0.050	0.060	0.068	0.072	0.076	0.079
105	0	0.010	0.021	0.032	0.042	0.053	0.063	0.071	0.076	0.079	0.082
108	0	0.011	0.022	0.033	0.044	0.055	0.066	0.074	0.079	0.083	0.086
111	0	0.011	0.023	0.034	0.046	0.058	0.068	0.077	0.082	0.086	0.089
114	0	0.012	0.024	0.036	0.048	0.060	0.071	0.080	0.086	0.089	0.092
117	0	0.012	0.025	0.037	0.050	0.062	0.074	0.083	0.089	0.093	0.096
120	0	0.013	0.026	0.039	0.052	0.065	0.077	0.086	0.092	0.096	0.099
123	0	0.013	0.026	0.040	0.053	0.067	0.079	0.090	0.096	0.099	0.102
126	0	0.013	0.027	0.041	0.055	0.069	0.082	0.093	0.099	0.103	0.106

(Continued)

TABLE Q.4.2 (*Continued*)

Flux Linkage Characteristics of a 12/8 SRM

Electrical Angle [°]	0 A	2 A	4 A	6 A	8 A	10 A	12 A	14 A	16 A	18 A	20A
129	0	0.014	0.028	0.043	0.057	0.071	0.085	0.096	0.102	0.106	0.109
132	0	0.014	0.029	0.044	0.059	0.074	0.088	0.099	0.105	0.109	0.112
135	0	0.015	0.030	0.045	0.061	0.076	0.090	0.102	0.109	0.112	0.116
138	0	0.015	0.031	0.047	0.063	0.078	0.093	0.105	0.112	0.116	0.119
141	0	0.016	0.032	0.048	0.064	0.081	0.096	0.108	0.115	0.119	0.122
144	0	0.016	0.033	0.050	0.066	0.083	0.098	0.111	0.118	0.122	0.125
147	0	0.017	0.034	0.051	0.068	0.085	0.101	0.113	0.121	0.125	0.128
150	0	0.017	0.035	0.052	0.070	0.087	0.104	0.116	0.124	0.127	0.130
153	0	0.017	0.035	0.054	0.072	0.090	0.106	0.118	0.126	0.130	0.133
156	0	0.018	0.036	0.055	0.073	0.092	0.109	0.121	0.128	0.132	0.135
159	0	0.018	0.037	0.056	0.075	0.094	0.111	0.123	0.130	0.135	0.138
162	0	0.019	0.038	0.057	0.077	0.096	0.113	0.124	0.131	0.137	0.140
165	0	0.019	0.039	0.059	0.078	0.098	0.115	0.126	0.133	0.138	0.141
168	0	0.019	0.040	0.060	0.080	0.100	0.117	0.127	0.134	0.139	0.142
171	0	0.020	0.040	0.061	0.082	0.102	0.119	0.129	0.135	0.140	0.143
174	0	0.020	0.041	0.062	0.083	0.103	0.120	0.129	0.136	0.140	0.143
177	0	0.020	0.042	0.063	0.084	0.105	0.121	0.130	0.136	0.141	0.144
180	0	0.021	0.042	0.063	0.085	0.105	0.122	0.131	0.137	0.141	0.144
183	0	0.020	0.042	0.063	0.084	0.105	0.121	0.130	0.136	0.141	0.144
186	0	0.020	0.041	0.062	0.083	0.103	0.120	0.129	0.136	0.140	0.143
189	0	0.020	0.040	0.061	0.082	0.102	0.119	0.129	0.135	0.140	0.143
192	0	0.019	0.040	0.060	0.080	0.100	0.117	0.127	0.134	0.139	0.142
195	0	0.019	0.039	0.059	0.078	0.098	0.115	0.126	0.133	0.138	0.141
198	0	0.019	0.038	0.057	0.077	0.096	0.113	0.124	0.131	0.137	0.140
201	0	0.018	0.037	0.056	0.075	0.094	0.111	0.123	0.130	0.135	0.138
204	0	0.018	0.036	0.055	0.073	0.092	0.109	0.121	0.128	0.132	0.135
207	0	0.017	0.035	0.054	0.072	0.089	0.106	0.118	0.126	0.130	0.133
210	0	0.017	0.035	0.052	0.070	0.087	0.104	0.116	0.124	0.127	0.130
213	0	0.017	0.034	0.051	0.068	0.085	0.101	0.113	0.121	0.125	0.128
216	0	0.016	0.033	0.050	0.066	0.083	0.098	0.111	0.118	0.122	0.125
219	0	0.016	0.032	0.048	0.064	0.081	0.096	0.108	0.115	0.119	0.122
222	0	0.015	0.031	0.047	0.063	0.078	0.093	0.105	0.112	0.116	0.119
225	0	0.015	0.030	0.045	0.061	0.076	0.090	0.102	0.109	0.112	0.116
228	0	0.014	0.029	0.044	0.059	0.074	0.088	0.099	0.105	0.109	0.112
231	0	0.014	0.028	0.043	0.057	0.071	0.085	0.096	0.102	0.106	0.109
234	0	0.013	0.027	0.041	0.055	0.069	0.082	0.093	0.099	0.103	0.106
237	0	0.013	0.026	0.040	0.053	0.067	0.079	0.090	0.096	0.099	0.102
240	0	0.013	0.026	0.039	0.052	0.065	0.077	0.086	0.092	0.096	0.099
243	0	0.012	0.025	0.037	0.050	0.062	0.074	0.083	0.089	0.093	0.096
246	0	0.012	0.024	0.036	0.048	0.060	0.071	0.080	0.086	0.089	0.092
249	0	0.011	0.023	0.034	0.046	0.058	0.068	0.077	0.082	0.086	0.089
252	0	0.011	0.022	0.033	0.044	0.055	0.066	0.074	0.079	0.083	0.086
255	0	0.010	0.021	0.032	0.042	0.053	0.063	0.071	0.076	0.079	0.082

(*Continued*)

TABLE Q.4.2 (*Continued*)

Flux Linkage Characteristics of a 12/8 SRM

Electrical Angle [°]	0 A	2 A	4 A	6 A	8 A	10 A	12 A	14 A	16 A	18 A	20A
258	0	0.010	0.020	0.030	0.040	0.050	0.060	0.068	0.072	0.076	0.079
261	0	0.009	0.019	0.029	0.038	0.048	0.057	0.065	0.069	0.073	0.076
264	0	0.009	0.018	0.027	0.037	0.046	0.054	0.061	0.066	0.069	0.073
267	0	0.008	0.017	0.026	0.035	0.043	0.052	0.058	0.062	0.066	0.069
270	0	0.008	0.016	0.024	0.033	0.041	0.049	0.055	0.059	0.063	0.066
273	0	0.008	0.015	0.023	0.031	0.039	0.046	0.052	0.056	0.059	0.063
276	0	0.007	0.014	0.022	0.029	0.036	0.043	0.049	0.053	0.056	0.059
279	0	0.007	0.013	0.020	0.027	0.034	0.040	0.045	0.049	0.053	0.056
282	0	0.006	0.012	0.019	0.025	0.031	0.037	0.042	0.046	0.050	0.053
285	0	0.006	0.012	0.017	0.023	0.029	0.035	0.039	0.043	0.046	0.050
288	0	0.005	0.011	0.016	0.021	0.027	0.032	0.036	0.040	0.043	0.046
291	0	0.005	0.010	0.015	0.019	0.024	0.029	0.033	0.037	0.040	0.043
294	0	0.004	0.009	0.013	0.018	0.022	0.026	0.030	0.033	0.037	0.040
297	0	0.004	0.008	0.012	0.016	0.020	0.023	0.027	0.031	0.034	0.037
300	0	0.003	0.007	0.010	0.014	0.017	0.021	0.024	0.028	0.031	0.034
303	0	0.003	0.006	0.010	0.013	0.016	0.019	0.022	0.026	0.029	0.032
306	0	0.003	0.006	0.009	0.012	0.015	0.018	0.021	0.024	0.027	0.030
309	0	0.003	0.006	0.008	0.011	0.014	0.017	0.020	0.022	0.025	0.028
312	0	0.003	0.005	0.008	0.011	0.013	0.016	0.019	0.021	0.024	0.027
315	0	0.003	0.005	0.008	0.010	0.013	0.015	0.018	0.021	0.023	0.026
318	0	0.002	0.005	0.007	0.010	0.012	0.015	0.017	0.020	0.022	0.025
321	0	0.002	0.005	0.007	0.010	0.012	0.014	0.017	0.019	0.022	0.024
324	0	0.002	0.005	0.007	0.009	0.012	0.014	0.016	0.019	0.021	0.024
327	0	0.002	0.005	0.007	0.009	0.012	0.014	0.016	0.018	0.021	0.023
330	0	0.002	0.005	0.007	0.009	0.011	0.014	0.016	0.018	0.020	0.023
333	0	0.002	0.004	0.007	0.009	0.011	0.013	0.016	0.018	0.020	0.022
336	0	0.002	0.004	0.007	0.009	0.011	0.013	0.015	0.018	0.020	0.022
339	0	0.002	0.004	0.006	0.009	0.011	0.013	0.015	0.017	0.020	0.022
342	0	0.002	0.004	0.006	0.009	0.011	0.013	0.015	0.017	0.019	0.021
345	0	0.002	0.004	0.006	0.008	0.011	0.013	0.015	0.017	0.019	0.021
348	0	0.002	0.004	0.006	0.008	0.011	0.013	0.015	0.017	0.019	0.021
351	0	0.002	0.004	0.006	0.008	0.010	0.013	0.015	0.017	0.019	0.021
354	0	0.002	0.004	0.006	0.008	0.010	0.013	0.015	0.017	0.019	0.021
357	0	0.002	0.004	0.006	0.008	0.010	0.013	0.015	0.017	0.019	0.021

TABLE Q.4.3

Terminal Voltage Characteristics of a 12/8 SRM at 1000 rpm

Electrical Angle [°]	0 A	2 A	4 A	6 A	8 A	10 A	12 A	14 A	16 A	18 A	20 A
0	0	0.4	0.9	1.3	1.7	2.1	2.6	3.0	3.4	3.8	4.3
3	0	0.4	0.9	1.3	1.8	2.2	2.7	3.1	3.6	4.0	4.5
6	0	0.5	1.0	1.5	2.0	2.5	3.0	3.5	4.0	4.5	5.0
9	0	0.6	1.1	1.7	2.2	2.8	3.3	3.9	4.4	5.0	5.5
12	0	0.6	1.2	1.8	2.4	3.0	3.6	4.2	4.8	5.4	6.0
15	0	0.7	1.3	2.0	2.6	3.3	3.9	4.6	5.3	5.9	6.6
18	0	0.7	1.4	2.1	2.9	3.6	4.3	5.0	5.7	6.4	7.2
21	0	0.8	1.6	2.3	3.1	3.9	4.7	5.5	6.3	7.1	7.9
24	0	0.8	1.7	2.5	3.4	4.3	5.1	6.0	6.8	7.7	8.5
27	0	0.9	1.8	2.8	3.7	4.6	5.6	6.5	7.4	8.4	9.3
30	0	1.0	2.0	3.0	4.0	5.1	6.1	7.1	8.1	9.1	10.2
33	0	1.1	2.2	3.4	4.5	5.6	6.7	7.9	9.0	10.1	11.2
36	0	1.2	2.5	3.7	4.9	6.2	7.4	8.7	9.9	11.2	12.4
39	0	1.4	2.7	4.1	5.5	6.9	8.2	9.6	11.0	12.4	13.8
42	0	1.5	3.1	4.6	6.2	7.7	9.3	10.8	12.4	13.9	15.5
45	0	1.8	3.5	5.3	7.1	8.9	10.7	12.5	14.3	16.1	17.9
48	0	2.0	4.1	6.2	8.2	10.3	12.4	14.5	16.6	18.6	20.7
51	0	2.4	4.9	7.4	9.8	12.3	14.8	17.3	19.8	22.3	24.7
54	0	3.0	6.0	9.1	12.2	15.2	18.3	21.4	24.4	27.5	30.5
57	0	3.9	7.8	11.8	15.8	19.8	23.8	27.8	31.7	35.5	37.8
60	0	5.4	10.9	16.4	21.9	27.4	32.9	37.8	40.6	42.4	44.5
63	0	7.2	14.6	22.0	29.5	36.8	42.1	44.1	46.5	48.1	49.4
66	0	7.7	15.6	23.6	31.5	39.4	45.1	48.8	49.9	51.4	52.3
69	0	7.9	16.0	24.2	32.4	40.5	47.5	50.6	52.6	52.8	53.0
72	0	7.8	15.9	24.0	32.2	40.3	47.8	52.3	53.3	54.1	55.2
75	0	7.8	15.8	23.9	31.9	40.0	47.7	53.0	54.1	55.2	55.6
78	0	7.7	15.7	23.7	31.7	39.7	47.6	53.2	54.8	55.7	55.8
81	0	7.9	16.0	24.2	32.4	40.5	48.0	53.7	55.8	55.8	56.2
84	0	7.8	15.9	24.0	32.1	40.2	47.8	53.9	55.7	56.3	56.4
87	0	7.7	15.7	23.8	31.9	39.9	47.7	54.0	55.8	56.4	56.9
90	0	7.7	15.6	23.6	31.6	39.6	47.5	53.9	56.0	56.7	57.0
93	0	7.8	15.9	24.1	32.3	40.4	47.8	54.1	56.3	56.9	57.3
96	0	7.7	15.8	23.9	32.0	40.1	47.6	54.0	56.6	56.9	57.2
99	0	7.7	15.7	23.8	31.8	39.8	47.5	53.8	56.5	57.0	57.3
102	0	7.6	15.6	23.5	31.5	39.4	47.3	53.5	55.8	57.1	57.2
105	0	7.8	15.9	24.0	32.2	40.2	47.6	53.5	57.1	57.3	57.5
108	0	7.7	15.8	23.8	31.9	39.9	47.3	53.3	56.6	57.2	57.5
111	0	7.6	15.6	23.6	31.6	39.6	47.2	53.2	56.2	57.1	57.4
114	0	7.6	15.5	23.4	31.3	39.2	46.9	52.7	57.0	57.0	57.4
117	0	7.7	15.8	23.9	31.9	39.9	47.2	52.8	56.5	57.1	57.5
120	0	7.6	15.6	23.7	31.6	39.5	46.9	52.6	56.3	56.9	57.3
123	0	7.6	15.5	23.4	31.3	39.2	46.7	52.4	56.3	56.9	57.5

(Continued)

TABLE Q.4.3 (*Continued*)

Terminal Voltage Characteristics of a 12/8 SRM at 1000 rpm

Electrical Angle [°]	0 A	2 A	4 A	6 A	8 A	10 A	12 A	14 A	16 A	18 A	20 A
126	0	7.5	15.3	23.2	31.0	38.8	46.4	52.1	56.2	56.7	57.1
129	0	7.6	15.7	23.7	31.6	39.5	46.6	52.1	56.1	56.6	56.9
132	0	7.5	15.5	23.4	31.3	39.0	46.2	51.4	55.7	56.1	56.4
135	0	7.5	15.3	23.2	31.0	38.6	45.9	51.1	55.3	55.7	55.9
138	0	7.4	15.2	22.9	30.6	38.2	45.6	50.6	52.5	55.1	54.9
141	0	7.5	15.5	23.3	31.2	38.8	45.5	50.0	54.4	54.0	53.3
144	0	7.4	15.3	23.1	30.8	38.3	44.9	48.9	53.9	52.4	51.7
147	0	7.4	15.1	22.8	30.4	37.8	44.4	47.3	49.6	50.5	49.6
150	0	7.2	14.9	22.5	30.0	37.2	43.6	45.4	46.7	48.3	47.5
153	0	7.4	15.1	22.8	30.4	37.6	43.1	42.6	40.9	45.9	45.6
156	0	7.2	14.8	22.4	29.8	36.9	41.9	39.2	35.8	41.8	43.5
159	0	7.1	14.6	22.0	29.3	36.1	40.2	35.3	31.8	40.5	40.5
162	0	6.9	14.2	21.5	28.6	35.1	38.1	31.5	28.1	32.9	36.0
165	0	6.9	14.3	21.5	28.6	34.9	35.9	27.7	25.0	25.0	31.6
168	0	6.7	13.7	20.7	27.4	33.4	32.6	24.5	21.9	19.3	20.8
171	0	6.3	12.9	19.5	25.9	31.2	28.5	21.2	19.1	16.9	14.6
174	0	5.7	11.7	17.6	23.3	27.9	23.9	17.8	16.0	14.1	12.2
177	0	4.9	10.0	15.1	20.0	23.7	19.3	14.3	12.8	11.4	10.0
180	0	2.5	5.1	7.7	10.3	12.2	10.2	8.4	8.0	7.3	6.8
183	0	−1.7	−3.5	−5.2	−6.9	−8.0	−5.1	−2.5	−1.2	0.3	1.6
186	0	−4.0	−8.3	−12.6	−16.6	−19.5	−14.2	−8.4	−6.0	−3.9	−1.5
189	0	−4.8	−10.0	−15.0	−19.9	−23.7	−18.9	−11.8	−9.2	−6.5	−3.7
192	0	−5.5	−11.2	−16.9	−22.5	−27.0	−23.4	−15.2	−12.2	−9.2	−6.3
195	0	−5.8	−12.0	−18.2	−24.1	−29.1	−27.4	−18.6	−15.2	−11.8	−11.9
198	0	−6.1	−12.6	−19.0	−25.2	−30.7	−30.8	−21.8	−18.3	−17.4	−23.3
201	0	−6.1	−12.5	−18.9	−25.2	−30.9	−33.0	−25.5	−21.2	−25.2	−27.5
204	0	−6.3	−12.9	−19.5	−25.9	−31.9	−35.2	−29.5	−24.9	−31.9	−32.2
207	0	−6.4	−13.1	−19.9	−26.5	−32.7	−36.7	−33.2	−29.0	−35.4	−35.1
210	0	−6.5	−13.4	−20.3	−27.0	−33.3	−38.1	−36.7	−34.1	−38.1	−37.0
213	0	−6.4	−13.2	−19.9	−26.6	−33.0	−38.5	−39.4	−39.8	−40.6	−39.1
216	0	−6.5	−13.4	−20.2	−27.0	−33.6	−39.3	−41.4	−43.3	−42.8	−41.1
219	0	−6.6	−13.6	−20.6	−27.4	−34.1	−39.9	−43.0	−45.0	−44.9	−43.2
222	0	−6.7	−13.8	−20.8	−27.8	−34.6	−40.4	−44.1	−47.1	−46.4	−44.9
225	0	−6.5	−13.5	−20.4	−27.2	−34.0	−40.5	−44.7	−48.0	−47.5	−46.4
228	0	−6.6	−13.7	−20.7	−27.6	−34.4	−40.9	−45.1	−50.6	−48.1	−47.4
231	0	−6.7	−13.8	−20.9	−27.9	−34.8	−41.1	−45.5	−46.8	−48.5	−47.9
234	0	−6.8	−14.0	−21.1	−28.2	−35.2	−41.5	−46.2	−50.9	−49.0	−48.4
237	0	−6.6	−13.7	−20.7	−27.6	−34.5	−41.4	−46.2	−47.8	−49.1	−48.7
240	0	−6.7	−13.8	−20.9	−28.0	−34.9	−41.6	−46.4	−49.6	−49.3	−49.0
243	0	−6.8	−13.9	−21.1	−28.2	−35.3	−41.8	−46.6	−49.5	−49.3	−48.8
246	0	−6.9	−14.1	−21.3	−28.6	−35.7	−42.1	−46.9	−49.7	−49.5	−49.1
249	0	−6.7	−13.8	−20.9	−27.9	−34.9	−41.9	−46.8	−49.7	−49.4	−48.9

(Continued)

TABLE Q.4.3 (*Continued*)

Terminal Voltage Characteristics of a 12/8 SRM at 1000 rpm

Electrical Angle [°]	0 A	2 A	4 A	6 A	8 A	10 A	12 A	14 A	16 A	18 A	20 A
252	0	−6.8	−13.9	−21.1	−28.2	−35.3	−42.1	−47.2	−49.9	−49.5	−48.9
255	0	−6.8	−14.1	−21.3	−28.5	−35.6	−42.3	−47.4	−49.7	−49.6	−49.0
258	0	−6.9	−14.2	−21.5	−28.8	−36.0	−42.5	−47.5	−49.9	−49.7	−49.0
261	0	−6.8	−13.9	−21.0	−28.1	−35.2	−42.2	−47.5	−49.6	−49.5	−48.7
264	0	−6.8	−14.0	−21.2	−28.4	−35.5	−42.4	−47.9	−49.7	−49.4	−48.9
267	0	−6.9	−14.1	−21.4	−28.6	−35.8	−42.6	−48.2	−49.8	−49.3	−48.8
270	0	−7.0	−14.3	−21.6	−28.9	−36.1	−42.7	−48.1	−49.5	−49.3	−48.9
273	0	−6.8	−13.9	−21.1	−28.2	−35.4	−42.4	−48.0	−49.2	−49.0	−48.5
276	0	−6.9	−14.0	−21.3	−28.5	−35.7	−42.6	−48.1	−49.0	−48.8	−48.5
279	0	−6.9	−14.2	−21.5	−28.7	−36.0	−42.7	−48.0	−49.0	−48.7	−47.9
282	0	−7.0	−14.3	−21.6	−29.0	−36.3	−42.9	−47.8	−49.0	−48.2	−47.7
285	0	−6.9	−14.0	−21.1	−28.3	−35.5	−42.5	−47.3	−48.0	−48.1	−47.3
288	0	−6.9	−14.1	−21.3	−28.6	−35.8	−42.7	−47.1	−47.4	−47.6	−47.1
291	0	−7.0	−14.2	−21.5	−28.8	−36.0	−42.7	−46.4	−46.6	−46.5	−46.8
294	0	−7.1	−14.3	−21.7	−29.0	−36.3	−42.4	−44.7	−45.9	−45.2	−44.5
297	0	−6.9	−13.9	−21.0	−28.2	−35.2	−40.0	−42.8	−43.1	−43.7	−43.8
300	0	−6.4	−12.9	−19.5	−26.1	−32.6	−37.0	−38.1	−39.7	−40.4	−41.0
303	0	−4.5	−9.2	−13.8	−18.5	−23.2	−27.8	−31.9	−33.8	−34.8	−36.0
306	0	−3.0	−6.2	−9.3	−12.4	−15.6	−18.7	−21.9	−25.0	−27.8	−29.3
309	0	−2.2	−4.3	−6.6	−8.8	−11.0	−13.2	−15.4	−17.7	−19.9	−22.1
312	0	−1.6	−3.2	−4.8	−6.5	−8.1	−9.7	−11.4	−13.0	−14.6	−16.3
315	0	−1.2	−2.4	−3.6	−4.9	−6.1	−7.3	−8.6	−9.8	−11.0	−12.3
318	0	−0.9	−1.8	−2.8	−3.7	−4.7	−5.6	−6.5	−7.5	−8.4	−9.4
321	0	−0.7	−1.4	−2.1	−2.8	−3.5	−4.2	−4.9	−5.6	−6.3	−7.0
324	0	−0.5	−1.0	−1.6	−2.1	−2.6	−3.2	−3.7	−4.2	−4.8	−5.3
327	0	−0.4	−0.8	−1.2	−1.5	−1.9	−2.3	−2.7	−3.1	−3.5	−3.9
330	0	−0.3	−0.5	−0.8	−1.1	−1.4	−1.6	−1.9	−2.2	−2.5	−2.8
333	0	−0.2	−0.3	−0.5	−0.7	−0.8	−1.0	−1.2	−1.3	−1.5	−1.7
336	0	−0.1	−0.2	−0.2	−0.3	−0.4	−0.5	−0.6	−0.7	−0.7	−0.8
339	0	0.0	0.0	0.0	0.0	0.0	0.0	0.0	0.0	−0.1	−0.1
342	0	0.1	0.1	0.2	0.3	0.3	0.4	0.4	0.5	0.6	0.6
345	0	0.1	0.3	0.4	0.5	0.7	0.8	0.9	1.0	1.2	1.3
348	0	0.2	0.4	0.6	0.8	1.0	1.1	1.3	1.5	1.7	1.9
351	0	0.2	0.5	0.7	1.0	1.2	1.5	1.7	1.9	2.2	2.4
354	0	0.3	0.6	0.9	1.2	1.5	1.8	2.1	2.3	2.6	2.9
357	0	0.4	0.7	1.0	1.4	1.7	2.1	2.4	2.8	3.1	3.5

1. Plot the torque and voltage characteristics of the 12/8 SRM as a function of electrical angles. The results should look similar to the ones given in Figure Q.4.1. How do the torque and terminal voltage profiles change with position and current? What are the similarities and differences between the torque and voltage profiles?

2. Plot both the static flux linkage characteristics versus electrical angle (for different currents) and versus current (for different electrical angles) respectively. The results should look similar to the ones given in Figure Q.4.2. How does the flux linkage profile change with position and current?

3. In order to calculate the dynamic current profile, we need to invert the flux linkage look-up table. Please calculate the current look-up table from the flux linkage look-up table using the algorithm in Figure 4.18. After the inversion, the look-up

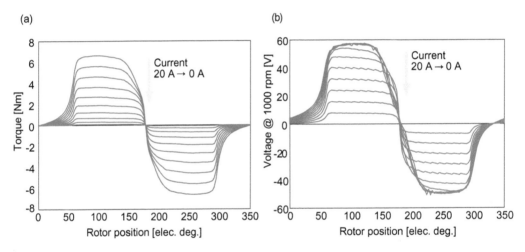

FIGURE Q.4.1
Torque and terminal voltage profiles of the 12/8 SRM: (a) torque and (b) terminal voltage.

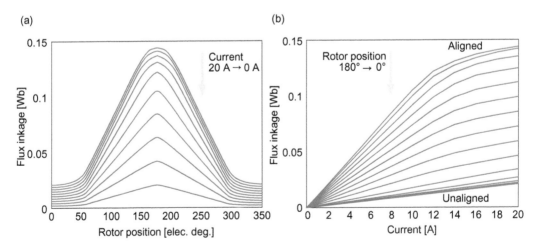

FIGURE Q.4.2
Flux linkage profiles of the 12/8 SRM: (a) flux linkage as a function of rotor position at different currents and (b) flux linkage as a function of current at different rotor positions.

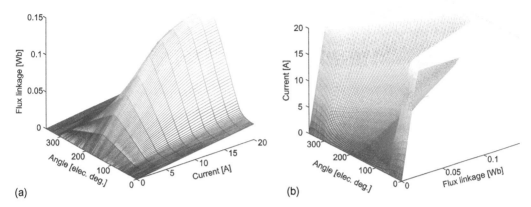

(a)   (b)

FIGURE Q.4.3
Flux linkage and current look-up tables: (a) flux linkage as a function of current and electrical angle and (b) current as a function of flux linkage and electrical angle.

table should look similar to the one given in Figure Q.4.3. You will use the current look-up table to calculate the dynamic performance in the next step.

4. Now let's start putting the dynamic model together. We will develop the dynamic model in MATLAB/Simulink and use scripts in MATLAB (2016a or later). Please save the Simulink file in the same folder where the SRM characterization files were saved. As shown in Figure Q.4.4, use a fixed-step discrete solver and set the

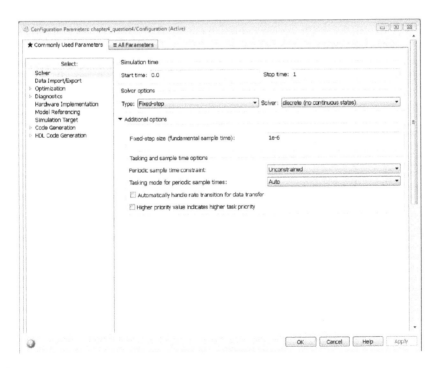

FIGURE Q.4.4
Configuration of Simulink parameters for the SRM dynamic model.

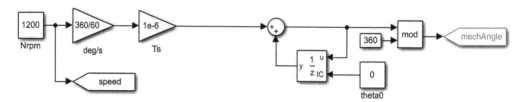

FIGURE Q.4.5
Implementation of mechanical angle model in Simulink.

fixed-step size to 1 micro-second. If the motor speed is 1200 RPM and the simulation stop time is 1 second, how many mechanical revolutions we would simulate?

5. Using (4.30), please add the mechanical angle model. Figure Q.4.5 shows a sample mechanical angle model developed in Simulink. Please run the simulation and plot the mechanical angle as a function of time. If the simulation stop time is 1 second, how many mechanical cycles did you observe? Did the result match with your calculations in the previous question?

6. Using (4.33), please develop the electrical angle model. Figure Q.4.6 shows a sample model for the electrical angle calculation. How is the k_{ph} parameter calculated for the 12/8 SRM for counterclockwise rotation? Plot the rotor position and the electrical angle of each phase together. How many electrical cycles do you see? What is the relationship between the number of electrical cycles with the number of stator poles and number of rotor poles?

7. Now we can start implementing the electrical model using the integration method shown in Figure 4.15. In question 2, we inverted the flux linkage look-up table to calculate the current look-up table $i_{LUT}(\lambda_{ph}, \theta_{elec})$. In order to easily validate the current control models that we will develop, let's first assume a linear relationship between flux linkage and current: $\lambda = Li$. We only make this assumption to validate our models, because this is not representative of SRM behavior under saturation. With this assumption, the value of L will be a constant and it will not be dependent on current or rotor position. Hence, the relationship between the

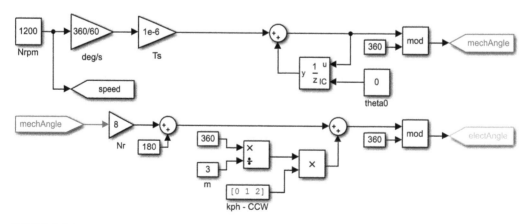

FIGURE Q.4.6
Implementation of electrical angle model in Simulink.

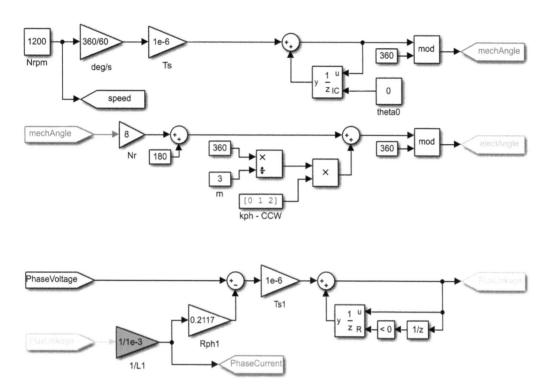

FIGURE Q.4.7
Implementation of phase electric model using integration method.

current and the flux linkage reduces from $i_{LUT}\left(\lambda_{ph}, \theta_{elec}\right)$ to $i = \lambda/L$. Once we validate the current controller model, we will replace this model with $i_{LUT}\left(\lambda_{ph}, \theta_{elec}\right)$.

Figure Q.4.7 shows an example for the electrical model using the integration method. Here, $L = 1\,\text{mH}$ and the phase resistance $R_{ph} = 0.2117\,\Omega$. Apply 10 V as the phase voltage and plot the phase current as a function of time. What is the mathematical expression for the current waveform?

8. In our simulation model, the simulation step time is 1 micro-second. This provides a good model of the physical system. However, when we want to implement controllers, we need to consider the limits of the sampling frequency of the digital controllers. The phase current in our model is also sampled every micro-second. This means that we would need a digital controller with 1 MHz sampling frequency, which might be difficult to achieve for some applications. Figure Q.4.8 shows the implementation of current sampling based on (4.43). The current sampling frequency is set at 1 kHz; hence, the counter counts up to 1 MHz/1 kHz = 1000. When the counter hits this value, it resets, and the current sampling block is triggered. Therefore, every 1000 steps, the phase current value at that time step is applied as the sampled current.

   a. Plot the phase current and sampled current for 1 kHz. What is the difference in the waveforms?

   b. What should the counting number be if the current sampling frequency is 22 kHz?

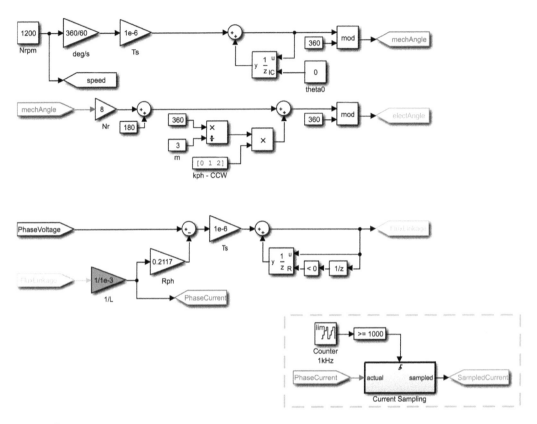

FIGURE Q.4.8
Implementation of current sampling.

    c. What is the difference between the sampled current waveform at 1 kHz and 22 kHz?

    d. Which sampling frequency would be better for control purposes?

9. Now we can model the phase excitation signals shown in Figure 4.10. A phase is turned on at θ_{ON} and turned off at θ_{OFF}. These angles are expressed in terms of electrical angles. Therefore, if θ_{ON} and θ_{OFF} are greater than zero, the phase excitation signal is applied when $\theta_{OFF} \geq \theta_{elec} \geq \theta_{ON}$. When θ_{ON} is greater than zero, it means that the phase is energized after the unaligned position for motoring mode of operation. Especially at high-speed condition, the phase excitation signal might be applied before the unaligned position so that the current rises before the induced voltage starts building up. In this case, θ_{ON} is expressed with a negative value to signify the phase advancement. With negative θ_{ON}, the phase excitation signal in motoring mode can be calculated as $(\theta_{OFF} \geq \theta_{elec} \geq 0) + (360 \geq \theta_{elec} \geq (360 + \theta_{ON}))$, provided that $\theta_{ON} < 0$. Figure Q.4.9 shows a sample model for the phase excitation signal calculation in motoring mode. Plot the phase excitation signal for $\theta_{ON} = 30°$, $\theta_{OFF} = 150°$ and $\theta_{ON} = -20°$, $\theta_{OFF} = 120°$. What is the difference between the phase excitation signals?

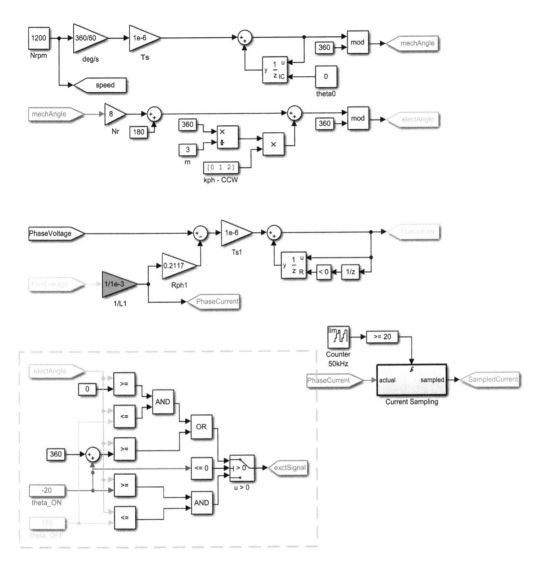

FIGURE Q.4.9
Implementation of phase excitation signal model in Simulink.

10. Now we can apply the hysteresis controller model in (4.36) and (4.37). Figures Q.4.10 and Q.4.11 show an example of how to implement the hysteresis control in the existing model. Please change the speed to $N_{rpm} = 600$, turn-on angle $\theta_{ON} = 30°$ and $\theta_{OFF} = 150°$. Set the current sampling to 20 kHz, current reference to 5 A and hysteresis band as 2%. Plot the phase currents and the switching signals with soft and hard switching. What is the difference between these two modulation techniques?

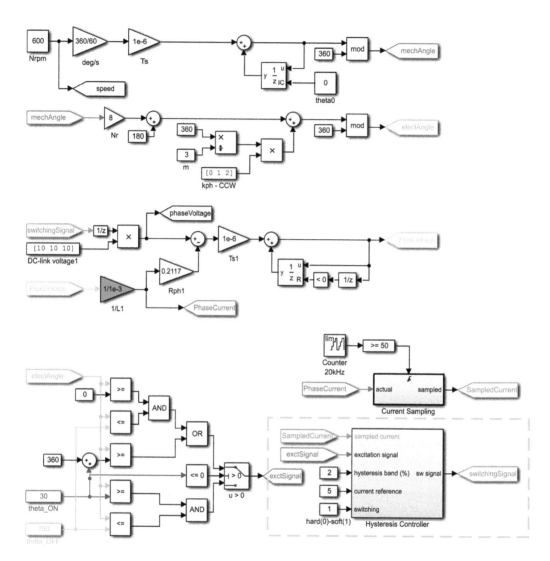

FIGURE Q.4.10
Implementation of hysteresis control.

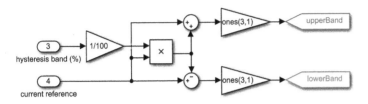

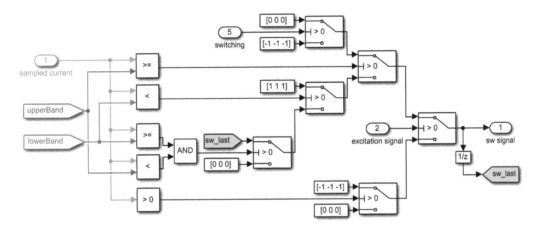

FIGURE Q.4.11
Details of hysteresis control.

11. Now we have a switched reluctance motor drive model. However, the phase flux linkage characteristics do not yet represent a practical SRM. We used constant inductance (linear) model to validate the current control model. Now, we can apply the current look-up table we developed in question 3 and the torque look-up table for the 12/8 SRM. Change the constant inductance model to the current look-up table model as shown in Figure Q.4.12. Set the current sampling to 50 kHz, speed to 1200 RPM, and current reference to 15 A. For $\theta_{ON} = 30°$ and $\theta_{OFF} = 150°$, the current and torque waveforms are given in Figure Q.4.13.

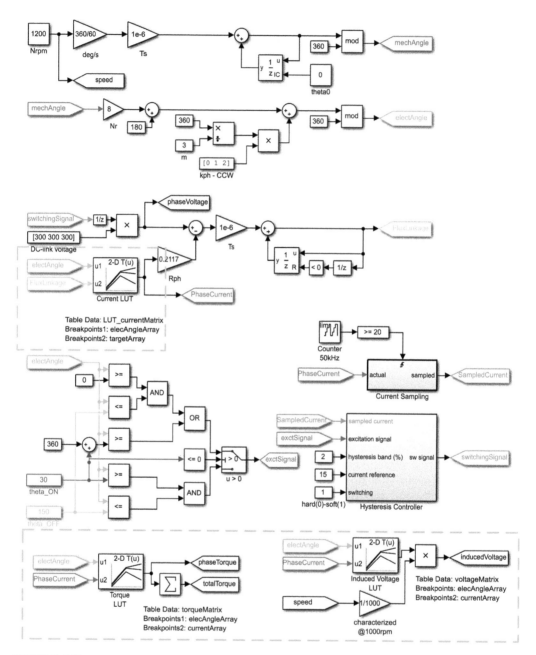

FIGURE Q.4.12
Implementation SRM current and torque look-up tables in the dynamic model.

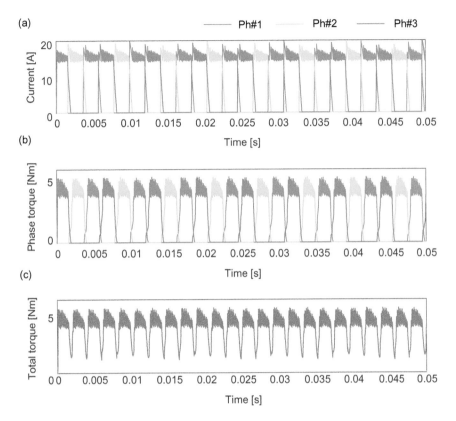

FIGURE Q.4.13
Dynamic results at 1200 rpm, 15A and for $\theta_{ON} = 30°$ and $\theta_{OFF} = 150°$.

In order to validate the dynamic model, we can apply the current waveforms calculated in the dynamic model to the FEA model of the motor. Then, we can compare the torque waveforms from the dynamic model and the FEA model. As shown in Figure Q.4.14, torque waveforms match closely. Please note that we haven't modeled mutual coupling in the dynamic model, but the FEA simulations take mutual coupling into account. The matching results show that mutual coupling is negligible when modeling this switched reluctance motor drive at these firing angles. It should be noted that the FEA validation takes significantly longer time to run than the dynamic model we just developed; outlining the importance of dynamic modelling.

If you apply the current waveform calculated from the dynamic model in the FEA model and measure voltage across the phase windings, which voltage would you observe: modulated phase voltage in the dynamic model or the induced voltage from the look-up table, why?

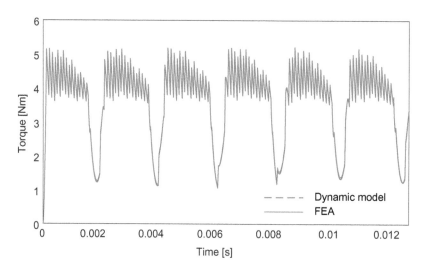

FIGURE Q.4.14
Comparison of torque calculated from the dynamic model and FEA model.

12. Change the speed to 10000 RPM, $\theta_{ON} = -10°$ and $\theta_{OFF} = 130°$. Figure Q.4.15 shows the current and torque waveforms. Due to wider conduction angle and high-speed operation, the period where two phases conduct at the same time is longer than the 1200 RPM case. Therefore, the effect of mutual coupling should be more noticeable. Figure Q.4.15 also shows the torque waveform from the FEA, when the calculated current waveforms are used in the FEA model. It can be seen that the torque waveform from FEA closely matches with the one from the dynamic model.

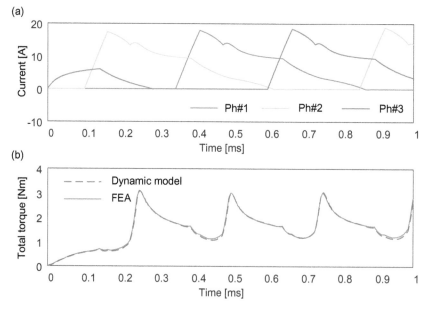

FIGURE Q.4.15
Dynamic results at 10000 rpm, 15A and for $\theta_{ON} = 10°$ and $\theta_{OFF} = 130°$.

For the given speed and conduction angles, please plot the phase voltage and induced voltage waveforms. What has changed as compared to the waveforms at 1200 rpm? Can you explain the relationship between the phase voltage, induced voltage, and current waveforms?

References

1. J. Dong, J. W. Jiang, B. Howey, H. Li, B. Bilgin, A. Callegaro, and A. Emadi, "Hybrid acoustic noise analysis approach of conventional and mutually coupled switched reluctance motors," *IEEE Transactions on Energy Conversion*, vol. 32, no. 3, pp. 1042–1051, 2017.
2. J. Dong, B. Howey, B. Danen, J. Lin, J. W. Jiang, B. Bilgin, and A. Emadi, "Advanced dynamic modeling of three-phase mutually-coupled switched reluctance machine," *IEEE Transactions on Energy Conversion*, vol. 33, no. 1, pp. 146–154, 2018.

5

Switched Reluctance Machines in Generating Mode

Berker Bilgin

CONTENTS

5.1 Introduction

When a switched reluctance machine (SRM) operates in motoring mode, it draws current from the source connected to the DC link. The inverter regulates the current, and torque is generated as a result of electromechanical energy conversion. Then, the torque applied to the shaft rotates the mechanical load at a certain speed. Hence, electrical energy is converted to mechanical energy.

Electromechanical energy conversion is bi-directional: mechanical energy can also be converted into electrical energy. In wind turbines, the wind power drives the shaft of the generator. By regulating the phase currents through the inverter, mechanical energy is converted into electrical energy. Hence, the output voltage and the current supplied to the load can be controlled.

In electric drives, the generating capability of electric machines is heavily used for regenerative braking. With conventional braking systems, the excess kinetic energy is converted to heat by friction brakes. In electric drives, the excess kinetic energy is converted to electrical energy by drawing mechanical power from the shaft. Therefore, the energy that would be wasted with frictional brakes can be recovered and the overall efficiency of the drive system is improved.

Due to its salient pole structure and lack of rotor excitation, SRM has a unique operation in generating mode. When a prime mover rotates the rotor of an SRM, the coils are open circuit and the SRM will not generate voltage. Therefore, Switched Reluctance Generator (SRG) requires an excitation source for the initial magnetization at the start-up [1].

5.2 Generating Mode of Operation

When a phase of the SRM is excited with current, the magnetic flux circulates on itself over the smallest reluctance path. This path is through the closest rotor pole, which becomes magnetized. Due to the magnetic force between the stator and rotor poles, the rotor pole tends to come in alignment with the stator pole. As the rotor pole moves towards the stator pole, reluctance of the magnetic circuit reduces with the decreasing air gap length resulting in higher stored energy in the magnetic circuit. This is the motoring mode of operation, which takes place within the positive slope of inductance profile as discussed in Chapter 3.

As the rotor pole moves towards the stator pole, it will tend to reach and stay at the aligned position (if the stator pole is excited with constant current, the rotor pole might slightly pass the stator pole due to the rotor inertia, but after a few oscillations, it will stay at the aligned position). At the aligned position, the reluctance of the magnetic circuit is at the minimum value. The rotor pole will tend to stay in this equilibrium. Anywhere before or beyond this point is a higher reluctance path. A small perturbation in position will apply a force on the rotor to keep it in alignment. Therefore, the rotor pole will not move beyond the aligned position without an external force.

If the rotor pole is forced to move beyond the aligned position by applying external torque to the shaft, the stator pole tends to decrease the reluctance in the circuit and tries to keep the rotor pole at the aligned position. This applies an opposing torque on the prime mover. The stored energy in the magnetic circuit increases. When the circuit is switched off, the stored energy in the magnetic circuit is converted into electrical energy and supplied back to the source.

At the end of the electrical cycle (aligned to the unaligned position), mechanical power drawn from the prime mover is converted into electrical energy. In simple terms, if the phase winding of SRG is connected to a load and the load current increases, in order to keep the voltage constant (to rotate the rotor at the same speed), the prime mover will have to apply more torque to the shaft. Thus, more power will be drawn from the prime mover.

If you followed the above description carefully, you probably noticed that we were moving the rotor pole away from the aligned position while the phase was connected to a voltage source. You might have questioned whether it was the prime mover or the excitation source, which increases the stored energy. Of course, it is the prime mover in generating mode. When the prime mover rotates the rotor, it overcomes the opposing torque, and the induced voltage, due to the rate-of-change-of flux linkage, increases the phase current. However, as mentioned earlier, an SRM will need excitation from the source for initial magnetization. After that, torque from the prime mover generates the motion, the rate-of-change-flux linkage induces voltage and creates the phase current.

5.3 Negative Torque and Negative Voltage

Figure 5.1 shows the phase flux linkage, terminal voltage, and electromagnetic torque of an SRM when the rotor pole completes one electrical cycle at constant speed and when the stator coils were excited with constant positive current. In this case, we assume that the rotor pole is moving in a counterclockwise direction.

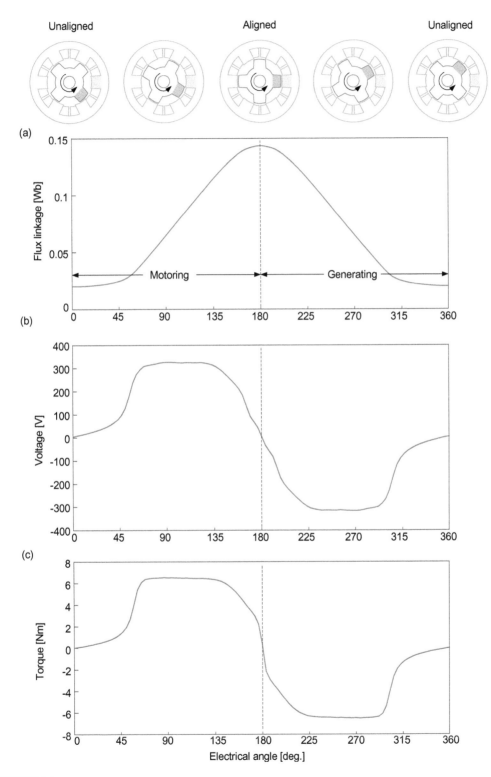

FIGURE 5.1
Electromagnetic profiles of SRM for one electrical cycle: (a) phase flux linkage, (b) voltage, and (c) torque.

When the rotor pole moves from unaligned to aligned position, electromagnetic torque is positive. This means that the forces acting on the rotor pole are in the same direction with the rotation. Terminal voltage is also positive. Since the coils are excited with constant positive current, this means that the terminal voltage and the phase current are in the same direction; therefore, the current will be drawn from the supply.

As shown in Figure 5.1, phase flux linkage decreases between aligned and unaligned position. This means that the reluctance is higher after the aligned position and the rotor pole will not move beyond this point without an external force. Here, negative torque shows that the prime mover needs to apply torque on the shaft to move the rotor pole from aligned to unaligned position.

In Figure 5.1, the coils are still energized with constant positive current after the aligned position. Then, negative voltage means that the current starts from the negative terminal of the supply and ends up at the positive terminal. This shows that the current is supplied back to the voltage source. As we will see later, the direction of the phase current doesn't change in generating mode. Since torque is a function of square of the phase current, it is independent of the direction of the phase current. However, the supply current changes its direction due to the negative voltage.

5.4 Four-Quadrant Operation

Industrial motor drives, such as electric traction systems and position and velocity control systems, require driving (motoring) and braking (generating) capability when the rotor rotates either in clockwise or counterclockwise direction. Figure 5.2 shows the four-quadrant operation in electric motor drive systems. In the first quadrant, the phases of SRM are excited in a certain order in the positive slope of the inductance profile, so that the rotor rotates in a counterclockwise direction. When the firing angles are changed to excite the phases in the negative slope of the inductance profile, braking torque is applied. Then the operation moves to the second quadrant. The rotor slows down due to the negative torque, but it keeps rotating in the counterclockwise direction. The excessive energy is supplied back to the source.

Figure 5.3 depicts this scenario in a 6/4 SRM. Assume that the initial position of the rotor is as shown in Figure 5.3a. The first rotor pole is aligned with the first stator pole; hence, phases are at 180°, 300°, and 60° electrical for counterclockwise rotation. If the conduction angles in motoring mode are 30° and 150° electrical, then Ph#3 should be energized and the rotor starts rotating counterclockwise. After 22.5° mechanical, Ph#3 is at 150° electrical and it is turned off. As shown in Figure 5.3b, Ph#2 is now at 30° electrical and it is excited to keep the motion in the same direction. After 30° mechanical, the rotor will be at a position given in Figure 5.3c, where Ph#1 should be turned on to produce torque in counterclockwise direction. This would keep the motor operating in the first quadrant in Figure 5.2.

Imagine that at this point, we would like to slow down the rotor. We need to apply breaking torque so the conduction angles should be after the aligned position. For this case, we assume that the conduction angles for the braking mode are 210° and 330° electrical. If we energize Ph#3 instead of Ph#1, the torque applied to the rotor will be in the clockwise direction. But due to the rotor's inertia, the direction of rotation cannot change instantaneously. The rotor slows down while moving in the counterclockwise direction and the excess kinetic energy is supplied back to the electrical source. Now

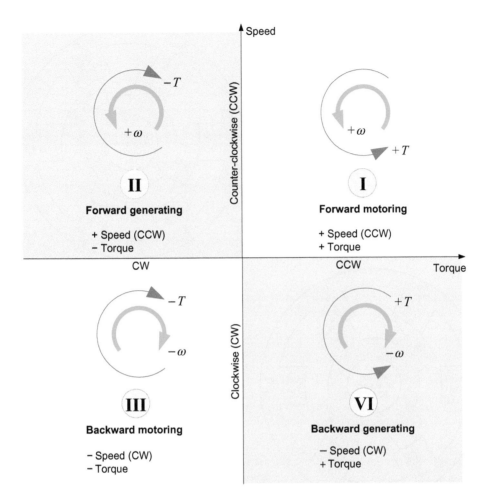

FIGURE 5.2
Four-quadrant operation of electric motor drives.

the torque and velocity are in opposite directions and, hence, the motor operates in the second quadrant in Figure 5.2.

From the position in Figure 5.3c, if the rotor rotates 15° mechanical in the counterclockwise direction, it comes to the position in Figure 5.3d. Imagine that we applied enough breaking torque and stopped the rotor at this position. Now phases are in 90°, 210°, and 330° electrical if the rotor was rotating in a counterclockwise direction. However, since the rotor speed is zero, any phase we energize will tend to get the closet rotor pole in alignment and torque will be in the same direction as the speed. At the position in Figure 5.3d, if we energize Ph#2, the rotor will start rotating in clockwise direction, but the electrical position of each phase will be different for clockwise position. Please remember from Chapter 4 that for a given stator and rotor pole configuration and number of phases, the electrical angle of each phase is dependent on the direction of rotation.

Figure 5.3e shows the electrical angles for clockwise rotation. Since the conduction angles for motoring mode are 30° and 150° electrical, we should energize Ph#3 in order to rotate the rotor in clockwise direction. Now we are in the third quadrant in Figure 5.2. After 30° mechanical, the rotor ends up in a position given in Figure 5.3f. At this position, the phases are at 30°, 270°, and 150° electrical. At this point, if we energize Ph#1, the torque

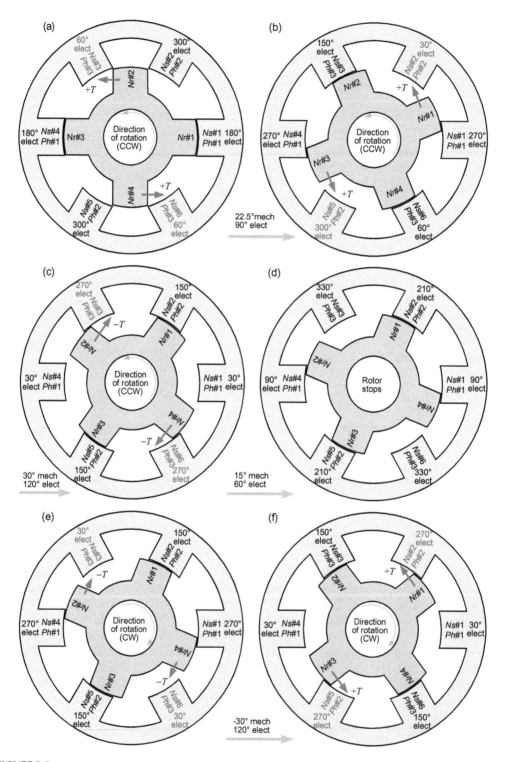

FIGURE 5.3
Four-quadrant operation of SRM: (a) initial position, (b) first quadrant operation, (c) second quadrant operation, (d) rotor stops, (e) third quadrant operation, and (f) fourth quadrant operation.

will be in the same direction as the motion, and the motor will keep running in the third quadrant. If we energize Ph#2, breaking torque will be applied while the rotor rotates in clockwise direction (due to inertia). Then, the operation moves to the fourth quadrant in Figure 5.2 where the speed is in clockwise direction, and the torque is in a counterclockwise direction.

5.5 Operating Modes in Switched Reluctance Generators

A conventional asymmetric bridge converter can be used to control the switched reluctance machine in generating mode as well. Figures 5.4 and 5.5 show the current waveforms in motoring and generating modes for hard- and soft-switching. In hard-switching, both switches are turned on and off at the same time to control the current. As shown in Figure 5.4a and b, current waveforms in motoring and generating modes are similar, but mirrored when hard-chopping is used. In generating mode, rise time at the turn-on is longer since the phase is energized close to aligned position where the inductance is higher. In motoring mode the phase is turned off close to aligned position; therefore, the fall time is longer.

As shown in Figure 5.4c, when both switches are on, the current flows into the phase; therefore, the source current is positive. When both switches are off, the diodes conduct due to the stored energy in the phase, and the current goes back to the supply as shown in Figure 5.4d.

Figure 5.5 shows the current waveforms for soft-switching. Please note that the waveforms in motoring and generating modes do not look alike anymore. In both modes, when both switches are on, the current is drawn from the source and the phase current starts building up. This is the phase excitation period in generating mode.

When the top switch (S_1) is turned off and the bottom switch (S_2) is kept in an on-state (or vice versa) as shown in Figure 5.5d, the phase winding is short circuited, and the phase current starts freewheeling. In motoring mode, the stored energy will be dissipated on the phase resistance, and phase current will reduce. The rate of change of current is lower than that of hard-chopping because the voltage applied to the phase is zero. In hard-chopping, when the current reduces by turning both switches off, as shown in Figure 5.4d, the voltage applied to the phase is $-V_{DC}$.

Unlike motoring mode, in generating mode, the phase current does not decrease when the current starts freewheeling, but it increases as shown in Figure 5.5b. Please also notice that the rate of change of current is higher as compared to excitation. This is one of the important aspects of generating mode in switched reluctance machines. Unlike motoring mode, with soft-switching, the rate-of-change-of-flux linkage does not oppose the current build up in generating mode [2]. Therefore, the negative voltage further boosts the current with a higher rate of change. When both switches are turned off at the end of the conduction period, the current is supplied back to the source. This is the generating period. When the current waveforms in hard- and soft-switching are compared, it can be observed that soft-switching can provide a smoother operation, since the excitation occurs once within the conduction period.

At higher speeds, the rate of change of flux linkage or induced voltage becomes higher than the DC link voltage. Therefore, similar to the motoring mode, switched reluctance

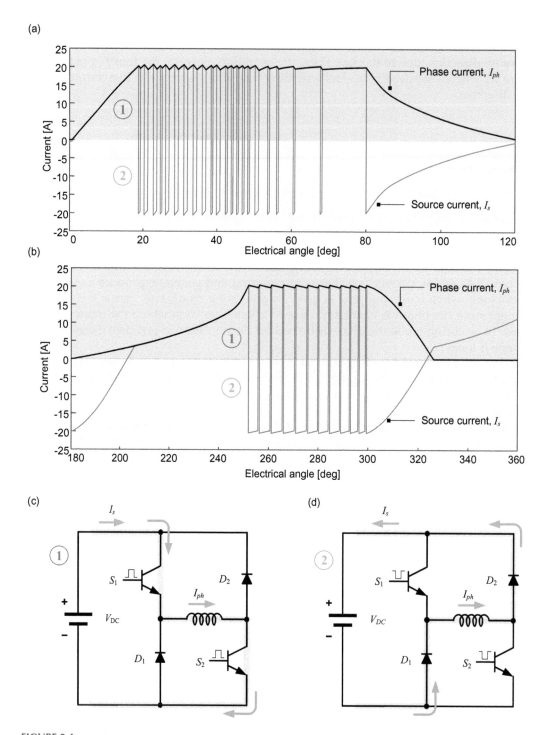

FIGURE 5.4
Phase and source current waveforms in hard-switching controlled switched reluctance machine: (a) motoring mode, (b) generating mode, (c) current flow when both switches are on, and (d) current flow when both switches are off.

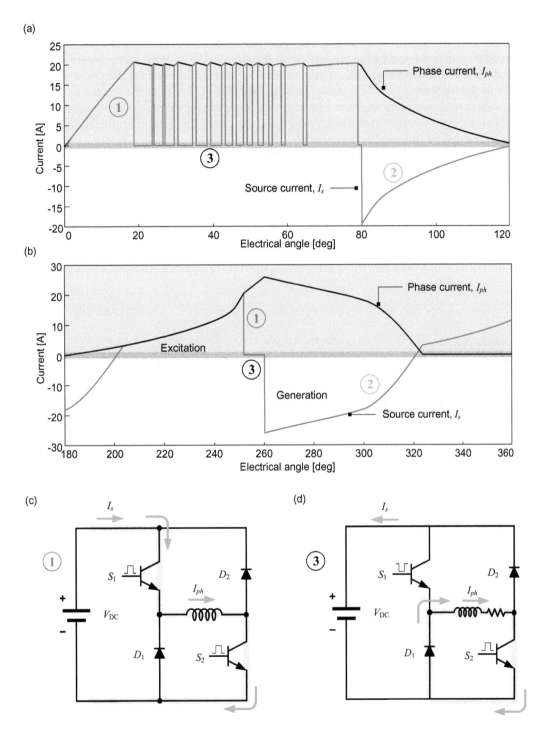

FIGURE 5.5
Phase and source current waveforms in soft-switching controlled switched reluctance machine: (a) motoring mode, (b) generating mode, (c) current flow when both switches are on, and (d) current flow when only one switch is off.

generator also works in single pulse mode at high speeds. In single pulse control, switching devices are turned on and off once per electrical cycle. When both switches are on, SRG draws the excitation current from the source. When both switches are off, the phase demagnetizes and supplies current to the source. Please note that the phase current both in motoring and generating modes are positive. This is because the torque in SRM is independent of direction of current. However, in generating mode, negative torque draws power from the prime mover, induces voltage and the phase current is supplied back to the source at the end of the conduction period.

Figures 5.6 and 5.7 show the phase current waveforms for single pulse operation for different conduction angles in motoring and generating mode, respectively. As shown in Figures 5.6a and 5.7a, the peak value of the phase current changes in both operating modes when the turn-on angle changes. As the turn-on angle decreases, the conduction angle gets larger and the peak current increases. As shown in Figure 5.6b, in motoring mode, the peak value of the phase current doesn't change for different turn-off angles if the phase is energized at the same turn-on angle. Since the DC link voltage opposes the induced voltage in motoring mode, the peak value of the

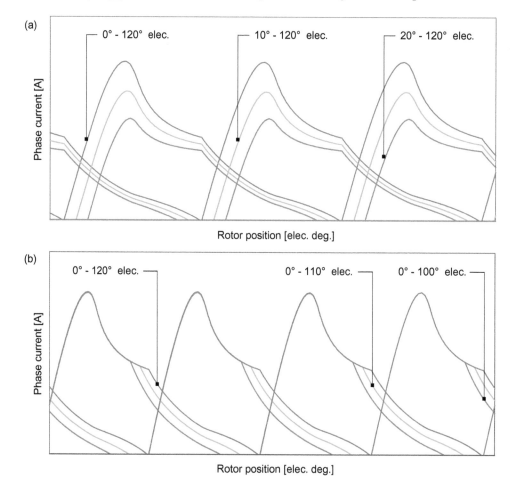

FIGURE 5.6
Phase current waveforms in motoring mode of operation: (a) different turn-on angles and (b) different turn-off angles.

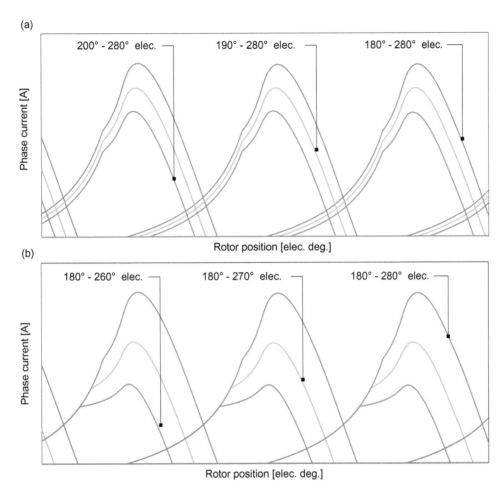

FIGURE 5.7
Phase current waveforms in generating mode of operation: (a) different turn-on angles and (b) different turn-off angles.

phase current can be controlled by adjusting the turn-on angle. This is not the case in generating mode. As shown in Figure 5.7b, the peak value of the phase current changes as the turn-off angle changes. This is because the negative induced voltage assists the current in the demagnetization period. Therefore, the peak value of the phase current in single-pulse controlled SRG depends on both turn-on and turn-off angles [3]. This phenomenon shows that there can be many different combinations of turn-on and turn-off angles in SRGs for the same output power. This can cause challenges in the control of SRG at high speeds [4].

In a single pulse operation in an SRG, care should be given on regulating the current using the conduction angles. In generating mode, the machine draws its excitation current from the DC link it supplies power to. Therefore, for a fixed conduction angle, a small perturbation in the load may cause an increase in the DC link voltage. This might lead to an increase in the excitation current and, hence, SRG generates even more current to the source after the both switches of the converter are turned off. This might further result in an increase in the DC link voltage leading to instable operation [5].

Questions and Problems

Figure Q.5.1 shows four-quadrant operation and switching modes in separately excited DC machine. Please compare the motoring and generating mode of operation in SRM and separately excited DC motor. What are the similarities and differences? Please answer the following questions.

1. What are the differences between the converters used in SRM and separately excited DC machine for four-quadrant operation? Which additional control flexibility would full bridge converter bring in SRM? Would it be necessary?

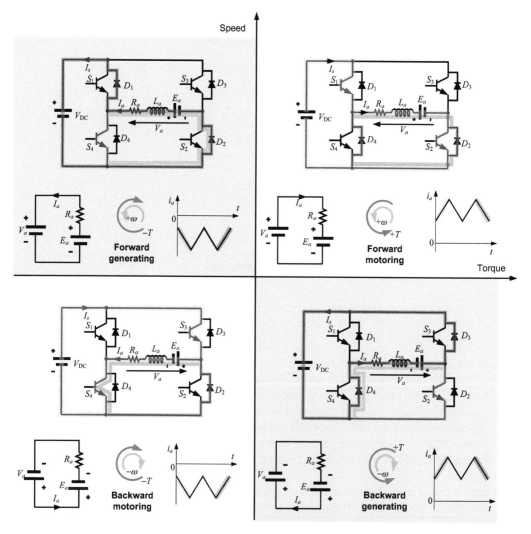

FIGURE Q.5.1
Conditions for four-quadrant operation in a separately excited DC machine.

2. Notice from Figure Q.5.1 that the armature current of separately excited DC machine is changing its direction in generating mode. Does the direction of phase current in SRM change in generating mode? Why?

3. Please compare how SRM and separately excited DC machine change direction of rotation. What is the difference?

4. Please compare the switching modes in SRG and separately excited DC motor when generating. What are the similarities?

5. In forward motoring mode, when the armature current increases, both S_1 and S_2 are on. When the armature current decreases, S_1 is off and the current freewheels through S_2 and D_4 (see Figure 5.5). Are these modes similar to operational modes of SRM? In this mode, what happens if we turn both S_1 and S_2 off?

References

1. B. Bilgin, A. Emadi, and M. Krishnamurthy, Switched reluctance generator with high higher number of rotor poles than stator poles, in *Proceedings of the IEEE Transportation Electrification Conference and Expo*, Dearborn, MI, June 2012, pp. 1–6.

2. Y. C. Chang and C. M. Liaw, Development and voltage feedback control for a switched reluctance generator, in *Proceedings of the IEEE International Conference on Electric Machines and Systems*, Seoul, Korea, October 2007, pp. 392–397.

3. P. Asadi, M. Ehsani, and B. Fahimi, Design and control characterization of switched reluctance generator for maximum output power, in *Proceedings of the IEEE Applied Power Electronics Conference and Exposition*, Dallas, TX, April 2006, pp. 1639–1644.

4. Y. Sozer and D. A. Torrey, Closed loop control of excitation parameters for high speed switched-reluctance generators, *IEEE Transactions on Power Electronics*, 19(2), 355–362, 2004.

5. D. A. Torrey, Switched reluctance generators and their control, *IEEE Transactions on Industrial Electronics*, 49(1), 13–14, 2002.

6

Materials Used in Switched Reluctance Machines

Elizabeth Rowan and James Weisheng Jiang

CONTENTS

6.1 Introduction

When describing an object, it is often easiest to start by describing what it is made from. Someone chose the material your chair or desk was made out of for specific reasons; perhaps for cost, sturdiness and strength, or aesthetics. How are the materials chosen for a motor?

Table 6.1 lists examples of materials used in an SRM and some of the material properties that are examined when selecting them. The list is not long, and indeed there are not many components in a switched reluctance machine which is part of what adds to its simplicity and robustness. When choosing a material, it is important to identify the critical properties that determine its selection. For instance, most of the properties listed could be examined for lamination steel; however, the critical properties evaluated when first selecting a lamination steel are generally gauge, flux density, loss density, and magnetic permeability. As a result, this chapter will focus on these critical properties. Figure 6.1 illustrates where these materials are used in an SRM. Components like the shaft, bearings, housing, and slot liners help hold the motor together under operation, while the lamination steel, wire coating, and encapsulation govern the performance of the machine [1].

An electric motor converts electromagnetic energy to mechanical energy. During this conversion there are losses in the form of thermal energy. The most heat sensitive components are in close proximity to heat generating sources such as the copper in the wires and the steel in the laminations. The performance of polymers used in insulation material is affected at much lower temperatures than the performance of metals. Thermal class ratings are assigned to magnetic wire insulation based on the stability of the coating over 20,000 hours at a given temperature. The coating can exist above this temperature for increasingly short durations with the possibility of irreversible aging.

The materials that go into the fundamental operation of the motor have been highly engineered – their properties pored over and tweaked to understand how to attain the

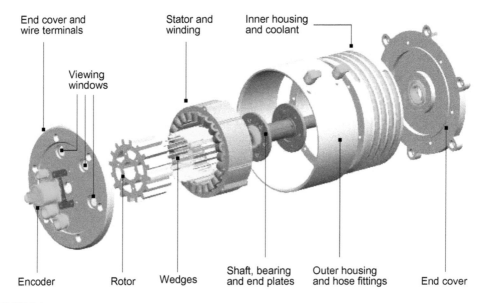

FIGURE 6.1
Exploded view of an SRM. (From Jiang, W., Three-phase 24/16 switched reluctance machine for hybrid electric powertrains: Design and optimization, McMaster University Ph. D. Thesis, Hamilton, ON, 2015.)

TABLE 6.1

Materials Used in Switched Reluctance Machines

Major component	Material		Mechanical and Manufacturing						Thermal			Magnetic			Electrical			Chemical	
		Density [kg/m³]	Hardness [HB]	Gauge, grade, AWG, or class	Poisson's ratio	Young's modulus [MPa]	Shear strength [MPa]	Yield strength [MPa]	Specific heat [kJ/kg/K]	Thermal conductivity [W/m/K]	Temperature constraint [°C]	Flux density [T]	Loss density [W/m³ or W/kg]	Permeability [H/m]	Resistivity [Ω·m]	Conductivity [S/m]	Current density [A/mm²]	Corrosion resistance	pH value, reactivity
Stator and rotor	Lamination steel			●								●	●	●	●				
	Coating	●		●														●	
Coil	Copper or aluminum			●							●		●		●	●	●		
	Polymer coating			●							●								
Housing	Aluminum (or steel)		●		●	●	●	●	●	●					●			●	●
	Coolant (if applicable)								●	●								●	●
Shaft	High strength steel		●		●	●	●	●											
Bearings	High strength steel		●	●														●	
Encapsulation	Various polymers				●	●	●	●		●								●	
Slot liners	Nomex by Dupont			●						●	●				●				
Wedge	Polyetheretherketone	●		●						●	●				●				

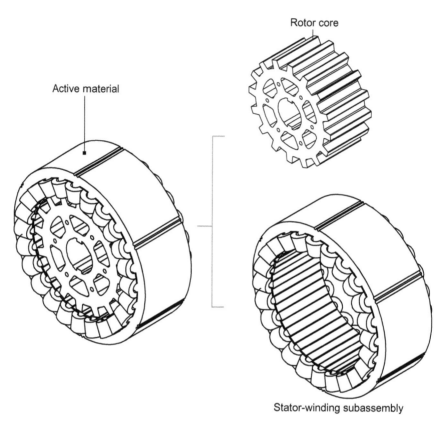

FIGURE 6.2
Active materials in an SRM.

greatest performance from them. The "active" components list can be boiled down to a number of simple roles: a magnetic flux carrier (lamination steel), an electrical conductor (copper or aluminum) and a variety of electrical insulators (encapsulation, wire coatings and lamination coatings). Figure 6.2 illustrates the active materials in an SRM.

Lamination steel, so named because it is cut from thin sheets being stacked or laminated together, is used primarily for its magnetic properties. SRMs are salient pole machines that generate torque based on the variance of magnetic reluctance. Salient poles protrude from the stator and rotor back irons, causing the rotor pole to experience a variation in air gap length when the rotor position changes. Both the rotor and stator are composed of steel lamination stacks. As the stator coils are excited, magnetic flux begins to flow creating a magnetic circuit.

Magnetic flux prefers to flow though materials with high magnetic permeability. Steel has much higher permeability than air, making it the preferred route. As the stator and rotor poles come into alignment, a flux pathway forms through the lamination steel in the salient rotor pole, salient stator pole, and back iron. The rotor will move towards the energized stator pole to minimize reluctance in the magnetic circuit. Reluctance decreases as more of the flux pathway passes through lamination steel instead of air. As the reluctance decreases from unaligned to partially aligned and towards aligned positions, electromagnetic forces produce torque. The unaligned position occurs when the rotor pole is not immediately adjacent to the energized stator pole, inductance is minimum, and the stator pole is not likely to be saturated. When the rotor pole and stator poles are aligned, inductance is maximized, and the stator pole is more likely to saturate. Without the lamination steel in the salient rotor and stator poles acting as a guide for the flux path, there would not be an effective torque production. The properties of the steel core significantly influence the efficiency of the conversion of electromagnetic energy to mechanical energy.

In early electrical machines, solid wrought iron was used until it was found that electromagnetic induction caused the formation of eddy currents that caused excessive heat. In 1880, a patent was filed by Thomas Edison in which he detailed separating thin strips of iron with paper (although he was not the first to use layered strips themselves). This innovation further improved the electrical isolation between laminations, reducing the energy expended as heat due to eddy currents. At around the same time, the production of rolled steel was improving and it was beginning to replace iron in rotating machines.

Currently, lamination steel is produced as rolled steel sheets with a layer of insulating material applied. These sheets are cut or punched into the correct shape for the rotor or stator cores. A great deal of care goes into how these steels are produced in order to create the desired mix of physical and magnetic properties. In order to tune these properties, steelmakers must concern themselves with both the macroscopic and microscopic properties of the steel. Macroscopically, the thickness of the steel will change how much energy is lost due to eddy currents. Microscopically they must use the effects of heat, time, and pressure as the steel goes through a lengthy production line to adjust the arrangement of the steel's internal atomic positions. We will cover steel production and the resulting material properties in more detail in Section 6.2.

Current flowing through motor windings generates a magnetic field, allowing the motor to spring to life. By managing the supplied current through the windings, we can control the machine. Motor windings consist of two components: the conductor and the insulating coating. The electrical conductor used in windings is typically copper or aluminum, while the insulating coating can be any of a number of electrically insulating polymers. The insulation materials will be examined in Section 6.3.

There are a large number of options that allow a motor designer to customize the wire to the specific needs of their motor. If the designer needs lower eddy current loss in the windings, using thinner strands of conductor wound together could be the answer, but it can also lower the fill factor in the stator slot. If the designer needs enhanced environmental protection, the appropriate polymer coating can be chosen, but it may also have a number of trade-offs based on the chemical composition of the polymer. Wire scheme design considerations will be examined in Chapter 7, while material considerations will be investigated in Section 6.4.

Encapsulation of the stator in an SRM can transport heat away from the hottest areas in the motor. Air pockets are filled with a solid material providing greater thermal conductivity. It can also offer additional mechanical strength, a higher dielectric breakdown voltage, and chemical and heat resistance. Encapsulation can provide passive, simple and relatively inexpensive thermal management.

6.2 Lamination Steel

Steel is used extensively in our modern society. In 2008, global steel production was 1,343 million tons. By 2015, that number had increased to 1,621 million tons [2,3]. Iron is the primary alloying element in steel; the secondary alloying element generally being carbon. Steel can be categorized by the additional alloying elements it contains. Alloys are metals created by mixing two or more elements, whose atoms combine in an orderly fashion.

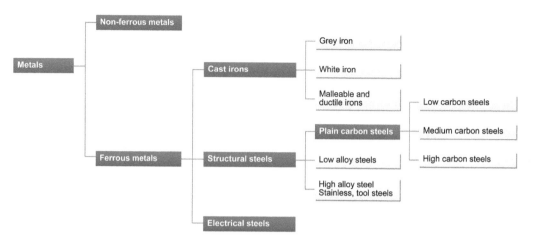

FIGURE 6.3
Classification of structural steels and other metals.

Elements may be chosen based on the specific properties they are able to impart to the steel when alloyed with it, or based on the properties of compounds that will form in the material. When purifying iron, residual unwanted elements may be left behind. Some elements are alloyed with steel to remove these impurities or render them harmless.

The element content added to steel depends on the application for which it will be used. Structural steels, shown in Figure 6.3, are produced and selected primarily for their mechanical properties. However, certain elements, like silicon or cobalt, can be added to steel to intentionally increase its magnetic properties. Often there are trade-offs when choosing which elements will be added. Electrical steel is a common type of lamination steel used in motors and is used throughout this chapter as the example material when discussing manufacturing and processing. However, although it is called steel, electrical steel lacks the characteristic carbon content of most structural steels. Figure 6.4 shows some key elements to be considered during the production of electrical steel. Table 6.2 provides generic details on the strengths of some of these alloying elements, not necessarily considering their electromagnetic performance. However, caution must be taken as the interactions between alloying elements can be complex, and the full effect of all alloying elements on each other must be considered.

6.2.1 Steel Production

Electrical steel production follows the same concept as traditional steelmaking. Traditional steelmaking generally follows one of two routes: production of steel from primarily iron ore, or production from primarily steel scrap, as shown in Figure 6.6. In the first route, liquid iron is extracted from iron ore, mainly hematite (Fe_2O_3) and magnetite (Fe_3O_4), using a blast furnace and coke, a solid product obtained from coal. The oxygen in the ore is removed by undergoing a reduction reaction with the high energy of the blast furnace and the carbon in the coke. Approximately 4% of the carbon remains in the hot metal after reduction.

High levels of impurities can be very detrimental to the final quality of the steel produced. In order to reduce impurities and inclusions as much as possible, steel can undergo a hot metal pretreatment. Hot metal pretreatments vary depending on what impurities the steelmaker is attempting to remove. Desiliconization, dephosphorization, and desulfurization can all take place at this stage. Reactants such as lime (CaO), hematite (Fe_2O_3), and

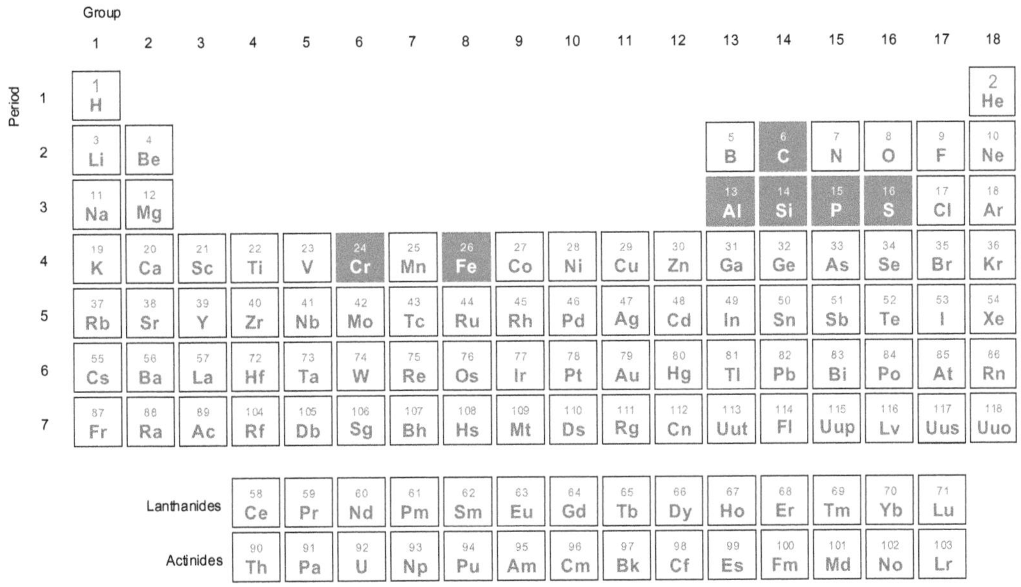

FIGURE 6.4
Major elements in electrical steel, highlighted in the periodic table.

TABLE 6.2

Generic Properties of Alloying Elements

Chemical Element	Properties
Carbon (C)	A common component in steelmaking and not generally considered an "additional" alloying element. In most rolled steels, carbon increases hardness, strength, and abrasion resistance, and causes the steel to respond more readily to heat treatment. However, it reduces ductility, toughness, impact properties and machinability. In electrical steel carbon is not desirable and is generally kept as low as 0.002%. If the carbon content is high, carbides are formed in the steel through the precipitation of carbon and other elements. The precipitates make it difficult to demagnetize the steel, and thus increase hysteresis losses.
Chromium (Cr)	Contributes to the strength and hardness of the steel. Forms stable carbides, and increases resistance to corrosion and abrasion. Helps steel maintain strength at elevated temperatures.
Aluminum (Al)	A useful deoxidizer. It restricts grain growth, which has specific implications to magnetic properties in electrical steels
Phosphorus (P)	Increases strength but reduces ductility
Silicon (Si)	Electrical steel is commonly referred to as silicon steel because silicon is one of the primary alloying elements. Silicon increases the resistivity with increasing weight percentage, as shown in Figure 6.5. Steel with a silicon content of 6% is considered high silicon steel and requires special treatment to produce. Silicon also acts as a deoxidizer and increases the hardenability.
Sulphur (S)	Increases machinability; however, it can decrease ductility in low carbon steels. Sulphur should be minimized in electrical steel since it can impede the magnetic performance.

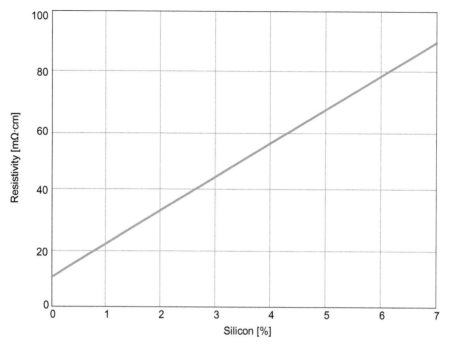

FIGURE 6.5
Relationship between electrical resistivity and silicon content in electrical steel. (From Beckley, P., Electrical steels for rotating machines, *IEEE Power and Energy Series*, Institute of Electrical Engineers, London, UK, Vol. 37, 2002.)

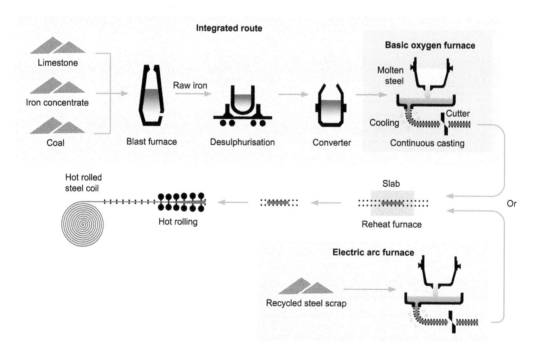

FIGURE 6.6
Integrated steelmaking for structural steel. (From Peacey, J.G., and Davenport, W.G., *The Iron Blast Furnace*, Pergamon Press, New York, 1979.)

fluorspar (CaF_2) may all be added at this stage to react with silicon, phosphorus or sulfur to be removed from the hot metal. Sodium carbonate (Na_2CO_3), calcium carbide (CaC_2), and magnesium (Mg) may be added specifically to reduce sulfur by greater amounts for low sulfur steels in a separate injection.

After undergoing hot metal pretreatment, the hot metal is added to a basic oxygen furnace (BOF), which oxidizes the carbon in the hot metal to lower its carbon content. Oxygen is added either through pure oxygen gas (O_2) or Fe_2O_3. Some oxygen once again remains in the steel and must be removed with subsequent treatments. It may seem odd that the process works back and forth between oxidation and reduction reactions; however, each time the hot metal undergoes one of these reactions fewer unwanted elements remain. Recycled steel scrap is also added to the BOF at this stage; normally 20%–30% is accepted into the furnace. This process is known as integrated steelmaking, and a simplified illustration of this process can be seen in Figure 6.6.

In the second route, recycled steel scrap is melted using electrical power and gas-oxygen burners in an electric arc furnace (EAF). This process is known aptly as electric arc furnace steelmaking, and it uses 90%–100% recycled steel scrap to produce steel. EAF requires only 7 MJ/kg of energy, compared to the 22 MJ/kg the integrated route requires. EAF needs the energy to melt steel while the integrated route requires the energy to melt the metal and also to participate in the reduction reaction than produces liquid iron. For comparison, aluminum requires 210 MJ/kg of energy to produce.

Regardless of which process is used to arrive at molten steel, alloying elements will be added prior to being cast into slabs for further processing. Coiled steel tends to be produced from slabs. Hot rolling will reduce the slab thickness to 2–4 mm. Thickness control and maintaining a good surface finish are challenging during hot rolling, so smaller gauges are generally continued via cold rolling.

Electrical steel can be produced using the integrated steelmaking route shown in Figure 6.6. After molten steel is produced as described above, silicon is added using ferrosilicon—an alloy of iron and silicon specifically produced for this purpose. A high purity ferrosilicon is used for electrical steel, with very small weight percentages (wt-%) of contaminants and approximately 70 wt-% silicon. The ferrosilicon is added using a steelmaking ladle after the oxygen blowing process has occurred and the carbon has been reduced. If available, the scrap added to the BOF will be electrical steel scrap from processing. Any scrap added must be high quality with very low carbon content. This is the preferred method to produce electrical steel in much of the world.

The magnetic properties of lamination steel are highly dependent on how the steel is processed, and these processes are carefully controlled. New methods of retaining the best properties are under continual research. In order to understand why these processes are chosen, the next section introduces the basics of metallurgy, followed by the continued discussion of the steelmaking process follows.

6.2.2 Introduction to Metallurgical Concepts

6.2.2.1 Crystallography

Atoms in metals are arranged in an orderly manner, and can be alloyed with metal and non-metal elements. Metals are considered crystalline, where the atoms are situated in a repeating array that extends throughout the material. The inset in Figure 6.7 shows the type of repeating pattern present in iron, where each dot is an atom, and each bold line is an atomic bond. This repetitive, three-dimensional pattern positions atoms in such a

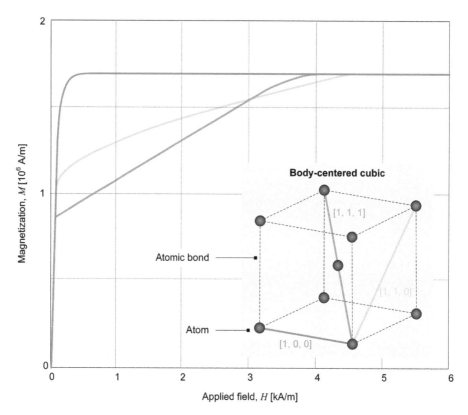

FIGURE 6.7
Body-centered cubic magnetization based on directions. (From Beckley, P., Electrical steels for rotating machines, *IEEE Power and Energy Series*, Institute of Electrical Engineers, London, UK, Vol. 37, 2002.)

fashion that long-range atomic order exists, which is to say over long atomic distances the pattern remains the same. This cubic pattern and the network of repeating units that spread throughout the material is known as a crystal lattice. Some properties depend on this spatial arrangement of atoms, and the long range atomic order that comes with it. For instance, in the iron-silicon crystal lattice, known as a body-centered cubic (BCC) lattice, certain directions are easier to magnetize. This has to do with how the electron spins of each atom have spontaneously aligned themselves.

Figure 6.7 shows the difference in magnetization for these different directions. The direction [100] is the easiest to magnetize and is therefore called the "easy magnetization direction." When a metal solidifies, atoms begin to arrange themselves in the repeating patterns of the crystal lattice. Figure 6.8 shows the growth of solid crystals from liquid metal, called "the melt." Here, the squares represent BCC units while the background is the melt. As the crystals grow, more liquid is turned into solid, until eventually the crystals impinge on each other. Each solidified crystal is then known as a grain, and where they touch, they form a grain boundary. During nucleation, these new crystals are usually randomly oriented with respect to each other.

Figure 6.8 shows an example of a growing grain (with thick, orange border lines) through the stages of its growth. As the grains get close to each other they impinge on each other, meaning that the melt between them disappears until only a grain boundary is left with

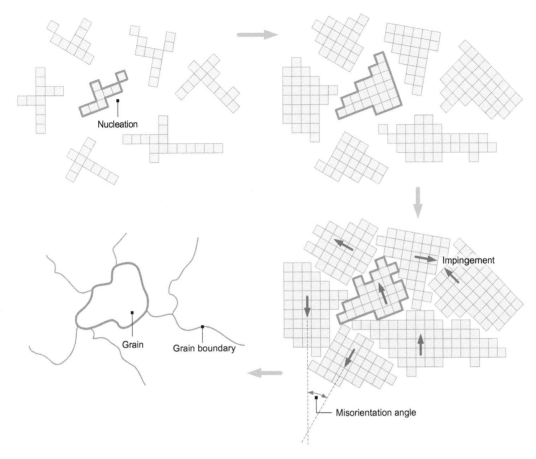

FIGURE 6.8
Solidification of a polycrystalline material. (From Callister, W.D., *Materials Science and Engineering: An Introduction*, John Wiley & Sons, New York, 1997.)

solid grains on each side. Since these grains formed in isolation, the orientation of their BCC crystals can be offset from each other. This difference in directionality is called the misorientation angle.

Defects and impurities can occur in the crystal lattice. During solidification, plastic deformation, or as the result of thermal stresses, interatomic bonds in a lattice can rupture and reform. When this happens, the lattice doesn't always line up correctly again, introducing a type of defect in the lattice called a dislocation. Due to the misorientation angle (see Figure 6.8) between grains, the lattice directions at a grain boundary do not match up.

Figure 6.9 shows an example of strained region around a dislocation. This figure can be thought of as a zoomed in version of a point in the grain boundary in Figure 6.8. The atoms and atomic bonds make up the body-centered cubic lattice in Figure 6.7. Without a strained region, the lattice would have a repeating symmetric pattern over the entire region shown in Figure 6.9. It can be observed that, due to the interruption in the lattice, a strained region occurs around the dislocation as areas of compression and tension. In a grain boundary, the grains are mismatched with each other, causing strain along its length. Therefore, a grain boundary can be viewed as a series of dislocations along its length, each causing

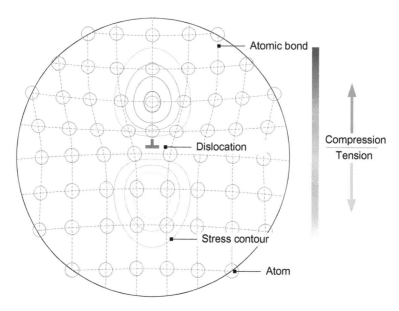

FIGURE 6.9
Dislocation causing compression and tension of crystal lattice. (From Callister, W.D., *Materials Science and Engineering: An Introduction,* John Wiley & Sons, New York, 1997.)

a region of stress. A higher misorientation angle, meaning a greater difference in the lattice between grains, can be thought of as a greater number of dislocations along its length.

6.2.2.2 Magnetic Qualities of Steel

Total core loss and permeability are important measures of magnetic performance used to evaluate lamination steels. Relative permeability (μ_r) refers to how easily a magnetic field can form inside a material:

$$\vec{B} = \mu_r \mu_0 \vec{H} \tag{6.1}$$

where B is the magnetic flux density, H is the applied magnetic field and μ_0 is the permeability of free space. Table 6.3 outlines which physical quantities are important to obtain

TABLE 6.3

Desirable Qualities for Electrical Steels

High Permeability	Low Losses
Small grain size	High alloying content
Good crystal texture	Good crystal texture
Reduced deformation, defects, or internal stresses	Reduced deformation, defects, or internal stresses
Low impurity level (Carbon, Sulphur, Nitrogen)	Low impurity level (Carbon, Sulphur, Nitrogen)
Avoid 'high energy' grain boundaries (can occur for example at high misorientation angle)	Able to be rolled to a low thickness (essential for thin laminations)

Source: Petrovic, D. S., Non-oriented electrical steel sheets, *Materials and Technologies,* 44(6), 317–325, 2010.

the best magnetic performance from steel. Trade-offs must be made for each application, whether they be to cost or material properties. For instance, steel thickness can be reduced to reduce eddy currents at the expense of increased cost for the precision and time required to make more laminations.

Lamination steel plays an important role in torque production and loss generation, thus greatly impacting the machine's overall performance. In an SRM, when the laminated steel core is saturated, the torque production capability increases. More flux will be generated when steel is present since the total reluctance of the magnetic circuit will be lower. However, while current is applied, the magnetic flux will cause changes in the magnetic structure of the material that generates heat. Permeability affects the torque density of an SRM, since it determines the reluctance the magnetic circuit will see as it comes into alignment. Flux density increases with increasing external field, while permeability decreases.

6.2.2.3 Heat Treatment

Annealing is a form of heat treatment in which a metal is heated to a high temperature, held there for some length of time, before cooling at a specific rate. As a metal is plastically deformed, microstructure and physical properties change. Grain shapes, strain hardening, and dislocation densities are all affected. Some of the energy that is used to deform the steel is trapped in the metal as strain energy. Annealing can relieve these stresses, increase softness, ductility or toughness, or promote a specific microstructure to form in the steel. Annealing raises the metal to an elevated temperature that is below its melting point. The temperature to which the metal is raised will depend on the properties the annealing is intended to produce. A phase diagram determines the type of steel that can be produced given a specific temperature and composition [6]. Heating and cooling rates as well as time spent at a given temperature will all impact the final material properties.

A typical annealing process will have three stages:

1. heating to a recommended temperature,
2. holding, or soaking at that temperature until the piece has attained a uniform temperature
3. cooling to room temperature.

Annealing takes time since the processes involved relies on the movement of atoms. Atomic movement may be accelerated by increasing the temperature; however, too large a gradient between the exterior and interior temperatures of a piece may cause it to warp. Many factors affect the temperature to which the steel should be brought in order to get the desired effect. In general, a heat treatment is ~700°C or greater. Processes initiated at these temperatures "unfreeze" the microstructure and allow stress gained during manufacturing to be relieved. The recovery process is hot enough to allow the movement and annihilation of dislocations and release some stress but not reform the microstructure. Recovery occurs in stress-relieving annealing; performed after lamination cutting to relieve the stress introduced through plastic deformation or temperature warping. A higher temperature process known as recrystallization allows new stress-free grains to grow, while allowing old grains to shrink. This process is used during continuous annealing, after hot rolling as we will see in Section 6.2.5. Large grains lower the hysteresis loss experienced during operation, but eddy current loss can increase if the grains get too big. Annealing can be useful in electrical steel for:

- Control of grain size
- Influencing of grain orientation
- Removal of stress
- Flattening
- Decarburization
- Control of hardness

Ultimately, the goal of annealing is to lower the internal stresses of the metal and to improve the magnetic performance.

6.2.3 Magnetic Domains and Hysteresis Loops

The source of magnetic properties in ferromagnetic materials is quantum effects at the atomic level. An understanding of magnetism at this depth is not required in order to get a macroscopic view of engineering systems. Magnetic properties can be described with reasonably accurate empirical models describing static and dynamic behavior for engineering applications. However, the physical problem of magnetic hysteresis engages a wide multi-disciplinary research effort, including mathematicians, physicists, material scientists, and engineers.

A typical hysteresis loop is illustrated in Figure 6.10 for a soft magnetic material like iron. This hysteresis loop can be created by applying a magnetic field to a strip of iron, for instance, and then reversing it repeatedly. What is found is that the magnetic induction (the B-field) of the iron strip will increase in strength along the external field direction and then decrease, but not along the same lines. Instead, the strength of the B-field decreases down a different path. The area traced out by these curves represents the hysteresis loss experienced by the iron strip.

Hysteresis loss is the energy from the external magnetic field being transformed into thermal energy. This transformation occurs because of the interaction between the external magnetic field and the internal magnetic structure of the iron. In order to begin understanding this curious behavior, we will need to examine concepts such as magnetic domains and domain walls. These terms describe the organization of the iron's internal magnetic field. Domains are composed of many individual magnetic moments, and their interaction with an applied external magnetic field allows domains to grow or shrink. The study of magnetic domain structure is fascinating, but one that requires a great deal of diverse background knowledge. An introduction will be presented here. If this explanation leaves the reader desiring more, there are several very good books that bring these topics to a deeper level [4,6,7].

As was introduced in the last section, grains are regions of similar crystal orientation. Where grains touch, they form grain boundaries—chains of dislocations that incur higher strain in the crystal lattice around them. Grains form an important analogy to domains; however, it is important to keep the concepts separate. Complicating the explanation of these two phenomena is the effect of grains on magnetic domains. Nonetheless, a great deal of insight can come from comparing and contrasting these two ideas.

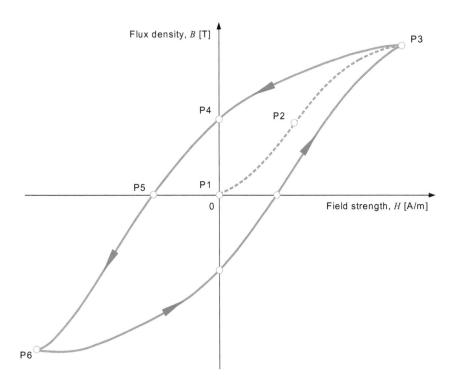

FIGURE 6.10

Magnetic hysteresis loop. (From Callister, W.D., *Materials Science and Engineering: An Introduction*, John Wiley & Sons, New York, 1997.)

TABLE 6.4

Comparison between Magnetic Domains and Crystal Grains

Comparable Features	Domains	Grains
Describes a region of:	Like magnetization direction	Like crystal direction
Is it a physical phenomenon (i.e., made of matter)	No, it is composed of a changing magnetic field	Yes, it is a concentration of atoms, arranged in a specific fashion
Building Blocks	Magnetic moments	Crystal lattice orientation

As shown in Table 6.4, a grain is the region with like crystal-lattice direction. The domain will still have regions of like direction; however, instead of a crystal lattice direction it will be a region of like magnetization direction. Domains in a demagnetized state are likely to be randomly oriented. What is meant by the phrase, "like magnetization direction?" To answer that, we must examine the idea of magnetic moments.

Electrons are governed by the laws of quantum mechanics and as such, a description using classical mechanics doesn't quite fit. However, classical descriptions provide a simple and instructional first step in understanding this complex behavior. It is important to remember that here we are using an analogy.

A magnetic moment describes the tiny magnetic field generated by the movement of electrons. Figure 6.11 illustrates the generation of the orbital spin and electron spin. Electrons orbit around their atomic nuclei, but also have their own spin, both of which generate magnetic fields. Some magnetic contributions from electrons are cancelled out by other electrons with similar, but opposite spins (e.g., one electron spins up and the other

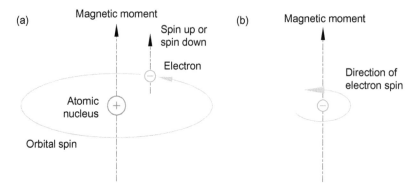

FIGURE 6.11
Illustration of magnetic moment generation caused by the movement of electrons: (a) electron spin and (b) magnetic moments from a classical perspective. (From Callister, W.D., *Materials Science and Engineering: An Introduction,* John Wiley & Sons, New York, 1997.)

one spins down). Together, the sum of these fields over all the electrons forms the overall magnetic moment of the atom. Magnetic moments, in turn will interact with each other and orient in similar directions to form the domains.

The spins of many electrons are cancelled out by their neighbors. However, ferromagnetic materials have uncancelled electron spins as a consequence of their electron structure. The spatial organization of electron spins affects the overall ferromagnetic characteristic of the material. The magnetic moments of ferromagnetic atoms couple with adjacent atoms to produce a much stronger internal field then for a material without mutual interaction. Only a small number of elements are ferromagnetic. These elements follow a curve of magnetization energy as a function of the ratio of atomic separation to unfilled shell diameter, as shown in Figure 6.12. This sweet spot of energy shells to atomic spacing means that electron spins could influence each other to align.

There is a relationship between the alignment of magnetic domains and the hysteresis curve. A magnetic domain is a region where there is a dominant magnetization direction, produced from the coupling of the individual magnetic moments within it. There may be a large number of domains in a polycrystalline material, and they are generally

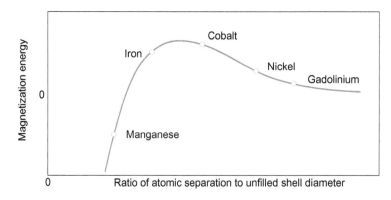

FIGURE 6.12
Bethe curve comparing magnetization energy for several element species. (From Beckley, P., Electrical steels for rotating machines, *IEEE Power and Energy Series,* Institute of Electrical Engineers, London, UK, Vol. 37, 2002.)

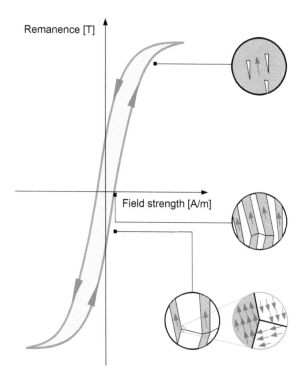

FIGURE 6.13

Change in domains with changing remanence and field strength. (From Bertotti, G., *Hysteresis in Magnetism: for Physicists, Material Scientists, and Engineers*, Academic Press, New York, 1998.)

microscopic in size. Magnetic domains can be differentiated by their directionality. The change between one magnetization direction to another forms a domain wall.

Figure 6.13 illustrates the response of magnetic domains with a changing external magnetic field. When an external field is applied, magnetic moments will be pulled in that direction. Similarly aligned moments will change their direction sooner than dissimilarly aligned moments. Since domain walls are the intersections of differently aligned domains, if the magnetic moments making up those domains shift alignment, the wall will move. Figure 6.14 shows a region of changing local magnetic moment direction over a domain wall. A domain wall can be thought of to have a thickness, represented by the change in magnetization direction.

As the magnetic field is applied, magnetic domains that are in line with the direction of the external field will grow, while domains that are opposed will shrink. As more and more of the domains become aligned to the externally imposed direction, they add a greater contribution to the overall magnetic field, which is the additional flux density provided by the steel. After the hardest domains to rotate have been moved, the flux density saturates. Saturation is the maximum magnetization possible in a material when all magnetic moments within the material are aligned in the direction of external magnetic field. In other words, saturation is equal to the combined net contributions of all the magnetic moments of all the atoms in the material.

6.2.3.1 Hysteresis Loss

Considering the nature of magnetic domains, how does the characteristic hysteresis loop manifest? Figure 6.15 organizes the phenomenon just discussed, with some expectation

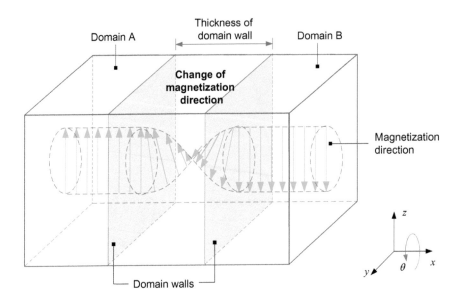

FIGURE 6.14
An illustration in the change in local magnetization direction over a domain wall. (From Bertotti, G., *Hysteresis in Magnetism: for Physicists, Material Scientists, and Engineers,* Academic Press, New York, 1998.)

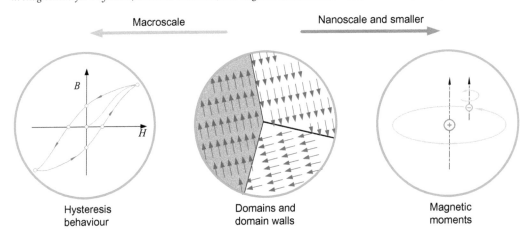

FIGURE 6.15
Comparison of how the constituent models describing hysteresis loss fit into each other.

of what length scale they may be found on. When an external magnetic field is applied, domains aligned with the external field grow and the ones unaligned shrink until all the magnetic moments are aligned. In a saturated material there are a small number of large domains. Theoretically a saturated material would contain only one domain; however, due to underlying crystal defects it may be difficult to apply the required field for domains to continue growing until only one was left.

What happens to a saturated material if the field is reversed? The magnetic moments of each domain have been rotated to accommodate a strong field in the first direction. When the field is reversed, some moments will be more inclined to begin moving in the other direction. All of the magnetic moments will begin to rotate; however, as some rotate more than others, new domains will form with domain walls at their borders. Eventually these new, more aligned domains will grow as we reach the bottom left region of the hysteresis loop until the metal saturates in the reverse field. The loop traces a different path every time because the growth of new aligned domains and shrinking of old unaligned domains is different every time. The hysteresis loops in Figures 6.10 and 6.13 are idealized averages if the system were to be brought back and forth between saturation at point P3 and saturation at point P6 in Figure 6.10. Smaller loops can be made by reversing the field direction at any point along the hysteresis curve below saturation. In this case the smaller loops will be different as well. This property is called "memory-dependent" or "history-dependent" behavior. In order to understand what the system will evolve into, we must understand where it has been. To answer why the loop is different every time, we need to reinvestigate hysteresis loss, and describe it in terms of crystal defects and magnetic domain walls.

6.2.3.2 Crystal Defects and Magnetic Domain Walls

Previously we spoke of the analogy between grains and domains: grains are composed of organized atoms and the domains are composed of organized magnetic moments. There is a relationship between defects in the crystal lattice and the movement of domain walls (recall the discussion on the dislocation in the crystal lattice and grain boundary in Figure 6.9). At points of dislocation, the domain walls appear to be "pinned," which is to say they resist movement.

The electrons of the atoms in the crystal lattice produce the magnetic moments. However, it's not immediately clear how changes in the structure of the crystal lattice would affect the magnetic moments and subsequently the domain walls. Unfortunately a complete description of this relationship is beyond the scope of this text. In fact a complete description connecting the quantum mechanical relationships arising from electron movement to the movement of domain walls and generation of hysteresis loss is still generally unknown. However, a simple description can be attained using general thermodynamic concepts of order and disorder. Thermodynamic potentials are equations of state that describe how a system will evolve based on constraints. Any constraints that cause a system to become more ordered will be favored, while any constraint that causes a system to contain more disorder will not be favored. This concept of order/disorder is a powerful tool when used to describe how the entire system will evolve to stimulus and how different aspects of a system can couple. When there are many constraints or stimuli, the system can become very difficult to predict.

Recall the tension and compression illustrated in Figure 6.9 due to the dislocation in the crystal lattice. The dislocation has introduced structural disorder to the crystal lattice system. We tend to say these areas of disorder are at a higher energy than the rest of the lattice. Since the lattice would prefer to be at the lowest energy possible, it seeks to release this energy. When the metal is annealed, the system is able to shift atoms and release the dislocation, lowering the energy and returning the area to a state closer to a perfect lattice.

Prior to annealing, the stress on the lattice causes a higher energy area due to the dislocation that "catches" the domain wall when it tries to move past. Consider a domain wall as a surface that requires energy to maintain. The lattice further away from a dislocation has lower energy than the dislocation at a higher energy creating a kind of hill or obstacle in the energy "landscape." Figure 6.16 illustrates how this so-called energy landscape maps to the grains. If that surface is interrupted by an obstacle, which could be any kind of defect, then when the domain wall pulls away there will be a "hole" left in the wall. Additional energy will be needed to fill the hole and maintain the surface once more. This extra energy is lost in the form of heat and contributes to the hysteresis loss.

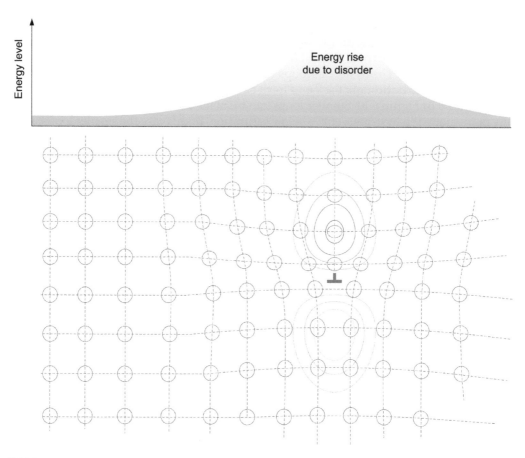

FIGURE 6.16
Energy landscape due to disorder of dislocation. (Adapted from Callister, W.D., *Materials Science and Engineering: An Introduction*, John Wiley & Sons, New York, 1997.)

Since grain boundaries are composed of dislocations, they are considered a type of defect and disrupt the movement of magnetic domain walls. By increasing the grain size, hysteresis loss can be decreased since there will be fewer grain boundaries per unit area. However, increased grain sizes also increases eddy current loss, analogous to how thicker laminations impose more losses. Smaller grains also increase the strength of the material due to the higher density of grain boundaries. The trade-off between hysteresis and eddy current losses creates an optimal grain size that varies with magnetic field strength.

6.2.3.3 Hard and Soft Magnetic Materials

The discussion so far has been geared towards soft magnetic materials, like electrical steel, which are ideal for the requirements of motors and transformers. As seen in Figure 6.17, soft magnetic materials have a smaller hysteresis loop area proportional to smaller hysteresis losses. The hysteresis loop area is kept thin and tall by having a low coercivity, and a high permeability allowing the material to saturate at high levels of magnetization in a relatively low external magnetic field.

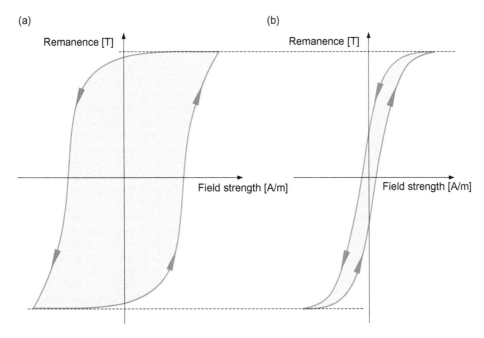

FIGURE 6.17

Comparison of B-H curves: (a) large magnetization and field strength, desirable for permanent magnets and magnetic recording and memory devices, and (b) large magnetization and small field strength is desirable for transformer and motor cores to minimize the energy dissipation.

Hard magnetic material has a large hysteresis loop area owing its shape to a high remanence, coercivity, and saturation flux density. These materials are used in permanent magnets because once they are magnetized, it is difficult to cause the orientation of their domains to change. This allows for a magnetic field to be retained in the permanent magnet long after the external field that magnetized it has been removed. We would expect the microstructure of a good hard magnetic material to contain many defects to impede domain wall movement. If a large enough field to demagnetize a hard magnetic material was applied we would expect large hysteresis losses. Electrical steels (Fe-Si) have a saturation magnetization less than pure iron but still high enough to be useful for most electric machines. In addition, electrical steels have a coercivity that allows a small magnetic field to magnetize them. This is important for SRM since a higher saturation magnetization means that more torque can be generated and a lower coercivity means a lower input current is required. While low coercivity materials can be created, such as nanocrystalline and amorphous alloys, their saturation magnetization tends to be low.

6.2.4 Types of Steels

Electrical steels are commonly used in electric machine manufacturing both for their role as high-quality flux magnifiers and for their moderate cost. There are several categories of electrical steels. The material selected greatly depends on the application and cost considerations. Figure 6.18 illustrates a comparison of flux density characteristics for several core materials with varying compositions and Figure 6.19 shows the core losses. The selection of the electrical steel for a given application generally start from these important properties.

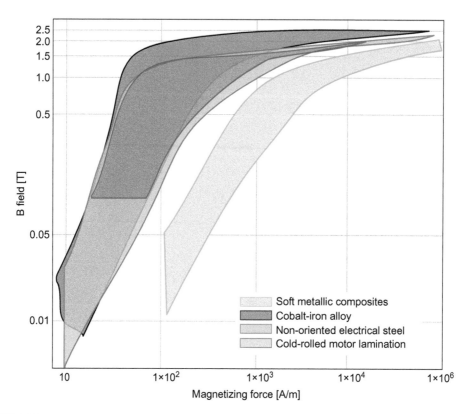

FIGURE 6.18
Illustration of relative Magnetic Flux Densities between lamination steel types and soft magnetic composites, reproduced from EMERF, Lamination Steels, Third Edition. (From Motion Control and Motor Association (MCMA), *Electric Motor Education and Research Foundation (EMERF)*, EMERF Lamination Steels, Third Edition, Arbor, Michigan, USA, 2009.)

6.2.4.1 Electrical Steels/Silicon Steels

The addition of silicon to steel increases the resistivity and lowers hysteresis losses by shrinking the hysteresis loop area but lowers the magnetization saturation. The addition of silicon also aids in the development of larger grain sizes. High silicon steels are used in machines where losses are the primary concern, such as high frequency applications, whereas low silicon steels are used in more torque-dense applications. High silicon content steel is more abrasive to cut and can compromise tool life during manufacturing. If silicon content increases beyond 3%, the rollability of the steel falls and special processes must be used.

Figure 6.20 shows the classification of electrical steels. Grain oriented steels are used in transformers, transducers and magnetic amplifier cores. Non-oriented steels are used in rotating machines such as motors, generators and alternators.

6.2.4.2 Grain-Oriented Electrical Steel (GOES)

Electrical steel sheets can be processed in a specific way causing the easy magnetization direction to align to a particular direction. This directionality causes lower core losses when magnetic flux is oriented in the preferred direction. However, in any other direction core losses increase. Grain-Oriented Electrical Steel (GOES) finds its primary use in transformers and other non-rotating applications.

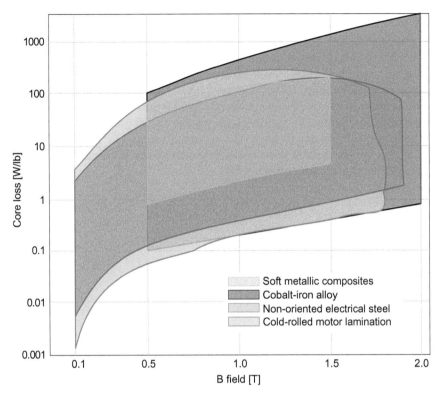

FIGURE 6.19
Illustration of total core loss for a variety of core materials, reproduced from EMERF, Lamination Steels, Third Edition. (From Motion Control and Motor Association (MCMA), *Electric Motor Education and Research Foundation (EMERF)*, EMERF Lamination Steels, Third Edition, Arbor, Michigan, USA, 2009.)

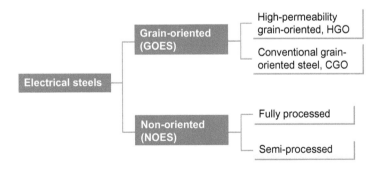

FIGURE 6.20
Classification of electrical steels. (From Callister, W.D., *Materials Science and Engineering: An Introduction*, John Wiley & Sons, New York, 1997.)

Relative properties of GOES: (in the preferred direction)

- Core loss: Better than non-oriented electrical steel (NOES)
- Permeability: Better than NOES
- Processing: Fair, more complicated than standard gauge NOES
- Cost: Higher than standard gauge NOES

6.2.4.3 Non-Oriented Electrical Steel (NOES)

NOES typically contain less silicon than GOES. NOES are processed to reduce core losses in rotating machines. Since rotating machines will have a changing magnetic flux in many directions, NOES maintain uniform magnetic properties with the grain and across it. For the best magnetic performance, fully processed NOES are annealed and coated prior to leaving the steel manufacturer. Fully processed grades have been processed to remove plastic deformation and stress. Very thin gauge grades (<0.25 mm) are becoming popular, with either a standard chemistry or high silicon content around 6.5%. Thin-gauge laminations have low core losses and are used in high frequency applications.

Relative properties of standard gauge NOES:

- Core loss: Good
- Permeability: Fair
- Processing: Easy, requires some processing
- Cost: Less than GOES, much less than Cobalt-Iron (Co-Fe) alloy

Relative properties of thin gauge NOES:

- Core loss: Better than standard NOES
- Permeability: Fair
- Processing: Fair, worse than standard gauge NOES. A Higher silicon content will require extra care
- Cost: More than GOES, less than Co-Fe

6.2.4.4 Powder Metal/Soft Magnetic Composites (SMC)

Soft Magnetic Composites (SMC) consist of metal particles coated in an insulating film. The metal particles are warm-formed into molds of the desired shape in order to maintain the integrity of the insulation. Since the particles are isolated from each other, the eddy current losses are much lower than steel laminations; however, SMC also have lower magnetic flux density. High temperature operation can be an issue for SMC that are coated with an insulating material made of polymers (similar to the insulation of a magnet wire). As such, special care must be taken to ensure that the material doesn't prematurely age or break down. Manufacturing of SMC requires completely separate procedures than lamination steels, which can cause the properties and cost to vary significantly.

6.2.4.5 Cobalt-Iron (Co-Fe) Alloys

Co-Fe alloys are able to reach a saturation flux density of 2.3 T, and possess similar losses to electrical steels. A typical composition is 49% cobalt, 49% iron, and 2% vanadium. However, with such high cobalt content the price of Co-Fe alloys can be more than ten times the price of electrical steels. Unlike silicon, cobalt is also ferromagnetic. Co-Fe alloys have high strength properties that can be enhanced during annealing, at the cost of its magnetic properties. Grain size determines the trade-off between strength and magnetic properties. Larger grains favor magnetic properties, while small grains favor strength. An annealing step is required for these materials before and after lamination punching. Before punching, the alloy is too soft in its unannealed state for laminations to be produced (a chemical treatment can be performed instead of annealing at this stage). After punching, the alloy contains too much stress to reveal its extraordinary magnetic properties. The annealing process must be carefully controlled for these alloys at high temperatures. Specialists in this area should be consulted if cobalt steel is to be considered in an application. Cobalt-steels are used in high performance, highly power-dense applications like aerospace and performance automotive.

Relative properties of standard gauge Co-Fe Alloys:

- Permeability: Better than NOES
- Processing: Hard, requires challenging processing
- Cost: Much higher than GOES, NOES or CRML

6.2.4.6 Other Types

Other lamination materials exist such as amorphous alloys and nickel-iron alloys. These materials have lower saturation flux densities, which might limit their use in SRM applications with high torque density requirements. Semi-processed NOES, otherwise known as cold-rolled motor laminations have largely been phased out of operation in much of the world. North America still produces these grades; however, losses tend to be quite high, and any cost-savings tends to be negated by additional annealing processes undertaken to reduce the losses.

6.2.5 Processing of Steel

In Figure 6.6, we looked at the processes in the integrated steelmaking for structural steel. Then, in order to understand those processes, basic metallurgy was introduced. Now, we will continue the discussion of the steelmaking process. Figure 6.21 illustrates the process required to create rolled structural steel. At the end of the processes in Figure 6.6, hot-rolled steel was brought to a desired thickness and it was rolled into a coil to continue down the line. Due to the limitations of the process, the thickness of the steel cannot be reduced further by hot rolling. Coils are large rolls of steel that are used as a convenient means of moving large quantities of thin steel sheet without damaging them. Coiling is done during processing to transport the coils from one processing line to another.

Figure 6.22 illustrates the processing the coil goes through before being cold rolled. Hot band from the hot roller is cooled and coiled to be moved to the annealing furnace. Annealing can be achieved in either continuous annealing furnace or in a bell furnace. In a continuous annealing process, a sheet is uncoiled and fed through an assembly that includes a long furnace and a cooling zone. Continuous annealing offers good control over heating and cooling rates, and thus is most commonly used. A bell-type furnace instead requires the coil to be cut into pieces and annealed in a stack. It takes longer to heat all the way through, and can create uneven heating distributions. A continuous annealing furnace can be attached to a manufacturing line that will shot-blast and pickle the steel to remove oxide scale formed during hot-rolling and annealing. To shot-blast the coil of steel, high pressure air fires abrasive steel particles at the sheet, while the pickling process runs steel through strong acid to remove contaminants such as oxide scale. Oxide scale is an electrical insulator; however it is partially magnetic and can carry flux that is unwanted for electrical steel. Under high flux levels, it will begin to increase the total hysteresis loss. Most or all the scale is removed during pickling creating a surface suitable for cold rolling.

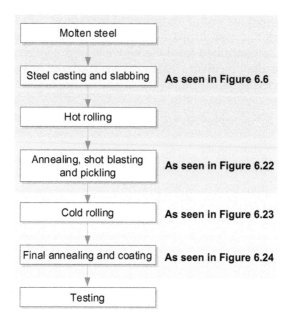

FIGURE 6.21
Process for plant schematic. (From Beckley, P., Electrical steels for rotating machines, *IEEE Power and Energy Series*, Institute of Electrical Engineers, London, UK, Vol. 37, 2002.)

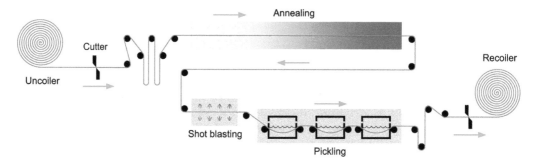

FIGURE 6.22
Preparations before cold rolling: annealing and descaling. (From Buswas, A.K., *Principles of Blast Furnace Ironmaking: Theory and Practice*, Cootha Publishing House, Brisbane, Australia, 1981; Cogent Power, Altogether more powerful—Non-oriented electrical steel. [Online]. Available: http://cogent-power.com/. Accessed: August 23, 2017.)

Figure 6.23 shows a schematic of a reversible cold rolling mill with a four-high roll stand configuration. Cold rolling controls the thickness of steel by applying a large amount of pressure normal to the sheet surface through a series of rolls called a roll stand. Mills can be configured to run a sheet forward through a series of roll stands (tandem) or back and forth through a single roll stand (reversing). For the four-high configuration, in each roll stand there are two, small "working" rolls at the steel surface and two larger "back-up" rolls in contact with the working roll. The working roll is smaller to achieve effective reduction in thickness down to lighter gauges, but this leaves the working roll susceptible to bending. The back-up rolls support the working roll so that it will not buckle. Cold rolling provides a smoother surface than hot rolling, and can produce more exact thicknesses. Surface roughness can be controlled to give an optimal smooth finish that allows for the

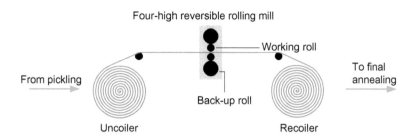

FIGURE 6.23
Process of cold rolling in a reversible cold rolling mill. (From Beckley, P., Electrical steels for rotating machines, *IEEE Power and Energy Series*, Institute of Electrical Engineers, London, UK, Vol. 37, 2002.)

best results during a final round of annealing. A uniform thickness is important for calculating stack height in motors, and for the best surface contacts between laminations.

Cold rolled steel has undergone significant plastic deformation during the thinning process and must be annealed to recover the correct microstructure. A "final" continuous annealing process releases the strain again, in a similar method to the hot band but with likely a different heating, soaking and cooling profile. The term "final annealing" could be misleading since it refers to the last annealing step on the production line (fully-processed lamination steel). However, annealing can be performed again after laminations have been stamped or assembled (semi-processed lamination steel), although this anneal is likely to only include a recovery process and not a recrystallization process. For laminations that will not have a final annealing after cutting, a long, hot (1,000°C) annealing cycle provides softer steel with large grain growth. The extended cycle gives time for grains to grow, and for the number of grain boundaries to reduce, decreasing hysteresis losses.

Generally the production line will add an electrically insulating coating to the steel before finishing the manufacturing process, as seen in Figure 6.24. Organic coatings must be applied to the steel after final annealing on the production line since they cannot withstand a final annealing after cutting. Non-organic coatings can withstand elevated temperatures used in stress-relief annealing.

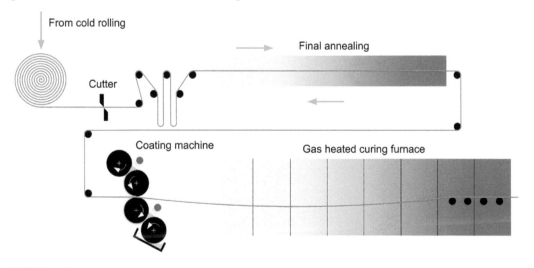

FIGURE 6.24
Process of coating curing. (From Cogent Power, Altogether more powerful—Non-oriented electrical steel. [Online]. Available: http://cogent-power.com/. Accessed: August 23, 2017.)

6.2.6 Grades and Losses

Thinner gauge steel will have fewer eddy current losses, as shown in Figure 6.25, since eddy current losses are inversely proportional to the square of thickness. However, a thinner gauge might not always be the best material selection. The application requirements of the machine being designed impacts the selection of steel grade. A small machine may prioritize price and opt for thicker gauge steel, with low to no silicon content and a low temperature, natural oxide coating. A high-power-density machine might require steel with high silicon content and a high temperature coating.

Standard measurement units for thickness include inch and millimeter. The Electrical Steel Standard Gauge is also still in use; however, it has been outdated in current published standards [8]. Gauge does not scale with either millimeters or inches, and does not provide a value thinner than 29 gauge (0.36 mm). While generally there are directly comparable thicknesses between measuring systems it is not always the case. It is important to consider the effect of rounding errors when converting between measurement units. Care should be taken when considering the equivalency of different manufacturers' published nominal measurements, as well as indicated manufacturing tolerances

There are several standards for specifying electrical steel designations. The European Committee for Iron and Steel Standardization is responsible for producing the European Standards (ENs) for structural steels. For example, EN 10025 : 2004 is the new European standard for structural steel. International Electrotechnical Commission (IEC) and EN standards designate electrical steels with a code containing information regarding the maximum allowable loss, steel thickness, and whether it is a fully or semi processed sheet. Both standards start with an M, indicating magnetic, followed a number indicating the maximum value for loss measured at 1.5 T and 50 Hz expressed in [W/100 kg]. Next, the thickness of the steel is indicated in units of 0.01 mm, and finally a designator describing the type of processing completed, where A denotes fully processed and K denotes semi-processed. For instance, Cogent steel produces grade M270-35A steel, a fully processed 0.35 mm thickness electrical steel. The thickness for this steel could also be expressed as 0.014 inches simply by converting to imperial or approximately 29 gauge since it's the closest measured gauge at 0.36 mm.

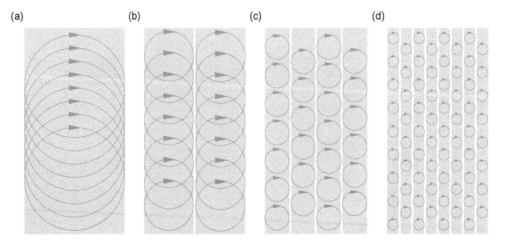

FIGURE 6.25
Comparison of eddy currents in steel: (a) without laminations, (b) with thick laminations, (c) with medium laminations, and (d) with thin laminations. (From Callister, W.D., *Materials Science and Engineering: An Introduction*, John Wiley & Sons, New York, 1997.)

TABLE 6.5

Partial List of Thin Gauge Products by Cogent Steel

Grade	Gauge [mm]	Max P10/400[a] [W/kg]	Typ P10/400 [W/kg]	Density [kg/dm³]
NO10	0.1	13.0	12.1	7.65
NO12	0.127	13.5	11.8	7.65
NO15	0.15	14.0	12.1	7.65
NO18	0.178	14.3	12.2	7.65
NO20-1200	0.2	12.0	11.4	7.60
NO20	0.2	13.5	12.3	7.65
NO25-1400	0.25	14.0	12.9	7.60
NO27-1500	0.27	15.0	13.7	7.60
NO30-1600	0.3	16.0	15.1	7.60
NO30	0.3	19.0	17.0	7.65

Source: Cogent Power, Altogether more powerful—Non-oriented electrical steel. [Online]. Available: http://cogent-power.com/. Accessed: August 23, 2017.

[a] P10/400 means iron losses at 1 T and 400 Hz.

The naming convention would indicate that this steel has a maximum allowable loss of 2.70 W/kg at 1.5 T and 50 Hz. The manufacturer's data sheet should always be consulted when applying values for electrical steel.

Thin gauge non-oriented electrical steels may be named differently, since they tend to be tested at higher frequencies. For instance, Cogent Steel has a thin gauge steel called NO20-1500. This steel is 0.20 mm thick; with a maximum loss of 15 W/kg at 1.0 T and 400 Hz. Table 6.5 shows a list of thin gauge non-oriented steels and some sample values.

Core loss values are usually measured by steel manufacturers under restricted conditions, necessitating care when comparing and choosing steel based on published loss values. Common frequencies like 50 or 60 Hz using sinusoidal excitation are used to determine the core loss, for instance. If core loss is listed as "W15/50" this indicates that the core loss was measured at 1.5 T and 50 Hz. A material listed as "W10/400" would have had its core loss measured at 1.0 T and 400 Hz. However, the stated core loss may vary significantly at other operating frequencies. Steels that have the same core loss at 50 Hz could have very different losses at higher frequencies due to their unique alloy composition, grain structure and grain orientation.

Figure 6.26a shows the loss values for several grades of steel at 1.5 T and 400 Hz and Figure 6.26b shows the loss values at 1.0 T and 2500 Hz. The designations for each grade of steel were tested at 1.5 T and 50 Hz, but show distinct changes under the two different operating conditions. Accurate materials data is needed over a wide range to evaluate steel's performance for the entire operating envelope of an SRM.

Figure 6.27 shows the losses produced by electrical steel under sinusoidal excitation at a constant frequency. However, an electric motor runs at variable speed meaning the frequency of the excitation is variable as well. In order to accommodate this, Finite Element Analysis (FEA) programs, like JMAG, take the loss curves and interpolate between them. Currently, analytic solutions are not yet comprehensive enough to predict the hysteresis loss of steel under varying excitation. Work has been done in physics and material science to determine the fundamental expressions of hysteresis loss. The stress in the lattice also needs to be addressed when modeling hysteresis loss. Manufacturing methods introduce a great deal of stress that can greatly change the B-H curve. Predications from materials science fundamentals require a complete understanding of the whole system.

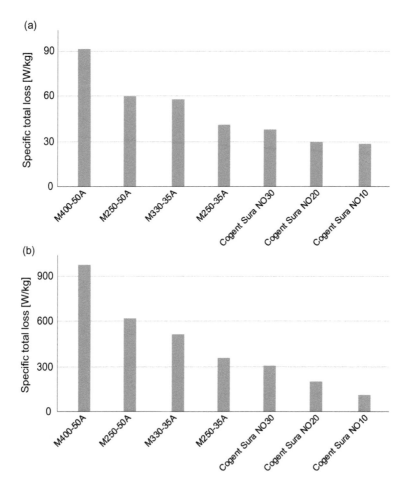

FIGURE 6.26
Comparison of losses: (a) 1.5 T, 400 Hz and (b) 1.0 T, 2500 Hz. (From Cogent Power, Altogether more powerful— Non-oriented electrical steel. [Online]. Available: http://cogent-power.com/. Accessed: August 23, 2017.)

6.2.7 Coating for Lamination

Coatings are used to electrically insulate laminations from each other in the stack. In addition to improving the resistance between individual laminations, coatings can provide lubricating effect for better punching performance, avoid metal welds between laminations, and provide a source of interlaminar gluing.

Coatings are rated based on an ASTM standard "C" designation, which can describe what the coating is made of, and what environmental conditions it can withstand. Table 6.6 describes the coating types specified by the ASTM standard. Melamine, polyesters, and epoxies/resins are examples of materials used in coatings today. Steel forms a natural oxide layer that is electrically insulating. The formation of this film can be controlled during an annealing process by adjusting the atmosphere in the furnace. These oxide coatings alone (C-0) are considered a "mill coat" and are only appropriate for small size machines. As the power of the machine increases, the resistance between laminations needs to increase. C-3 is the only organic varnish typically used and cannot withstand the same temperature ranges as other designations. However, it is an excellent electrical insulator and it is less

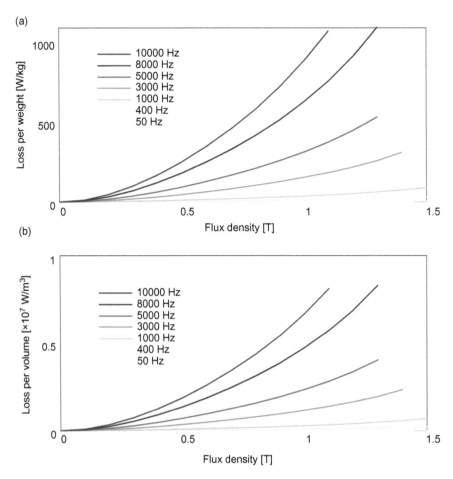

FIGURE 6.27
Losses at different frequencies, Cogent NO10: (a) loss per weight and (b) loss per volume. (From Cogent Power, Altogether more powerful—Non-oriented electrical steel. [Online]. Available: http://cogent-power.com/. Accessed: August 23, 2017.)

demanding on lamination cutting tools allowing for improved tool life when compared to other coatings.

Coatings may be required to withstand high temperatures. A great deal of stress can be introduced into lamination stacks during assembly, which compromises the magnetic performance of the steel. In order to relieve some of this stress, the lamination steel can be annealed multiple times during manufacturing. For instance, a C-3 organic coating is not able to withstand the temperatures of stress-relieve annealing without damage. If a lamination manufacturer wanted to anneal the lamination stack after punching to relieve stress, an organic coating could not be used. Corrosive atmospheres and vibrations may also be of concern to the integrity of coatings since any crack or fracture in the coating can allow for increased eddy current losses. Vibration and environmental effects can be of particular concern for traction motors. It should also be noted that, depending on the stack assembly method, there could be movement between laminations.

Since coatings are applied on the surfaces of a steel sheet after cutting, the cut edges are exposed to the atmosphere. The laminations can be deburred and then recoated to

TABLE 6.6

List of Standard Electrical Steel Coatings Specified in ASTM A976-03

C-0	The steel has a natural, ferrous oxide surface. The thin oxide layer provides sufficient insulating quality for small cores. It can withstand normal stress-relieving temperatures. The condition of the oxidized surface can be enhanced by controlling the atmosphere during anneal. Also known as a mill coat.
C-1	An oxide is formed during contact with an oxidizing furnace atmosphere at the end of heat-treatment. Looks bluish-grey.
C-2	An inorganic coating of magnesium oxide and silicates that reacts with the surface of the steel during high-temperature annealing. Principally used in distribution transformer cores composed of grain oriented electrical steels; not for stamped laminations due to its abrasiveness.
C-3	Enamel or varnish coating that enhances punchability and is resistant to normal operating temperatures. Will not withstand stress-relief annealing. Suitable for operating temperatures up to about 180°C.
C-4	Coating formed by chemically treating the steel surface followed by a high temperature curing treatment. Can withstand stress-relieving temperatures, but some reduction in insulation resistivity may occur. Suitable for moderate levels of insulation resistance.
C-4-AS	An anti-stick treatment that provides protection against lamination adhesion during annealing of semi-processed grades.
C-5	An inorganic coating similar to C-4 with added filler such as aluminum phosphate, or similar ceramics for additional insulation. C-5 is used when high levels of inter-laminar resistance between laminations are required. The coating will withstand stress-relief annealing up to 840°C, but some reduction in insulation resistivity may occur. Can be applied overtop of C-2 coatings for additional insulation.
C-5 AS	The same coating as C-5, but with a thinner coating thickness. Anti-stick for semi-processed grades.
C-6	A combined organic/inorganic coating that improves the punchability of steel, and is suitable for punched laminations. The coating can withstand burn-off treatments, at 320°C–540°C used to rebuild motors by removing stator winding insulation. It is not considered able to withstand stress-relief annealing.

Source: The American Section of the International Association or Testing Materials (ASTM), Standard Classification of Insulating Coatings for Electrical Steels by Composition, Relative Insulating Ability Application (ASTM A976). [Online]. Available: http://www.astm.org/. Accessed: August 23, 2017.

reduce the risk of a short circuit. This is typically an option selected by large equipment manufacturers since the power ranges are much higher. Table 6.7 shows examples of several lamination coatings, including Suralac® 9000—a self-bonding coating applied to thin gauge laminations.

Tool quality when punching is directly related to the magnetic properties of the finished steel. Worn tooling may cause more plastic deformation leading to higher stress and, hence, higher core losses. Therefore, it is important to maintain tool quality to achieve better magnetic performance after cutting the laminations. The coating on the steel will affect how the tools wear over time.

6.2.8 Cutting Methods and Core Building

Cutting the motor lamination shape out of a steel sheet introduces stress into the crystal lattice and, as a result, degrades the magnetic performance of the laminations. Stress in the lattice causes magnetic permeability to decrease and iron losses to increase, specifically the hysteresis component. The degradation of the magnetic properties is affected by the amount of steel cut, the type of cutting technique used, and the cutting angle relative to the rolling direction. The simple design of the rotor and stator in an SRM allows for economic cutting by limiting the path length of the cut. As the magnetizing properties in

TABLE 6.7

Specs of SURALAC Coatings from Cogent Steel

Designation	Suralac 1000	Suralac 3000	Suralac 5000	Suralac 7000	Suralac 9000
Type	Organic	Organic with fillers	Semi-organic	Inorganic	Organic
Thickness per side [μm]	0.7–6	3.5–6	0.7–1.2	0.7–3.5	3.5–5.5
Standard thickness [μm]	2.5	6	1.2	1.5	4.5
Colour	Clear to brown	Grey	Brown to grey	Grey	Clear
Temperature capability, in air, continuous [°C]	180	180	200	270	[a]
Stamping lubricant	Yes	Yes	Yes	Yes	[b]
Typical pencil hardness [H]	8–9	8–9	8–9	9	5–7
Typical thickness [μm]	0.7 / 2.5	3.5 / 6	0.7 / 1.2	0.7 / 1.5 / 3.5	N/A
Typical welding[c]	Good / Spec	Spec / Spec	Exc / Exc	Exc / Good / Mod	N/A
Typical punching[c]	Exc / Good	Good / Mod	Good / Exc	Good / Good / Mod	Good
Surface insulation resistance, per lamination [Ωcm²]	10 / 50	>200	3 / 7	20 / 40	>200

Source: Cogent Power, Altogether more powerful—Non-oriented electrical steel. [Online]. Available: http://cogent-power.com/. Accessed: August 23, 2017.

[a] At 125°C the bonding strength is 50% of that at room temperature, strength is regained when the temperature is lowered. After 20,000 hours at about 170°C, the room temperature bonding strength is reduced by 50%.

[b] Lubricants used at stamping may impact on the bonding strength and are not recommended.

[c] exc = excellent, good = good, mod = moderate, spec = special precautions/techniques needed.

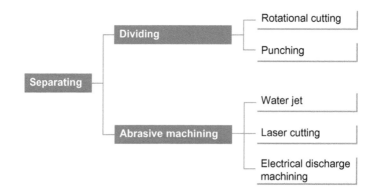

FIGURE 6.28
Manufacturing process separating according to DIN 8580 applied to electrical engineering.

the steel degrade, the core will need more magnetizing field to develop the same induction level, which means more current will be required. Maintaining good magnetic properties is important for the long-term efficiency of the motor.

Figure 6.28 shows some common types of cutting techniques for rotor and stator laminations. Each method introduces different levels of stress and will affect the permeability and losses in the final core differently. Punching is an efficient high-volume cutting technique. A press uses a punch and die to cut the entire lamination at once, allowing for high-speed operation as well as the ability to cut multiple pieces simultaneously, such as the rotor and stator together. However, punching can introduce burrs to the edges of the cut piece, which can form electrical paths between laminations after assembly. Burr formation can be controlled by keeping dies sharp, either through the use of long-lasting die materials like carbide or through frequent sharpening. Tool design is another important aspect to consider, ensuring that the tool was designed to work with the mechanical properties of the steel being used. Any burrs that are introduced during the punching process must be removed prior to assembly. Burrs can be removed through grinding or chemical means. Mechanical cutting techniques like punching and rotational cutting cause plastic deformation along the edge of the piece, distorting the grain structure of the metal.

Laser cutting is a low-production volume method of lamination cutting. However, it requires no dies or punches and can therefore be more flexible than punching when prototyping or for cutting large pieces that might not fit in a press. Laser cutting doesn't cause plastic deformation of the steel like mechanical cutting does; however, the heat generated by the laser can affect the magnetic properties. Low stress cutting methods, such as water jet cutting and electrical discharge machining, allow for precise and flexible cutting, with very little degradation of the steel's magnetic properties. These cutting techniques usually give the best results; however, these techniques are slow. For instance, water jet can cut at a speed of 800 mm/min for 0.5 mm thick laminations. Figure 6.29 illustrates the relative lot speed to flexibility of production using several cutting techniques. After the laminations are cut, the stressed region extends in from the cut edge. Heat treatment can be used to relieve this effect, if required. Laminations are not always annealed after cutting, generally to save costs.

When comparing the cutting techniques, water jet cutting and electrical discharge machining provide the best magnetic performance. However, when considering volume manufacturing, punching is the best overall for high lot size, low losses, and high permeability. Despite the lack of gross plastic deformation in the laser sample, the magnetic

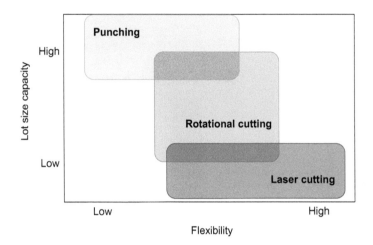

FIGURE 6.29

Methods for lamination production. (From Beckley, P., Electrical steels for rotating machines, *IEEE Power and Energy Series*, Institute of Electrical Engineers, London, UK, Vol. 37, 2002.)

properties could still suffer from the induced thermal stresses at the edges. The losses incurred by these stresses in laser cutting can be controlled by speed, power, laser frequency, and gas pressure [12–16].

Lamination cutting introduces the greatest stresses during core building. However, when the core is assembled, pressure is added to hold the laminations tightly together to avoid vibration during motor operation. The compressive strain caused when pressing the lamination stack together also affects the magnetic performance of the motor, which affects the hysteresis loss. Damage to the insulation can cause additional eddy current losses.

As with the core assembly, inserting the shaft and adding the housing also present challenges to the condition of the magnetic performance of the steel. For interior rotor designs, the rotor can be heated—allowing the shaft to be inserted—and then cooled. However, this imposes thermal stresses on the steel in the rotor core, as well as compressive forces. A similar procedure can be performed for the stator frame, with similar challenges.

6.3 Insulation Materials

There are two primary roles of insulation: to provide safe and efficient operation of the machine and to not impede the performance of the machine by its presence. Insulation is very important to the safe operation of an electric machine. Insulation systems must contend with the environmental and operational requirements of the machine and retain a safety margin to avoid failure. However, insulation materials also take up space and add weight and cost, and can thermally insulate temperature sensitive components. Insulation should be considered from a system perspective to avoid waste, while protecting against failure, and increasing the lifetime of the motor. Figure 6.30 shows insulation systems for round and rectangular wires. For SRM, coils exist only on the stator and, therefore, the majority of insulation material exists in the stator slots.

Insulation materials can be found around wires in coils, slot wedges, and slot liners. Sometimes, slot liners are also called groundwall or main wall insulation. Wire insulation

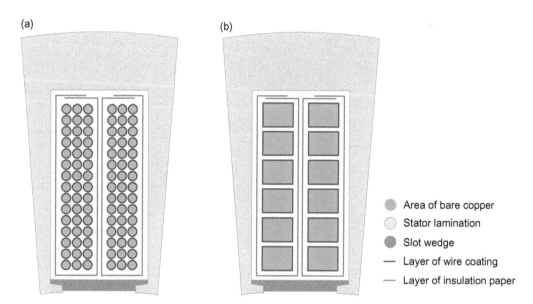

FIGURE 6.30
A closer look of the slot and coils: (a) round wire and (b) rectangular wire.

prevents short circuits between wires in a coil and can be exposed to high transient voltages. Slot wedges help restrict the movement of the coils in the stator slot. Polyetheretherketone (PEEK), an organic thermoplastic is a common material used for slot wedges. Slot liners provide a barrier between potential sharp edges on the lamination stack and the soft wire coating. Without the liner, vibration could cause the two parts to rub together, damaging the wire coating and causing a short circuit. In addition, the slot liner transfers heat from the coil to the stator core and should therefore be thermally conductive. Air gaps should be avoided for thermal reasons and to avoid partial discharges (PD). A common slot liner material is NOMEX 410 paper.

6.3.1 Polymer Materials

Many insulation components involve polymers, a class of materials that tend to have high electrical resistance. Figure 6.31 summarizes the molecular characteristics that separate different types of polymers. These molecular characteristics manifest as a wide range of physical properties. Table 6.8 presents a list of insulation materials and their electrical resistivity. All are polymers or contain crucial polymer components. Most polymers, though not all, are insulating at room temperature. Insulating polymers are the ones in which the filled valence band in their electron energy band structure is separated from the empty conduction band by a rather large band gap. In order for the insulation material to conduct, electrons must be excited by more than 2eV. However, as the temperature rises, this band gap will shrink, and electrical conductivity will increase.

A polymer is a giant molecule, sometimes called a macromolecule, comprised of monomers (*mer* meaning part, *mono* meaning one, and *poly* meaning many) connected by covalent bonds. Monomers simply represent the repeating groups of atoms that form the building blocks of polymers, and polymerization is the chemical reaction (or reactions) necessary to join them. Covalent bonds are the chemical bonds formed between non-metals where each atom contributes a number of electrons to be shared;

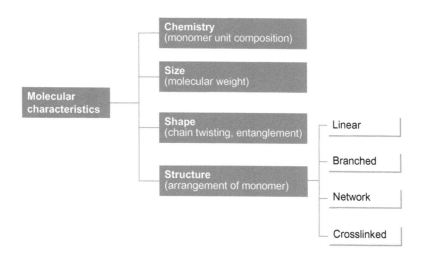

FIGURE 6.31

Categorization of polymers by molecular characteristics.

TABLE 6.8

Insulation Materials Used in Electric Machines

Material	Thermal Conductivity [W/m/K]	Electric Resistivity [ohm × cm]	Manufacturer
Elan-tron® MC4260	0.60–0.70	8×10^{14}	Elantas
Kapton® FN	0.12	$1.4–2.3 \times 10^{17}$	DuPont
Nomex® Paper	0.12–0.15	$8 \times 10^{11}–8 \times 10^{16}$	DuPont
Victrex® PEEK 450G	0.25	4.9×10^{16}	Victrex
LORD Circalok™ 6006	1.1	1×10^{16}	Lord
ECCTreme™ ECA 3000	0.18	$>10^{18}$	DuPont
Teflon® PTFE	0.22	$>10^{18}$	DuPont

carbon dioxide (CO_2) is an example. While carbon dioxide is three atoms in size, polymers can be long enough that it bridges the length scales between the atomic world and macroscopic objects. The behavior of polymers largely depends on this "mesoscale" size, which is why for many polymer characteristics, only the topology of the chain is described instead of its detailed chemical composition.

Figure 6.32 is a representation of what a single polymer chain could look like, where r is the chain end-to-end length. Since polymers are long chains, there are certain properties that are self-evident: they are flexible to varying degrees, other chains cannot pass through them, and their length is determined by the number of "links" (monomers) in the chain. The inset in Figure 6.32 illustrates the repeating connected monomers. Assuming the monomers are all the same, they must be able to bond to at least two other monomers to form a chain, for instance have two or more bonding sites. The number of bonds a monomer can form is known as its functionality. Polymers can bend at each chain link, allowing limited movement, and arrangement in two and three dimensions. The range of movement is restricted by the type of bond between monomers (double versus single covalent bond), among other properties.

Polymers used for insulation typically have high-molecular weight. Polymer chains can have a variety of structures that describe how monomers connect together.

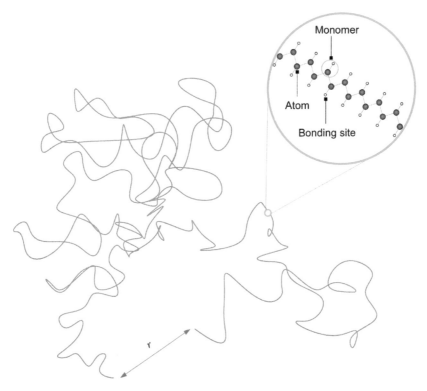

FIGURE 6.32
Schematic representation of polymer chain. (From Callister, W.D., *Materials Science and Engineering: An Introduction,* John Wiley & Sons, New York, 1997.)

Figure 6.33 shows examples of different polymer structures. So far, Figure 6.32 has shown linear chains. Linear chains in Figure 6.33a can tangle around each other and strong intermolecular forces can influence their behavior. Branched polymers in Figure 6.33b are similar to linear chains, but side-chains occasionally connect to the main branch. Crosslinked polymers in Figure 6.33c have linear chains and branches chemically bonded to each other, allowing large segments of the polymer to be free, while tethered at certain points. Networked polymers in Figure 6.33d form from monomers with high functionality, allowing them to bond to multiple units. However, the line between networked and crosslinked polymers is somewhat flexible, since highly crosslinked polymers can be considered networks.

Two types of polymers are found frequently in insulating materials: thermoplastics and thermosets. Each is a family of polymers distinguished by their structures and consequently their mechanical properties under elevated temperatures. Thermosets are highly crosslinked network structure polymers formed by heating. Thermosets are generally harder and stronger than thermoplastics. Some examples of thermosets include:

- Rubber (natural, butyl, silicone)
- Polyamide (Nylon)
- Polyester (Mylar)
- Polypropylene (PP)
- Polystyrene (PS)

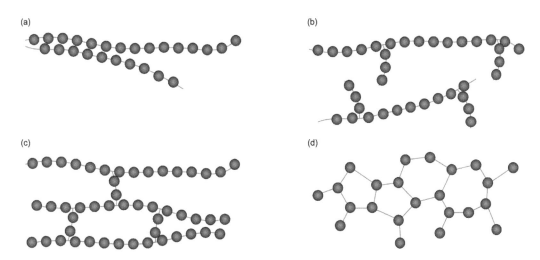

FIGURE 6.33
Polymer chain structures: (a) linear, (b) branched, (c) crosslinked, and (d) network. (From Callister, W.D., *Materials Science and Engineering: An Introduction,* John Wiley & Sons, New York, 1997.)

- Polyvinyl chloride (PVC)
- Polymethylmetacrylate (PMMA)
- Polycarbonate (PC)
- Polytetrafluoroethylene (PTFE)
- Epoxy, phenolic, silicon, and polyester resins

Thermoplastics are typically linear chains, with some branching. As temperature increases they will soften until liquid, and as temperature decreases they will harden. Some thermoplastics examples include:

- Polyethylene (PE, LDPE, MDPE, HDPE, XLPE)
- Ethylene-Propylene (EPR)
- Polyimide
- Polyetheretherketone (PEEK)

Due to atomic mechanisms, polymer chains are in constant tiny motion, moving or buckling based on the laws of thermodynamics and statistical mechanics. As the temperature increases, the motion of the polymer chain increases, causing previously tangled chains to come loose. Thermoplastics have temporary links between chains through tangling while thermoset chains are chemically bonded to each other. As temperature rises the thermoplastic chains will come loose, causing the material to soften. However, when cooled they can regain their previously tangled state. Under temperature increase, the thermoset chains will experience greater tension, but remain bonded until breaking. These properties of thermoplastics and thermosets affect the lifetime of the motor and they should be considered when selecting the insulating material for the given application.

6.3.2 Dielectric Breakdown and Partial Discharge

In an electric machine, dielectric materials can be found in wire insulation and slot liners. A dielectric material is an insulating material that can be polarized with a displacement current. Polymer materials tend to be dielectric materials, although there are notable exceptions. Thermoplastics and thermosets are types of polymers that act as organic dielectric materials.

Positive and negative charges can build up between insulation components, and given a strong enough electric field, the opposing surfaces will act like the plates of a capacitor. When the dielectric material temporarily loses its property of non-conductivity due to a high electric field, this is known as a dielectric breakdown. Table 6.9 shows a list of dielectric materials used as insulators.

A partial discharge (PD) can occur when a high electric stress (measured in kV/mm) causes the electrical breakdown of air, causing a spark as shown in Figure 6.34. The dielectric constant of air is much smaller than that of the insulating materials. Any electric field close enough to the dielectric breakdown voltage of the insulators is already enough to ionize the air. The spark is referred to as a partial discharge because only a small area of

TABLE 6.9

Solid Dielectrics for Electrical Insulation

Inorganic	Ceramics (AlN, HBN)
	Glass, Quartz
	Cements and Minerals (Mica)
Organic	Thermoplastics
	Thermosets
Composites	Kevlar
	Carbon
	Fiberglass

Source: Stone, G.C. et al., *Electrical Insulation for Rotating Machines: Design, Evaluation, Aging, testing, and Repair,* IEEE Press, Wiley, Hoboken, NJ, 2014.

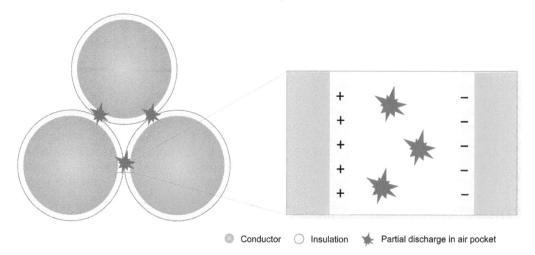

● Conductor   ○ Insulation   ✹ Partial discharge in air pocket

FIGURE 6.34

Partial discharge in an air pocket. (From Stone, G.C. et al., Electrical *Insulation for Rotating Machines: Design, Evaluation, Aging, Testing, and Repair,* IEEE Press, Wiley, 2014.)

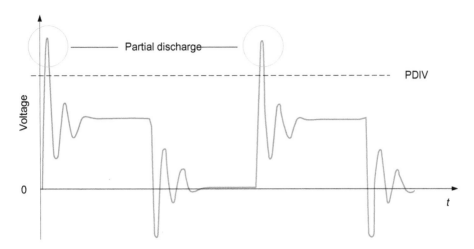

FIGURE 6.35
A voltage surge passing the partial discharge inception voltage. (From Stone, G.C. et al., Electrical *Insulation for Rotating Machines: Design, Evaluation, Aging, Testing, and Repair,* IEEE Press, Wiley, 2014.)

the insulation has been affected. Over time, the PD will damage the insulation material, causing aging. The discharges will break chemical bonds in the insulation material, and eventually cause a ground fault to occur.

Partial discharges can occur due to voltage surges from the inverter, or due to environmental effects. Moisture in the environment or any dirt and debris that might make its way into the motor could cause damage to a number of systems. For windings, the compounding of these effects could cause wear in the thin coating, allow cracks to come in contact with each other, allow water to infiltrate them, or allow a partial discharge. PD can be mitigated by minimizing environmental effects and air trapped between insulation layers, as well as by selecting insulation designed for high voltage environments.

Figure 6.35 illustrates conditions for partial discharge to occur during an inverter surge. The voltage has reached a value above the partial discharge inception voltage (PDIV), which is the voltage above which partial discharge may occur, increasing the likelihood of partial discharge event. This likelihood is impacted by environmental conditions. White powder appearing on coils could be a sign of insulation deterioration due to partial discharge. Filler materials can be added to the polymers in film coating to add extra resistance against partial discharges. These fillers are generally inorganic nanoparticles that increase the PDIV—the voltage below which partial discharge does not occur. Partial discharges can also occur in covered wire, with a similar cause and effect. Mica paper is often used to disrupt the formation of PD.

6.4 Magnet Wire

The wire used in electrical machines is generally called magnet wire. Magnet wire consists of a conductor and an insulator. The National Electrical Manufacturers Association (NEMA) Magnet Wire (MW) 1000 standard governs the material requirements and testing

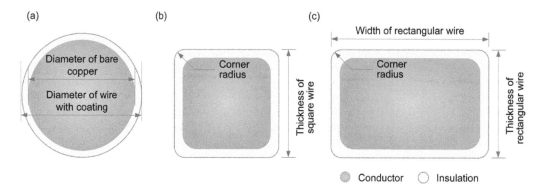

FIGURE 6.36
Schematic of Wire Shapes: (a) round, (b) square, and (c) rectangular.

details for all types of magnet wire. As shown in Figure 6.36, for the purposes of the NEMA standard, wire conductors have three main shapes:

- round, with a continuous bare or insulated diameter,
- square, with equal nominal width and thickness
- rectangular, with thickness being the smaller of the two dimensions and width being the larger.

Conductors are generally aluminum or copper. Square and rectangular shaped wires have rounded edges. The arcs of the radii of the rounded edges must merge smoothly into the flat surface of the wire. This is to avoid any sharp edges that could pierce the insulation. Square and rectangular wire types can increase the slot fill factor compared to round wires, since there is less space between wires. Wire sizes are given in the American Wire Gauge (AWG) sizes, where larger gauge numbers indicate smaller diameters of wire.

Under NEMA, insulation is "a dielectric medium applied to a conductor." Wire can be film (also called enamel) coated, which is generally a polymer coating, or a covered insulator like glass or paper (also called covering). According to NEMA, a film coating is a "continuous barrier of polymeric insulation." while a covering is "a fibrous or tape insulating material that is wound, wrapped, or braided around bare or film coated conductor." A covering could be a paper or glass insulator in the form of fibrous wrapping or tape. The tape is wrapped around the wire with a degree of overlap between winds. In both the film and covering cases, there is a focus on the integrity of the insulation, so that the conductor is not exposed along the wire length. A crack in the film coating is defined as exposing the bare conductor at 6x–10x magnification for the transformer oil resistance and hydrolytic stability test.

NEMA defines the term called "build," which denotes the increase in diameter, width, or thickness of magnet wire due to film insulation. Covered insulation does not have builds; instead a percentage value is given for what percentage of tape is overlapped on subsequent winds around the conductor. Builds scale in terms of how much increase there is over a "single build," and are named: heavy build for two times single build, triple build for three times, and quadruple build for four times. These builds are typically multiple layers of poly- mer coating. For instance, MW 36 rating details a heavy build film insulation that has two layers: a polyester basecoat and a polyamideimide topcoat, which allow for the combined properties of both polymer types. Table 6.10 gives examples for the thickness by build for standard AWG wire types and the resistance per unit length experienced at a given current.

TABLE 6.10

Specifications of Round Copper Wires, AWG 15 ~ 30

		Diameter [mm]					Resistance Per unit Length [Ω/km or mΩ/m]
			With Wire Insulation, d_w				
AWG	Current [A]	Bare Copper, d_c	Single	Heavy	Triple	Quadruple	
15	15.5	1.45034	1.49098	1.52908	1.57226	1.59512	10.66
16	12.3	1.29032	1.33096	1.36906	1.40716	1.43002	13.45
17	9.8	1.15062	1.19126	1.22428	1.26238	1.28524	16.95
18	7.75	1.02362	1.06172	1.09474	1.1303	1.15824	21.37
19	6.14	0.91186	0.94742	0.98044	1.01346	1.0414	27
20	4.89	0.8128	0.84836	0.87884	0.90932	0.93472	33.86
21	3.87	0.7239	0.75692	0.78486	0.81534	0.84328	42.79
22	3.04	0.64262	0.67564	0.70104	0.73152	0.75692	54.44
23	2.44	0.57404	0.60706	0.63246	0.65786	0.6858	67.81
24	1.93	0.51054	0.54102	0.56642	0.59182	0.61722	85.92
25	1.53	0.45466	0.4826	0.50546	0.53086	0.5588	108.6
26	1.2	0.40386	0.4318	0.45212	0.47752	0.50292	138
27	0.968	0.361	0.389	0.409	0.42926	0.45212	171.1
28	0.761	0.320	0.348	0.366	0.38608	0.4064	217.8
29	0.611	0.287	0.312	0.330	0.35052	0.37084	271.2
30	0.477	0.254	0.277	0.295	0.31496	0.3353	347.2

Adapted from: National Electrical Manufacturers Association (NEMA) Standard, ANSI/NEMA MW 1000–2015, Rosslyn, Virginia, USA, March 31, 2015.

Self-bonding wire coatings have adhesive coatings, when set, which allow turns to bond together and avoid slippage. Self-bonding wire also follows the build nomenclature; however, instead of being rated as single, heavy, triple, or quadruple build, these insulations are described by type. Since self-bonding is an outer coating, type 1 self-bonding describes a heavy build wire, type 2 describes a triple build wire, and type 3 describes a quadruple build wire.

Polymer materials, as a family of materials, are generally the most vulnerable to environmental effects like temperature and chemicals. As they begin to break down and age, the motor is at a greater risk of a number of failure mechanisms. As electrical insulators, polymer materials generally stand in the way between vital componentry and the outside world (i.e. dirt and moisture). A short circuit could occur if the insulation on the stator windings develops cracks.

Polymer materials behave fundamentally different from metals in many ways, generally having lower melting points and more plastic mechanical properties. However, polymer materials themselves are a diverse group, and can have widely varying properties. As a point of comparison, copper has a melting point of 1,085°C while its class-H insulator has a temperature rating (thermal class) of 200°C. Yet despite this vulnerability, polymers play a very important role in the lifetime of a machine.

6.4.1 Properties of Conductors

Aluminum and copper are common choices for magnet wire conductor, and each has its own benefits. The use of aluminum or copper is highly dependent on the application.

TABLE 6.11

Examples of Conductors in Magnet Wire

Material	Example Alloy	Price
Copper	ETP (electrolytic tough pitch -UNS C11040)	$2.91 USD/lb
Aluminum	Aluminum 1350	$0.92 USD/lb

Source: Investment Mine, Aluminum Prices and Aluminum Price Charts. [Online]. Available: http://www.infomine.com/. Accessed: August 23, 2017.

TABLE 6.12

Properties of Copper and Aluminum Wire

Material	Electric Resistivity @ 20°C [ohm × m]	Thermal Conductivity @ 20°C [W/m/K]	Density [kg/m³]	Specific Heat @ 20°C [kJ/kg/K]
Copper	1.724×10^{-8}	386	8890	0.385
Aluminum	2.826×10^{-8}	205	2700	0.833

Source: Bertotti, G., *Hysteresis in Magnetism: for Physicists, Material Scientists, and Engineers*, Academic Press, New York, 1998.

Aluminum has a lower mass density than copper, and is also cheaper. Table 6.11 gives the commodity prices of copper and aluminum as of August 2017. Copper dominates when attempting to get the maximum electrical conductivity by volume (aluminum has 61% the conductivity of copper on a volume basis). This factor could be important, for instance, when considering the conductor fill factor in a stator slot.

As shown in Table 6.12, aluminum is less dense than copper. The decrease in electrical conductivity of aluminum is not proportional to the density. Therefore, in weight restricted applications, using aluminum wires can be a feasible solution even though the volume of the aluminum wire could be higher than copper wires for the same current density.

6.4.2 Thermal Class of Magnet Wire

The operation of a motor produces losses, which are heat sources that raise the temperature of the motor. Resistive losses in the conductors of a coil can be significant, depending on the operation and performance requirements of the motor. Table 6.13 gives examples of some common magnet wire types and their associated thermal class. A thermal class for film-coated wires is determined based on the thermal endurance and heat shock capability of a motor. Thermal endurance (NEMA WM 1000 3.58.1) is measured on round, heavy build magnet wire, from which rectangular and square wire are based. AWG 18 wire is preferred to be used during this test. Thermal classes are reported as a numeric value; however, a historic letter designation still exists. Table 6.13 also gives examples of materials used for each thermal class. Polyamides and polyimides tend to be used in high temperature applications, while epoxies have good chemical resistance. Nanometer-sized particles can be combined with these base resins to create the highest thermal rating films.

Figure 6.37 shows thermal performance to price increase over a range of AWG sizes for comparison purposes. It should be noted that the prices of magnet wires are subject to fluctuations in commodity prices. Thermal class does not strictly relate to price. While the highest thermal class wire may be more expensive this is not always the case. As seen in Figure 6.37, the difference in the price of Class A and Class H wires is negligible. Besides thermal class, wire selection should take into account other properties, such as windability and dielectric

TABLE 6.13

Examples of Thermal Classes and Insulating Materials of Round Copper Magnet Wires according to ANSI/NEMA MW1000-2015

Magnet Wire Thermal Class	Letter Designation	Standard	Insulating Material	
			Underlying Coating	**Superimposed Coating**
105	A	MW15-C	Polyvinyl acetal-phenolic	
130	B	MW28-C	Polyurethane	Polyamide
155	F	MW79-C	Polyurethane	
		MW80-C	Polyurethane	Polyamide
		MW41-C	Glass fiber covered	
180	H	MW76-C	Polyester (amide) (imide)	Polyamide
		MW77-C	Polyester (imide)	
		MW78-C	Polyester (imide)	Polyamide
		MW82-C	Polyurethane	
		MW83-C	Polyurethane	Polyamide
		MW50-C	Glass fiber covered	
200	K	MW74-C	Polyamide (amide) (imide)	
		MW35-C	Polyester (amide) (imide)	Polyamide-imide
		MW44-C	Glass fiber covered	
220	M	MW61-C	Aromatic polyamide	
		MW37-C	Polyester (amide) (imide)	Polyamide-imide
		MW81-C	Polyamide-imide	
240	C	MW16-C	Aromatic polyimide	

Source: National Electrical Manufacturers Association (NEMA) Standard, ANSI/NEMA MW 1000-2015, Rosslyn, Virginia, USA, March 31, 2015.

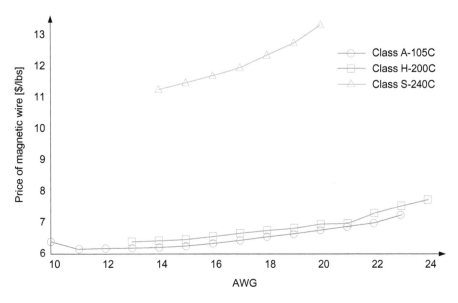

FIGURE 6.37
Price difference between heavy build magnet wire types with different thermal classes.

breakdown strength. While thermal class is an important measure of the performance of a wire over its lifetime, in a high temperature environment thermoplastic flow should also be considered. Thermoplastic flow, also known as "Cut-Through Temperature," is the temperature at which the coating softens and, under a given pressure, will deform to allow electrical contact between conductors. The cut-through temperature is well above a wire's rated thermal class temperature. Thermal class is determined by its stability over a long time period; however thermoplastic flow is tested over a much shorter time period.

Figure 6.38 reports the temperature at which the average life of the tested wire crosses 20,000 hours for aluminum and copper conductors. In order for the thermal class to be determined, the extrapolated life for the wire at the rated temperature, must meet or exceed 20,000 hours. The thermal class must be at least 20°C lower than the heat shock capability temperature. Heat shock is used to evaluate the resistance of the wire to cracking under rapid temperature changes after being physically stressed. This is a pass/fail measurement determined by whether cracks are visible after the test. In a motor, the wire is stressed during winding and forming, and then heated over the course of its lifetime. For covered wire, such as glass or paper covered wire, the Institute of Electrical and Electronics Engineers (IEEE) has not established test procedures for determining the thermal class and, as a result, it is established through experience.

A magnet wire coating can have a different class for aluminum or copper, and is representative of the wire alone—not the total machine. Aluminum tends to have a higher thermal index and heat shock values than copper wire. At temperatures above 200°C, copper will oxidize at an increased rate, becoming brittle and pitted. Protecting the copper with silver or nickel can extend life above 200°C, if compatible with the covering or film coating.

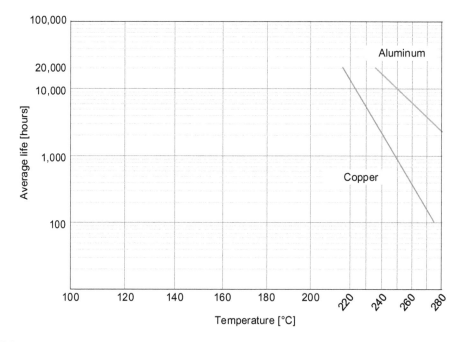

FIGURE 6.38
18 AWG heavy build thermal aging. (From Superior Essex, Essex GP/MR-200 Magnet Wire /Winding Wire. [Online]. Available: http://www.superioressex.com/. Accessed: August 23, 2017.)

6.4.3 Windability

The highest mechanical stresses tend to occur while winding the coils. During fast, automated winding, wire will experience the greatest deformation from bending, friction, and strong tension. Damage to insulation films from winding decreases the reliability of the coil. However, by increasing tension, the slot fill factor increases. Adding lubrication and using abrasion resistant film coatings can aid in winding. While coefficient of friction is not a NEMA standard test, it can be used to determine how successfully the wires will slide past each other while winding. With too much friction, the coil may become uneven; for instance, bunching on one side. Without enough friction, the coil may slide off each other instead of forming a compact helix. The term "Scrape Resistance" is used to give an indication of coating resistance to physical damage. A load is reported in grams that was required (under specific conditions) to remove the coating from the wire due to scraping in one direction.

Adhesion and flexibility is used to determine how much mechanical stress the magnet wire can undergo. The magnet wire coating is elongated and bent around a mandrel to determine failure. Figure 6.39 shows a wire being wrapped around a mandrel whose diameter is k times the diameter of uninsulated copper, d_c. Therefore, $1d_c$ refers to a mandrel of same diameter as the bare conductor, and $3d_c$ refers to a mandrel of three times the diameter of the bare conductor being tested. Failure is determined either by cracking or loss of adhesion to the wire. Failure for this test can indicate the quality of the cure for the insulation, or the surface conditions of the conductor. For motor design, this property, as well as heat shock, is important for evaluating end turn lengths. Tables 6.14 and 6.15 show the bendability of several types of wires given different mandrel sizes for a standard test. Polyester resin gives the best adhesion between film coating and conductor, followed by polyesterimide resin and polyamideimide. Cracks in wire coatings that don't extend down to the conductor (or substrate) can eventually compromise the performance during operation. A small crack can propagate to the conductor or create other cracks. Tables 6.14 and 6.15 illustrate whether the crack is small, medium, or down to the substrate so that the likelihood of failure can be evaluated. For instance, a Class H fiberglass covered wire (H-DGC) is likely to create a medium crack if used on a mandrel less than $8d_c$ and is likely

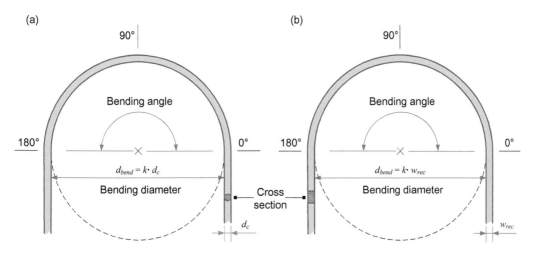

FIGURE 6.39
Illustration of bend diameter for copper wires: (a) round and (b) rectangular. (From National Electrical Manufacturers Association (NEMA) Standard, ANSI/NEMA MW 1000-2015, Rosslyn, Virginia, USA, March 31, 2015.)

TABLE 6.14

Characteristics of Round Covered Wire

Wire Type	Coating Thickness [mm]	Breakdown Voltage [V]	Breakdown Voltage after Heating and 6d Bend [V]	Bendability					
				$2d_c$	$4d_c$	$6d_c$	$8d_c$	$10d_c$	$12d_c$
Class F glass fiber covered wire (F-DGC)	0.150	880 (590)	720 (480)	▲	▲	▲	▲	▲	●
Class H glass fiber covered wire (H-DGC)	0.145	920 (630)	760 (520)	▲	▲	▲	●	●	○
Polyamide imide/Polyester glass fiber wound wire (1AIW-SGTC)	0.107	4,600 (4,300)	4,500 (4,210)	●	○	○	○	○	○
Polyester (PEW)	0.068	5,900 (8,680)	5,700 (8,380)	○	○	○	○	○	○
Remarks		Metal foil method (Electrical stress values in V/mm)	59 Metal Foil Method F-DGC: 180°C for 6 hours H-DGC: 210°C for 6 hours Other: 150°C for 6 hours	○ Good ● Small Crack ▲ Medium Crack × Crack from which substrate is visible Bending Angle = 180°					

Source: Hitachi, Magnet Wire Selection and Use Directions for Magnet Wire. [Online]. Available: http://www.hitachi-metals.co.jp/. Accessed: August 23, 2017.

TABLE 6.15

Characteristics of Rectangular Covered Wire

Wire Type	Coating Thickness (mm)	Breakdown Voltage (V)			Bendability			
		Ordinary State	150°C	200°C	$2w_{rec}$	$4w_{rec}$	$6w_{rec}$	$8w_{rec}$
Class F glass fiber covered wire (F-DGC)	0.191 × 0.134	940	950	940	×	▲	●	○
Class H glass fiber covered wire (H-DGC)	0.196 × 0.143	830	840	830	×	▲	●	○
Aromatic polyamide paper covered wire	0.138 × 0.13	2,300	2,200	2,200	×	●	●	○
Mica tape covered wire	0.154 × 0.156	6,400	6,300	6,000	●	○	○	○
Aromatic polyimide tape-covered wire	0.105 × 0.088	9,340	9,800	9,600	○	○	○	○
Polyester (PEW)	0.064 × 0.065	5,200	5,400	4,750	○	○	○	○
Remarks		Metal Foil Method After heating for 24 hours						

○ Good
● Small Crack
▲ Medium Crack
× Crack from which substrate is visible
w_{rec} is conductor width
Bending Angle = 180°

Source: Hitachi, Magnet Wire Selection and Use Directions for Magnet Wire. [Online]. Available: http://www.hitachi-metals.co.jp/. Accessed: August 23, 2017.

to create a small crack on a mandrel less than $12d_c$. Therefore, this wire has restrictions on how much it can bend without damage in a coil. If the end-turns require the wire to bend at a diameter more than 12 times the diameter of the bare conductor then the wire is in danger of forming cracks at its insulation. Over the lifetime of the motor, even small cracks can grow large.

6.5 Shaft Materials

The shaft is made from high strength steel to avoid failure from high stresses experienced during rotation. Shaft material selection depends on the desired properties in the required motor application such as mechanical and thermal properties, rigidity, hardness, wear resistance, machinability, noise absorption, manufacturing process, and cost. Material selection also depends on the motor size, application, and environment. A mechanical analysis should be performed to determine what stress loading the shaft will be under.

The material can be heat-treated to increase material hardness and yield, and ultimate tensile strength using processes such as spheroidizing, annealing, normalizing, carburizing, and quenching. The most common material for electric motor shafts is low to medium carbon steel, and can be hot- or cold-rolled. A comparison of the yield strength, ultimate strength, and other material properties for some common shaft materials can be observed in Table 6.16.

Questions and Problems

1. Suppose you are selecting the lamination steel for the switched reluctance motor shown in Figure Q.6.1. You are interested in thin gauge electrical steel for the rotor and stator, and have a thin gauge lamination and a thicker gauge lamination to investigate.

    a. Consider the gauges NO30-1600 (Table Q.6.1) and M470-50A (Table Q.6.2). Using the information given in Section 6.2.6, "Grades and Losses," and Table 6.5, what are the thicknesses of these electrical steels? Plot and compare the B-H curves. What is the difference between the magnetization characteristics of NO30-1600 and M470-50A?

    b. Tables Q.6.3 and Q.6.4 show the loss density characteristics of NO30-1600 and M470-50A. Plot the loss density on the y-axis and the magnetic flux density on the x-axis at 50 Hz for these two gauges. What relationship do you expect to see?

    c. Using the data in Tables Q.6.3 and Q.6.4, please plot the loss density on the y-axis and the magnetic field density on the x-axis, for NO30-1600 and M470-50A at 50, 400, 2000, and 5000 Hz. Include each frequency in the same plot. What do you expect to observe for each curve as the frequency increases? If you wanted to fit a polynomial to these curves, could a single function fit all the curves with good accuracy?

TABLE 6.16

Properties of Some Shaft Materials

Material	Yield Tensional Strength (MPa)	Ultimate Tensional Strength (MPa)	Shear Strength (MPa)	Fatigue Strength (MPa)	Elongation (%)	Hardness Brinell (HB)	Machinability Rating (%)	Vibration Damping
AISI 1008 cold rolled	285	340	196	170	20	95	55	OK
AISI 1020 cold rolled	350	420	242	193[a]	15	121	65	OK
AISI 1045 cold rolled, annealed	505	585	338	268[a]	12	170	65	OK
304 SS	241	586	334	241[b]	55	149	45	A little better than steel
Ductile iron 65-45-12	310	448	336	179	12	131–220	160	Good
Gray cast iron 40	–	293	393	128[b]	<1	235	70	Best
A7075-T6	503	572	331	159[c]	11	150	120	Poor

Source: Tong, W., *Mechanical Design of Electric Machines*, CRC Press, Boca Raton, FL, 2014.

[a] based on 107 cycles.

[b] based on 108 cycles.

[c] based on 5 × 108 cycles.

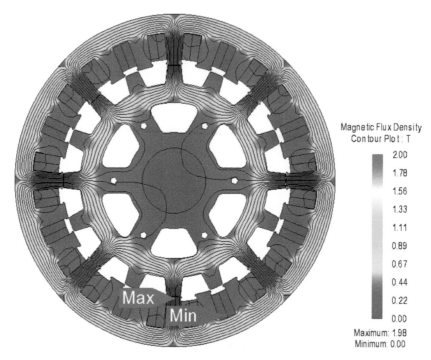

FIGURE Q.6.1
Magnetic flux density contour plot of a 24/16 SRM.

TABLE Q.6.1

Magnetization Characteristics of NO30-1600 at 50 Hz

H (A/m)	B (T)	H (A/m)	B (T)
10	0.0260	400	1.3520
20	0.0550	450	1.3630
30	0.1120	500	1.3720
40	0.2180	750	1.4100
50	0.4900	1000	1.4380
60	0.6510	1250	1.4530
70	0.7870	1500	1.4700
80	0.8730	2000	1.4920
90	0.9520	2500	1.5230
100	1.0190	5000	1.6240
125	1.1000	7500	1.6900
150	1.1770	10,000	1.7520
175	1.2310	25,000	1.8820
200	1.2580	50,000	1.9220
250	1.3020	75,000	1.9400
300	1.3180	100,000	1.9480
350	1.3370		

TABLE Q.6.2

Magnetization Characteristics of M470-50A at 50 Hz

H (A/m)	B (T)	H (A/m)	B (T)
47.9697	0.1	140.4117	1.1
59.6417	0.2	165.6715	1.2
67.5242	0.3	210.6460	1.3
74.8249	0.4	314.8213	1.4
81.8074	0.5	678.4804	1.5
88.9717	0.6	2011.7	1.6
95.7895	0.7	4875.9	1.7
104.1299	0.8	9079	1.8
113.2770	0.9	14,926	1.9
124.3332	1	23,962	2.0

TABLE Q.6.3

Loss Characteristics of NO30-1600

B (T)	50 Hz (W/m³)	400 Hz (W/m³)	2000 Hz (W/m³)	5000 Hz (W/m³)
0.01	0.760	15.20	235.60	1216
0.02	3.80	59.280	912	4560
0.03	9.120	129.20	1900	9880
0.04	16.720	243.20	3344	15,428
0.05	26.60	380	5168	25,384
0.06	39.520	554.80	7448	34,580
0.07	54.720	760	10,032	48,836
0.08	72.960	988	13,072	59,660
0.09	91.20	1292	16,416	77,520
0.10	114	1596	19,988	92,720
0.20	456	6232	74,024	316,920
0.30	912	13,224	156,560	656,640
0.40	1444	22,268	263,720	1,117,200
0.50	2128	32,984	396,720	1,717,600
0.60	2812	45,524	559360	
0.70	3648	59,812	753,920	
0.80	4484	76,000	988,000	
0.90	5472	94,240	1,261,600	
1.00	6612	114,760	1,588,400	
1.10	7904	138,320	1,968,400	
1.20	9500	164,920	2,424,400	
1.30	11,476	198,360	2,964,000	
1.40	13,984	242,440	3,655,600	
1.50	16,644	298,680	4,590,400	
1.60	19,076			
1.70	21,280			
1.80	23,180			

TABLE Q.6.4

Loss Characteristics of M470-50A

B (T)	50 Hz (W/m³)	400 Hz (W/m³)	2000 Hz (W/m³)	5000 Hz (W/m³)
0.10	268.2653	4504.2	54,652	212,230
0.20	945.4069	15,747	173,610	706,160
0.30	1903.1	32,122	362,660	
0.40	3079.5	53,234	637,030	
0.50	4447.4	79,761	1,031,005	
0.60	5995.4	112,170	1,584,442	
0.70	7700.4	152,350	2,333,147	
0.80	9629.9	200,870	3,306,570	
0.90	11,765	259,350	4,533,003	
1.00	14,125	329,580	6,040,893	
1.10	16,732	413,300		
1.20	19,629	511,579		
1.30	22,922	626,810		
1.40	26,779	760,960		
1.50	31,474	898,710		
1.60	36,255	1,034,712		
1.70	39,844	1,161,904		
1.80	42,613			
1.90	45,035			
2.00	47,986			

TABLE Q.6.5

Magnetization Characteristics of Stainless Steel

H (A/m)	B (T)	H (A/m)	B (T)
0	0	1172.8	0.9516
10	0.0033	1589.2	1.0850
79.4230	0.0280	1846.7	1.1400
153.9893	0.0680	2234.5	1.1952
218.8420	0.1161	2672.9	1.2373
269.1245	0.1600	3343.7	1.2830
301.6181	0.1919	4879.5	1.3594
366.3327	0.2720	6576.9	1.4178
453.2889	0.3880	8355.7	1.4602
604.0939	0.5614	11,409	1.5250
751.9661	0.7089	13,921	1.5658
959.0397	0.8495	19,525	1.6137

TABLE Q.6.6

Magnetization Characteristics of Cast Iron

H (A/m)	B (T)	H (A/m)	B (T)
0	0	1038.6	1.0250
2.9817	0.0038	1408.0	1.100
6.4981	0.0083	1784.5	1.1490
11.0837	0.0145	1995.3	1.1715
17.2731	0.0230	2625.8	1.2220
25.4541	0.0348	3105.9	1.2500
35.4269	0.0499	5440.0	1.3250
46.8447	0.0683	6884.8	1.3617
59.3609	0.0898	9127.7	1.4033
72.6132	0.1139	10,539	1.4263
86.1775	0.1403	15,323	1.4900
99.6141	0.1679	21,719	1.5500
112.4832	0.1958	27,452	1.5875
135.7459	0.2499	34,628	1.6250
157.3738	0.3000	43,626	1.6624
198.5561	0.3917	58,218	1.7106
221.5804	0.4374	73,137	1.7441
301.1807	0.5624	95,813	1.7900
419.0198	0.6950	101,560	1.8008
563.7012	0.8125	109,710	1.8144
619.1451	0.85	112,870	1.8190
767.8853	0.9285	116,400	1.8238
954.6943	1.00	123,030	1.8323

d. Tables Q.6.5 and Q.6.6 shows the typical magnetization characteristics for stainless steel and cast iron. Please plot them together with the magnetization characteristics of NO30-1600 and M470-50A. What do you observe between electrical steel and other iron alloys?

e. If you were asked to design an SRM requiring high torque density at low speed, which material(s) would you use? How does your selection change if you were asked to design an SRM that would mostly run at high speed?

2. Please define the terms listed below as they relate to magnet wire. Please describe how you would use them to choose between magnet wire types for your motor.

- Thermal class
- Adhesion and flexibility
- Breakdown voltage

References

1. Jiang W., Three-Phase 24/16 Switched reluctance machine for hybrid electric powertrains: Design and optimization, *McMaster University Ph. D. Thesis*, Hamilton, ON, 2015.
2. World Steel Association, World Steel in Figures 2016. [Online]. Available: http://www.worldsteel.org/. Accessed: August 23, 2017.
3. Irons, G. Steel industry, *Berkshire Encyclopedia of Sustainability: Vol. 2 The Business of Sustainability*, Berkshire Publishing Group, Great Barrington, MA, 2009.
4. Beckley, P., Electrical steels for rotating machines, *IEEE Power and Energy Series*, Vol. 37, Institute of Electrical Engineers, London, UK, 2002.
5. Peacey, J. G., Davenport, W. G., *The Iron Blast Furnace*, Pergamon Press, New York, 1979.
6. Callister, W. D., *Materials Science and Engineering: An Introduction*. John Wiley & Sons, New York, 1997.
7. Bertotti, G., *Hysteresis in Magnetism: For Physicists, Material Scientists, and Engineers*. Academic Press, New York, 1998.
8. Motion Control and Motor Association (MCMA), Electric Motor Education and Research Foundation (EMERF), EMERF Lamination Steels Third Edition, Arbor, Michigan, USA, 2009.
9. Buswas, A. K., *Principles of Blast Furnace Ironmaking: Theory and Practice*, Cootha Publishing House, Brisbane, Australia, 1981.
10. Cogent Power, Altogether more powerful—Non-oriented electrical steel. [Online]. Available: http://cogent-power.com/. Accessed: August 23, 2017.
11. The American Section of the International Association or Testing Materials (ASTM), Standard Classification of Insulating Coatings for Electrical Steels by Composition, Relative Insulating Ability Application (ASTM A976). [Online]. Available: http://www.astm.org/. Accessed: August 23, 2017.
12. Emura, M., Landgraf, F. J. G., Ross, W., Baretta, J. R., The influence of cutting technique on the magnetic properties of electrical steels, *Journal of Magnetism and Magnetic Materials*, 255, 358–360, 2003.
13. Siebert, R., Schneider, J., Beyer, E., Laser cutting and mechanical cutting of electrical steels and its effect on the magnetic properties, *IEEE Transactions on Magnetics*, 50(4), 1–4, 2014.
14. Belhadj, A., Baudouin, P., Houbaert, Y., Simulation of the HAZ and magnetic properties of laser cut non-oriented electrical steels, *Journal of Magnetism and Magnetic Materials*, 248, 34–44, 2002.
15. Evine, N. E., Petrovi, D. S., Non-oriented electrical steel sheets, *Materials and Technology*, 44(6), 317–325.
16. Belhadj, A., Baudouin, P., Breaban, F., Deffontaine, A., Dewulfd, M., Houbaertb, Y., Effect of laser cutting on microstructure and on magnetic properties of grain non-oriented electrical steels, *Journal of Magnetism and Magnetic Materials*, 256, 20–31, 2003.
17. Stone, G. C., Culbert, I., Boulter, E. A., Hussein, D., *Electrical Insulation for Rotating Machines: Design, Evaluation, Aging, Testing, and Repair*. IEEE Press, Wiley, Hoboken, NJ, 2014.
18. National Electrical Manufacturers Association (NEMA) Standard, ANSI/NEMA MW 1000-2015, Rosslyn, Virginia, USA, March 31, 2015.
19. Investment Mine, Aluminum Prices and Aluminum Price Charts. [Online]. Available: http://www.infomine.com/. Accessed: August 23, 2017.
20. Superior Essex, Essex GP/MR-200 Magnet Wire/Winding Wire. [Online]. Available: http://www.superioressex.com/. Accessed: August 23, 2017.
21. Hitachi, Magnet Wire Selection and Use Directions for Magnet Wire. [Online]. Available: http://www.hitachi-metals.co.jp/. Accessed: August 23, 2017.
22. Tong, W., Mechanical Design of *Electric Machines*. CRC Press, Boca Raton, FL, 2014. Accessed: August 23, 2017.

7

Design Considerations for Switched Reluctance Machines

James Weisheng Jiang

CONTENTS

There are various geometrical and operational parameters that need to be taken into account during the design process of an SRM. In this chapter, we will start with motor performance metrics, such as output torque, torque ripple, and torque-speed profiles. Then, we will discuss the different motor operations, for instance, constant torque, constant power, duty cycles, and drive cycles. Winding parameters and SRM dimensions, especially stator and rotor pole arc angles, are also discussed in detail in this chapter.

7.1 Motor Performance Parameters

7.1.1 Major Specifications of an SRM

The power of electric motors can be rated in watts (W) or horsepower (hp). The relationship between the units can be expressed as

$$\text{Power rating [hp]} = \frac{\text{Power rating [kW]}}{0.746}. \tag{7.1}$$

For instance, the traction motor used by the 2010 Prius has a rated power of 60 kW at its peak load, equal to about 80 hp. When comparing and assessing electric motors, power-to-weight ratio (kW/kg) is used, which is also called the specific power (SP). The 2010 Prius traction motor has a mass of 36.7 kg with casing and cooling system included giving a specific power of 1.6 kW/kg.

The definition of volumetric power density (PD) of a motor is similar to SP. It is the ratio of the rated power of the motor to its volume. The definition of motor volume can differ. If only the active material, within a motor is considered, the volume of the motor is the volume that contains the motor's rotor, stator, and winding, which can be calculated as:

$$V_{active} = \frac{\pi D_s^2}{4} L_{total} \tag{7.2}$$

where D_s is the outer diameter of the stator core and L_{total} is the total axial length of the active volume. Similarly, two other related parameters are nominal-torque-to-active-weight ratio and nominal-power-to-active-weight ratio. If the active weight is replaced with the total weight of a motor, two new parameters are created: nominal-torque-to-total-weight ratio and nominal-power-to-total-weight ratio. In this case, the volume of the entire motor is used with casing and cooling system included. For instance, the volume of the entire 2010 Prius traction motor with casing and cooling system is 12.5 L. Thus, the power density (PD) would be 4.8 kW/L.

7.1.2 Torque and Torque Quality

The torque in each phase of an SRM is dependent on the relative position between the stator and rotor poles, and the level of excitation current. During phase commutation, phase torques, as shown in Figure 7.1a, are added together, and the overall profile ends up with a pulsated waveform as shown in Figure 7.1b.

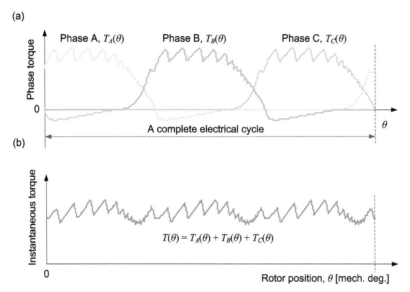

FIGURE 7.1
Waveforms of torque: (a) phase torque and (b) instantaneous torque.

The instantaneous torque waveform $T(\theta)$ can be decomposed as:

$$T(\theta) = T_{ave} + T_{ripple}(\theta) \tag{7.3}$$

where $T(\theta)$ is the instantaneous torque at each rotor position θ, T_{ave} is the average torque, and $T_{ripple}(\theta)$ is the periodic component of the instantaneous torque waveform. The average torque can be calculated using the following equation:

$$T_{ave} = \frac{1}{\theta_2 - \theta_1} \int_{\theta_1}^{\theta_2} T(\theta)d\theta \tag{7.4}$$

where $(\theta_1 - \theta_2)$ is equal to a complete electrical cycle θ_{cycle}. For an SRM, θ_{cycle} is defined when a rotor pole moves from the unaligned position to the next similar position. Please note that T_{ave} is not always equal to the average of T_{max} and T_{min}. Figure 7.2 shows the process of decomposing an instantaneous torque waveform. $T(\theta)$ can be a function of the rotor angular position or time. The torque quality can be measured in a number of ways, for instance:

$$Ripple_{Normalized} = \frac{T_{max} - T_{min}}{T_{ave}} \tag{7.5}$$

$$Ripple_{Percentage} = \frac{T_{max} - T_{min}}{T_{ave}} \times 100\% \tag{7.6}$$

$$T_{Peak-to-peak} = T_{max} - T_{min} \tag{7.7}$$

$$\Delta T_{RMS} = \sqrt{\frac{1}{\theta_2 - \theta_1} \int_{\theta_1}^{\theta_2} \left(T(\theta) - T_{ave}\right)^2 d\theta} = \sqrt{\frac{1}{\theta_2 - \theta_1} \int_{\theta_1}^{\theta_2} \left(T_{ripple}(\theta)\right)^2 d\theta} \tag{7.8}$$

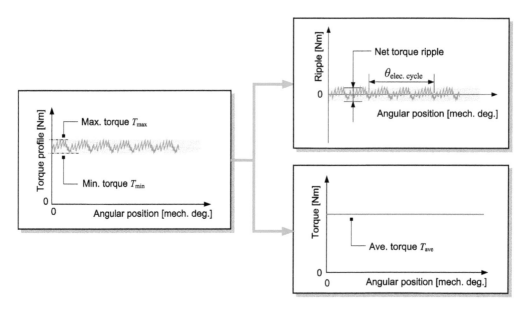

FIGURE 7.2
Decomposition of instantaneous torque waveform into average torque and torque ripple.

$Ripple_{Normalized}$, as in (7.6), is the ratio between the net torque ripple and the average torque. This value is unit-less and can be easily converted to $Ripple_{Percentage}$ by multiplying it by 100%. These two values are commonly used to measure the torque ripple. The merit of $T_{Peak-to-peak}$ in (7.7) is that it gives the magnitude of the torque pulsation in Nm. One common drawback of these three measurements is that only the maximum and minimum values of the torque ripple are used to evaluate the torque ripple over a certain period. For instance, if a sinusoidal and square waveform have the same maximum and minimum values, it would not be possible to tell the differences between these waveforms by these three metrics.

ΔT_{RMS} can measure the deviation of all the data points on a periodic torque waveform from its average value. Furthermore, ΔT_{RMS} is in Nm and will always have a positive value. This makes ΔT_{RMS} a better objective for the optimization of firing angles, which will be discussed in Chapter 9.

Rotational speed of the rotor can be expressed in revolutions per minute (rpm) n_m, revolutions per second (rev/s, Hz) f_m (or f_{mech}), or radians per second (rad/s) ω_m. Revolutions per minute (rpm) is frequently used to express the rotational speed of a motor. The relationship between angular speed (also called angular frequency), ω_m, and n_m is:

$$\omega_m = \frac{\pi}{30} n_m \tag{7.9}$$

The relationship between angular speed ω_m, and revolutions per second f_m is:

$$\omega_m = 2\pi f_m \tag{7.10}$$

7.1.3 Torque-Speed Profile

Maximum average torque values at different speed points define the shape of an electric motor's torque-speed profile. The rated torque of an electric motor is the peak average torque in the constant torque region. The torque-speed profile is limited by the maximum allowable current (the root mean square (RMS) value of phase current constraint), and the DC link voltage. A typical torque-speed profile for an SRM has three regions: constant-torque region, constant-power region, and falling power region, as shown in Figure 7.3.

In the constant-torque region, the peak current can be reached under the given voltage constraints and the current waveform can be shaped by chopping, as shown in Figure 7.4. At the base speed, the peak torque can be achieved with the maximum current and rated voltage. Even though the first region is called the constant-torque region, for an SRM, the torque-speed envelope in the constant-torque region is not always flat (discussed in Chapter 9). In the constant-power region, the peak torque is inversely proportional to the speed.

As shown in Figure 7.4, at the constant power region, even though the commanded current is the same as for the base speed, the current can never reach this value due to the limited DC link voltage. Hence, the energy-conversion loop is much smaller compared to the base speed, which results in a lower output torque. As the motor speed increases further away from the constant-power region, the output torque drops rapidly. In the falling power region, the output torque is inversely proportional to the square of the speed. Figure 7.4 shows that the energy-conversion area for very high-speed operation shrinks even further.

In the case of acceleration, depending on the system requirements, the motor only operates at the peak-torque condition for a limited time due to thermal constraints. In this case, the torque-speed profile for transient operation is used, as shown in Figure 7.5. For a longer-term operation, the maximum temperature rise constraint has to be considered and the torque-speed profile of continuous operation will be used.

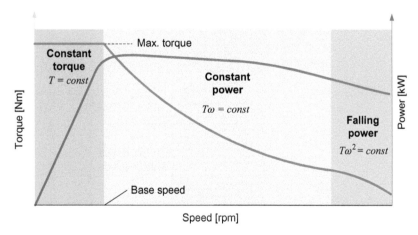

FIGURE 7.3
Typical torque-speed and power-speed profiles for an electric motor.

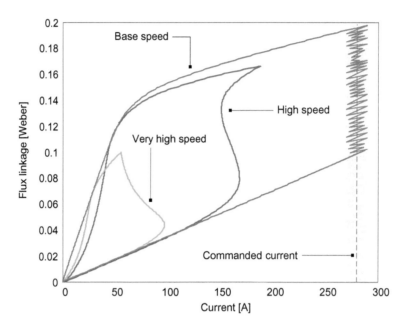

FIGURE 7.4
Energy-conversion loops for base speed, high speed, and very high speed. (From Jiang, W.J., Three-phase 24/16 switched reluctance machine for hybrid electric powertrains: Design and optimization, *Ph.D. Dissertation, Department of Mechanical Engineering*, McMaster University, Hamilton, ON, 2016.)

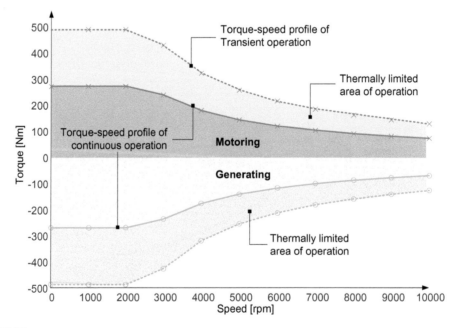

FIGURE 7.5
Comparison of continuous and transient operations for electric machines. (From Emadi, A. (Ed.), *Advanced Electric Drive Vehicles*, CRC Press, Boca Raton, FL, 2014.)

7.2 Motor Operations

7.2.1 Basic Types of Operations

Depending on the application, an electric motor can operate under different load conditions. Some basic types of motor operations are shown in Figure 7.6: (a) constant torque with increasing speed, (b) constant power, (c) torque increases linearly with speed, and (d) torque increases quadratically with speed. More complicated motor duty cycles can be created by combining these operations.

For the constant torque operation, the torque is more or less constant with speed. The power in this type of operation increases linearly as the motor speed increases. The applications of this operation can be conveyors, hoists, and crushers. For the constant power operation, the torque is inversely proportional to the motor speed. For instance, if the motor is employed in a winder, as the material builds up, the diameter of the winder increases, which requires a gradual decrease of the motor speed. In displacement pumps, torque increases linearly with speed as shown in Figure 7.6c. Centrifugal pumps have the load type shown in Figure 7.6d. Some operating types for different applications are summarized in Table 7.1.

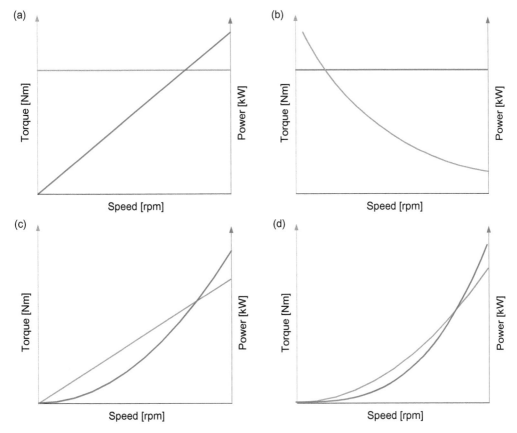

FIGURE 7.6
Four basic motor operations: (a) constant torque, (b), constant power, (c) linear increase of torque with speed, and (d) increase of torque with square of speed.

TABLE 7.1

Operation Types for Different Applications

Application	Required Torque or Power
Conveyor, screw compressor, hoist	Constant torque
Helical displacement pump	Torque increases linearly with speed
Fans, blowers, centrifugal pump	Torque increases with speed squared
Winder, unwinder	Constant power

7.2.2 Duty Cycles

The IEC (International Electrotechnical Commission) defines eight common duty cycles for electric motors. Figure 7.7 shows the first two duty cycles. For instance, in the case of S1, as shown in Figure 7.7a, the motor can operate at a rated load for an unlimited period of time to reach its temperature limit, T_{limit}. During this process, the motor reaches its thermal equilibrium. For S2, as shown in Figure 7.7b, the on-load period is too short for the motor to reach a thermal equilibrium. The subsequent off-load is long enough for the motor to cool down to the ambient temperature.

7.2.3 Priority Operating Regions

As stated earlier, an electric machine has its specific application, which means that there will be some priority operating regions where the motor operates more frequently than the rest of the points in its torque-speed map. Figure 7.8 shows an example of the torque-speed profile used for a washing machine. The motor only has two cycles: washing and spinning. In the washing cycle, the output torque required is 20 Nm at 47 rpm. The motor is required to deliver 4 Nm at 1350 rpm in the spinning cycle [4]. Furthermore, the motor spends nearly 80% of its time in the washing cycle. When an SRM is designed for the washing machine, these two operating points will be priority design targets.

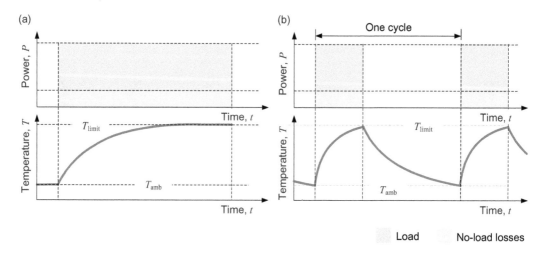

FIGURE 7.7

Two common types of motor duty cycles defined by IEC: (a) S1 and (b) S2. (From Agrawal, K.C., *Industrial Power Engineering Handbook*, Newnes, Boston, MA, 2001.)

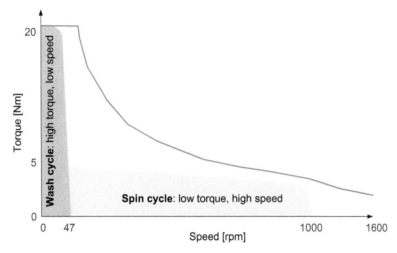

FIGURE 7.8
Two major operating regions for an electric motor used for a washing machine. (From Castano, S. M., Yang, R., Mak, C., Bilgin, B., and Emadi, A. External-rotor switched reluctance motor for direct-drive home appliances, in *Proceedings of the Annual Conference of the IEEE Industrial Electronics Society (IECON)*, Washington, DC, October 2018.)

A more sophisticated case can be seen in Figure 7.9. It shows the priority operating points of a traction motor for a hybrid electric powertrain. The motor must meet all these requirements. In order to have an even better understanding of the usage of the motor, the operating frequencies for a traction motor under different driving cycles will be needed. A driving cycle is used for this purpose which is a series of data points for the speed of a vehicle versus time, which is used to assess the performance of the vehicle.

Figure 7.10 shows two typical driving cycles: UDDS and US06. The Urban Dynamometer Driving Schedule (UDDS) is used for light duty vehicle testing. US06 is a high acceleration, aggressive driving schedule. Figure 7.11 shows the usage of a traction motor for a hybrid electric powertrain under UDDS and US06, respectively. The motor's efficiency should be

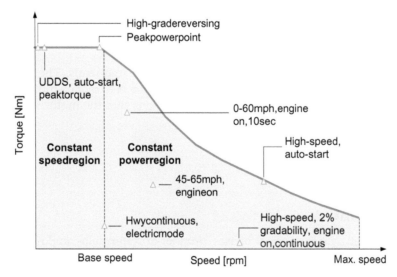

FIGURE 7.9
Typical torque-speed characteristics and priority operating points of a traction motor in a hybrid electric powertrain. (From Bilgin, B. et al., *IEEE Trans. Transport. Electrification*, 1, 4–17, 2015.)

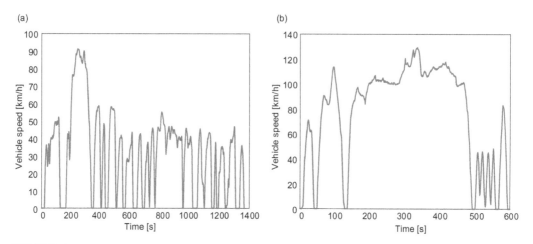

FIGURE 7.10
Two typical drive cycles: (a) UDDS and (b) US06. (From Emadi, A. (Ed.), *Advanced Electric Drive Vehicles*, CRC Press, Boca Raton, FL, 2014.)

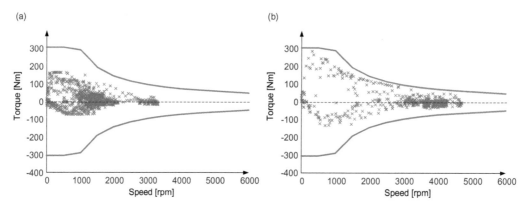

FIGURE 7.11
Usage of the traction motor under a certain drive cycle used in a hybrid electric vehicle powertrain: (a) UDDS and (b) US06.

enhanced in the regions where the motor operates more frequently. The driving cycles represent a close approximation to estimate the load profile of a traction motor. More information might still be needed such as ambient conditions that the motor will experience during operation.

7.3 Motor Configurations

7.3.1 Number of Phases

The relationship between number of phases, number of stator poles, and number of rotor poles has been discussed in Chapter 3. The common industry practice is to use three phases. At least three phases are needed to maintain smooth torque both in clockwise and counterclockwise rotation. Figure 7.12 shows four conventional SRMs with 2, 3, 4, and 5 phases, respectively.

FIGURE 7.12
SRM geometries with different numbers of phases: (a) two-phase 4/2 SRM, (b) three-phase 6/4 SRM, (c) four-phase 8/6 SRM, and (d) five-phase 10/8 SRM.

As the number of phases increases, the RMS value of current of each phase reduces. In high-power applications, this is an advantage for the sizing of the motor. Higher number of phases also means higher number of poles. For the given stator outer diameter, the space available for the coils reduces as the number of stator poles increases. Therefore, a higher number of phases might not be practical for low- and medium-power motors. A higher number of phases can provide a considerable improvement in torque quality. In addition, the DC link current ripple will be lower in higher phase machines, which helps reduce the DC link capacitance requirement in the converter.

Motors with a higher number of phases also require converters to drive them. In low- and medium-power motors, this can increase the cost of the motor drive system. In high-power motors, depending on the application, motor drive systems with a higher number of phases might be a viable option considering the torque quality, and the size of the motor and the converter.

7.3.2 Number of Poles

Chapter 3 quantifies how to define the number of rotor poles as a function of the number of stator poles and the number of phases. Figure 7.13 shows six configurations of SRMs all with three phases. For a three-phase 24/16 SRM, each phase has eight stator poles.

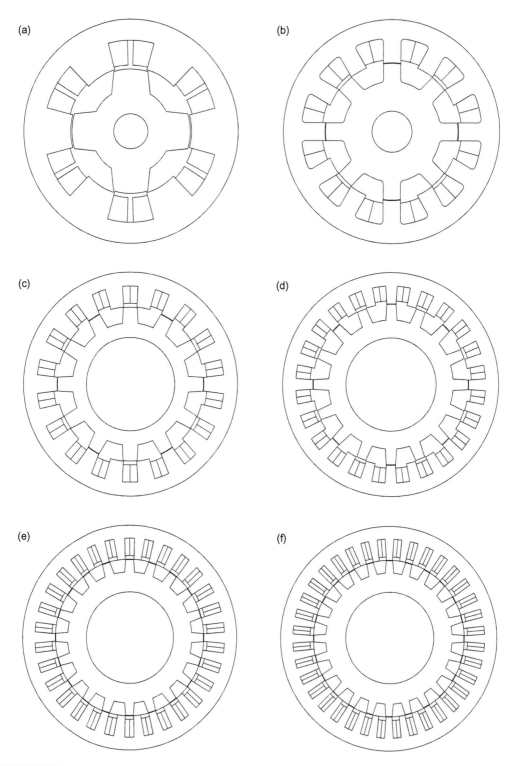

FIGURE 7.13
Three-phase conventional SRM examples: (a) 6/4, (b) 12/8, (c) 18/12, (d) 24/16, (e) 30/20, and (f) 36/24.

When the rotational speed of the motor, ω_r is fixed, increasing the number of stator and rotor poles will cause the excitation frequency to increase.

An increase in the number of stator and rotor poles does not necessarily lead to an improvement in the torque quality. For the given stator outer diameter constraint, as the number of poles increase, the width of the stator teeth and the slot area reduces. This situation can be observed in Figure 7.13e and f in three-phase 30/20 and 36/24 SRMs. A smaller slot area makes it difficult to include enough coil strands to reduce the copper loss and achieve the required current density. Besides, as the number of slots increases, the coil connections become more complicated. In addition, as the stator pole width reduces, the stator core might oversaturate at the peak current which might lead to a reduction in torque production capability. As discussed in Chapter 2, torque production capability and power factor of an SRM improves when it operates in the saturation region of the electrical steel. In SRM design, it is important to balance the appropriate amount of saturation with other design tradeoffs and constraints by properly defining the dimensions and control parameters.

7.3.3 Fundamentals of SRM Dimensions

Figure 7.14 shows the design parameters for a conventional SRM illustrated on a 6/4 configuration. N_S and N_R are the number of stator poles and the number of rotor poles, respectively. The symbols and corresponding definitions are summarized in Table 7.2.

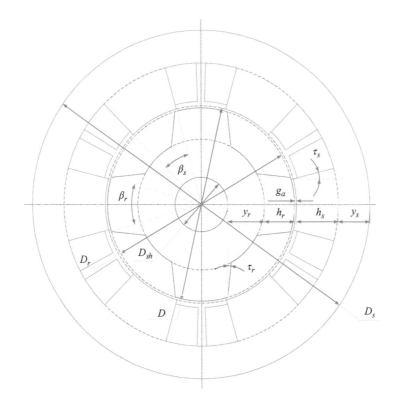

FIGURE 7.14
Design parameters for an SRM illustrated using a sample 6/4 SRM. (From Jiang, W.J., Three-phase 24/16 switched reluctance machine for hybrid electric powertrains: Design and optimization, *Ph.D. Dissertation, Department of Mechanical Engineering*, McMaster University, Hamilton, ON, 2016.)

TABLE 7.2

Major Dimensions of 6/4 SRM

Symbol	Definition
D	Bore diameter [mm]
D_s	Stator outer diameter [mm]
D_r	Rotor outer diameter [mm]
D_{sh}	Rotor shaft diameter [mm]
h_r	Rotor pole height [mm]
h_s	Stator pole height [mm]
N_{Ph}	Number of phases
N_s	Number of stator poles
N_r	Number of rotor poles
β_r	Rotor pole arc angle [mech. deg. or radian]
β_s	Stator pole arc angle [mech. deg. or radian]
y_s	Stator back iron thickness [mm]
y_r	Rotor back iron thickness [mm]
τ_r	Rotor taper angle [mech. deg. or radian]
τ_s	Stator taper angle [mech. deg. or radian]

Source: Jiang, W.J., Three-phase 24/16 switched reluctance machine for hybrid electric powertrains: Design and optimization, *Ph. D Dissertation, Department of Mechanical Engineering*, McMaster University, Hamilton, ON, 2016.

The effect of these parameters on the motor performance will be discussed throughout this chapter.

L_S and L_R are the stack lengths of the stator and the rotor of an SRM, respectively, as shown in Figure 7.15. As shown in Figure 7.16, the total length L of the stator is the summation of the lengths of end turns L_{end}, and L_S.

$$L_{total} = 2L_{end} + L_S \tag{7.11}$$

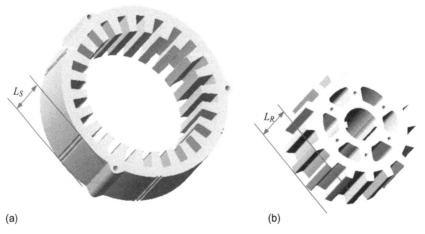

(a)   (b)

FIGURE 7.15
Stack lengths: (a) stator core and (b) rotor core.

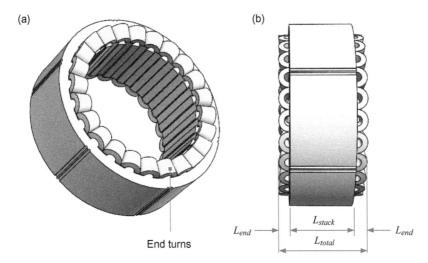

FIGURE 7.16
Winding end turns: (a) end turns and (b) total length of a stator-winding system.

Please note that the stack length of the stator, L_S does not necessarily have to be equal to the stack length of the rotor, L_R. There are many cases where the rotor stack length L_R is larger than the stator stack length L_S to take advantage of the stator winding end turns to generate higher torque.

The stator and rotor of an SRM are generally made of individually insulated electrical steel laminations to reduce the eddy current losses. The applied insulation coating as well as imperfect manufacturing and assembly processes bring in another key parameter; the stacking factor. The stacking factor for either the stator or the rotor of an SRM is the ratio of the total thickness of the lamination sheets to the axial length of the iron core, which can be calculated as:

$$S_f = \frac{N_{lam}t_{lam}}{L} \tag{7.12}$$

where N_{lam} is the number of lamination sheets for either the stator or the rotor, t_{lam} is the thickness for lamination sheets without the insulation coating, and L is either the stator stack length L_S or the rotor stack length L_R. The stacking factor is always less than 1 and can vary from 0.8 to 0.95. The typical stacking factor for a lamination thickness of 0.1 mm is 89.6%, while for 0.2 mm, it is 92.8% and for 0.5 mm, it is 95.8% [2]. As the lamination thickness, t_{lam}, increases, the stacking factor approaches 1.

7.3.4 Diameters and Air Gap Length

There are four major diameters in a switched reluctance machine: stator outer diameter D_s, bore diameter (also stator inner diameter) D, rotor outer diameter D_r, and shaft diameter (also rotor inner diameter) D_{sh}. D_s defines the radial boundary of the motor core, which is normally a geometric constraint. The stator inner diameter, D defines the outer boundary of the air gap when the rotor and the stator poles are aligned as shown in Figure 7.17.

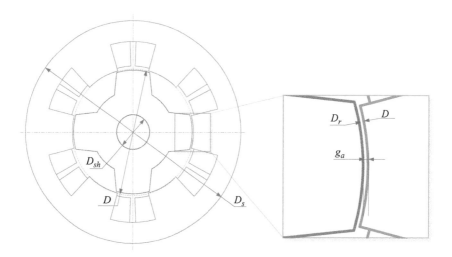

FIGURE 7.17
Diameters and air gap length for an SRM illustrated using a sample 6/4 SRM.

D_r defines the inner boundary of the air gap. Therefore, the air gap length g_a, can be calculated as:

$$g_a = \frac{1}{2}(D - D_r)$$ (7.13)

At the aligned position, the stator and rotor teeth have the minimum air gap as shown in Figure 7.18a. At this position, the reluctance is minimum, and inductance is maximum. At the unaligned position, the effective air gap between the stator and rotor teeth reaches its maximum, as shown in Figure 7.18b. At this position, the reluctance is maximum, and the inductance is minimum. If a smaller air gap length is used, the reluctance will be lower, and more flux can be generated with the same magneto-motive force, which can increase the output torque of the motor. But meanwhile, radial force also increases, which can affect a motor's noise and vibration behavior. Furthermore, air gap length is limited by mechanical constraints, such as concentricity, circularity, and so on.

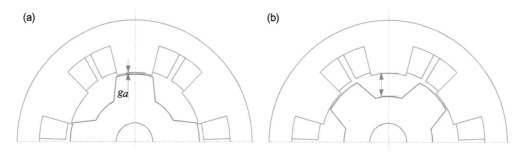

FIGURE 7.18
Gap between rotor and stator: (a) air gap, aligned position and (b) unaligned position.

7.4 Winding Schemes

Conventionally, an SRM has concentrated coils wound around each stator pole. But, there are different ways in which these coils can be connected together to create the phase windings. In this section, windings for SRMs will be discussed. A number of definitions related to windings will be introduced, such as conductors, strands, turns, coils, series path, parallel path, phase winding, and motor winding.

7.4.1 Slot and Coil

If an SRM is cut in the radial direction, you will see the cross section of the motor as shown in Figure 7.19. On the stator, a number of slots can be seen, and within each slot, wires fill up certain areas. Take the 12/8 SRM, for example. It has 12 slots and 12 stator poles. The wires are wound around each stator pole, and, thus, there are two groups of wires belonging to two different phases in each slot. Figure 7.20 shows a simplified version of slot area and wires.

Wires wound around a stator pole make up a coil, and the coil can have one turn, two turns, and more turns, as shown in Figure 7.21, depending on the design. Coil side, also called coil leg, is the part of the coil that is totally immersed in the stator slot, as shown in

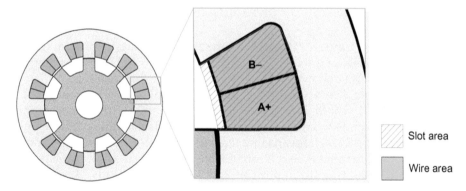

FIGURE 7.19
Illustration of slot area and wire area.

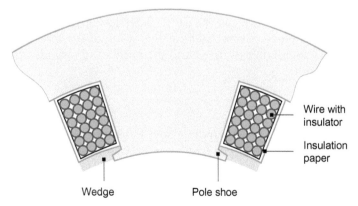

FIGURE 7.20
Calculation of slot fill factor.

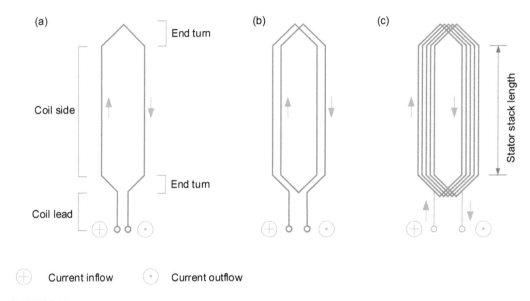

FIGURE 7.21
Illustration of number of turns per coil: (a) one-turn coil, (b) two-turn coil, and (c) multi-turn coil.

Figure 7.21a. The coil in Figure 7.21a only has one turn. The parts of the coil that can be seen at both ends of the stator are called end turns (also called coil end, overhang, or end winding). Figure 7.21b and c shows that a coil can have more than one turn.

Figure 7.22 shows four different configurations of coils. If only observed by the cross sections of the slots, it would not be possible to differentiate these configurations, because all of them have 24 wires in one side of the slot. The coil might have only one turn, as shown in Figure 7.22a. It means that the coil has 24 strands. These 24 strands are connected in a parallel manner. The coil might have two turns, as shown in Figure 7.22b. In this case, the coil has two turns per slot and 12 parallel strands. As compared to Figure 7.22a, the coils in Figure 7.22b can carry half of the current, because the effective cross section of the copper is made of 12 wires instead of 24. But the flux generated in the magnetic circuit links with two turns instead of one. Hence, assuming the magnetic characteristics of the core is the same for these two cases, the total magneto-motive force (number of turns times current) stays the same. Selection of the number of turns and number of stands is related to current and voltage rating of the machine. Later in this section, the effects of the number of turns and the number of strands on motor performances will be discussed.

7.4.2 Wires and Slot Fill Factor

Magnet wire is widely used in SRM windings and has two major components: the bare copper and the insulation material. The insulation around the copper conductor enables contact between the wires without causing electrical short circuit between turns. The details of magnet wire insulation have been discussed in Chapter 6.

In North America, the American National Standards Institute (ANSI) standardizes magnet wires [6]. Magnet wire has standardized diameters, which are categorized using American Wire Gauge (AWG). Different standards offer different AWG values, but for the

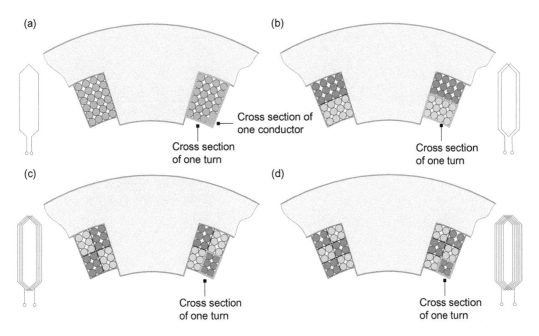

FIGURE 7.22
Examples of winding configurations with the same slot fill factor: (a) 1 turn per slot, 24 parallel strands per turn, (b) 2 turns per slot, 12 parallel strands per turn, (c) 4 turns per slot, 6 parallel strands per turn, and (d) 6 turns per slot, 4 parallel strands per turn.

same AWG number, the wire size should be the same. For instance, AWG 15 has a nominal diameter of 1.45 mm for its bare copper.

Only a certain portion of the slot area can contain the wires, which brings another important definition called slot fill factor. Figure 7.23a shows the slot area A_s. Two important dimensions of the slot are the height of the slot l_{slot}, and the width of slot opening w_{slot}.

There are two slot fill factors: bare copper slot fill factor and wire slot fill factor. The bare copper slot fill factor is the ratio of bare copper area over the slot area. As for one slot:

$$ff_{copper} = \frac{2N_{turn}N_{str}\pi\left(\dfrac{d_c}{2}\right)^2}{A_s} = \frac{N_{turn}N_{str}\pi d_c^2}{2A_s} \tag{7.14}$$

where d_c is the diameter for bare copper. For instance, ff_{copper} of 0.4 means that 40% of the slot area is occupied by bare copper. Please also note that each slot has two coils that belong to two different phases. This is why N_{turn} was multiplied by 2 in (7.14). The wire slot fill factor is the ratio of wire area over the slot area for one slot:

$$ff_{wire} = \frac{2N_{turn}N_{str}\pi\left(\dfrac{d_w}{2}\right)^2}{A_s} = \frac{N_{turn}N_{str}\pi d_w^2}{2A_s} \tag{7.15}$$

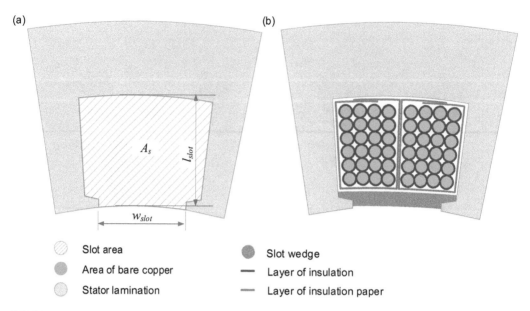

(a)   (b)

A_s l_{slot}

w_{slot}

⧄ Slot area   ● Slot wedge

● Area of bare copper   — Layer of insulation

○ Stator lamination   ⸺ Layer of insulation paper

FIGURE 7.23
A closer look of the slot and round wires: (a) slot area and (b) cross section of slot area.

where d_w is the diameter for wire with insulation. d_w is always greater than d_c for the same wire gauge. Thus for the same slot with the same number of conductors, ff_{copper} is always smaller than ff_{wire}. Wire insulation and insulation thickness based on thermal class have been discussed in Chapter 6.

A high slot fill factor means that more wires are inserted into the slot. However, it should be noted that the slot should have enough space available not only for the conductors, but also the insulation layers. In addition, depending on the winding technique, it is also important to ensure there is enough space available to insert the coils into the slots. In practice, when using round conductors, the slot fill factor is selected between 35% and 60%. For hand-wound coils, the slot fill factor can be between 35% and 40%. Lower fill factor enables hand windings, and higher fill factor might require special winding tooling or an automated winding process. Selection of the slot fill factor is an important parameter in machine design process. It has to be selected considering practical constraints such as manufacturability, cost, torque density, and thermal constraints. It should also be noted that higher slot fill factor can lead to larger end-turn length. If the SRM design has an axial length constraint, it might be necessary to target a smaller slot fill factor to reduce coil axial length.

A higher slot fill factor can be achieved with bar windings, which are applied in traction motors and in applications where phase-to-phase voltage is above 690 V. Rectangular or bar wires, as shown in Figure 7.24, can be formed in a certain shape, so that the turn-to-turn voltage does not exceed the dielectric breakdown of the insulation system. The form of the rectangular wires enable higher fill factor. In traction applications, bar wires are used to reduce the copper losses at low-speed operation.

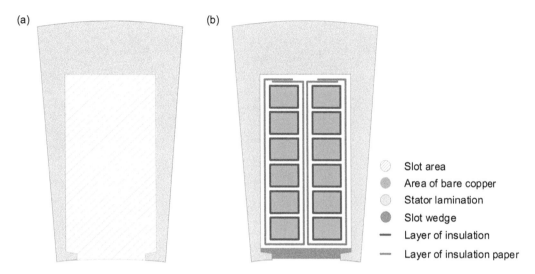

FIGURE 7.24
A closer look of the slot and bar wires: (a) slot area and (b) cross section of slot area.

7.4.3 Current and Current Density

If the current waveform of one phase of an SRM over a complete electrical cycle is represented by the function $i(t)$ over a complete electrical cycle $[T_1, T_2]$, then the RMS value of the waveform is calculated as:

$$I_{RMS} = \sqrt{\frac{1}{T_2 - T_1} \int_{T_1}^{T_2} [i(t)]^2 \, dt} \tag{7.16}$$

If the waveform can be discretized by n number of points $\{i_1, i_2, \ldots, i_n\}$ over an electrical cycle, the RMS value of the current sequence can be calculated as:

$$I_{RMS} = \sqrt{\frac{1}{n} \left(i_1^2 + i_2^2 + \cdots + i_n^2 \right)} \tag{7.17}$$

Figure 7.25 shows two current waveforms from constant-torque and constant-power regions. The RMS values of each waveform are shown in Figure 7.25 as well. For an SRM, there is a constraint for the RMS value of phase current and it helps determine the torque-speed profile, especially in the constant torque region. In an electric motor, the RMS value of the phase current is limited by thermal and power supply constraints. As shown in (7.18), current density is defined as the ratio of RMS value of the current waveform over

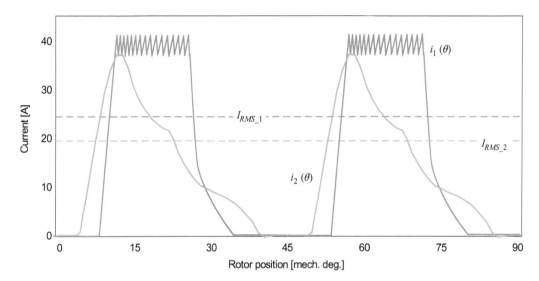

FIGURE 7.25
Current waveforms at different speed and corresponding RMS value of currents.

the bare copper area. It is an important parameter in electric machine design. It defines the torque production capability, but it is also limited by the thermal constraints.

$$J = \frac{I_{RMS\_coil}}{A_s ff_{copper}/(2N_{turn})} = \frac{2I_{RMS\_coil}N_{turn}}{A_s ff_{copper}} \ [\text{A/mm}^2, \text{ or A/inch}^2] \tag{7.18}$$

Here, I_{RMS_coil} is the RMS value of the current flowing through the coil and A_s is the slot area. The current density is limited by the slot area, the slot fill factor, the number of turns, and the number of strands. The constraint of RMS phase current and cooling method are correlated. I_{RMS} can also be used to estimate copper loss. If the RMS current of one coil is I_{RMS}, and the resistance of the coil is R_{coil}, the copper loss for that coil can be estimated as [7]:

$$P_{copper\_coil} = I_{RMS\_coil}^2 R_{coil} \tag{7.19}$$

7.4.4 Combinations of Turns and Strands

For the given stator and rotor geometries, the motor performance can be affected by changing the number of turns and the number of strands. For a given 24/16 SRM stator geometry, we will analyze the performance of the motor when the number of turns varies between 12 and 20, and the number of strands varies between 7 and 12. The slot fill factor is an important constraint that can be used to choose the number of turns and the number of strands. In this case, the slot fill factor is limited at 54%; hence, some of the combinations of number of turns and strands are eliminated. With the number of turns fixed, decreasing the number of strands leads to a reduction in the current density, which means the slot fill factor cannot be too low either. Therefore, the slot fill factor for this case is between 0.48 and 0.54.

Figure 7.26 compares the output torque of these combinations at 3000 rpm, the base speed of the motor. It is noteworthy that as the number of turns increases from 12 to 17, the peak output torque increases. However, after it peaks at 17 turns, the increase in the number of turns leads to a reduction of the peak torque due to high-induced voltage. This phenomenon will be discussed in detail in Chapter 9.

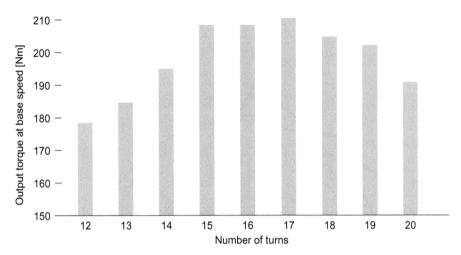

Comparison of peak torque at 3000 rpm for motors with different number of turns. (From Jiang, W.J., Three-phase 24/16 switched reluctance machine for hybrid electric powertrains: Design and optimization, *Ph.D. Dissertation, Department of Mechanical Engineering*, McMaster University, Hamilton, ON, 2016.)

7.4.5 Coil Connection

Each phase of an SRM has multiple coils. For instance, 12/8 SRM has three phases and each phase has four stator poles, thus four coils. As shown in Figure 7.27, on one side of the stator pole, current either flows in or out. Now, let's discuss how to connect the four coils of Phase A for the 12/8 SRM.

As it can be seen in Figure 7.28a, all four coils of Phase A can be connected in series. In this manner, the current in each coil is the same. Assuming balanced operation, the flux linkage and, hence, the induced voltage of each coil will be the same. Therefore, the power supply has to overcome a total voltage of 4u to apply the current i as shown in Figure 7.28b. This type of coil connection is denoted as 4S, where four means the number of coils for one phase and S means series connection.

Figure 7.29 shows one possible connection of coils with two parallel paths. In Figure 7.29a, the adjacent coils (coils A1-A2 and coils A3-A4) are connected in series. These two series

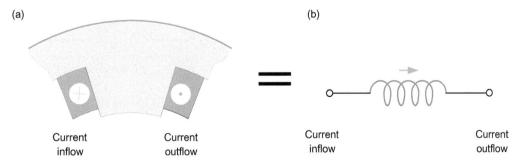

Phase coil: (a) stator coil and (b) its electrical schematic equivalence.

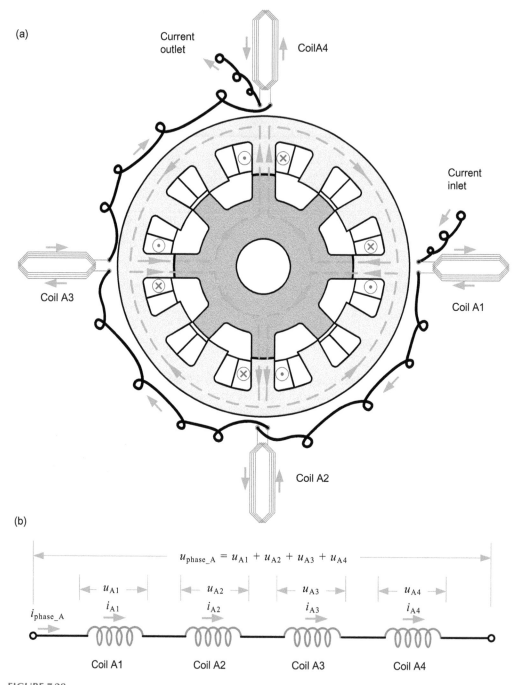

FIGURE 7.28
Series connection of four coils for phase A, 4S: (a) coil connection for phase A and (b) electrical schematic equivalence.

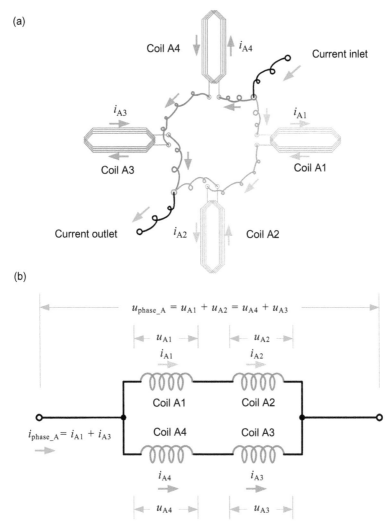

FIGURE 7.29
Series and parallel connection of four coils for phase A, 2S2P_A: (a) coil connection for phase A and (b) electrical schematic equivalence.

paths are then connected in parallel. This type of coil connection is called 2S2P_A. Figure 7.30a shows the 2S2P_B connection in which the coils in the diagonal positions are connected in series [8]. This type of connection will not help reduce unbalanced magnetic pull, which will be discussed in Chapter 8. Coils can also be connected in parallel first and then in series as shown in Figure 7.30b. In all of these configurations, the voltage rating of the supply reduces to 2u and the current rating increases to 2i as shown in Figure 7.29b. Please note that only the input and output terminals of the phase winding are connected to the converter. Therefore, in case of parallel paths, such as in Figure 7.29, the converter will regulate i_{phase_A}; however, i_{A1} and i_{A4} might be different if the coil flux linkages are different due to eccentricity.

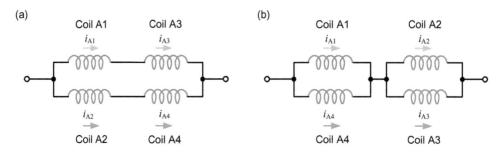

FIGURE 7.30
Other two possible coil connections with two parallel paths: (a) 2S2P_B and (b) 2P2S.

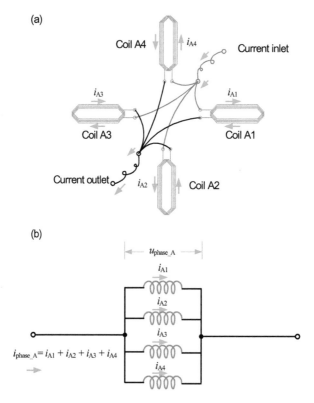

FIGURE 7.31
Parallel connection of all coils: (a) coil connection for phase A and (b) electrical schematic equivalence.

Finally, Figure 7.31 shows the case in which all the coils are connected in parallel, which is called 4P. As shown in Figure 7.31b, the phase voltage is u and the phase current is 4i for this connection.

7.4.6 The Whole Winding

If the coil connection for one phase is determined, then the winding scheme for all phases can be put together. As shown in Figure 7.32, three-phase 12/8 SRM has 12 coils and each phase has four coils.

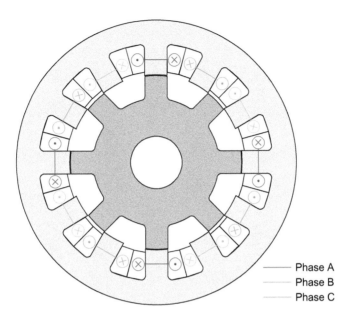

FIGURE 7.32
All phase coils for 12/8 SRM.

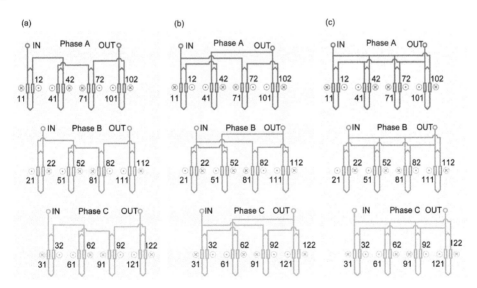

FIGURE 7.33
Whole winding scheme for 12/8 SRM: (a) 4S, (b) 2S2P, and (c) 4P.

Figure 7.33 shows the windings diagram of the 12/8 SRM for series (4S), series/parallel (2S2P), and parallel (4P) connections.

Figure 7.34 shows an example of the winding scheme and coil connections for a three-phase 24/16 SRM. The number of turns per coil, N_{turn}, for this motor is 15. The number of strands, N_{str}, is 10. The wire size used in this motor is AWG 20. In this example, for each phase, eight coils in different slots are connected in series. Table 7.3 summarizes the specs related to the winding scheme.

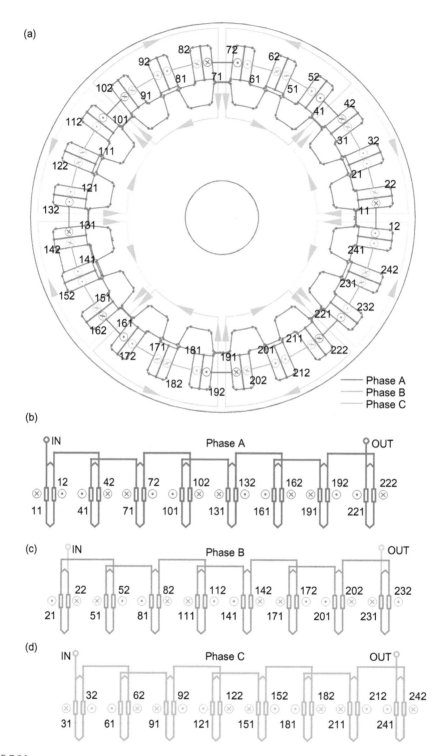

FIGURE 7.34
Winding scheme for all three phases for 24/16 SRM. Eight coils of each phase are connected in series. Each coil has 15 of turns with 10 parallel strands: (a) whole winding scheme, (b) Phase A, (c) Phase B, and (d) Phase C. (From Jiang, J.W. et al., *IEEE Trans. Trans. Electrification.*, doi: 10.1109/TTE.2017.2664778, 2017.)

TABLE 7.3
Summary of Wining Configuration for 24/16 SRM

Description	Symbol	Value
Number of phase	m	3
Number of stator slot	N_s	24
Number of turns	N_{turn}	15
Number of strands	N_{str}	10
Number of poles	N_p	8
Number of pole pairs	pp	4
Wire gauge	—	20 AWG
Wire insulation	—	Heavy
Slot depth [mm]	l_{slot}	23.56 mm
Width of slot opening [mm]	w_{slot}	10.88 mm
Slot area for 1 coil [mm²]	A_s	161.49
Copper area for 1 coil [mm²]	A_c	77.83
Wire area (1 coil): [mm²]	A_w	90.98
Bare copper slot fill factor	ff_{bare}	0.48
Wire slot fill factor	ff_{wire}	0.56
RMS current constraint [A]	I_{RMS}	143
Max. current density [A/mm²]	J	21.7
Coil connection	—	8S

Source: Jiang, W.J., Three-phase 24/16 switched reluctance machine for hybrid electric powertrains: Design and optimization, *Ph.D. Dissertation, Department of Mechanical Engineering*, McMaster University, Hamilton, ON, 2016.

7.5 Stator and Rotor Pole Arc Angles

7.5.1 Pole Arc Angles

Stator and rotor pole arc angles are among the most important geometrical parameters in SRMs and they have significant effect on the torque quality. The definitions of stator pole arc angle, β_s and rotor pole arc angle, β_r are given in Figures 7.35 and 7.36, respectively.

7.5.2 Maximum Values for Pole Arc Angles

In an SRM, torque production is based on the change in the magnetic reluctance. To ensure this, the length of the arc between two consecutive rotor poles should be larger than the length of the stator pole arc. If this condition is not satisfied, there won't be a fully unaligned position and at any rotor position, rotor pole will partially or fully align with the stator pole.

The circumference of the rotor can be calculated as:

$$C_r = \pi D_r \tag{7.20}$$

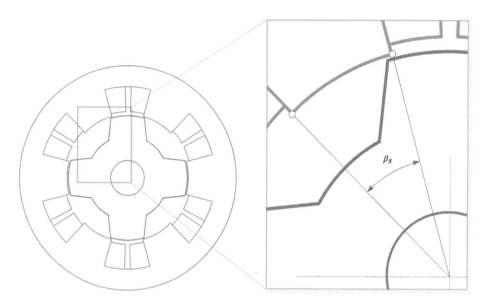

FIGURE 7.35
Illustration of stator pole arc angle using an example of 6/4 SRM.

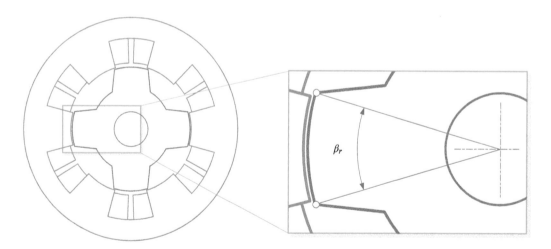

FIGURE 7.36
Illustration of rotor pole arc angle using an example of 6/4 SRM.

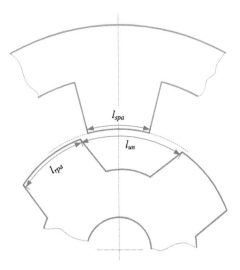

FIGURE 7.37
Arc lengths of rotor pole, stator pole, and unaligned position.

Assuming that the rotor pole arc angle is in degrees, the arc length of the rotor pole, as shown in Figure 7.37, can be calculated as:

$$l_{rpa} = \frac{D_r}{2} \frac{\pi \beta_r}{180} \tag{7.21}$$

There are N_r number of rotor poles and N_r number of gaps between the rotor poles. Therefore, the total of the arc lengths of the rotor poles:

$$L_{rpa} = N_r l_{rpa} = N_r \frac{D_r}{2} \frac{\pi \beta_r}{180} \tag{7.22}$$

The total arc length available between the rotor poles:

$$L_{un} = \pi D_r - L_{rpa} = \pi D_r - N_r \frac{D_r}{2} \frac{\pi \beta_r}{180} \tag{7.23}$$

Since there are N_r number of spaces between the rotor poles, the arc length available for the unaligned position is [10]:

$$l_{un} = \frac{\pi D_r}{N_r} - \frac{D_r}{2} \frac{\pi \beta_r}{180} \tag{7.24}$$

Similarly, the length of the stator pole arc can be calculated as:

$$l_{spa} = \left(\frac{D_r}{2} + g_a\right)\frac{\pi\beta_s}{180}$$ (7.25)

Since l_{un} in (7.24) should be greater than or equal to l_{spa} in (7.25), the following inequality is obtained:

$$\left(\frac{\pi D_r}{N_r} - \frac{D_r}{2}\frac{\pi\beta_r}{180}\right) - \left(\frac{D_r}{2} + g_a\right)\frac{\pi\beta_s}{180} \geq 0$$ (7.26)

Since the air gap g_a is very small, the inequality in (7.26) can be further simplified into:

$$\frac{(\beta_r + \beta_s)\pi}{180} \leq \frac{2\pi}{N_r}$$ (7.27)

If β_r and β_s are in radian, the maximum value of the sum of the stator and rotor pole arc angles is expressed as:

$$\beta_s + \beta_r < \frac{2\pi}{N_r}$$ (7.28)

7.5.3 Minimum Value of Pole Arc Angles

We have derived a maximum value for the stator and rotor pole arc angles. The derivation was related to geometrical parameters to maintain unaligned position. The minimum value of the stator and rotor pole arc angles is related to the self-starting capability of an SRM. The condition of self-starting states that, when one phase is at the end of the conduction period, the successive phase should be at least at the start of its conduction period to enable torque production at any position. In order to quantify the minimum value of pole arc angles and the condition for self-starting, first the relationship between β_s and β_r needs to be explained.

The derivation of the minimum value of the pole arc angles is based on ideal inductance profiles. In order to simplify the discussion on the relationship between β_s and β_r, the circular geometry of a 6/4 SRM is converted to a linear geometry, as shown in Figure 7.38. At the initial position, we assume that the stator pole S1 and the rotor pole R1 are aligned. Imagine that the motor is cut through the center of S1 and R1, and all the arcs are bent to straight-line segments. In the horizontal view, the rotor outer diameter and the stator bore diameter become irrelevant. Since the air gap length is small, the rotor outer diameter and stator bore diameter can be assumed to be equal.

The linear geometry in Figure 7.38 will be used to explain the relationship between β_s and the rotor β_r. Since the stator is stationary and the rotor is rotating, the relative position between the stator and the rotor is θ. It is assumed that at the initial rotor position, θ is zero. In the stationary polar system of the stator, the angular position is denoted as γ. In Figure 7.38, it is also assumed that the centers of the first stator pole S1

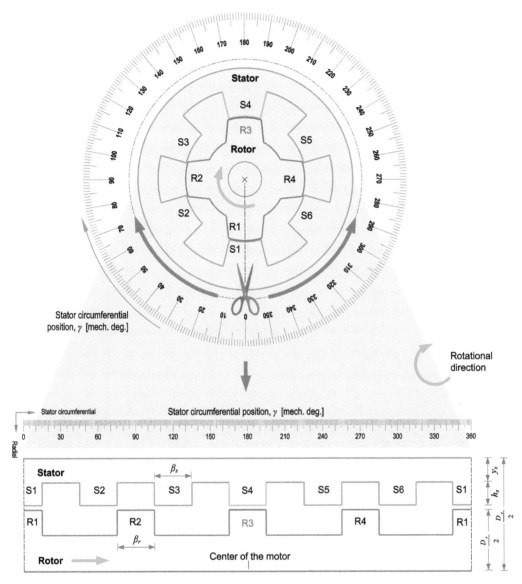

FIGURE 7.38
Conversion of a circular motor geometry to a linear geometry.

and the first rotor pole R1 are on $\gamma = 0°$ at $\theta = 0°$. For a 6/4 SRM, since there are two stator poles per phase, S4 is also aligned with R3.

7.5.3.1 Relationship between Pole Arc Angles

Figure 7.39 shows a three-phase 6/4 SRM at $\theta = 0°$ with β_s and β_r both equal to 30°. The reason of this selection will be explained later.

Since the center of the first stator pole is at $\gamma = 0°$, the angle of the left corner of the first stator pole is $(360 - \beta_s/2)$ and the right corner of the first stator pole is $\beta_s/2$. Similarly, the angle of the left corner of the first rotor pole is $(360 - \beta_r/2)$ and the left corner is $\beta_r/2$.

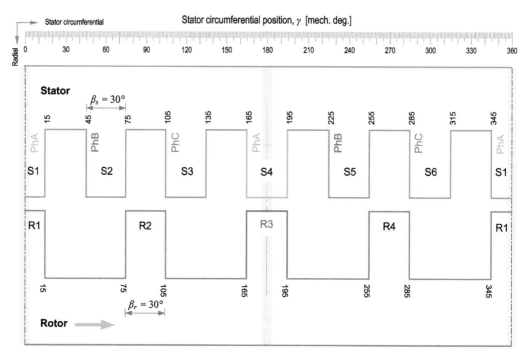

FIGURE 7.39
Relative position between rotor and stator, 6/4 SRM, $\theta = 0°$, aligned position for PhA, $\beta_s = 30°$, $\beta_r = 30°$.

For symmetric stator and rotor geometries, the angles of each corner are apart by the stator pole pitch $360/N_s$. Since the linear geometry is a cut-out version of the circular geometry, the angles should not be larger than $360°$. Therefore, the left and right corners of the stator poles can be expressed as:

$$\gamma_{s\_left} = \text{mod}\left[\left(360° - \frac{\beta_s}{2}\right) + \frac{360°}{N_s}(k-1), 360°\right] \tag{7.29}$$

$$\gamma_{s\_right} = \text{mod}\left[\frac{\beta_s}{2} + \frac{360°}{N_s}(k-1), 360°\right] \tag{7.30}$$

where $k = 1, 2, ..., N_s$

Similarly, the left and right corners of the rotor poles can be expressed as:

$$\gamma_{r\_left} = \text{mod}\left[\left(360° - \frac{\beta_r}{2}\right) + \frac{360°}{N_r}(j-1), 360°\right] \tag{7.31}$$

$$\gamma_{r\_right} = \text{mod}\left[\frac{\beta_r}{2} + \frac{360°}{N_r}(j-1), 360°\right] \tag{7.32}$$

where $j = 1, 2, ..., N_r$

To simplify the calculations, we can ignore the mod function for the first few pole positions. Since the angles are increasing when the pole numbers increase, we can assume that the angles of the left corners of the stator and rotor poles are $-\beta_s/2$ and $-\beta_r/2$. Therefore, the (7.29)–(7.32) can be simplified for the first few poles as:

$$\begin{cases} \gamma_{s\_left} = -\dfrac{\beta_s}{2} + \dfrac{360°}{N_s}(k-1) \\[2mm] \gamma_{s\_right} = \dfrac{\beta_s}{2} + \dfrac{360°}{N_s}(k-1) \\[2mm] \gamma_{r\_left} = -\dfrac{\beta_r}{2} + \dfrac{360°}{N_r}(j-1) \\[2mm] \gamma_{r\_right} = \dfrac{\beta_r}{2} + \dfrac{360°}{N_r}(j-1) \end{cases} \tag{7.33}$$

At $\theta = 0°$, the first stator pole (S1) and the first rotor pole (R1) are aligned. When the rotor moves to the right by half of the rotor pole pitch, this can be expressed as:

$$\theta_{1/2 \text{ rotor pole pitch}} = \frac{1}{2}\frac{360°}{N_r} \tag{7.34}$$

S1 and, hence, S4 will be at the unaligned position as shown in Figure 7.40. For the 6/4 SRM, this corresponds to $\theta = 45°$. It can be observed from Figure 7.40 that the difference between the angle of the left corner of S1 and the right corner of R4 is the same as the difference between the angle of the right corner of S1 and the angle of the left corner of R1 at this rotor position.

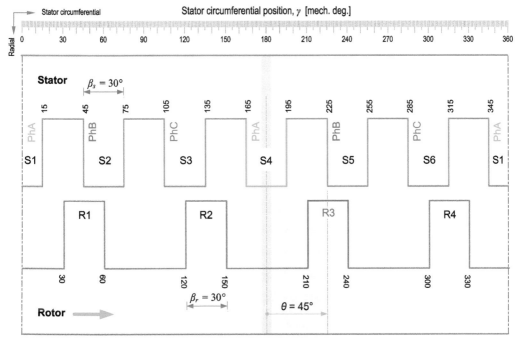

FIGURE 7.40
Relative position between rotor and stator, 6/4 SRM, $\theta = 45°$, unaligned position for PhA, $\beta_s = 30°$, $\beta_r = 30°$.

At the rotor position given in Figure 7.40, the right corner of R4 is at 330°. The right corner of R1 is at 60°. Therefore, the distance between these two corners is 90° (360 − 330 + 60°), which equals to the rotor pole pitch 360/N_r. The distance between these two points is covered by the rotor pole arc angle of R1, stator pole arc angle of S1, and the space between R4-S1 and S1-R1. Therefore, the distance (mech. deg.) between the right corner of R4 and the left corner of S1 is the half of the distance that is not covered by the stator and rotor poles under one rotor pole pitch, which equals:

$$\theta = \frac{1}{2}\left[\frac{360°}{N_r} - \left(\beta_s + \beta_r\right)\right] \qquad (7.35)$$

If the rotor moves to the right by this value, the left corner of R4 will be aligned with the right corner of S1. For a 6/4 SRM, this angle is 15°, if $\beta_s = 30°$ and $\beta_r = 30°$. Now, from $\theta = 45°$ in Figure 7.40, if the rotor moves to the right by 15° ($\theta = 60°$), the position of the stator and rotor poles will be as in Figure 7.41.

After this position, if the rotor moves to right by β_s (or β_r since they are the same), R4 will be aligned with S1. This corresponds to $\theta = 90°$ and the position of the stator and rotor pole at this angle are shown in Figure 7.42.

For an ideal inductance profile shown in Figure 7.43, it can be assumed that the torque generation happens when the inductance profile has a positive slope between the start of the alignment and the aligned position. Before the start of the alignment at $\theta = 60°$, we assume that the unaligned inductance is constant. Since the rate of change of inductance is zero, no torque will be generated between the unaligned position and the start of the alignment. At $\theta = 90°$ in Figure 7.42, R4 is fully aligned with S1. For the idealized inductance profile in Figure 7.43, Phase A will see a constant aligned inductance. Therefore, no torque would be generated if Phase A is excited at that position.

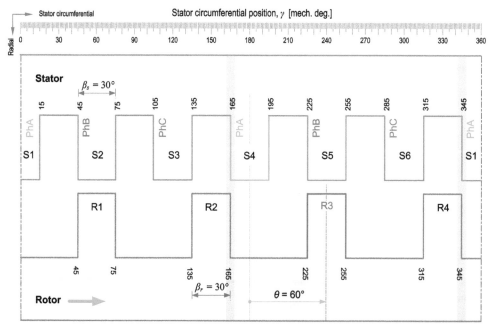

FIGURE 7.41
Relative position between rotor and stator, 6/4 SRM, $\theta = 60°$, $\beta_s = 30°$, $\beta_r = 30°$.

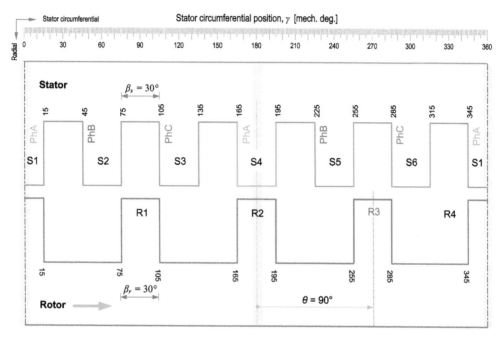

FIGURE 7.42
Relative position between rotor and stator, 6/4 SRM, $\theta = 90°$, $\beta_s = 30°$, $\beta_r = 30°$.

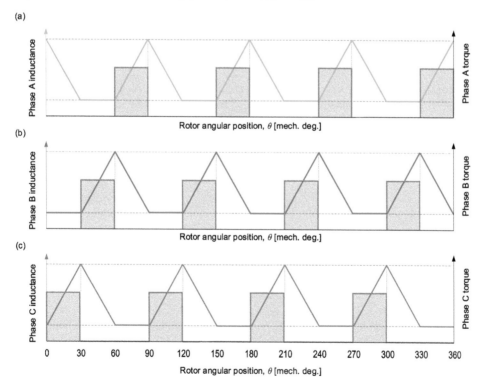

FIGURE 7.43
Stator inductance profiles and phase torque for three successive phases, $\beta_s = 30°$, $\beta_r = 30°$: (a) Phase A, (b) Phase B, and (c) Phase C.

At $\theta = 90°$ in Figure 7.42, R4 is aligned with S1 and the conduction period for Phase A is completed. Since the rotor is moving to the right, the next phase to energize is Phase C, because R1 is at the start of alignment with S3 (same condition can be observed between S6 and R3, since S3 and S6 belong to the same phase). This condition explains the reason why β_s and β_r were selected as 30° at the beginning. If these angles were smaller, at $\theta = 90°$ in Figure 7.42, the right corner of R1 would not align with the left corner of S3 in Figure 7.42. In this case, Phase C would still be having the constant unaligned inductance, which results in zero torque generation. Phase A is at the aligned position in Figure 7.42, which is also no-torque region.

In a starting condition, if the initial position of the rotor was at $\theta = 90°$ as in Figure 7.42 and the pole arc angles were smaller than 30°, the motor could not start by itself. This is the condition of self-starting in SRMs and it defines the minimum values of the pole arc angles. We will use this relationship later to quantify the minimum values of pole arc angles.

7.5.3.2 Self-Starting Condition

In order to satisfy the self-starting condition, when one phase is at the end of the conduction period (aligned position for an idealized inductance profile), the successive phase should be at least at the start of its conduction period (start of alignment for an idealized inductance profile) to enable torque production at any position, as shown in Figure 7.43.

The self-starting condition will be examined in three different configurations: three-phase 6/10 SRM, five-phase 10/12 SRM, and four-phase 16/12 SRM. The positions of the poles at $\theta = 0°$ are given in Figures 7.44 through 7.46, respectively.

For the 6/10 SRM in Figure 7.44, 10/12 SRM in Figure 7.45, 16/12 SRM in Figure 7.46 SRMs, at $\theta = 0°$, the first rotor poles are aligned with the first stator poles, which corresponds to the end of the conduction period for Phase A. For the 6/10 SRM in Figure 7.44, if the rotor is moving to the right, Phase C should be energized. The closest rotor pole to S3, which belongs to Phase C, is R4. In order to maintain the self-starting condition, at $\theta = 0°$, the right corner of R4 should have the same angle as the left corner of S3. This relationship can be quantified using (7.33):

$$\frac{\beta_r}{2} + \frac{360°}{N_r}(4-1) = -\frac{\beta_s}{2} + \frac{360°}{N_s}(3-1) \tag{7.36}$$

If (7.36) is solved for β_s and β_r, the result is:

$$\beta_r + \beta_s = \frac{720°}{N_s N_r} \times (2N_r - 3N_s) \tag{7.37}$$

For the 6/10 SRM, since S6 also belongs to Phase C, the same condition can be derived using R9 and S6:

$$\beta_r + \beta_s = \frac{720°}{N_s N_r} \times (5N_r - 8N_s) \tag{7.38}$$

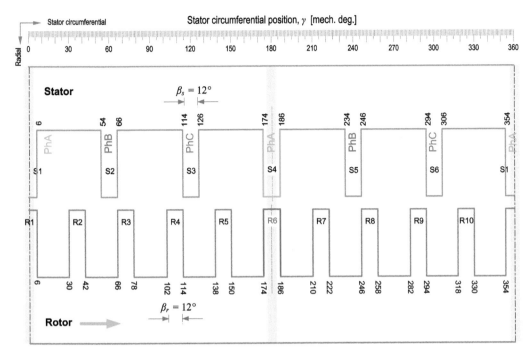

FIGURE 7.44

Relative position between rotor and stator, three-phase 6/10 SRM, $\theta = 0°$ aligned position for PhA, $\beta_s = 12°$, $\beta_r = 12°$.

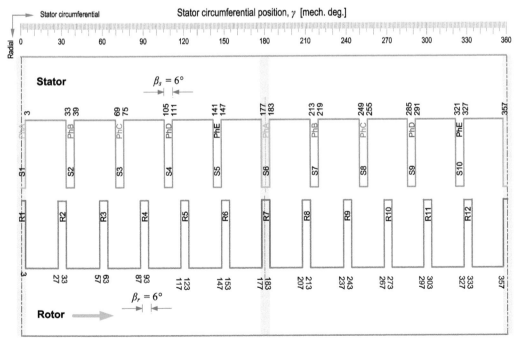

FIGURE 7.45

Relative position between rotor and stator, five-phase 10/12 SRM, $\theta = 0°$, $\beta_s = 6°$, $\beta_r = 6°$.

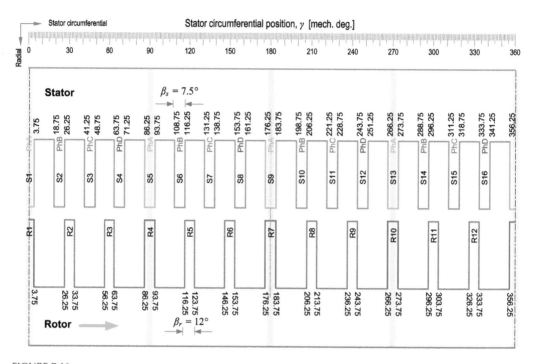

FIGURE 7.46

Relative position between rotor and stator, 4-phase 16/12 SRM, $\theta = 0°$, aligned position for PhA, $\beta_s = 7.5°$, $\beta_r = 7.5°$.

For the five-phase 10/12 SRM in Figure 7.45, the successive phase to energize is Phase B. The closest rotor pole to S2 is R2. Hence,

$$\frac{\beta_r}{2} + \frac{360°}{N_r} \times (2-1) = -\frac{\beta_s}{2} + \frac{360°}{N_s} \times (2-1) \tag{7.39}$$

Therefore,

$$\beta_r + \beta_s = \frac{720°}{N_s N_r} \times (N_r - N_s) \tag{7.40}$$

Since S7 also belongs to Phase B as shown in Figure 7.45, for R8 and S7, we can express the condition as:

$$\beta_r + \beta_s = \frac{720°}{N_s N_r} \times (6N_r - 7N_s) \tag{7.41}$$

For the four-phase 16/12 SRM in Figure 7.46, the successive phase to energize is Phase D. The closest rotor pole to S4 is R3. Hence,

$$\frac{\beta_r}{2} + \frac{360°}{N_r} \times (3-1) = -\frac{\beta_s}{2} + \frac{360°}{N_s} \times (4-1) \tag{7.42}$$

Therefore,

$$\beta_r + \beta_s = \frac{720°}{N_s N_r} \times (3N_r - 2N_s) \tag{7.43}$$

For S8, which belongs to the same phase, R6 is the closest rotor pole in Figure 7.46. Hence, the relationship between these poles can be written as:

$$\frac{\beta_r}{2} + \frac{360°}{N_r} \times (6-1) = -\frac{\beta_s}{2} + \frac{360°}{N_s} \times (8-1) \tag{7.44}$$

Therefore,

$$\beta_r + \beta_s = \frac{720°}{N_s N_r} \times (7N_r - 5N_s) \tag{7.45}$$

Similarly, if S12 and R9 are studied, the equation will be

$$\beta_r + \beta_s = \frac{720°}{N_s N_r} \times (11N_r - 8N_s) \tag{7.46}$$

One more equation can be obtained from S16 and R12:

$$\beta_r + \beta_s = \frac{720°}{N_s N_r} \times (15N_r - 11N_s) \tag{7.47}$$

As shown in Table 7.4, the results of these expressions are all equal to the number of poles, which is (N_s/m). Therefore, the expression for the self-starting condition can be derived as:

$$\beta_s + \beta_r \geq \frac{720°}{N_s N_r} \frac{N_s}{m} \tag{7.48}$$

TABLE 7.4

Generalization of Equations for Self-Starting Condition

SRM	m	N_s	N_r	Equation		Number of Poles, N_s/m
Three-phase 6/10	3	6	10	$\beta_r + \beta_s = \frac{720°}{N_s N_r} \times (2N_r - 3N_s) = \frac{720°}{N_s N_r} \times 2$	(7.37)	
				$\beta_r + \beta_s = \frac{720°}{N_s N_r} \times (5N_r - 8N_s) = \frac{720°}{N_s N_r} \times 2$	(7.38)	$N_s/m = 6/3 = 2$
Five-phase 10/12	5	10	12	$\beta_r + \beta_s = \frac{720°}{N_s N_r} \times (N_r - N_s) = \frac{720°}{N_s N_r} \times 2$	(7.40)	
				$\beta_r + \beta_s = \frac{720°}{N_s N_r} \times (6N_r - 7N_s) = \frac{720°}{N_s N_r} \times 2$	(7.41)	$N_s/m = 10/5 = 2$
Four-phase 16/12	4	16	12	$\beta_r + \beta_s = \frac{720°}{N_s N_r} \times (3N_r - 2N_s) = \frac{720°}{N_s N_r} \times 4$	(7.43)	
				$\beta_r + \beta_s = \frac{720°}{N_s N_r} \times (7N_r - 5N_s) = \frac{720°}{N_s N_r} \times 4$	(7.45)	
				$\beta_r + \beta_s = \frac{720°}{N_s N_r} \times (11N_r - 8N_s) = \frac{720°}{N_s N_r} \times 4$	(7.46)	$N_s/m = 16/4 = 4$
				$\beta_r + \beta_s = \frac{720°}{N_s N_r} \times (11N_r - 8N_s) = \frac{720°}{N_s N_r} \times 4$	(7.47)	

Equation (7.48) can be simplified as:

$$\beta_s + \beta_r \geq \frac{720°}{mN_r} \tag{7.49}$$

If β_r and β_s are in radian, (7.49) can be rewritten as:

$$\beta_s + \beta_r \geq \frac{4\pi}{mN_r} \tag{7.50}$$

Combined with the maximum value of the pole arc angles in (7.28), the complete inequality for the stator and rotor pole arc angles can be obtained as:

$$\frac{4\pi}{mN_r} \leq \beta_s + \beta_r < \frac{2\pi}{N_r} \tag{7.51}$$

The minimum value of the pole arc angles in (7.50) shows that self-starting condition could be maintained if $\beta_s > \beta_r$ or $\beta_r > \beta_s$. Conventionally $\beta_r > \beta_s$ is commonly applied. However, when the rotor pole count increases in configurations with a higher number of rotor poles, $\beta_s > \beta_r$ can provide better space utilization and, hence, higher torque output.

If we assume that $\beta_s = \beta_r$, (7.50) can be simplified as:

$$\beta_{s,r} \geq \frac{360°}{mN_r} \tag{7.52}$$

For the three-phase 6/4 SRM, $360°/mN_r$ equals to 30°. This is the same value that was defined in Figure 7.39. In a 6/4 SRM if the stator pole is selected as 30° and if the rotor pole is higher than or equal to that, the self-starting condition should be satisfied.

Equation (7.52) quantifies the minimum value of the pole arc angles in mechanical degrees. In one mechanical revolution, the same electrical angle repeats itself by N_r times. If (7.52) is multiplied by N_r, the minimum value of pole arc angles is expressed in electrical degrees that equals to $360°/m$. This is the electrical phase shift between the successive stator phases.

7.5.3.3 Nonlinear Effects

Note that the derivation of the minimum value of the pole arc angles was based on ideal inductance profiles. It was assumed that the torque production was only between the start of the alignment and the unaligned position. It was also assumed that the rate of change of inductance was zero between the unaligned position and the start of the alignment. In practice, the inductance profile is not linear. As shown in Figure 7.47, SRM can still generate torque before the alignment starts. There are many other factors that affect the unaligned inductance and, hence, the self-starting condition. Some of these factors are the taper angles on the stator and the rotor, pole shoes on the stator, and the nonlinear magnetization characteristics of electrical steel (e.g. saturation). The derivation of self-starting condition certainly gives an idea about the selection of the pole arc angles, but it is recommended to conduct electromagnetic finite element analysis to optimize the pole arc angles during SRM design process.

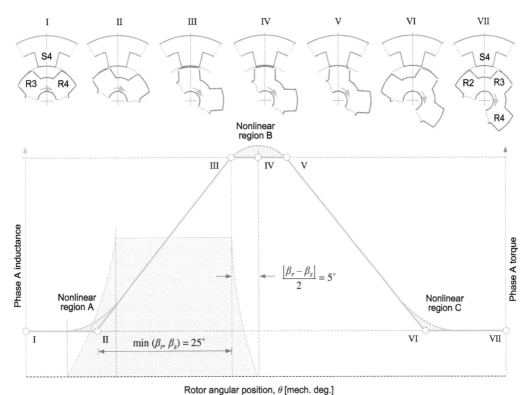

FIGURE 7.47
Nonlinear effect of inductance profile of SRM with $\beta_r > \beta_s$.

7.6 Other Stator and Rotor Dimensions

7.6.1 Pole Heights

The definition of the stator pole height is illustrated in Figure 7.48. The stator pole height is almost equal to the coil height. Some space at the tip of the stator pole has to be left for the pole shoes and wedges, which ensure that the coils can be kept in place. Figure 7.49 shows the illustration of the rotor pole height. The selection of the rotor pole height is related to how the stator flux penetrates into the rotor core. Due to the nonlinear characteristics of an SRM, it is recommended to analyze how different rotor pole heights affect the performance of the motor during a design process. It should be noted that, when the rotor is rotating, its poles work like a fan, which will generate windage loss. Therefore, if the rotor pole is too long, it will also create too much noise and vibration. In critical applications, such as in high-speed traction motors, the gaps between the rotor poles can be filled up with epoxy or similar material to improve the structural integrity of the rotor and reduce the windage loss. The rotor geometry should be well optimized to keep the filler material safe at high-speed operation.

Figure 7.50 shows an example of how the stator pole height affects the performances of 24/16 SRM at 2000 rpm. For these models, the stator pole height increases from 24 mm to 27 mm with 0.5 mm increment. Since the slot area increases, the copper slot fill factor

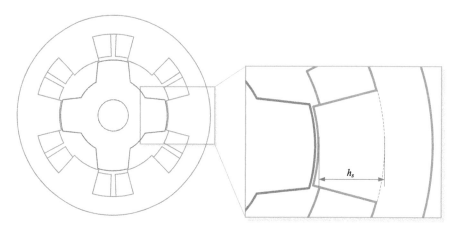

FIGURE 7.48
Illustration of stator pole height using an example of 6/4 SRM.

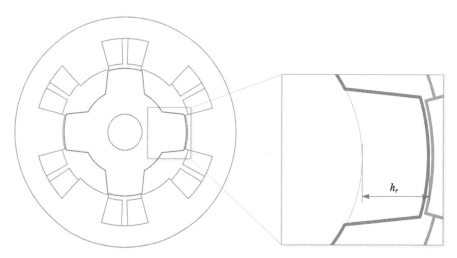

FIGURE 7.49
Illustration of rotor pole height using an example of 6/4 SRM.

decreases from 0.56 to 0.50. As shown in Figure 7.50a, output torque decreases as the stator pole height increases, because rotor outer diameter decreases to satisfy the stator outer diameter constraint. Hence, the lever arm from the shaft to the air gap decreases. In addition, longer stator poles increase the amount of iron that has to be saturated and affect the magnetic flux linkage characteristics. This impacts the torque production capability and power factor in SRMs. Figure 7.50b shows that the torque ripple tends to increase and decrease as the stator pole height increases from 24 to 27 mm for the given 24/16 SRM. The torque ripple varies in a nonlinear way due to the change in the magnetic circuit and conduction angles at the same time.

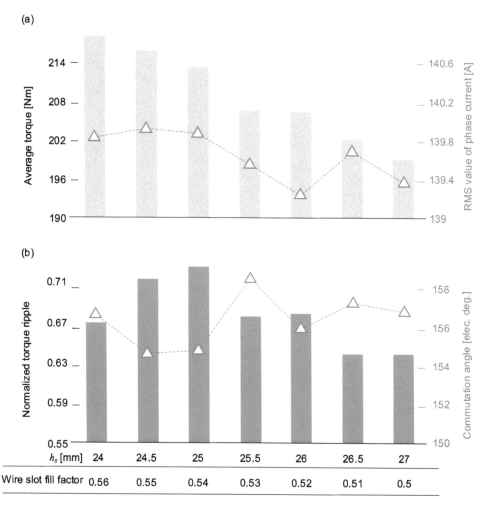

FIGURE 7.50
Comparison of motor performances at 2000 rpm and 240 A based on optimized conduction angles, variation in stator pole height, 24/16 SRM: (a) average torque and (b) normalized torque ripple. (From Jiang, J.W. et al., *IEEE Trans. Trans. Electrification*, 3(1), 76–85, March 2017.)

7.6.2 Taper Angles

Stator and rotor taper angles can be added to the poles to change the motor performances and also support the poles as shown in Figures 7.51 and 7.52 respectively. Figure 7.53 shows a series of models with different stator taper angles varying from 0° to 7°. For the sake of analysis, the stator pole height and stator back-iron thickness are adjusted accordingly to maintain that the slot fill factor of 0.54. The sum of stator pole height and stator back iron thickness stays constant to guarantee that the location of the air gap is fixed. For the given 24/16 SRM, as the stator taper angle increases, the output torque increases almost linearly in the 24/16 SRM. Meanwhile, the higher value of stator taper angle leads to the reduction of normalized torque ripple. It should be noted that, due to the nonlinear characteristics of the motor, different trends can be observed depending on the saturation level and conduction angles.

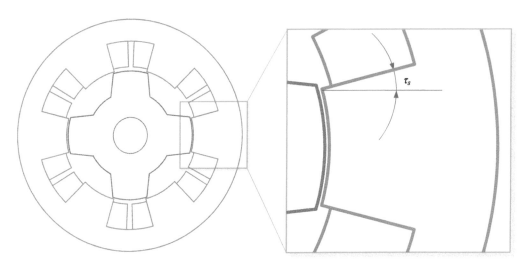

FIGURE 7.51
Illustration of stator taper angle using an example of 6/4 SRM.

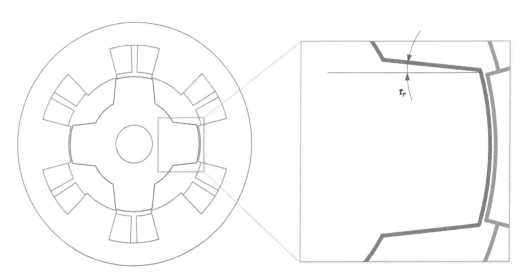

FIGURE 7.52
Illustration of rotor pole height using an example of 6/4 SRM.

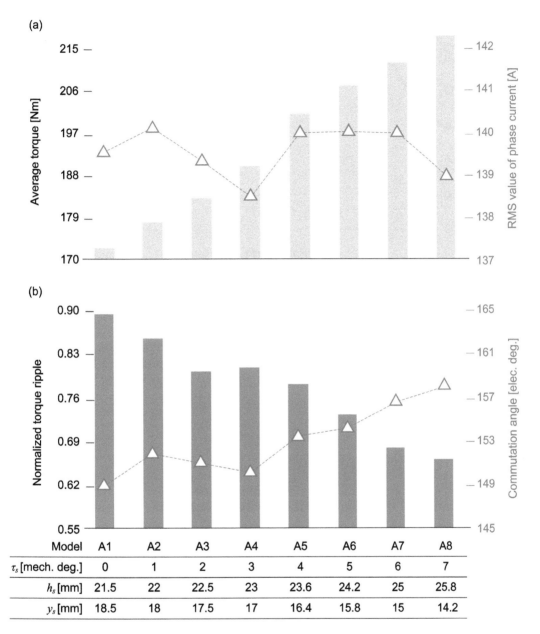

FIGURE 7.53
Comparison of motor performances at 2000 rpm and 240 A based on optimized conduction angles, variation in stator taper angle, 24/16 SRM: (a) average torque and (b) normalized torque ripple. (From Jiang, J.W. et al., *IEEE Trans. Trans. Electrification*, 3(1), 76–85, March 2017.)

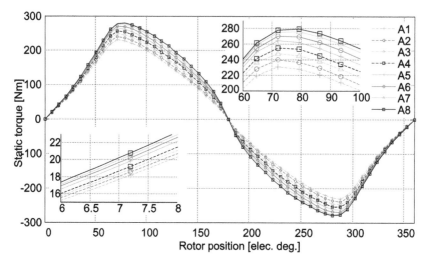

FIGURE 7.54
Comparison of static phase torque waveforms, Cases A1-A8, at peak reference current, 240 A. (From Jiang, J.W. et al., Design optimization of switched reluctance machine using genetic algorithm, *Proceedings of the 2015 IEEE International Electric Machines and Drives Conference (IEMDC)*, Coeur d'Alene, ID, May 2015, pp. 1671–1677, 2015.)

Figure 7.54 compares the static phase torque waveforms for these cases. It shows that enlarging the stator taper angle causes the static torque to rise faster and have a higher peak value in the 24/16 SRM. A8 has the highest phase flux linkage because it has the largest taper angle among Cases A1–A8.

7.6.3 Back Iron Thicknesses

Figures 7.55 and 7.56 show the definition of the stator and rotor back iron thicknesses, respectively. A thicker stator back iron provides more support for the stator poles, can help reduce acoustic noise. If the stator back iron thickness is too thin, the deformation will occur more easily. The thickness of the rotor back iron depends on the shaft diameter.

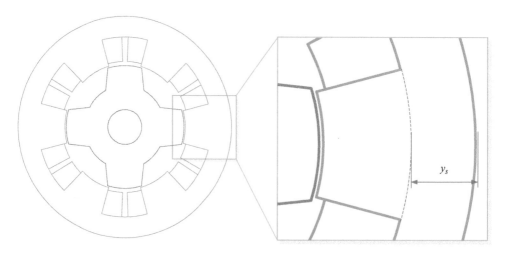

FIGURE 7.55
Illustration of stator back iron thickness using an example of 6/4 SRM.

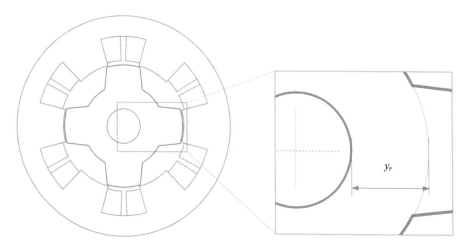

FIGURE 7.56
Illustration of rotor back iron thickness using an example of 6/4 SRM.

In the design of an SRM, the maximum flux density in the back irons has to be carefully studied. If the back irons are too thin, they saturate quickly and the torque production capability of the SRM can drop significantly.

7.6.4 Stator Pole Shoe and Slot Wedges

Pole shoes are affixed to the stator poles, as shown in Figure 7.57. The pole shoes are used to hold the pole wedges. The wedges are inserted in each slot and hold the windings in place. The wedges can also compress the coils in the slots and affect the slot fill factor. Dimensions of the pole shoe, such as height, width, and fillet radius, affect the output torque and torque ripple.

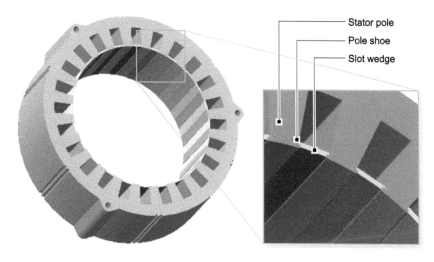

FIGURE 7.57
Shape and placement of pole shoes and slot wedges. (From Jiang, J.W. et al., *IEEE Trans. Trans. Electrification*, 3(1), 76–85, March 2017.)

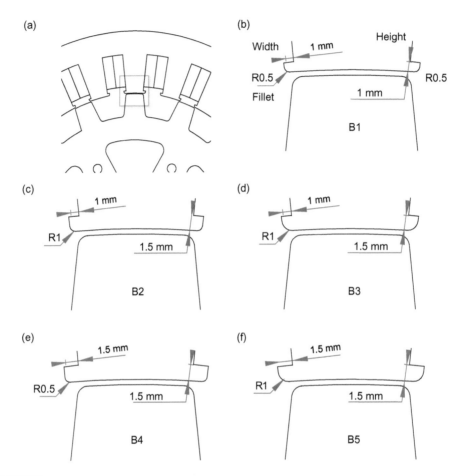

FIGURE 7.58
Five design candidates for stator pole shoes: (a) close up of the air gap, (b) Case B1, (c) Case B2, (d) Case B3, (e) Case B4, and (f) Case B5. (From Jiang, W.J., Three-phase 24/16 switched reluctance machine for hybrid electric powertrains: Design and optimization, *Ph.D. Dissertation, Department of Mechanical Engineering*, McMaster University, Hamilton, ON, 2016.)

Figure 7.58 shows five designs of stator poles, which differ in size and shape. Figure 7.59 compares the motor performances for these five motor models. Generally, adding the pole shoe reduces the output torque. The amplitude of average torque tends to decrease as the size of the pole shoe increases, as shown in Figure 7.59a. Figure 7.59b shows that pole shoes help reduce torque ripple and, commonly, the torque ripple improves more as the size of the pole shoe increases.

7.6.5 Fillets

The fillet, a rounding of an interior or exterior corner, can be used on both the rotor and stator poles. They can eliminate sharp edges to relieve the stress concentrations at these points and make the components less prone to fatigue failure. Figure 7.60 shows the example of fillets applied to the stator. The fillet radius at the stator pole tip is 1 mm and the fillet radius at the bottom of the stator pole is 2 mm. Figure 7.61 shows the fillets on the rotor. The fillet radius at the rotor pole tips is 1 mm and the fillet radius at the bottom of the rotor

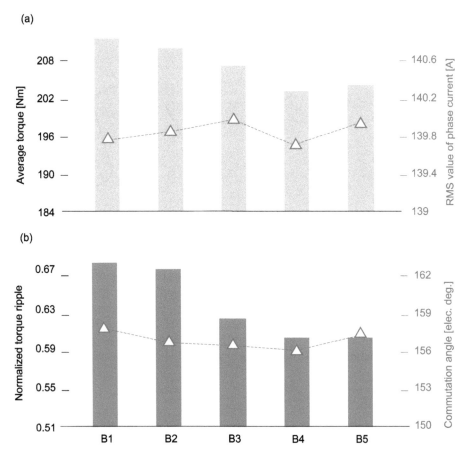

FIGURE 7.59
Comparison of motor performances at 2000 rpm and 240 A based on optimized conduction angles, variation in pole shoes: (a) average torque and (b) normalized torque ripple. (From Jiang, J.W. et al., *IEEE Trans. Trans. Electrification*, 3(1), 76-85, March 2017.)

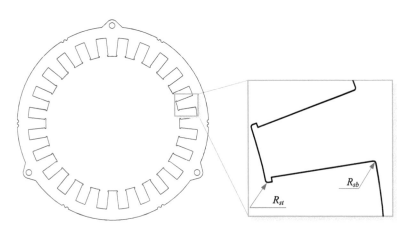

FIGURE 7.60
Illustration of stator fillets using an example of 24/16 SRM.

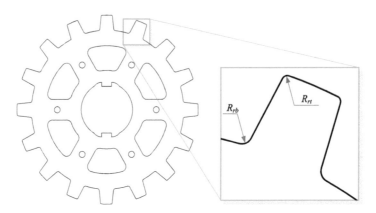

FIGURE 7.61
Illustration of rotor fillets using an example of 24/16 SRM.

pole is 2 mm. If the fillet is comparable to the size of either the stator pole or the rotor pole, finite element analysis (FEA) simulation based on detailed mesh has to be performed to study the effects of the fillet on motor performances.

Fillets and pole shoes should be carefully designed analyzed in an SRM. Since they are close to the air gap, they can have significant effect on the reluctance of the magnetic circuit and, hence, the shape of the static torque profile. This can affect the induced voltage and the peak torque.

Questions and Problems

1. Given a stator stack length is 50 mm and a stacking factor is 0.95, please estimate the number of lamination sheets when the nominal thickness of each lamination is 0.30, 0.25, and 0.1 respectively.

2. Figure 7.34 shows the winding scheme for all three phases for 24/16 SRM. For each phase, eight coils are connected in series. Each coil has 15 turns and each turn has 10 parallel strands. Plot two alternate winding diagrams of three phases for the same 24/16 SRM: (a) each phase has two parallel paths, and (b) each phase has four parallel paths.

3. Consider a coil with 4 turns and 5 strands per turn, and a slot fill factor of 0.5. The slot fill factor is required to be within the range of 0.4 (included) to 0.6 (included). The slot geometry and wire gauge are fixed. Suppose that numbers of strands and turns are less than or equal to 12. What are the possible combinations of numbers for turns and strands? Calculate the slot fill factor for each case.

4. Given that the constraints for stator and rotor pole arc angles are:

$$\begin{cases} \beta_r > \beta_s \geq \dfrac{2\pi}{mN_r} \\ \dfrac{4\pi}{mN_r} \leq \beta_s + \beta_r < \dfrac{2\pi}{N_r} \end{cases}$$

and x-axis is the rotor pole arc angle and y-axis is the stator pole arc angle, the feasible region can be plotted for selecting the stator and rotor pole arc angles for a three-phase 6/4 SRM on the x-y plane. The shaded area in Figure Q.7.1 shows the region in which $\beta_r > \beta_s$.

Similarly, $\beta_s \geq 2\pi/mN_r$ can be added to the previous plot. The overlapping region, as shown in Figure Q.7.2, is the solution set that satisfies both inequalities. Add the inequality conditions in the second equation and show the feasible region.

5. The figures showing relative position between the rotor poles and stator poles in Section 7.5.3 of this chapter were produced primarily using the MATLAB® scripts given in this question. Reproduce the scripts first in MATLAB. Save these five functions in the same folder.

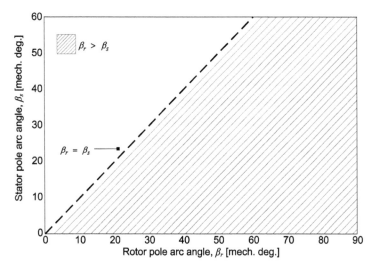

FIGURE Q.7.1
Feasible region for $\beta_r > \beta_s$.

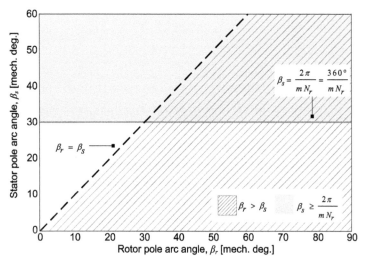

FIGURE Q.7.2
Feasible region for $\beta_r > \beta_s$ and $\beta_s \geq 2\pi/mN_r$.

The code has one main function, which calls four sub-functions:

- **vertical_lines,**
- **stator_rotor_angles,**
- **lines_between_poles,** and
- **lines_between_pole_corners.**

The main function is given as:

```
%% Update parameters
Ns=XX;              % number of stator poles, to be updated
Nr=XX;              % number of rotor poles, to be updated
m=XX;               % number of phases, to be updated
rotor_position=XX;  % rotor position in mech. deg., to be updated
beta_s=XX;          % stator pole arc angle, to be updated
beta_r=XX;          % rotor pole arc angle, to be updated

%% Initialization
theta=0:0.5:360;
stator_pole=1;   stator_back_iron=stator_pole+0.5;
rotor_pole=0.9;    rotor_back_iron=rotor_pole-0.5;

temp=find(theta==rotor_position); %%%%%%%%%%%%%%%%%% PARAMETERS

% calculate the angle of the corners of the rotor and stator poles at
% different rotor positions
[stator_angles,rotor_angles]=stator_rotor_angles(Ns,Nr,beta_s,beta_r,theta);


%% Plotting the stator poles

% locate and plot the corners
for i=1:length(stator_angles);
    plot(stator_angles(i),stator_pole,'ob'); xlim([0 360]);
    ylim([0 2.0]); hold on

text(stator_angles(i),stator_back_iron,num2str(stator_angles(i)),'Rotation',9
0);
    xlabel ('Stator circumferential position [mech. deg.]')
    set(gca, 'YTickLabelMode', 'manual', 'YTickLabel', []);
    set(gca,'YTick',[])
    set(gca,'XTick',[0:60:360])
    set(gca,'xaxislocation','top');
end

for i=1:Ns
    % plotting lines between the corners of the stator poles
    point1=stator_angles((2*i)-1);  point2=stator_angles((2*i));
    lines_between_pole_corners(point1,point2,50,stator_pole,1);
    % vertical lines
    vertical_lines(point1,point2,stator_pole,stator_back_iron,50,i,1,m);
end

% connecting the poles
for i=1:Ns
    point1=stator_angles(2*i);
    if ((2*i)+1)>length(stator_angles)
        point2=stator_angles(((2*i)+1)-length(stator_angles));
    else
        point2=stator_angles((2*i)+1);
    end
    lines_between_poles(point1,point2,stator_back_iron,50,1);
end

% plotting the rotor poles
for i=temp:temp %length(theta)
```

```
    text(20,0.2,['Theta = ',num2str(theta(i)),' deg']);
    rotor_angles_to_analyze=rotor_angles(i,:);
    for j=1:length(rotor_angles_to_analyze);
        plot(rotor_angles_to_analyze(j),rotor_pole,'ok');

text(rotor_angles_to_analyze(j),rotor_back_iron,num2str(rotor_angles_to_analy
ze(j)),'Rotation',-90);
    end
    for j=1:Nr
        % plotting lines between the corners of the rotor poles
        point1=rotor_angles_to_analyze((2*j)-1);
point2=rotor_angles_to_analyze((2*j));
        lines_between_pole_corners(point1,point2,50,rotor_pole,2);
        % vertical lines
        vertical_lines(point1,point2,rotor_pole,rotor_back_iron,50,j,2,m);
    end
    % connecting the poles
    for j=1:Nr
        point1=rotor_angles_to_analyze(2*j);
        if ((2*j)+1)>length(rotor_angles_to_analyze)
            point2=rotor_angles_to_analyze(((2*j)+1)-
length(rotor_angles_to_analyze));
        else
            point2=rotor_angles_to_analyze((2*j)+1);
        end
        lines_between_poles(point1,point2,rotor_back_iron,50,2);
    end
end
```

The first sub-function is given as:

```
function
[]=vertical_lines(point1,point2,PolePos,YokePos,NumOfPoints,NsNum,Component,P
hases)

if abs(Component-1)<=1e-6
    colorCode='-b';
else
    colorCode='-k';
end

ycoordinates=linspace(PolePos,YokePos,NumOfPoints);
xcoordinates=point1*ones(length(ycoordinates),1);
plot(xcoordinates,ycoordinates,colorCode,'LineWidth',2);

ycoordinates=linspace(PolePos,YokePos,NumOfPoints);
xcoordinates=point2*ones(length(ycoordinates),1);
plot(xcoordinates,ycoordinates,colorCode,'LineWidth',2);

if abs(Component-1)<=1e-6
    comp='S';
    multp=1;
else
    comp='R';
    multp=-1;
end

if point1>point2
    location=point1+abs(point1-(360+point2))/2;
    if location>=360
        location=location-360;
    end
    if abs(Component-1)<=1e-6
        text(location,PolePos+multp*(1*abs(YokePos-
PolePos)/4),[comp,num2str(NsNum)],'Rotation',90);
        phaseNum=mod(NsNum,Phases);
        if phaseNum<=0
            phaseNum=phaseNum+Phases;
        end
        text(location,PolePos+multp*(3*abs(YokePos-
PolePos)/4),['Ph',num2str(phaseNum)],'Rotation',90);
    else
        text(location,PolePos+multp*(1*abs(YokePos-
PolePos)/4),[comp,num2str(NsNum)],'Rotation',90);
    end
else
    if abs(Component-1)<=1e-6
        text(point1+abs(point2-point1)/2,PolePos+multp*(1*abs(YokePos-
PolePos)/4),[comp,num2str(NsNum)],'Rotation',90);
        phaseNum=mod(NsNum,Phases);
        if phaseNum<=0
            phaseNum=phaseNum+Phases;
        end
        text(point1+abs(point2-point1)/2,PolePos+multp*(3*abs(YokePos-
PolePos)/4),['Ph',num2str(phaseNum)],'Rotation',90);
    else
        text(point1+abs(point2-point1)/2,PolePos+multp*(1*abs(YokePos-
PolePos)/4),[comp,num2str(NsNum)],'Rotation',90);
    end
end
end
```

The second sub-function is given as:

```
function
[stator_angles,rotor_angles]=stator_rotor_angles(Ns,Nr,beta_s,beta_r,theta)

tau_r=360/Nr;   tau_s=360/Ns; % stator and rotor pole pitch

stator_angles(1,1)=360-beta_s/2;  stator_angles(1,2)=beta_s/2;
% the angle of the corner points of the first stator pole
% it is assumed that the center of the first stator pole is aligned with
% the x-axis
% there are no negative angles. Minus 6 degrees is (360-6) = 354 degrees

for i=2:Ns % since the first stator pole is calculated above, it starts from
the second one
    for j=1:2 % since we have to calculate the angle for each corner, the
length of the array is 2*Ns
        stator_angles((2*i)-(2-j))=mod(stator_angles(((2*i)-(2-j))-
2)+tau_s,360); % the difference between the angles of the corners of each
stator pole is the stator pole pitch.
    end
end

rotor_angles(1,1)=360-beta_r/2;  rotor_angles(1,2)=beta_r/2;
% the angle of the corner points of the first rotor pole
% it is assumed that the center of the first rotor pole is aligned with the
% x-axis
% there are no negative angles

% the loop below is the same as the stator. This is valid only for the
% first rotor position
for i=2:Nr
    for j=1:2 % since we have to calculate the angle for each corner, the
length of the array is 2*Nr
        rotor_angles((2*i)-(2-j))=mod(rotor_angles(((2*i)-(2-j))-
2)+tau_r,360);
    end
end

% now we calculate the angles of each rotor pole corner at different rotor
% positions.
dim=length(rotor_angles); % depending on the rotor position array, its length
might end up higher 2*Nr. We fix this length before extending it for other
rotor positions.
for i=2:length(theta)
    for j=1:dim
        rotor_angles(i,j)=mod(rotor_angles(i-1,j)+(theta(i)-theta(i-1)),360);
% basically we add the increase in rotor position
    end
end
```

The third sub-function is given as:

```
function []=lines_between_poles(point1,point2,YokePos,NumOfPoints,Comp)

if abs(Comp-1)<=1e-6
    colorCode='-b';
else
    colorCode='-k';
end

if point1>point2
    xcoordinates=linspace(point1,360,NumOfPoints);
ycoordinates=YokePos*ones(length(xcoordinates),1);
    plot(xcoordinates,ycoordinates,colorCode,'LineWidth',2);
    xcoordinates=linspace(0,point2,NumOfPoints);
ycoordinates=YokePos*ones(length(xcoordinates),1);
    plot(xcoordinates,ycoordinates,colorCode,'LineWidth',2);
else
    xcoordinates=linspace(point2,point1,NumOfPoints);
ycoordinates=YokePos*ones(length(xcoordinates),1);
    plot(xcoordinates,ycoordinates,colorCode,'LineWidth',2);
end

end
```

The last sub-function is given as:

```
function
[]=lines_between_pole_corners(point1,point2,NumOfPoints,PoleHeight,Comp)

if abs(Comp-1)<=1e-6
    colorCode='-b';
else
    colorCode='-k';
end

if point1>point2 % the pole is at the center of the x-axis. In this case we
need two lines. From the first corner to the end and from the beginning to
the second corner
    xcoordinates=linspace(point1,360,NumOfPoints);
ycoordinates=PoleHeight*ones(length(xcoordinates),1);
    plot(xcoordinates,ycoordinates,colorCode,'LineWidth',2);
    xcoordinates=linspace(0,point2,NumOfPoints);
ycoordinates=PoleHeight*ones(length(xcoordinates),1);
    plot(xcoordinates,ycoordinates,colorCode,'LineWidth',2);
else
    xcoordinates=linspace(point2,point1,NumOfPoints);
ycoordinates=PoleHeight*ones(length(xcoordinates),1);
    plot(xcoordinates,ycoordinates,colorCode,'LineWidth',2);
end
```

To plot the relative position between the rotor and stator for 3-phase 6/4 SRM ($\beta_s = 30°$, $\beta_r = 30°$) when $\theta = 0°$, the parameters in the main function should be updated accordingly:

```
Ns=6;                % number of stator poles, to be updated
Nr=4;                % number of rotor poles, to be updated
m=3;                 % number of phases, to be updated
rotor_position= 0;   % rotor position in mech. deg., to be updated
beta_s=30;           % stator pole arc angle, to be updated
beta_r=30;           % rotor pole arc angle, to be updated
```

After the codes are finished running, the figure will be plotted as shown in Figure Q.7.3:

If the rotor position θ is changed to 15°, the results in Figure Q.7.4 should be observed. In this case, the rotor moved by 15.

```
Ns=6;                % number of stator poles, to be updated
Nr=4;                % number of rotor poles, to be updated
m=3;                 % number of phases, to be updated
rotor_position= 15;  % rotor position in mech. deg., to be updated
beta_s=30;           % stator pole arc angle, to be updated
beta_r=30;           % rotor pole arc angle, to be updated
```

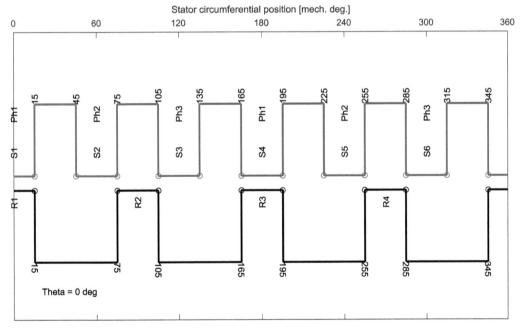

FIGURE Q.7.3
Snapshot of MATLAB plot, relative position between rotor and stator, 3-phase 6/4 SRM, $\theta = 0°$, $\beta_s = 30°$, $\beta_r = 30°$.

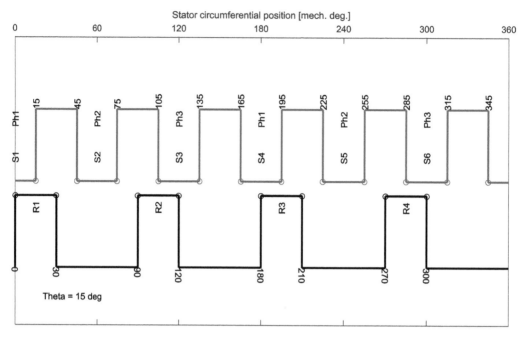

FIGURE Q.7.4

Snapshot of MATLAB plot, relative position between rotor and stator, 3-phase 6/4 SRM, $\theta = 15°$, $\beta_s = 30°$, $\beta_r = 30°$.

In order to plot the relative position between the rotor and stator for 3-phase 12/8 SRM ($\beta_s = 15°$, $\beta_r = 15°$) when $\theta = 0°$, the parameters in the main function should be updated accordingly:

```
Ns=12;                  % number of stator poles, to be updated
Nr=8;                   % number of rotor poles, to be updated
m=3;                    % number of phases, to be updated
rotor_position= 0;      % rotor position in mech. deg., to be updated
beta_s=15;              % stator pole arc angle, to be updated
beta_r=15;              % rotor pole arc angle, to be updated
```

MATLAB should produce the results shown in Figure Q.7.5.

After tuning the scripts, please analyze the following configurations for $\beta_s = \beta_r$.

- 3-phase 6/4 SRM (Figure 7.13a)
- 3-phase 12/8 SRM (Figure 7.13b)
- 3-phase 18/12 SRM (Figure 7.13c)
- 3-phase 24/16 SRM (Figure 7.13d)

First calculate the minimum values of pole arc angles for all the geometries. Plot the relative position between the rotor and stator for all four geometries for the minimum values of pole arc angles when $\theta = 0°$. The MATLAB plots for 6/4 and 12/8 SRMs have already been given. Fill the blanks in Table Q.7.1. For three-phase 6/4 SRM, the minimal β_s and β_r, the number of poles, and the expression for $\beta_s + \beta_r$ have already been calculated and included in Table 7.5.

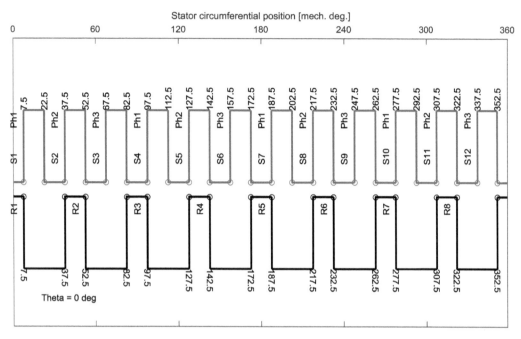

FIGURE Q.7.5

Snapshot of MATLAB plot, relative position between rotor and stator, 3-phase 12/8 SRM, $\theta = 0°$, $\beta_s = 15°$, $\beta_r = 15°$.

TABLE Q.7.1

The Relationship between Stator and Rotor Pole Arc Angles

Geometry	m	N_s	N_r	$\beta_{r\_min} = \beta_{s\_min} =$ [mech. deg.]	Number of Poles	Expression for $\beta_s + \beta_r$
Three-phase 6/4 SRM	3	6	4	30	2	$\beta_r + \beta_s = 720°/N_sN_r \times (2N_r - N_s)$
						$\beta_r + \beta_s = 720°/N_sN_r \times (5N_r - 3N_s)$
Three-phase 12/8 SRM						
Three-phase 18/12 SRM						
Three-phase 24/16 SRM						

References

1. Jiang, J. W. Three-phase 24/16 switched reluctance machine for hybrid electric powertrains: Design and optimization, Ph.D. Dissertation, Department of Mechanical Engineering, McMaster University, Hamilton, ON, 2016.
2. Emadi, A. (Ed.) *Advanced Electric Drive Vehicles*. CRC Press, Boca Raton, FL, 2014.
3. Agrawal, K. C. *Industrial Power Engineering Handbook*. Newnes, Boston, MA, 2001.
4. Castano, S. M., Yang, R., Mak, C., Bilgin, B., and Emadi, A. External-rotor switched reluctance motor for direct-drive home appliances, in *Proceedings of the Annual Conference of the IEEE Industrial Electronics Society (IECON)*, Washington, DC, October 2018..
5. Bilgin, B., P. Magne, P. Malysz, Y. Yang, V. Pantelic, M. Preindl, A. Korobkine, W. Jiang, M. Lawford, and A. Emadi. Making the case for electrified transportation. *IEEE Transactions on Transportation Electrification*, 1(1), 4–17, 2015.
6. National Electrical Manufacturers Association Standard, ANSI/NEMA MW 1000-2015, Rosslyn, Virginia, USA, March 31, 2015.
7. Miller, T. J. E. (Ed.). *Electronic Control of Switched Reluctance Machines*. Newnes, Oxford, UK, 2001.
8. Jiang, J. W., B. Bilgin, A. Sathyan, H. Dadkhah, and A. Emadi, Analysis of unbalanced magnetic pull in eccentric interior permanent magnet machines with series and parallel windings, *IET Electric Power Application*, 10(6), 526–538, 2016.
9. Jiang, J. W., B. Bilgin, and A. Emadi, Three-phase 24/16 switched reluctance machine for hybrid electric powertrains: Design and optimization, *IEEE Transactions on Transportation Electrification*, 3(1), 76–85, March 2017.
10. Bilgin, B., A. Emadi, and M. Krishnamurthy, Design considerations for switched reluctance machines with a higher number of rotor poles, *IEEE Transactions on Industrial Electronics*, 59(10), 3745–3756, 2012.
11. Jiang, J. W., B. Bilgin, B. Howey, and A. Emadi, Design optimization of switched reluctance machine using genetic algorithm, *Proceedings of the 2015 IEEE International Electric Machines and Drives Conference* (IEMDC), Coeur d'Alene, ID, May 2015, pp. 1671–1677.

8

Mechanical Construction of Switched Reluctance Machines

Yinye Yang, James Weisheng Jiang, and Jianbin Liang

CONTENTS

8.1 Introduction

Upon first inspection of an electric motor, the motor housing and drive shaft are observed from outside. The housing is used as the main structure and supports many other components which cannot be seen unless the housing is opened. Figure 8.1 shows the major components for a switched reluctance machine (SRM). Three major subassemblies can be identified in the exploded view: shaft-rotor, stator-winding, and housing (also called casing or frame).

Even though the structure of an SRM is relatively simple, a good design that considers practical manufacturing requires a broad expertise and experience across various disciplines. During the design phase of a motor, a large number of parameters must be carefully considered and determined. This chapter covers the components and assembly of SRMs. It also provides discussions on SRM design and manufacturing. Assembly considerations with respect to lamination alignment, rotor balance, and shaft alignment are discussed to provide guidance on mechanical design of SRMs.

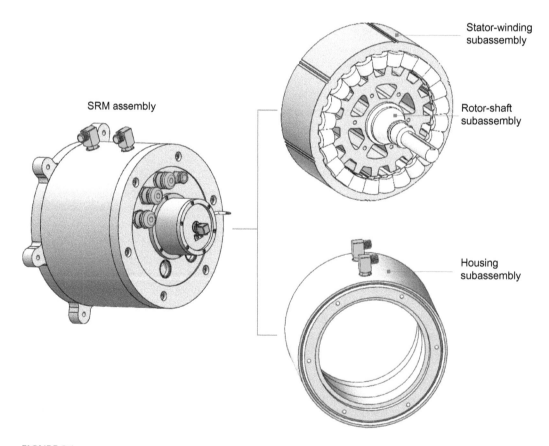

FIGURE 8.1
Major construction of a switched reluctance machine. (From Jiang, W.J., Three-phase 24/16 switched reluctance machine for hybrid electric powertrains: Design and optimization, Ph. D Dissertation, Department Mechanical Engineering, McMaster University, Hamilton, ON, 2016).

8.2 Frame

The frame of an SRM is the main physical boundary of the motor. Two major dimensions of a motor frame are the motor casing diameter and the motor casing axial length, which basically determine the physical volume of the motor. There are many occasions that the motor casing mass can be a design constraint. There are a number of dimensions used in determining the motor frame size and mounting elements. The major dimensions of horizontally foot-mounted electric motor frame according to National Electrical Manufacturers Association (NEMA) are shown in Figure 8.2.

There are a number of other mounting types, for instance wall mounted, ceiling mounted, pedestal mounted, face mounted, and flange mounted. Figure 8.3 shows some common mounting arrangements. The corresponding codes of NEMA and the International Electrotechnical Commissions (IEC) are also indicated. The mounting arrangement affects the physical dimensions of the motor and it is determined based on the application.

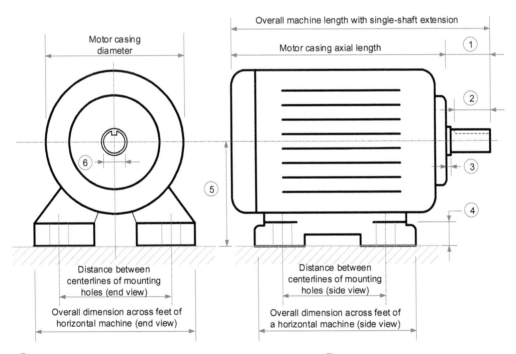

①	Length of shaft from the end of housing to the end of shaft drive end	④	Thickness of the mounting foot
②	Length of the shaft available for coupling on the drive end	⑤	Centerline of the shaft to bottom of the feet
③	Distance between the end of housing to the shaft shoulder	⑥	Shaft extension diameter

FIGURE 8.2
Typical dimensions of foot-mount electric motor frames. (From Chalmers, B.J. (Ed.), *Electric Motor Handbook*, Amsterdam, the Netherlands: Elsevier, 2013.)

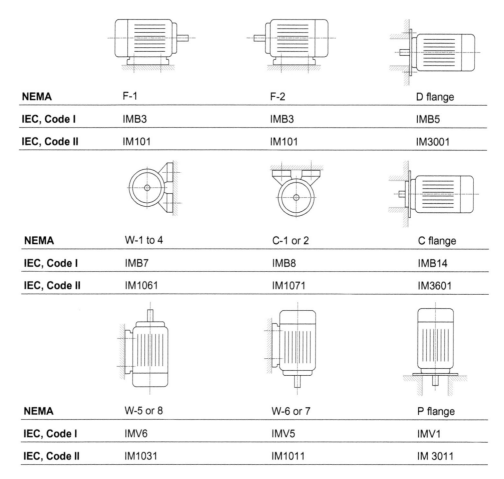

NEMA	F-1	F-2	D flange
IEC, Code I	IMB3	IMB3	IMB5
IEC, Code II	IM101	IM101	IM3001
NEMA	W-1 to 4	C-1 or 2	C flange
IEC, Code I	IMB7	IMB8	IMB14
IEC, Code II	IM1061	IM1071	IM3601
NEMA	W-5 or 8	W-6 or 7	P flange
IEC, Code I	IMV6	IMV5	IMV1
IEC, Code II	IM1031	IM1011	IM 3011

FIGURE 8.3

Common mounting arrangements for electric motors. (From Chalmers, B.J. (Ed.), *Electric Motor Handbook*, Amsterdam, the Netherlands: Elsevier, 2013.)

The sizing of the motor frame depends mainly on the volume of the stator-winding subassembly, and the type of enclosure, the method, and the type of mounting. The enclosure (the housing or frame) should meet some specific environmental requirements. The cooling method will also be added into the design of the motor enclosure. Selection of the cooling method is based on the power, current density, losses, and the allowable temperature rise.

Figure 8.4 shows the number of layers for an SRM when a liquid jacket is used for a water or oil cooling system. The very inner layer is the shaft-rotor subassembly. Then there is the air gap layer between the rotor and the stator. The next layer is the stator-winding subassembly. The coolant flows between the inner and outer frame layers. The inner frame layer separates the stator from the coolant. For air-cooled motors, fins are added to the outer layer to facilitate the heat dissipation. Figure 8.5 shows an example of an SRM

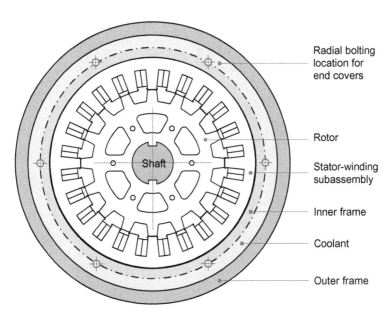

FIGURE 8.4
Layers of an SRM with water jacket. (From Jiang, W.J., Three-phase 24/16 switched reluctance machine for hybrid electric powertrains: Design and optimization, Ph.D. Dissertation, Department Mechanical Engineering, McMaster University, Hamilton, ON, 2016).

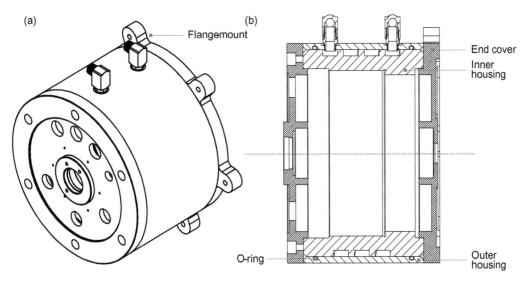

FIGURE 8.5
SRM housing with coolant channel: (a) 3D geometry and (b) cross section view. (From Jiang, W.J., et al., *IEEE Trans. Trans. Electrification*, 3(1), 76-85, March 2017.)

FIGURE 8.6
The housing of 24/16 traction SRM.

housing subassembly with a water jacket for cooling. The inner housing and outer housing form the coolant channel, which has a spiral shape. The end covers are bolted to the frame on both sides. Figure 8.6 shows the housing with water jacket for a 24/16 traction SRM.

8.3 Enclosures

There are a number of considerations that need to be made when designing the enclosure for a motor. The most important consideration should be the application of the motor. An SRM can be used in the automotive, chemical, food, mining, or many other industries. In each application, certain industry standards have to be met. The environment in which the motor will operate needs to be taken into account as well, because the enclosure has to ensure the protection and safety of the machine and operators. The enclosure might also need to prevent the motor from exposure to water, dust, chemicals, or other outside elements. In many noise-sensitive applications, the frame of the motor has to provide the noise attenuation to meet the requirements. As discussed earlier, the frame also serves as an important component for heat dissipation. The NEMA provides the definitions for various motor enclosures, as summarized in Table 8.1. Generally speaking, there are two major

TABLE 8.1

Common NEMA Electric Motor Enclosures

	Description	Definition
Open	Drip-proof (ODP)	Operate with dripping liquids up to 15-degree angle from vertical.
	Splash-proof	Operate with splashing liquids up to 100-degree angle from vertical.
	Guarded	Guarded by limited size openings (less than 3/4 in.)
	Semi-guarded	Only top half of motor guarded.
	Drip-proof fully guarded	Drip proof motor with limited size openings.
	Externally ventilated	Ventilated with separate motor driven blower, can have other types of protection.
	Pipe ventilated	Openings accept inlet ducts or pipe for air cooling.
	Weather protected type 1	Ventilating passages minimize entrance of rain, snow, and airborne particles. Passages are less than 3/4 in. in diameter.
	Weather protected type 2	In addition to type 1, motors have passages to discharge high-velocity particles blown into the motor.
Totally enclosed	Non-ventilated (TENV)	Not equipped for external cooling.
	Fan-cooled (TEFC)	Cooled by external integral fan.
	Explosion-proof (TEXP)	Withstands internal gas explosion. Prevents ignition of external gas.
	Dust-ignition-proof	Excludes ignitable amounts of dust and amounts of dust that would degrade performance.
	Waterproof	Excludes leakage except around shaft.
	Pipe-ventilated	Openings accept inlet ducts or pipe for air cooling.
	Water-cooled	Cooled by circulating water.
	Water to air-cooled	Cooled by water-cooled air.
	Air-to-air cooled	Cooled by air-cooled air.
	Guarded TEFC	Fan cooled and guarded by limited size openings.
	Encapsulated	Has resin-filled windings for severe operating conditions.

Source: Chalmers, B.J. (Ed.), *Electric Motor Handbook*, Amsterdam, Netherlands: Elsevier, 2013.

categories: totally enclosed and open. For an open motor, the winding is exposed to the external air to provide cooling. The cooling of a totally enclosed motor is done by a fan or coolant jacket. The IEC standards provide a more detailed description of motor enclosures.

8.4 Stator and Rotor

All electric machines are composed of two essential components: the stationary part—the stator and the movable or rotational part—the rotor. For a typical radial flux machine, the stator is fixed to the machine frame or the housing, the rotor is placed on either inside or outside the stator and is mounted with bearings to deliver the mechanical power to the load.

8.4.1 Laminations

The stator and rotor in an SRM are typically made of non-oriented soft magnetic electrical steel. The properties of the core materials are subject to numerous factors including purification of steel, controlling of alloying elements, grain orientation, and grain sizes [4]. Since the stator and rotor cores of an SRM can be exposed to high frequencies, it is often desirable to use electrical steel with lower losses. In addition, it is always desired to use a core material with high saturation flux density to improve the torque density and power factor [5]. High-saturation alloys and soft magnetic powder composites are also available for different SRM applications. Table 8.2 lists some of the materials and manufacturers for SRM cores.

Similar to other types of electric machines, the stator and rotor in SRMs are typically laminated to reduce eddy current losses. Various thicknesses are available from lamination manufacturers for different applications and types of machines. In general, thin laminations generate lower core losses and, therefore, achieve higher efficiency. Various grades of electrical steel also exist with different levels of volumetric core losses.

As discussed in Chapter 6, an insulation surface coating separates lamination steels. Proper lamination insulation prevents electrical shorting within the lamination stack. It increases the resistance of the core in the axial direction, which reduces the eddy current losses. Eddy currents are generated within each lamination by the alternating magnetic field. Poor insulation or no insulation between laminations will lead to inter-laminar eddy currents and, hence, additional core losses. In addition, lamination insulation provides resistance to corrosion and rust, and acts as a lubricant during stamping. Thus, it is necessary to apply the insulation to at least one side of the lamination sheets. If only one side of the lamination is coated with insulation, then all the lamination sheets should be oriented so that the insulated sides face the same direction during the lamination stacking process. In addition, local saturation can be caused from manufacturing techniques, such as lamination interlocks and cut outs, which results in local eddy current concentration and, hence, additional stator losses.

Once the lamination material, grade, and thickness are selected, the lamination steel will be cut into shapes to form the stator and rotor. Stamping is typically used for high-volume and high-speed production. The stator and rotor laminations can be punched simultaneously within the same sheet of steel. The laminations are stamped concentrically to produce the external stator and internal rotor, or vice versa for internal stator and external rotor. Laser cutting is another method to produce laminations. It is more suitable for large-size laminations where it is difficult to use a stamping machine and it provides a flexible solution for low volume prototypes.

TABLE 8.2

Sample Core Materials for Switched Reluctance Machine

Carbon Steel	Silicon Steel	Soft Magnetic Composite	High Saturation Alloys	Amorphous Alloys	Nano-crystalline Alloys
ASTM 1000 series	AK-Steel M series ArcelorMittal iCARe™ series Cogent Power™ M series JFE JNE series	HoganasAB Somaloy™ series Kobe Steel MH series	Carpenter Hiperco™ series Hitachi Metals YEP-2V VAC Vacoflux series	Hitachi Metglas™ amorphous metal Hitachi Metals Powerlite™ series VAC VITROVAC™ series	Hitachi Metals FT series VAC VITROPERM™ series

8.4.2 Stacking

Lamination stacking is the next process to assemble the stator and rotor cores respectively. Laminations are stacked axially on a mandrel with keys to align the salient slots and poles. For machines with a large stack length, the laminations are rotated periodically to mitigate asymmetry or misbalance in the manufacturing process. It is a common practice to provide clamping pressure to the lamination stacks in order to increase the stacking factor and prevent relative movement between laminations. End caps or clamp rings are typically used for these purposes. Figure 8.7 presents a prototyped double-rotor SRM stack with end clamps.

Several methods are commonly used for securing the lamination cores. Riveting and bolting offer reliable and economical solutions and, therefore are most commonly implemented. However, the rivet heads may interfere with the end windings and the holes punched in the laminations may introduce local stress concentration and flux concentration. Welding is another economical and reliable approach. However, weld joints electrically connect laminations, which creates shorting paths for eddy currents and increases iron losses. Weld joints should be carefully designed and preferably applied to low flux concentration areas, commonly at the stator outer diameter as shown in Figure 8.8. Finishing processes are also required after welding in some occasions to clean and smooth the weld joints, for instance, to achieve a good contact between the laminations and housing or structural components. Alternatively, lamination bonding applies adhesives to secure the lamination cores without introducing additional components or changing lamination geometries. The drawback of this method is that it can provide lower mechanical strength than riveting or welding.

Another stacking method is lamination interlocking. Dimples are placed on each lamination during the punching process. Each lamination then nests inside the neighboring lamination during the stacking process to form a solid lamination core. The interlocking process requires a complex and expensive die. The dimples create extra iron losses, and, therefore are typically placed at the back iron where the flux is lower.

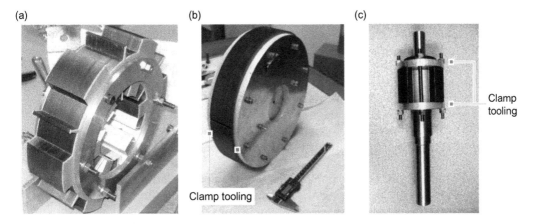

(a) (b) (c)

Clamp tooling

Clamp tooling

FIGURE 8.7
Lamination stacks: (a) stator, (b) exterior rotor, and (c) interior rotor. (From Yang, Y. et al., *IEEE Trans. Energy Convers.*, 30, 671–680, 2015; Yang, Y., *Double-Rotor Switched Reluctance Machine for Integrated Electro-Mechanical Transmission in Hybrid Electric Vehicles*, Ph.D. Dissertation, McMaster University, Hamilton, ON, February 2014.)

Weld joints on the stator

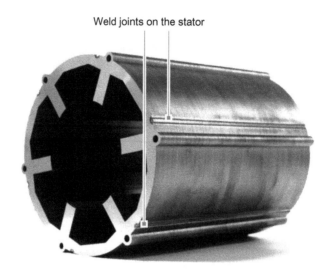

FIGURE 8.8
Weld joints on the stator of a 6/10 switched reluctance machine. (Courtesy of EDEC Laboratory, Illinois Institute of Technology, Chicago, IL). (From Bilgin, B. et al., Switched reluctance generator with higher number of rotor poles than stator poles, in *Proceedings of the IEEE Transportation Electrification Conference and Expo*, Dearborn, MI, pp. 1–6 June 2012; Bilgin, B. et al., *IEEE Trans. Ind Electron*, 60, 2564–2575, 2013; Bilgin, B. et al., *IEEE Trans. Ind. Electron.*, 59, 3745–3756, 2012.).

8.4.3 Design Features

Various slot and pole shapes exist for different switched reluctance machines. The number of poles largely determines the pole angles with certain degrees of design flexibility. The pole height can be adjusted to optimize output performance and efficiency. As discussed in Chapter 7, the pole teeth can be designed with parallel, tapered, or curved shapes depending on the application. Pole shoes can be added on the pole tips to support the slot wedges and to keep the stator coils in place.

Many design details can be added to the switched reluctance machine laminations. To reduce the rotor mass, cut-outs can be placed at the shaft side of the rotor back iron, where flux density is typically low. The location of the cut-outs shall be carefully designed to avoid either local stress concentration or excessive plastic deformation. Figure 8.9a shows the cut-outs applied on the rotor of a 24/16 SRM designed for a traction application. Figure 8.9b shows the cuts-outs on the rotor of a 6/14 SRM designed for an HVAC application. Similarly, small holes can be designed at low flux density areas for bolts or rivets to secure the cores. Furthermore, weld joints may be created at the outer circumferential surface of the cores. Keyways or knurls can be added at the inner surface of the rotor cores to guide the integration of motor and the shaft. Air barriers can be placed inside the laminations to guide or separate the flux lines [11].

Stator cores can be bolted to the housing, as shown in Figure 8.10a. For this type of connection, the stator cores can be easily disassembled from the housing. The stator lamination sheets are also bolted together. For the stator geometry without bolt holes, as shown in Figure 8.10b, the stator lamination sheets can be either glued or welded together. However, welding the stator lamination might affect the motor performance. Interference fit can be used to secure the stator core to the housing. The housing can be heated before installing the stator core. Bonding coating, for example SURALAC™ 9000 introduced by Cogent, can

(a) (b)

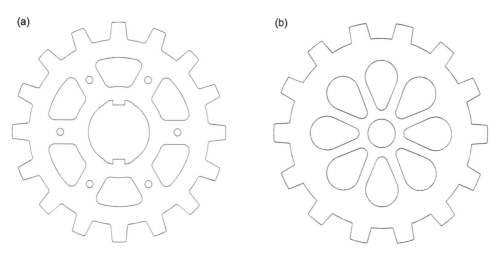

FIGURE 8.9
Cut-outs on the rotor of (a) 24/16 SRM and (b) 6/14 SRM.

(a) (b)

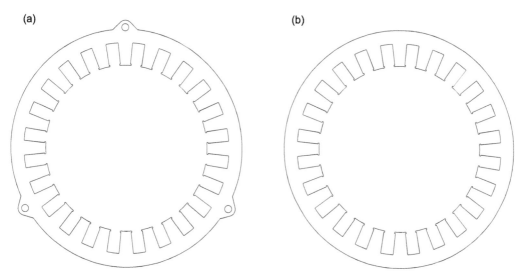

FIGURE 8.10
Stator lamination: (a) with bolt holes and (b) without bolt holes.

be used to bond lamination sheets together which also reduces noise and vibration in the stator core. The bonding coating also has a lubricating effect, which can prolong the life of the stamping tools [12].

8.5 Windings

Windings are the set of conducting coils embedded in the stator slots that are wound around each stator pole of an SRM. When excited with current, coils generate the magnetic field, which creates the magnetic interaction between the stator and rotor poles.

(a)   (b)   (c)

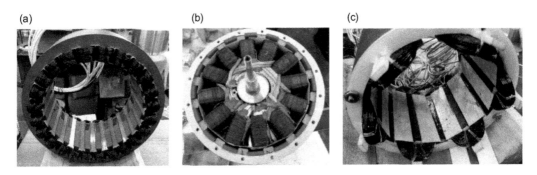

FIGURE 8.11
Actual windings in SRMs: (a) 24/16 traction SRM, (b) external-rotor e-bike SRM, and (c) HVAC SRM.

As discussed in Chapter 7, switched reluctance machines employ the simplest winding construction compared to other types of electric machines. The coils are simply wound around each stator pole. This is called concentrated winding. The coils can be connected in series, parallel, or in combination of series/parallel to create the phase winding. There is no need for any winding on the rotor. This simplifies the manufacturing and it improves thermal reliability and high-speed capability of the motor. Figure 8.11 shows some examples for SRMs designed for traction, e-bike, and HVAC applications.

8.5.1 Magnet Wires

For switched reluctance machines, coils are typically made of magnet wires, which are electrical conductors coated with insulation layers. Copper and aluminum are the most commonly used materials for conductors because of their high electrical conductivity, strong tensile strength, and relatively low price. Copper has higher density and cost, but it offers higher electrical conductivity than aluminum. This means that for the same ampere rating, copper wires require a smaller cross section area. In addition, copper also has the advantage of a higher melting point, higher tensile strength, and lower thermal expansion coefficient. Based on these advantages copper is more often used in machine windings, and it is almost exclusively used for low voltage, small and medium size switched reluctance machines.

Several cross-sectional shapes for magnet wires are available as shown in Figure 8.12. Round cross-sectional shapes of magnet wires are most commonly used in SRMs. This

(a)   (b)   (c)

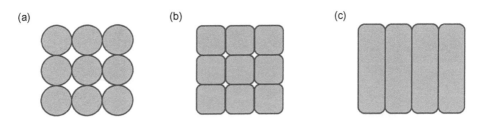

FIGURE 8.12
Magnet wire cross section variations: (a) round wire, (b) square wire, and (c) rectangular wire.

type of coil is generally called a stranded coil. They are easy to manufacture, flexible to bend during the assembly process, and provide good mechanical strength due to their lack of sharp corners or edges between neighboring wires. The wires can either be manually wound turn-by-turn or automatically layered by winding machines. However, due to the fundamental geometry limit of the round cross-sectional shape, such as the gaps between round wires, the practical slot fill factor is limited to 60% also considering the thickness of insulation layers and slot liners. The practical slot fill factor for round wires is between 35% and 40% for hand-wound coils to include enough space for the insulation layers and the winding process.

Thicker magnet wires with square or rectangular cross-sectional shapes can be implemented to achieve a higher slot fill factor. These are generally called form coils since they can be designed in a certain form by placing each conductor at a specific location. However, square and rectangular magnet wires are seldom used in small or medium size switched reluctance machines. Smaller motors usually have a short axial length and small stator poles which requires a high end-winding bend radius that could cause high mechanical stress on the corners of square or rectangular wires. Generally, the cost of form coils is higher than stranded coils. Round magnet wires form a random-wound machine winding, which are commonly used on applications with 690 V phase-to-phase voltage or below. Beyond this point, rectangular wires are used and are formed so that the turn-to-turn voltage does not exceed the dielectric break-down of the insulation system.

8.5.2 Winding Insulation

Insulation layers are necessary around the copper conductor to enable the contact between the wires without causing any electrical short circuit. As discussed in Chapter 6, magnet wire insulation is generally made of organic material, which softens at a lower temperature than copper or electrical steel. The mechanical strength of the insulation is also much lower than copper or steel. Therefore, electrical insulation around the conductors is generally the limit for the lifetime of the stator winding.

In addition, winding-to-stator insulation is necessary in SRMs to prevent possible winding-to-ground short circuit faults. For prototype machines, slot liner and slot wedges are typically applied. These insulation materials are usually made of aramid or Mylar layered paper. Nomex is an example of insulation paper supplied by DuPont for slot insulation. It is often used in industrial applications due to its high mechanical strength, flexibility, high chemical and moisture resistance, and dielectric breakdown strength at high temperatures. The thickness of the slot insulation material ranges between 0.1 and 0.65 mm. Figure 8.13 illustrates a switched reluctance machine prototype with insulations.

For large volume production machines, injection molded plastic insulation is typically used to form a structure that surrounds the stator lamination to prevent contact between the winding and the lamination steel. Figure 8.14 shows how injection molded plastic insulation is applied in an SRM. The injection molded plastic insulation structure provides various advantages compared to slot liners. The formed structure has consistent insulation thickness, is more reliable and durable, faster to manufacture, and often enables an integrated way to place and route the end wires. The slot wedge can also be a part of injection molded plastic insulation; therefore, in those applications pole shoes would not be necessary which simplifies the stamping of the laminations.

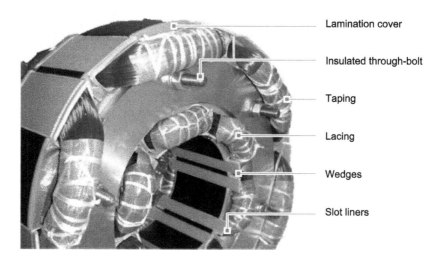

FIGURE 8.13
Switched reluctance machine prototype with insulations.

FIGURE 8.14
Switched reluctance machine injection mold plastic insulation.

Figure 8.15 shows how slot wedges are applied on the stator of an SRM without the injection molded plastic insulation.

Typical plastic injection molding materials for stator and rotor include polyamide (PA), such as Nylon PA 6, PA 66; polyethylene (PE), and polybutylene terephthalate (PBT). Glass fiber reinforced polyphenylene sulfide (PPS), such as PPS-GF30, PPS-GF40, and polysulfone (PSU), can be used for harsh, high temperatures and contaminative

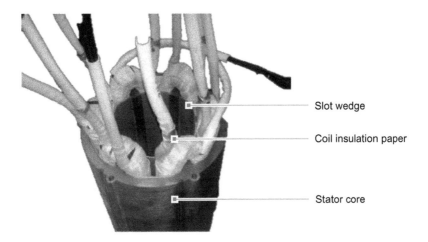

FIGURE 8.15
Coils and slot insulation shown on the prototype of a 6/10 SRM. (Courtesy of EDEC Laboratory, Illinois Institute of Technology, Chicago, IL). (From Bilgin, B. et al., Switched reluctance generator with higher number of rotor poles than stator poles, in *Proceedings of the IEEE Transportation Electrification Conference and Expo*, Dearborn, MI, pp. 1–6 June 2012; Bilgin, B. et al., *IEEE Trans. Ind Electron*, 60, 2564–2575, 2013; Bilgin, B. et al., *IEEE Trans. Ind. Electron.*, 59, 3745–3756, 2012.)

environments, for instance, where the motor is in direct contact with transmission oil or engine coolant. In addition, encapsulation of the stator via vacuum pressure impregnation (VPI) process helps to improve the transmission of heat from the winding to the stator core by replacing the air pockets with varnish or thermally conductive epoxy. It also provides higher dielectric and mechanical strength, and chemical and heat resistance.

8.5.3 Insulation Test

It is important to test the winding insulation integrity during off-line motor testing, which typically happens during machine assembly check, maintenance, or prototype building. The purpose of these tests is to ensure that all the phases are balanced, and well insulated at turn-to-turn, phase-to-phase, and phase-to-ground. Multiple tests are commonly conducted including the Meg-Ohm test, polarization index test, high-potential test, step-voltage test, and surge test [13]. An advanced off-line motor test analyzer is capable of testing all these within one set.

Meg-Ohm test measures the stator winding insulation resistance by applying a constant DC voltage to the winding phase terminals while connecting the housing to the ground. A low Meg-Ohm value indicates a poor insulation system and, hence, the likelihood for the machine to have phase-to-ground shortage and eventually failure. It is recommended by the Institute of Electrical and Electronics Engineers (IEEE) 43-2013 [14] that the minimum insulation resistance for a machine rated below 1 kV line-to-line voltage is 5 MΩ with the temperature corrected to 40°C.

Polarization index (PI) is another common test to examine the conditions of the stator winding insulation. A test analyzer applies a DC voltage to the winding terminals and records the insulation resistance variation over a 10-minute period as the molecules in the insulation system get polarized. A PI ratio is defined as the insulation resistance ratio between the

10-minute mark and the 1-minute mark. According to IEEE 43-2013, the minimum PI value for a thermal class A rated machine is 1.5 while it needs to be 2.0 for thermal class B, F, and H.

The step-voltage test is similar to polarization index test, except that the input voltage increases in steps at various voltage levels. In fact, a step-voltage test is typically performed right after the polarization index test. The reason for applying increasing voltage steps is to ensure that the insulation system can withstand the normal day-to-day voltage spikes the machine will typically see during starting and stopping.

High-potential test, often called hi-pot test, is conducted to verify the dielectric voltage withstand capability of the stator winding insulation. A high DC voltage is applied to the winding terminals and leakage current is measured over time to determine the insulation integrity. According to the IEEE Standard 95-2002 [15], the hi-pot test voltage is two times the machine rated voltage plus 1,000 V, for instance $2 \times U + 1000$, where U is the rated voltage.

The final test is the surge test, which is an important test to detect potential winding faults in turn-to-turn, coil-to-coil, and phase-to-phase insulations. Surge test works on the principle that the oscillating frequency of the winding changes when the inductance or the capacitance changes. During a surge test, a short, high-voltage pulse generated by a charged capacitor is released into the stator winding, creating voltage difference between turns within one coil. Depending on the coil inductance, a corresponding waveform is recorded in the surge test analyzer. If all the turns, coils, and phases are well insulated, the waveform for each phase would be identical. Otherwise, there may be weak insulation or shortages in the winding that shortens the current path and lead to a form of leakage.

8.5.4 Winding Connections and Unbalanced Magnetic Pull

Depending on the machine drive constraints, such as current, and performance objectives, such as radial force balance and noise and vibration reduction, switched reluctance machine coils can be connected in different ways depending on the number of coils per phase. As discussed in Chapter 7, in a three-phase 24/16 SRM, six coils per phase can be connected in series or parallel or in a combination of series/parallel configuration. Examples of different coil connections can be seen in Figure 8.16. If the motor configuration can employ only two coils per-phase such as in three-phase 6/4 or four-phase 8/6 SRM, the coils can be connected either in series or parallel. The majority of SRMs employ series connected coils. In this section, the effect of winding connection on the unbalanced magnetic pull will be discussed.

The unbalanced magnetic pull (UMP) is generated when eccentricity between the rotor and the stator occurs. Eccentricity between the rotor and the stator can be generated either during the manufacturing stages, assembly process, operation, or maintenance period. Factors including manufacturing tolerances, displaced bearings, rotor lamination assembling problem, and uneven load all contribute to the eccentricity and, thus, create UMP. UMP can further lead to the bending of the shaft, reduction of the mechanical stiffness of the shaft and even serious dry friction between the rotor and the stator.

Three types of eccentricities are often observed: static eccentricity, dynamic eccentricity, and mixed eccentricity. To illustrate the mechanism for eccentricity, three axis have been identified, as shown in Figure 8.17. l_{sc} is the geometric center axis for stator-case assembly,

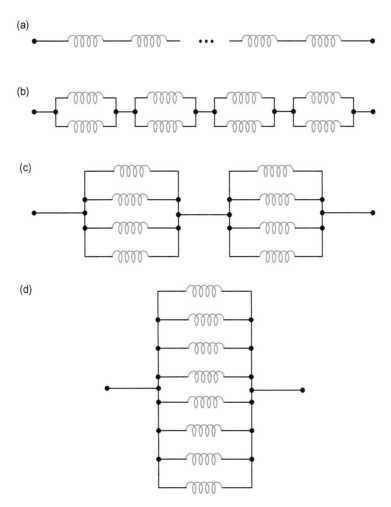

FIGURE 8.16
Winding connections for one phase, 24/16 SRM: (a) series winding, (b) 4S2P (combination of series and parallel winding), (c) 2S4P, and (d) 8P (all parallel winding). (From Jiang, J.W. et al., *IET Electr. Pow. Appl.*, 10, 526–538, 2016.)

l_r represents the rotor's geometric center axis, and l_{sb} is the geometric axis for the shaft-bearing assembly, which is also the rotation axis. It is assumed that bearings fit tightly on the shaft and the motor's back covers. Problems in stator lamination stack assembly, stator and case misalignment, or lose stator stack can lead to a tilted stator-case assembly geometric center axis l_{sc} with respect to the shaft-bearing assembly geometric center axis l_{sb}. Shaft misalignment, bent shaft, or bearing displacement all result in a tilted shaft-bearing axis l_{sb} with respect to the rotor geometric center axis l_r. The spatial relationship between l_{sc}, l_r, and l_{sb} determines the type of spatial eccentricity. We'll focus on 2D analysis, which means C_{sc} (stator-case geometric center), C_r (rotor geometric center), and C_{sb} (shaft-bearing geometric center, also referred as rotation center) can be analyzed regarding their planetary relationships.

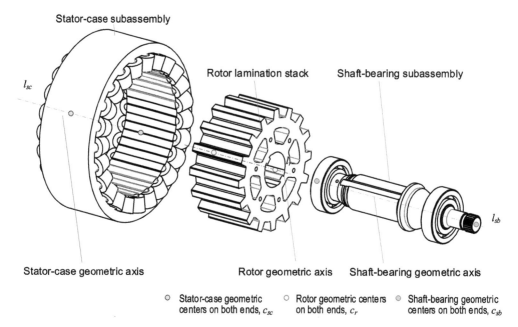

FIGURE 8.17
Spatial relationship between shaft-bearing assemble axis, rotor geometric center axis, and stator-case assemble axis. (From Jiang, J.W. et al., *IET Electr. Pow. Appl.*, 10, 526–538, 2016.)

The ideal scenario is when the three centers are all concentric with each other, as shown in Figure 8.18a. For static eccentricity, C_r is concentric to C_{sb} but eccentric to C_{sc}, as shown in Figure 8.18b. As the name suggests, for static eccentricity, the rotor center remains unchanged while the rotor rotates. However, for dynamic eccentricity, the rotor center is rotating as the rotor spins. For dynamic eccentricity shown in Figure 8.18c, C_{sb} is concentric regarding C_{sc}. Mixed eccentricity in Figure 8.18d is more like a combination of static eccentricity and dynamic eccentricity. A special case of mixed eccentricity can be observed when the stator-shaft displacement, l equals r. In practice, static eccentricity, dynamic eccentricity, and mixed eccentricity occur the most often.

For the static eccentricity, the rotor geometric center C_r is concentric with respect to the rotor rotation center C_{sb}; however, a displacement exists between the rotor rotation center C_{sb} and the stator-case geometric center C_{sc}. The displacement between the stator geometric center and the rotor geometric center is along the positive X-axis, as shown in Figure 8.18b. Since the relative position between the stator lamination and the rotor rotation center is fixed, the air gap lengths between the rotor and the stator are unchanged around the entire bore when the rotor rotates.

Figure 8.19 describes the relative position between the rotor and the stator with dynamic eccentricity applied. The rotor spins from its initial position in Figure 8.19a around its rotation center in counter-clockwise direction by θ degrees. Figure 8.19a–d shows four consecutive positions for the machine with dynamic eccentricity as the rotor rotates from the initial position for a complete revolution. For the dynamic eccentricity, the total air gap difference is fixed as the rotor rotates and the air gap differences along X- and Y-axis change periodically. Figure 8.20a–d shows four consecutive positions for the machine with mixed eccentricity, in which the total air gap difference is varying periodically.

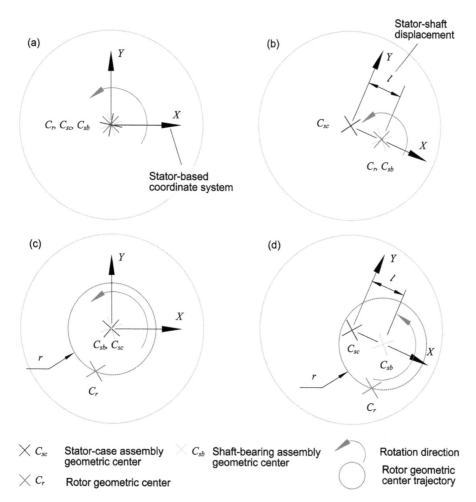

FIGURE 8.18
Spatial relationships between C_r, C_{sc}, and C_{sb}: (a) all concentric, (b) static eccentricity, (c) dynamic eccentricity, and (d) mixed eccentricity. (From Jiang, J.W. et al., *IET Electr. Pow. Appl.*, 10, 526–538, 2016.)

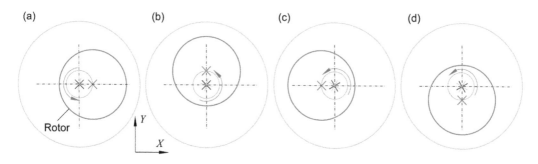

FIGURE 8.19
Rotor center trajectory analysis for dynamic eccentricity: (a) 0 mechdeg., initial position, (b) 90 mech. deg., (c) 180 mech. deg., and (d) 270 mech. deg. (From Jiang, J.W. et al., *IET Electr. Pow. Appl.*, 10, 526–538, 2016.)

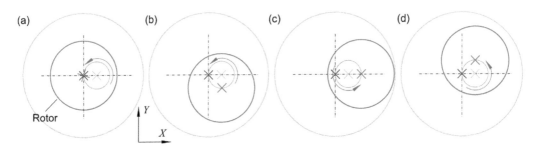

FIGURE 8.20
Rotor center trajectory analysis for mixed eccentricity: (a) initial position, (b) 90 deg., (c) 180 deg., and (d) 270 deg. (From Jiang, J.W. et al., *IET Electr. Pow. Appl.*, 10, 526–538, 2016.)

Figure 8.21 compares the UMP waveforms for four winding connections, 8S (all series), 4S2P (two parallel paths), 2S4P (four parallel paths), and 8P (all parallel), for 24/16 SRM with dynamic eccentricity. It can be seen that with more parallel paths in the winding, a higher reduction of UMP can be achieved when eccentricity exists. The UMP waveform for 8S with dynamic eccentricity is fluctuating around a fixed value. The UMP waveforms can also be plotted in a radial coordinate, as shown in Figure 8.22. As mentioned earlier, for motors with dynamic eccentricity, since the total air gap difference is fixed as the rotor rotates, the UMP distributes evenly in the radial coordinate.

A parallel connection also helps reduce UMP when mixed eccentricity exists, as shown in Figure 8.23. As discussed earlier, in the case of mixed eccentricity, the total air gap difference varies periodically as the rotor rotates. The total air gap difference

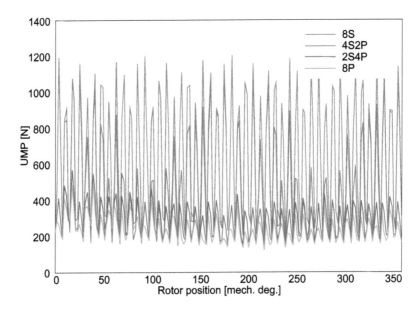

FIGURE1 8.21
UMP waveforms for 8S, 4S2P, 2S4P, and 8P, dynamic eccentricity (0.1 mm on positive X-axis).

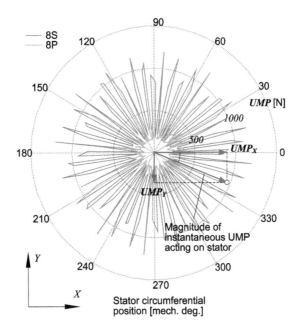

FIGURE 8.22
Comparison between UMP distribution patterns for 8S and 8P, dynamic eccentricity (0.1 mm on positive X-axis).

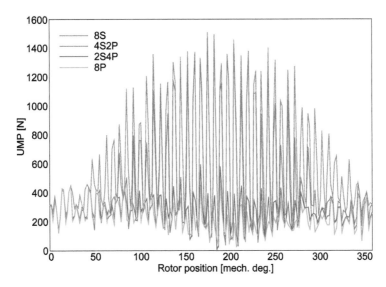

FIGURE 8.23
Comparison between UMP waveforms for 8S, 4S2P, 2S4P, and 8P, mixed eccentricity (0.1 mm on positive X-axis).

is also reflected on the distribution of the UMP in a polar plane, as shown in Figure 8.24. Since the rotation center is shifted along the positive X-axis and the rotor center's trajectory lies in the left half plane, the UMP distribution pattern also tends to act in the same area. It can be seen that the UMP distribution for 8P is compressed, compared with 8S.

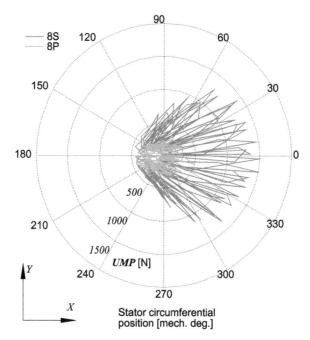

FIGURE 8.24

Comparison between UMP distribution patterns for 8S and 8P, mixed eccentricity (0.1 mm on positive X-axis).

8.6 Bearings

Bearings and shafts are critical components in SRMs to support the rotating rotor and align the rotor both radially and axially to the stator, as shown in Figure 8.25. Since the air gap of SRMs are relatively small compared to that of other machine types, it is important to have a rigid machine structure to ensure a uniform air gap. An uneven air gap is often a cause for high torque ripple, unbalanced radial pull, and unwanted vibration and noise. Poorly conditioned bearings also lead to reduced machine performance and increased loading due to excessive friction. It is reported that more than half of all motor failures can be attributed to bearing failures [17]. Bearings can be categorized into two primary types: hydrodynamic journal bearings and anti-friction rolling bearings.

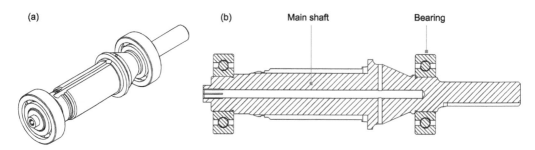

FIGURE 8.25

Shaft and bearing subassembly: (a) 3D geometry and (b) cross section view.

8.6.1 Journal Bearings

Journal bearing, also called solid bearing or plain bearing, is a type of bearing that comprises only bearing surface with no rolling elements. The design and construction of journal bearings are relatively simple. They are compact, lightweight, and can withstand high load. The working principle of a journal bearing is based on using grease, oil, or gases to form a thin film layer to separate the rotating shaft from the stationary journal bearing surface. Since shaft diameter is typically slightly smaller than bearing surface diameter, an eccentricity between the shaft center and the bearing surface center results due to gravity, as the shaft tends to rest on the bearing bottom dead center (BDC). This is illustrated in Figure 8.26a. During shaft rotation, a convergent wedge clearance is developed due to the relative surface movement, creating hydrodynamic pressure in the film to carry the shaft as shown in Figure 8.26b. This hydrodynamic pressure lifts the shaft and allows the journal bearing to tolerate moderate vibration and dynamic load variations [18].

Journal bearings are advantageous for low-cost switched reluctance machines operating at light load conditions in lubricated applications, such as electric drives for water pumps and oil pumps. The primary advantages of journal bearings include:

- Light weight
- Low cost manufacturing
- Long life under normal load operation
- Low operation noise
- Less transmitted vibration
- Shock load tolerant capability
- Electrical isolation from rotor to ground
- Less electromagnetic interference
- Less sensitivity to contamination
- Less mounting accuracy requirement
- Ease of maintenance

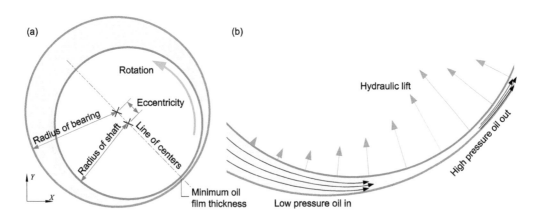

FIGURE 8.26
Journal bearing working principle: (a) journal bearing cross section model and (b) journal bearing operation principle. (From Leader, M.E., *Understanding Journal Bearings*, [Online]. Available: http://edge.rit.edu/edge/ P14453/public/Research/2-_LEADER_-_Understanding_Journal_Bearings.pdf. [Accessed January 23, 2017].)

- Good heat dissipation with lubrication
- Good wear dust clearance with lubrication

A wide variety of materials can be used for journal bearings including iron, copper, bronze, carbon, graphite, ceramics, plastics, and composite materials [18]. Lead-based alloys have been virtually eliminated in modern applications due to environmental concerns and lack of strength. Different materials have different properties and the material should be carefully chosen to suit the application. Properties such as material strength, hardness, surface roughness, temperature rating, chemical resistance, electrical conductivity, and dry run capability should all be considered in the design process. It is important to ensure smooth operation and longevity of journal bearings by proper material selection, installation, and lubrication.

Side grooves can be added in the inner journal surface for better heat dissipation and material wear clearance. Figure 8.27 illustrates examples of journal bearings with clearance grooves. In addition, shaft load, speed, and lubrication viscosity should also be taken into account in material selection. These properties have significant impact on the journal bearing's performance and life. Figure 8.28 illustrates the Stribeck curve, which was proposed by German engineer Richard Stribeck in 1902. It shows the friction factor as a function of the shaft load, speed, and lubrication viscosity [19]. It can be observed that friction in the journal bearings start high at very low speeds, decrease to a minimum where metal-to-metal contact is eliminated, and increase thereafter due to fluid dynamic viscosity as shaft speed increases. It is thus ideal to locate the friction factor somewhere in the center region of the Stribeck curve. Since speed and load are typically determined by the bearing application, lubrication viscosity is important for design optimization.

8.6.2 Rolling Bearing

The other common bearing type is the anti-friction rolling bearing. Rolling bearings are typically composed of an inner ring, an outer ring, multiple rolling elements (such as balls and rollers), and a cage that holds these together. The rings guide the rolling elements to roll with very little rolling resistance and little sliding in the race. The rings and the rolling elements take load while the cage sustains no direct load. The cage functions to reduce

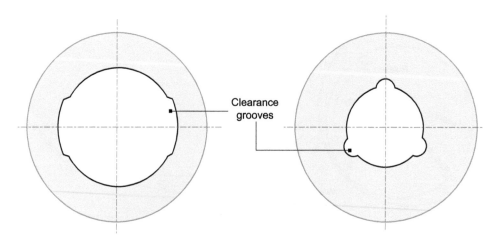

Clearance grooves

FIGURE 8.27
Journal bearings with clearance grooves.

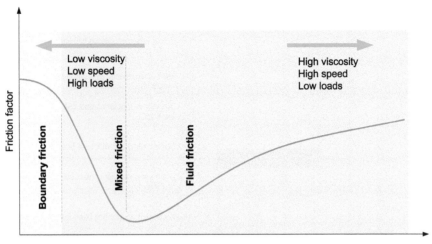

FIGURE 8.28
Stribeck curve showing friction factor as a function of viscosity, speed, and load.

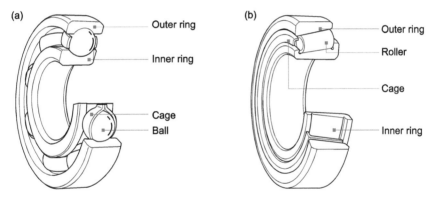

FIGURE 8.29
Ball bearing and roller bearing construction: (a) ball bearing and (b) roller bearing. (From NTN Global, *Classification and Characteristics of Rolling Bearings*. Available: http://www.ntnglobal.com/en/products/catalog/pdf/2202E_a01.pdf. [Accessed: March 5, 2017].)

friction and wear, to hold the rolling elements at equal distance from each other, to guide them into load race, and to prevent them from falling out.

Rolling bearings are typically classified by the type of the rolling element, such as ball bearing versus roller bearing, as shown in Figure 8.29. Ball bearings are designed to handle both radial and thrust loads, to some extent. In comparison, roller bearings provide higher capability to withstand heavy load by distributing it over a large cylindrical area. However, roller bearings are sensitive to the direction of the load. This will be further discussed when we talk about different types of roller bearings.

Rolling bearings are also classified by the type of load that the bearing is designed to withstand (such as radial bearing versus thrust bearing). Radial bearings accommodate loads that are predominantly perpendicular to the shaft, while thrust bearings accommodate loads that are predominantly in the axial direction of the shaft. Rolling bearings can be further classified as shown in Figure 8.30. The classification is based on

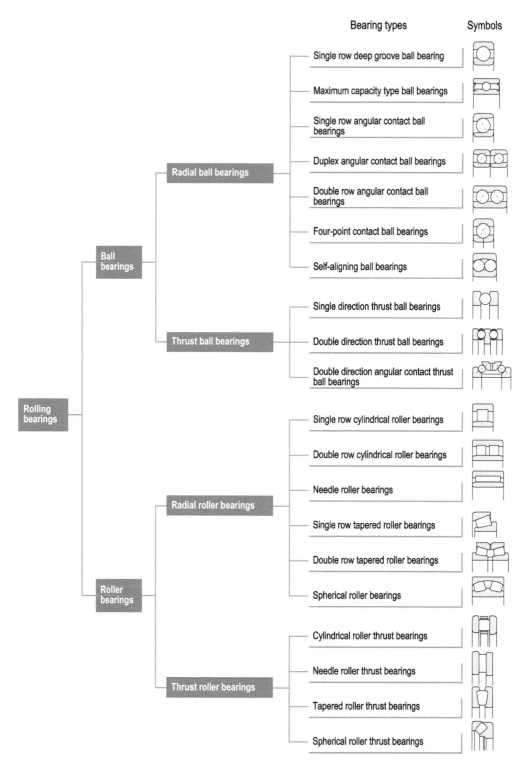

FIGURE 8.30

Rolling bearing classification. (From Nachi, *Ball and Roller Bearings*, Machi-Fujikoshi, Tokyo, Japan, 2014.)

(i) the shape of the rollers (ii) the number and the shape of the raceways (iii) the relationship between the bearing rings (inner and outer rings) and the rolling element, and (iv) the use of accessories. Simplified bearing symbols are illustrated next to the corresponding bearing types.

Table 8.3 further summarizes the typical characteristics of different rolling bearing types, where the arrows show the allowable load direction and capacity, and the dots mark the performance level. A longer arrow means the bearing is capable of handling higher loads in that particular direction, while the more dots a bearing gets, the better performance it has in that category.

TABLE 8.3

Typical Rolling Bearing Characteristics

Bearing Type	Symbols	Load Carrying Capacity (Axial load ← / Radial load ↑)	High Speed Rotation	Accuracy	Low Noise	Rigidity
Deep groove ball bearings			••••	•••	•••	•
Angular contact ball bearings			••••	•••	•••	•
Double row angular contact ball bearings			•••	•	•	•
Self-aligning ball bearing			••	•	•	•
Cylindrical roller bearings			•••	•••	••	••
Double row cylindrical roller bearings			•••	•••	••	•••
Tapered roller bearings			••	•••	•	••
Double-row tapered roller bearings			••	•	•	••••
Spherical roller bearings			••	•	•	•••
Needle roller bearings			••	•	•	••
Single-row direction thrust ball bearings			•	•	••	•
Double-row thrust angular contact ball bearings			•••	•••	••	••

(*Continued*)

TABLE 8.3 (*Continued*)

Typical Rolling Bearing Characteristics

Bearing Type	Symbols	Load Carrying Capacity		High Speed Rotation	Accuracy	Low Noise	Rigidity
		Axial load	Radial load				
Thrust cylindrical roller bearings		←		•	•	•	•••
Thrust tapered roller bearings		←		•	•	•	•••
Spherical roller thrust bearings		←	↑	•	•	•	•••

Source: Nachi, *Ball and Roller Bearings*, Machi-Fujikoshi, Tokyo, Japan, 2014.

As rolling bearings must be able to maintain high reliability and rotational precision under repeated high stress, materials with high strength, hardness, and fatigue-resistant characteristics are required for bearing rings and rolling elements. Standard materials are high carbon chromium bearing steel or case hardening steel with a hardened carburized outer layer. The steel is heat-treated to enhance hardness and attain optimum resistance to rolling fatigue. Bearing surfaces are fine polished to a very high accuracy by using special machine tools. Ceramic materials can also be used for special applications, such as those that require a high level of hardness, electrical insulation, and chemical resistance. In addition, cold-rolled steel, polyamide, and brass can be used for bearing cages. Compared to journal bearings, rolling contact bearings have lower starting and running resistance, consume less amount of lubricant, and are less prone to wear, thus require far less maintenance. Figure 8.31 demonstrates a typical bearing selection procedure. Multiple factors and selection criteria should be considered when making the final decision.

8.6.3 Axial Positioning and Bearing Fit

The axial positioning of the bearing is important in the mechanical design of electric machines. As shown in Figure 8.32a, one end of the bearing is usually in contact with the shaft shoulder, and the other end should be axially positioned. A wave spring washer can be used to axially fix the position of the bearing. The wave spring washer provides a preload on the outer ring of the bearing. This helps compensate the dimensional variations due to the assembly tolerance stack up errors and change in temperature. The wave spring washer can help to reduce acoustic noise and vibration in electric machines by absorbing the impulse force.

Figure 8.32b shows a retaining ring that can be used for the axial positioning of the inner and outer rings. The snap ring shown in Figure 8.32c is assembled in the groove of the bearing, and it helps to fix the axial position of the bearing, since it is in contact with the housing or the end cover. A snap ring is usually specified for heavy duty or impact loading applications.

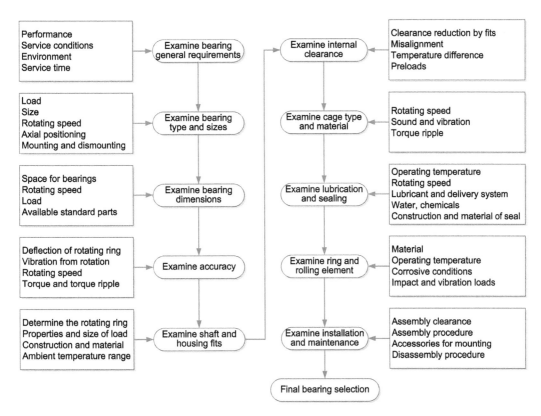

FIGURE 8.31
Bearing selection procedure.

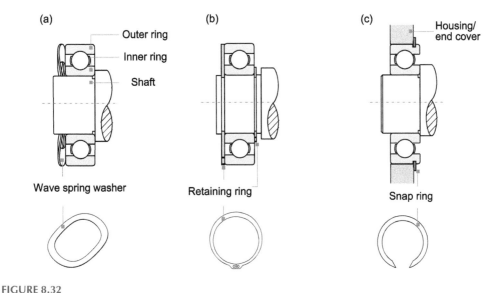

FIGURE 8.32
Axial positioning of a bearing: (a) bearing with wave spring washer and shaft shoulder, (b) bearing with retaining ring, and (c) bearing with snap ring.

8.7 Shafts

The rotor shaft is another critical component in SRMs to support the machine structures and transmit torque and speed to the external load. Typically, a machine shaft is supported by two bearings on each end of the shaft, and the rotor is mounted to the center of the shaft. It is crucial for the shaft to be designed so that it can handle the combined tension, bending, and torsional effects through its lifetime. Some motors undergo frequent start-stops and change of direction, which causes cyclic torsional stress on the shaft [17]. The distance between the two bearings is designed to be as short as possible to enhance the shaft stiffness and reduce shaft stress and deflection. It is also not desired to have long shaft overhang as this would increase the shaft bending stress and may lead to a rotor unbalance issue.

8.7.1 Shaft Loading

The shafts in SRMs typically endure various combined load types, including radial and axial loads, steady and cyclic loads, torsional and transverse loads, fitting preloads and shock loads, and so on. Descriptions of each load type are as follows:

- Radial loads are caused by unbalanced magnetic pull due to rotor saliency and non-uniform air gap distance along the motor radial direction.
- Axial loads are the loads perpendicular to the shaft cross section. They are typically smaller than radial loads and are a result of end winding unbalanced magnetic pull. Shafts can also be subject to axial loads resulted from axial pre-loading or a misaligned shaft coupling.
- Steady loads are primary design loads that occur continuously on a non-rotating shaft such as transverse loads due to gravity, and tension loads by pulley or chain tensioner. The shaft in this case does not transmit output torque, and functions to support mechanical structures and serve as a pivot axis while rotor assembly rotates around the shaft via bearings; for example the shaft used for external rotor in-wheel motor in electric bicycle applications.
- Cyclic loads apply repeated alternating stresses and strains on the shaft. For example, gravity causes bending stress in the shaft between the two bearings. The stress varies with shaft position from tension to compression and back to tension in every shaft rotation.
- Torsional load is the main load on the cross section of the shaft due to torque transmission. As a result, torsional shear stress occurs throughout the shaft from the shaft coupling end to the rotor lamination end. The maximum torsional shear stress is located on the surface of the shaft as the shear stress is proportional to the shaft diameter.
- Transverse loads are the loads perpendicular to the shaft axis. For a horizontally mounted motor, the motor weight acts as a transverse load between the supporting bearings. If there are gears or sprocket wheels mounted on the shaft, the shaft is also subjected to transverse loading from pulling tension or compression to transmit power.

- Shaft preloading can result from interference fit of rotor cores, bearings, clamps, etc. Preloads can also be applied between gear members that are mounted on the shafts to properly maintain gear mesh tooth contact.
- Shock loads are intermittent loads that occur non-consistently on the shaft. It happens when loads start, stop, suddenly change, or experience interruption in a rapid pace due to kinetic energy stored in the motor inertial system.

These combined loads vary with loading cycles, road conditions, shaft speed, acceleration, and exert tension, torsion, bending, and compression on the shaft cross section. As the rotating motion may be continuous, intermittent, uni-directional, or bi-directional, the shaft should be designed to tolerate the maximum stress under the worst-case scenario with a defined safety margin.

8.7.2 Shaft Geometry

The shaft geometry changes depending on the application. Figure 8.33 shows the shaft of an SRM designed for a traction application. The simplest shaft is a solid round straight shaft with rotor cores pressed on or shrunk on it. Hollow shafts are common for large motors and generators. The shaft outer diameter may vary to suit for different mounting features such as bearings, clamps, rotor cores, drive gears, and speed sensors. Different steps are thus created so that the central shaft part has the largest outer diameter and the end has the smallest. Figure 8.34 illustrates a shaft geometry example with different features on the shaft surface. The output end of the main shaft can either have keyway or spline to transmit torque to the load.

Shaft seats manufacturing tolerances designed for bearings are critical to ensure good bearing fit, smooth motor operation, and long bearing life. The amount of interference between the mating shaft surface and the mounting bearing can be divided into three categories: loose fit (clearance fit), slide fit (transition fit), and press fit (interference fit). Loose fit allows for easy bearing installation but too loose of a fit may lead to excessive wear, vibration, and bearing ring creep and fracture. Slide fit is a transition fit that is neither loose nor tight. It applies slight press fit to help prevent creep but also allows certain degrees of assembly flexibility. Press fit helps to secure the bearings on the shaft without internal clearance. Although press fit naturally makes installation of bearings more difficult and susceptible to damage, problems such as noise, vibration, and loose assembly can be avoided with proper calculations and careful installation. Table 8.4 presents ISO-preferred system of limits and types of fits for reference. Please refer to [22] to interpret the information given in the table.

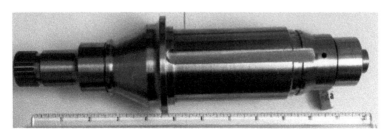

FIGURE 8.33
Shaft of a traction SRM.

(a)

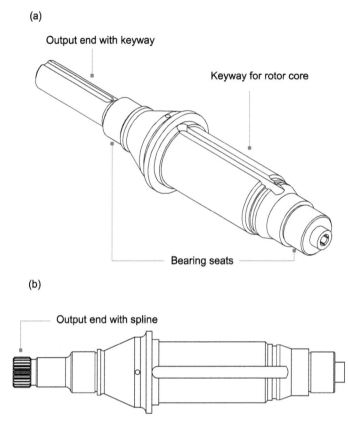

Output end with keyway

Keyway for rotor core

Bearing seats

(b)

Output end with spline

FIGURE 8.34

Shaft geometry and features: (a) output end with keyway and (b) output end with spline.

TABLE 8.4

ISO Preferred System of Limits and Fits

	Hole Basics	Shaft Basics	Descriptions	Notes
Clearance Fits	H11/c11	C11/h11	Loose running fit for wide commercial tolerances or allowances on external members.	↑ More clearance
	H9/d9	D9/h9	Free running fit not for use where accuracy is essential, but good for large temperature variations, high running speeds, or heavy journal pressures.	
	H8/f7	F8/h7	Close running fit for running on accurate machines and for accurate location at moderate speeds and journal pressures.	
	H7/g6	G7/h6	Sliding fit not intended to run freely, but to move and turn freely and locate accurately.	

(Continued)

TABLE 8.4 (*Continued*)

ISO Preferred System of Limits and Fits

	Hole Basics	Shaft Basics	Descriptions	Notes
	H7/h6	H7/h6	Locational clearance fit provides snug fit for locating stationary parts; but can be freely assembled and disassembled.	
Transition Fits	H7/k6	K7/h6	Locational transition fit for accurate location, a compromise between clearance and interference.	
	H7/n6	N7/h6	Locational transition fit for more accurate location where greater interference is permissible.	
Interference Fits	H7/p6	P7/h6	Locational interference fit for parts requiring rigidity and alignment with prime accuracy of location but without special bore pressure requirements.	
	H7/s6	S7/h6	Medium drive fit for ordinary steel parts or shrink fits on light sections, the tightest fit usable with cast iron.	More interference
	H7/u6	U7/h6	Force fit suitable for parts which can be highly stressed or for shrink fits where the heavy pressing forces required are impractical.	

Source: International Organization for Standardization, ISO Standards Handbook—Limits, fits and surface properties, in ISO, 1999.

There are many parameters that need to be considered when selecting the fit for the bearing and shaft, as well as for the bearing and the housing/end cover. Some of these parameters are conditions of rotation, magnitude of the load, bearing internal clearance, temperature differences, running accuracy, design and material of the shaft and housing, ease of mounting and dismounting, and displacement of the bearing in the non-locating position. Figure 8.35 shows fitting types and the tolerances for the connection of the bearing with the shaft and the housing. Please refer to Table 8.4 for the description of the tolerances.

8.7.3 Shaft Assembly

Figure 8.36 shows an example of shaft-rotor assembly. Figure 8.37 shows the actual shaft-rotor assembly. The most economical way to assemble a shaft into a rotor core is press fit. However, if the press fit is too tight, the shaft may be bent, or the rotor laminations might be damaged. On the other hand, if it is too loose, the rotor core may slide on the shaft in operation. Shaft knurling is sometimes applied to facilitate the rotor press-fit assembly process and secure the rotor core in operation. In addition, motor shafts usually carry keyed or splined components to enhance power transmission. Shaft keys offer a reliable solution to connect the shaft to the external loads, which is also a cost-effective approach as the keys are easy to install and disassemble. In addition, shaft keys can serve to protect shafts and motor components from overload conditions. In comparison, splines are directly machined on shafts in which there is no relative movement between the spline teeth and the shafts. As the stress is uniformly distributed in the spline teeth, splined shafts are capable of transmitting more torque and power. They provide better rotor balance due to symmetry in geometry as compared to the shafts with keys. Furthermore, it is common to apply heat treatment to splined shafts to improve the surface hardness, strength, and wear resistance.

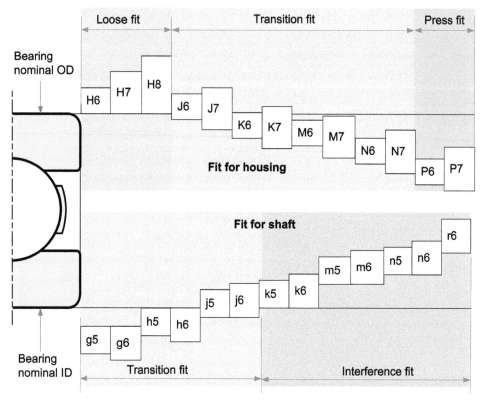

FIGURE 8.35
Fit for the bearing with shaft and the housing.

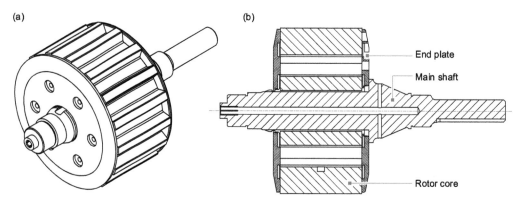

FIGURE 8.36
Shaft-rotor subassembly: (a) 3D geometry and (b) cross section view.

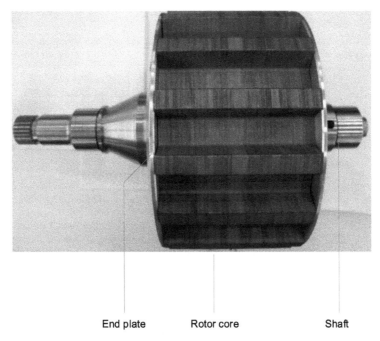

FIGURE 8.37
Shaft-rotor subassembly.

8.7.4 Circumferential Positioning of the Shaft

Keys and keyways are the most commonly used method for circumferential positioning of the rotor core on the shaft. Figure 8.38 shows two- and four-key structures. The keyways are cut-out on the shaft, which reduces the strength of the shaft. Therefore, they need to be designed carefully. A shaft might require a larger diameter when more keyways are adopted.

Table 8.5 shows the number of keys on the rotor laminations in several types of electric vehicle (EV) and hybrid electric vehicle (HEV) traction motors. Two keyways are more widely used for the connection between the shaft and rotor laminations. However, the

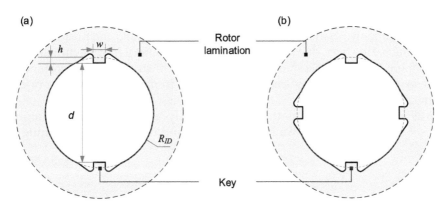

FIGURE 8.38
Rotor and shaft connected with keys: (a) two-key and (b) four-key.

TABLE 8.5

Number of Keyways on the Shafts for HEV/EV Traction Motors

Number of Shaft Keyway	EV/HEV Model
4	2003 Prius
2	2004 Prius,
	2008 LS 600h,
	2007 Camry,
	2010 Prius,
	2012 LEAF
1	Tesla Model S 60
0 (press fit is used)	2016 Audi A3 e-tron

Source: Ricardo Knowledge, Tesla Model S 60 Benchmarking Overview. [Online]. Available: http://estore.ricardo.com/; Burress, T. et al., *Benchmarking of Competitive Technologies 2012 Nissan LEAF*, Oak Ridge National Laboratory, 2012; Ricardo Knowledge, xEV Benchmarking and Competitive Analysis Database, 2017. [Online]. Available: http://estore.ricardo.com/; Rogers, S.A., Annual progress report for advanced power electronics: 2012, U.S. Department of Energy, USA, Technical Report, 2012. [Online]. Available: https://energy.gov/; Burress, T., *Benchmarking of Competitive Technologies*, Oak Ridge National Laboratory, 2012. [Online]. Available: https://energy.gov/; Rogers, S.A., Annual progress report for advanced power electronics: 2010, U.S. Department of Energy, USA, Technical Report, 2010. [Online]. Available: https://energy.gov/; Olszewski, M., *Evaluation of 2004 Toyota Prius Hybrid Electric Drive System*, Oak Ridge National Laboratory, 2012. [Online]. Available: https://info.ornl.gov/; Olszewski, M., *Evaluation of 2005 Honda Accord Hybrid Electric Drive System*, Oak Ridge National Laboratory, 2005. [Online]. Available: https://digital.library.unt.edu/. Accessed: October 24, 2017; Burress, T.A. et al., *Evaluation of the 2008 Lexus LS 600H Hybrid Synergy Drive System*, Oak Ridge National Laboratory, 2009. [Online]. Available: https://digital/library.unt.edu/. Accessed: October 34, 2017; Burress, T., *Benchmarking State-of-the-art Technologies*, Oak Ridge National Laboratory, 2013 U.S. DOE Hydrogen and Fuel Cells Program and Vehicle Technologies Program Annual Merit Review and Peer Evaluation Meeting, 2013. [Online]. Available: https://energy.gov/.

trend is using less keyway to reduce the shaft diameter and obtain a compact motor design. For example, in the Tesla Model S 60, only one keyway is used. In the 2016 Audi™ A3 e-tron traction motor, a press fit is used for the connection between the shaft and rotor core, and, thus, no keyways are used on the shaft.

8.7.5 Axial Positioning of the Shaft

Rotor core, end plates, end covers, and bearings should be axially fixed on the shaft in a reliable way. Commonly, shaft shoulders are used to axially hold the positions of mounting parts and hubs. However, due to certain design constraints it may be difficult to design a shaft with more than a couple shaft shoulders. Axial constraints are important because an unstable axial connection may lead to degradation in motor performance or even the failure of the motor.

Figure 8.39 shows some options that are commonly used for axial positioning on a shaft. The set screw in Figure 8.39a has full thread and no head. The retaining ring in

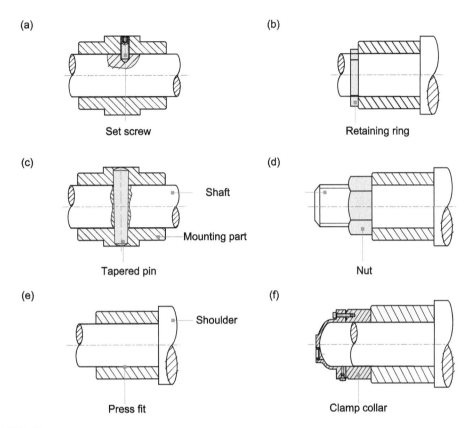

FIGURE 8.39
Various methods for the axial positioning between the shaft and mounting parts: (a) set screw, (b) retaining ring, (c) tapered pin, (d) nut, (e) press fit, and (f) clamp collar. (From Ugural, A.C., *Mechanical Design of Machine Components*, Second ed., Vol. 18., CRC Press, Boca Raton, FL, January 2015.)

Figure 8.39b is fixed by the shaft groove and it can be used to fix the position of the mounting parts. In Figure 8.39c, tapered pin is knocked in the aligned holes of the shaft and mounting parts. A nut uses the thread to axially fix the position of the mounting parts, as shown in Figure 8.39d. Press fit and clamp collar methods are shown in Figure 8.39e and f. Press fits require good tolerance design and accurate manufacturing for the shaft and mounting parts. A clamp collar uses friction force for the axial positioning.

The axial position restraining method should be selected by considering the torque requirement of the motor. Clamp collar, set screw, retaining ring, and tapered pin are suitable for low torque applications. For medium load applications, nut, tapered pin, and clamp collar could be preferred. Press fit (interference fit) is the most suitable option for axial positioning in high torque applications.

8.8 Shaft Design for Traction Motors

There are many important parameters for shaft design, such as shaft rigidity, critical speed, and maximum tolerated bending and torsional shear stress. In this section, we will present the design of the shaft of a 24/16 SRM for a traction application. Readers who are interested in shaft design can reference to [17] and [34] for more details.

Different HEV and EV manufacturers have various preferences for their traction motors. New versions of traction motors are being developed for higher power, higher torque, higher maximum speed, and higher power density. These requirements can pose challenges on the shaft design. Higher power and torque usually require a larger shaft diameter and longer axial length. Higher power density means a more compact design, which might require a shorter shaft with a smaller diameter. For higher rotational speeds, the shaft should be designed to avoid the critical speed to reduce shaft vibration.

Figure 8.40 shows the shafts for the induction motor in the Tesla Model S 60 and 2012 Nissan Leaf traction motors. Some similarities can be observed between these two designs. Both are stepped shafts and have a larger diameter in the middle section. Shaft shoulders are used in both designs for axial positioning. In addition, a spline is used for torque output in both traction motors. It can be noted that the keyway of the shaft in the

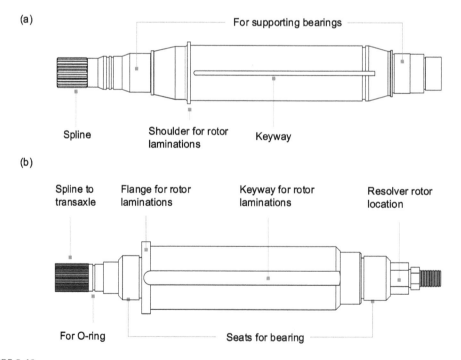

FIGURE 8.40
Examples of shafts for traction motors used in electrified vehicles: (a) Tesla Model S 60 and (b) 2012 Nissan Leaf. (From Ricardo Knowledge, Tesla Model S 60 Benchmarking Overview. [Online]. Available: http://estore.ricardo.com/.; Burress, T. et al., *Benchmarking of Competitive Technologies 2012 Nissan LEAF*, Oak Ridge National Laboratory, 2012. [Online]. (Accessed: October 24, 2017.))

traction motor of the Tesla Model S 60 is narrower than that of the 2012 Nissan Leaf. In both designs, various shaft sections are available for O-rings, laminations, the resolver rotor, and bearings.

8.8.1 Design Procedures for the Shaft

The flow chart of the shaft design process is shown in Figure 8.41. After the mounting parts and the shaft material have been determined, structural design and shaft strength examination take place. In the structural design process, the bearings and their locations are selected. The designer also has to determine whether a key or spline will be used for the torque output. In the shaft examination process, the strength, critical speed, and stress concentrations of the shaft are examined.

8.8.2 Circumferential Positioning

As described in Section 8.7.4, keys are commonly used for the connection between the rotor core and the shaft. Figures 8.42 and 8.43 show two kinds of key-keyway structure on a 24/16 SRM designed for a traction application. In the design in Figure 8.42, a separate key is used for the connection between the shaft and the rotor core. The main drawback of this method is that the key should be axially positioned and fixed, which might add complexity to the rotor assembly.

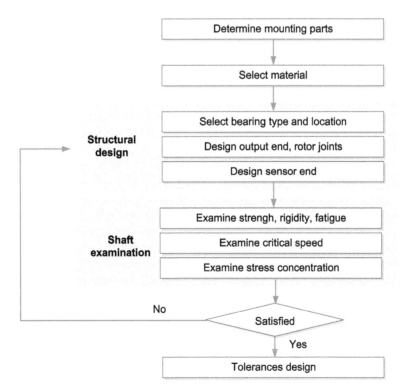

FIGURE 8.41
Flow chart for the shaft design process.

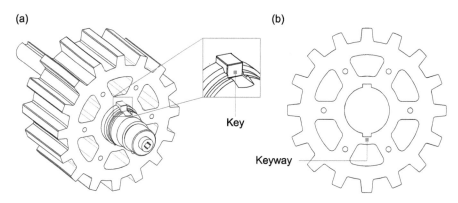

FIGURE 8.42
Keyway on rotor core: (a) 3D geometry and (b) rotor with keyways.

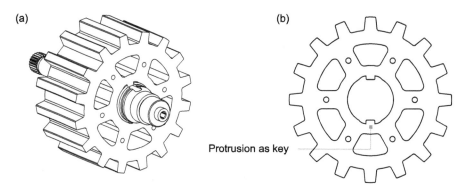

FIGURE 8.43
Protrusion on rotor as key: (a) 3D geometry and (b) rotor.

In the structure shown in Figure 8.43, the key is a part of the rotor lamination. This configuration can reduce the parts required for the axial positioning, since the position of the rotor core is axially fixed by the end plates. This key-keyway structure in Figure 8.43 is utilized in several commercial traction motors, such as the 2003 Prius, 2008 LS 600h, 2007 Camry, 2010 Prius, 2012 LEAF, and Tesla Model S 60.

8.8.3 Bearing Positioning

The type and the position of the bearings should be selected and designed carefully to maintain the position of the shaft during operation. This is vital to maintain the operating life of the traction motor. Bearings also have an important function in maintain a consistent air gap throughout the operating life of the motor.

Table 8.6 shows the comparison of the bearings in the 2012 Nissan Leaf, Tesla Model S 60, and the 24/16 traction SRM, along with their positions, which is based on the assembly shown in Figure 8.44. The 24/16 traction SRM and Tesla Model S 60 use the same bearings. BB1-3763 SKF™ can provide excellent insulation to reduce the electric currents induced in the bearings because the balls bearings are made of silicon nitride.

TABLE 8.6

Comparison of Bearings and Their Positions of Different Traction Motor Designs

	24/16 SRM	2012 Nissan Leaf	Tesla Model S 60
Type of bearing	BB1-3793 SKF	NSK 6206V	BB1-3793 SKF
Bearing width, w [mm]	21	16	21
Bearing OD [mm]	80	62	80
Bearing ID [mm]	35	30	35
Stack length, L_1 [mm]	92	151	152
Bearing span, L_2 [mm]	185.72	230[a]	280[a]
Shaft length, L [mm]	292.22	360[a]	365[a]
Length ratio, L_1: L_2: L	1:2:3.18	1:1.53:2.38	1:1.84:2.4

Note:
[a] The data is estimated based on published reports [23] and [24].

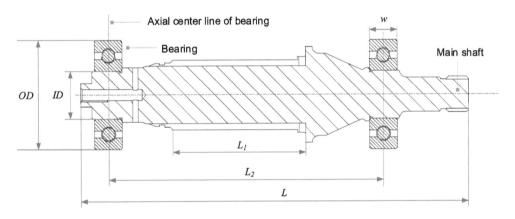

FIGURE 8.44
Shaft and bearing subassembly.

The position of the bearings should be carefully selected because it affects the lateral rigidity and the deflection of the shaft. For traction motors, the maximum acceptable deflection is 0.003 times the bearing span [33]. Hence, bearing positions should be based on the shaft length, stack length, and bearing span. It can be noticed from Table 8.6, that the ratio of the stack length to the bearing span in the 24/16 traction SRM is 1:2.3, while the ratio of the bearing span to the shaft total length is 2:3.18. These two ratios are similar to that of the 2012 Nissan Leaf and Tesla Model S 60 traction motors.

8.8.4 Output End of the Shaft

Splines are commonly used for the connection between the shaft and the load. The manufacturing cost of splines is usually higher than a key-keyway design; however, the torque transmitted by a spline is more stable. Involute type splines and straight teeth splines are used in electric motor applications. An involute spline is simpler to manufacture and it is also more commonly used than a straight tooth spline, because its teeth are stronger with less stress concentration. As shown in Figure 8.45, an external spline is incorporated in the main shaft of the 24/16 traction SRM.

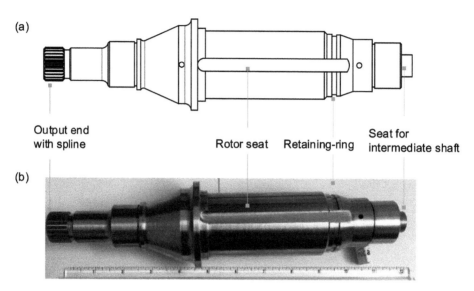

FIGURE 8.45
Main shaft with spline: (a) drawing of the main shaft and (b) actual main shaft.

8.9 Assembly Considerations

Since SRMs typically employ relatively a smaller air gap to enhance power and torque density, it is important to have uniform air gaps. Hence, careful mechanical design and stringent tolerance control are necessary during the assembly process.

A complete SRM assembly includes the rotor assembly, stator assembly, then followed by the motor casing assembly. The rotor assembly is integrated with the stator assembly via bearings and housing. The rotor assembly of an SRM is relatively simple since there are no magnets or winding on the rotor. Figure 8.46 shows the assembly drawing of a 24/16 SRM.

In addition to providing better thermal management due to simple rotor construction, SRMs also have the advantage of a more balanced rotor structure since there is no magnet insertion misalignment or winding unbalance. These issues are commonly found in permanent magnet machines and wound-rotor machines as a result of magnet manufacturing tolerance deviations, magnet shifting within slots caused by resin varnish, shifts caused by injection over molding pressure, or resin varnish cavities within rotor windings.

For an interior rotor configuration, the rotor laminations are stacked and aligned, and then pressed fitted into the shaft. The rotor laminations can also be heated up so that a shrink fit can be applied to secure the rotor assembly. This helps to reduce the internal stress caused by the press fitting process; however, it increases the assembly process time and cost as a trade-off. End clamps can be applied to press against the rotor laminations to tighten the stack assembly. Insulation washers are also typically applied to prevent any leakage current as well as to protect the end laminations. Rotor lamination stamping typically provides fairly good accuracy control; however, for machines that require tight

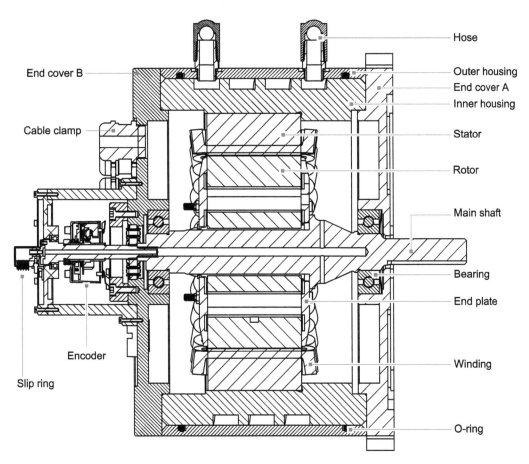

End cover B

Cable clamp

Encoder

Slip ring

Hose

Outer housing
End cover A
Inner housing

Stator

Rotor

Main shaft

Bearing

End plate

Winding

O-ring

FIGURE 8.46
Assembly drawing of a 24/16 SRM.

tolerance air gaps, rotor outer surface machining can be applied to minimize the rotor run out. Figure 8.47 illustrates a typical SRM rotor assembly [11].

The stator assembly is comprised of stator laminations and windings. SRM windings are typically wound on each stator tooth. Using conventional techniques, each coil is wound after the entire stator lamination stack is assembled. As an alternative, a segmented stator can be applied, where the coils are wound on each pole separately then the stator is assembled. Segmented stator designs can be commonly found in large SRMs due to the stamping size limits of the stamping tools. Segmented stators have recently been trending in small machine designs to increase the copper fill factor by allowing more turns to achieve higher power density. Segmented stator can achieve 20%–30% more copper fill factor than the conventional one-piece ring stators that have the same cross-sectional area [17]. The smaller lamination size also achieves better utilization of the lamination material. The disadvantage of using segmented stator configuration is the difficulty in the assembly process, such as stator assembly concentricity control and lead terminal connection. Another disadvantage is higher eddy current losses that are

(a)

Insulation washer   Rotor   Shaft

Shaft collar clamp

(b)

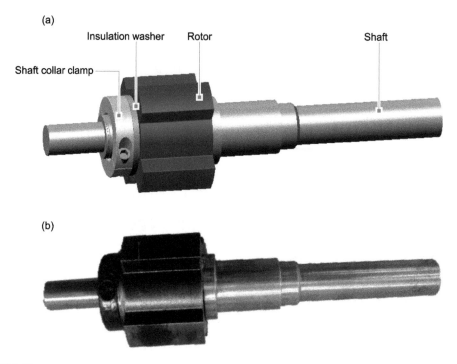

FIGURE 8.47
Switched reluctance machine rotor assembly: (a) general assembly 3D model and (b) actual parts. (From Yang, Y., Double-rotor switched reluctance machine for integrated electro-mechanical transmission in hybrid electric vehicles, Ph.D. Dissertation, McMaster University, Hamilton, ON, February 2014.)

generated as a result of the gaps in the punched edges between neighboring lamination segments and residual stress from the cuts. Figure 8.48 illustrates a prototype segmented stator tooth with individual-wound coil.

In an ideal situation, the rotor assembly and the stator assembly would be concentric and the air gap distance between the rotor outer diameter and stator inner diameter is uniform. However, it is inevitably impossible for any machine to achieve this in reality due to deviations from the manufacturing tolerances and assembly stack up misalignment. As a result, rotor eccentricity is typically found in electric machines. For switched reluctance machines, this is particularly critical as most of the switched reluctance machines utilize a small air gap to enhance torque and power density. Thus, any deviation in air gap length caused by assembly eccentricity will have impact on the machine output parameters, notably the vibration and noise caused by the unbalanced radial force pull.

Rotor balance is another critical aspect for rotor assembly. It is especially important for switched reluctance machines to ensure safety and normal operation as many of them are designed for high speed applications. An unbalanced rotor is typically caused by a mismatched rotor weight distribution. Improper manufacturing tolerances, material defects, unsymmetrical structures such as keys and key ways, all lead to rotor unbalance. Rotor unbalance is a significant source of noise and vibration. It is also likely to cause excess wear and stress on the bearings and other supporting components. Furthermore, machine efficiency and lifetime could be negatively affected by an unbalanced rotor. Fortunately, switched reluctance machines have the advantage of simple rotor structure. This eliminates

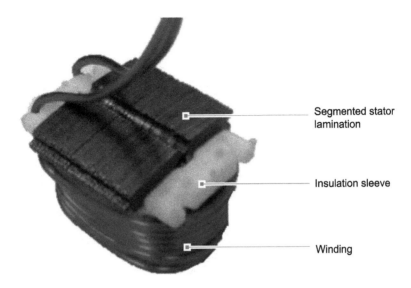

FIGURE 8.48
Segmented stator tooth with individual-would coil.

the balance issues caused from non-uniformly weighed magnets or windings as well as the potential mismatched weight distribution caused by these extra components.

Finally, the machine output shaft misalignment could lead to significant machine noise and vibration issues and results in excessive loading and reduction in machine life. It occurs when the centerline of the motor shaft does not perfectly align with the driven shaft. Figure 8.49 illustrates various types of shaft misalignment. Parallel misalignment happens when the centerlines of the motor shafts are offset by a distance. This can either be horizontal misalignment, vertical misalignment, or the combination of the two. When the motor is running, both the motor shaft and the driven shaft have to be bent and they both rotate with an eccentric load. This leads to component stress, wear, and fatigue. Similar to parallel misalignment, angular misalignment can also be categorized as horizontal

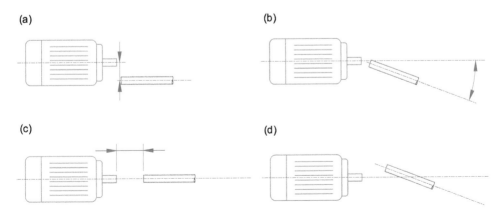

FIGURE 8.49
Various types of shaft misalignments: (a) parallel, (b) angular, (c) axial, and (d) combined.

angular misalignment and vertical angular misalignment. An axial misalignment occurs when the distance between the drive and driven shafts are too far. It can be corrected by moving one of the motors in parallel. In practice, motor shaft misalignment could be the combination of all these three types.

8.10 Balancing of Rotor Subassembly

8.10.1 Unbalance in Rotor Subassembly

Rotor balancing is a necessary and important step of the motor manufacturing process for SRM rotors. Rotor unbalance is one of the sources of vibration and noise in SRMs. The centrifugal forces caused by rotor unbalance also leads to excessive loading on the bearing which will shorten the operation lifetime of the bearings. The rotor unbalance becomes even more challenging if the SRMs are designed for high-speed application where the centrifugal forces are much larger. The discussion of rotor balance presented in this section is for a rigid rotor. A rotor subassembly is considered as rigid if its unbalance can be corrected in any two correction planes (e.g., end surfaces). For a rigid rotor, the residual unbalance of the rotor subassembly does not change significantly at any speed below the maximum operating speed of the SRM. Flexible rotors are not included in the discussion of this section because elastic deflection or deformations occurs in the flexible rotor when it is spinning. A rigid rotor can also be distinguished from a flexible rotor by its operational spinning speed. The maximum spinning speed of a rigid rotor is less than 50% of the first critical speed of the rotor subassembly [35]. The critical speed of the rotor is defined as the rotational speed at which the resonance occurs on rotor.

Rotor unbalance appears because of the uneven mass distribution in the rotor subassembly. As discussed previously, there are three kinds of rotor unbalance, including static unbalance, moment unbalance, and dynamic unbalance. The static unbalance of a thin circular plate is shown in Figure 8.50. There is an eccentricity, e, between the center of mass and the rotational center. This eccentricity leads to centripetal forces, which is applied to the center of mass when the rotor is rotating. The static unbalance can be reduced by

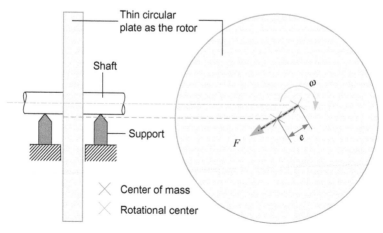

FIGURE 8.50
Static unbalance in a thin circular plate.

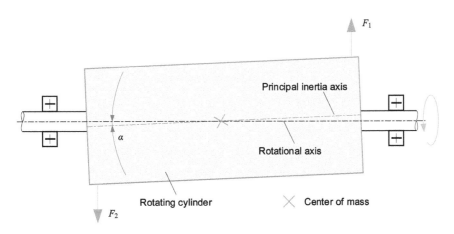

FIGURE 8.51
Moment unbalance in a rotating cylinder.

changing the mass distribution, including removing or adding mass on the rotor subassembly. For a rotor whose diameter is 7–10 times larger than its axial length, the correction of rotor unbalance can be performed by adding or removing mass on a single plane of the rotor, for example, on one end plane of the rotor.

Moment unbalance can also appear for rotors whose diameter is less than 7 to 10 times of the rotor axial length [36]. The moment unbalance in a rotating cylinder is shown in Figure 8.51. Unlike the static unbalance, the rotational axis goes through the center of mass. Therefore, there is no eccentricity between the rotational axis and the center of mass. However, due to uneven mass distribution with regard to the rotational axis, which are located on both ends of the cylinder, the principal inertia axis is not aligned with the rotational axis. When a rotor is free to rotate around an axis, torque must be applied to change its angular momentum. This axis is the principal inertia axis. The uneven distribution of mass along the rotational axis leads to vibration and extra loading on the bearings. Dynamic unbalance is a combination of static unbalance and moment unbalance. Dynamic unbalance appears in the actual rotor subassembly and should be corrected.

8.10.2 Permissible Residual Unbalance

With a spinning test, the magnitude of the rotor unbalance can be measured using a rotor balance machine. Then the rotor unbalance should be corrected by adding or removing mass to change the mass distribution of the rotor subassembly. Correcting the rotor unbalance perfectly would be every costly. Therefore, the permissible residual unbalance, U_{per}, is defined as the tolerance for the unbalance of the rotor after balancing. The calculation of the permissible residual unbalance, U_{per}, is given as follows:

$$U_{per} = e_{per} \times m \tag{8.1}$$

where e_{per} is the permissible specific unbalance, m is the mass of the rotor subassembly. e_{per} can be determined by the balance quality grades and the maximum speed of the SRM. The balance grades, shown in Table 8.7, are recommended by ISO1940-1. The magnitude of the balance quality grade is defined as: $e_{per} \times n/6$, where n is the speed of the rotor in rpm. As shown in Figure 8.52, the selection of permissible specific unbalance is based on the balance quality grade and the maximum spinning speed of the rotor.

TABLE 8.7

Guidance for Balance Quality Grades for Rotors

Application	Balance Quality Grade	Magnitude of Balance Quality Grade, $e_{per} \times n/6$ [mm/s]
Agricultural machinery, crushing machines	G 16	16
Electric motors and generators of at least 80 mm shaft height and with maximum rated speed up to 950 rpm; Electric motors of shaft heights smaller than 80 mm;	G 6.3	6.3
Electric motors and generators of at least 80 mm shaft height and of maximum rated speed above 950 rpm	G 2.5	2.5
Audio and video drives, Grinding machine drives	G 1	1
Gyroscopes, Spindles and drives of high-precision systems	G 0.4	0.4

Source: ISO 1940/1, *Balance Quality Requirements of Rigid Rotors*, International Organization for Standardization, Geneva, Switzerland, 1986.

Note: The shaft height is the distance from the center of the shaft to the mounting surface.

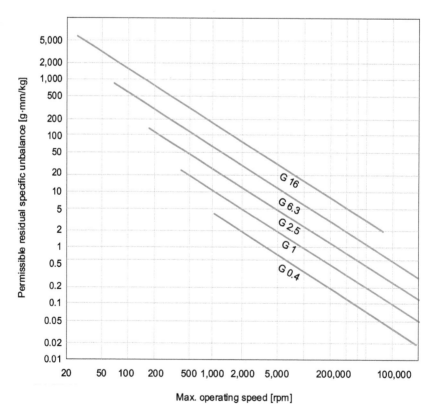

FIGURE 8.52

Permissible residual specific unbalance based on balance quality grade, G and spinning speed, *n*. (From ISO 1940/1, *Balance Quality Requirements of Rigid Rotors*, International Organization for Standardization, Geneva, Switzerland, 1986.)

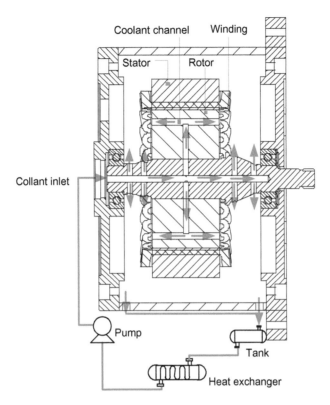

FIGURE 8.53
Mechanical construction of an SRM with internal oil cooling.

The cost for rotor balancing and the mechanical construction of SRMs will affect the selection of balance grades. In the mass production of SRMs, a lower balance quality grade results in a higher cost for rotor balancing. For SRMs that use an internal oil cooling system, the coolant flows through the coolant channel in the shaft and the rotor, and the coolant is in contact with the rotor (see Figure 8.53). The rotor subassembly cannot be balanced with the coolant in the balancing machine; therefore, it is not advantageous to balance the rotor to a small magnitude of balance quality.

8.10.3 Correction of Rotor Unbalance

All the rotating elements of the rotor should be included in the measurement and correction of unbalance. Figure 8.54 shows the rotor-shaft-endplate subassembly of a 24/16 traction SRM for the measurement and correction of unbalance. For an external-rotor SRM, as shown in Figure 8.55, the rotor and the endcaps are in a subassembly for the unbalance measurement and correction.

The measurement of unbalance for the external-rotor SRM is shown in Figure 8.56a. The rotor subassembly is spun up using the belt drive. The laser tachometer is used to measure the rotating speed of the rotor. The magnitude of the vibration, which is measured by the accelerometers, is converted into the magnitude of unbalance. In the measurement

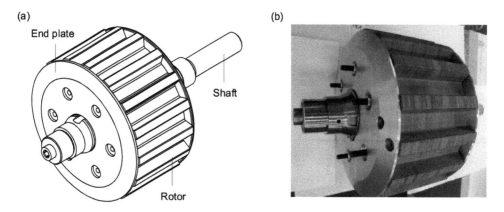

FIGURE 8.54
Rotor-shaft-endplate subassembly in a 24/16 SRM: (a) CAD drawing and (b) actual subassembly.

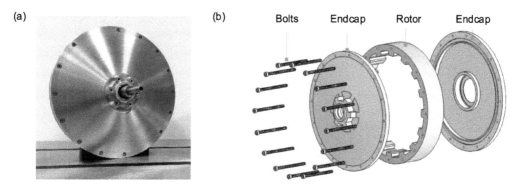

FIGURE 8.55
Mechanical construction of an external-rotor SRM: (a) actual assembly and (b) exploded view of rotor-endcaps subassembly.

of unbalance, it is not necessary to conduct the unbalance measurement at the maximum operating speed of the rotor. Generally speaking, the existence of unbalance does not change with the change of speed. The spinning speed in the test only changes the centripetal forces and thus the magnitude of the vibration. If the measurement of the vibration is accurate enough, during the measurements, the rotor subassembly can spin at a speed which is much lower than its maximum operating speed.

Figure 8.56b shows the measurement of unbalance in the rotor of an SRM designed for an HVAC application. As shown in Figure 8.57, the unbalance is corrected by adding mass in the cut-out of the rotor. For this SRM, since there is no endplate in the rotor subassembly, removing mass on the rotor could reduce the motor performance. For this reason, adding mass is the preferred balancing technique. The added material is steel-filled epoxy, which has good adhesion characteristics with an operating temperature range of −30°C ~ 121°C.

(a)

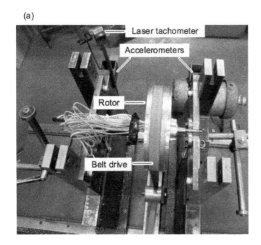

(b)

FIGURE 8.56
Rotor unbalance measurement: (a) external-rotor SRM and (b) rotor of an HVAC SRM. (Courtesy of Delstar, Montreal, QC.)

FIGURE 8.57
Correction of unbalance for the rotor of an HVAC SRM.

Questions and Problems

1. Describe the advantages of mechanical construction of an SRM compared to other types of machines? Compare the rotor configurations between switched reluctance machine, induction machine, interior permanent magnet machine, and brushless permanent magnet machine. What are the differences in terms of machine construction and what are the benefits?

2. List all the necessary insulations in switched reluctance machine construction. Describe the function of each insulation system and material used.

3. Compare journal bearings and rolling bearings. What are the differences? What are their preferred applications?

4. What are the mechanical construction concerns that should be considered during the machine design phase? What can be done to avoid these issues?

5. Figure Q.8.1 shows an SRM designed for an e-bike application, which uses an exterior rotor. Please refer to the exploded view of the motor as shown in Figure Q.8.2, and fill the blanks in Table Q.8.1.

(a)

(b)

(c)

(d)

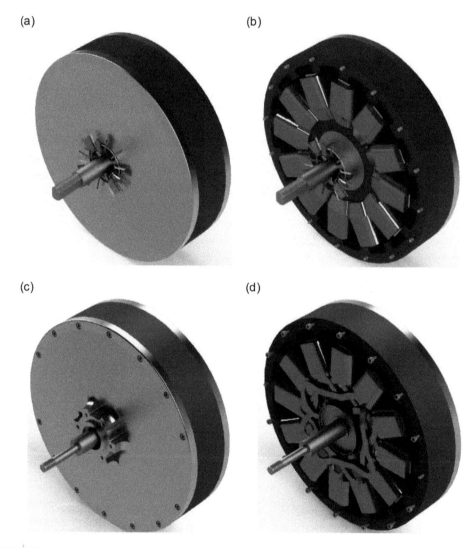

FIGURE Q.8.1
SRM with exterior rotor for E-bike: (a) with end cover A, (b) without end cover A, (c) with end cover B, and (d) without end cover B.

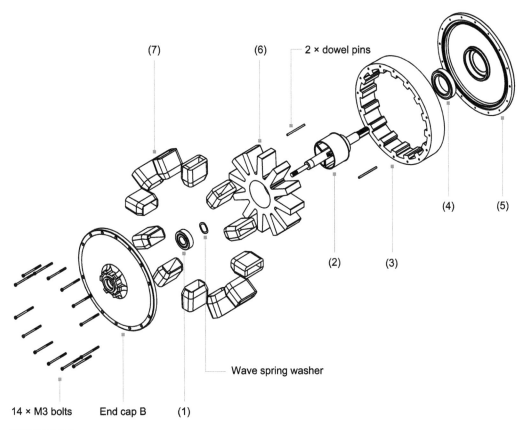

FIGURE Q.8.2
Exploded view of the SRM.

TABLE Q.8.1

Partial Part List with Blanks to Be Filled

Part #	Name	Possible Material or Model	Possible Connection With	Fitting Type
(1)			End cap B	Transition fit
(3)			(1), (4)	
(4)			(6)	
(5)			Rotor slot fillers	
(6)			(5)	
(7)			(3)	Bolts
(8)			(2)	
			(6)	

References

1. Jiang, J. W., Three-phase 24/16 switched reluctance machine for hybrid electric powertrains: Design and optimization, Ph.D. Dissertation, Department Mechanical Engineering, McMaster University, Hamilton, ON, 2016.
2. Chalmers, B. J. (Ed.) *Electric Motor Handbook*. Amsterdam, Netherlands: Elsevier, 2013.
3. Jiang, J. W., B. Bilgin, and A. Emadi, Three-phase 24/16 switched reluctance machine for hybrid electric powertrains: Design and optimization, *IEEE Transactions on Transportation Electrification*, 3(1), 76–85, March 2017.
4. Bilgin, B. and A. Sathyan, Fundamentals of electric machines, in *Advanced Electric Drive Vehicles*, Boca Raton, FL, CRC Press, 2015, pp. 107–186.
5. Sgobba, S., Physics and measurements of magnetic materials, in *CERN Accelerator School CAS 2009: Specialised Course on Magnets*, Bruges, Belgium, June 2009.
6. Yang, Y., N. Schofield, and A. Emadi, Double-rotor switched reluctance machine (DRSRM), *IEEE Transactions on Energy Conversion*, 30(2), 671–680, 2015.
7. Yang, Y., Double-rotor switched reluctance machine for integrated electro-mechanical transmission in hybrid electric vehicles, Hamilton, ON, Ph.D. Dissertation, McMaster University, February 2014.
8. Bilgin, B., A. Emadi, and M. Krishnamurthy, Switched reluctance generator with higher number of rotor poles than stator poles, in *Proceedings of the IEEE Transportation Electrification Conference and Expo*, Dearborn, MI, June 2012, pp. 1–6.
9. Bilgin, B., A. Emadi, and M. Krishnamurthy, Comprehensive evaluation of the dynamic performance of a 6/10 SRM for traction application in PHEVs, *IEEE Transactions on Industrial Electronics*, 60(7), 2564–2575, 2013.
10. Bilgin, B., A. Emadi, and M. Krishnamurthy, Design considerations for switched reluctance machines with a higher number of rotor poles, *IEEE Transactions on Industrial Electronics*, 59(10), 3745–3756, 2012.
11. Yang, Y., N. Schofiled, and A. Emadi, Double-rotor switched reluctance machine design, simulations, and validations, *IET Electrical Systems in Transportation*, 6(2), 117–125, 2016.
12. Cogent Power, Altogether more powerful - Non-oriented electrical steel. [Online]. Available: http://cogent-power.com/. (Accessed: March 20, 2017).
13. Geiman, J., DC step-voltage and surge testing of motors, *Maintenance Technology*, 20(3), 32, 2007.
14. 43-2013, *IEEE Recommended Practice for Testing Insulation Resistance of Electric Machinery*, IEEE Standard, 2013.
15. 95-2002, *IEEE Recommended Practice for Insulation Testing of AC Electric Machinery (2300 V and Above) With High Direct Voltage*, IEEE Standard, 2002.
16. Jiang, J. W., B. Bilgin, A. Sathyan, H. Dadkhah, and A. Emadi, Analysis of unbalanced magnetic pull in eccentric interior permanent magnet machines with series and parallel windings, *IET Electrical and Power Application*, 10(6), 526–538, 2016.
17. Tong, W., *Mechanical Design of Electric Motors*, Boca Raton, FL: CRC Press, 2014.
18. Leader, M. E., *Understanding Journal Bearings*, [Online]. Available: http://edge.rit.edu/edge/P14453/public/Research/2-_LEADER_-_Understanding_Journal_Bearings.pdf. (Accessed January 23, 2017).
19. Alford, L., *Bearings and Their Lubrication*, New York: McGraw Hill, 1912.
20. NTN Global, *Classification and Characteristics of Rolling Bearings*. Available: http://www.ntn-global.com/en/products/catalog/pdf/2202E_a01.pdf. (Accessed: March 5, 2017).
21. Nachi, *Ball and Roller Bearings*, Tokyo, Japan: Machi-Fujikoshi, 2014.
22. International Organization for Standardization (ISO), Limits, fits, and surface properties, Geneve, Switzerland: International Organization for Standardization, 1999.
23. Ricardo Knowledge, Tesla Model S 60 Benchmarking Overview. [Online]. Available: http://estore.ricardo.com/. (Accessed: October 24, 2017).

24. Burress, T. et al., *Benchmarking of Competitive Technologies 2012 Nissan LEAF*, Oak Ridge National Laboratory, 2012. [Online]. (Accessed: October 24, 2017).

25. Ricardo Knowledge, xEV Benchmarking and Competitive Analysis Database, 2017. [Online]. Available: http://estore.ricardo.com/. (Accessed: March 20, 2017).

26. Rogers, S. A., Annual progress report for advanced power electronics: 2012, U.S. Department of Energy, USA, Technical Report, 2012. [Online]. Available: https://energy.gov/. (Accessed: October 24, 2017).

27. Burress, T. Benchmarking of competitive technologies, Oak Ridge National Laboratory, 2012. [Online]. Available: https://energy.gov/. (Accessed: October 24, 2017).

28. Rogers, S. A., Annual progress report for advanced power electronics: 2010, U.S. Department of Energy, USA, Technical Report, 2010. [Online]. Available: https://energy.gov/. (Accessed: October 24, 2017).

29. Olszewski, M., *Evaluation of 2004 Toyota Prius Hybrid Electric Drive System*, Oak Ridge National Laboratory, 2012. [Online]. Available: https://info.ornl.gov/. (Accessed: October 24, 2017).

30. Olszewski, M., *Evaluation of 2005 Honda Accord Hybrid Electric Drive System*, Oak Ridge National Laboratory, 2005. [Online]. Available: https://digital.library.unt.edu/. (Accessed: October 24, 2017).

31. Burress, T. A. et al., *Evaluation of the 2008 Lexus LS 600H Hybrid Synergy Drive System*, Oak Ridge National Laboratory, 2009. [Online]. Available: https://digital/library.unt.edu/. (Accessed: October 24, 2017).

32. Burress, T., *Benchmarking State-of-the-art Technologies*, Oak Ridge National Laboratory, 2013 U.S. DOE Hydrogen and Fuel Cells Program and Vehicle Technologies Program Annual Merit Review and Peer Evaluation Meeting, 2013. [Online]. Available: https://energy.gov/. (Accessed: October 24, 2017).

33. Ugural, A. C., *Mechanical Design of Machine Components*, Second ed., Vol. 18., Boca Raton, FL: CRC Press, January 2015.

34. Erik, O., *Machinery's Handbook*, Norwalk, CT: Industrial Press, 2012.

35. MacCamhaoil, M., *Static and Dynamic Balancing of Rigid Rotors*, Bruel and Kjaer, application note, Naerum, Denmark, 2016.

36. ISO 1940/1, *Balance Quality Requirements of Rigid Rotors*, Geneva, Switzerland: International Organization for Standardization, 1986.

9

Control of Switched Reluctance Machines

Jin Ye, Haoding Li, and James Weisheng Jiang

CONTENTS

9.1 Introduction

While SRMs share many similarities with other electric machines, the principles governing their control separate them from other machine topologies. SRMs differs from most machine topologies in their unique geometry and torque production mechanism, which results in challenges in modeling the motor. The machine's doubly salient structure creates large variations in the air gap, which means that the machine flux linkage is a nonlinear function of rotor position. Additionally, SRMs generally operate in the magnetically saturated region. This can make the modeling and control of the machine challenging.

Due to its robustness, SRMs feature a wide speed range, which makes it suitable for traction applications. However, the control mechanism across this wide speed range will vary, due to the different operating conditions and limitations that arise at different speeds. The control for SRMs must be able to account for operating under these different conditions, thus increasing the difficulty in designing the controller.

Due to these challenges, SRM controls require some level of sophistication and sensitivity on the part of the designer. In this chapter, an overview of the underlying principles of SRM control will be provided and state-of-the-art control strategies will be examined. Furthermore, this chapter will familiarize the reader with the common current control strategies in use, as well as extensions to these strategies that have been developed more recently such as the torque sharing function.

In order to design a controller for any system, it is paramount that the designer has an intimate understanding of the dynamics of the system itself. Control for SRMs is no exception. Some understanding of the dynamics of SRMs will provide insight regarding the significance of various controls strategies that must be applied in different situations.

9.2 System Dynamics

Due to its simple construction, the inherent dynamics of an SRM is modeled based on its magnetic characteristics, as discussed in Chapter 4. When the machine geometry is defined, the shape of the air gap determines the inductance profile, which will have an effect on the machine's dynamic behavior. Due to the doubly salient structure, this inductance profile varies with position, which has a significant impact on torque production. Torque production in SRMs depends on the rate of change of inductance. The equation for torque under linear operation conditions is given as:

$$\tau = \frac{1}{2}i^2\frac{\partial L(\theta)}{\partial \theta} \tag{9.1}$$

where L is the phase inductance and θ is the rotor position. It can be seen from (9.1) that the amount of torque produced scales with current. However, as with any real system, there is a finite time associated with increasing current to the required level. Since each phase of the machine can be reduced to an inductor with an iron core for analysis purposes, the behavior of the current may be likened to that of a first order electrical system. When some constant voltage, V is applied, the current will behave in accordance with (9.2), which was also provided in Chapter 4.

$$V = iR + L(\theta)\frac{\partial i}{\partial t} + i\frac{\partial L(\theta)}{\partial \theta}\omega_r \qquad (9.2)$$

On the right-hand side of (9.2), the last term represents the motional electro-motive force (EMF), which is generated due to the time-varying magnetic field induced from the phase current. This voltage opposes the rate of change of current, which has implications on the control capability of the dynamic system. Like the torque, it can be noted that the induced voltage term is dependent on the rate of change of inductance, which shows the fundamental role the machine inductance plays on the dynamics of the system, and indirectly, how much the machine design will affect the control limitations of the system. From (9.2), it can be seen that the induced voltage term scales with the speed of rotation. Therefore, at higher speeds, the induced voltage will have a bigger impact on current control in the system, which will reflect the control strategies available for the SRM.

9.3 Control Overview for SRM

The heart of the control scheme for SRM is current control, in which the phase current is shaped based on a series of predetermined parameters in order to meet different objectives for speed, torque, power, and efficiency across a wide range of operating conditions, subject to the dynamics described in (9.2). By having the capabilities to control the current, the torque-speed characteristics of the machine can be derived. This then allows for an outer closed loop with which the machine speed is controlled. Figure 9.1 shows a complete overview of the conventional control paradigm used in SRMs.

At the center of the control structure shown in Figure 9.1, it can be seen that a current controller feeds switching signals to the asymmetric bridge converter, which controls the phase current based on some current reference, i_{ref}. The current reference, as well as other control parameters like the conduction angles (to be discussed in detail later in

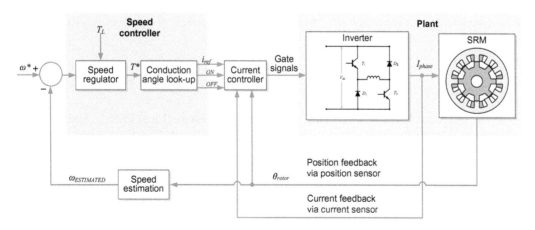

FIGURE 9.1
Overview of the control structure for SRMs which incorporates closed loop control for phase current, shaft torque, and rotor speed.

the chapter), are calculated based on the torque and speed requirements of the machine. Finally, the machine speed, ω, is estimated by differentiating the machine position θ_{rotor}, which is measured using a position sensor. This is compared to the speed reference, ω, to determine the necessary torque required to accelerate or decelerate the machine to drive the load torque, T_L.

In order to build the complete control structure described by Figure 9.1, parameters such as θ_{on} and θ_{off}, amongst many others, must be found through simulation. The following sections give a detailed description of each of the three controllers shown in Figure 9.1, as well as the necessary control parameters.

9.4 Current Control

9.4.1 Conditions for Current Control

Due to the salient structure of SRMs, there is high variance in the inductance over its electrical cycle, and so its natural dynamics is highly non-linear, as described by (9.2). As the machine rotates, a motional EMF can be quantified to characterize the rate of change of current in one phase of an SRM. This relationship is given in (9.3).

$$\varepsilon = i \frac{\partial L(\theta)}{\partial \theta} \omega_r \tag{9.3}$$

The EMF cannot be directly measured across the phase windings of the SRM. In a brushed direct current (DC) motor, the back EMF can be seen by rotating the armature in a constant magnetic field using a prime mover and measuring the voltage on the terminals. However, with the SRM, since the EMF is also dependent on the phase current, simply rotating the rotor will not produce any voltage. Should the phase be excited, only the phase voltage from the DC-link can be seen. Therefore, for the SRM, the motional EMF is fundamentally different from the back EMF on a brushed DC motor. As the motor rotates, the motional EMF is induced in the phase. This limits the growth of phase current and makes the motional EMF (induced voltage) relevant to the discussion of the dynamics of phase current in SRMs, as is outlined in (9.4).

$$\frac{V - \varepsilon}{L} = \frac{\partial i}{\partial t} \tag{9.4}$$

In (9.4), the relationships between the DC-link voltage, V, the motional EMF, ε, and the rate of change of current are shown by rearranging (9.2) and neglecting the effects of phase resistance. Since the EMF is a linear function of rotor speed, the rate of change of current becomes influenced by the machine speed as well.

Without any current control, the dynamic current profiles for the three-phase SRM are given in Figure 9.2. Figure 9.2a shows that the excitation occurs on the rising inductance period of the electrical cycle, while Figure 9.2b shows the dynamic current profile at three different speeds. Figure 9.2c shows the induced voltage profile that is responsible for the shape of the corresponding current profiles in Figure 9.2b. While the induced

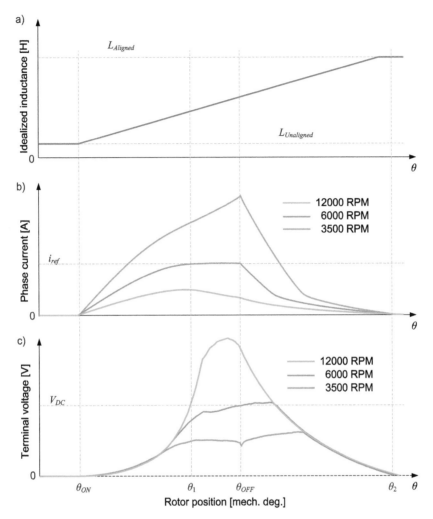

FIGURE 9.2
Dynamic profiles at various operating speeds excited with constant conduction angles: (a) phase inductance, (b) phase current, and (c) phase voltage.

voltage is not directly measurable, it can be characterized by simulating the terminal voltage at different rotor positions under constant current excitation, as discussed in Chapter 4. In a dynamic profile such as Figure 9.2c, the induced voltage profile can be thought of as the voltage that the DC-link voltage has to overcome to generate the current profiles shown in Figure 9.2b. Equation (9.3) shows that this quantity is a function of speed, and so the induced voltage has a more drastic effect at higher speeds. From Figure 9.2, it can be seen that at each speed, the phase is excited at θ_{on} and positive DC-link voltage is applied across the phase windings. At the turn-on, energy will continually be injected into the phase, and the current will behave based on the dynamics outlined in (9.2), until the time θ_{off} when negative DC-link voltage is applied across the phase. At the turn-on, energy is removed from the phase until current is depleted, which occurs at θ_2.

At 3500 RPM, the current will continue to rise when the DC-link voltage is applied to the phase because the induced voltage has a low amplitude due to the low speed of operation. At this speed, the induced voltage does not restrict the rate of change of current. Naturally, the high current will incur significant copper losses, and if left uncontrolled, it may exceed the thermal limit of the copper wires, which might result in damage to the coils. In actual operation, this current is chopped by the converter so that it follows the current reference, i_{ref} to guarantee torque production that would meet the torque command. This results in trapezoidal current profiles.

At 6000 RPM, it can be seen that the current naturally tapers off after the phase is excited as the induced voltage will rise to the DC-link voltage value. Since the amplitude of the induced voltage is close to the DC-link voltage, there is little change in the current profile until the turn-off time occurs.

In order to fully appreciate the effect of the induced voltage, the current profile when the machine is running at 12,000 RPM must be observed. In this case, it can be seen that at θ_1, the current naturally decays even when positive DC-link voltage is still being applied. This is because the induced phase voltage becomes higher than the value of the DC-link voltage due to the high speed, causing the current to decay, as shown in Figure 9.2c. From this example, it can be seen that at sufficiently high speeds, controlling the peak value of current through switching might not be possible due to the influence of the induced voltage that naturally limits the rate of change of current. Thus, to achieve the necessary average torque at high speeds, the conduction angles θ_{on} and θ_{off} are controlled. However, worse torque quality will typically result due to the pulsated nature of the current profiles.

9.4.2 Hysteresis Control

The peak current at low speeds must be controlled to be at some commanded value i_{ref}, in order to prevent the current from exceeding the thermal limitations and to maintain smooth operation of the machine. This is accomplished by switching the DC-link voltage that is fed into the phase. The current command is set by the torque controller, as shown in Figure 9.1, and is tracked using a current controller. The current controller generates switching signals, which are fed to the inverter to switch the DC-link voltage and to regulate the phase current.

To maintain the phase current at some constant value i_{ref}, a hysteresis band is defined by I_{upper} and I_{lower}. These values are calculated using the current reference, i_{ref}, and a tolerance β, as shown in (9.5). Generally, β is given as a percentage of the current reference, i_{ref}.

$$\begin{aligned} I_{upper} &= i_{ref} \times (1 + \beta) \\ I_{lower} &= i_{ref} \times (1 - \beta) \end{aligned} \tag{9.5}$$

This hysteresis band is imposed on the 3500 RPM current profile shown in Figure 9.2, to limit the current amplitude to the level set by i_{ref}. The resulting profile is shown in Figure 9.3.

By comparing the voltage profile of Figure 9.3 to 9.2, it can be seen that during the conduction period, the voltage of Figure 9.3 alternates between the DC-link voltage

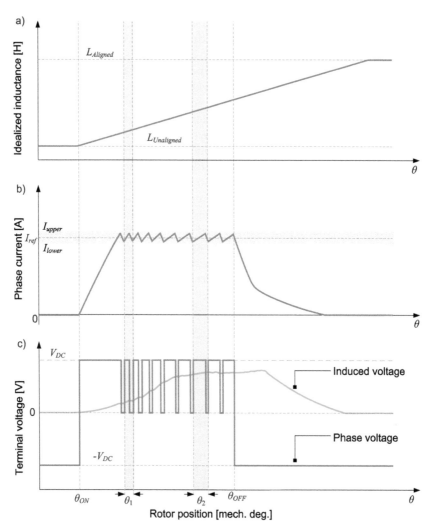

FIGURE 9.3
Current control for SRM demonstrating current chopping with (a) phase inductance, (b) phase current, and (c) phase voltage.

and zero to maintain the current within the hysteresis band, using the feedback from the current sensor. Notably, the induced voltage amplitude remains below the DC-link value for the entire conduction period, which makes the current control necessary at this speed.

The hysteresis current controller has a variable switching frequency, which at every instant in time is a function of the phase inductance as well as the width of the hysteresis band. Fundamentally, if the rate of change of current is higher at a particular instant, it would take less time for the current to go outside the hysteresis band, and so the current controller must respond by switching the device faster. Alternately, a lower rate of change of current would mean the switching will occur slower in time.

To this effect, the switching frequency will be higher near the unaligned position since there is higher rate of change of current due to lower induced voltage and lower phase inductance.

For this reason, the selection of the conduction period will have an impact on the switching frequency. From the inductance profile in Figure 9.3a, it can be seen that the phase excitation starts closer to the unaligned position. As the induced voltage builds up, the rate of change of current reduces throughout the excitation period, and so the switching period at θ_2 is much longer than the period at θ_1, as shown in Figure 9.3b.

When controlling the current within the conduction period, phase voltage can be switched between the positive or negative DC-link voltage, or it can switch between zero and DC-link voltage as shown in Figure 9.3c, to keep the current within the hysteresis band. While the former would result in a faster response, it would require a much higher switching frequency as well as causing higher harmonics in the current.

Furthermore, current tracking can be improved and the current ripple may be reduced by decreasing the size of the hysteresis band, which would improve the torque ripple and acoustic properties of the machine. However, selecting the width of this band is, in reality, limited by the capabilities of the switching devices used, the current sensor, and the motor controller. Finally, the conduction angles must be optimized, which would also have an effect on the dynamic performance of the motor.

9.4.3 Effect of Conduction Angles

When designing the controls for SRMs, the conduction angles θ_{on} and θ_{off} should be selected to maximize the performance of the motor at various operating conditions. At low speeds, while hysteresis control will deliver the required torque by selecting the proper current reference, the torque quality and efficiency can be improved by carefully selecting the conduction angles. Meanwhile at high speeds, conduction angles are the only parameters for control as peak current control is not possible. Figure 9.4 shows the effect of varying conduction angles on the dynamic current and torque profiles of an SRM.

In Figure 9.4, three dynamic profiles are shown with three sets of conduction angles. In profile 1, a short conduction period is applied which utilizes less current and, thus, creates lower copper losses. However, the small conduction period results in low torque production and high torque ripple due to the lack of commutation between phases. A wider conduction period is used in profiles 2 and 3, and it can be seen in Figure 9.4c that the torque quality is improved in these cases while also improving average torque. However, by delaying the θ_{off}, some negative torque production is introduced, as shown in the phase torque in Figure 9.4b. Profile 3 has a more delayed θ_{off} than profile 2, resulting in more negative torque production due to the prolonged phase excitation period. Despite both profiles 2 and 3 are turned off at the aligned position, tail current remains in the phase due to the residual stored energy in the magnetic circuit.

Practically, optimization is used to determine the conduction angles that are suitable for each operating point on the torque speed map. As can be seen from Figure 9.4, many conflicting objectives can be considered when choosing the optimal conduction angles. When selecting the conduction angles, the dynamics of the phase current, as well as the speed, are important factors. As the speed increases, the rate of change of current is reduced due to the effect of increased induced voltage. Consequently, the conduction angles are advanced so that each phase will have enough time to energize prior to reaching positive torque production.

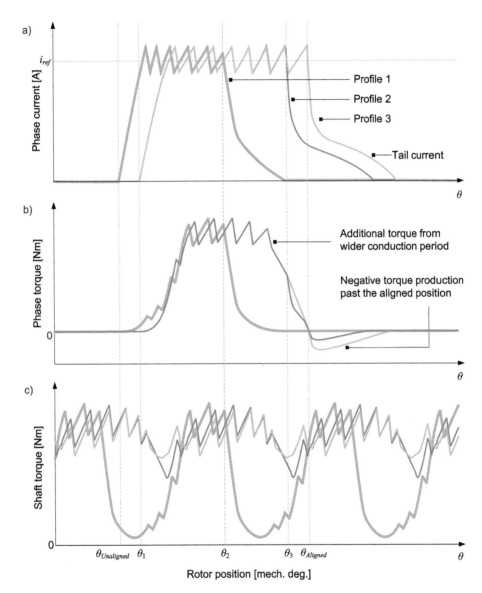

FIGURE 9.4
Dynamic profiles for SRM demonstrating the effect of varying conduction angles on (a) phase current, (b) phase torque, and (c) shaft torque.

9.5 Optimization of the Conduction Angles

Proper selection of the conduction angles can significantly affect the average torque and torque ripple. In this section, the effects of turn-on and turn-off angles on average torque, torque ripple, and other objectives will be discussed. Genetic Algorithm (GA) will be used to find the optimized firing angles for SRMs in both single- and multi-objective optimization. In this text, the terms firing angles and commutation angles both refer to conduction angles, and are used interchangeably.

9.5.1 Effect of Conduction Angles

Figure 9.5 shows the effects of conduction angles on six parameters, including torque ripple, average output torque, root mean square (RMS) value of the phase current, loss, and efficiency. The conduction angles affect the torque quality. As shown in Figure 9.5a, at a fixed speed and current, the torque ripple – defined as the ratio of the difference between maximum torque and minimum torque to the average torque – is unacceptable in a wide search domain of firing angles. Here unacceptable torque ripple means that the torque ripple is over 120%. In order to improve the torque quality, the torque ripple has to be minimized. The minimum torque ripple is not always achievable because there are constraints on phase current and conduction angles. Furthermore, the objective of minimizing torque ripple has to be balanced with the objective of maximizing output torque. These two objectives are often conflict as it will be discussed later in this section.

As shown in Figure 9.5b, the RMS value of phase current, I_{RMS}, is almost proportional to the conduction angle, which is the difference between the turn-on and turn-off angles. I_{RMS} represents the thermal limitation and is an important constraint in the optimization of conduction angles. For instance, if 130 A is used as the constraint of I_{RMS}, this means that any combination of θ_{ON} and θ_{OFF} resulting in an I_{RMS} higher than 130 A will be ruled out. Reference speed and the commanding current also affect the contours of I_{RMS}.

Figure 9.5c shows the contours of average torque as θ_{ON} and θ_{OFF} vary, with the motor operating at 2000 RPM. The maximum output torque is not always attainable because of the constraints on I_{RMS} and the firing angles. The objective of maximizing output torque can be used as the only objective in a single-objective optimization. By mixing it with one or more of other objectives, a multi-objective optimization can be formulated.

Other objectives can be formulated to represent certain features of SRMs. For instance, the ratio of average output torque over the RMS value of phase current, T_{ave}/I_{RMS}, is often used to indicate efficiency. Figure 9.5d shows an example of the contours of T_{ave}/I_{RMS}, as firing angles vary. If we put Figure 9.5c and d side-by-side, you will see that both contours have similar trends; however, the peaks of the two are different. This happens because of the trend where I_{RMS} decreases towards the bottom right corner as the commutation decreases. Figure 9.5e and f show the contours of total loss and efficiency of a certain SRM at 2000 RPM and 200 A. At a specific speed, decreasing the total loss does not always necessarily mean that the efficiency will be improved. The output torque value has to be taken into account. The peak of T_{ave}/I_{RMS} does not correspond to the peak of efficiency if you are comparing Figure 9.5d and f. SRMs are highly nonlinear systems and special attention needs to be paid if any simplification is applied to model the system. The objective of minimizing normalized torque ripple improves torque quality; however, it does not necessarily mean that the noise and vibration of the motor is improved. Please note that Figure 9.5 shows the contours of certain parameters for a specific motor at a certain operating point. The trends of the contours are not generic.

9.5.2 Optimization of Commutation Angles

Tuning the turn-on and turn-off angles affects the output torque, the motor efficiency, and the torque ripple. However, it can be extremely time-consuming to manually tune these angles for each operating point. In order to speed up the tuning process for the conduction angles, a GA based optimization will be discussed in this section using the MATLAB® Global Optimization Toolbox [2].

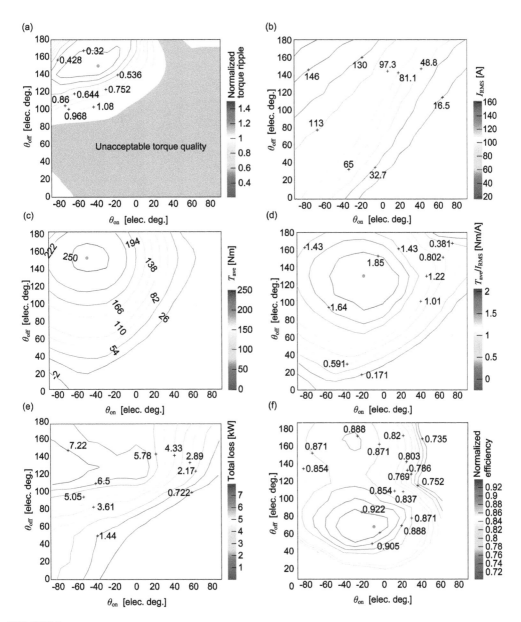

FIGURE 9.5

An example of effects of firing angles, 2000 rpm (reference speed), 200 A (commanded current): (a) normalized torque ripple, (b) RMS phase current, (c) average torque, (d) ratio of average torque over RMS phase current, (e) total loss, and (f) efficiency. (From Jiang, J.W., Three-phase 24/16 switched reluctance machine for hybrid electric powertrains: Design and optimization, Ph.D. Dissertation, Department Mechanical Engineering, McMaster University, Hamilton, ON, 2016.)

To optimize the conduction angles, a model is needed that calculates the dynamic torque profile of the machine for the given conduction angles. The dynamic model of a switched reluctance motor drive was introduced in Chapter 4. This model can be built in the MATLAB/Simulink environment for use with the MATLAB Global Optimization Toolbox. This model is utilized to calculate the objectives for the given variables, operating

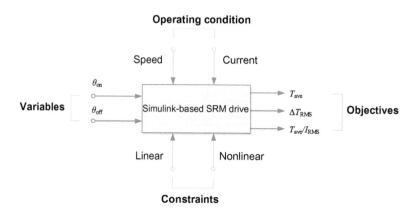

FIGURE 9.6
Description of the formulation of the optimization problem for conduction angles.

conditions, and constraints. The model is called by a GA optimizer in MATLAB® using either [x, fval,...] = ga(fitnessfcn, nvars,...) (single-objective) or [x, fval, ...] = gamultiobj(fitnessfcn, nvars,...) (multi-objective) functions. Please refer to the MATLAB® Global Optimization Toolbox manual for details concerning these functions [2]. In this model, firing angles will be the variables, as shown in Figure 9.6. Reference speed and commanding current represent the operating conditions, which are inputs for the SRM drive model. The constraints of the optimization problem can be grouped into linear and nonlinear constraints. A number of objectives can be formulated and used in either single- or multi-objective optimization.

9.5.2.1 Optimization Flowchart

Figure 9.7 shows the flowchart used in the optimization problem. It can be used both for the selected operating points and for the entire operating range. The flowchart can be divided into the four following sections. In the first section (I), a specific SRM is characterized (the model was discussed in Chapter 4). The flux linkage and torque lookup tables can be obtained and used in the SRM drive model. Then three loops are used in the flowchart. In the innermost loop (II), the SRM drive model built in Simulink is called by the optimizer. The turn-on θ_{ON} and the turn-off θ_{OFF} angles are variables fed into the SRM drive. The objective functions used in the optimization problem vary in different scenarios.

Until the stop criteria constraints are met, the optimizer continues its search for a global optimum in the second loop (III). If it is a multi-objective optimization problem, the optimized results for a specific operating point are plotted in the Pareto-front and decision-making logic is involved to choose a single solution for the front. In the third loop (IV), either randomly or evenly selected operating points, as in the combination of reference speed and commanded current, are fed into the second loop. Each operating point corresponds to one specific optimization problem.

9.5.2.2 Objectives

The three most commonly used objectives are discussed in this section. They are maximizing average output torque in (9.6), minimizing the RMS value of the net torque ripple

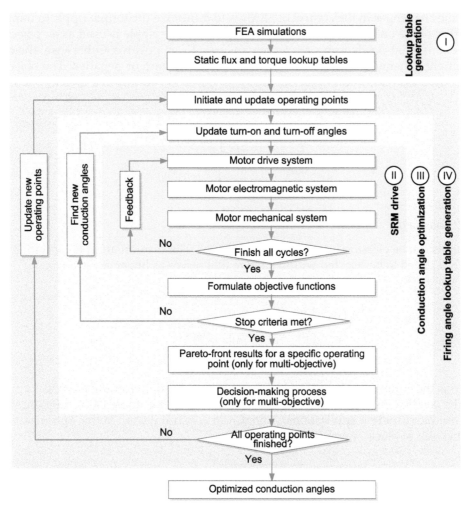

FIGURE 9.7
Flowchart for optimization of conduction angle. (From Jiang, J.W., Three-phase 24/16 switched reluctance machine for hybrid electric powertrains: Design and optimization, Ph.D. Dissertation, Department Mechanical Engineering, McMaster University, Hamilton, ON, 2016.)

in (9.7), and maximizing the ratio of average output torque to the RMS value of the phase current in (9.8). The objective f_1 is used to evaluate the average torque:

$$f_1 = T_{ave} = \frac{1}{\theta_1 - \theta_2} \int_{\theta_1}^{\theta_2} T(\theta)\, d\theta \tag{9.6}$$

where $T(\theta)$ is the torque waveform as a function of rotor position and $(\theta_1 - \theta_2)$ is equals to a complete electrical cycle. The objective f_2, which is the RMS value of net torque ripple, addresses the concern over the quality of the output torque:

$$f_2 = \Delta T_{RMS} = \sqrt{\frac{1}{\theta_1 - \theta_2} \int_{\theta_1}^{\theta_2} \left(T(\theta) - T_{ave}\right)^2 d\theta} \tag{9.7}$$

One of the challenges in the control of SRMs is to minimize the torque ripple to improve the quality of the output torque. If minimizing the torque ripple is used as an objective, another constraint needs to be set, which ensures T_{ave} is positive. Otherwise, the optimizer might generate conduction angles that make T_{ave} zero or negative. The objective f_3 can be used as an indicator of the efficiency for the SRM running at a specific operating point:

$$f_3 = \frac{T_{ave}}{I_{RMS}} = \frac{\dfrac{1}{\theta_1 - \theta_2} \displaystyle\int_{\theta_1}^{\theta_2} T(\theta)\, d\theta}{\sqrt{\dfrac{1}{\theta_1 - \theta_2} \displaystyle\int_{\theta_1}^{\theta_2} I(\theta)^2\, d\theta}} \tag{9.8}$$

9.5.2.3 Constraints

Constraints can be classified as linear and nonlinear. The turn-on angle θ_{on} and the turn-off angle θ_{off} need to be selected within a range that serves as linear constraints:

$$\begin{bmatrix} 1 & -1 \\ -1 & 1 \end{bmatrix} \begin{bmatrix} \theta_{on} \\ \theta_{off} \end{bmatrix} < \begin{bmatrix} -\dfrac{360}{m} \\ \dfrac{360k}{m} \end{bmatrix} \tag{9.9}$$

where m is the number of phases, and k is a constant value determining the upper boundary of the conduction angle. For instance, the value for k is set to be 1.4. A different value can be used for different applications. A nonlinear constraint used in this optimization is the RMS value of phase current:

$$I_{RMS} = \sqrt{\frac{1}{\theta_2 - \theta_1} \int_{\theta_1}^{\theta_2} I^2(\theta)\, d\theta} \leq I_{RMS\_constraint} \tag{9.10}$$

Figure 9.8 shows the domains of linear and nonlinear constraints for a certain SRM operating at 2000 RPM and 200 A. In Figure 9.8, the two blue triangles represent two regions that violate the nonlinear constraints. The lower triangle means the firing angle is below the lower boundary. The upper triangle indicates that the firing angle is above the upper boundary of the constraint. The green triangle stands for a region in which I_{RMS} is greater than its constraint. The change in I_{RMS} as the firing angle varies can be seen in Figure 9.5b.

Two single-objective optimization problems can be formulated as:

- Case I: max. T_{ave} (f_1 in (9.6)), I_{RMS} as constraint
- Case II: max. T_{ave}/I_{RMS} (f_3 in (9.8)), I_{RMS} as constraint

f_2 in (9.7) can also be used as the objective in the single-objective optimization. In this case, another constraint of the required torque at certain operating points has to be employed. This means that the ΔT_{RMS} will be minimized while satisfying the constraints of I_{RMS} and the required torque. A lookup table of the required values of output torques has to be provided,

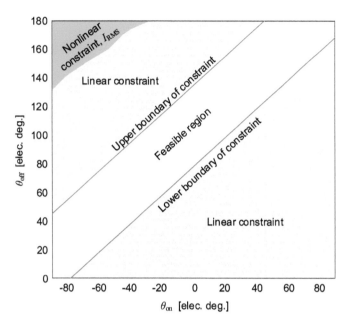

FIGURE 9.8
An example of constraints, 2000 rpm, 200 A.

which might make the optimization more complicated. The lookup table has three dimensions: speed (input), current reference (input), and required torque (output).

9.5.2.4 Brief Introduction to Genetic Algorithm (GA) Optimizer

Figure 9.9 is a short explanation of the GA optimizer. GA optimization is based on the idea of natural selection (survival of the fittest), first raised by Charles Darwin, and it may be applied to solve optimization problems. A population represents a set of possible solutions for a given problem. The population number, that is, the number of solutions, stays the same throughout this exploration. Each individual, analogous to a chromosome, possesses several variables, just like genes. Each individual has a fitness value, which represents its ability to compete with others. The fitness value could be determined by an objective function. Based on each individual's fitness value, parents are selected to produce better offspring that combine their genes. The process of parent election is analogous to the idea of survival of the fittest. The definition of crossover represents mating between the selected parents. Mutation means that random modification would be introduced, which could be used to inhibit premature convergence. At last, when a new generation of offspring vary little from those in previous generations, the algorithm has converged into a set of solutions, according to this algorithm.

The GA could be generalized into the following steps. First randomly choose a population and determine their fitness values. Then the following procedures would be repeated in a cycle until the new generation is not noticeably different from the previous: (1) evaluate the fitness of the population, (2) select the best parents from the population, (3) perform crossover on the selected parents to create a new generation, and (4) perform a mutation of the new population. A more detailed explanation of GA is included in the appendix of this chapter.

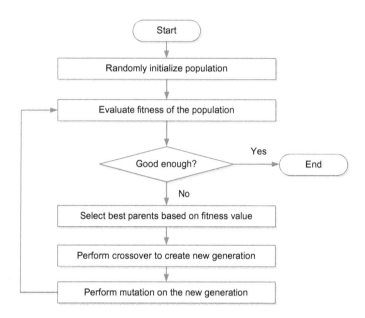

FIGURE 9.9
Flowchart of GA optimizer. (From Mathworks. Global Optimization Toolbox User's Guide, 2016 [Online]. Available: http://www.mathworks.com/help/pdf_doc/gads/gads_tb.pdf. (Accessed: February 27, 2017).)

9.5.3 Multi-Objective Optimization

Single-objective optimization for commutation angles, for instance with the objective of maximizing average torque, has its own limitation. The objective of maximizing output torque is important; however, there are some other objectives that need to be considered as well, such as, minimizing torque ripple. The objectives must be balanced, and compromises have to be made based on specific operating points. For example, at some operating regions, output torque has to be given more emphasis; while, at other operating regions, either torque quality or efficiency is more important. The method of transforming objectives into one single objective, for instance, by summing them up with assigned weights, is not recommended because the range of each cost function may not be well known beforehand and the correlations between objectives and variables are ignored. Multi-objective optimization for commutation angles is needed because a number of optimized results will be given for one optimization problem instead of just one result as in single-objective optimization. In this section, the three objectives are employed in the multi-objective optimization problems individually. Two multi-objective optimization problems are:

- Case III: max. T_{ave} (f_1 in (9.6)), while min. ΔT_{RMS} (f_2 in (9.7)), I_{RMS} as constraint
- Case IV: max. T_{ave}/I_{RMS} (f_3 in (9.8)), while min. ΔT_{RMS} (f_2 in (9.7)), I_{RMS} as constraint

9.5.3.1 Pareto Front Results

The multi-objective optimizer produces more than one optimized result for a specific operating point, which can be plotted on a Pareto front. For instance, each green bubble in Figure 9.10 represents an optimized solution of Case III, a multi-objective optimization. The values of conduction angles for all optimized results are also plotted to observe its

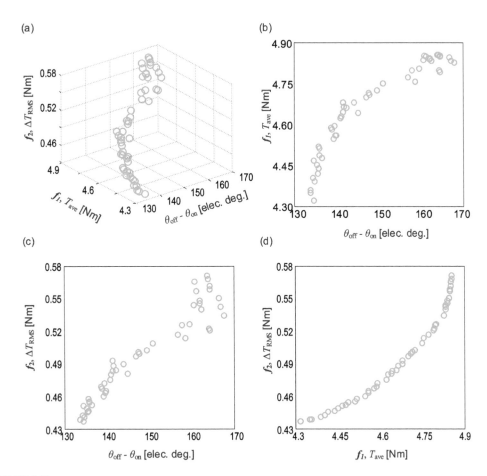

FIGURE 9.10
Results of multi-objective optimization, Case III (max. T_{ave} (f_1) and min. ΔT_{RMS} (f_2)), 1000 rpm: (a) relationship between f_1, f_2 and $\theta_{off} - \theta_{on}$, (b) relationship between f_1 and $\theta_{off} - \theta_{on}$, (c) f_2 and $\theta_{off} - \theta_{on}$, and (d) Pareto front results. (From Jiang, J.W., Three-phase 24/16 switched reluctance machine for hybrid electric powertrains: Design and optimization, Ph.D. Dissertation, Department Mechanical Engineering, McMaster University, Hamilton, ON, 2016.)

correlation with the optimization objectives. Figure 9.10b shows that T_{ave} tends to increase as the conduction angle increases, because multiphase conduction intensifies. Meanwhile, the increase in the conduction angle, from 130 to 165 electrical degrees, leads to an increase of ΔT_{RMS}, as shown in Figure 9.10c. It is clear upon observing Figure 9.10d that trade-offs are necessary to ensure the multi-objective optimizer delivers acceptable results.

Figure 9.11 shows the optimized results of Case IV, for the SRM operating at 1,000 RPM. As shown in Figure 9.11b, as the conduction angle increases, ΔT_{RMS} tends to decrease, which indicates that the torque quality improves. Figure 9.11c shows that T_{ave}/I_{RMS} decreases as the conduction angle increases. This happens because within the given range of conduction angles, I_{RMS} increases more significantly than T_{ave}. It is also worth mentioning that the conduction angle span of Case IV is smaller than that of Case III. On each Pareto front, as shown in Figures 9.10d and 9.11d, only one solution will be selected as the single solution to the optimization problem. It's interesting to note that the Pareto front results for Case III and IV experience opposite trends: one convex (Figure 9.10d) and the other is concave (Figure 9.11d). Different objectives and different types of linear and nonlinear constraints

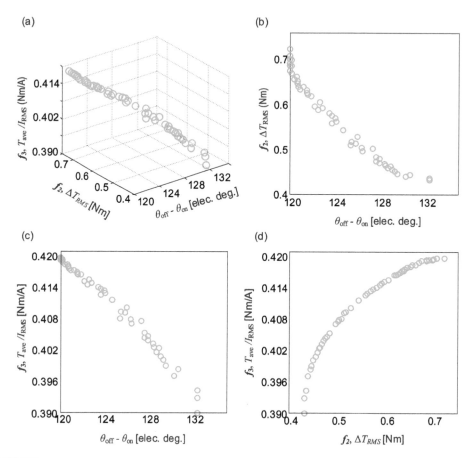

FIGURE 9.11

Results of multi-objective optimization, Case IV (max. T_{ave}/I_{RMS} (f_3) and min. ΔT_{RMS} (f_2)), 1000 rpm: (a) relationship between f_1, f_2, and $\theta_{off} - \theta_{on}$, (b) relationship between f_2 and $\theta_{off} - \theta_{on}$, (c) f_3 and $\theta_{off} - \theta_{on}$, and (d) Pareto front results. (From Jiang, J.W., Three-phase 24/16 switched reluctance machine for hybrid electric powertrains: Design and optimization, Ph.D. Dissertation, Department Mechanical Engineering, McMaster University, Hamilton, ON, 2016.)

all affect the shapes of Pareto front results. A decision-making process should be used to achieve this, and it will be discussed in the next section.

9.5.3.2 Decision-Making Process

For a multi-objective optimization problem, the user's preferences are needed for choosing the optimized solution from the Pareto front results. For a small number of operating points, the decision-maker can handpick solutions from Pareto front results; however, this process can be time-consuming if optimization needs to be completed for a large number of points over the entire operating range. A set of algorithms is designed to accelerate the decision-making process.

Figure 9.12 shows the flowchart for the decision-making process in the multi-objective optimization problems (Case III and IV). For a specific operating point, the Pareto front results among other relevant results are first extracted from the optimizer and may include, (a) T_{ave}, (b) torque ripple (c) I_{RMS}, and (d) conduction angle ($\theta_{off} - \theta_{on}$). It should

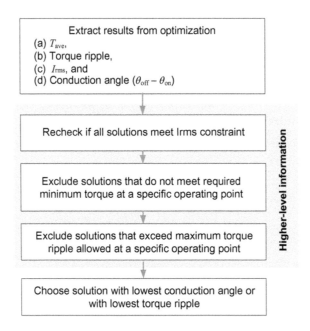

FIGURE 9.12

Flowchart for decision-making process of multi-objective optimization in conduction angles. (From Jiang, J.W., Three-phase 24/16 switched reluctance machine for hybrid electric powertrains: Design and optimization, Ph.D. Dissertation, Department Mechanical Engineering, McMaster University, Hamilton, ON, 2016.)

be noted that higher-level information is needed in choosing one solution from multiple trade-off solutions.

In this process, the constraint placed on I_{RMS}, the required maximum torque, and the maximum torque ripple at different speeds is the higher-level information. The value of I_{RMS} could be rechecked for all solutions to see if the current constraint is met. In many applications, a specific range of average output torque is pre-determined, from a given torque-speed profile. Then, the solutions from the Pareto front, which do not satisfy the required maximum torque, may be excluded.

For instance, T_{ave_r} represents the required torque at a specific operating point as shown in Figure 9.13. Any solution on the Pareto front with values higher than T_{ave_r} as indicated by green and red circles, is acceptable considering the average torque. In the decision-making process, T_{ave_r} works more like a constraint instead of an objective. If maximizing torque ripple is the other objective, several solutions, indicated by red circles in Figure 9.13, are excluded from consideration. More ranking criteria, such as torque ripple and conduction angle $\theta_{off} - \theta_{on}$ can be included in the decision-making process. If data regarding the required torque does not exist, the highest torque $T_{highest}$ on the Pareto front can be used as a reference. At this operating point, 5% of $T_{highest}$ can be compromised to improve the torque quality. The subset of the Pareto front comprises the results with the output torque ranging from $T_{highest}$ to $0.95\ T_{highest}$. Within the subset, the result with the lowest torque ripple will be picked as the final optimized result. As stated earlier, for some operating points, torque quality will be given a priority. In this case, an even higher compromise of output torque has to be made; for instance 10% of the peak torque, can be sacrificed.

Multi-objective optimization has its merits. As stated earlier, the multi-objective optimization can take more than one objective and generate a number of optimized results.

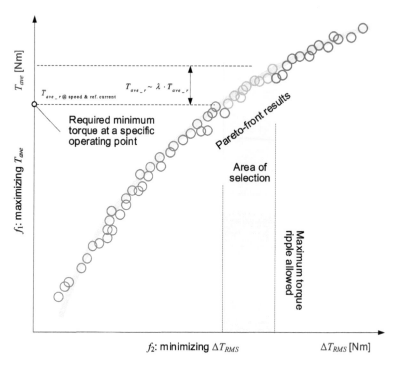

FIGURE 9.13
Decision-making algorithm in ranking Pareto-front results with maximizing T_{ave} and maximizing torque ripple as two objectives. (From Jiang, J.W., Three-phase 24/16 switched reluctance machine for hybrid electric
powertrains: Design and optimization, Ph.D. Dissertation, Department Mechanical Engineering, McMaster
University, Hamilton, ON, 2016.)

In addition, the relationship between objectives can be analyzed. However, decisions have
to be made in selecting one final result from the Pareto front and compromises have to
be made between objectives. Multi-objective optimization is also more computationally
expensive than single-objective optimization.

9.5.4 Optimization for the Entire Operating Range

Figure 9.14 shows a typical torque-speed map for an SRM. Each cross on the chart represents an operating point that is required to undergo optimization in order to determine
the conduction angles. Thus, to characterize an SRM's performance over the entire operating range, a number of points need to be selected over the operating envelop and optimization for conduction angles has to be completed. As we discussed earlier, both single- and
multi-objective optimizations can be used to solve this problem. Under the same set of
constraints, if maximizing average torque is used as the objective in the single-objective
optimization as in Case I, the torque-speed envelop will enlarge and cover higher torque
values. If a multi-objective optimization method is used in Case III, for instance, since the
output torque at each speed point will have to be balanced with the torque quality, then
the torque-speed envelop might be narrower than in Case I.

Figure 9.15 shows an example of optimized turn-on and turn-off angles. For each combination of reference speed and current commands, a unique pair of conduction angles is
determined based on either single- or multi-objective optimization. The surfaces of turn-
on and turn-off angles are highly nonlinear. As shown in Figure 9.15a, for the same current,

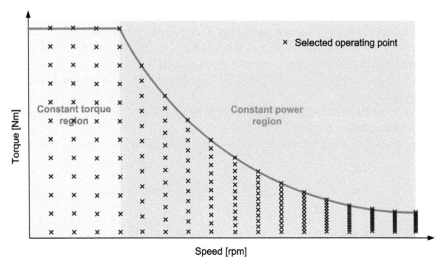

FIGURE 9.14
Selected operating points used to characterize SRM over the entire operating range.

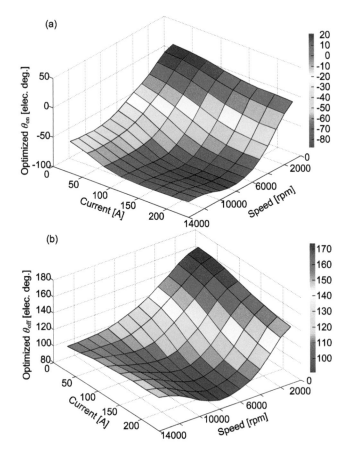

FIGURE 9.15
Lookup tables of optimized firing angles: (a) turn-on angles and (b) turn-off angles.

the turn-on angle tends to advance as the speed increases. For instance, when 200 A is the reference current, the turn-on angle changes from 0 electrical degree to –70 electrical degrees gradually, as the speed increases from 0 RPM to 14,000 RPM. A similar surface is observed for the turn-off angles as shown in Figure 9.15b. At very low speeds and low currents, the turn-off angle is higher. These two lookup tables will be used to determine the torque-speed envelope.

Figure 9.16 shows the characterization of an SRM based on the lookup tables for optimized conduction angles. Figure 9.16a shows the contour map for the RMS phase current. It can be seen that the constraint of RMS phase current determines the maximum

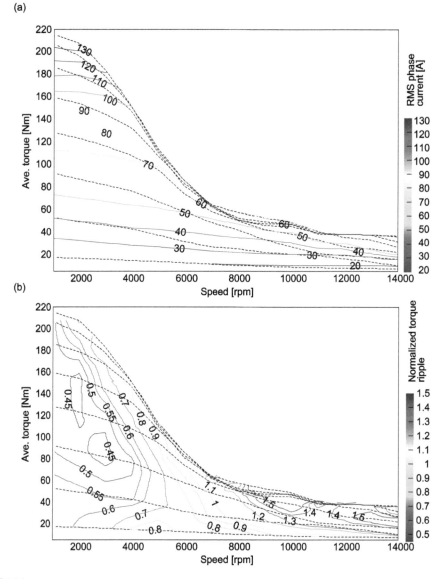

FIGURE 9.16

Characterization of an SRM's performance based on lookup tables for firing angles using Simulink-based SRM drive: (a) contours of RMS value of phase current and (b) contours of normalized torque ripple. (From Jiang, J.W. et al., *IEEE Trans. Transport. Electrification*, 3(1), 76–85, March 2017.)

output torque in the constant torque region. As the constraint increases, for instance from 70 A to 120 A, the maximum output torque increases, and the constant torque region becomes narrower. Figure 9.16b shows the contours of normalized torque ripple. It can be seen that within the high-speed region, the torque quality worsens as the speed increases. Other contour maps, for peak-to-peak torque, copper loss, iron loss, efficiency, and so on, can all be plotted on the torque-speed profile to visualize the performance of the motor.

9.6 Torque Sharing Functions

So far, in this chapter conduction angle control has been introduced as the primary control method for SRMs. Using conduction angle control, the only parameters considered are the turn-on angle, the turn-off angle, and the peak current, which is set to achieve the required torque output. Using this control method, rectangular current profiles are created, and it has been shown that they can be tracked using a hysteresis current controller. Lastly, it has been shown that by adjusting these parameters using optimization methods, the torque ripple and efficiency of the machine can be optimized.

However, while conduction angle control is very simple to implement, its performance is limited due to the fact that there are only three parameters available for adjustment. As an example, for torque ripple minimization, only the timing of the commutation region can be defined, while the current dynamics within the commutation region are not controlled.

Alternatively, more advanced techniques have been proposed in SRM literature where the entire current reference can be defined using a more generalized set of parameters without being constrained to simply conduction angles and peak current. Motivation for the development of these techniques came from the need to improve the torque ripple in SRMs beyond the capabilities of conduction angle control. These current profile defining algorithms are called torque sharing functions.

9.6.1 Fundamentals of Torque Sharing Functions

A torque sharing function (TSF) intelligently divides a constant torque reference amongst different phases by defining a reference current profile for each phase. Ideally, when this current reference profile is tracked by a current controller, the total torque contribution from all of the phases will add up to the constant torque, thus eliminating torque ripple. A very simple way of defining a torque sharing function is through using analytical expressions to model the dynamic behavior of the phase torque contribution. In light of this, the torque reference of the kth phase is defined as in (9.11).

$$T_{e\_ref}(k) = \begin{cases} 0 & 0 \le \theta < \theta_{ON} \\ T_{e\_ref} f_{rise}(\theta) & \theta_{on} \le \theta < \theta_{on} + \theta_{ov} \\ T_{e\_ref} & \theta_{on} + \theta_{ov} \le \theta < \theta_{off} \\ T_{e\_ref} f_{fall}(\theta) & \theta_{off} \le \theta < \theta_{off} + \theta_{ov} \\ 0 & \theta_{off} + \theta_{ov} \le \theta < \theta_{p} \end{cases} \qquad (9.11)$$

where $T_{e\_ref(k)}$ is the reference torque for k^{th} phase, T_{e_ref} is total torque reference, $f_{rise}(\theta)$ is the rising function for the incoming phase that increases from zero to one, and $f_{fall}(\theta)$ is the decreasing function for the outgoing phase that decreases from one to zero, for some rotor position, θ. In (9.11), note that when $f_{rise}(\theta)$ and $f_{fall}(\theta)$ are selected carefully to add to unity for all θ, the phase torque contributions of the outgoing phase and the incoming phase will add up to the constant torque command, T_{e_ref}. Other parameters, such as θ_{on}, θ_{off}, θ_{ov}, and θ_p, are the turn-on angle, turn-off angle, overlapping angle, and pole pitch, respectively. The pole pitch, θ_p, is defined in (9.12) using the number of rotor poles N_r.

$$\theta_p = \frac{2\pi}{N_r} \tag{9.12}$$

An example of the torque reference of (9.11) is shown in Figure 9.17. The figure shows that outside the commutation region, only one phase conducts. Inside the commutation region, the phase torques are defined by $f_{rise}(\theta)$ and $f_{fall}(\theta)$, which must be chosen carefully to add up to unity for all θ.

A phase current command can then be interpolated from the torque profiles generated by the TSF. One way to achieve this is to invert the static torque characteristics shown in Chapter 4 in a similar way that the flux linkage characteristics is inverted. Then the current references can be obtained from the resulting look-up table. An example of the inverted torque look-up table used to obtain the current references for the 12/8 SRM is shown in Figure 9.18.

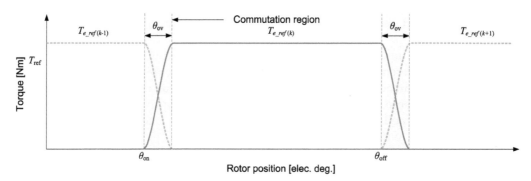

FIGURE 9.17
An example of the torque reference produced by the torque sharing function.

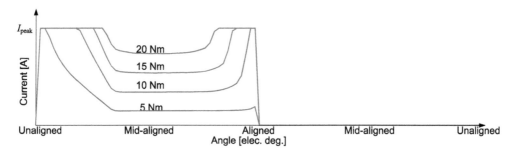

FIGURE 9.18
The inverted static torque characteristic showing the amount of current needed.

Figure 9.18 shows the amount of current command needed to create a given torque command, in one electrical cycle. This can be thought of as the amount of current needed to create the desired torque to draw the rotor pole towards the aligned position. Another interesting feature about these inverted torque look-up tables is that from the inversion process, saturation on the current can be seen near the aligned and unaligned positions. At these positions, torque producing capability of the machine is low. Therefore, in order to make a given torque command, an excessive amount of current is needed, which may not be realizable in practice. Hence, a peak current constraint is set within the inversion process to limit the range of current in the look-up tables.

Using these look-up tables, the torque command created by the TSF can be converted into a current command. The resulting current reference is a function dependent on the rotor position. For the torque reference in Figure 9.17, the corresponding current reference is shown in Figure 9.19. Note that by choosing different values for θ_{ON}, θ_{OFF}, and θ_{ov}, different current profiles can be obtained as illustrated in Figure 9.19. The parameters used for each case is given in Table 9.1. For real-time implementation, these current references are stored as look-up tables in the SRM drive and tracked by the current controller. This is shown in Figure 9.20.

It can be seen so far that these current profiling techniques involve a much more complex way of defining a current reference compared to the three parameters required for conduction angle control, but with the benefit of more freedom in the range of possible control actions. Using torque sharing functions, current is shaped to specifically reduce torque ripple in the commutation region of the phase excitation. Interestingly, Figure 9.19 shows that the shape of the current profile can vary dramatically even just by shifting the

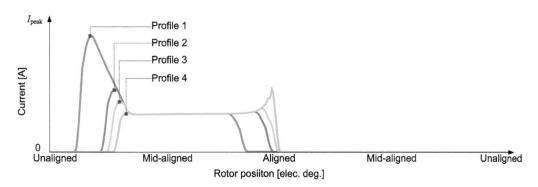

FIGURE 9.19
The phase current reference produced by the TSF for four different set of excitation parameters.

TABLE 9.1

TSF Parameters and the Respective RMS Current Values of the Current References in Figure 9.19

	Turn-on Angle [°e]	Turn-off Angle [°e]	Overlap Angle [°e]	RMS Current [A]
Profile 1	20	140	15	11.42
Profile 2	40	160	15	8.12
Profile 3	45	165	15	7.83
Profile 4	50	170	15	7.86

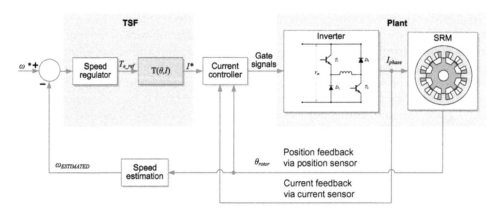

FIGURE 9.20
Overview of control structure for SRM with TSF.

conduction period, and not limited to simple rectangular waveforms. However, there are advantages and disadvantages to using one specific current profile compared to another. Therefore, the choice of parameters for the TSF must be carefully made. Some metrics to quantify the performance of a given TSF are described in the following section.

9.6.2 Evaluation Metrics of Torque Sharing Function

The selection of a suitable TSF is challenging when considering both the torque-speed range and drive efficiency. A current reference that requires less current usage is better than one that requires more current. Also, the performance of each TSF depends on how well the current reference it produces is tracked by the current controller. As the speed of the SRM increases, phase current is not able to track the reference due to the higher induced voltage. This makes the TSF less effective at high speeds. In light of this, two metrics are commonly used to characterize the performance of any TSF.

9.6.2.1 Copper Losses

The copper loss associated with phase current is defined by (9.13)

$$P_{cu} = RI_{RMS}^2 \tag{9.13}$$

where R is the phase resistance. The copper losses depend on the RMS value of the phase current measured over one electrical period, which is defined in (9.14).

$$I_{RMS} = \sqrt{\frac{1}{\theta_p} \int_0^{\theta_p} i_k^2(t) d\theta} \tag{9.14}$$

Therefore, when defining a torque sharing function, the RMS value of the phase current should be considered, as it affects the drive efficiency. In Figure 9.19, for example, four current references have been created using one particular TSF with varying conduction periods. While these current profiles have all been created to meet the same torque command, their RMS current values vary. In particular, in profile 1, the conduction period is most advanced, where the turn-on angle is near the unaligned position. In this case, higher current is required to compensate for the lack of torque producing capabilities near the unaligned position, leading to high RMS current. On the other hand, the RMS current in profile 3 is noticeably lower since the phase is excited in a range where the torque production capabilities are much higher. Therefore, when considering the copper losses for this particular example, it is more desirable to excite the phase between 45° electrical and 165° electrical, than to excite the phase much earlier.

9.6.2.2 Rate of Change of Flux Linkage

For the torque sharing function to properly reduce torque ripple, the current reference that it produces must be properly tracked. However, as seen so far in this chapter, current tracking capabilities are limited by the voltage dynamics in the phase, which is speed dependent. Therefore, depending on how well the TSF can cater to the natural dynamics of the phase of the SRM, the effective speed range for the given TSF may be limited. To understand this better, the following relationship can be derived using from the phase voltage equation of an SRM, in which the voltage drop on the phase resistance is neglected:

$$V_{DC} = \frac{d\lambda(\theta,i)}{dt} \tag{9.15}$$

Assuming constant current excitation for simplicity, the right-hand side can be broken down as:

$$V_{DC} = \frac{d\lambda(\theta)}{d\theta}\frac{d\theta}{dt} \tag{9.16}$$

From which it can be seen that

$$\frac{V_{DC}}{\omega} = \frac{d\lambda(\theta)}{d\theta} \tag{9.17}$$

Equation (9.17) suggests that, assuming that the DC-link voltage is constrained to be constant, the rate of change of flux linkage in the machine decreases as the machine speeds up. Therefore, if a torque sharing function can return a current reference that maintains a low maximum rate of change of flux linkage, that TSF can be effective across a wider speed range.

At a given speed, the rate of change of flux linkage due to the current reference must satisfy (9.18) for the current to be successfully tracked.

$$\frac{d\lambda(\theta)}{d\theta} \le \frac{V_{dc}}{\omega} \tag{9.18}$$

The maximum absolute value of the rate of change of flux linkage (ARCFL) M_λ is defined as:

$$M_\lambda = max\left\{ \frac{d\lambda_{rise}}{d\theta}, -\frac{d\lambda_{fall}}{d\theta} \right\} \tag{9.19}$$

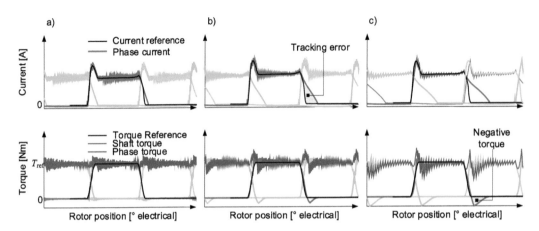

FIGURE 9.21
Simulation results of the cubic TSF at (a) 1000 RPM, (b) 2000 RPM, and (c) 3000 RPM showing the effect of increase in speed on the performance of the TSF for the 12/8 SRM using $\theta_{on} = 40°$e, $\theta_{off} = 160°$e, $\theta_{ov} = 20°$e.

where λ_{rise} is the rising flux linkage for the incoming phase, and λ_{fall} is the decreasing flux linkage for the outgoing phase. The theoretical maximum ripple-free speed (TRFS) ω_{max} can then be derived as:

$$\omega_{max} = \frac{V_{DC}}{M_\lambda} \tag{9.20}$$

The effect of speed on the rate of change of flux linkage is shown in Figure 9.21 to demonstrate the limitation of the speed range of the TSF. In Figure 9.21, the same current reference profile produced by the cubic TSF is tracked in three constant speed simulations. Analytical torque sharing functions including the cubic TSF will be explained in the next section.

In Figure 9.21a, the simulation has been run at 1000 RPM. In this case, there is a small tracking error in the current waveform, leading to a smooth torque profile with 0.18 Nm RMS of torque ripple. In Figure 9.21b, the simulation has been run at 2000 RPM. It may be observed that in this case, the current tracking error becomes more significant. Therefore, current in the outgoing phase cannot decrease as fast as the current command. This is due to the limitations in the rate of change of flux linkage at higher speeds. As a result, negative torque is produced in the phase leading to higher torque ripple. At 2000 RPM, 0.27 Nm RMS of torque ripple is produced. Finally, in Figure 9.21c, the current tracking error is more pronounced at 3000 RPM, as the rate of change of flux linkage is even more limited. In this case, significant negative torque is produced resulting in 0.38 Nm RMS of torque ripple, as well as a decrease in the average torque compared to the other cases. Table 9.2 summarizes the results from these simulations.

TABLE 9.2
TSF Parameters and the Respective RMS Current, Average Torque, and Torque Ripple Values for the Cases in Figure 9.21

	Speed [RPM]	Torque Command [Nm]	RMS Current [A]	Average Torque [Nm]	RMS Torque Ripple [Nm]
Simulation 1	1000	3	8.07	3.00	0.18
Simulation 2	2000	3	8.21	2.99	0.27
Simulation 3	3000	3	7.98	2.77	0.38

9.6.3 Analytical Torque Sharing Functions

In this section, some classical TSFs found in literature are presented to the reader. In these families of TSF, the torque contribution of each phase in the commutation region, denoted by $f_{rise}(\theta)$ and $f_{fall}(\theta)$, is approximated using mathematical functions. The most common among these functions are the linear, cubic, and exponential TSFs.

9.6.3.1 Linear TSF

The linear TSF is shown in (9.21) and its waveform is shown in Figure 9.22. During commutation, the TSF for the incoming phase is increasing linearly from 0 to 1, whereas the TSF for the outgoing phase is decreasing from 1 to 0.

$$f_{rise}(\theta) = \frac{1}{\theta_{ov}}(\theta - \theta_{on})$$

$$f_{fall}(\theta) = 1 - f_{rise}(\theta + \theta_{on} - \theta_{off}) \tag{9.21}$$

9.6.3.2 Cubic TSF

The cubic TSF approximates the phase torque contribution in the commutation region as a cubic function. This is shown in Figure 9.23, and the expressions for $f_{rise}(\theta)$ and $f_{fall}(\theta)$ are represented as (9.22) with coefficients α_0, α_1, α_2, and α_3.

$$f_{rise}(\theta) = \alpha_0 + \alpha_1(\theta - \theta_{on}) + \alpha_2(\theta - \theta_{on})^2 + \alpha_3(\theta - \theta_{on})^3 \tag{9.22}$$

In order to make sure that this TSF is continuous for all rotor positions, the constraints are set as follows:

$$f_{rise}(\theta) = \begin{cases} 0, & (\theta = \theta_{on}) \\ 1, & (\theta = \theta_{on} + \theta_{ov}) \end{cases} \quad \frac{df_{rise}(\theta)}{d\theta} = \begin{cases} 0, & (\theta = \theta_{on}) \\ 0, & (\theta = \theta_{on} + \theta_{ov}) \end{cases} \tag{9.23}$$

Then, by substituting (9.22) for (9.23), the coefficients of the cubic TSF can be derived as (9.24) and the cubic TSF can be obtained as (9.25).

$$\alpha_0 = 0; \ \alpha_1 = 0; \ \alpha_2 = \frac{3}{\theta_{ov}^2}; \ \alpha_3 = \frac{-2}{\theta_{ov}^3} \tag{9.24}$$

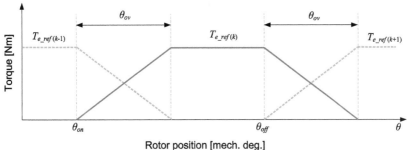

FIGURE 9.22
Linear TSF.

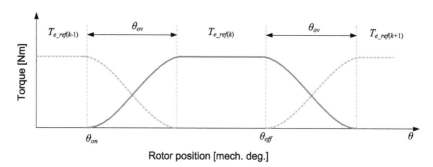

FIGURE 9.23
Cubic TSF.

$$f_{rise}\left(\theta\right) = \frac{3}{\theta_{ov}^2}\left(\theta - \theta_{on}\right)^2 - \frac{2}{\theta_{ov}^3}\left(\theta - \theta_{on}\right)$$

$$f_{fall}\left(\theta\right) = 1 - f_{rise}\left(\theta + \theta_{on} - \theta_{off}\right)$$
(9.25)

9.6.3.3 Exponential TSF

The exponential TSF is defined as (9.26), where an exponential function is used to define the torque in the commutation region. Figure 9.24 shows an example of the exponential TSF.

$$f_{rise}\left(\theta\right) = 1 - exp\left(\frac{-\left(\theta - \theta_{on}\right)^2}{\theta_{ov}}\right)$$

$$f_{fall}\left(\theta\right) = 1 - f_{rise}\left(\theta + \theta_{on} - \theta_{off}\right)$$
(9.26)

9.6.3.4 Advanced Formulations of Torque Sharing Functions

More recently, different methods of formulating TSFs offline have been proposed in literature. These methods differ from the analytical methods in that they do not require explicit functions to define the torque reference in the commutation region. Instead, they

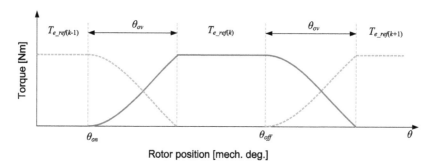

FIGURE 9.24
Exponential TSF.

rely on optimization techniques to define the current profile to minimize torque ripple, using the TSF metrics as optimization objectives. In this way, additional freedom is available for the design of the current reference, allowing for an increased ability to control the machine. Some of these methods are then modified to calculate the current reference in real time. These modified algorithms are referred to as online algorithms.

In order to improve either torque-speed performance or efficiency, several offline TSFs have been proposed [4,5]. Offline TSFs cannot be adjusted during the operation of the machine; however, their parameters can be optimized according to one or two evaluation criteria. By optimizing the control parameters, TSF shows greater potential in reducing torque ripples at higher speed or in improving the efficiency. In [4], the turn-on and overlap angles of TSFs discussed above are optimized in order to minimize both the maximum ARCFL and copper losses by using genetic algorithm. In [5–7], other shapes of TSFs are proposed that either improve torque-speed performance or efficiency. In [8], an offline TSF for torque ripple reduction of SRM drives over a wide speed range is proposed. In [9–13], torque sharing functions based on optimization of the current waveform have been proposed.

Appendix

A.9.1 Brief Introduction of Genetic Algorithm

A.9.1.1 Selection Process Introduction

The main role of the selection function is to give preference to more healthy individuals and allow them to pass their genes to the next generations as shown in Figure A.9.1. The selection function determines the number of children an individual could reproduce. In other words, the selection function determines the probability for an individual to be chosen as a potential candidate for production, as a function of their level of fitness, that is, the numerical value of the objective function.

MATLAB's Genetic Algorithm Toolbox offers five options for the selection function:

- **selectionremainder** (Stochastic remainder sampling without replacement)
- **selectionroulette** (Choose parents using roulette wheel)
- **selectionstochunif** (Choose parents using stochastic universal sampling)
- **selectiontournament** (Each parent is the best of a random set)
- **selectionuniform** (Choose parents at random)

Individuals selected in a specific manner from a generation form a "mating pool", the size of which is generally the same as that of the population. In other words, on average each individual would have a chance to reproduce. It is not unusual that some healthy individuals might receive more than one chance of production while unfit individuals may not have a chance to be incorporated into the mating pool at all. However, different selection functions apply their own method in terms of selecting candidates for the mating pool and allotting chances for production.

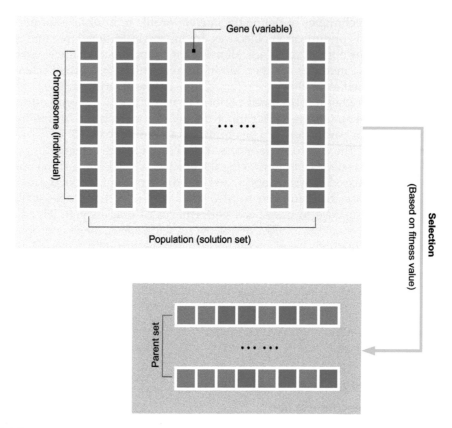

FIGURE A.9.1
Selection function mechanism.

A.9.1.1.1 Stochastic Sampling with Replacement Method

The Stochastic Sampling with Replacement Method is one of the most popular methods for filling the mating pool. Each individual is first assigned a value P_f, which could be obtained by

$$P_f = \frac{F_i}{F_A} \tag{A.9.1}$$

where F_i is the fitness value for a particular individual and F_A is the sum of every individuals' fitness values. For instance, an individual with $P_f = 1.36$ directly places one copy of itself into the mating pool. The remaining fractional part 0.36 indicates that this instance still has a 0.36 chance of being the selected once. Another individual, for instance, with $P_f = 0.54$ has a 0.54 chance of being selected once.

The probabilities of being chosen as a potential parent may be based on the numerical value of the objective function. All individuals from one certain generation can be ranked according to their fitness values. Then this ranking could be used to calculate the probability of being chosen. Therefore, the probability of an individual being chosen as a parent can be obtained by

$$P_n = \frac{N_{total} - N_{n,rank} + 1}{\sum_{k=1}^{N_{total}} k}$$
(A.9.2)

where N_{total} is the size of the population, and $N_{n,rank}$ is the rank for a particular individual. The Roulette Wheel Selection Method is commonly used to select parent candidates from a generation. The sum of the fitness values for all individuals constitutes the pie. The size of each individual corresponds to their fitness value, which means a healthy individual actually occupies a larger space. For instance, Individual 1, as shown in Figure A.9.2, has the highest fitness value and consequently takes the largest area in the pie. This indicates that this individual has the highest chance of being selected as a parent. The odds of any segment being selected remains the same. The function stops when the desired number of parents is chosen. In this manner, the area that each individual occupies represents the likelihood that it will be selected for the mating pool. The Basic Roulette Wheel Selection Method functions similarly to the Stochastic Sampling with Replacement Method.

A.9.1.1.2 Stochastic Universal Sampling

Stochastic Universal Sampling, as shown in Figure A.9.3, functions differently than the Roulette Wheel Method. N equally-spaced pointers are employed by the Stochastic Universal Sampling method while a single pointer is used by the Roulette Wheel Selection Method. N is equal to the number of the parents that will be selected for the mating pool. In this example, 10 individuals ($N = 10$) will be selected. The comb-like pointers start from a random number in the range [0, F_A/N]. As the example illustrates, individuals # 1, 1, 2, 2, 3, 4, 4, 5, 7, 9 are selected for the mating pool. It can be seen that individuals #1 and #4 are selected twice because they possess higher level of fitness. Compared with the Stochastic Universal Sampling Method, the Roulette Wheel Method might perform poorly when an individual possesses a much higher level of fitness than the rest.

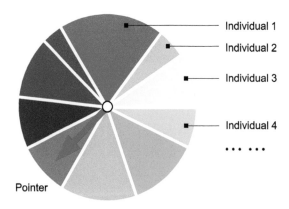

FIGURE A.9.2
Roulette wheel selection method.

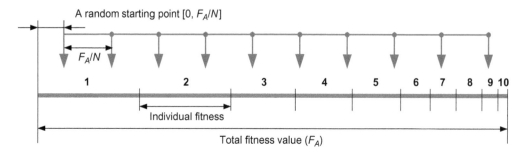

FIGURE A.9.3
Stochastic universal sampling method.

A.9.1.2 Introduction to Crossover Operation

The crossover operation, also referred as recombination, is applied to any pair of parents that are chosen to "produce offspring", as shown in Figure A.9.4. The "fitness" of the individuals from the current generation will be evaluated based on the objective function, from which parents will be selected to reproduce and generate a new generation of descendants. In this way, the "fit genes" from the current generation will be preserved and handed down to the next generation. During this process, less healthy individuals will be ignored and the evolution for the solution set proceeds towards increasing the fitness value based on the objective function. For instance, the population would be evolving towards decreasing the numerical value of the objective function, if the optimization problem is to seek a minimization.

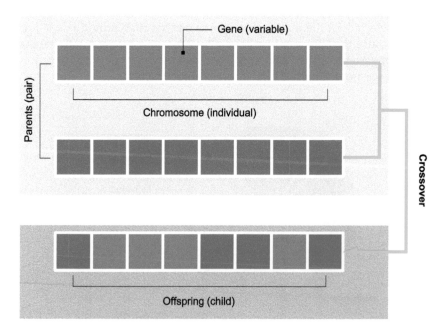

FIGURE A.9.4
Crossover function mechanism.

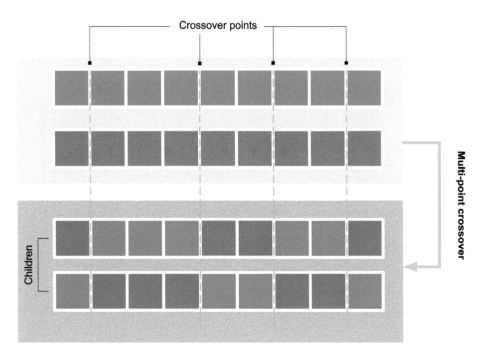

FIGURE A.9.5
Multi-point crossover mechanism.

For the Multi-Point Crossover method, as shown in Figure A.9.5, a certain number of the crossover points would be determined randomly with no duplication. Subsequently, the equivalent segments of both parents would be exchanged, and two children will be generated as a consequence.

A.9.1.2.1 Uniform Crossover Method

The Uniform Crossover Method employs a "crossover mask", which works like an on-off switch, as shown in Figure A.9.6. The crossover mask, which is a binary random number generator and has the same length as the chromosome structure, determines which parent will supply the offspring with chromosomes. In comparison with the previous method, the Uniform Crossover Method requires no understanding of the significance of any particular bit and assigns each bit with the same probability of being inherited by offspring.

A.9.1.2.2 Intermediate Crossover Method and Linear Crossover Method

The Intermediate Crossover Method (ICM) and Linear Crossover Method (LCM) are quite similar. One important difference is that the ICM could produce offspring in a larger area as shown with the shaded area in Figure A.9.7a. The children generated by the LCM perturb less than with the ICM. The method in which parents produce children using the ICM method may be described by

$$O_1 = P_1 + \alpha(P_2 - P_1) \tag{A.9.3}$$

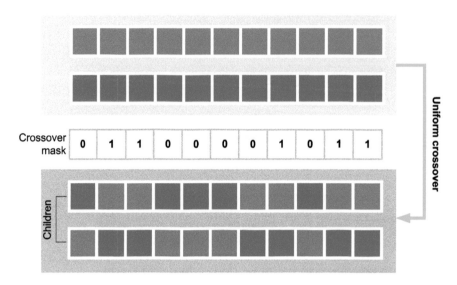

FIGURE A.9.6
Mask crossover mechanism.

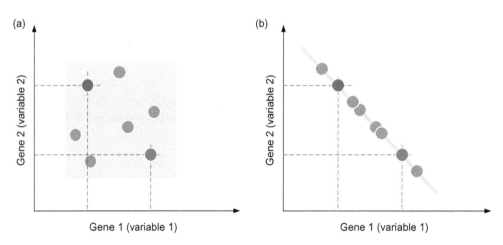

FIGURE A.9.7
Crossover method comparison: (a) intermediate crossover method and (b) linear crossover method.

where P_2, P_1 represent the two parents that will be mated, and α is a scaling factor chosen randomly for different variables over some interval, [−0.25, 1.25]. α remains the same for all genes within the LCM. Figure A.9.7 illustrates the differences between the two methods. The figure describes a scenario in which both parents have a chromosome with two genes, such as an individual with two variables. The LCM will only generate offspring that exist of the green line as illustrated by Figure A.9.7b, while the ICM would generate offspring in a much larger area as described by the blue square in Figure A.9.7a.

A.9.1.3 Introduction to Mutation

Figure A.9.8 shows that mutation would be performed on the genes of the new generation and will generate different chromosomes, i.e., individuals in the solution set. The mutation process prevents premature convergence and ensures that a global solution set would be attained. The mutation function modifies genes in the chromosomes randomly with a low level of probability over a range such as [0.001, 0.01]. The relationship between the selection process and crossover process is shown in Figure A.9.9.

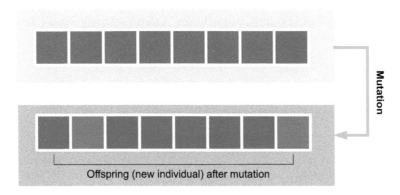

FIGURE A.9.8
Mutation function mechanism.

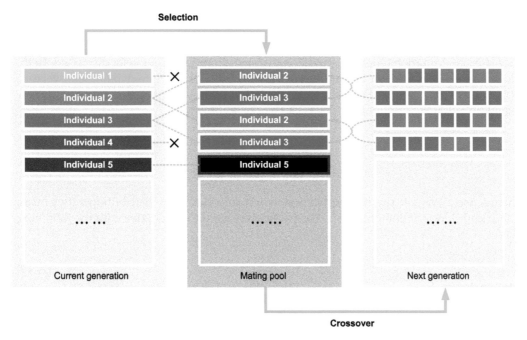

FIGURE A.9.9
Selection and crossover relationship.

The MATLAB Genetic Algorithm toolbox offers two options for the mutation function:

- **mutationuniform** (Uniform multi-point mutation)
- **mutationgaussian** (Gaussian mutation)

A.9.1.3.1 Uniform Multi-Point Mutation

Uniform Multi-Point Mutation uses the Random Mutation Operator (RMO). In RMO, a mutated individual would be generated in the neighborhood of the parent set for a particular generation. $X^{(t,j)} = (x_1^{(t,j)},...,x_n^{(t,j)})$ represents the jth individual in the tth generation. Within the same population, each gene would be in an interval $(x_k^{(t,L)}, x_k^{(t,U)})$, which represent the lower and upper bounds, respectively. A mutated gene would then be created according to

$$x_k^{(t,jm)} = x_k^{(t,j)} + u\Delta(x_k^{(t,U)} - x_k^{(t,L)}) \tag{A.9.4}$$

where u is a random value in $[-1,1]$ and Δ is the mutation step size. For the Polynomial Mutation Operator, the equation above would be modified as follows

$$x_k^{(t,jm)} = x_k^{(t,j)} + \delta_k(x_k^{(t,U)} - x_k^{(t,L)}) \tag{A.9.5}$$

where:

$$\delta_k = \begin{cases} (2u)^{1/(\eta+1)} - 1 & \textbf{if } u < 0.5 \\ 1 - [2(1-u)]^{1/(\eta+1)} & \textbf{if } u \geq 0.5 \end{cases} \tag{A.9.6}$$

u is a random number from within the range $[0,1]$ and η is a parameter to control the variability of the perturbation.

A.9.1.3.2 Gaussian Mutation Operator

The Gaussian Mutation Operator (GMO) is different from the previously discussed methods as it introduces a vector of standard deviations called the "strategy parameter vector". $(x_1^{(t,j)},...,x_n^{(t,j)},\sigma_1^{(t,j)},...,\sigma_n^{(t,j)})$ represents the jth individual in the tth generation and its corresponding strategy parameters. The GMO works according to

$$\sigma_n^{(t,jm)} = \sigma_n^{(t,j)} \cdot e^{\tau'z + \tau z_k} \qquad k = 1,...,n \tag{A.9.7}$$

$$x_k^{(t,jm)} = x_k^{(t,j)} + \sigma_n^{(t,jm)} \cdot z_k' \qquad k = 1,...,n \tag{A.9.8}$$

where z_k and z_k' denote random numbers drawn from a Gaussian distribution with a mean of zero and standard deviation of 1. The parameters would work better with the following choices:

$$\tau = \frac{1}{\sqrt{2\sqrt{n}}} \tag{A.9.9}$$

$$\tau' = \frac{1}{\sqrt{2n}} \tag{A.9.10}$$

The GMO is commonly applied to all of the elements of the solution.

Questions and Problems

1. The effect of hysteresis control on the phase current is described in great detail in this chapter. Provide pseudocode to model the hysteresis controller in a dynamic modeling environment. Assuming the rest of the motor is modeled by a current source, define the necessary variables and assume that the conduction period is run in a for-loop.

2. Out of the three phase currents shown in Figure 9.2, which one has the steepest slope from θ_{on} to θ_1 and which one is flattest? Why is that?

3. Hysteresis control is a variable-frequency switching strategy. Switching occurs to regulate the current within the tolerance band. It can be seen in Figure 9.3 that the switching frequency decreases over time. Why is the switching frequency at the start of the conduction period higher than that at the end of the conduction period?

4. Figure 9.2 shows the phase current profile in one phase of an SRM at three different speeds. If current control is applied at 6000 RPM, would the maximum switching frequency over the conduction period be higher or lower than the case at 3000 RPM? Why?

5. In practice, the switching frequency of the hysteresis controller should be within some rated value, due to the thermal limitations of the power electronics. This is not reflected in the simulation for Figure 9.2. Without limiting the switching frequency directly, what can be done to make sure that the switching frequency is within the rated value of the power electronics? How would this affect the phase current and the machine torque?

6. The speed of the SRM is governed by the first-order differential equation

$$J\frac{d\omega}{dt} = -B\omega + T - T_L \qquad \text{(Q9.1)}$$

where J is the motor inertia, B is the friction coefficient, T is the motor torque, T_L is the load torque, and ω is the machine speed. The open-loop mechanical system transfer function, $\omega(s)/T(s)$ can be derived by taking the Laplace transform of (Q9.1) and neglecting the load torque as it can be considered as a disturbance. Derive the expression for the time constant of the mechanical system.

   A Proportional-Integral (PI) controller can be implemented to regulate the speed error, $E(s)$, to zero, such that the speed ω is equal to the speed command, ω^*. The transfer function of a PI controller is given as:

$$C(s) = \frac{T(s)}{E(s)} = k_p + \frac{k_i}{s} \qquad \text{(Q9.2)}$$

where:

$$E(s) = \omega^*(s) - \omega(s) \qquad \text{(Q9.3)}$$

Using a PI controller to regulate the machine speed to its commanded value, derive the closed-loop system transfer function assuming a feedback gain $H(s) = 1$. Neglect T_L.

7. The linear TSF is provided by (9.11), where the values of $f_{rise}(\theta)$ and the values of $f_{fall}(\theta)$ are given by (9.16). A piece of pseudocode is shown below to realize this TSF in MATLAB.

```
// User defined inputs, these are constants
define T_ref; //reference current
define theta_OV; // overlap angle
define theta_ON; // turn-on angle
define theta_OFF; // turn-off angle

        myTorqueRef = array(361); //set up empty array

for angle = 0:360
        if angle<theta_ON
myTorqueRef(angle) = 0;
        elseif theta_ON<angle<theta_ON+theta_OV
myTorqueRef(angle) = (1/theta_OV)*(angle-theta_ON);
        elseif theta_ON+theta_OV<angle<theta_OFF
myTorqueRef(angle) = 1;
        elseif theta_OFF<angle<theta_OFF+theta_OV
                myTorqueRef(angle) =1-(1/theta_OV)*(angle-theta_OFF);
        else
myTorqueRef(angle) = 0;
        end
end

myTorqueRef = myTorqueRef*T_ref;
return 0:360, myTorqueRef
```

Using this pseudocode, draw the torque reference generated by the linear TSF for a three-phase SRM using the following values. (Note that all angles are in degrees electrical).

1. Tref = 1, $\theta_{on} = 0°$, $\theta_{off} = 120°$, $\theta_{ov} = 20°$
2. Tref = 1, $\theta_{on} = 0°$, $\theta_{off} = 120°$, $\theta_{ov} = 10°$

What happened to the commutation region when the overlap angle was reduced from 20° to 10°? When the phase torque references were summed together, what kind of function is the resulting waveform? This is the total torque reference. Would a torque profile like this be desirable? How would the waveforms from part a) and b) change if a torque command Tref = 2 was set?

8. Figure Q.9.1 shows the phase current and torque profile at 3000 RPM, along with the torque and current references generated by the linear TSF. As can be seen, the dynamic torque profile features significant torque ripple in this specific case and demonstrates that the TSF is not effective in this scenario. What is the reason for this? Highlight critical elements in the current reference during the commutation regions that lead to the poor performance of the TSF in this specific scenario. How can the performance of the TSF be improved?

9. When the operating speed of the SRM increases, does TSF perform better or worse? Why?

10. SRMs feature a nonlinear model and, therefore, the Simulink model is used as an objective function to optimize the conduction angles. In this question, we will practice how to solve an optimization problem using a Genetic Algorithm and Simulink.

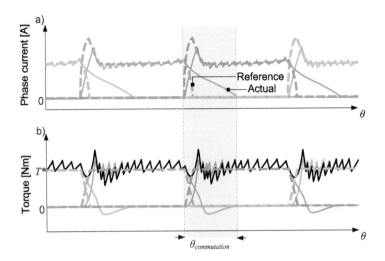

FIGURE Q.9.1
Application of linear TSF at 3000 RPM.

$$\text{Given that} \begin{cases} y_1 = x_1^2 + x_2^2 \\ y_2 = (x_1 - 10)^2 + 4(x_2 - 10)^2 \end{cases}$$

For all the following optimization problems, the constraints to apply are:

$$\begin{cases} -5 \le x_1 < 25 \\ -10 < x_2 \le 20 \\ x_1 - x_2 < -10 \\ (x_1 - 15)^2 + (x_2 - 15)^2 \le 400 \end{cases}$$

First, we will solve the single-objective optimization: minimize $y = 0.4y_1 + 0.6y_2$. We will formulate the objective function using Simulink and solve the optimization problem using GA in MATLAB.

Figure Q.9.2 shows the Simulink model for the objective function. Fixed-step simulation with two steps per simulation is recommended for this problem.

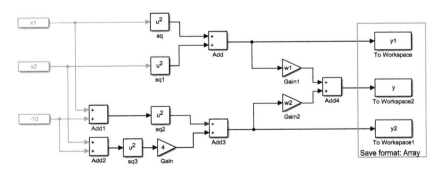

FIGURE Q.9.2
Simulink model.

Sample code for the optimization problem is provided below. Please save the Simulink model and the MATLAB scripts in the same folder.

```
%% Save this code a separate m-file
%%%%%%%%%%%%% initialization %%%%%%%%%%%%%%%

% Written in Matlab 2016a

global x1 x2 y1 y2 y w1 w2 simulink_obj
simulink_obj = 'y1_y2.slx'; % Name of the Simulink file.

%%
%%%%%%%%%%%%%% single-obj optimization %%%%%%%%%%%%%%%

%%%% Weights of objectives

w1 = 0.4; % weight function for y1 in the objective function
w2 = 0.6; % weight function for y2 in the objective function

%%%% constraints:

LB = [-5, -10]; % lower boundary
UB = [25, 20]; % upper boundary

ConstraintFunction = @non_linear_cons; % nonlinear constraint to be called

A = [1, -1]; b = [10]; % Linear inequality constraints
Aeq = []; beq = []; % No linear equality constraints

%%%% GA settings

options = gaoptimset('PlotFcns',{@gaplotbestf,@gaplotbestindiv},...
'PopulationSize',20,'EliteCount', 10,'TolFun',1e-6,'StallGenLimit',20);
% In GA options, please try to adjust the values of 'PlotFcns',
% 'PopulationSize', 'EliteCount', 'TolFun', 'StallGenLimit', etc. and see how
% they affect the optimized results.


numberOfVariables = 2;

[solution, eval] = ga (@single_obj,numberOfVariables,A,b,Aeq,beq,...
LB,UB,ConstraintFunction,options);
```

```
%% save this code as a separate m-file titled same as the name of the
function
%%%%%%%%%%%%%% objective function %%%%%%%%%%%%%%%

function eval = single_obj(x)
global x1 x2 y1 y2 y w1 w2 simulink_obj
x1 = x(1);
x2 = x(2);
model=simulink_obj;
load_system (model) % Two functions, load_system and sim, will be used to
call the Simulink model. To reduce the computational time, please reset the
simulation time, solver option, step size, etc. for your Simulink model.
sim(model)
ave_y = mean(y);
eval=ave_y;
```

```
%% save this code as a separate m-file titled same as the name of the
function
%%%%%%%%%%%%%%%% nonlinear constraint %%%%%%%%%%%%%%%%

function [c, ceq] = non_linear_cons(x)
% GA optimizer will call it by ConstraintFunction = @non_linear_cons;

c = [(x(1)-15)^2 + (x(2)-15)^2 - 400];
ceq = [];
```

Before discussing the results, the optimization problem must be investigated in more detail. Three groups of constraints are used in the optimization problem: (1) upper and lower boundaries (UB, LB), as shown in Figure Q.9.3a and b, (2) the linear inequality constraint (a, b), as shown in Figure Q.9.3c, and (3) nonlinear constraint, as shown in Figure Q.9.3d.

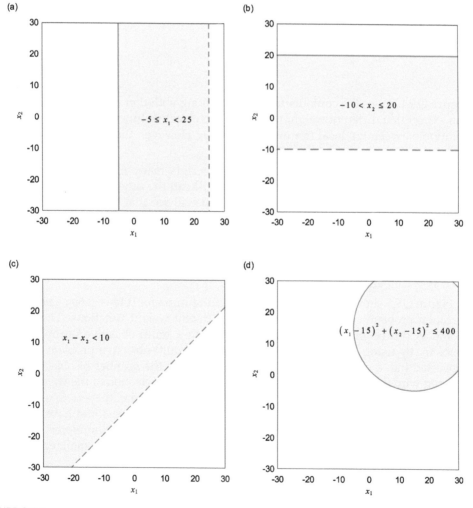

FIGURE Q.9.3
Plots of constraints: (a) boundary for x_1, (b) boundary for x_2, (c) linear constraint, and (d) nonlinear constraint.

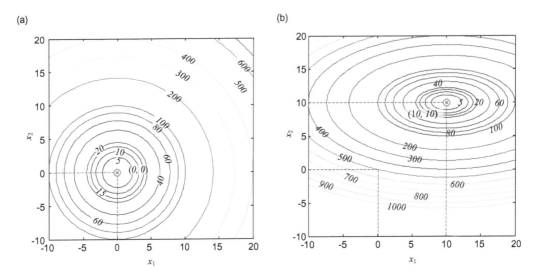

FIGURE Q.9.4
Plots of objective functions: (a) y_1 and (b) y_2.

Figure Q.9.4 shows the contours for y_1 and y_2 along with their global minima. You can expect that different weight factors for the objective function will change the solution between the local minima of y_1 and y_2. However, the change will not be linear due to the nonlinear constraint.

We now have all of the models and the scripts ready for the optimization. Execute the first m-file that you created. It will call the other scripts during the optimization. If you keep Simulink open, you will see that the Simulink model runs each time to calculate the value of the objective function. After the optimization has completed, the results for x_1 and x_2 are stored in a variable called *eval*. You will see that the optimized values are $x_1 = 6$ and $x_2 = 8.57$. The value of the objective function is saved in the variable called *solution*, which is calculated as $y = 0.4y_1 + 0.6y_2$ and is equal to 58.2857.

Figure Q.9.5 shows the contours and global minima for $0.4y_1 + 0.6y_2$. The minimization problem can be calculated with different weight functions. Therefore, y_1 and y_2 can be minimized individually using a multi-objective optimization problem. By using the weighted sum method, a multi-objective problem can be converted into a single-objective problem. Although the number of objectives is reduced with this method, the result is sensitive to the selection of the weights. In the given problem, the objectives are just numbers; therefore, they were unitless. In an SRM application, the two objectives might not be unitless and have different physical properties; for instance torque in Nm and current in amperes. In that case, the objectives cannot be just summed up. They should be normalized based on reasonable scales.

Now, for the same functions and constraints, solve the optimization problems:

a. **Single-objective optimization**: minimize $y = 0.1y_1 + 0.9y_2$

b. **Multi-objective optimization**: minimize y_1 and minimize y_2

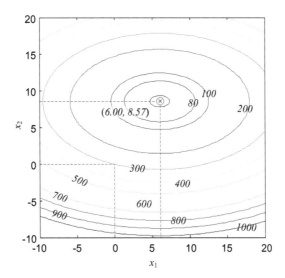

FIGURE Q.9.5
Plot of single-objective function $0.4y_1 + 0.6y_2$.

For all the questions, formulate the objective function using Simulink. Solve all the optimization problems using GA in MATLAB. For each of the questions, plot the feasible region, contours of the objective function(s), and solution(s) on its $x_1 - x_2$ plane. For b, please also plot the Pareto front on a $y_1 - y_2$ plane.

Hint: For the multi-objective optimization make the following modifications in the code:

```
Main function:
options =
gaoptimset('PlotFcns',{@gaplotpareto,@gaplotscorediversity},'PopulationSize',
20,'EliteCount', 10,'TolFun',1e-6,'StallGenLimit',20);

[eval, solution] =
gamultiobj(@multi,numberOfVariables,A,b,Aeq,beq,LB,UB,ConstraintFunction,opti
ons);

Objective Function:
ave_y1 = mean(y1);

ave_y2 = mean(y2);

eval (1)=ave_y1;
eval (2)=ave_y2;
```

11. A highly-simplified SRM model built in Simulink can be seen in Figure Q.9.6. In the model, it was assumed that phase torque can be calculated as $I \sin(\theta_{elec} 2\pi/360)$ where I is the phase current and θ_{elec} is the electrical angle of the phase in degrees.

The counter in the Electrical Angle Calculation block counts up to 360°. For a three-phase machine, the initial electrical angle for each phase can be assumed as 0°, 120°, and 240° electrical. According to the torque expression, zero degree electrical is the unaligned position. Please note that the output of Electrical Angle

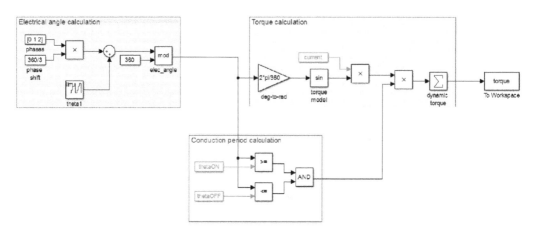

FIGURE Q.9.6
Model of highly-simplified SRM built in Simulink.

Calculation block is a vector at each time step and it gives the electrical angle for each phase. This is accomplished by entering the array [0 1 2] in the "constant" block named "phases".

The Conduction Angle Calculation block compares the electrical angle with the turn-on and turn-off angles. Within the conduction period, the model calculates the phase torque for the given current. Here it is assumed that rectangular current is applied for simplification.

Please reproduce this model and use it in this question. Please use a fixed-step solver. The inputs for the SRM model are thetaON θ_{on}, thetaOFF θ_{off} and current. The output of the model is an array that includes the data for the torque waveform. Use the GA optimizer in MATLAB® to solve the single-objective optimization problem: maximize the average torque, T_{ave} when the current is at 20 A. What are the optimized conduction angles? What is the average torque?

The constraints for all problems in this question $\begin{cases} -90 \leq \theta_{on} \leq 90 \\ 0 \leq \theta_{off} \leq 180 \\ \theta_{off} - \theta_{on} < 160 \end{cases}$

12. Now consider the SRM dynamic model that was built in Chapter 4. It can be used as an objective function providing the torque ripple and current waveform for the given input parameters to optimize the conduction angles. Figure Q.9.7 shows the updated Simulink model of the SRM for the optimization. The simulation timestep and the total simulation time are updated as parameters to analyze the motor operation at different speeds. The "To Workspace" blocks send the torque and phase current waveforms to the MATLAB workspace for the optimization.

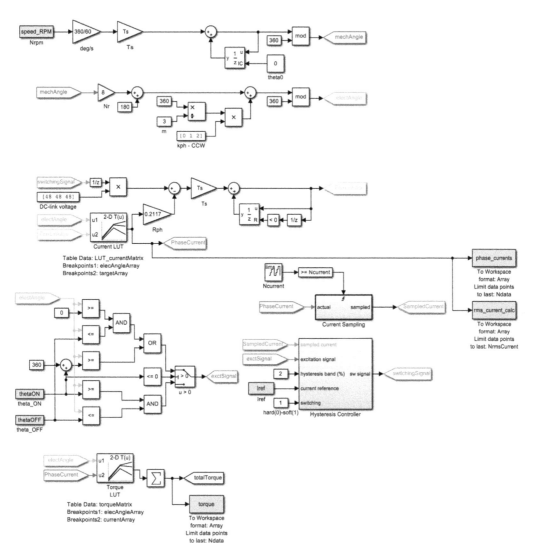

FIGURE Q.9.7
SRM dynamic model for optimization.

The "rms_current_calc" block sends only the current for the last electrical cycle to calculate the RMS value. Please change the DC-link voltage to 48 V.

The sample code below shows the single objective optimization algorithm to maximize the average torque for the given RMS current constraint. Please note that the algorithms and the Simulink model are not optimized; they are provided for guidance. Please save the algorithms as separate m-files. Please name the Simulink model as "SRM_model" and ensure that the model and m-files are saved in the same folder.

```matlab
%%%%%%%%%%%%%% Optimize_1_RunOptimization %%%%%%%%%%%%%%%
function [thetaON, thetaOFF, torque_ave, torqueRipplePerct,...
        torqueRippleRMS, RMS_Current, torque_data, current_data]...
    = Optimize_1_RunOptimization (speed_RPM, Iref, P_rotor,...
    elecCycleTransient, elecCycleToRetrieve, simStepsPerCycle, fcurrent,...
    GA_PopSize, GA_EliteCount, GA_TolFun, GA_StallGenLimit,...
    simulink_model_name, I_RMS_constraint)

    load_system(simulink_model_name);

    % setup GA optimization settings:
    [GA_options, LB, UB, A, b, Aeq, beq, numberOfVariables]...
    = setOptParams(GA_PopSize, GA_EliteCount, GA_TolFun, GA_StallGenLimit);

    % set optimization dynamic model settings (speed dependant):
    [Tsim, Ts, Ndata, Ncurrent, NrmsCurrent] = Optimize_setSimParams...
    (speed_RPM, P_rotor, elecCycleTransient, ...
    elecCycleToRetrieve, simStepsPerCycle, fcurrent);

    % Run the optimizer:
    [firing_final, torque_ave_final, torque_ripple_perc_final, ...
        torque_ripple_RMS_final, RMS_current_final, torque_results, ...
        current_results] = Optimize_2_RunOptimizer(GA_options, LB, UB,...
        A, b, Aeq, beq, numberOfVariables, simulink_model_name,...
        Iref, speed_RPM, Tsim, Ts, Ndata, NrmsCurrent, Ncurrent,...
        I_RMS_constraint);

    thetaON = firing_final(1);  thetaOFF = firing_final(2);
    torque_ave = torque_ave_final;
    torqueRipplePerct = torque_ripple_perc_final*100;
    torqueRippleRMS = torque_ripple_RMS_final;
    RMS_Current = RMS_current_final;
    torque_data = torque_results;
    current_data = current_results;

    close_system(simulink_model_name);
    clc;
end

%%
function [GA_options, LB, UB, A, b, Aeq, beq, numberOfVariables]...
    = setOptParams(GA_PopSize, GA_EliteCount, GA_TolFun, GA_StallGenLimit)

    % turn ON = LB(1):UB(1)
    % turn OFF = LB(2):UB(2)
    LB = [-90, 90];
    UB = [90, 180];

    % single objective optimization
    GA_options = gaoptimset('PlotFcns',{@gaplotbestf,@gaplotbestindiv},...
        'PopulationSize',GA_PopSize,'EliteCount',GA_EliteCount,'TolFun',...
        GA_TolFun,'StallGenLimit',GA_StallGenLimit);

    % Linear inequality constraints:
    A = [1, -1];
    b = 0;

    % Linear equality constraints:
```

```
    Aeq = [];
    beq = [];

    numberOfVariables = 2;
    % Number of decision variables (#1: turnON, #2: turnOFF)
end

%%
function [Tsim, Ts, Ndata, Ncurrent, NrmsCurrent] = ...
        Optimize_setSimParams(speed_RPM, P_rotor,...
        elecCycleTransient, elecCycleToRetrieve, simStepsPerCycle,...
        fcurrent)
    % Number of Electrical Cycles to Calculate:
    elecCycleToCalc = elecCycleTransient + elecCycleToRetrieve;
    % total number of electrical cycles to simulate, including the
    % transients
    mechDegPerCycle = 360/P_rotor; % mechanical angle the rotor takes
    % at each stroke
    mechRadToCalc = deg2rad(mechDegPerCycle)*elecCycleToCalc; % total
    % mechanical angles in radians to be simulated including the transients
    speed_rad_s=speed_RPM*(pi/30);
    Tsim = mechRadToCalc/speed_rad_s; % total simulation time

    % simulation step time:
    Ts=Tsim/(simStepsPerCycle*elecCycleToCalc); % simStepsPerCycle is an
    % input parameter. It defines how many points should be simulated in
    % the dynamic model at each electrical cycle. When this value is...
    % multiplied with the
    % elecCycleToCalc, the result is the total number of points to be
    % simulated. When this value divides Tsim, the result is the simulation
    % step time.

    Ndata=simStepsPerCycle*elecCycleToRetrieve; % Ndata is applied in the
    % To Workspace blocks in the Simulink model. So only the data at
    % the steady state is used in the optimization.

    NrmsCurrent=simStepsPerCycle; %NrmsCurrent is applied in another To...
    % Workspace block in the simulink model to capture the phase current...
    % in the last electrical cycle. This data will be used to calculate...
    % the RMS value of the phase current.

    % current sampling:
    Tcurrent = (1/fcurrent);
    Ncurrent = round(Tcurrent/Ts);
end
```

```
%%%%%%%%%%%%%%%% Optimize_2_RunOptimizer %%%%%%%%%%%%%%%%
function [firing_final, torque_ave_final, torque_ripple_perc_final,...
        torque_ripple_RMS_final, RMS_current_final, torque_results,...
        current_results] = Optimize_2_RunOptimizer(GA_options, LB, UB,...
        A,b, Aeq, beq, numberOfVariables, simulink_model_name,...
        Iref, speed_RPM, Tsim, Ts, Ndata, NrmsCurrent, Ncurrent,...
        I_RMS_constraint)

    set(0,'DefaultFigureVisible','off');
    % prevent GA algorithm from stealing focus

    % Initialize Vars:
    torque_ave = [];
    current_rms = [];
    fire_last = [];
    FF=@Optimize_FitnessFunction;        % fitness function
    CF=@Optimize_ConstraintFunction;     % nonlinear constraint function

    % single objective optimization
    [firing_final, ~, ~, ~] = ga(FF, numberOfVariables,A,b,Aeq,beq,LB,...
        UB,CF,GA_options);

    % after final firing angles are obtained, make sure the results
    % are updated properly:
    [torque_ave_final, torque_ripple_perc_final, ...
        torque_ripple_RMS_final, RMS_current_final, torque_results, ...
        current_results] = Optimize_3_RunOptModel (simulink_model_name,...
        firing_final(1), firing_final(2),Iref, speed_RPM, Tsim, Ts,...
        Ndata, NrmsCurrent, Ncurrent);

    set(0,'DefaultFigureVisible','on');

    %-------------------------------------------------------------------------
    %%
    function f = Optimize_FitnessFunction (fire)

        if ~isequal(fire,fire_last) % only run dynamic model if necessary
            [torque_ave, ~, ~, ~, ~, ~] = ...
                Optimize_3_RunOptModel (simulink_model_name,...
                fire(1), fire(2),Iref, speed_RPM, Tsim, Ts, Ndata,...
                NrmsCurrent, Ncurrent);
            fire_last = fire;
        end

        % maximize average torque fitness function:
        f = -torque_ave;

    end
```

```
%-------------------------------------------------------------------------
%%
function [c, ceq] = Optimize_ConstraintFunction (fire)

    if ~isequal(fire,fire_last) % only run dynamic model if necessary
        [torque_ave, ~, ~, current_rms, ~, ~] = ...
            Optimize_3_RunOptModel (simulink_model_name, fire(1),...
            fire(2),Iref, speed_RPM, Tsim, Ts, Ndata, NrmsCurrent,...
            Ncurrent);
        fire_last = fire;
    end

    % nonlinear inequality constraint:
    c = current_rms - I_RMS_constraint;

    % nonlinear equality constraint
    ceq = [];
end
end
```

```
%%%%%%%%%%%%%%% Optimize_3_RunOptModel %%%%%%%%%%%%%%%%%
function [torque_ave, torque_ripple_percent, torque_ripple_rms, ...
    current_rms, torque_results, current_results] = ...
    Optimize_3_RunOptModel (simulink_model_name, thetaON, thetaOFF,...
    Iref, speed_RPM, Tsim, Ts, Ndata, NrmsCurrent, Ncurrent)

    % run the dynamic model:
    results = sim(simulink_model_name, 'SrcWorkspace', 'current',...
        'SimulationMode', 'normal');

    % get results
    torque_results = results.get('torque'); % get the torque data
    torque_ave = mean(torque_results); % calculate the average value
    torque_max=max(torque_results); % calculate the maximum torque
    torque_min=min(torque_results); % calculate the minimum torque
    torque_ripple_percent=(torque_max-torque_min)/torque_ave; % calculate
    %the percentage torque ripple
    torque_ripple_rms=rms(torque_results-torque_ave); % calculate
    % the rms value of the torque ripple
    current_results = results.get('phase_currents'); % get the phase
    % currents
    rms_current_calc = results.get('rms_current_calc'); % get the phase
    % currents in the last electrical cycle

    % we can estimate the rms value of the phase current by ignoring the
    % commutation. This approach will slightly over estimate the rms
    % current.
    current_rms=mean(rms_current_calc);
    temp=rms_current_calc(:,1); % calculate the rms value using only one
    % phase
    temp1=find(temp>0);
    if length(temp1)>3
        temp2=temp(temp1(1):temp1(end));
        current_rms = mean(temp2);
    else
        current_rms = 0; % if the conduction angle is zero
    end
end
```

[Run the optimization by calling the 'Optimize_1_RunOptimization" function from the workspace as follows:

```
speed_RPM=1000; Iref=10; P_rotor=8; elecCycleTransient=3;
elecCycleToRetrieve=5;
simStepsPerCycle=5000; fcurrent=80e3; GA_PopSize=20;
GA_EliteCount=10; GA_TolFun=1e-2; GA_StallGenLimit=40;
simulink_model_name='SRM_model'; I_RMS_constraint=12;

[thetaON, thetaOFF, torque_ave, torqueRipplePerct,...
      torqueRippleRMS, RMS_Current, torque_data, current_data]...
    = Optimize_1_RunOptimization (speed_RPM, Iref, P_rotor,...
    elecCycleTransient, elecCycleToRetrieve, simStepsPerCycle, fcurrent,
GA_PopSize,...
    GA_EliteCount, GA_TolFun, GA_StallGenLimit, simulink_model_name,...
    I_RMS_constraint);
```

What are the optimized conduction angles?

References

1. J. W. Jiang, Three-phase 24/16 switched reluctance machine for hybrid electric powertrains: Design and optimization, Ph.D. Dissertation, Department of Mechanical Engineering, McMaster University, Hamilton, ON, 2016.
2. Mathworks. Global Optimization Toolbox User's Guide, 2016 [Online]. Available: http://www.mathworks.com/help/pdf_doc/gads/gads_tb.pdf. (Accessed: February 27, 2017).
3. J. W. Jiang, B. Bilgin, and A. Emadi, Three-phase 24/16 switched reluctance machine for hybrid electric powertrains: Design and optimization, *IEEE Trans. Transport. Electrification*, 3(1), 76–85, March 2017.
4. X. D. Xue, K. W. E. Cheng, and S. L. Ho, Optimization and evaluation of torque sharing function for torque ripple minimization in switched reluctance motor drives, *IEEE Trans. Power Electron.*, 49(1), 28–39, 2002.
5. V. P. Vujičić, Minimization torque ripple and copper losses in switched reluctance drive, *IEEE Trans. Power Electron.*, 27(1), 388–399, 2012.
6. A. C. Pop, V. Petrus, C. S. Martis, V. Iancu, and J. Gyselinck, Comparative study of different torque sharing functions for losses minimization in switched reluctance motors used in electric vehicles propulsion, in *Proceedings of the International Conference on Optimization of Electrical and Electronic Equipment* (OPTIM), Brasov, Romania, May 2012, pp. 356–365.
7. P. Kiwoo, X. Liu, and Z. Chen, A non-unity torque sharing function for torque ripple minimization of switched reluctance generators, in *Proceedings of the European Conference on Power Electronics and Applications* (EPE), Lille, France, September 2013, pp. 1–10.
8. J. Ye, B. Bilgin, and A. Emadi, An offline torque sharing function for torque ripple reduction of switched reluctance motor drives, *IEEE Trans. Ener. Conv.*, 30(2), 726–735, 2015.
9. J. Ye, B. Bilgin, and A. Emadi, An extended-speed low-ripple torque control of switched reluctance motor drives, *IEEE Trans. Power Electron.*, 30(3), 1457–1470, 2015.
10. H. Li, B. Bilgin, and A. Emadi, *Torque Ripple Reduction in Switched Reluctance Machines*, McMaster Technology ID 15-050, Discovery Disclosure, McMaster University, December 12, 2017, Hamilton, ON.

11. J. Ye and A. Emadi, *Torque Ripple Reduction in Switched Reluctance Motor Drives*, McMaster Technology ID 14-024, (Patent Application No. US 14/599,838 and CDN 2,878,561, January 17, 2014), US 9,742,320, August 22, 2017.

12. J. Ye and A. Emadi, *Extended-Speed Low-Ripple Torque Control of Switched Reluctance Motor Drives*, McMaster Technology ID 14-050, (Patent Application No. US 14/565,940 and CDN 2,874,157, December 10, 2014), US 9,641,119, May 2, 2017.

13. H. Li, B. Bilgin, and A. Emadi, An Improved Torque Sharing Function for Torque Ripple Reduction in Switched Reluctance Machines, *IEEE Trans. Power Electron.*, 2018, Early Access. doi: 10.1109/TPEL.2018.2835773.

10

Power Electronic Converters to Drive Switched Reluctance Machines

Jin Ye

CONTENTS

10.1 Introduction

In AC motor drives, bipolar phase current is required to generate torque, and four-quadrant converters are applied. Moreover, the switches in each phase are connected in series, leading to possible device damage under the shoot-through fault condition. In contrast to the AC motor drives, the direction of the SRM torque is not dependent on the direction of the current. Therefore, unipolar phase current is required. Besides, SRM has fault-tolerant operation capability because of the independent windings. If one of the windings has a fault, an SRM can still operate with a degraded performance. This chapter introduces several converter topologies for the SRM drives and then gives comparative evaluation of power converters.

10.2 Asymmetric Bridge Converter

The asymmetric bridge converter for a three-phase SRM shown in Figure 10.1 is the most widely applied converter in SRM drives. It allows independent current control of different phases and, therefore, it maintains good torque control. However, it includes two switches and two diodes per phase, which is costly for low-cost applications.

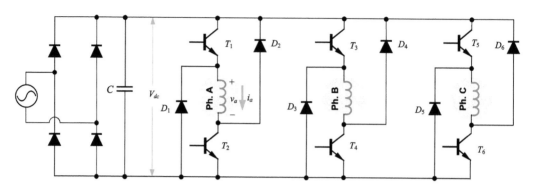

FIGURE 10.1
The asymmetric bridge converter for a three-phase SRM.

There are two switching patterns: bipolar modulation (hard switching) and unipolar modulation (soft switching). In hard switching, when the switches T_1 and T_2 are turned on, the phase voltage v_a is equal to the DC-link voltage V_{dc}, and the phase current i_a increases. When the phase current hits the upper limit of the current reference, the switches T_1 and T_2 are turned off. The phase current i_a circulates through the diodes D_1, D_2, and DC-link capacitors, and the stored energy in the magnetic circuit is transferred back to DC-link capacitors. In this condition, the phase voltage is equal to $-V_{dc}$ and, therefore, the current decreases until it hits the lower limit of the hysteresis band. Then, the phase current i_a rises up again by turning on the switches T_1 and T_2. When the phase A current needs to be turned off completely, the switches T_1 and T_2 are both turned off. The phase current will decrease to zero until the stored energy in the winding is depleted. The operational waveforms are shown in Figure 10.2. The required breakdown voltages of devices T_1, T_2, D_1, and D_2 should be higher than the DC-link voltage V_{dc}.

With the hard switching pattern, phase current circulates between the DC link and the phase winding in a switching cycle, which increases the power loss dissipation caused by switching losses as well as the current ripple in the DC-link capacitors. Soft switching can be applied to improve the converter's efficiency. The operational waveforms with soft switching are shown in Figure 10.3. During the interval θ_1 and θ_2, the switch T_1 is always on, and T_2 is turned on and off to track the current reference. When the switch T_2 is on, the phase voltage v_a is V_{dc}, and the phase current i_a rises up. When i_a hits the upper limit of the current reference, the switch T_2 is turned off. Then, the phase current freewheels through the diode D_2 and the switch T_1. The phase winding voltage v_a is equal to the sum the voltage drops of D_2 and T_1 during conduction. Since the voltage drop is very low, the phase current i_a decreases slowly. After θ_2, the current reference becomes zero. Switches T_1 and T_2 are both turned off to reduce the phase current i_a rapidly. Compared to the hard switching shown in Figure 10.2, the switching losses are reduced with the soft switching.

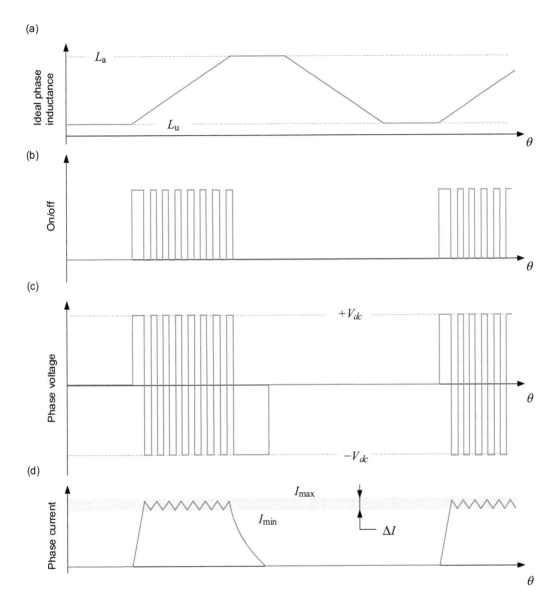

FIGURE 10.2
The operational waveforms of the asymmetric bridge converter using bipolar (hard) switching pattern: (a) ideal phase inductance, (b) on/off signal, (c) phase voltage, and (d) phase current.

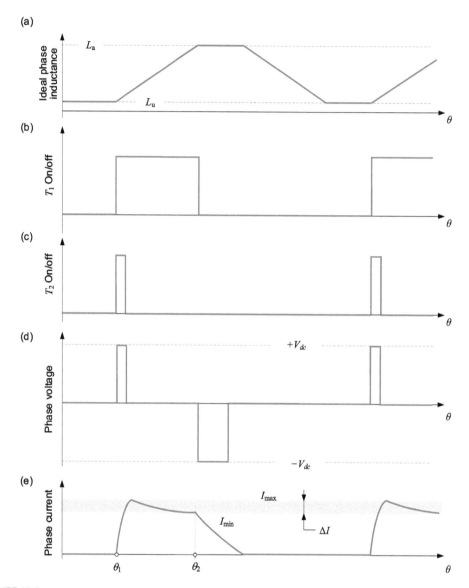

FIGURE 10.3
The operational waveforms of the asymmetric bridge converter using unipolar (soft) switching pattern: (a) ideal phase inductance, (b) T_1 on/off, (c) T_2 on/off, (d) phase voltage, and (e) phase current.

10.3 (N + 1)-Switch Converters

In SRM drives, the phase winding does not conduct during the whole electrical cycle. Therefore, some of the power devices can be shared between different phase windings and the number of the switches can be reduced. Several converters with shared switches will be discussed in this chapter. Since the total number of switches in these topologies is one plus the number of phases, these converters are categorized as (N + 1)-switch converters.

A (N + 1)-switch converter for a three-phase SRM is shown in Figure 10.4. Three-phase windings are connected to four switches. The reduction of the number of switches is

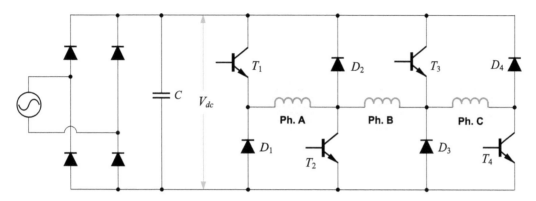

FIGURE 10.4
(N + 1)-switch converter for the three-phase SRM.

achieved by sharing a switch between two phase windings. Compared to the asymmetric bridge converter, the current ratings of the shared devices (T_1, D_1, T_2, and D_2) increase. When SRM operates at low speed, the phase current can decrease rapidly because of the low induced voltage. Therefore, no overlapping between different phase currents occurs. The current of each phase can be independently controlled. However, when SRM operates at high speed, phase currents may overlap between different windings because of the higher induced voltage. This converter will lose the independent phase current control capability due to the shared power devices. For instance, when phase A is required to be turned off, both T_1 and T_2 should be off to decrease the phase current to zero. If phase B needs to be excited before phase A current falls to zero, T_2 and T_3 should be on. Therefore, there is a contradiction for the required switching status of T_2. In order to avoid the current overlap at high speed operation, the conduction angle of phase B needs to be delayed during the commutation from phase A to phase B. By delaying the conduction angle, the generated torque is reduced. Therefore, the drawback of this converter is that it cannot independently control the phase currents, especially at high speed.

In order to improve the current control performance, a converter with shared devices for the four-phase SRM is shown in Figure 10.5. Compared to the (N + 1)-switch converter shown in Figure 10.4, there is one more switch and one more diode for this converter shown in Figure 10.5. The phase windings A and B share the power devices D_2 and T_2. The phase winding C and D share the power devices D_5 and T_5. As these devices are not shared by the adjacent phase windings, this converter can guarantee that only one phase current

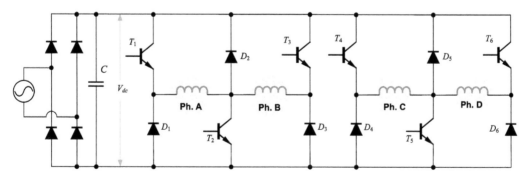

FIGURE 10.5
A converter with shared devices for the four-phase SRM.

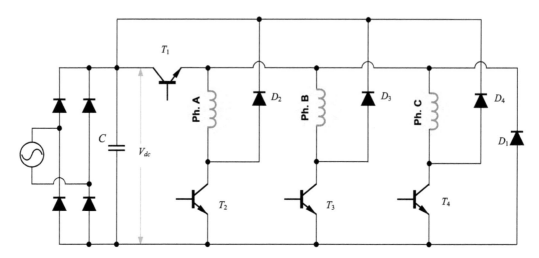

FIGURE 10.6
An alternative (N + 1)-switch converter for the three-phase SRM.

flows through each device anytime. Therefore, each phase current can be independently controlled. The torque control performance is better than that of the (N + 1)-switch converter in Figure 10.4. However, this converter is only suitable for the SRM with an even number of phases.

Another example of a (N + 1)-switch converter for the three-phase SRM is shown in Figure 10.6. The switch T_1 and diode D_1 are shared by three-phase windings and, therefore, the current ratings of T_1 and D_1 are higher than other switches. Similar to the converter shown in Figure 10.4, the phase currents cannot be controlled independently when they overlap at high speed.

10.4 C-Dump Converter

A C-dump converter is shown in Figure 10.7. This converter contains both the buck and boost converters. The operational waveforms of C-dump converter are shown in Figure 10.8, and the illustration of five modes of C-dump converter is depicted in Figure 10.9.

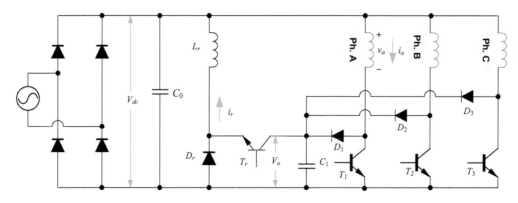

FIGURE 10.7
The conventional C-dump converter.

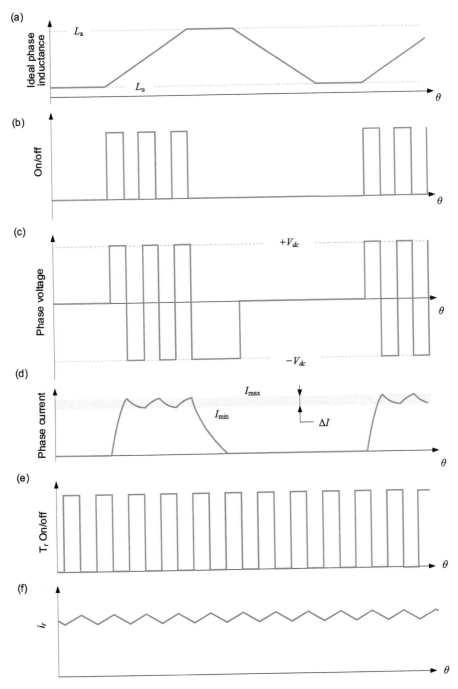

FIGURE 10.8
The operational waveforms of the conventional C-dump converter: (a) ideal phase inductance, (b) on/off, (c) phase voltage, (d) phase current, (e) T_r on/off, and (f) i_r.

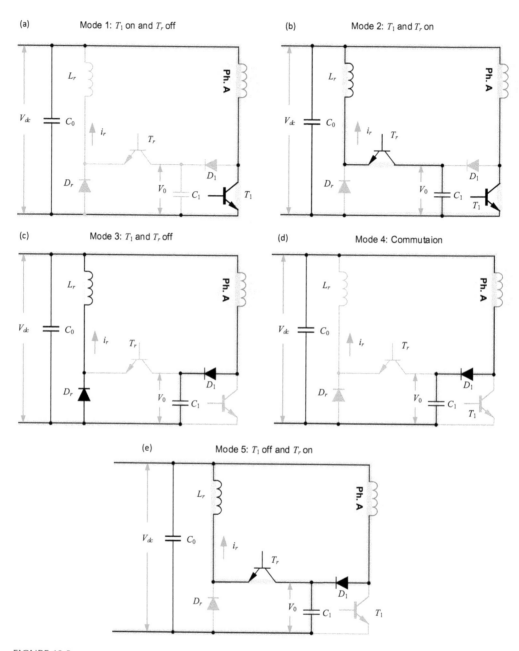

FIGURE 10.9
Five Modes of C-dump converter: (a) Mode 1, (b) Mode 2, (c) Mode 3, (d) Mode 4, and (e) Mode 5.

By taking phase A as an example, the DC-link voltage V_{dc}, the inductance of phase A winding, the switch T_1, the diode D_1, and the capacitor C_1 make up a boost converter. The capacitor C_1, the switch T_r, the diode D_r, the inductor L_r, and the DC-link voltage V_{dc} make up a buck converter. T_1 is used to control the phase current and T_r is used to control the voltage output of the capacitor C_1. When phase A current is below the reference, T_1 is turned on and V_{dc} is applied to the phase winding and phase A is working in mode 1 or 2. When the current of phase A rises above the reference, T_1 is turned off and $V_{dc} - V_0$ is applied to the phase winding. Phase A now works in one of the modes 3, 4, or 5, depending on V_0. Once the status of T_1 is defined by the phase current, the specific operational mode can be utilized by the status of T_r. For example, when phase A current is below the reference and T_r is on, phase A is working in mode 2. When the current of phase B needs to be built up, T_2 is turned on and T_1 is turned off in Figure 10.7. The current of phase B flows through C_0 and T_2, and the phase A current flows through D_1, C_0, and C_1, and charges C_1. Thus, the C-dump converter allows independent control of different phases. T_r is used to control the voltage V_0 to provide higher demagnetization voltage during the commutation, leading to the higher voltage ratings for power devices. The additional inductor and switches/diodes are needed, which may increase the costs. Also, the C-Dump converter does not achieve freewheeling, which may increase the acoustic noise and switching losses [1].

10.5 N-Switch Converters

Several converters with a single switch per phase are available. For an N-phase SRM, if the number of the switches of the converter is N, then these converters are categorized as N-switch converters. An N-switch converter using bifilar windings for the three-phase SRM is shown in Figure 10.10. The bifilar phase winding contains two closely spaced, parallel windings. For each phase, one of the bifilar windings is connected to a switch and the other one to a freewheeling diode.

The operational waveforms of this converter are shown in Figure 10.11. The voltage v_a is equal to the DC-link voltage V_{dc} when the switch T_1 is on, leading to an increase in i_a. The voltage across the D_1 is $(1 + n_1/n_2)*V_{dc}$ when T_1 is on, where n_1 and n_2 are the number of turns for the bifilar windings. When T_1 is off, the diode D_1 is forward biased. The voltage of the phase winding v_a is equal to $-(n_1/n_2)*V_{dc}$. Therefore, the phase current i_a decreases, and the stored energy of this winding is sent back to the DC link. The voltage across the switch T_1 is equal to $(1 + n_1/n_2)*V_{dc}$ when it is turned off. Although this converter has only one switch for each phase, the voltage ratings of power devices are higher compared to the asymmetric bridge converter [1].

Another single switch per phase converter with a split DC-link capacitor is shown in Figure 10.12. The operational waveforms are shown in Figure 10.13. When the switch T_1 is

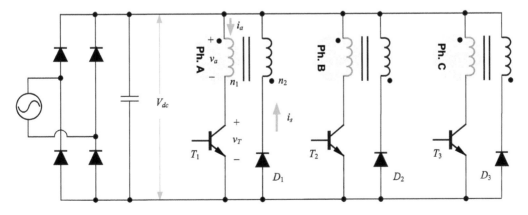

FIGURE 10.10
A converter with bifilar windings for three-phase SRM.

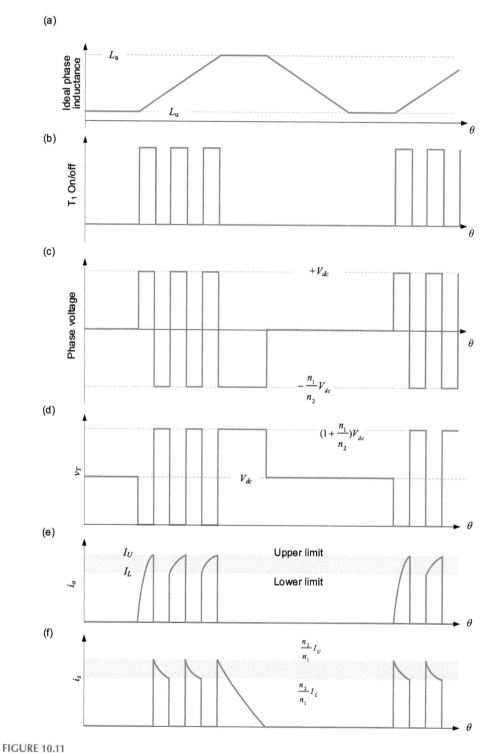

FIGURE 10.11
The operational waveforms of the converter with bifilar windings: (a) ideal phase inductance, (b) T_1 on/off, (c) phase voltage, (d) v_T, (e) i_a, and (f) i_s.

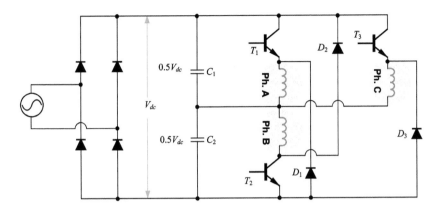

FIGURE 10.12
Split DC converter for three-phase SRM.

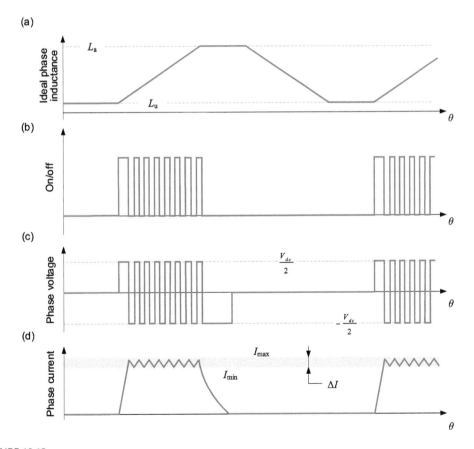

FIGURE 10.13
The operational waveforms of the split DC converter: (a) ideal phase inductance, (b) on/off, (c) phase voltage, and (d) phase current.

on, the phase voltage v_a is equal to $V_{dc}/2$ and, therefore, the phase current i_a increases. When the switch T_1 is off, the phase current i_a freewheels through diode D_1, and the stored energy is sent back to the capacitor C_2. The breakdown voltages of the switch T_1 and the diode D_1 need to be higher than the DC-link voltage V_{dc}. It should be noted that the DC-link voltage utilization is low for this converter. Only half of the DC-link voltage can be applied to the phase windings. Although this converter only needs one switch for each phase, this advantage is outweighed by the low DC-link voltage utilization.

In order to improve the usable DC-link voltage for the phase windings, a split AC converter can be used as shown in Figure 10.14. A split AC converter has three switches, three diodes, and two capacitors. When the current of phase A is below the reference, T_1 is turned on and V_{dc} is applied. The phase A current flows through C_1 and T_1. When the current of phase A rises above the reference, T_1 is turned off and $-V_{dc}$ is applied. The current of phase A flows through D_1 and C_2. When the current of phase B needs to be built up, T_2 is turned on and T_1 is turned off. The current of phase B flows through C_2 and T_2. The demagnetization voltage of phase A is $-V_{dc}$. The split AC power converter also allows partial independent control of different phases. Compared with the split DC power converter, the split AC converter provides higher magnetization/demagnetization voltage, which decreases the commutation torque ripple. Moreover, no efforts have to be made to balance the voltage of C_1 and C_2. The usable voltage for the windings is the DC-link voltage V_{dc} in this converter. The power devices need to have a voltage rating two times the DC-link voltage.

An alternative single switch per phase converter is shown in Figure 10.15. When phase A needs to carry the current, the switches T_2 and T_3 are turned on and off. When phase B needs to carry the current, the switches T_1 and T_3 are operating. Because of the shared switch T_3 between phases A and B, the currents of these two phases cannot be controlled independently when the phase currents overlap. Similarly, the currents of phase C and phase D cannot be controlled independently. Also, the efficiency of this converter is lower as compared to the asymmetric bridge converter because of the extra diodes in series with the phase windings. Moreover, this converter is only suitable for the electric machines with an even number of phases. These drawbacks limit the applications of this converter.

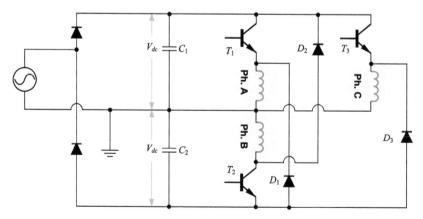

FIGURE 10.14
Split AC converter for three-phase SRM.

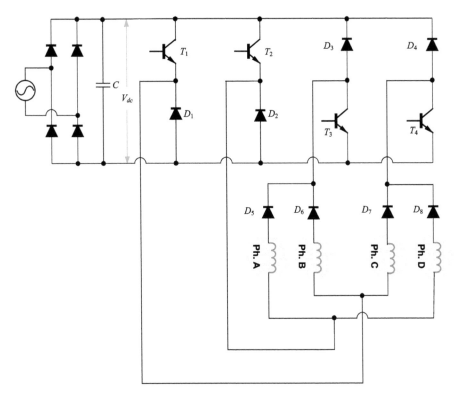

FIGURE 10.15
An alternative single switch per phase converter for a four-phase SRM.

10.6 Asymmetric Neutral Point Diode Clamped (NPC) Three-Level Power Converter

In the unipolar (soft switching) mode, an asymmetric bridge converter can apply two different levels of voltages to a phase of SRM. As shown in Figure 10.3, when both switches are on, the voltage across the terminals of a phase winding is V_{dc}. When one of the switches is turned off, the phase voltage equals zero. When both switches are off, phase voltage is $-V_{dc}$, which is the same voltage level as before, but with opposite polarity. This is why an asymmetric bridge converter is called a two-level converter.

If a converter is capable of applying more than two voltage levels, it is called a multi-level converter. As compared to conventional two-level converters, multi-level converters have several advantages, such as lower magnitude of current ripple, lower power losses at high switching frequency, lower common-mode voltage, and lower electromagnetic interference (EMI) [2]. Due to these advantages, multi-level converters are popular in medium voltage applications.

The asymmetric three-level neutral point diode clamped (NPC) converter for three-phase SRM drives is shown in Figure 10.16 [3]. The asymmetry of this topology is similar to that in the asymmetric bridge converter. The switch pairs T_{1A}-T_{2A} and T_{3A}-T_{4A}, and the diodes D_{1A} and D_{2A} are diagonally positioned to control phase A current. A similar structure can be observed in the asymmetric bridge converter in Figure 10.1. It can be observed that two DC-link capacitors are connected to the neutral point u_n and the voltage across them is

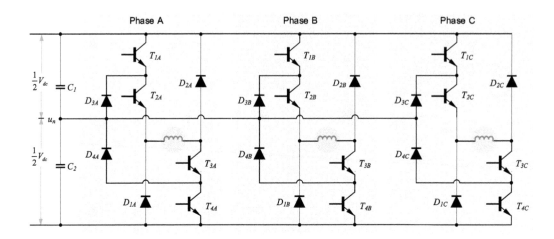

FIGURE 10.16
Asymmetric neutral point diode clamped (NPC) three-level power converter. (From Peng, F. et al., *IEEE Trans. Power Electron.*, 32(11), 8618–8631, November 2017.)

the half of the DC-link voltage. The term diode clamped comes from the fact that diodes D_{3A} and D_{4A} in Phase A are used to connect the phase terminals to the neutral point. This enables applying different voltage levels across the phase winding.

In the asymmetric three-level neutral diode clamped (NPC) converter in Figure 10.16, the required blocking voltages of the main switches (T_{1X}, T_{2X}, T_{3X}, and T_{4X}, $X = A$, B, or C) and clamping diodes (D_{3X}, D_{4X}) are half of the DC-link voltage ($V_{dc}/2$). The required blocking voltages of D_{1X} and D_{2X} are the DC-link voltage (V_{dc}). There are nine operational modes for each phase leg, which are illustrated in Figure 10.17. States of the four switches in one phase are illustrated in Table 10.1. The phase voltage (u_w) and the neutral point voltage (u_n) under different modes are also listed in Table 10.1. The on- and off-state of the switches are denoted as 1 and 0, respectively. The up arrow ↑ represents the increase and down arrow ↓ represents the decrease in the corresponding voltage. The cross × represents no changes.

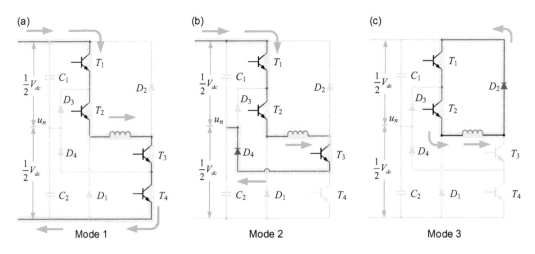

FIGURE 10.17
Operational modes of the asymmetric NPC three-level power converter: (a) Mode 1, (b) Mode 2, (c) Mode 3, (d) Mode 4, (e) Mode 5, (f) Mode 6, (g) Mode 7, (h) Mode 8, and (i) Mode 9.   (*Continued*)

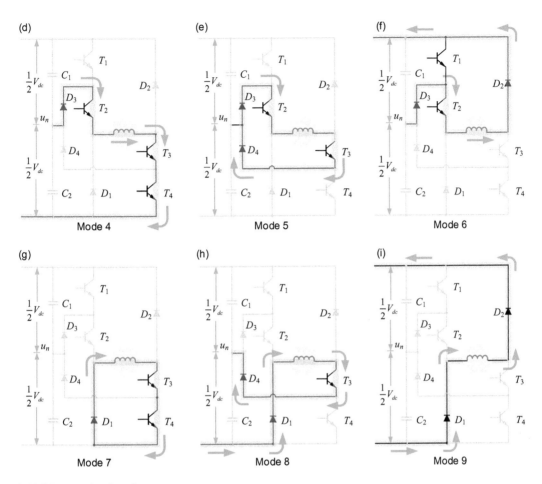

FIGURE 10.17 (Continued)
Operational modes of the asymmetric NPC three-level power converter: (a) Mode 1, (b) Mode 2, (c) Mode 3, (d) Mode 4, (e) Mode 5, (f) Mode 6, (g) Mode 7, (h) Mode 8, and (i) Mode 9.

TABLE 10.1

States of the Switches in Different Operation Modes

Mode Number	T_1	T_2	T_3	T_4	u_w	u_n
1	1	1	1	1	V_{dc}	×
2	1	1	1	0	$0.5\,V_{dc}$	↑
3	1	1	0	0	0	×
4	0	1	1	1	$0.5\,V_{dc}$	↓
5	0	1	1	0	0	×
6	0	1	0	0	$-0.5\,V_{dc}$	↓
7	0	0	1	1	0	×
8	0	0	1	0	$-0.5\,V_{dc}$	↑
9	0	0	0	0	$-V_{dc}$	×

A modified modulation scheme is applied to determine which operating mode is used [3]. The detailed operational modes of the asymmetric NPC three-level converter are explained below.

Mode 1: All switches are on. The DC-link voltage V_{dc} is applied to the phase winding and the load current flows through T_1, T_2, T_3, and T_4. The diodes D_1, D_2, D_3, and D_4 are blocked. No current flows into the neutral point. Therefore, the neutral point potential is unchanged in this mode. The phase voltage is V_{dc}.

Mode 2: T_1, T_2, and T_3 are on; T_4 is off. Diodes D_1, D_2, and D_3 are blocked. The load current flows through T_1, T_2, T_3, and D_4. Current flows into the neutral point. Therefore, the neutral point potential increases in this mode. The phase voltage is around $V_{dc}/2$.

Mode 3: T_1 and T_2 are on; T_3 and T_4 are off. Load current flows through T_1, T_2, and the freewheeling diode D_2. Diodes D_1, D_3, and D_4 are blocked. No current is injected into the neutral point. Therefore, the neutral point potential is unchanged in this mode. The phase voltage is zero.

Mode 4: T_1 is off; T_2, T_3, and T_4 are on. D_1, D_2, and D_4 are blocked. Current flows through D_3, T_2, T_3, and T_4. The neutral point potential decreases in this mode because the current flows out from the neutral point. The phase voltage is around $V_{dc}/2$.

Mode 5: T_1 and T_4 are off; T_2 and T_3 are turned on. Current flows through T_2, T_3, D_3, and D_4. Diodes D_1 and D_2 are blocked. No current is injected into the neutral point. Therefore the neutral point potential is unchanged in this mode. The phase voltage is zero.

Mode 6: T_1, T_3, and T_4 are off; T_2 is on. Current flows through D_3, T_2, and D_2. Diodes D_1 and D_4 are blocked. Current flows out from the neutral point. Therefore, the neutral point potential decreases in this mode. The phase voltage is around $-V_{dc}/2$.

Mode 7: T_1 and T_2 are off; T_3, and T_4 are on. Current flows through T_3, T_4, and the freewheeling diode D_1. Diodes D_2, D_3, and D_4 re blocked. No current is injected into the neutral point. Therefore, the neutral point potential is unchanged in this mode. The phase voltage is zero.

Mode 8: T_1, T_2, and T_4 are off; T_3 is on. Current flows through D_1, T_3, and D_4. Diodes D_2 and D_3 are blocked. Current is injected into the neutral point. Therefore, the neutral point potential increases in this mode. The phase voltage is around $-V_{dc}/2$.

Mode 9: All the switches are off. Current flows through the freewheeling diodes D_1 and D_2. Diodes D_3 and D_4 are blocked. No current is injected into the neutral point. Therefore the neutral point potential is unchanged in this mode. The phase voltage is zero.

10.7 Other SRM Converters

A converter with higher demagnetization voltage is proposed in [4] to achieve faster commutation as well as higher torque generation. The circuit diagram of this converter is shown in Figure 10.18. This converter is derived by adding a buck-boost converter to the asymmetric bridge converter. The switch T_m, the inductor L_m, and the diode D_m compose the buck-boost converter. The voltage of the capacitor C_1 can be controlled by adjusting the

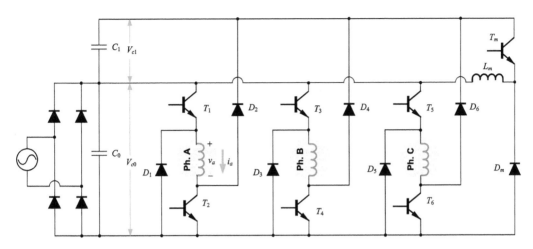

FIGURE 10.18
A converter with higher demagnetization voltage for the three-phase SRM.

duty cycle of the switch T_m. When the switches T_1 and T_2 are on, the voltage V_{c0} is applied to the winding of phase A. When the switches T_1 and T_2 are off, the diodes D_1 and D_2 are forward biased. The summation of the voltages V_{c0} and V_{c1} is applied to the winding, and the phase voltage v_a is equal to $-(V_{c0} + V_{c1})$. Therefore, the current falls down faster compared to the conventional asymmetric bridge converter. The commutation between the adjacent phases is faster, and the higher torque can be generated especially at high-speed operation. The voltage ratings of the power devices are equal to $V_{c0} + V_{c1}$.

Another converter with variable DC-link voltage is proposed in [5], as shown in Figure 10.19. The converter is similar to the C-dump converter. It also includes buck and boost converters. The components T_r, D_r, and L_r compose a buck converter. The boost converter is composed of the phase winding, T_1, and diode D_1. The voltage V_{c1} can be controlled by adjusting the duty cycle of the switch T_r. The advantage of this converter is that it can independently control the phase currents, and the DC-link voltage is adaptable to different speed ranges. However, this converter includes an extra buck converter, and it introduces extra power losses. The voltage ratings of the denoted devices are equal to V_{c0}.

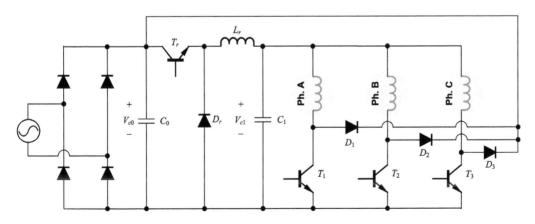

FIGURE 10.19
A variable DC-link converter by inserting a buck converter for the three-phase SRM.

10.8 Comparison of Power Electronic Converters for Three-Phase SRMs

The five power electronic converters including the asymmetric bridge converter, (N + 1)-switch converter, C-dump converter, and split DC and split AC converters are compared in terms of device rating, switching losses, conduction loss, and control performance in this section for three-phase SRMs [6,7]. The maximum root mean square (RMS) phase current of the motor is I_d and the DC-link voltage is V_{dc}. The volt-ampere (VA) rating of the converter is an important criterion to evaluate the cost of the converter. The converter VA rating is defined as $N \times V \times I$ where V and I are the voltage rating and RMS current rating of the switches, and N is the number of the switches. For easier comparison, the normalized power losses are used. The switching loss of a single switch with V_{dc} voltage stress and I_d current stress is assumed to be 1 per-unit (pu). The conduction loss of a single switch or diode with I_d current is also assumed as 1 pu.

The asymmetric bridge converter shown in Figure 10.1 has two switches and two diodes per phase. In the asymmetric bridge converter, the voltage rating and maximum RMS current are V_{dc} and I_d and, therefore, the asymmetric bridge converter has a 6VA rating in three-phase SRMs. The switch in the asymmetric bridge converter has V_{dc} voltage stress and I_d current stress and, therefore, switching loss of the switching device is 1 pu. In the hard (bipolar) switching pattern, the asymmetric bridge converter has 6 pu for both switching and conduction losses of the switch, and a 6 pu for diode conduction loss. In the soft (unipolar) switching pattern, the asymmetric bridge converter has a 3 pu switching loss, a 9 pu switch conduction loss, and a 3 pu diode conduction loss. In the soft switching mode, when the phase current decays, it flows through the switch instead of the diode. Therefore, the conduction loss of switches will be higher, but the total conduction loss stays the same.

The (N + 1)-switch converter shown in Figure 10.6 has four switches and four diodes per phase. This topology cannot achieve independent control of the current, which might be undesirable for high-performance motor drives. The cost may be reduced with the lower number of switches and diodes. However, the power ratings of T_1 and D_4 are much higher than those of other switches and diodes due to the repeated switching. The shared switch T_1 in (N + 1)-switch converter has $3I_d$ current stress and V_{dc} voltage stress. Other three switches in (N + 1)-switch converter have I_d current stress and V_{dc} voltage stress. Therefore, (N + 1)-switch converter has 6 VA rating in total. For (N + 1)-switch power converter, T_1 has three times switching power loss due to repeated switching and, therefore, total switching loss is also 6 pu. Similarly, the (N + 1)-switch power converter has a 6 pu switch and diode conduction losses.

The C-dump converter shown in Figure 10.7 has four switches, four diodes, one inductor, and two capacitors. When the voltage V_o of the C-dump converter is controlled at $2V_{dc}$ and $3V_{dc}$, the maximum voltage for switches is also $2V_{dc}$ and $3V_{dc}$. The VA rating of the C-dump converter is increased to $4 \times 2V_{dc} \times I_d$ and $4 \times 3V_{dc} \times I_d$, respectively. The switching loss is 8 pu and 12 pu with these two voltages. With the same current rating of the switch and the diode, the conduction loss of the switch and the diode in C-dump converter is decreased to 4 pu since the number of switches and diodes are decreased.

The split DC converter shown in Figure 10.12 allows partial independent control of different phases. Moreover, efforts have to be made to balance the voltage of C_1 and C_2. In addition, the increased number of capacitors adds additional cost to the system and lowers the power density considering the bulky size of the capacitors. The voltage rating of three switches in split DC converter is V_{dc} and, therefore, the split DC converter has $3 \times V_{dc} \times I_d$ VA rating. The total switching loss of split DC converter is 3 pu considering V_{dc} voltage stress and I_d current stress. The conduction loss of the transistor and diode in a split DC converter are both 3 pu.

The split AC converter shown in Figure 10.14 has three switches, three diodes, and two capacitors. The split AC power converter also allows partial independent control of different phases. Compared with the split DC power converter, the split AC converter provides higher magnetization/demagnetization voltage, which might decrease the commutation torque ripple. Moreover, no efforts have to be made to balance the voltage of C_1 and C_2. The voltage rating of three switches in split AC converter is $2V_{dc}$ and, therefore, the split AC converter has $3 \times 2V_{dc} \times I_d$ VA rating. The VA rating of split AC converter is the same as the asymmetric bridge converter. The total switching loss of split AC converter is increased to 6 pu considering $2V_{dc}$ voltage stress. Since the current rating of split AC converter and split DC converter is the same, the split AC converter has the same conduction loss as the split DC converter.

A detailed comparison of the five studied power electronic converters in terms of VA rating, power loss, and control performance is listed in Table 10.2. The (N + 1)-switch converter and split AC converter have the same VA rating as the asymmetric bridge converter despite the reduced number of switches and diodes. Therefore, these three converters have similar costs. The C-dump converter has the same magnetization/demagnetization voltage as split AC and asymmetric bridge converters when the voltage V_o of the capacitor is controlled at $2V_{dc}$. However, it has a higher VA rating and switching losses. Therefore, split AC and asymmetric bridge converters are more effective in terms of cost and control performance.

10.9 Comparison of Control Performance of Power Electronic Converters

In order to evaluate the performance of different converters in a switched reluctance motor drive, a 12/8 SRM simulation model is built in MATLAB®/Simulink as introduced in Chapter 4. Instantaneous torque control using the linear torque sharing function (TSF) introduced in Chapter 9 is applied. The turn-on angle θ_{on}, turn-off angle θ_{off} and overlapping angle θ_{ov} of linear TSF are set to 5°, 20°, and 2.5°, respectively. Hysteresis current control with 0.5A hysteresis band is applied and the DC-link voltage V_{dc} is set to 300 V. The torque reference is set to 1.5 Nm. The RMS current and torque ripple are expressed as (10.1) and (10.2), respectively.

$$I_{rms} = \sqrt{\frac{1}{\theta_p} \int_0^{\theta_p} i_k^2 d\theta} \tag{10.1}$$

$$T_{rip} = \frac{T_{max} - T_{min}}{T_{av}} \tag{10.2}$$

where i_k is k^{th} phase current; T_{av}, T_{max}, and T_{min} are the average torque, maximum torque, and minimum torque, respectively.

The comparison of torque ripple, the average torque, and RMS current of the power converters is shown in Figures 10.20 through 10.22, respectively. As shown in Figure 10.20, C-dump converter shows the lowest torque ripple up to 6000 rpm due to higher demagnetization voltage. Since split AC and asymmetric power converters have the same magnetization/demagnetization voltage, they have the similar torque ripple, average torque,

TABLE 10.2

Comparison of Power Electronic Converters for a Three-Phase SRM

Converter	Asymmetric Converter		(N + 1)-Switch Converter	Split DC Converter	Split AC Converter	C Dump Converter	
	Bipolar switching	Unipolar Switching				Capacitor Voltage $2V_{dc}$	Capacitor Voltage $3V_{dc}$
Number of switches	6	6	4	3	3	4	4
Number of diodes	6	6	4	3	3	2	2
Number of capacitors	1	1	1	2	2	2	2
Number of inductors	0	0	0	0	0	1	1
Magnetization voltage	V_{dc}	V_{dc}	V_{dc}	$0.5V_{dc}$	V_{dc}	V_{dc}	V_{dc}
Demagnetization voltage	$-V_{dc}$	$-V_{dc}$	$-V_{dc}$	$-0.5V_{dc}$	$-V_{dc}$	$-V_{dc}$	$-2V_{dc}$
Total power ratings (VA)	$6V_{dc}I_d$	$6V_{dc}I_d$	$6V_{dc}I_d$	$3V_{dc}I_d$	$6V_{dc}I_d$	$8V_{dc}I_d$	$12V_{dc}I_d$
Transistor switching losses (pu)	6	3	6	3	6	8	12
Transistor conduction losses (pu)	6	9	6	3	3	4	4
Diode conduction losses (pu)	6	3	6	3	3	4	4
Phase independence	Yes	Yes	No	Partial	Partial	Yes	Yes
Freewheeling	No	Yes	Yes	No	No	No	No

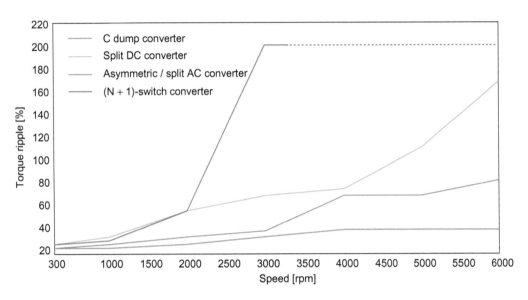

FIGURE 10.20
Comparison of torque ripples for different SRM converters.

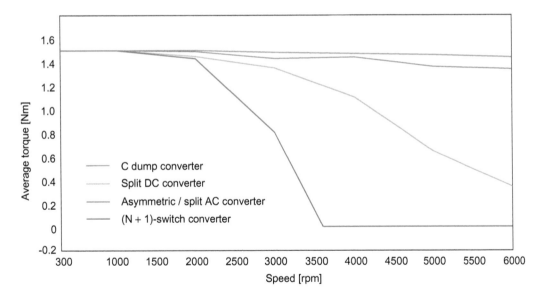

FIGURE 10.21
Comparison of average torque for different SRM converters.

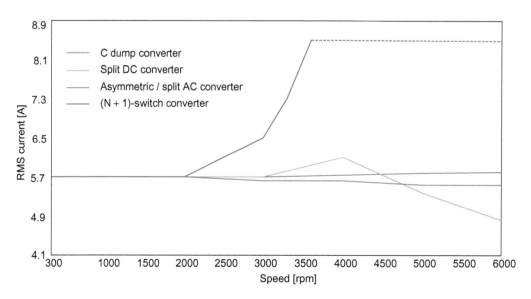

FIGURE 10.22
Comparison of RMS current for different SRM converters.

and RMS current. Split AC has the same VA rating, switching losses as the asymmetric bridge converter, but lower conduction losses. Therefore, a split AC power converter can be preferred in terms of efficiency. The phase current is not independently controlled in a (N + 1)-switch converter, which leads to a very high torque ripple and much lower average torque at higher speed. Although the (N + 1)-switch converter has the same VA rating, and conduction and switching losses as the asymmetric bridge converter, it is worse in terms of torque-speed performance. A (N + 1)-switch converter is not a feasible choice due to these reasons. Split DC converters produce around one-fifth of the average torque of the split AC and C-dump converters at 6000 rpm, while having one half of the VA rating, and conduction and switching losses. Therefore, split AC converter and C-dump converters could be the possible candidates for low-cost and high-performance SRM drives. They have similar RMS current and average torque up to 6000 rpm. However, the split AC converter shows almost twice as much torque ripple over the wide speed range with one half of VA rating and switching losses compared with the C-dump converter. Also, compared to split AC and C-dump converters, the asymmetric bridge converter has a reduced number of capacitors and inductors.

Questions and Problems

1. Asymmetric bridge converter is the most widely used power converter in SRM drives. How do you compare the bipolar (hard) switching and unipolar (soft) switching pattern?

2. Compared to the asymmetric bridge converter, (N + 1)-switch converter has a lower number of switches? Does it mean that (N + 1)-switch converter is more cost-effective?

3. What is the drawback of the non-independent control? Why do we need to improve the magnetization/demagnetization voltage?

4. The table below shows the switching states for different operational modes for the three-level NPC converter. Please fill out the rest of the table.

Mode Number	T_1	T_2	T_3	T_4	D_1	D_2	D_3	D_4	u_w	u_n
1	1	1	1	1	0	0	0	?	?	X
2	1	1	1	0	0	?	0	?	?	↑
3	1	?	0	0	?	1	0	0	0	X
4	0	?	1	?	0	0	1	0	$0.5\,V_{dc}$	↓
5	0	1	1	0	0	0	0	?	0	X
6	0	1	?	?	?	1	1	?	?	↓
7	0	0	1	?	?	0	0	?	0	X
8	0	0	1	0	1	?	0	1	$-0.5\,V_{dc}$	↑
9	0	0	0	0	1	?	0	0	?	X

5. The table below shows the comparison of VA ratings of the asymmetric bridge (two-level) and asymmetric NPC three-level converters. The maximum RMS phase current of the motor is I_d and the DC-link voltage is V_{dc}.

	Two-Level (Asymmetric Bridge)			NPC Three-Level		
	Rating	Amount	Total	Rating	Amount	Total
Switches	?	6	?	?	12	?
Switches total			?			$6 \times V_{dc} \times I_d$
D1X, D2X	?	6	?	?	6	?
D3X, D4X	N/A	N/A	N/A	?	6	?
Diodes total			?			?

Please fill out the rest of the table. What can you say about the VA ratings of the switching devices and diodes in these converters? The NPC three-level converter needs two DC-link capacitors. The capacitance of each capacitor has to be doubled to maintain the same voltage on the DC-link. However, compared to conventional asymmetric bridge converter, the capacitor voltage is divided in half. What does it say about the size and cost of the capacitors in these converters? What could be the other components, which might affect the cost of the NPC three-level converter?

6. Three-phase bridge inverter is commonly used to drive permanent magnet and induction machines. Figure Q.10.1 below shows the circuit diagram of a three-phase bridge inverter. Can we use this converter to drive an SRM? How should we modify the SRM windings to utilize the conventional inverter?

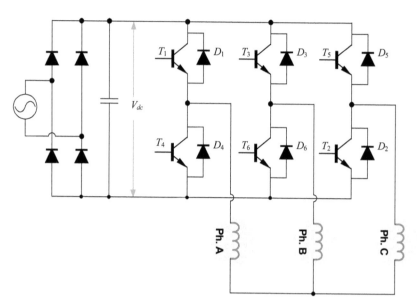

FIGURE Q.10.1
Standard three-phase inverter.

7. Figure Q.10.2 shows the circuit diagram of a single-bus star-connected SRM converter. It is also known as Miller converter. This converter benefits from a star connection for the motor phases. It reduces the required number of leads to (N + 1) and it needs (N + 1) switches and (N + 1) diodes. This topology reduces the number of wires and semiconductor switches [8]. What could be the challenges in controlling this converter?

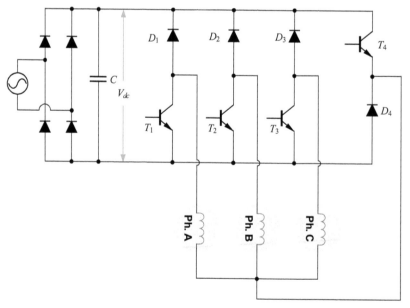

FIGURE Q.10.2
Circuit diagram for a single-bus star-connected SRM converter.

References

1. R. Krishnan, *Switched Reluctance Motor Drives: Modeling, Simulation, Analysis, Design, and Applications*. Boca Raton, FL: CRC Press, June 2001.
2. J. Rodriguez, Multilevel inverters: A survey of topologies, controls, and applications, *IEEE Trans. Ind. Electron.*, 49(4), 724–738.
3. F. Peng, J. Ye, and A. Emadi, An asymmetric three-level neutral point diode clamped converter for switched reluctance motor drives, *IEEE Trans. Power Electron.*, 32(11), 8618–8631, November 2017.
4. A. K. Jain and N. Mohan, SRM power converter for operating with high demagnetization voltage, *IEEE Trans. Ind. Appl.*, 41(5), 1224–1231, 2005.
5. Y. Kido, N. Hoshi, A. Chiba, S. Ogasawara, and M. Takemoto, Novel switched reluctance motor drive circuit with voltage boost function without additional reactor, in *Proceedings of the IEEE European Conference on Power Electronics and Applications*, Birmingham, UK, August-September 2011, pp.1–10.
6. K. Ha, C. Lee, J. Kim, R. Krishnan, and S. G. Oh, Design and development of low cost and high efficiency variable speed drive system with switched reluctance motor, *IEEE Trans. Ind. Appl.*, 43(3), 703–713, May 2007.
7. J. Ye and A. Emadi, Power electronic converters for 12/8 switched reluctance motor drives: A comparative analysis, in *Proceedings of the IEEE Transportation Electrification Conference and Expo (ITEC)*, Dearborn, MI, June 2014.
8. P. Shamsi and B. Fahimi, Single-bus star-connected switched reluctance drive, *IEEE Trans. Power Electron.*, 28(12), 5578–5586, December 2013.

11

Position Sensorless Control of Switched Reluctance Motor Drives

Jin Ye

CONTENTS

11.1 Introduction

In general, for torque or speed control of a switched reluctance machine (SRM), position feedback from an encoder or resolver is necessary. However, in order to reduce the cost and volume of the motor drive, position sensorless control of SRM becomes an alternative solution to obtain the rotor position. Since some magnetic characteristics of SRM such as flux and self-inductance are rotor position dependent, rotor position can be obtained by estimating these magnetic characteristics online.

A diagram of classification of sensorless methods for SRMs is shown in Figure 11.1. Passive rotor position estimation based on the measurement of terminal voltage and phase current of active phases does not require additional hardware and it does not generate additional power losses. Self-inductance-based position estimation and the flux-based

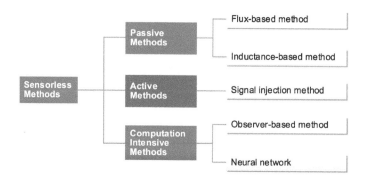

FIGURE 11.1
Classification of position sensorless control schemes for SRM.

rotor position estimation are two approaches of passive position sensorless techniques [1–5]. Signal injection method is a method for low speed operation [1,6–9]. High-frequency signal is injected to the inactive phase to obtain the inductance, which is later converted to rotor position. Computation intensive methods such as observer-based estimation, and neural network are also presented in [10–15]. They show some robustness or model independence; however, they require high computational complexity. Methods to improve the accuracy of the position estimation in switched reluctance machines have been presented in [16].

As the speed increases, the overlapping region of the active phases becomes significant and mutual flux cannot be neglected anymore. The accuracy of both inductance-based and flux linkage-based rotor position estimation methods reduce at higher speed due to the mutual flux between active phases. In addition, torque sharing function (TSF) is widely used in instantaneous torque control to reduce commutation torque ripple [17,18]. When TSF is applied, overlapping areas of incoming and outgoing phases are significant even at low speed. Therefore, mutual flux has to be considered to achieve accurate estimation of rotor position over a wide speed range. The position estimation including the mutual flux will also be discussed in this chapter [19,20].

11.2 Inductance-Based Position Estimation

In SRMs, hysteresis control is applied for phase current control, as shown in Figure 11.2. Upper and lower current references of the kth phase are denoted as i_{k_up} and i_{k_low}, respectively. The hysteresis band is represented as:

$$\Delta i_k = i_{k\_up} - i_{k\_low} \tag{11.1}$$

When the switches T_1 and T_2 are turned on as shown in Figure 11.3a, DC-link voltage is applied to the phase; hence, the phase current slope is positive. When the switches T_1 and T_2 are turned off, as shown in Figure 11.3b, the phase current slope is negative. The voltage equations neglecting mutual coupling are then derived as (11.2) and (11.3) when switches are on and off, respectively.

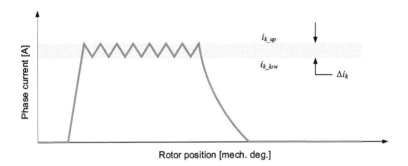

FIGURE 11.2
Hysteresis control of phase current in SRM.

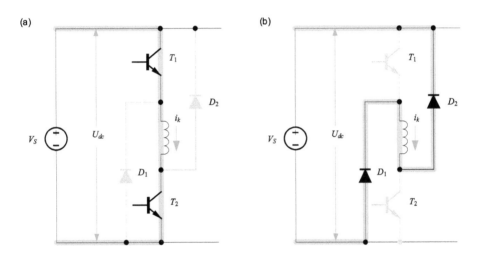

FIGURE 11.3
Two switching states of an SRM driver: (a) ON state and (b) OFF state.

$$V_{dc} = Ri_k + L_{inc\_k,k} \frac{di_k(t_{k\_on})}{dt} + \frac{\partial L_{k,k}}{\partial \theta} i_k \omega_m \qquad (11.2)$$

$$-V_{dc} = Ri_k + L_{inc\_k,k} \frac{di_k(t_{k\_off})}{dt} + \frac{\partial L_{k,k}}{\partial \theta} i_k \omega_m \qquad (11.3)$$

where $L_{inc\_k,k}$ is the incremental self-inductance of the kth phase, which can be defined as:

$$L_{inc\_k,k} = L_{k,k} + i_k \frac{dL_{k,k}}{di} \qquad (11.4)$$

where t_{k_on} and t_{k_off} are time instants when the kth phase switching states are ON and OFF during a switching period, respectively; $di_k(t_{k\_on})/dt$ and $di_k(t_{k\_off})/dt$ are the slope of kth phase current at t_{k_on} and t_{k_off}, respectively; $L_{k,k}$ is the self-inductances of the kth phase; θ and ω_m are rotor position and angular speed of SRM, respectively; and V_{dc} is the DC-link voltage. Equation (11.4) expresses how the inductance changes with current. Therefore, the incremental self-inductance $L_{inc\_k,k}$ represents how the phase inductance changes with current at a certain rotor position.

The switching period is usually short enough and, therefore, variation of the mechanical speed, inductance, back EMF and resistance can be neglected. Then, the incremental self-inductance can be derived as in (11.5) by combining (11.2) and (11.3). For a given DC-link voltage, unsaturated self-inductance can be estimated by using the phase current slope difference between ON and OFF states.

$$L_{inc\_k,k} = \frac{2V_{dc}}{\dfrac{di_k\left(t_{k\_on}\right)}{dt} - \dfrac{di_k\left(t_{k\_off}\right)}{dt}} \tag{11.5}$$

where $L_{inc\_k,k}$ is estimated k^{th} phase incremental self-inductance without considering the mutual flux. By neglecting the variation of the speed, induced voltage, and ohmic resistance in a switching period, the self-inductance is estimated by measuring the change in the slope of the phase current [1]. Incremental inductance can be estimated by using phase current slope difference and, therefore, the application of this method can be extended to magnetic saturation region [3]. The influence of the variation of resistance is eliminated. Hence, this method is capable of operating at low speeds.

The incremental self-inductance is a function of rotor position and phase current as shown in (11.6). Therefore, the rotor position can be estimated based on the inductance-current-position characteristics shown in (11.7), by neglecting the mutual coupling.

$$L_{inc\_k,k} = f(i_k, \theta) \tag{11.6}$$

$$\theta = f^{-1}(i_k, L_{inc\_k,k}) \tag{11.7}$$

Alternatively, the position can be obtained based on the analytical expression of the incremental self-inductance. The analytical expression for the incremental self-inductance can be given by (11.8) in terms of a truncated Fourier series [21]. Once the phase current is measured using the current sensor and the incremental self-inductance is estimated, the rotor position can be determined.

$$L_{inc\_k,k}(i_k, \theta) = \sum_{n=0}^{2} L_n(i_k)\cos(n(N_r\theta + \theta_0)) \tag{11.8}$$

$$= L_0(i_k) + L_1(i_k)\cos(N_r\theta + \theta_0) + L_2(i_k)\cos(2N_r\theta + 2\theta_0)$$

where n represents the order of harmonics, N_r represents the number of rotor poles, and θ_0 represents the phase shift. For an 8/6 SRM, the analytical expression of the incremental self-inductance can be given by (11.9).

$$L_{inc\_k,k}(i_k, \theta) = L_0(i_k) + L_1(i_k)\cos(6\theta + \theta_0) + L_2(i_k)\cos(12\theta + 2\theta_0) \tag{11.9}$$

In [21], the analytical expression of self-inductance is also given in terms of the aligned, midway-to-aligned, and unaligned inductances. If 0° is assumed as the aligned position and one stroke is 60° mechanical in 8/6 SRM, the midway position is at 15° and the unaligned position is at 30°. Inserting these values for θ in (11.9) and using trigonometric periodicity rules, the aligned, midway, and unaligned inductances of 8/6 SRM can be obtained as in (11.10)–(11.12):

$$L_a(i_k, \theta) = L_0(i_k) + L_1(i_k)\cos(\theta_0) + L_2(i_k)\cos(2\theta_0)$$

$$L_m(i_k, \theta) = L_0(i_k) - L_1(i_k)\sin(\theta_0) - L_2(i_k)\sin(2\theta_0) \qquad (11.10\text{--}11.12)$$

$$L_u(i_k, \theta) = L_0(i_k) - L_1(i_k)\cos(\theta_0) + L_2(i_k)\cos(2\theta_0)$$

where L_a, L_u, and L_m are the self-inductance at aligned position, unaligned position, and midway position, respectively.

Phase self-inductance is estimated only in the active region by using the phase current slope difference at rotating shaft condition. In a three-phase machine, each phase takes up one third of rotor period and three-phase inductance estimation will cover the total rotor period. The classification of the phase self-inductance estimation region is shown in Figure 11.4. Self-inductance estimation is classified into phase A, B, and C self-inductance estimation. Phase self-inductance estimation regions are selected to avoid the inductance estimation near unaligned rotor position. Because, near unaligned position, the change of phase self-inductance with rotor position is relatively low and, therefore, slight error in inductance estimation might lead to a much higher error in rotor position estimation.

The rotor position is estimated based on the corresponding phase self-inductance estimation region. When the estimated inductance of a phase reaches the maximum value, the self-inductance estimation is transferred to the next region. For example, when phase-A self-inductance estimation region is selected, estimated phase-A self-inductance is converted to the rotor position at each switching period by using position-inductance characteristics. Once the estimated phase-A inductance reaches L_{max}, phase inductance estimation is changed from phase-A self-inductance estimation region to phase-B self-inductance estimation region and the rotor position is updated based on phase B self-inductance. The flowchart of the rotor position estimation algorithm using phase current slope self-inductance estimation is shown in Figure 11.5.

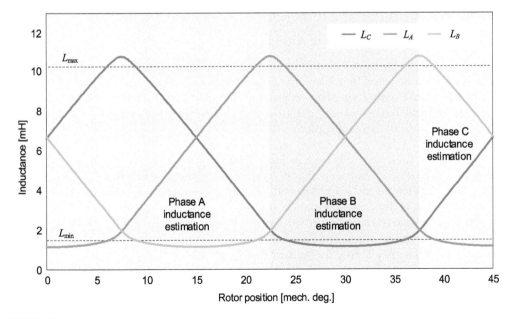

FIGURE 11.4
Classification of self-inductance estimation region. (From Ye, J. et al., *IEEE Trans. Pow. Electron.*, 20(3), 1499–1512, 2015.)

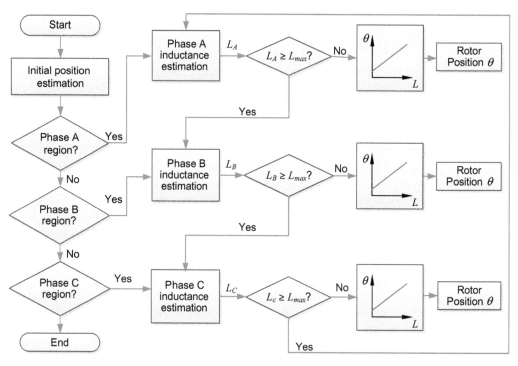

FIGURE 11.5
Flowchart of the rotor position estimation algorithm using phase current slope self-inductance estimation. (From Ye, J. et al., *IEEE Trans. Pow. Electron.*, 30(3), 1499–1512, 2015.)

11.3 Flux-Based Position Estimation

The flux-based rotor position estimation obtains the self-flux linkage by integrating the terminal voltage subtracted by the voltage across the ohmic resistance as shown in (11.13). In (11.13), the mutual coupling is neglected and the resistance is measured in advance. The accuracy could be deteriorated due to the variation of the ohmic resistance and accumulation error due to integration.

$$\lambda_k = \int (v_k - i_k R)\, dt \tag{11.13}$$

where v_k and i_k are the measured voltage and current respectively; λ_k is the estimated self-flux linkage; and R is the phase resistance.

The self-flux linkage is a function of rotor position and phase current as shown in (11.14). The rotor position can be estimated based on flux linkage-current-position characteristics as shown in (11.15) by neglecting the mutual coupling.

$$\lambda_k = f(i_k, \theta) \tag{11.14}$$

$$\theta = f^{-1}(i_k, \lambda_k) \tag{11.15}$$

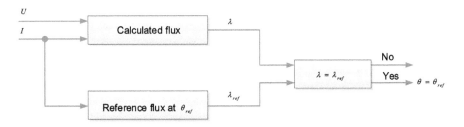

FIGURE 11.6
Diagram of an improved flux-current method. (From Zeng, W. et al., A new flux/current method for SRM rotor position estimation, in *Proceeding of International Conference on Electrical Machines and Systems*, Tokyo, Japan, pp. 1–6, 2009.)

Similar to the inductance-based position estimation, this method requires a 3D lookup table, which needs to be measured in advance and takes up a significant amount of memory space.

Different methods have been proposed to improve the flux linkage-based position estimation in switched reluctance machines. In [5], flux linkage is estimated based on the amplitude of the first order switching harmonics of the phase voltage and current. The influence of resistance variation and low induced voltage for rotor position estimation is reduced and, hence, both accuracy and applicable speed range are improved. In [22], the reference flux and corresponding reference position are predefined. By comparing the calculated flux linkage with the reference flux linkage at these particular positions, the rotor position can be obtained. The advantage of this approach is that fewer rotor positions need to be detected. This significantly reduces memory space and computational capacity. The diagram of this method is shown in Figure 11.6.

In [23], a high-resolution position estimator is proposed to improve the accuracy of the position estimation by adding a correcting variable to the primary flux-based position estimator. This method shows poor accuracy at low speed when induced voltage is small. Also, the accuracy is deteriorated by the variation of the ohmic resistance and accumulation error due to integration.

11.4 Signal Injection Method

The signal injection method is applicable for low-speed operation. When SRM operates at a low speed, the rotor position is estimated by injecting voltage/current signal to inactive phases. The signal injection methods can be classified into two categories: high-frequency pulse injection method and the low-amplitude sinusoidal voltage method.

High-frequency voltage pulses are injected into an inactive phase to obtain the information of phase inductance. At low speeds, a voltage pulse of V_{dc} is applied to the inactive phase for a short period of time Δt, leading to a current rise Δi. Therefore the voltage equation can be expressed as in (11.16) by neglecting the induced voltage and mutual coupling.

$$V_{dc} \approx L_{kk} \frac{\Delta i}{\Delta t} \qquad (11.16)$$

Then the self-inductance of the inactive phases can be estimated based on (11.17) by measuring the DC-link voltage V_{dc} and the current rise Δi.

$$L_{kk} \approx V_{dc} \frac{\Delta t}{\Delta i} \qquad (11.17)$$

High-frequency voltage pulses are injected into an inactive phase rather than the active phases. For a three-phase SRM, no more than two phases are active at the same time. In order to determine the inactive phase for pulse injection, the three-phase self-inductance is divided into 6 sectors and each sector has corresponding estimation and active phases. Assume that at 0° (electrical degrees), phase A at the aligned position and at 180° (electrical degrees), phase A is at the unaligned position. The self-inductance profile of phase A, phase B, and phase C are shown in Figure 11.7. When one phase is

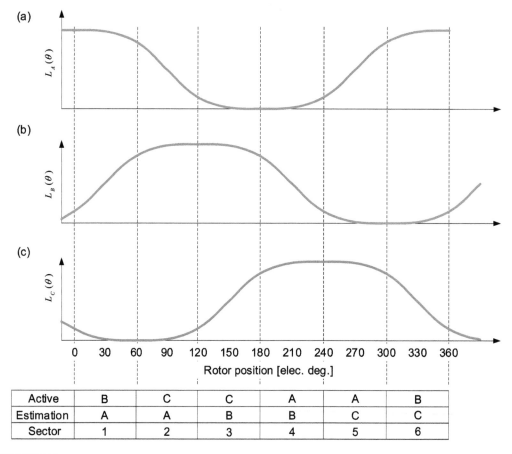

Active	B	C	C	A	A	B
Estimation	A	A	B	B	C	C
Sector	1	2	3	4	5	6

FIGURE 11.7
Sector definition for self-inductance estimation: (a) phase A, (b) phase B, and (c) phase C.

conducting, one of two inactive phases is used to estimate the rotor position. In sector 1, phase B is the active phase, and phase A (inactive phase) is used for position estimation. Similar analysis can be applied to other regions as well.

The rotor position can be obtained by using either the phase inductance-current-rotor position characteristics or the analytical expression as discussed in the inductance-based position estimation. However, the injected pulse may produce negative torque and additional power losses. In addition, it takes time for the injected current to reduce to zero. The next pulse cannot be injected to the phase before the previous pulse reduces to zero. This leads to poor performance at high-speed operation. Signal injection method is more useful at low-speed operation.

Different methods have been proposed to improve the high-frequency pulse injection-based rotor position estimation in SRMs. In [6], the working sector of the SRM is selected by comparing amplitude of the current response of the inactive phases with two predefined thresholds. This method is simple but it is not promising for instantaneous torque control. In [7], voltage pulse is injected to determine the speed range of SRM, and then the corresponding dynamic model is defined for rotor position estimation. In [1], a rotor position estimation scheme based on phase inductance vector is proposed. A pulse is injected to get the inductance of inactive phases and full cycle inductance is obtained. This method has advantages of no need of a *priori* knowledge on magnetic characteristics of the machine. Some pulse injection methods with additional circuit [9] have also been reported, which may increase the cost and complexity of implementation.

The low-amplitude sinusoidal voltage method is an alternative method for rotor position estimation in SRM. Based on the phase and amplitude variation, the phase modulation (PM) and the amplitude modulation (AM) techniques were discussed in [24]. When a sinusoidal voltage is applied to an inactive phase, either phase or amplitude of the current response can be measured and then the inductance information can be extracted. In general, the PM technique uses zero crossing detectors for voltage and current to extract inductance information, and the AM technique uses the envelope of the current response. This method is robust to the noise due to the switching of the inverter; however, both techniques need an additional voltage source. In summary, voltage injection methods often suffer from either additional power losses or low speed constraint.

11.5 Computation Intensive Methods

11.5.1 Observer-Based Methods

The observer-based methods are alternative ways to estimate the rotor position in SRM [7,22,25,26]. Compared to flux and inductance-based methods, the observer-based methods are less sensitive to measurement noise and numerical residual errors. The control diagram of an SRM with observer-based position estimation method is shown in Figure 11.8. In this figure, $\hat{\theta}$ and θ are the estimated and real rotor positions, respectively. $\hat{\lambda}_k$ and λ_k are the estimated flux linkage and real flux linkage, respectively. $\hat{\omega}$ and $\hat{\alpha}$ are the estimated speed and angular acceleration. The observer is used to estimate the position, speed, and acceleration.

We assume that the parameters of the SRM, including the electromagnetic characteristics and motion dynamics are known. The flux linkage of SRM is a function of both the rotor position and the phase current, and it can be stored in a look up table. When the position is

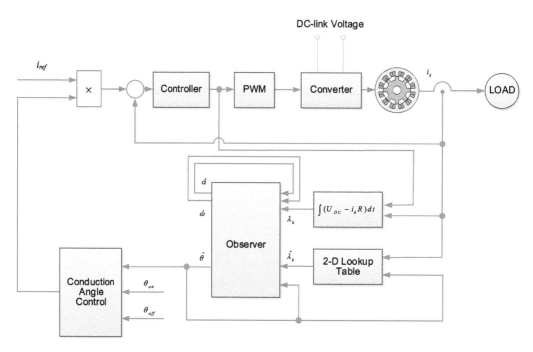

FIGURE 11.8
The control diagram of the SRM based on the observer method.

estimated by the observer, the flux linkage can be estimated based on the predefined look up tables. The estimation error can be developed from the difference between the real flux λ_k and the estimated flux $\hat{\lambda}_k$, and this estimation error can then be used to tune the SRM model until the estimated flux is equal to the real value. This can be accomplished by the observer.

The observers may take different forms, including flux observer [7,22] and sliding mode observer [26]. In [7] and [22], instead of comparing the estimated rotor position with the commutation angles, the estimated flux is compared with the commutation flux. The selection of the parameters of the observer determines its stability and dynamics. This method may suffer from computational complexity and large initial errors.

The sliding mode observer (SMO) is gaining interest in the rotor position estimation of the SRM due to its robustness and high reliability, especially in high-speed operation [26–33]. A brief introduction to sliding mode observer is presented in Appendix B of this chapter. In [26], a novel sliding-mode observer was introduced considering the impact of the magnetic saturation, load disturbance, parameter variation, and model errors. The sliding mode surface was selected as the difference between the estimated and measured phase currents. Switching gains, including the flux-linkage gain, the speed gain, and the position gain, were analyzed and optimized to ensure the stability and robustness to the system disturbance and parameter variations. In [27–29], the sliding mode surface was selected as the difference between the measured flux linkage and the estimated flux linkage. In [27], the sliding mode observer was improved by considering the discrete-time formulation of the observer; the advantages and limitations of fixed-point and floating point computations were discussed. In [28], a torque-ripple-minimization controller was introduced based on the sliding-mode observer. A detailed guideline for a comprehensive evaluation of the sliding mode observer was introduced in [29]. A hybrid observer of CSMO (current sliding mode observer) and FSMO (flux-linkage sliding mode observer) was proposed in [30]. By combining the CSMO at low speed and FSMO at high speed, the proposed hybrid observer shows better performance over wide speed range. Practical implementation of four-quadrant sensorless schemes based on the sliding-mode observer was discussed in [31,32].

11.5.2 Phase Lock Loop

The phase lock loop (PLL) method was developed to reduce the impact of the measurement noise and numerical residual errors in order to improve the position estimation accuracy of SRM drives [25]. In addition, the PLL method has better estimation accuracy during acceleration and deceleration compared to the observer based methods [25]. PLL has been used widely in field of the phase and frequency detection. In particular, PLL method has been used to control power electronic converters and electric machines. PLL can be regarded as a control system as shown in Figure 11.9, where the error between estimated position and real position works as the input to the control system. The controller should be designed to guarantee the stability and dynamics of the control system. The position estimation error detection is determined by either inductance based or flux based method. When the PLL system becomes stable, the estimated position would be ideally equal to the real position and it is utilized in the closed-loop control of switched reluctance motors. Please see Appendix A of this chapter for more information of phase lock loop based position estimation.

11.5.3 Fuzzy Logic and Neural Network

The fuzzy logic-based position estimation method has the advantage of no model dependency. The concept of fuzzy logic stems from the self-learning system. The self-learning system mainly includes both the static motor characteristics and real-time operating conditions. The real-time operating conditions, including saturation effects, mutual inductance, eddy currents, and temperature variation, can be considered to improve the accuracy of the rotor position estimation method.

Artificial neural network (ANN) is an alternative solution considering the high nonlinearity of the SRM system. In general, the inputs of ANN training data are current and flux; the outputs are the corresponding rotor position and speed. Interested readers can refer to [11] for more details. In [34], a novel, single neural

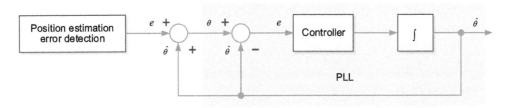

FIGURE 11.9
Block diagram of the phase lock loop-based position estimation.

proportional-integral-derivative (PID) control was proposed to estimate the rotor position based on the radial basis function (RBF) neural network, which utilizes both the offline and online training to enhance the adaptability and robustness. These methods show some robustness or model independence; however, they suffer from the computational complexity.

11.6 Initial Position Estimation

The initial rotor position of the SRM needs to be estimated to ensure a smooth and reliable start-up [35,36]. Several methods of the initial rotor position estimation are discussed in this section.

11.6.1 Two Current Threshold Method

The two current threshold method has been widely adopted to estimate the initial position of SRM. The three phase self-inductance profiles are divided into different sectors by two thresholds of each phase [35]. As shown in Figure 11.7, for a three-phase SRM, the self-inductance is divided into six sectors. Current thresholds for each sector, including the low current threshold and high current threshold, are defined in Table 11.1. Voltage pulses are injected to the three phases simultaneously and position sector can be determined by comparing the current amplitude of three phases within two thresholds. For example, if the current amplitude of phase A is below the low threshold, the current amplitude of phase C is above the high threshold, and the current amplitude of phase B is between the two thresholds, the rotor is located in Sector 1. Once the sector is defined, active phase can be determined, and then the initial position can be estimated.

11.6.2 One Current Threshold Method

One current threshold method is proposed in [36]. This method utilizes the time required for the current to reach the threshold and self-inductance profile to determine the initial rotor position sector. As shown Figure 11.10, the inductance profile of three-phase machine in one electrical period can be divided into six sectors by comparing the magnitude of self-inductances. The summary of the one current thresholds method is given in Table 11.2.

TABLE 11.1

Summary of the Two Current Thresholds Method

Sector	Phase A	Phase B	Phase C
1	$i_A <$ Low Threshold	Low Threshold $< i_B <$ High Threshold	$i_C >$ High Threshold
2	Low Threshold $< i_A <$ High Threshold	$i_B <$ Low Threshold	$i_C >$ High Threshold
3	$i_A >$ High Threshold	$i_B <$ Low Threshold	Low Threshold $< i_C <$ High Threshold
4	$i_A >$ High Threshold	Low Threshold $< i_B <$ High Threshold	$i_C <$ Low Threshold
5	Low Threshold $< i_A <$ High Threshold	$i_B >$ High Threshold	$i_C <$ Low Threshold
6	$i_A <$ Low Threshold	$i_B >$ High Threshold	Low Threshold $< i_C <$ High Threshold

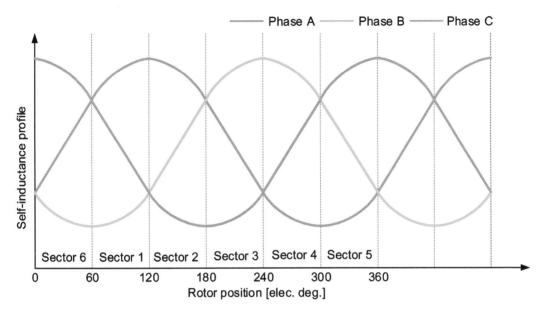

FIGURE 11.10
Self-inductance profile of a 6/4 SRM in the linear region.

TABLE 11.2

Summary of One Current Threshold Method

Time Required	Self-Inductance	Rotor Position	Sector	Active Phase
$t_1 > t_3 > t_2$	$L_A > L_C > L_B$	0°–60°	6	C
$t_3 > t_1 > t_2$	$L_C > L_A > L_B$	60°–120°	1	A
$t_3 > t_2 > t_1$	$L_C > L_B > L_A$	120°–180°	2	B
$t_2 > t_3 > t_1$	$L_B > L_C > L_A$	180°–240°	3	C
$t_2 > t_1 > t_3$	$L_B > L_A > L_C$	240°–300°	4	A
$t_1 > t_2 > t_3$	$L_A > L_B > L_C$	300°–360°	5	B

11.6.3 Inductance Based Initial Rotor Position Estimation Method

One current threshold method utilizes the time required for the current to reach to the threshold value to estimate the inductance. In the inductance-based initial rotor position estimation, instead of comparing the phase current with the current threshold, the three-phase self-inductance can be directly compared to determine the initial rotor position sector [33]. The same pulses are injected into three phases to calculate the self-inductance and then sector can be determined. Table 11.3 summarizes the principle of inductance based method. Compared to current threshold method, it is unnecessary to select the value of threshold. The flowchart of this initial position detecting method is shown in Figure 11.11.

TABLE 11.3

Initial Position Detection by Comparing Phase Inductance

Inductance Comparisons	Rotor Position	Estimation Phase
$L_A > L_B \geq L_C$	$0° \leq \theta_{initial} < 60°$	A
$L_B > L_A \geq L_C$	$60° \leq \theta_{initial} < 120°$	A
$L_B > L_C \geq L_A$	$120° \leq \theta_{initial} < 180°$	B
$L_C > L_B \geq L_A$	$180° \leq \theta_{initial} < 240°$	B
$L_C > L_A \geq L_B$	$240° \leq \theta_{initial} < 300°$	C
$L_A > L_C \geq L_B$	$300° \leq \theta_{initial} < 360°$	C

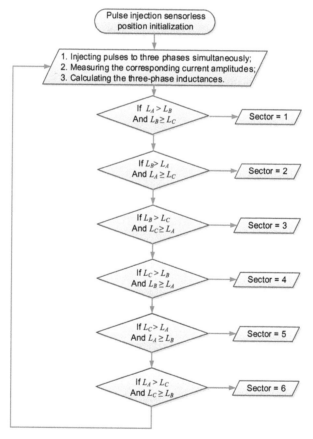

FIGURE 11.11
Inductance based initial rotor position estimation method.

Appendix A: Phase Lock Loop (PLL) Based Method

In a small neighborhood of an operational point, λ_k can be written as (A.11.1).

$$\Delta\lambda_k = \left.\frac{\partial\lambda_k}{\partial i_k}\right|_{\theta=const}\Delta i + \left.\frac{\partial\lambda_k}{\partial\theta}\right|_{i=const}\Delta\theta \tag{A.11.1}$$

Equation (A.11.1) can be written as in (A.11.2) by neglecting the variation of the phase current.

$$\theta - \hat{\theta} = \left.\frac{\partial\theta}{\partial\lambda_k}\right|_{i=const}\left(\lambda_k - \hat{\lambda}_k\right) \tag{A.11.2}$$

where $\hat{\theta}$ and θ are the estimated and real rotor position, respectively. The $\hat{\lambda}_k$ and λ_k is the estimated flux linkage and real flux linkage, respectively.

The motion equations of the SRM can be expressed as:

$$\dot{\theta} = \omega \tag{A.11.3}$$

$$\dot{\omega} = \alpha \tag{A.11.4}$$

where ω is the angular speed and α is the angular acceleration. In a digital implementation, the estimated position at the current sampling time step (n) comes from the previous time step $(n-1)$, and it is denoted as $\hat{\theta}(n\,|\,n-1)$. Then the estimated flux is obtained as (A.11.5).

$$\hat{\lambda}_k(n\,|\,n-1) = \psi\left(\hat{\theta}(n\,|\,n-1), i_k(n)\right) \tag{A.11.5}$$

According to (A.11.2), the estimation error can be obtained as (A.11.6).

$$e(n) = \hat{\theta}(n\,|\,n) - \hat{\theta}(n\,|\,n-1) = \left.\frac{\partial\theta}{\partial\lambda_k}\right|_{i_k(n)}\left(\lambda_k(n) - \hat{\lambda}_k(n\,|\,n-1)\right) \tag{A.11.6}$$

where $\hat{\theta}(n\,|\,n)$ is the estimated rotor position at the current time step, $e(n)$ is the estimation error, and $g = \left.\partial\theta/\partial\lambda_k\right|_{i_k(n)}$ is the iteration gain of the numerical calculation.

A PLL is designed based on motion equations [27], which shows some promise in reducing the impact of measurement noise and numerical residual error,

$$\hat{\dot{\theta}} = \hat{\omega} + k_\theta e \tag{A.11.7}$$

$$\hat{\dot{\omega}} = \hat{\alpha} + k_\omega e \tag{A.11.8}$$

$$\hat{\dot{\alpha}} = k_\alpha e \tag{A.11.9}$$

where $\hat{\theta}$, $\hat{\omega}$ and $\hat{\alpha}$ are the estimated position, angular speed and angular acceleration respectively; e is the estimation error. The transfer function of the error dynamics of the system can be expressed as (A.11.10). This is a third-order system and it is stable if the gains k_θ, k_ω and k_α are positive.

$$\frac{e}{\dot{\alpha}} = \frac{1}{s^3 + k_\theta s^2 + k_v s + k_\alpha} \tag{A.11.10}$$

In the digital implementation, (A.11.7)–(A.11.9) can be expressed in the discrete domain as (A.11.11)–(A.11.13).

$$\hat{\theta}(n+1|n) = \hat{\theta}(n|n-1) + \hat{\omega}(n|n-1)T + k_\theta e(n)T \tag{A.11.11}$$

$$\hat{\omega}(n+1|n) = \hat{\omega}(n|n-1) + \hat{\alpha}(n|n-1)T + k_\omega e(n)T \tag{A.11.12}$$

$$\hat{\alpha}(n+1|n) = \hat{\alpha}(n|n-1) + k_\alpha e(n)T \tag{A.11.13}$$

Based on (A.11.6) and (A.11.11) through (A.11.13), the position can be estimated with a three-order PLL, which will reduce the impact of measurement noise and numerical residual errors. The diagram of the third order PLL is shown in Figure A.11.1.

If $\partial\theta/\partial\lambda_k$ item is removed from (A.11.6), and the PLL is reduced to the second order, the position estimation is reformatted as (A.11.14) through (A.11.16). This method is known as the flux observer-based position estimation.

$$e(n) = \left(\lambda_k(n) - \hat{\lambda}_k(n|n-1)\right) \tag{A.11.14}$$

$$\hat{\theta}(n+1|n) = \hat{\theta}(n|n-1) + \hat{\omega}(n|n-1)T + k_\theta e(n)T \tag{A.11.15}$$

$$\hat{\omega}(n+1|n) = \hat{\omega}(n|n-1) + k_\omega e(n)T \tag{A.11.16}$$

which is a typical flux observer based position estimator. If the third-order PLL is removed, position estimation can be reformatted as (A.11.17). This method is known as a numerical method based direct calculation method.

$$\hat{\theta}(k) = \hat{\theta}(k-1) + \frac{\partial\theta}{\partial\lambda_k}\bigg|_{i_k(n)} \left(\lambda_k(n) - \hat{\lambda}_k(k-1)\right) \tag{A.11.17}$$

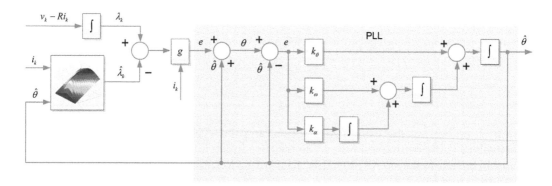

FIGURE A.11.1
Block diagram of the third-order PLL.

Therefore, the third-order PLL method in (A.11.11) through (A.11.13) is a combination of the observer-based method and direct calculation method.

Appendix B: Sliding Mode Observer (SMO) Based Method

In this section, sliding mode observer will be briefly introduced. The control diagram of an SRM controller with SMO is shown in Figure A.11.2. The control diagram of SRM based on SMO is similar to the one shown in Figure 11.8. The observer block will be the sliding mode observer.

First of all, an appropriate sliding-mode surface should be selected. The flux-linkage estimation error e_k is defined as the difference between measured flux linkage and estimated flux linkage:

$$e_k(t) = \lambda_k(t) - \hat{\lambda}_k(t) \tag{A.11.18}$$

The sliding-mode surface is selected as the total flux-linkage estimation error.

$$s = p\, e_k(t) \tag{A.11.19}$$

where p is a factor that keeps s being monotonically increasing with θ for a certain current. p is defined as

$$p = \begin{cases} 1, \hat{\theta} \in (180,360] \\ -1, \hat{\theta} \in (0,180] \end{cases} \tag{A.11.20}$$

Then the observer can be designed as follows.

$$\dot{\hat{\theta}}(t) = \hat{\omega}(t) + k_\theta\, \mathrm{sgn}(s) \tag{A.11.21}$$

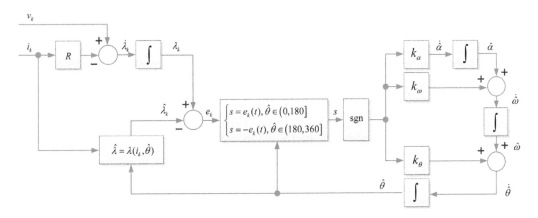

FIGURE A.11.2
Block diagram of the sliding-mode observer.

$$\dot{\hat{\omega}}(t) = \hat{\alpha}(t) + k_\omega \, \mathrm{sgn}(s) \qquad\qquad\qquad (A.11.22)$$

$$\dot{\hat{\alpha}}(t) = k_\alpha \, \mathrm{sgn}(s) \qquad\qquad\qquad (A.11.23)$$

where $\hat{\theta}$ and $\hat{\omega}$ are the estimated rotor position and speed, respectively; k_ω, k_θ, and k_α are SMO gains. The detailed selection guides of SMO gains can be found in [33]. To reduce the chattering, the saturation function is used to replace the signum (sgn) function in the observer. Assuming the boundary layer of the linear region is μ, a modified observer can be expressed as in (A.11.24) through (A.11.26).

$$\dot{\hat{\theta}}(t) = \hat{\omega}(t) + k_\theta \mathrm{sat}(s) \qquad\qquad\qquad (A.11.24)$$

$$\dot{\hat{\omega}}(t) = \hat{\alpha}(t) + k_\omega \mathrm{sat}(s) \qquad\qquad\qquad (A.11.25)$$

$$\dot{\hat{\alpha}}(t) = k_\alpha \mathrm{sat}(s) \qquad\qquad\qquad (A.11.26)$$

where "sat" function is defined as

$$\mathrm{sat}(s) = \begin{cases} \mathrm{sgn}(s), & |s| > \mu \\ s/\mu, & |s| \le \mu \end{cases} \qquad\qquad\qquad (A.11.27)$$

Questions and Problems

1. What are the three types of position sensorless methods for SRM drives?
2. What are the advantages and disadvantages of inductance based and flux based passive sensorless methods?
3. How does pulse injection method work?
4. How do we estimate the initial position of the SRM drives?
5. An SRM is designed to have a torque-speed curve in Figure Q.11.1. How can we estimate the rotor position when the SRM is running in the operational point A? Can we use the flux-based method?
6. As shown in Figure 11.4, the self-inductance estimation is classified to three regions when the SRM is working the motoring mode. Now we run the same SRM in generating mode. Can we use the same position estimation method?
7. Does the inductance-based estimation method in (11.5) work in the entire speed range?

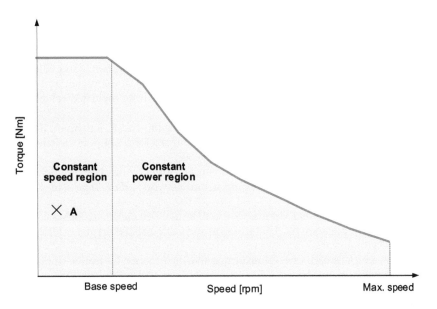

FIGURE Q.11.1
Block diagram of the third-order PLL.

References

1. J. Cai and Z. Deng, Sensorless control of switched reluctance motor based on phase inductance vectors, *IEEE Trans. Power Electron.*, 27(7), 3410–3423, 2012.
2. H. Gao, F. R. Salmasi, and M. Ehsani, Inductance model-based sensorless control of the switched reluctance motor drive at low speed, *IEEE Trans. Power Electron.*, 19(6), 1568–1573, 2004.
3. J. Cai and Z. Deng, A sensorless starting control strategy for switched reluctance motor drives with current threshold, *Electr. Pow. Compo. Sys.*, 41(1), 1–15, 2013.
4. I. H. Al-Bahadly, Examination of a sensorless rotor position measurement method for switched reluctance drive, *IEEE Trans. Ind. Electron.*, 55(1), 288–295, 2008.
5. K. Ha, R. Kim, and R. Krishnan, Position estimation in switched reluctance motor drives using the first switching harmonics through Fourier series, *IEEE Trans. Ind. Electron.*, 58(12), 5352–5359, 2012.
6. G. Pasquesoone, R. Mikail, and I. Husain, Position estimation at starting and lower speed in three-phase switched reluctance machines using pulse injection and two thresholds, *IEEE Trans. Ind. Appl.*, 47(4), 1724–1731, 2013.
7. B. Fahimi, A. Emadi, and R. B. Sepe, Four quadrant position sensorless control in SRM drives over the entire speed range, *IEEE Trans. Power Electron.*, 20(1), 154–163, 2005.
8. E. Ofori, T. Husain, Y. Sozer, and I. Husain, A pulse injection based sensorless position estimation method for a switched reluctance machine over a wide speed range, in *Proceedings of the IEEE Energy Conversion Congress and Expo*, Denver, CO, September 2013, pp. 518–524.
9. L. Shen, J. H. Wu, and S. Y. Yang, Initial position estimation in SRM using bootstrap circuit without predefined inductance parameters, *IEEE Trans. Power Electron.*, 26(9), 2449–2456, 2011.
10. A. Khalil, S. Uderwood, I. Husain, H. Klode, B. Lequesne, S. Gopalakrishnan, and A. Omekanda, Four-quadrant pulse injection and sliding-mode-observer-based sensorless operation of a switched reluctance machine over entire speed range including zero speed, *IEEE Trans. Ind. Appl.*, 43(3), 714–723, 2007.

11. E. Mese and D. A. Torrey, An approach for sensorless position estimation for switched reluctance motors using artificial neural networks, *IEEE Trans. Power Electron.*, 17(1), 66–75, 2002.

12. C. A. Hudson, N. S. Lobo, and R. Krishnan, Sensorless control of single switch-based switched reluctance motor drive using neural network, *IEEE Trans. Ind. Electron.*, 55(1), 321–329, 2008.

13. L. Henriques, L. Rolim, W. Suemitsu, J. Dente, and P. Branco, Development and experimental tests of a simple neuro-fuzzy learning sensorless approach for switched reluctance motors, *IEEE Trans. Power Electron.*, 26(11), 3330–3344, 2011.

14. S. Paramasivam, S. Vijayan, M. Vasudevan, R. Arumugam, and R. Krishnan, Real-time verification of AI based rotor position estimation techniques for a 6/4 pole switched reluctance motor drive, *IEEE Trans. Magn.*, 43(7), 3209–3222, 2007.

15. E. Estanislao, D. X. Juan, C. Roberto, and P. Ruben, Sensorless control for a switched reluctance wind generator, based on current slope and neural networks, *IEEE Trans. Ind. Electron.*, 56(3), 817–825, 2009.

16. J. Ye and A. Emadi, *Systems and Methods for Rotor Position Determination*, McMaster Technology ID 14-059, (Patent Application No. US 14/676,110 and CDN 2,887,080, April 1, 2015), US 9,722,517, August 1, 2017.

17. J. Ye, B. Bilgin, and A. Emadi, An offline torque sharing function for torque ripple reduction of switched reluctance motor drives, *IEEE Trans. Ener. Conv.*, 30(2), 726–735, 2015.

18. J. Ye, B. Bilgin, and A. Emadi, An extended-speed low-ripple torque control of switched reluctance motor drives, *IEEE Trans. Power Electron.*, 30(3), 1457–1470, 2015.

19. J. Ye, B. Bilgin, and A. Emadi, Elimination of mutual flux effect on rotor position estimation of switched reluctance motor drives considering magnetic saturation, *IEEE Trans. Pow. Electron.*, 30(2), 532–536, 2015.

20. J. Ye, B. Bilgin, and A. Emadi, Elimination of mutual flux effect on rotor position estimation of switched reluctance motor drives, *IEEE Trans. Pow. Electron.*, 30(3), 1499–1512, 2015.

21. C. S. Edrington, B. Fahimi, and M. Krishnamurthy, An auto calibrating inductance model for switched reluctance motor drives, *IEEE Trans. Ind. Electron.*, 54(4), 2165–2173, 2007.

22. W. Zeng, C. Liu, Q. Zhou, J. Cai, and L. Zhang, A new flux/current method for SRM rotor position estimation, in *Proceeding of International Conference on Electrical Machines and Systems*, Tokyo, Japan, 2009, pp. 1–6.

23. G. Gallegos-Lopez, P. C. Kjaer, and T. J. E. Miller, High-grade position estimation for SRM drives using flux linkage/current correction model, *IEEE Trans. Ind. Appl.*, 35(4), 859–869, 1999.

24. M. Ehsani, I. Husain, S. Mahajan, and K. R. Ramani, New modulation encoding techniques for indirect rotor position sensing in switched reluctance motors, *IEEE Trans. Ind. Appl.*, 30(1), 85–91, 1994.

25. F. Peng, J. Ye, and A. Emadi, Position sensorless control of switched reluctance motor based on numerical method, in *Proceedings of the IEEE Energy Conversion Congress and Expo (ECCE)*, Milwaukee, WI, September 2016, pp. 1–6.

26. Y. J. Zhan, C. C. Chan, and K. T. Chau, A novel sliding-mode observer for indirect position sensing of switched reluctance motor drives, *IEEE Trans. Ind. Electron.*, 46(2), 390–397, 1999.

27. R. A. McCann, M. S. Islam, and I. Husain, Application of a sliding-mode observer for position and speed estimation in switched reluctance motor drives, *IEEE Trans. Ind. Appl.*, 37(1), 51–58, 2001.

28. S. Mir, M. E. Elbuluk, and I. Husain, Torque-ripple minimization in switched reluctance motors using adaptive fuzzy control, *IEEE Trans. Ind. Appl.*, 35(2), 461–468, 1999.

29. M. S. Islam, I. Husain, R. J. Veillette, and C. Batur, Design and performance analysis of sliding-mode observers for sensorless operation of switched reluctance motors, *IEEE Trans. Control Syst. Technol.*, 11(3), 383–389, 2003.

30. M. Divandari, A. Koochaki, M. Jazaeri, and H. Rastegar, A novel sensorless SRM drive via hybrid observer of current sliding mode and flux linkage, in *Proceedings of the IEEE International Electric Machines & Drives Conference*, Antalya, Turkey, 2007, vol. 1, pp. 45–49.

31. G. Tan, Z. Ma, S. Kuai, and X. Zhang, Four-quadrant position sensorless control in switched reluctance motor drives based on sliding mode observer, in *Proceedings of the IEEE International Conference on Electrical Machines and Systems*, Tokyo, Japan, 2009, pp. 1–5.

32. S. A. Hossain, I. Husain, H. Klode, B. Lequesne, A. M. Omekanda, and S. Gopalakrishnan, Four-quadrant and zero-speed sensorless control of a switched reluctance motor, *IEEE Trans. Ind. Appl.*, 39(5), 1343–1349, 2003.

33. X. Wang, F. Peng, and A. Emadi, A position sensorless control of switched reluctance motor based on sliding-mode observer, in *Proceedings of the IEEE Transportation Electrification Conference (ITEC)*, Dearborn, MI, June 2016, pp. 1–6.

34. T. Shi, C. Xia, M. Wang, and Q. Zhang, Single neural PID control for sensorless switched reluctance motor based on RBF neural network, in *Proceedings of the 6th World Congress on Intelligent Control and Automation*, Dalian, China, 2006, pp. 8069–8073.

35. G. Pasquesoone, R. Mikail, and I. Husain, Position estimation at starting and lower speed in three-phase switched reluctance machines using pulse injection and two thresholds, *IEEE Trans. Ind. Appl.*, 47(4), 1724–1731, 2011.

36. M. Krishnamurthy, C. S. Edrington, and B. Fahimi, Prediction of rotor position at standstill and rotating shaft conditions in switched reluctance machines, *IEEE Trans. Power Electron.*, 21(1), 225–233, 2006.

12

Fundamentals of Vibrations and Acoustic Noise

James Weisheng Jiang and Jianbin Liang

CONTENTS

12.1 Introduction

SRM has a simple geometry and excitation-free rotor construction, which enables more robust operation at high-speed and high-temperature conditions. However, acoustic noise and vibration can be significant concerns in SRMs. Before we discuss the noise and vibration in SRMs, we need to equip ourselves with the fundamentals and tools for noise and vibration studies, which is the core of this chapter.

Time and space varying radial force density acting on the stator core causes vibration. Radial force density over time and space can be treated as a surface wave, or more accurately, a cylindrical wave. Thus, in this chapter, the fundamentals of waves will be discussed first.

The spring-mass models are used to introduce the concepts of natural frequency, damping ratio, forcing frequency, and resonance. As the level of complexity of the structure increases, more complicated mode shapes can appear. Complicated mode shapes can be created based on the mode shapes of simple structures. Therefore, the mode shapes for straight beam, square plate, cylindrical shell, and circular plate are discussed in this chapter.

Resonance can happen when the pattern of a certain forcing harmonic matches the mode shape, and the forcing frequency of the harmonic is close to the natural frequency of the mode shape. In order to obtain the shapes and forcing frequencies of forcing harmonics, fast Fourier transform (FFT) is needed. Towards the end of this chapter, 1D and 2D FFT will be discussed.

Finally, the interaction between forcing harmonics at different orders and mode shapes will be investigated. The fundamentals and the tools discussed in this chapter will be heavily used in the next chapter when analyzing acoustic noise and vibration in SRMs.

12.2 Waves

In an electric machine, radial force density acting on the stator core cause vibrations. The radial force density varies in time and space, and it can be treated as a time- and space-varying surface wave. Therefore, we need to discuss the fundamentals of waves first.

12.2.1 Temporal and Spatial Waves

A wave can vary in the spatial domain and it can also change over time. Figure 12.1a shows a wave traveling over time. Its frequency can be calculated based on the temporal period T:

$$f_{mech} = \frac{1}{T} \text{ [Hz]} \tag{12.1}$$

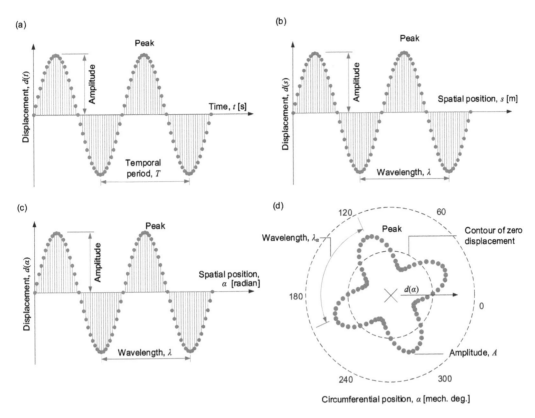

FIGURE 12.1
Examples of one dimensional (1D) wave: (a) time domain in seconds, (b) spatial domain in meters, (c) spatial domain in radians, and (d) polar system in mechanical degrees.

Figure 12.1b shows a snapshot of a wave over spatial domain. It can be seen that the wave is periodic over the wavelength of λ. In SRMs, the wave for radial force density occurs over the air gap circumference. Thus, the wave can be plotted in radians, as shown in Figure 12.1c. Figure 12.1d shows another form of space-varying wave plotted in the polar system in mechanical degrees. In this chapter and the next chapters, almost all the spatial positions are circumferential positions. Please note that in a polar system, counterclockwise is defined as the positive rotational direction.

As shown in Figure 12.2, a surface wave travels both in time and spatial domains. The figure shows four different three-dimensional surface waves. The spatial or circumferential position is expressed in mechanical degrees. The surface wave in Figure 12.2a is varying with space, but fixed over time domain. Two periods can be observed over 360 mechanical degrees. Figure 12.2b shows another surface wave, which is time-varying. Four periods can be seen over 1 second. Figure 12.2c shows a surface wave, which varies over both time and circumferential dimensions. Compared to Figure 12.2c, the surface wave shown in Figure 12.2d travels in the opposite direction. The direction of a wave will be discussed later in this section.

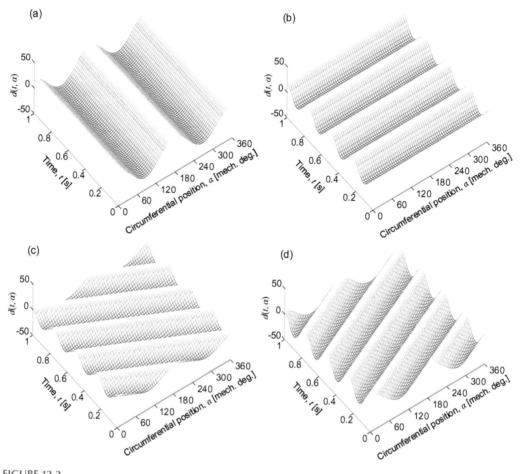

FIGURE 12.2
3D surface wave: (a) space-varying only, (b) time-varying only, (c) space-time-varying, positive direction, and (d) space-time-varying, negative direction.

In electric motors, the force density wave in the airgap also changes with time and space (circumference). Assuming a uniform distribution over the axial direction, Figure 12.3 shows a simple example of how the force wave changes with time and circumferential position. It can be seen that, the waves in Figure 12.3 rotate clockwise (in negative rotational direction). It is assumed here that the waves are uniform over the axial length. Therefore, the 3D version of the waves shown in Figure 12.3 would be as in Figure 12.4. In the case of Figure 12.3, the magnitude of the wave was constant. As shown in Figure 12.5, the magnitude of the wave can change as it rotates circumferentially. Again, since the counterclockwise direction is defined as positive rotational direction, the wave in Figure 12.5 is rotating in the negative direction over time.

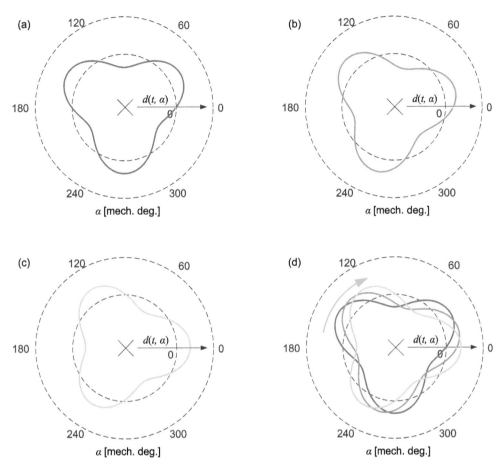

FIGURE 12.3
Surface wave with fixed amplitude over both time and circumference: (a) t_1, (b) t_2, (c) t_3, and (d) t_1, t_2, and t_3.

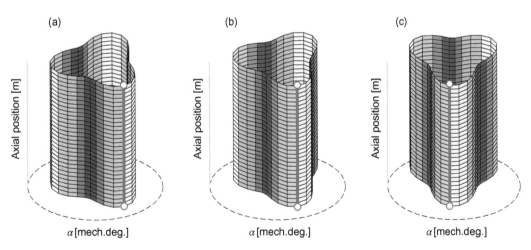

FIGURE 12.4
Surface wave with even axial distribution: (a) t_1, (b) t_2, and (c) t_3.

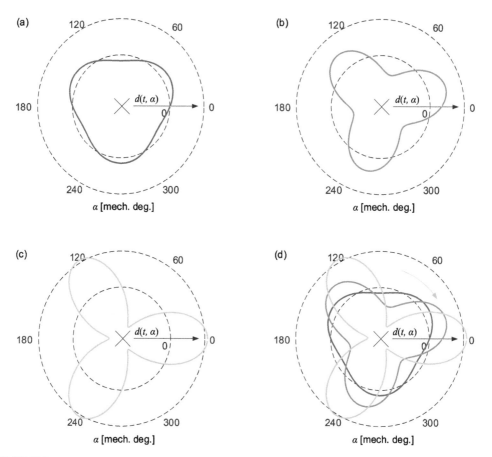

FIGURE 12.5
Progressive wave with time-varying amplitude over both time and circumference: (a) t_1, (b) t_2, (c) t_3, and (d) t_1, t_2, and t_3.

For SRMs, since the phases conduct in sequence, the forces acting on the stator can be seen as a wave traveling in both time and spatial (circumferential) domains. Figure 12.6 shows the combination of both radial and tangential force density at six different positions. Figure 12.6a–c show three positions when phase A is excited. It can be seen that as the rotor rotates, the radial force density grows. Figure 12.6d and e show how the radial and tangential force density grows for the phase B. Then it comes to the excitation of the phase C, as shown in Figure 12.6f.

The force density wave changes over time and circumferential domains, but it is more complicated than a sinusoidal wave. However, the force density wave can still be decomposed into a series of sine or cosine waves using Fast Fourier Transform (FFT). The FFT method will be discussed later in this chapter.

12.2.2 Analytical Expression of 1D Wave

A 1D wave over time domain can be expressed generally as:

$$d(t) = A\cos(\omega_u t + \phi) \tag{12.2}$$

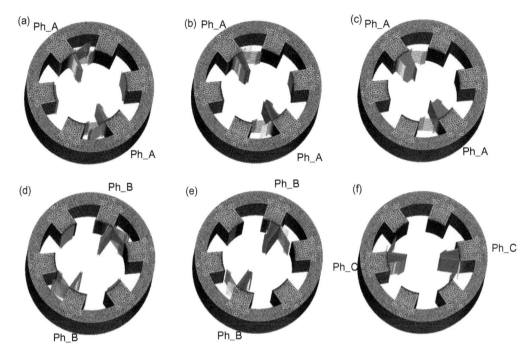

FIGURE 12.6
Nodal force, 6/4 SRM: (a) Ph_A is conducting, t_1, (b) Ph_A is conducting, t_2, (c) Ph_A is conducting, t_3, (d) Ph_B is conducting, (e) Ph_B is conducting, and (f) Ph_C is conducting.

where A is the amplitude of the wave, ω_u is the temporal angular frequency, and ϕ is the phase angle.

The amplitude A can be time-varying or space-varying. But, in this section, the amplitude is considered to have a constant positive value. A minus sign in front of the amplitude represents the wave rotating in the negative direction. The ω_u represents the angular frequency in radians per second. The relationship between the angular frequency ω_u and the mechanical frequency f_{mech} is:

$$\omega_u = 2\pi u f_{mech} \tag{12.3}$$

where u is the temporal order of the wave and f_{mech} is the mechanical frequency for $u = 1$. Please note that u can be negative or positive, or even zero. Hence, the equation for 1D wave can be rewritten as:

$$d(t) = A\cos(2\pi u f_{mech}t + \phi) \tag{12.4}$$

If T is the period of the fundamental ($u = 1$), the mechanical frequency can be expressed as:

$$f_{mech} = \frac{1}{T} \tag{12.5}$$

Figure 12.7 shows a few waves with varying temporal order (u = 1–6). As the temporal order increases, ω_u increases accordingly. By changing the phase angle, the wave shifts in the time domain, as shown in Figure 12.8. The wave for $d_2(t)$ is shifted in the time domain by $\pi/6$ compared to $d_1(t)$.

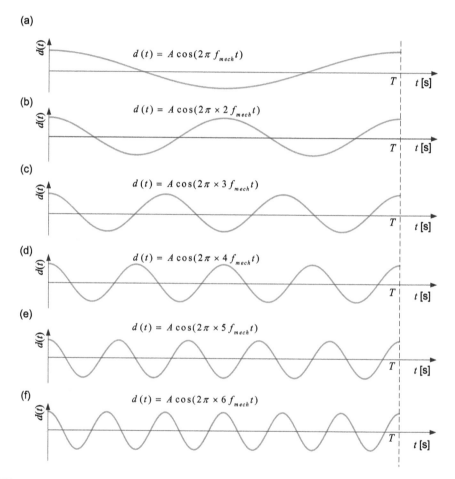

FIGURE 12.7
Waves with varying temporal order: (a) $d(t) = A\cos(2\pi \times 2f_{mech}t)$; (b) $d(t) = A\cos(2\pi \times 2f_{mech}t)$; (c) $d(t) = A\cos(2\pi \times 3f_{mech}t)$; (d) $d(t) = A\cos(2\pi \times 4f_{mech}t)$; (e) $d(t) = A\cos(2\pi \times 5f_{mech}t)$; and (f) $d(t) = A\cos(2\pi \times 6f_{mech}t)$.

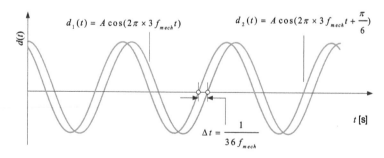

FIGURE 12.8
Wave shifted over time domain.

A 1D wave can also be expressed in the spatial domain. The radial force density is commonly plotted over $(0, 2\pi)$ as a surface wave or a cylindrical wave in the polar system. Thus, the fundamental wavelength for the first (base) circumferential or spatial order ($v = 1$) is 2π:

$$\lambda_{base} = \lambda_1 = 2\pi \tag{12.6}$$

The spatial frequency for $v = 1$ can be expressed as:

$$\xi_{base} = \frac{1}{\lambda_{base}} = \frac{1}{2\pi} \tag{12.7}$$

The wavelength of spatial order, v is:

$$\lambda_v = \frac{2\pi}{v} \tag{12.8}$$

Figure 12.9 shows a 1D waves with different spatial (circumferential) orders. Please note that the unit for the x-axis in Figure 12.9 is radians. As mentioned earlier, the circumferential waves shown in Figure 12.9 can be plotted in a polar system as shown in Figure 12.10. The relationship between the waveforms in Figure 12.9 and the ones in Figure 12.10 can be seen in Figure 12.11.

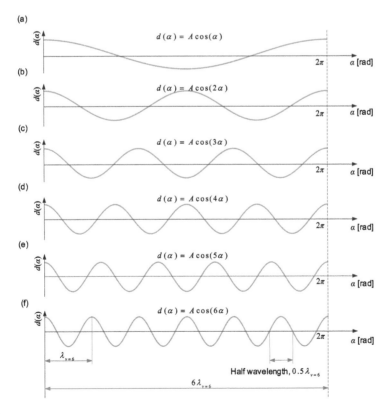

FIGURE 12.9
Waves with varying spatial or circumferential order: (a) $d(t) = A\cos(\alpha)$; (b) $d(t) = A\cos(2\alpha)$; (c) $d(t) = A\cos(3\alpha)$; (d) $d(t) = A\cos(4\alpha)$; (e) $d(t) = A\cos(5\alpha)$; and (f) $d(t) = A\cos(6\alpha)$.

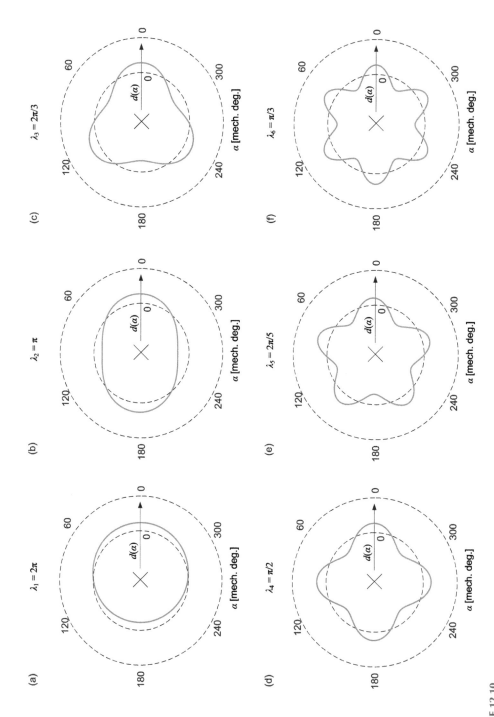

FIGURE 12.10
Six circumferential orders ($v = 1$–6) in a polar system: (a) $\lambda_1 = 2\pi$, (b) $\lambda_2 = \pi$, (c) $\lambda_3 = 2\pi/3$, (d) $\lambda_4 = \pi/2$, (e) $\lambda_5 = 2\pi/5$, and (f) $\lambda_6 = \pi/3$.

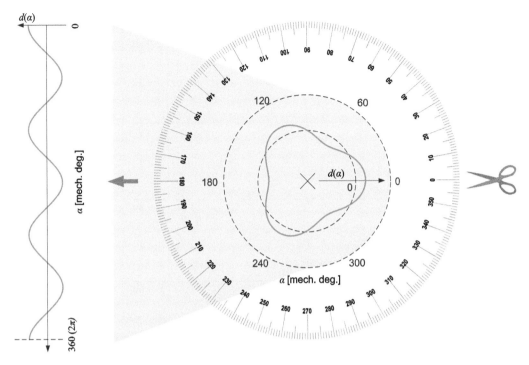

FIGURE 12.11
Conversion of a wave from polar system to Cartesian coordinate system.

For some applications, half waves are used, as shown in Figure 12.12. In this chapter, half waves are used to express the axial mode, which describes the pattern of axial deformation for straight beam or cylindrical shell. The axial order in this chapter and the next is denoted as ax, which is an integer similar to the temporal order u and spatial order v. If the total length of the first axial mode wave is l ($\lambda_{base} = \lambda_{ax=1} = l$), the wavelength for a certain axial mode, ax, can be expressed as:

$$\lambda_{ax} = \frac{l}{ax} \tag{12.9}$$

The analytical expression for axial mode shape with supported-supported constraint is given as:

$$d(x) = \sin\left(\frac{ax \cdot \pi \cdot x}{l}\right) \tag{12.10}$$

12.2.3 Analytical Expression of Surface Waves

Before the discussion of analytical expression for surface waves, the data structure of a surface wave needs to be defined. The data points for a surface wave are structured as shown in Figure 12.13. Each row stores space-varying data for a certain time point.

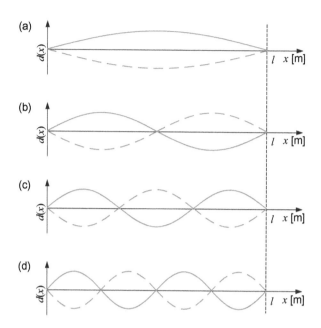

FIGURE 12.12
Half waves with axial order $ax = 1, 2, 3,$ and 4: (a) $ax = 1$; (b) $ax = 2$; (c) $ax = 3$; and (d) $ax = 4$.

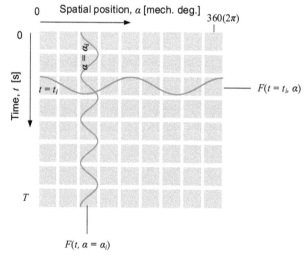

FIGURE 12.13
Data structure for a surface wave.

Each column represents time-varying data. From left to right, the circumferential position in mechanical degree (in radians) increases. From top to bottom, time increases. Then the time dimension can be replaced with rotor position in mechanical degree or electrical degree. The data in Figure 12.13 is used to plot 3D surface waveforms in the format as shown in Figure 12.14.

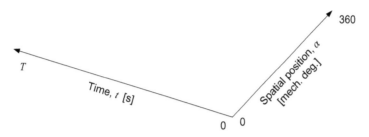

FIGURE 12.14
3D format of surface wave.

Traditionally, a surface wave is expressed as:

$$d(t,\alpha) = A\cos(\omega_u t - v\alpha + \phi) \tag{12.11}$$

The minus sign before $v\alpha$ terms represents that the wave rotates in the counter clockwise direction (positive rotational direction). To simplify the analysis, the version with a plus sign will be used, which shows that the wave is rotating in clockwise direction:

$$d(t,\alpha) = A\cos(\omega_u t + v\alpha + \phi) \tag{12.12}$$

The expression in (12.12) can be reorganized as:

$$d(t,\alpha) = A\cos(\omega_u t + v\alpha + \phi)$$
$$= A\cos(u\omega_{mech}t + v\alpha + \phi)$$
$$= A\cos(2\pi u f_{mech}t + v\alpha + \phi) \tag{12.13}$$

As discussed earlier, u is the temporal order, and v is the spatial or circumferential order. For a certain plane or cylindrical wave, the order of the wave can be expressed as (u, v). Since u and v can both be any integer (negative, zero, or positive), the order can be located in a v-u coordinate system, as shown in Figure 12.15. Please note that u comes before v in the expression of the order (u, v) because this arrangement makes the MATLAB® programming simpler. This notation is similar to that of a cell in a matrix, as shown in Figure 12.16.

For a certain surface wave, if its temporal order u equals 4, its spatial order v equals 2, and its phase angle is zero, its analytical expression can be written using (12.13):

$$d(t,\alpha) = A\cos(2\pi \times 4 f_{mech}t + 2\alpha) \tag{12.14}$$

If the amplitude of the surface wave is 2 and the mechanical frequency is 1 Hz, the analytical expression in (12.14) reduces to:

$$d(t,\alpha) = 2\cos(2\pi \times 4t + 2\alpha) \tag{12.15}$$

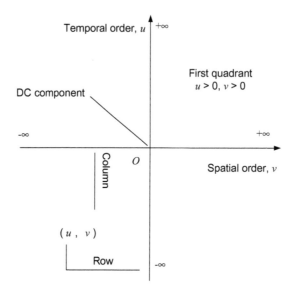

FIGURE 12.15
Expression of temporal order and spatial order in *v-u* coordinates.

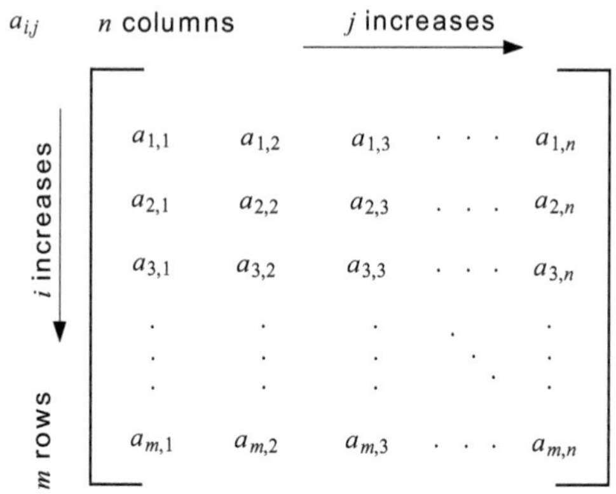

$a_{i,j}$ *n* columns   *j* increases

$$
\begin{bmatrix}
a_{1,1} & a_{1,2} & a_{1,3} & \cdots & a_{1,n} \\
a_{2,1} & a_{2,2} & a_{2,3} & \cdots & a_{2,n} \\
a_{3,1} & a_{3,2} & a_{3,3} & \cdots & a_{3,n} \\
\vdots & \vdots & \vdots & \ddots & \vdots \\
a_{m,1} & a_{m,2} & a_{m,3} & \cdots & a_{m,n}
\end{bmatrix}
$$

i increases

m rows

FIGURE 12.16
An m-by-n matrix.

The 3D figure and 2D contour map for this surface wave can be seen in Figure 12.17. At a certain time, two peaks and two troughs can be observed over the spatial domain. At a certain circumferential position, four peaks and four troughs can be observed over the time domain. It is apparent that *u* determines the number of peaks for the temporal dimension

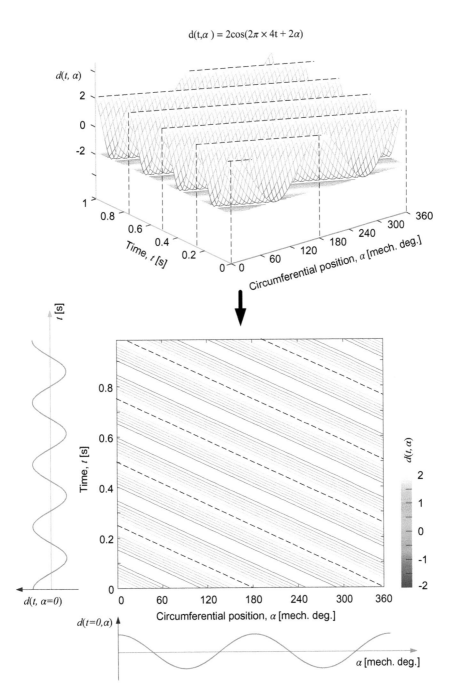

FIGURE 12.17
Illustration for surface wave $d(t,\alpha) = 2\cos(2\pi \times 4t + 2\alpha)$.

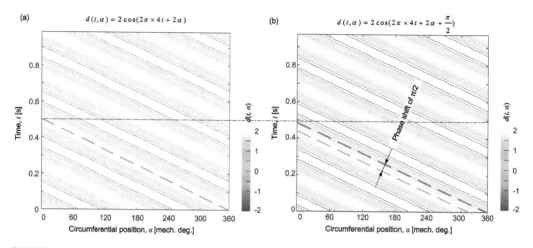

FIGURE 12.18

Effect of phase angle on the location of (progressive) surface wave: (a) before phase shift and (b) after phase shift.

and v determines the number of peaks for the spatial dimension. The signs of u and v also affect the rotational direction of the wave, which will be explained in the next section. Figure 12.18 shows that the phase angle only affects the location of the surface wave. As compared to the waveform in Figure 12.18a, the one in Figure 12.18b is shifted by $\pi/2$.

Figure 12.19 shows two special surface waves with either u or v equal to zero. As shown in Figure 12.19a, when u is zero, there is no variation on the wave in the time domain. When v equals zero, as shown in Figure 12.19b, at a specific time, the wave over the circumference is uniform.

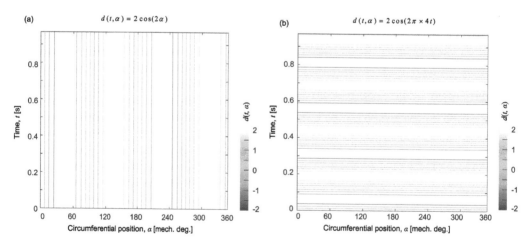

FIGURE 12.19

Two special surface waves: (a) $u = 0$, $v = 2$ and (b) $u = 4$, $v = 0$.

12.2.4 Rotational Direction of a Circumferential Wave

As explained in Figure 12.3, at each time step, the circumferential wave can be plotted in a polar coordinate system. If the peaks in the spatial or circumferential domain travel with time, the wave rotates in the polar system when time increases. This is illustrated in Figure 12.20. As the time increases from t_0 to t_3, the oval shape rotates in the clockwise direction (or in the negative rotational direction). The rotational direction can be determined by the sign of the temporal order v and spatial order ν: $\mathrm{sgn}(u)\,\mathrm{sgn}(v)$. If $\mathrm{sgn}(u)\,\mathrm{sgn}(v)$ is positive, the rotational direction is negative (CW). If $\mathrm{sgn}(u)\,\mathrm{sgn}(v)$ is negative, the rotational direction is positive (CCW). The rotational directions for orders in four quadrants of u-v system can be found in Figure 12.21.

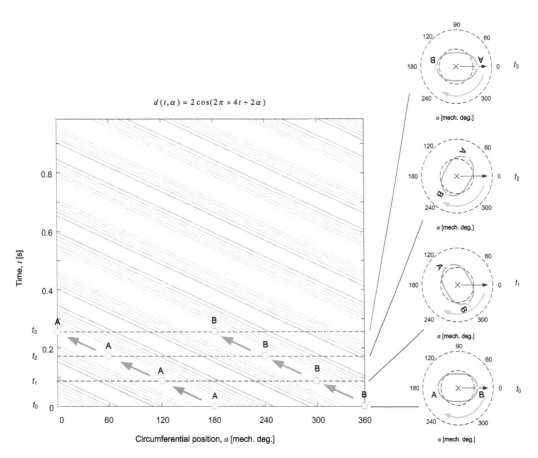

FIGURE 12.20
Illustration of rotational direction for a surface wave.

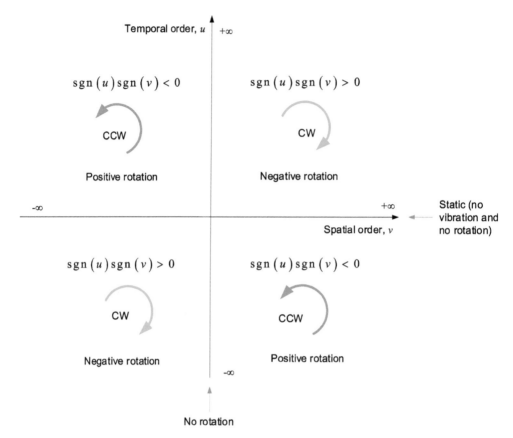

FIGURE 12.21
The rotational directions in four quadrants.

12.3 Fundamentals of Mechanical Vibrations

12.3.1 Undamped Spring-Mass Model

In order to study the interaction between forces and the motor structure, first we need to discuss some basic models for mechanical vibrations. Using these models, we will discuss the concept of oscillation, natural frequency, damping, forcing frequency, and resonance. Figure 12.22 shows three positions for an undamped spring-mass system with one degree of freedom. A mass m is attached to the end of a spring. The mass of the spring is neglected. Initially, the damping ratio of the spring is set to zero.

When the system is disturbed in the x direction, for instance by pulling slightly on the mass, the mass will start oscillating at a certain frequency. The equation of motion for this system can be expressed as:

$$m\ddot{x}(t) + kx(t) = 0 \tag{12.16}$$

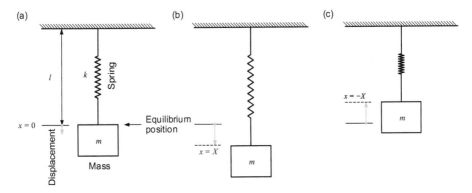

FIGURE 12.22
Undamped spring-mass system: (a) static equilibrium position, (b) lowest position, and (c) highest position.

where m is the mass, k is the stiffness of the spring, $x(t)$ is the displacement of the mass over time, and $\ddot{x}(t)$ is the acceleration of the mass. The term $kx(t)$ is the force exerted on the mass according to Hooke's law. Equation (12.16) can be reorganized as:

$$\ddot{x}(t) + \frac{k}{m}x(t) = 0 \tag{12.17}$$

The term k/m can be replaced by ω_n ($\omega_n = \sqrt{k/m}$), which is a constant for the system and will be discussed later. Thus, (12.17) can be expressed as:

$$\ddot{x}(t) + \omega_n^2 x(t) = 0 \tag{12.18}$$

The differential equation in (12.18) can be solved with two initial conditions:

$$\begin{cases} x(t) = X\cos(\omega_n t - \phi) \\ x(0) = x_0 \\ \dot{x}(0) = v_0 \end{cases} \tag{12.19}$$

where $x(0)$ is the initial displacement (it is the displacement of the initial pull), $\dot{x}(0)$ is the initial speed (the speed the moment when the mass is released), X is the amplitude, and ϕ is the phase angle of the oscillation. From the expression of $x(t)$ in (12.19), it can be seen that the motion of the mass is the projection of a circular motion on the x direction with constant angular frequency ω_n, as shown in Figure 12.23. The velocity of the mass can be obtained as [1]:

$$\dot{x}(t) = -\omega_n X\sin(\omega_n t - \phi) \tag{12.20}$$

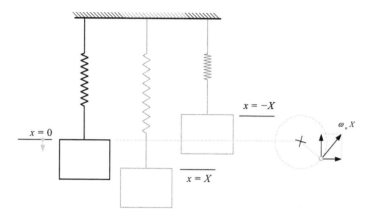

FIGURE 12.23
Relationship between motions for undamped spring-mass system and a constant-frequency circular motion.

It is noteworthy that ω_n [rad/s] is the angular frequency of the constant-frequency circular motion. Now, the time for the undamped spring-mass system to complete one vibrational cycle, or for the constant-frequency circular motion to finish one revolution, can be calculated as:

$$\tau = \frac{2\pi}{\omega_n} = \frac{2\pi}{\sqrt{\dfrac{k}{m}}} \tag{12.21}$$

Thus, the natural frequency, f_n [Hz], of the spring-mass system can be derived as:

$$f_n = \frac{1}{2\pi}\sqrt{\frac{k}{m}} \tag{12.22}$$

It can be noticed from (12.22) that, the system's natural frequency is independent of the magnitude and the phase angle. The natural frequency is determined by the mass and the stiffness of the spring. The amplitude X can be calculated from the constant ω_n and the initial conditions:

$$X = \sqrt{x_0^2 + \left(\frac{v_0}{\omega_n}\right)^2} \tag{12.23}$$

Figure 12.24 shows the change in the amplitude of the oscillation of the undamped spring-mass system when the mass is pulled in the x direction for $x_0 = X$, as shown in Figure 12.24. When the mass is released, the initial speed is zero, $v_0 = 0$. The phase angle can be calculated as:

$$\phi = \tan^{-1}\left(\frac{v_0}{x_0\omega_n}\right) \tag{12.24}$$

For $v_0 = 0$, the phase angle equals zero.

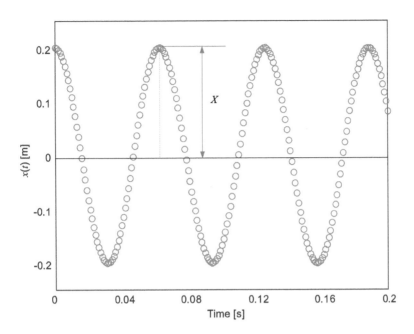

FIGURE 12.24
Motion of an undamped spring-mass system with phase angle equal to zero degree.

12.3.2 Spring-Mass System with Viscous Damping

In practice, the oscillation discussed in the previous section cannot be perpetual because losses will damp the system and, hence, the oscillation of the system will decay. After a disturbance, the mass will be bouncing up and down, but meanwhile the mass is trying to return to its static equilibrium position. Viscous damping force can be modeled as:

$$F = -c\dot{x} \tag{12.25}$$

where c is the damping coefficient and $\dot{x}$ is the velocity of the mass. The damping force is proportional to the velocity, but acting in the opposite direction. Thus, the differential equation for the damped spring-mass system shown in Figure 12.25, can be expressed as:

$$m\ddot{x}(t) + c\dot{x}(t) + kx(t) = 0 \tag{12.26}$$

Equation (12.26) can be rewritten as:

$$\ddot{x}(t) + 2\varsigma\omega_n\dot{x}(t) + \omega_n^2 x(t) = 0 \tag{12.27}$$

where $\omega_n = \sqrt{k/m}$, and ς is the damping ratio, which can be expressed as:

$$\varsigma = \frac{c}{2m\omega_n} = \frac{c}{2\sqrt{mk}} \tag{12.28}$$

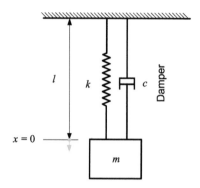

FIGURE 12.25
Damped spring-mass system.

There are four cases of damping depicted in Figure 12.26 and they are related to the value of ς. When $\varsigma = 0$, the system is the undamped spring-mass system, which was discussed in the previous section. When $0 < \varsigma < 1$, the system is underdamped and the oscillation decays over time. When $\varsigma > 1$, this case is referred to as an overdamped system where the system's response doesn't have an overshoot. When $\varsigma = 1$, the system is called critically damped, which is between the overdamped and underdamped cases.

When $0 < \varsigma < 1$, the solution of the damped spring-mass system can be expressed as [1]:

$$x(t) = Ce^{-\varsigma\omega_n t}\cos\left(\omega_d t - \phi\right) \tag{12.29}$$

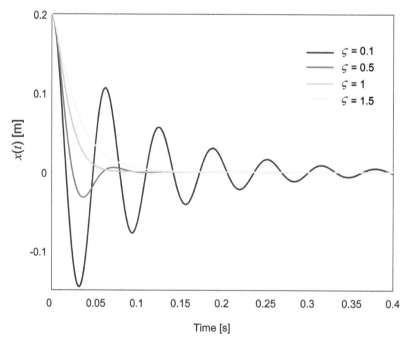

FIGURE 12.26
Damped spring-mass system with varying damping ratio, zero initial speed.

where C is the amplitude. ω_d is called the damped circular frequency and ϕ is the phase angle. $e^{-\varsigma\omega_n t}$ acts as the decaying factor. ω_d can be calculated as:

$$\omega_d = \sqrt{1-\varsigma^2}\,\omega_n \tag{12.30}$$

When $0 < \varsigma < 1$, ω_d is always smaller than or equal to ω_n. The amplitude and phase angle can be calculated as:

$$\begin{cases} C = \sqrt{x_0^2 + \left(\dfrac{\varsigma\omega_n x_0 + v_0}{\omega_d}\right)^2} \\ \phi = \tan^{-1}\left(\dfrac{\varsigma\omega_n x_0 + v_0}{\omega_d x_0}\right) \end{cases} \tag{12.31}$$

When $\varsigma = 1$, the displacement can be expressed as:

$$\begin{cases} x(t) = e^{-\omega_n t}\left[x_0 + (v_0 + \omega_n x_0)t\right] \\ x(0) = x_0 \\ \dot{x}(0) = v_0 \end{cases} \tag{12.32}$$

It can be seen from (12.32) that $x(t)$ has a decaying component but has no oscillating term. Figure 12.27 shows how the initial conditions affect the shape of the displacement for the

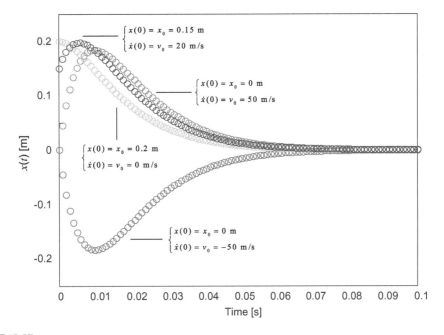

FIGURE 12.27
Response of damped spring-mass system with damping ratio equal to one and varying initial conditions.

system. It can be seen that whenever the initial speed is not equal to zero, there is only one clear undershoot of the response, but no subsequent overshoots.

When the system is overdamped ($\varsigma > 1$), the solution is given by [1,2]:

$$x(t) = C_1 e^{\left(-\varsigma+\sqrt{\varsigma^2-1}\right)\omega_n t} + C_2 e^{\left(-\varsigma-\sqrt{\varsigma^2-1}\right)\omega_n t} \tag{12.33}$$

where C_1 and C_2 can be expressed by using the initial conditions:

$$\begin{cases} C_1 = \dfrac{x_0\omega_n\left(\varsigma+\sqrt{\varsigma^2-1}\right)+v_0}{2\omega_n\sqrt{\varsigma^2-1}} \\[4mm] C_2 = \dfrac{-x_0\omega_n\left(\varsigma-\sqrt{\varsigma^2-1}\right)-v_0}{2\omega_n\sqrt{\varsigma^2-1}} \\[4mm] x(0) = x_0 \\[2mm] \dot{x}(0) = v_0 \end{cases} \tag{12.34}$$

When ς is much greater than 1, the system is heavily damped, and the solution can be approximated by [2,3]:

$$x(t) = x_0 + \frac{v_0}{2\varsigma\omega_n}\left(1-e^{-2\varsigma\omega_n t}\right) \tag{12.35}$$

12.3.3 Forced Damped Spring-Mass System

When a periodic force with a certain frequency acts on the damped spring-mass system, as shown in Figure 12.28, the system can be called as a driven damped harmonic oscillator or a forced damped harmonic oscillator. The periodic force can be expressed as:

$$F(t) = F_0 \sin\left(\omega_f t\right) \tag{12.36}$$

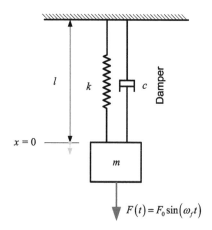

FIGURE 12.28
Forced damped spring-mass system.

where F_0 is the force amplitude and ω_f is the forcing frequency.

It is assumed that the external force acts on the mass when the object is at the equilibrium position. The motion of the forced damped spring-mass system can be expressed by:

$$\ddot{x}(t) + 2\varsigma\omega_n\dot{x}(t) + \omega_n^2 x(t) = \frac{F_0}{m}\sin(\omega_f t) \tag{12.37}$$

The solution of (12.37) is a combination of transient response and steady state response [2,3]:

$$x(t) = Xe^{-\varsigma\omega_n t}\sin(\omega_d t + \phi) + \frac{F_0}{k}\frac{\sin(\omega_f t - \phi_s)}{\sqrt{\left[1 - \left(\frac{\omega_f}{\omega_n}\right)^2\right]^2 + \left(2\varsigma\frac{\omega_f}{\omega_n}\right)^2}} \tag{12.38}$$

where ϕ_s is the phase angle at the steady state and it can be calculated as:

$$\phi_s = \arctan\left[\frac{2\varsigma\frac{\omega_f}{\omega_n}}{1 - \left(\frac{\omega_f}{\omega_n}\right)^2}\right] \tag{12.39}$$

r can be defined as the ratio of the forcing frequency to the natural frequency:

$$r = \frac{\omega_f}{\omega_n} \tag{12.40}$$

The amplitude or maximum displacement at the steady-state can be given as:

$$D = \frac{F_0/k}{\sqrt{\left[1 - \left(\frac{\omega_f}{\omega_n}\right)^2\right]^2 + \left(2\varsigma\frac{\omega_f}{\omega_n}\right)^2}} \tag{12.41}$$

It can be seen from (12.41) that even though the periodic force is a function of time, its maximum displacement is proportional to the force amplitude. Since $k = \omega_n^2 m$, (12.41) can be rewritten as:

$$D = \frac{F_0/m}{\sqrt{\left(\omega_n^2 - \omega_f^2\right)^2 + 4\varsigma^2\omega_n^2\omega_f^2}} \tag{12.42}$$

The ratio of the displacement at state-state in (12.41) to F_0/k is defined as the magnification factor, MF:

$$MF = \frac{1}{\sqrt{\left[1 - \left(\frac{\omega_f}{\omega_n}\right)^2\right]^2 + \left(2\varsigma\frac{\omega_f}{\omega_n}\right)^2}} \tag{12.43}$$

Equation (12.43) can be reorganized by using the definition of the ratio of frequencies in (12.40):

$$MF = \frac{1}{\sqrt{\left[1-r^2\right]^2 + \left(2\varsigma r\right)^2}}$$

(12.44)

Figure 12.29 shows the relationship between MF and r with varying damping ratio. It can be seen that as the damping ratio decreases, MF increases significantly when the frequency ratio gets closer to one. The magnification factor reaches its maximum when the following condition is met:

$$\omega_f = \omega_n \sqrt{1 - 2\varsigma^2}$$

(12.45)

This is when the resonance would happen to the system. When resonance occurs, small periodic driving force can lead to oscillations with much larger amplitude. When designing a motor, methods have to be used to avoid resonance. When the system is undamped ($\varsigma = 0$), the magnification factor reaches infinity when $\omega_f = \omega_n$. For an SRM, vibrational energy in the stator-frame system first builds up as the system resonates and then radiates out as sound. Transmissibility is defined as the ratio of the output to input. Transmissibility is greater than 1 when the forcing frequency and natural frequency of the system coincide, and resonance occurs. Therefore, accurate calculation of natural frequencies of different

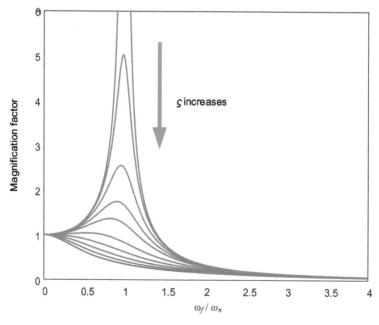

FIGURE 12.29
Relationship between MF and ω_f/ω_n with varying ς.

vibrational modes is essential for the acoustic noise estimation in SRM. In the next sections, we will discuss different vibrational modes and how to calculate their natural frequencies.

12.4 Mode Shapes

Previously, we discussed the simple harmonic oscillator, which has one natural frequency and vibrates in one direction. In reality, an object can vibrate in different manners and have a number of vibration modes. For a specific object, each mode of vibration has a distinct natural (modal) frequency and a mode shape. The shape and dimensions of the object directly affect the modes of vibration. Also, constraints on the object affect the mode shapes and the corresponding natural frequencies as well. The more complex the structure is, the more complicated the modes of vibration can be. In this section, we will explore the modes of vibrations for a few commonly used geometries, such as a straight beam, square plate, cylindrical shell, and circular plate.

12.4.1 Straight Beam

Generally, mode shapes of complex structures can be analyzed using the mode shapes of a straight beam. The mechanism for supporting the beam leads to the definition of the boundary conditions, or constraints, for the two ends of the beam. One end of the beam can be free, supported, or clamped. As the name suggests, if one end of the beam is free, then that end can move and rotate. If the end of the beam is supported, then that end cannot move but can be free to rotate. If the end is clamped, it cannot be moved or rotated. Figure 12.30 shows the combinations of the three types of constraints. If one end of the beam is supported and the other end free, as shown in Figure 12.30a, the end that is supported can rotate freely, but cannot move. If the constraint is changed to be clamped with the other end to be free as shown in Figure 12.30b, the clamped end cannot move or rotate. The function for the mode shapes of clamped-free beam is [4]:

$$d(x) = \cosh(\beta_n x) - \cos(\beta_n x) - \sigma_n \left[\sinh(\beta_n x) - \sin(\beta_n x) \right] \tag{12.46}$$

where n is the axial mode, l is the total length of the beam, and x ($0 \leq x \leq l$) is the axial location. β_n is axial-mode dependent and can be calculated as:

$$\beta_n = \frac{(2n-1)\pi}{2} \tag{12.47}$$

σ_n also depends on the axial mode n, the value of which can be found in Table 12.1. When both ends are supported, as shown in Figure 12.30c, the mode shapes of the beam are more sinusoidal and the function for its mode shape will be [4]:

$$d(x) = \sin\frac{n\pi x}{l} \tag{12.48}$$

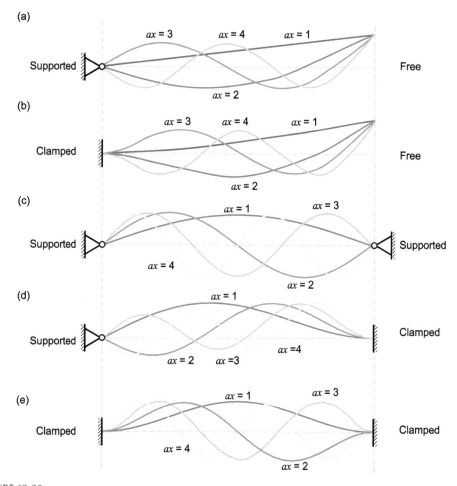

FIGURE 12.30
Mode shapes of a straight beam with different constraints at different axial modes, n: (a) supported-free (s-f),
(b) clamped-free (c-f), (c) supported-supported (s-s), (d) supported-clamped (s-c), and (e) clamped-clamped (c-c).
(From Wang, C.Y. and Wang, C.M., *Structural Vibration: Exact Solutions for Strings, Membranes, Beams, and Plates*,
CRC Press, London, UK, 2013.)

The case in Figure 12.30d is a mixture of two cases shown in Figure 12.30c and e. The
function for the mode shapes of clamped-clamped in Figure 12.30e is similar to the
mode shapes of clamped-free, which is shown in (12.46). The equations for β_n and δ_n

are different for mode shapes of clamped-clamped and clamped-free. For the clamped-clamped case, β_n is also axial-mode dependent [4]:

$$\beta_n = \frac{(2n+1)\pi}{2} \tag{12.49}$$

σ_n for the clamped-clamped case also depends on the axial mode n and the values can be found in Table 12.1. If the constraint shown in Figure 12.30e is used for an electric motor, it means that two ends of the motor are clamped. Figure 12.30b shows the case in which only one end of the motor is clamped, and this arrangement is frequently used when testing electric motors. We will discuss further about these two cases in the following sections.

TABLE 12.1

Functions of Mode Shapes

Type of Constraint	Function of Mode Shapes ($0 \leq x \leq l$)	β_n	σ_n
Clamped-free	$\cosh(\beta_n x) - \cos(\beta_n x)$ $-\sigma_n\left[\sinh(\beta_n x) - \sin(\beta_n x)\right]$	$\dfrac{(2n-1)\pi}{2}$	0.7341 ($n=1$) 1.0185 ($n=2$) 0.9992 ($n=3$) 1 ($n>3$)
Supported-supported	$\sin\dfrac{n\pi x}{l}$	N/A	N/A
Clamped-clamped	$\cosh(\beta_n x) - \cos(\beta_n x)$ $-\sigma_n\left[\sinh(\beta_n x) - \sin(\beta_n x)\right]$	$\dfrac{(2n+1)\pi}{2}$	0.982502 ($n=1$) 1.00078 ($n=2$) 0.999966 ($n=3$) 1 ($n>3$)

Source: Wang, C.Y., and C.M. Wang. *Structural Vibration: Exact Solutions for Strings, Membranes, Beams, and Plates,* CRC Press, London, UK, 2013.

12.4.2 Square Plate

The mode shapes in a two-dimensional plane will be discussed on a square plate example. Figure 12.31 shows a number of modes for a square plate. It is assumed that all four sides of the square plate are supported; hence, the mode shapes of both sides are more sinusoidal. When both dimensions are at mode 1, the square plate vibrates back and forth in the center of the geometry. When the vertical dimension is at its mode 2 and horizontal in mode 1, two vertical halves of the plate vibrate in opposite directions.

When the vertical mode increases to 3, it can be seen that the plate is divided into three almost equal pieces, each vibrating opposite to its adjacent piece. It can be seen from Figure 12.31 that each column has the same horizontal mode. The rows have the same vertical mode. The red and blue colors of the excitation waveforms in the horizontal and vertical dimensions show the pairs.

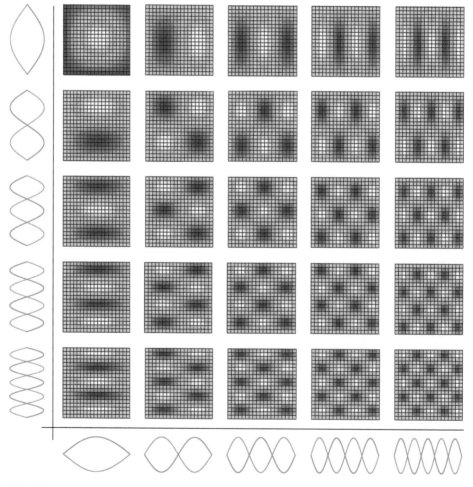

FIGURE 12.31
Mode shapes for square plate, supported-supported. (From Arthur, W.L. and Qatu, M.S., *Vibration of Continuous Systems*, McGraw Hill Professional, New York, 2011.)

12.4.3 Cylindrical Shell

In acoustic noise and vibration analysis, SRMs can be modelled as cylindrical shells. For cylindrical shells, we will use two dimensions: circumferential mode (*circ*) and axial mode (*ax*). Figure 12.32 shows how a cylindrical shell vibrates in different circumferential modes. The axial modes of the cylindrical shell are identical to the modes of a straight beam. Figure 12.33 shows the first three axial modes for the cylindrical shell when the constraint is supported-supported for both ends. If the constraint for the two ends is changed to

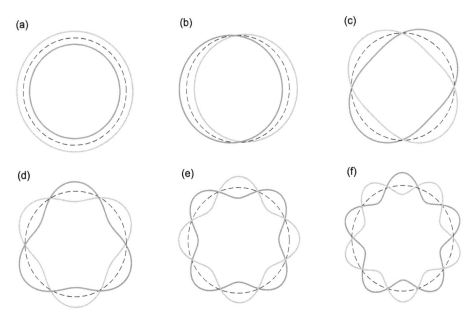

FIGURE 12.32
Circumferential mode shapes for a cylindrical shell: (a) *circ* = 0, (b) *circ* = 1, (c) *circ* = 2, (d) *circ* = 3, (e) *circ* = 4, and (f) *circ* = 5. (From Dunn, F. et al., *Springer Handbook of Acoustics*, Springer, New York, 2015.)

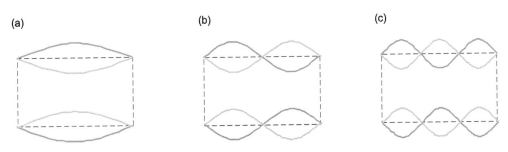

FIGURE 12.33
Axial mode shapes of a cylindrical shell, supported-supported: (a) *ax* = 1, (b) *ax* = 2, and (c) *ax* = 3.

clamped-clamped, the first three mode shapes can be seen in Figure 12.34. When the shell is supported on one end, the deflection of that end is constrained. When the shell is clamped on one end, the deflection and the rotation angle of that end are constrained at the same time. Figure 12.35 shows a closer view of the axial mode shapes for the clamped-clamped and supported-supported cases. It can be observed that in the supported-supported case, the mode shapes are more sinusoidal. If one end of the cylindrical shell is free to vibrate and the other end is clamped, the first three axial modes can be seen in Figure 12.36.

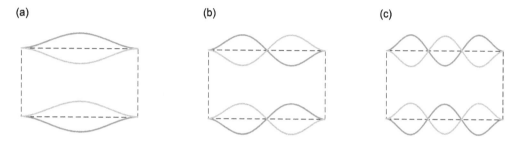

FIGURE 12.34
Axial mode shapes of a cylindrical shell, clamped-clamped: (a) $ax = 1$, (b) $ax = 2$, and (c) $ax = 3$.

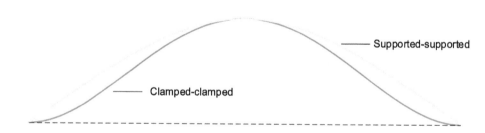

FIGURE 12.35
Comparison of mode shapes of clamped-clamped and supported-supported.

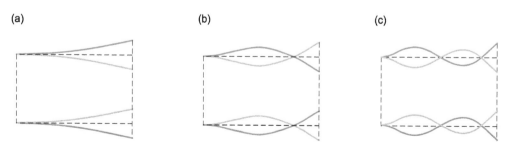

FIGURE 12.36
Axial mode shapes of a cylindrical shell, clamped-free: (a) $ax = 1$, (b) $ax = 2$, and (c) $ax = 3$.

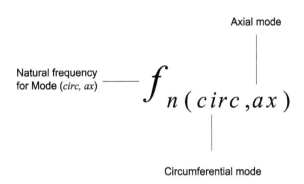

FIGURE 12.37
Notation for natural frequency of mode (*circ, ax*) for cylindrical shell.

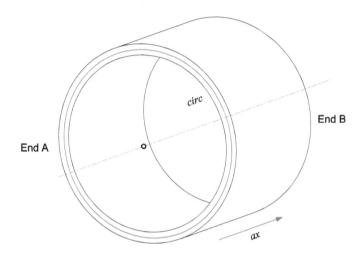

FIGURE 12.38
An example of cylindrical shell with diameter of 300 mm and length of 600 mm.

The notation of the natural frequency for the mode (*circ, ax*) of the cylindrical shell can be seen in Figure 12.37. We will use the cylindrical shell shown in Figure 12.38, as an example to illustrate the mode shapes under two different constraints: clamped-clamped and clamped-free, which are close to the actual support. The cylindrical shell has an outer diameter of 300 mm, a thickness of 10 mm, and an axial length of 600 mm. First, the mode shapes for the cylindrical shell with two ends clamped are studied. The mode shapes and their natural frequencies are calculated using ACTRAN™ software.

12.4.3.1 Clamped-Clamped Constraint

Figure 12.39 shows the first four mode shapes for the cylindrical shell with *circ* equal to 0 and 1, and *ax* equal to 1 and 2. The minimum value of *ax* is 1 rather than 0. This is because *ax* = 0 means that the shell will deform uniformly in the axial direction. However, due to the boundary conditions, the shell at the constraint end would not deform uniformly in the axial direction. Please note that the contour color of the deformation shows the

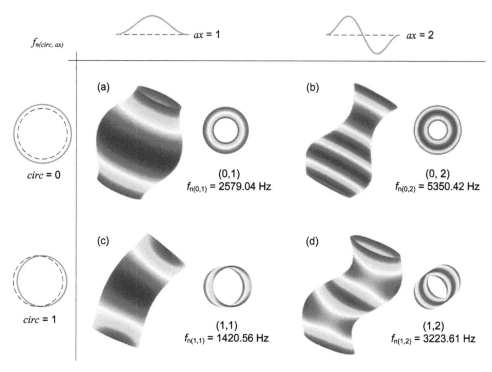

FIGURE 12.39
Mode shapes of a cylindrical shell, clamped-clamped: (a) (0,1), (b) (0,2), (c) (1,1), and (d) (1,2).

absolute value. Therefore, the deformation represented by a contour color can be a hill or a valley. In some figures, this might not be noticeable due to the angle the picture of the mode shapes was captured.

For mode (0,1), as shown in Figure 12.39a, the circumferential mode is zero and the axial mode is 1. The circumferential deformation is concentric when $circ = 0$. When the axial mode increases from 1 (Figure 12.39a) to 2 (Figure 12.39b), the number of peak and trough increases from one to two. When the circumferential mode is one (Figure 12.39c) and (Figure 12.39d), it can be seen that the deformation is eccentric at different axial locations. It should be noted that the results are from 3D Finite Element Analysis (FEA) simulation and they are the snapshot of the mode shapes. If the results were in the form of animation, it could be seen that the peaks and troughs were alternating. When axial mode further increases, as shown in Figure 12.40, the number of peaks and troughs increases accordingly.

When the circumferential mode increases, the number of radial deformations increases. Figure 12.41 shows the vibration modes for the cylindrical shell when $circ$ is 2 and 3. When $circ$ equals 2, two peaks and two troughs can be seen on the circumference of the cylindrical shape, as shown in Figure 12.41a and b. When $circ$ equals 2 and ax equals 2, the number of peak and trough in the axial direction is 2. This mode can be written as (2,2), as shown in Figure 12.41b. Modes (3,1) and (3,2), shown in Figure 12.41c and d respectively, both have 3 peaks and 3 troughs circumferentially. Please note again that the contour color is the absolute value of the deformation. That means if a peak and a tough have the same absolute value of deformation, but in different directions, the contour color for both can be

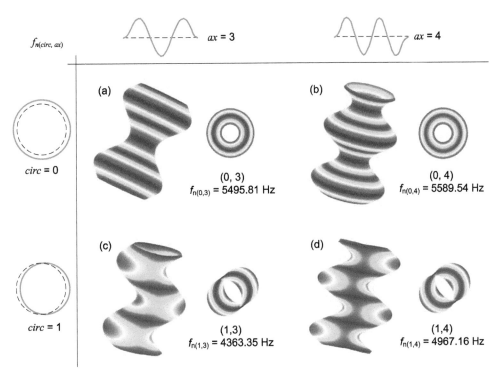

FIGURE 12.40
Mode shapes of a cylindrical shell, clamped-clamped: (a) (0,3), (b) (0,4), (c) (1,3), and (d) (1,4).

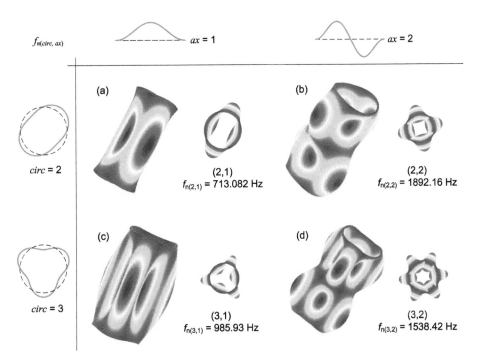

FIGURE 12.41
Mode shapes of a cylindrical shell, clamped-clamped: (a) (2,1), (b) (2,2), (c) (3,1), and (d) (3,2).

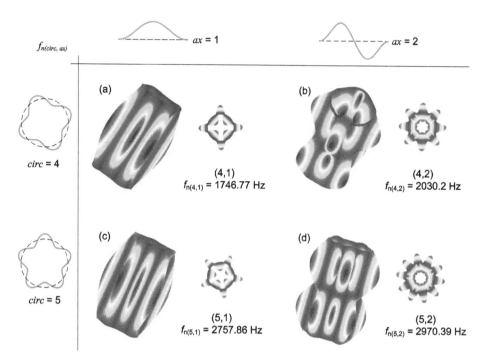

FIGURE 12.42
Mode shapes of a cylindrical shell, clamped-clamped: (a) (4,1), (b) (4,2), (c) (5,1), and (d) (5,2).

identical. Figure 12.42 shows a group of four mode shapes with higher number of circumferential modes. The number of circumferential peaks and troughs further increases with the circumferential mode.

When the axial order increases to 3 and 4, as shown in Figure 12.43, the number of axial peaks and troughs increases accordingly. Figure 12.44 shows the mode shapes when both the axial and circumferential modes are high. It should be noted that the frequency of mode (3,3) (2395.51 Hz) is lower than Mode (2,3) (3031.2 Hz), even though (3,3) has a higher number of circumferential order.

12.4.3.2 Clamped-Free Constraint

We have discussed a number of mode shapes for a cylindrical shell with two ends clamped. However, the motor might be clamped on one end only, like a suspension bridge. Now, we will show another group of mode shapes for the same cylindrical shell with one end clamped. You will notice that the natural frequency and the mode shape for the same order will be affected when the constraint is different.

Figure 12.45 shows the first few mode shapes for the cylindrical shell in Figure 12.38. The only difference is that the cylindrical shell is clamped on one end and the boundary condition for the other end is free. First, the axial mode is examined. It was shown in Figure 12.42a that, if the cylindrical shell is clamped on both ends, a trough or a peak were observed when $ax = 1$. When the cylindrical shell is clamped on only one end, the shell is either tilted upward or downward when $ax = 1$, as shown in Figure 12.45a and c. When $ax = 2$, one trough or peak is added to the mode shape, compared with the case $ax = 1$ (see Figure 12.45b and d). The clamped end of the cylindrical shell basically

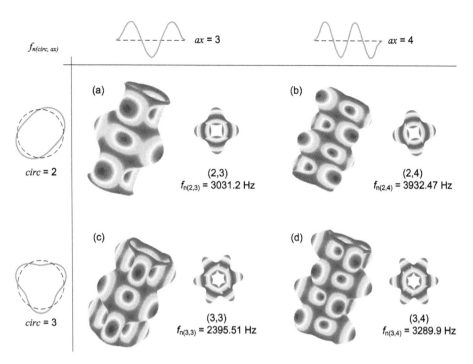

FIGURE 12.43
Mode shapes of a cylindrical shell, clamped-clamped: (a) (2,3), (b) (2,4), (c) (3,3), and (d) (3,4).

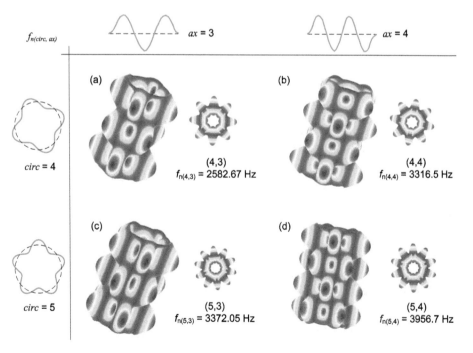

FIGURE 12.44
Mode shapes of a cylindrical shell, clamped-clamped: (a) (4,3), (b) (4,4), (c) (5,3), and (d) (5,4).

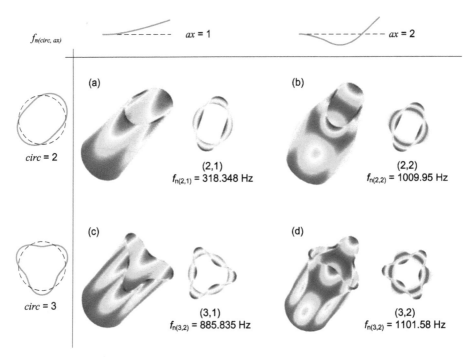

FIGURE 12.45
Mode shapes of a cylindrical shell, clamped-free: (a) (2,1), (b) (2,2), (c) (3,1), and (d) (3,2).

has no deformation. The other end of the cylindrical shell, which has a free boundary condition, is tilted inward or outward. Figure 12.46 shows the mode shapes with higher circumferential modes. The circumferential deformations are similar to that of clamped-clamped structure. Figures 12.47 and 12.48 show the mode shapes for higher circumferential and axial modes. When $ax = 3$, one peak, one trough and one tilting (or slope) can be seen in the axial direction. For $ax = 4$, another peak or trough is added compared to $ax = 3$.

For cylindrical shells, the two constraints discussed, clamped-clamped, and clamped-free, have many similarities. For instance, they both have axial and circumferential modes. The circumferential mode shapes for both constraints are identical. However, the natural frequencies of these constraints are different for the same mode shapes. Table 12.2 compares the natural frequencies for the same modes. It can be seen that the natural frequencies of clamped-clamped constraint are generally greater than clamped-free. This shows us the importance of boundary conditions in acoustic noise modeling and analysis.

SRM can have more vibrational dimensions than just circumferential and axial directions. For instance, the stator can bend, twist in different directions. In addition, the stator teeth can be modelled as suspension and have their own vibrational modes. This is why many vibrational modes can be observed from acoustic FEA simulations if the frequency range is high enough. Because the audible frequency range for human ears is between 20 Hz and 20 kHz, only mode shapes in this frequency range are taken into consideration in our analysis.

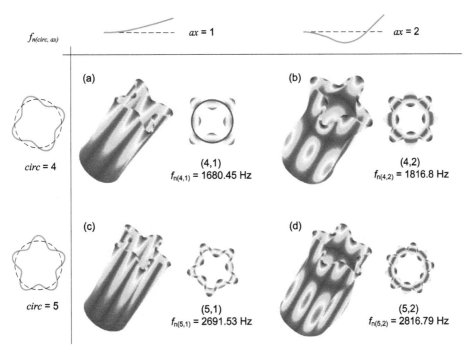

FIGURE 12.46
Mode shapes of a cylindrical shell, clamped-free: (a) (4,1), (b) (4,2), (c) (5,1), and (d) (5,2).

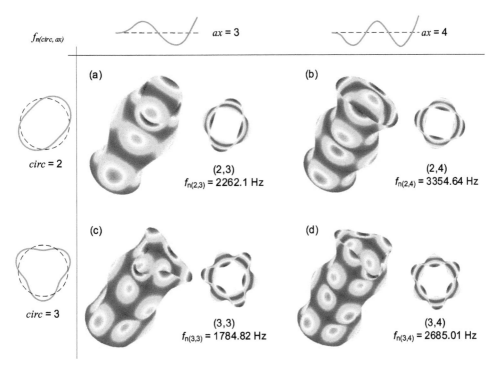

FIGURE 12.47
Mode shapes of a cylindrical shell, clamped-free: (a) (2,3), (b) (2,4), (c) (3,3), and (d) (3,4).

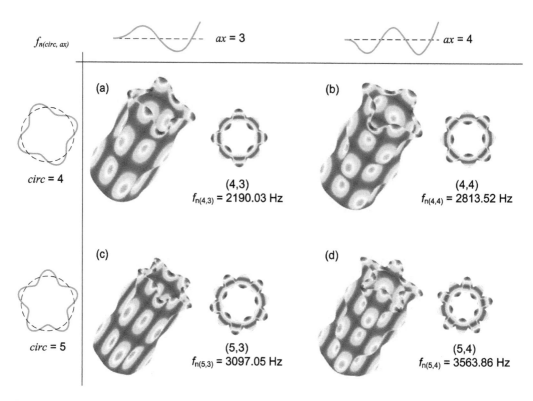

FIGURE 12.48
Mode shapes of a cylindrical shell, clamped-free: (a) (4,3), (b) (4,4), (c) (5,3), and (d) (5,4).

TABLE 12.2

Comparison of Natural Frequencies for Vibration Modes between
Clamped-Clamped (cc) and Clamped-Free (cf)

| Mode (*circ, ax*) | Natural Frequency $f_{n(circ, ax)}$ [Hz] | | Difference [%] (cc–cf)/ cc × 100% |
	Clamped-Clamped, cc	Clamped-Free, cf	
(2,3)	3031.20	2262.10	25.37
(2,4)	3932.47	3354.64	14.69
(3,3)	2395.51	1784.82	25.49
(3,4)	3289.90	2685.01	18.39
(4,1)	1746.77	1680.45	3.80
(4,2)	2030.20	1816.80	10.51
(4,3)	2582.67	2190.03	15.20
(4,4)	3316.50	2813.52	15.17
(5,1)	2757.86	2691.53	2.41
(5,2)	2970.39	2816.79	5.17
(5,3)	3372.05	3097.05	8.16
(5,4)	3956.70	3563.86	9.93

12.4.4 Circular Plate

The end covers of an SRM can be modelled as circular plates. If the diameter of the housing of a motor is greater than its axial length, the housing of the motor can be modelled as a circulate plate, as well. For a circular plate, it has two dimensions: circumferential order (*circ*) and radial order (*r*). A circular plate's circumferential order is similar to that of the cylindrical shell. If a line segment is drawn from the center of the circular plate to the point of the circumference, the deformation of the line segment is identical to a beam with supported-free constraint, as shown in Figure 12.49.

Figure 12.50 shows four mode shapes (*r* = 1, 2 and *circ* = 1, 2) for the circular plate with 300 mm diameter and 10 mm axial thickness. Mode (1,1) has one peak and one trough on its circumference, as shown in Figure 12.50a. Mode (1,2) has two peaks and two troughs in its circumference, as shown in Figure 12.50b. When the radial mode increases to 2, one trough or peak is added to the deformation on the radial direction of the circular plate, as shown in Figure 12.50c and d. Figure 12.51 shows that when the circumferential mode increases further, the number of circumferential troughs and peaks increase accordingly. The knowledge of mode shapes for a circular plate is important, because when analyzing a motor with an axial length much smaller than its diameter, the mode shapes of the circular plates can appear frequently in the low frequency range. Figure 12.52 shows a few mode shapes for the structure of an exterior rotor SRM. The outer diameter of the structure is much greater than its axial length. The end covers of the structure show a few patterns of the mode shapes of a circular plate.

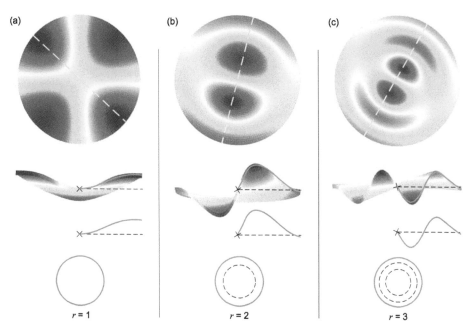

FIGURE 12.49
Illustrations of radial modes for circular plate: (a) *r* = 1, (b) *r* = 2, and (c) *r* = 3.

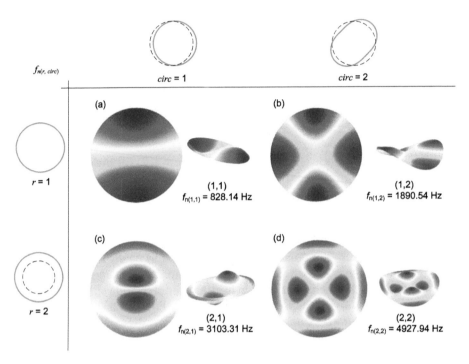

FIGURE 12.50
Mode shapes of a circular plate: (a) (1,1), (b) (1,2), (c) (2,1), and (d) (2,2).

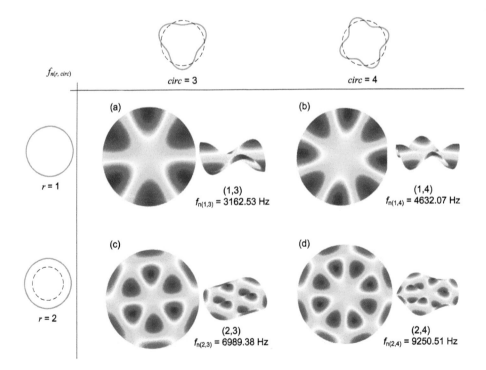

FIGURE 12.51
Mode shapes of a circular plate: (a) (1,3), (b) (1,4), (c) (2,3), and (d) (2,4).

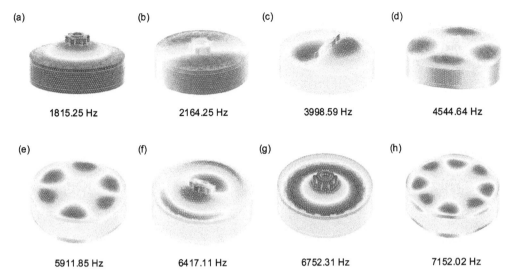

(a)   1815.25 Hz   (b)   2164.25 Hz   (c)   3998.59 Hz   (d)   4544.64 Hz

(e)   5911.85 Hz   (f)   6417.11 Hz   (g)   6752.31 Hz   (h)   7152.02 Hz

FIGURE 12.52
Selected mode shapes of an exterior rotor SRM: (a) (1,0), (b) (1,1), (c) (2,1), (d) (1,2), (e) (2,3), (f) (3,1), (g) (2,0), and (h) (2,4).

As mentioned earlier, as the complexity of the structure increases, the varieties of the mode shapes increases. For instance, if the stator pole of an SRM has high length to width ratio, the stator poles can vibrate significantly in the low frequency range and a number of mode shapes can appear on the stator poles.

12.5 FFT Decomposition

When radial forces act on the structure of an SRM, vibrations are generated resulting in radiated noise. The radial force waveform can be seen as a wave in the circumferential and time domains with a number of harmonics. In order to have a good understanding of the vibration generation, the features of each and every harmonic content have to be investigated. FFT decomposition is such a tool, which can be used to decompose the radial force wave into its harmonic components.

12.5.1 1D FFT

FFT decomposes a wave, periodic or not, into its harmonic components, as shown in Figure 12.53. The wave can be approximated by a collection of cosine and sine functions with varying amplitudes at different harmonic frequencies:

$$f(t) = a_0 + \sum_{u=1}^{\infty} \left[a_u \cos(2\pi f_{mech} u t) + b_u \sin(2\pi f_{mech} u t) \right]$$

$$= a_0 + a_1 \cos(2\pi f_{mech} t) + a_2 \cos(2\pi f_{mech} \times 2t) + a_3 \cos(2\pi f_{mech} \times 3t) + \dots \qquad (12.50)$$

$$+ b_1 \sin(2\pi f_{mech} t) + b_2 \sin(2\pi f_{mech} \times 2t) + b_3 \sin(2\pi f_{mech} \times 3t) + \dots$$

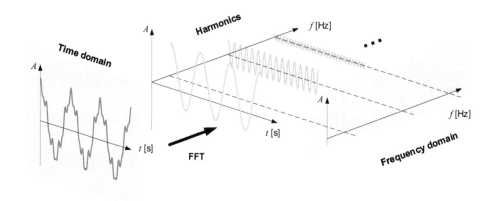

FIGURE 12.53
Mechanism of FFT.

where a_0 is the DC component, and u is the temporal harmonic order, a_u ($u = 1, 2, ...$) is the amplitude for cosine function, b_u ($u = 1, 2, ...$) is the amplitude for sine function, and f_{mech} is the mechanical frequency, which serves as the base frequency as introduced in Section 12.2.2. Since, the cosine and sine series can also be combined into a cosine series with phase angles, (12.50) can be rewritten as:

$$f(t) = C_0 + \sum_{u=1}^{\infty} \left[C_u \cos\left(2\pi f_{mech} u t + \phi_u\right) \right]$$

$$= C_0 + C_1 \cos\left(2\pi f_{mech} t + \phi_1\right) + C_2 \cos\left(2\pi f_{mech} \times 2t + \phi_2\right) + C_3 \cos\left(2\pi f_{mech} \times 3t + \phi_3\right) + ...$$

$$(12.51)$$

where C_0 equals a_0 in (12.50), ϕ_u ($u = 1, 2, ...$) is the phase angle, and C_n ($u = 1, 2, ...$) is the amplitude for cosine function. Phase angle ϕ_u can be calculated as:

$$\phi_u = \arctan\left(-\frac{b_u}{a_u}\right) \qquad (12.52)$$

The amplitude C_u can be obtained as:

$$C_u = \sqrt{a_u^2 + b_u^2} \qquad (12.53)$$

The function in (12.51) can also be expressed using Euler's equation, which is given as:

$$e^{i\theta} = \cos\theta + i\sin\theta \qquad (12.54)$$

Thus, sine and cosine functions can be written as:

$$\begin{cases} \cos\theta = \dfrac{e^{i\theta} + e^{-i\theta}}{2} \\[3mm] \sin\theta = \dfrac{e^{i\theta} - e^{-i\theta}}{2} \end{cases} \qquad (12.55)$$

Hence, the Fourier series in (12.51) can be reorganized using Euler function as:

$$f(t) = C_0 + \sum_{u=1}^{\infty} \left[\frac{C_u}{2} \left(e^{i(2\pi f_{mech}ut+\phi_u)} + e^{-i(2\pi f_{mech}u+\phi_u)} \right) \right]$$

$$= D_0 + \sum_{u=1}^{\infty} \left[D_u e^{i2\pi f_{mech}ut} + D_{-u} e^{-i2\pi f_{mech}ut} \right]$$

(12.56)

Please note that the FFT function in MATLAB renders results in this format. The coefficients in (12.56) are:

$$\begin{cases} D_0 = C_0 \\[2mm] D_u = \dfrac{C_u}{2} e^{i\phi_u} \\[2mm] D_{-u} = \dfrac{C_u}{2} e^{-i\phi_u} \end{cases}$$

(12.57)

Figure 12.54 shows the decomposition of a time-varying wave. The waveform, $f(t)$ is composed of two harmonic orders, 3rd and 15th. The third harmonic order has amplitude of 20. The 15th order has amplitude of 3. Using MATLAB, the harmonic orders and phase angles for all harmonic components of $f(t)$ can be found, as shown in Figure 12.55. Please note that the spectrums of the amplitudes and phase angles are both one-sided ($u \geq 0$). The spectrums can be two-sided, as shown in Figure 12.56. On the left side of the spectrums, u is negative. The amplitudes of the harmonics in two-sided spectrum are half of the amplitudes in the one-sided spectrum. In this chapter and the next chapter, both the one-sided and two-sided spectrums are used. There are a number of questions on this topic at the end of this chapter.

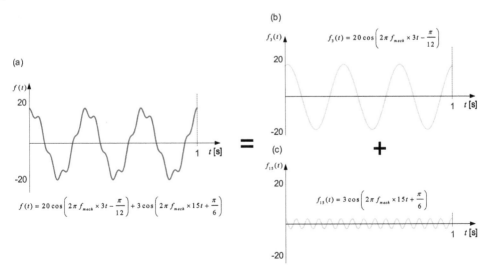

FIGURE 12.54
Decomposition of a time-varying wave: (a) $f(t)$, (b) $f_3(t)$, and (c) $f_{15}(t)$.

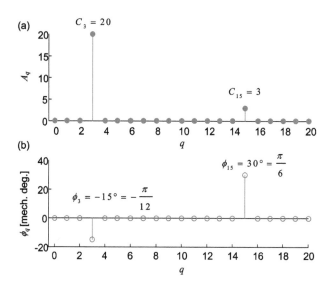

FIGURE 12.55
One-sided spectrum of harmonics: (a) amplitudes and (b) phase angles.

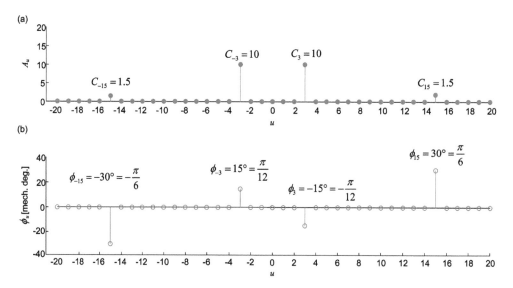

FIGURE 12.56
Two-sided spectrum of harmonics: (a) amplitudes and (b) phase angles.

The waves in the spatial domain, as in Figures 12.3 and 12.5, can also be studied in a similar manner. If the wave is observed at a certain time point, it can be decomposed into a certain number of cosine functions. Unlike the previous case, now spatial position (circumferential position) is the variable. Please note that, 2π is used as the fundamental wavelength for all the circumferential waveforms. The circumferential position, α, can be defined as:

$$\alpha \in [0, 2\pi) \tag{12.58}$$

If v is defined as the spatial harmonic order, FFT series as a function of the spatial position can be written as:

$$f(\alpha) = C_0 + \sum_{v=1}^{\infty} \left[C_v \cos(\alpha v + \phi_v) \right]$$

$$= C_0 + C_1 \cos(\alpha + \phi_1) + C_2 \cos(\alpha \times 2 + \phi_2) + C_3 \cos(\alpha \times 3 + \phi_3) + \dots$$

(12.59)

Figure 12.57 shows the decomposition of a circumferential wave. The fundamental wavelength is 2π. The wavelengths of the third and fifteenth orders are $2\pi/3$ and $2\pi/15$,

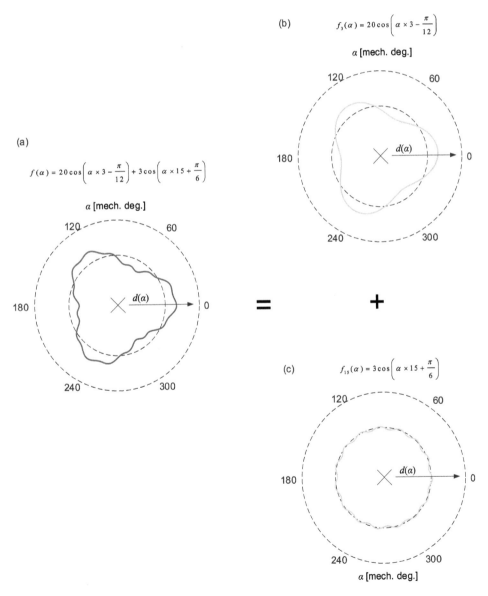

Decomposition of a circumference-varying wave: (a) $f(t)$, (b) $f_3(t)$, and (c) $f_{15}(t)$.

respectively. The amplitudes and phase angles are the same as the case in Figure 12.55. For SRMs, the circumferential radial force density wave is directly correlated with the deformation of the stator structures.

12.5.2 2D FFT

A surface wave in time and spatial domains can be treated as a function and decomposed into its harmonic components:

$$f(t,\alpha) = C_{0,0} + \sum_{u=1}^{\infty} \sum_{v=1}^{\infty} \left[C_{u,v} \cos\left(2\pi f_{mech} ut + v\alpha + \phi_{u,v}\right) \right] \qquad (12.60)$$

As discussed earlier, u is the harmonic order for time domain and v is the harmonic order for spatial domain. Figure 12.58 shows a surface wave that varies in both time and spatial domains. The surface wave can be decomposed first using 1D FFT in time domain at each spatial position. Then a second round of 1D FFT can be performed on the results from the first round of FFT analysis. The two rounds of 1D FFT analyses can be reversed, which means that spatial FFT can be applied before time FFT. MATLAB also provides functions for multi-dimensional FFT decomposition. For instance, a surface wave can be decomposed using `fft2` function in MATLAB. Some of the questions at the end of this chapter can help the reader understand this process better.

For a surface wave that is expressed analytically as:

$$f(t,\alpha) = C_{0,0} + f_{3,1}(t,\alpha) + f_{3,6}(t,\alpha) + f_{7,1}(t,\alpha) + f_{7,6}(t,\alpha) \qquad (12.61)$$

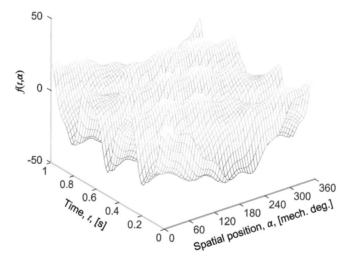

FIGURE 12.58
Time-spatial-varying surface $f(t,\alpha)$.

the functions of the harmonic components can be as follows:

$$
\begin{cases}
C_{0,0} = 10 \\
f_{3,1}(t,\alpha) = C_{3,1}\cos\left(2\pi f_{mech}\times 3\times t+\alpha\times 1\right) \\
f_{3,6}(t,\alpha) = C_{3,6}\cos\left(2\pi f_{mech}\times 3\times t+\alpha\times 6\right) \\
f_{7,1}(t,\alpha) = C_{7,1}\cos\left(2\pi f_{mech}\times 7\times t+\alpha\times 1\right) \\
f_{7,6}(t,\alpha) = C_{7,6}\cos\left(2\pi f_{mech}\times 7\times t+\alpha\times 6\right)
\end{cases}
\tag{12.62}
$$

If the individual functions in Figure 12.59 are added up together, the surface waveform in Figure 12.58 would be achieved. Figure 12.59a shows the surface for $f_{3,1}(t,\alpha)$, which has

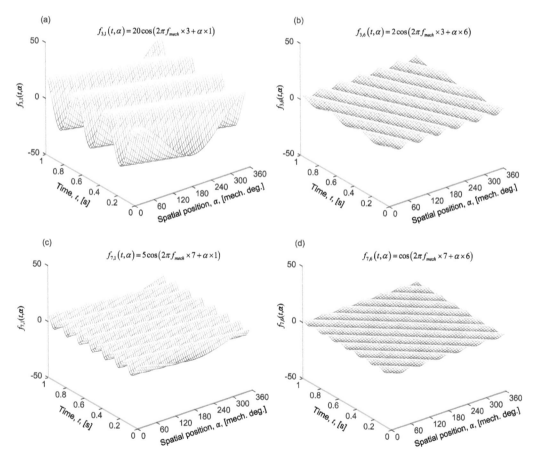

FIGURE 12.59
Harmonics: (a) $f_{3,1}(t,\alpha)$, (b) $f_{3,6}(t,\alpha)$, (c) $f_{7,1}(t,\alpha)$, and (d) $f_{7,6}(t,\alpha)$.

the third temporal order ($v = 3$) and the first spatial order ($u = 1$). One complete sinusoidal wave can be seen over the spatial dimension. Three complete sinusoidal waves can be seen over the time dimension. The amplitude of $f_{3,1}(t,\alpha)$ is 20. Similarly, $f_{3,6}(t,\alpha)$ has three complete sinusoidal waves over 1 second, as shown in Figure 12.59b. The spatial order of $f_{3,6}(t,\alpha)$ is 6, which leads to six complete sinusoidal waves over 360 mechanical degrees. Similar to $f_{3,1}(t,\alpha)$, $f_{7,1}(t,\alpha)$ has one complete wave over a complete mechanical revolution, as shown in Figure 12.59c. $f_{7,1}(t,\alpha)$ has a much higher temporal order at 7, which leads to a higher frequency signal over time. The last function $f_{7,6}(t,\alpha)$ is the noisiest over both the time and spatial domain, as shown in Figure 12.59d, but its amplitude is small.

If the amplitudes are arranged in a four-quadrant table, it would have the format shown in Table 12.3. Similar to two-sided spectrum in Figure 12.56, the amplitudes are reduced by half. The contrast of the cell color indicates the value of the amplitudes. A condensed format of the table can be found in Table 12.4. The heat map for the amplitudes of the harmonics can be seen in Figure 12.60. It can be seen that the harmonics are shown in only two quadrants because u and v are either positive or negative.

For this surface wave, all the phase angles are zero, thus the phase angle matrix is:

$$\Phi_{(u,v)} = \begin{bmatrix} \phi_{3,1} & \phi_{3,6} \\ \phi_{7,1} & \phi_{7,6} \end{bmatrix} = \begin{bmatrix} 0° & 0° \\ 0° & 0° \end{bmatrix} \tag{12.63}$$

If the phase angle matrix was changed to:

$$\Phi_{(u,v)} = \begin{bmatrix} \phi_{3,1} & \phi_{3,6} \\ \phi_{7,1} & \phi_{7,6} \end{bmatrix} = \begin{bmatrix} 0° & -90° \\ 180° & 60° \end{bmatrix} \tag{12.64}$$

TABLE 12.3

Table for Amplitudes of Harmonics in Four Quadrants

$C_{(u,v)}$		Spatial Order, v												
		−6	−5	−4	−3	−2	−1	0	1	2	3	4	5	6
Temporal order, u	7								2.5					0.5
	6													
	5													
	4													
	3								10.0					1.0
	2													
	1													
	0						10.0							
	−1													
	−2													
	−3	1.0					10.0							
	−4													
	−5													
	−6													
	−7	0.5					2.5							

TABLE 12.4

Condensed Table for Amplitudes of Harmonics

$C_{(u,v)}$		Circumferential Order, v				
		−6	**−1**	**0**	**1**	**6**
Temporal order, u	7				$C_{7,1} = 2.5$	$C_{7,6} = 0.5$
	3				$C_{3,1} = 10.0$	$C_{3,6} = 1.0$
	0			$C_{0,0} = 10.0$		
	−3	$C_{-3,-6} = 1.0$	$C_{-3,-1} = 10.0$			
	−7	$C_{-7,-6} = 0.5$	$C_{-7,-1} = 2.5$			

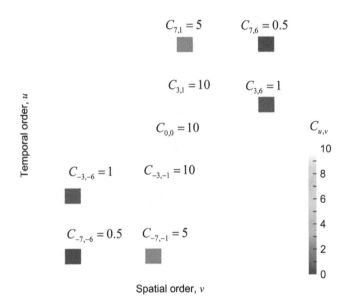

FIGURE 12.60

Heat map for amplitudes of the harmonics $f(t,\alpha)$ in four quadrants.

then the four-quadrant heat map for the phase angles would look like in Figure 12.61. It can be seen that the phase angles in the third quadrant are the opposites of the ones in the first quadrant.

If the harmonic orders were changed to the following:

$$O_{(u,v)} = \begin{bmatrix} (-3,1) & (-3,-6) \\ (7,1) & (7,-6) \end{bmatrix} \tag{12.65}$$

Then, the amplitudes of the harmonics would appear in all four quadrants, as shown in Figure 12.62. This example is to show that any half plane (upper (1st, 2nd), bottom (3rd, 4th), left (2nd, 3rd), or right (1st, 4th)) of the harmonics table (amplitudes and phase angles) contains all necessary information to represent a surface wave.

2D FFT analysis will be used heavily in decomposing radial force density waves for acoustic noise analysis in SRMs. The harmonics will appear in all four quadrants. Through

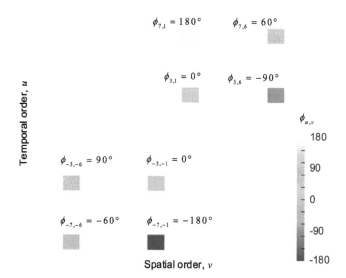

FIGURE 12.61
Heat map for phase angles of the harmonics $f(t,\alpha)$.

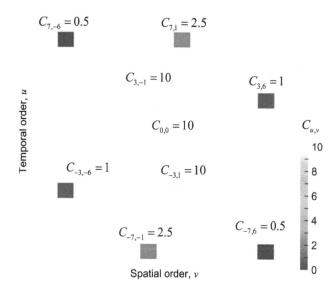

FIGURE 12.62
Heat map for amplitudes of the harmonics $f(t,\alpha)$ in four quadrants.

some simple mathematical operation, the harmonics can be grouped in two quadrants (first and second, i.e., planes where $u > 0$, or first and fourth, i.e., planes where $v > 0$). This will be discussed when we introduce superposition of harmonics in the upcoming sections.

As it was shown in Figure 12.20, $\mathrm{sgn}(u)\mathrm{sgn}(v)$ is used to determine the rotational direction of the harmonics with order (u, v). A plus or minus sign is added prior the amplitude of the harmonic to indicate the direction of rotation. If $\mathrm{sgn}(u)\mathrm{sgn}(v)$ is positive, the rotational direction is negative (clockwise) and a minus sign is added prior the amplitude of the harmonic. If $\text{sgn}(u)\text{sgn}(v)$ is negative, the rotational direction is positive (counter-clockwise), and a plus sign is added prior to the amplitude of the harmonic. This means that, the harmonic $C_{7,-6}$, can be written as $+C_{7,-6}$ because, $\text{sgn}(u)\text{sgn}(v)$ is negative, and the harmonic rotates in the positive direction. The harmonic $C_{7,1}$, rotates in the negative direction, thus it can be expressed as $-C_{7,1}$. The harmonics in Figure 12.62 can be expressed in a matrix, as shown in Table 12.5. This notation will be used heavily in the next chapter.

12.5.3 Shifting FFT Results

The results from a 1D FFT in MATLAB has the format shown in Figure 12.63a. The DC component is placed as the first element of the vector. When the number of original 1D data points, N, is an even number, the first half of the results (from 1st to $N/2$) have the positive mode numbers, u. For the first half of the results in Figure 12.63a, u increases moving from left to right. The last element of the vector has $u = -1$. For the second half (from $N/2 + 1$ to N), u is negative and, moving from left to right, the absolute value of u decreases. If the dominant harmonics have low orders, with the original format of the FFT results in Figure 12.63a, the dominant harmonics are distributed at both ends of the results. Using `fftshift` command in MATLAB, the dominant harmonics can be shifted to the center of the spectrum. If N is an even number, the centralized spectrum will have a format as shown in Figure 12.63b. When N is odd, the arrangement of `fftshift` results is as shown in Figure 12.64b.

If 2D FFT is applied in MATLAB, the results would look like as shown in Figure 12.65. It can be observed that the sign of the temporal and spatial orders in each quadrant do not match with the format presented in Figure 12.21. It can be seen that u equals zero in the first row. In the first column, v equals to zero. The rest of the results are grouped in four quadrants, but low harmonic orders are not centralized. This is why `fftshift` is needed

TABLE 12.5

Matrix of Harmonics with Plus and Minus Sign Indicating Rotational Direction

$C_{(u,v)}$		Circumferential Order, v				
		-6	**-1**	**0**	**1**	**6**
Temporal order, u	7	$+C_{7,-6}$			$-C_{7,1}$	
	3		$+C_{3,-1}$			$-C_{3,6}$
	0			$C_{0,0}$		
	-3	$-C_{-3,-6}$			$+C_{-3,1}$	
	-7		$-C_{-7,-1}$			$+C_{-7,6}$

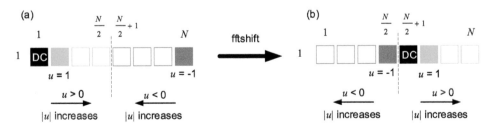

FIGURE 12.63
1D `fftshift`, N is an even number: (a) original format and (b) after shifting.

FIGURE 12.64
1D fftshift, N is an odd number: (a) original format and (b) after shifting.

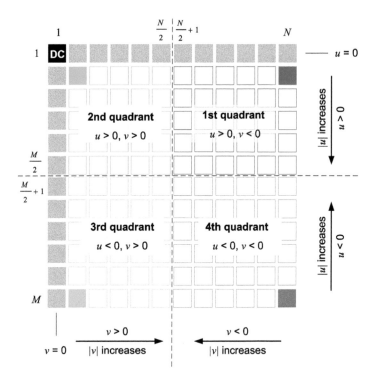

FIGURE 12.65
Arrangement of 2D FFT results.

for 2D FFT results. Figure 12.66 shows the mechanism of row and column fftshift. After the fftshift, the results should be flipped upside down so the quadrants are in the same format as in Figure 12.21. In this section, the fftshift will be discussed. We included a problem at the end of this chapter to practice how to apply this method.

When the number of columns, N and the number of rows, M in the 2D FFT analysis results in Figure 12.65 are both even numbers, fftshift rearranges the 2D FFT results in a manner as shown in Figure 12.66.

When N is an even number and M is an odd number, after fftshift, the DC component is located at $((M+1)/2,(N+2)/2)$, as shown in Figure 12.67.

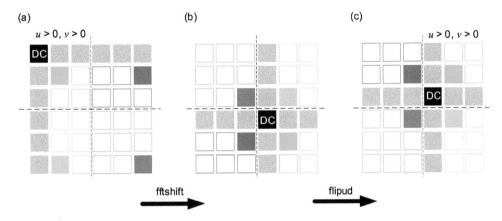

FIGURE 12.66
Functions of `fftshift` and `flipud` in MATLAB: (a) original, (b), after shifting, and (c) after flipping.

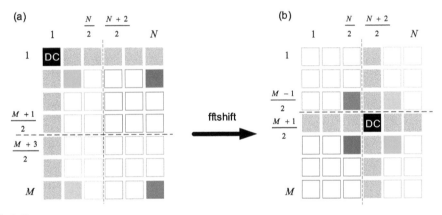

FIGURE 12.67
`fftshift` when N is even, and M is odd: (a) original format and (b) after shifting.

When N is an odd number and M is an even number, after `fftshift`, the DC component is at $((M+1)/2,(N+2)/2)$ as shown in Figure 12.68. When N and M are both odd numbers, after `fftshift`, the DC component is right at the center of the matrix $((M+1)/2,(N+1)/2)$, as shown in Figure 12.69.

In some special cases when the harmonics in the (i) 1st and 3rd, or (ii) 2nd and 4th quadrants are identical, then having the harmonics (i) in the 1st and 2nd quadrant, or (ii) in the 2nd and 3rd quadrant would be enough to analyze all the harmonics. In these cases, 2D `fftshift` would not be necessary. Instead, only column or row shift could be used. The `fftshift (matrix, 1)` in MATLAB performs row-only shift. When N and M are both even numbers, the rearrangement of results after performing `fftshift (matrix, 1)` can be seen in Figure 12.70. After row-only `fftshift`, the harmonics on

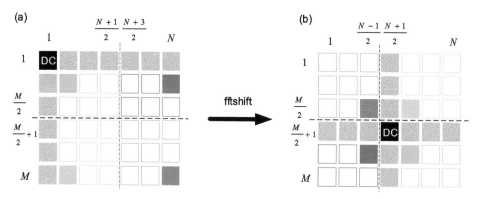

FIGURE 12.68
`fftshift` when N is odd, and M is even: (a) original format and (b) after shifting.

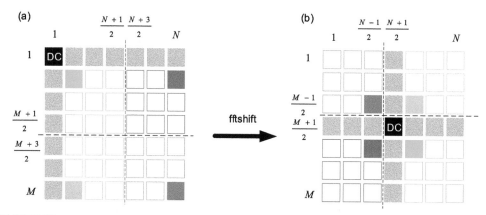

FIGURE 12.69
`fftshift` when both N and M are odd numbers: (a) original format and (b) after shifting.

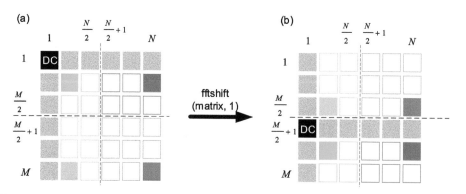

FIGURE 12.70
Row-only `fftshift`, `fftshift (matrix, 1)` when both N and M are even numbers: (a) original format and (b) after shifting.

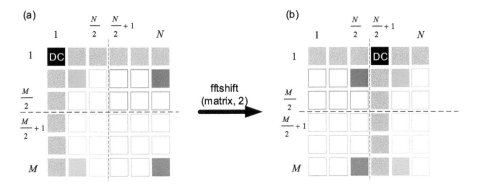

FIGURE 12.71

Column-only `fftshift`, `fftshift (matrix, 2)` when both N and M are even numbers: (a) original format and (b) after shifting.

either the left or right side of the matrix can be used to approximate the original surface wave. Similarly, `fftshift (matrix, 2)` performs column-only shift. Figure 12.71 shows the rearrangement for `fftshift (matrix, 2)` when N and M are both even numbers. After column-only `fftshift`, the harmonics on either the upper or lower side of the matrix can be enough to reconstruct the original surface wave.

12.5.4 Axial Decomposition

The 3D surface waveforms of the mode shapes for six circumferential orders ($circ = 1,..., 6$) are shown in Figure 12.72. Axial direction is another dimension that should be taken into account. Earlier, we have seen a number of examples with no axial variation. At a certain time and at a certain circumferential position, the radial force density can vary axially as well, as shown in Figure 12.73a. The axial modifying factor represents the flux leakage at the ends of the motor core. Besides, there could be SRM prototypes with a skewed rotor. In this case, a dimension, representing a skew angle, has to be added to the axial direction, as shown in Figure 12.73b. Figure 12.73c shows an example with both axial modifying factor and axial-twisting coefficient.

These considerations will make the analysis of radial force density more complicated because multi-dimensional FFT has to be used. This concern is raised because the data of radial force density is usually generated using 2D electromagnetic simulations. To use the 2D data in a 3D noise and vibration model, the data is extended evenly over the axial direction. Thus, the axial modifying and twisting effects are neglected. 3D FEA electromagnetic simulations can solve this problem. However, 3D FEA simulation is computationally expensive. A couple of 3D FEA simulations can be used to generate a set of calibrating factors in the axial direction. The set of calibrating factors can be used on extending the 2D FEA results into 3D.

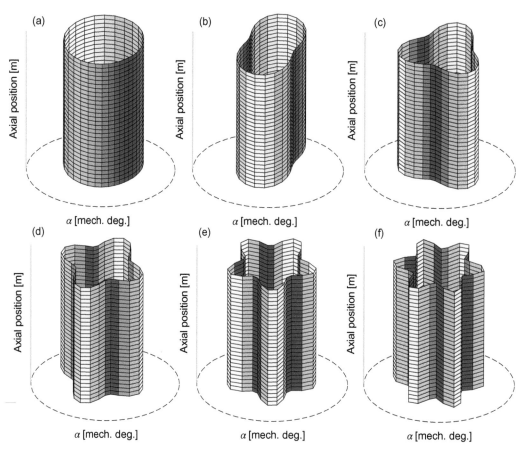

FIGURE 12.72
First six circumferential orders in 3D domain: (a) *circ* = 1, (b) *circ* = 2, (c) *circ* = 3, (d) *circ* = 4, (e) *circ* = 5, and (f) *circ* = 6.

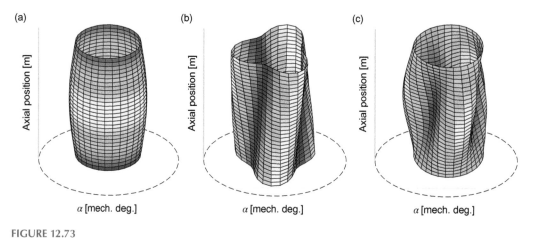

FIGURE 12.73
Variation over axial direction: (a) axial modifying factor, (b) axial-twisting coefficient, and (c) axial modifying factor + axial-twisting coefficient.

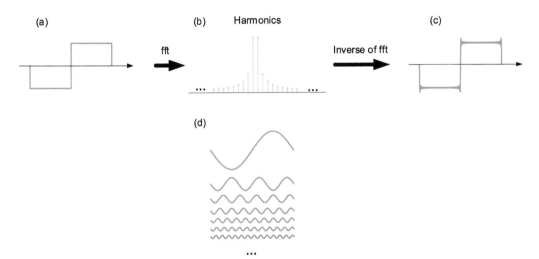

FIGURE 12.74
Decomposition of square wave using FFT: (a) original, (b) amplitudes of harmonics, (c) reproduced, and (d) waveforms of harmonics.

We will assume that the radial force density is evenly distributed along the axial direction. As shown in Figure 12.74, the square wave can be decomposed into a series of sinusoidal waves.

Similarly, a rectangular wave can also be approximated by using a series of sinusoidal waves, as shown in Figure 12.75. This means that an even distribution of radial force density can be decomposed into a series of sinusoidal distributions. It can be seen that a time-varying axially uniform distribution of radial force density wave can excite a number of axial modes. Higher axial modes, which have much higher natural frequencies, are harder to excite.

If only the first and the third axial mode ($ax = 1$ and 3) are considered, the rectangular wave can also be approximated, as shown in Figure 12.76 for the supported-supported constraint. As discussed earlier in this chapter, the mode shapes of supported-supported

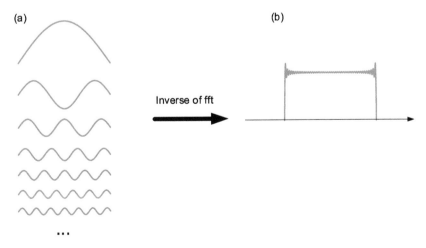

FIGURE 12.75
Reconstruction of rectangular wave using sinusoidal waves: (a) harmonics and (b) reproduced.

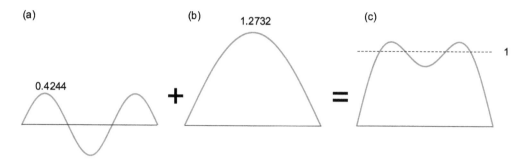

FIGURE 12.76
Reproduction of a rectangular function using harmonics with $ax = 1$ and 3, s-s constraint: (a) $ax = 3$, (b) $ax = 1$, and (c) reproduced.

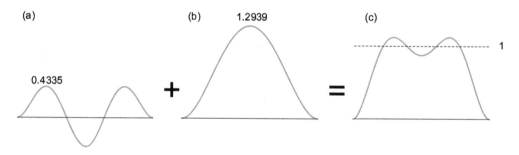

FIGURE 12.77
Reproduction of a rectangular function using harmonics with $ax = 1$ and 3, c-c constraint: (a) $ax = 3$, (b) $ax = 1$, and (c) reproduced.

type of constraints are sinusoidal. However, clamped-clamped constraint has its own mode shapes. Figure 12.77 shows the reconstruction of the rectangular wave using the first and the third axial modes for clamped-clamped constraint. To obtain accurate estimation of noise level for an electric machine, higher axial modes should not be ignored.

12.6 Superposition of Harmonics

In the last section, the procedures for the FFT decomposition have been discussed and presented. The harmonics of the radial force wave can be obtained and displayed in all four quadrants on a the u-v plane. To reduce the size for the matrix of the harmonics and accelerate the matrix computation, the harmonics with identical $|u|$ and $|v|$ (absolute values of temporal and spatial orders) can be superposed. The discussion in this section helps understand the relationship between mode shapes and the harmonics.

12.6.1 Procedures for Superposing Surface Waves

Figure 12.78 shows several scenarios for the superposing of harmonics that will be heavily used when analyzing acoustic noise and vibration in SRMs. The first scenario, which

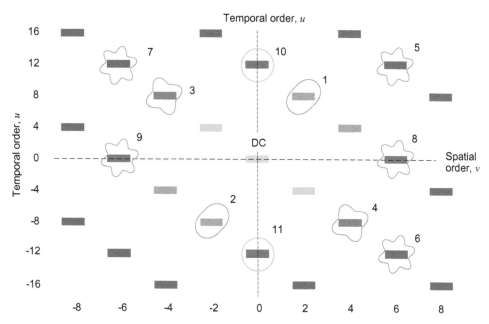

FIGURE 12.78
Scenarios for superposition of harmonics.

is more common, is the harmonics 1 and 2. The two harmonics are located in the first and the third quadrants and are symmetric with respect to the origin. The second pair, 3 and 4, is similar to the first pair, but they are in the second and the fourth quadrants. The third pair, 5 and 6, is symmetrical with respect to v axis. The fourth pair, 5 and 7, is symmetric with respect to the u axis. Harmonics 8 and 9 are the fifth pair. They are symmetrical with respect to u axis and located on the v axis. The sixth pair, 10 and 11, is symmetrical with respect to v axis and reside on the u axis. The last pair is 8 (u equals 0) and 5 (u is nonzero), and they both have the same spatial order. Please note that the line patterns around the harmonics in Figure 12.78 represent the mode shapes of those harmonics.

The following sections cover all the cases of superposition of surface waves that will be used in the next chapter. Each surface wave is represented by its temporal order, u, spatial order, v and its phase angle ϕ, denoted as (u, v, ϕ). Equation (12.66) will be used to explain the superposition of two harmonics with identical amplitudes:

$$A\cos\alpha + A\cos\beta = 2A\cos\frac{\alpha+\beta}{2}\cos\frac{\alpha-\beta}{2} \tag{12.66}$$

12.6.2 (u, v, ϕ_1) and $(-u, -v, \phi_2)$

Figure 12.79 shows the orders for the two surface waves that are located in the first and the third quadrants and are also symmetrical with respect to the origin. It is assumed that u and v are both greater than zero. The two surface waves shown in Figure 12.79 can be formulated as in (12.67):

$$\begin{cases} d_1(t,\alpha) = A\cos(u\omega_{mech}t + v\alpha + \phi_1) \\ d_2(t,\alpha) = A\cos(-u\omega_{mech}t - v\alpha + \phi_2) \end{cases} \tag{12.67}$$

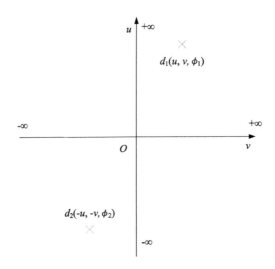

FIGURE 12.79
Superposition of surface waves (u, v, ϕ_1) and $(-u, -v, \phi_2)$.

These two orders are mirror-symmetric with respect to the origin and reside in the first and third quadrants. The sum of these two surface waves yields:

$$d_{sum}(t, \alpha) = d_1(t, \alpha) + d_2(t, \alpha)$$

$$= A\cos(u\omega_{mech}t + v\alpha + \phi_1) + A\cos(-u\omega_{mech}t - v\alpha + \phi_2)$$

$$= \left[2A\cos\left(\frac{\phi_1 + \phi_2}{2}\right)\right]\cos\left(u\omega_{mech}t + v\alpha + \frac{\phi_1 - \phi_2}{2}\right) \quad (12.68)$$

The amplitude of the sum of the two surface waves is $2A\cos((\phi_1 + \phi_2)/2)$. If $\phi_1 = -\phi_2$, (12.68) can be rewritten as:

$$d_{sum}(t, \alpha) = 2A\cos(u\omega_{mech}t + v\alpha + \phi_1) \quad (12.69)$$

It can be seen that if the two surface waves have opposite temporal and spatial orders, and opposite phase angles, the amplitude for the sum of the two surface waves is the sum of their individual amplitudes.

If $\phi_1 + \phi_2 = (2k+1)\pi$, the sum of the two surface waves is given as:

$$d_{sum}(t, \alpha) = A\cos(u\omega_{mech}t + v\alpha + \phi_1) + A\cos(-u\omega_{mech}t - v\alpha + \phi_2)$$

$$= 2A\cos\frac{\phi_1 + \phi_2}{2}\cos\left(u\omega_{mech}t + v\alpha + \frac{\phi_1 - \phi_2}{2}\right) \quad (12.70)$$

$$= 0$$

It shows that the two surfaces cancel each other in this case.

12.6.3 $(u, -v, \phi_1)$ and $(-u, v, \phi_2)$

Figure 12.80 shows the orders of the two surface waves that are located in the second and the fourth quadrants and are also symmetrical with respect to the origin. The two surface waves can be formulated as:

$$\begin{cases} d_1(t,\alpha) = A\cos(u\omega_{mech}t - v\alpha + \phi_1) \\ d_2(t,\alpha) = A\cos(-u\omega_{mech}t + v\alpha + \phi_2) \end{cases} \tag{12.71}$$

Similar to the previous case, the sum of these two surface waves is:

$$\begin{aligned} d_{sum}(t,\alpha) &= d_1(t,\alpha) + d_2(t,\alpha) \\ &= A\cos(u\omega_{mech}t - v\alpha + \phi_1) + A\cos(-u\omega_{mech}t + v\alpha + \phi_2) \\ &= \left[2A\cos\left(\frac{\phi_1 + \phi_2}{2}\right)\right]\cos\left(u\omega_{mech}t - v\alpha + \frac{\phi_1 - \phi_2}{2}\right) \end{aligned} \tag{12.72}$$

Identical to the previous case, the amplitude of the sum of the surface waves is related to the value of $((\phi_1 + \phi_2)/2)$.

12.6.4 (u, v, ϕ_1) and $(-u, v, \phi_2)$

Figure 12.81 shows the orders of the two surface waves that are mirror-symmetrical with respect to the v axis and residing on the first and the fourth quadrants. The sum of these two surface waves is given as:

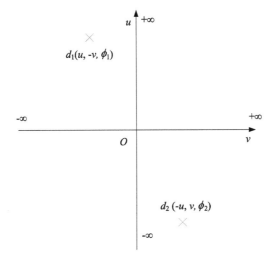

FIGURE 12.80
Superposition of surface waves $(u, -v, \phi_1)$ and $(-u, v, \phi_2)$.

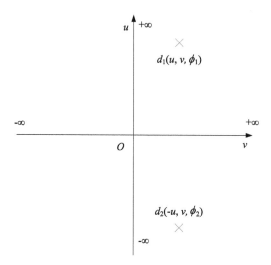

FIGURE 12.81
Superposition of surface waves (u, v, ϕ_1) and $(-u, v, \phi_2)$.

$$d_{sum}(t,\alpha) = d_1(t,\alpha) + d_2(t,\alpha)$$

$$= A\cos(u\,\omega_{mech}t + v\alpha + \phi_1) + A\cos(-u\,\omega_{mech}t + v\alpha + \phi_2) \qquad (12.73)$$

$$= 2A\cos\left(v\alpha + \frac{\phi_1 + \phi_2}{2}\right)\cos\left(u\,\omega_{mech}t + \frac{\phi_1 - \phi_2}{2}\right)$$

It can be seen that the amplitude of the surface wave after the summation reaches its peak or trough when the following conditions are met:

$$\begin{cases} v\alpha + \dfrac{\phi_1 + \phi_2}{2} = \pi k \\[3mm] u\omega_{mech}t + \dfrac{\phi_1 - \phi_2}{2} = \pi j \end{cases} \qquad (12.74)$$

where k and j are integers. The sum reaches its peak when,

$$\begin{cases} v\alpha + \dfrac{\phi_1 + \phi_2}{2} = \pi k \\[3mm] u\omega_{mech}t + \dfrac{\phi_1 - \phi_2}{2} = \pi(k + 2j) \end{cases} \qquad (12.75)$$

The sum reaches its trough when

$$\begin{cases} v\alpha + \dfrac{\phi_1 + \phi_2}{2} = \pi k \\[3mm] u\omega_{mech}t + \dfrac{\phi_1 - \phi_2}{2} = \pi(k + 2j + 1) \end{cases} \qquad (12.76)$$

This case is more complicated than the previous cases. Three examples will be used to illustrate the pattern of the sum of the two surface waves. In the first example, the two surface waves are defined as:

$$\begin{cases} d_1(t,\alpha) = \cos(2\omega_{mech}t + 2\alpha) \\ d_2(t,\alpha) = \cos(-2\omega_{mech}t + 2\alpha + \dfrac{\pi}{3}) \end{cases} \tag{12.77}$$

For the two surface waves in (12.77), the absolute values of the temporal and spatial orders equal to 2. The amplitudes of these surface waves are both 1. The second surface wave has a phase shift of $\pi/3$. The sum of these two surface waves is:

$$d_{sum}(t,\alpha) = d_1(t,\alpha) + d_2(t,\alpha)$$

$$= \cos(2\omega_{mech}t + 2\alpha) + \cos(-2\omega_{mech}t + 2\alpha + \frac{\pi}{3}) \tag{12.78}$$

$$= 2\cos\left(2\alpha + \frac{\pi}{6}\right)\cos\left(2\omega_{mech}t - \frac{\pi}{6}\right)$$

If the mechanical frequency is 1 Hz, the two surface waves in (12.77) and their sum in (12.78) can be seen in Figure 12.82. The behavior of the sum is similar to a standing wave. A standing wave, also known as a stationary wave, has constant amplitude at different spatial points [6]. In other words, standing wave can be seen as a wave in which its peaks or any other points on the wave do not move spatially. Figure 12.83 shows an example of a standing wave. As the wave vibrates from t_0 to t_4, the wave does not move in the spatial domain. If plotted in the polar system, the standing wave has a pattern, as shown in Figure 12.84. Please note that the standing wave does not rotate circumferentially.

In the second example, u is 4 and, hence, the sum of two surface waves are given as:

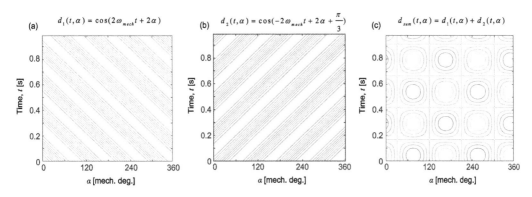

FIGURE 12.82

Superposition of surface waves: (a) $d_1(t,\alpha) = \cos(2\omega_{mech}t + 2\alpha)$, (b) $d_2(t,\alpha) = \cos(-2\omega_{mech}t + 2\alpha + \dfrac{\pi}{3})$, and (c) $d_{sum}(t,\alpha) = d_1(t,\alpha) + d_2(t,\alpha)$.

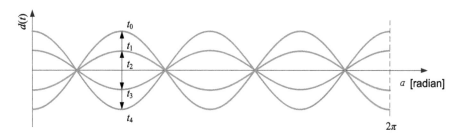

FIGURE 12.83
An example of a standing wave.

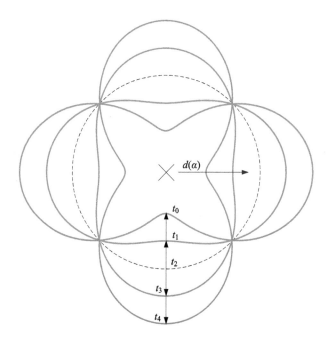

FIGURE 12.84
The standing wave plotted in a polar system.

$$d_{sum}(t,\alpha) = d_1(t,\alpha) + d_2(t,\alpha)$$

$$= \cos(4\omega_{mech}t + 2\alpha) + \cos(-4\omega_{mech}t + 2\alpha + \frac{\pi}{3})$$ (12.79)

$$= 2\cos\left(2\alpha + \frac{\pi}{6}\right)\cos\left(4\omega_{mech}t - \frac{\pi}{6}\right)$$

The contours for these two surface waves and their sum are shown in Figure 12.85. Compared with first example in Figure 12.82, the summed surface wave in Figure 12.85 changes faster in the time domain because of higher temporal frequency.

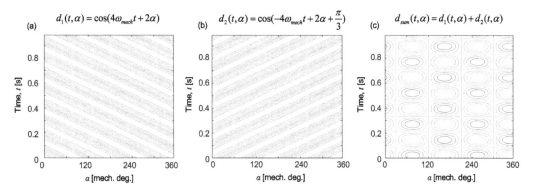

FIGURE 12.85

Superposition of surface waves: (a) $d_1(t,\alpha) = \cos(4\omega_{mech}t + 2\alpha)$, (b) $d_2(t,\alpha) = \cos(-4\omega_{mech}t + 2\alpha + \frac{\pi}{3})$, and (c) $d_{sum}(t,\alpha) = d_1(t,\alpha) + d_2(t,\alpha)$.

In the third example, the amplitudes of the two surface waves are different. The amplitude for the second surface wave is increased to two. Then, the sum of the two surface waves is:

$$d_{sum}(t,\alpha) = d_1(t,\alpha) + d_2(t,\alpha)$$

$$= \cos(4\omega_{mech}t + 2\alpha) + 2\cos(-4\omega_{mech}t + 2\alpha + \frac{\pi}{3})$$

(12.80)

As shown in Figure 12.86, the amplitude of the sum equals to 3 and is still the sum of the amplitudes of the two surface waves. Compared with the second example in Figure 12.85, it can be seen that the summed surface wave in Figure 12.86 is more similar to the surface wave of $d_2(t)$ because $d_2(t)$ has a bigger amplitude and thus has a larger effect.

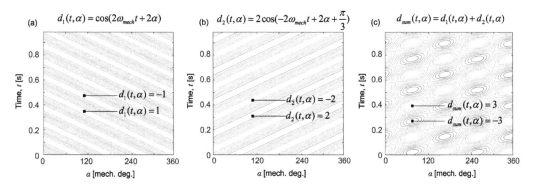

FIGURE 12.86

Superposition of surface waves: (a) $d_1(t,\alpha) = \cos(2\omega_{mech}t + 2\alpha)$, (b) $d_2(t,\alpha) = 2\cos(-2\omega_{mech}t + 2\alpha + \frac{\pi}{3})$, and (c) $d_{sum}(t,\alpha) = d_1(t,\alpha) + d_2(t,\alpha)$.

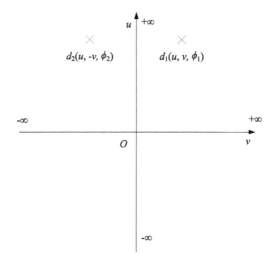

FIGURE 12.87
Superposition of surface waves (u, v, ϕ_1) and $(u, -v, \phi_2)$.

12.6.5 (u, v, ϕ_1) and $(u, -v, \phi_2)$

When two orders (u, v) and $(u, -v)$ of the two surface waves are mirror-symmetrical with respect to u, as shown in Figure 12.87, the sum of the two surface waves is given as:

$$
\begin{aligned}
d_{sum}(t,\alpha) &= d_1(t,\alpha) + d_2(t,\alpha) \\
&= A\cos(u\omega_{mech}t + v\alpha + \phi_1) + A\cos(u\omega_{mech}t - v\alpha + \phi_2) \\
&= A\cos\left(u\omega_{mech}t + \frac{\phi_1 + \phi_2}{2}\right)\cos\left(v\alpha + \frac{\phi_1 - \phi_2}{2}\right)
\end{aligned}
\tag{12.81}
$$

The pattern for the summed surface wave is identical to that in 12.6.4.

12.6.6 $(0, v, \phi_1)$ and $(0, -v, \phi_2)$

When the orders of surface waves are both on the v axis and mirror-symmetric with respect to u axis, as shown in Figure 12.88 the sum of two surface waves is given as:

$$
\begin{aligned}
d_{sum}(t,\alpha) &= d_1(t,\alpha) + d_2(t,\alpha) \\
&= A\cos(v\alpha + \phi_1) + A\cos(-v\alpha + \phi_2) \\
&= \left[2A\cos\left(\frac{\phi_1 + \phi_2}{2}\right)\right]\cos\left(v\alpha + \frac{\phi_1 - \phi_2}{2}\right)
\end{aligned}
\tag{12.82}
$$

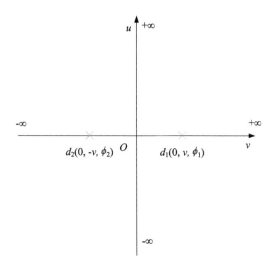

FIGURE 12.88
Superposition of surface waves $(0, v, \phi_1)$ and $(0, -v, \phi_2)$.

When $((\phi_1 + \phi_2)/2) = \pi k$, the equation for the summed surface wave is:

$$d_{sum}(t,\alpha) = d_1(t,\alpha) + d_2(t,\alpha)$$

$$= A\cos(v\alpha + \phi_1) + A\cos(-v\alpha + \phi_2) \tag{12.83}$$

$$= \pm 2A\cos\left(v\alpha + \frac{\phi_1 - \phi_2}{2}\right)$$

It can be seen that the sum of the surface waves is only a function of the circumferential position α. This means that, at a specific point on the circumference, it has a constant value over time. That means there is no oscillation.

When $((\phi_1 + \phi_2)/2) = \pi k + \pi/2$, the two surface waves cancel each other, and their sum is zero:

$$d_{sum}(t,\alpha) = d_1(t,\alpha) + d_2(t,\alpha)$$

$$= A\cos(v\alpha + \phi_1) + A\cos(-v\alpha + \phi_2)$$

$$= \left[2A\cos\left(\frac{\phi_1 + \phi_2}{2}\right)\right]\cos\left(v\alpha + \frac{\phi_1 - \phi_2}{2}\right) \tag{12.84}$$

12.6.7 $(u, 0, \phi_1)$ and $(-u, 0, \phi_2)$

When the two surface waves are located on the u axis and mirror-symmetrical with respect to v axis, as shown in Figure 12.89, the sum of the two surface waves is:

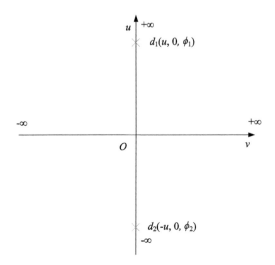

FIGURE 12.89
Superposition of (progressive) surface waves $(u, 0, \phi_1)$ and $(-u, 0, \phi_2)$.

$$d_{sum}(t,\alpha) = d_1(t,\alpha) + d_2(t,\alpha)$$

$$= A\cos(u\omega_{mech}t + \phi_1) + A\cos(-u\omega_{mech}t + \phi_2)$$

$$= \left[2A\cos\left(\frac{\phi_1 + \phi_2}{2}\right)\right]\cos\left(u\omega_{mech}t + \frac{\phi_1 - \phi_2}{2}\right)$$

(12.85)

When $((\phi_1 + \phi_2)/2) = \pi k$, the summed surface wave is expressed as:

$$d_{sum}(t,\alpha) = d_1(t,\alpha) + d_2(t,\alpha)$$

$$= A\cos(u\omega_{mech}t + \phi_1) + A\cos(-u\omega_{mech}t + \phi_2)$$

$$= \pm 2A\cos\left(u\omega_{mech}t + \frac{\phi_1 - \phi_2}{2}\right)$$

(12.86)

More specifically, if $\phi_1 + \phi_2 = 0$, the (12.86) can be rewritten as:

$$d_{sum}(t,\alpha) = d_1(t,\alpha) + d_2(t,\alpha)$$

$$= A\cos(u\omega_{mech}t + \phi_1) + A\cos(-u\omega_{mech}t + \phi_2)$$

$$= 2A\cos\left(u\omega_{mech}t + \phi_1\right)$$

(12.87)

From (12.87), it can be seen that the summed surface wave is only a function of time. At a certain time, it is uniform over the circumference. The waves at different time steps are all concentric.

12.6.8 (u, v, ϕ_1) and $(0, v, \phi_2)$

As shown in Figure 12.90, this scenario describes two harmonics, one located on v axis and both having the same circumferential order. The harmonic content on the v axis has an analytical expression:

$$d_2(t,\alpha) = A\cos\left(v\alpha + \phi_2\right) \tag{12.88}$$

Because the temporal order of this harmonic is 0, it does not vary with time. The spatial order v of $d_2(t, \alpha)$ is not zero. Hence, this harmonic has a certain circumferential shape. The surface wave $d_2(t, \alpha)$ serves as the base, or equilibrium position. The other harmonic $d_1(t, \alpha)$ changes with time with a circumferential order v. The summed surface wave of $d_1(t, \alpha)$ and $d_2(t, \alpha)$ is shown in (12.89).

$$
\begin{aligned}
d_{sum}(t,\alpha) &= d_1(t,\alpha) + d_2(t,\alpha) \\
&= 2\cos(4\omega_{mech}t + 2\alpha) + \cos(2\alpha)
\end{aligned}
\tag{12.89}
$$

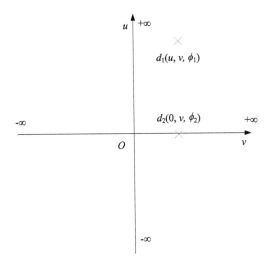

FIGURE 12.90
Superposition of surface waves (u, v, ϕ_1) and $(0, v, \phi_2)$.

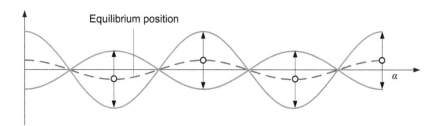

FIGURE 12.91
An example of uneven circumferential equilibrium position.

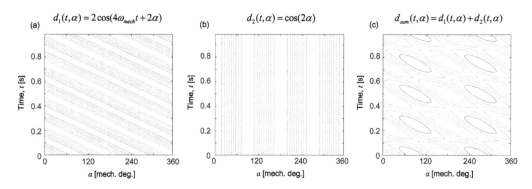

FIGURE 12.92
Superposition of surface waves: (a) $d_1(t,\alpha)=2\cos(4\omega_{mech}t+2\alpha)$, (b) $d_2(t,\alpha)=\cos(2\alpha)$, and (c) $d_{sum}(t,\alpha)=d_1(t,\alpha)+d_2(t,\alpha)$.

The summed surface wave is shown in Figure 12.92. It oscillates around the equilibrium position created by $d_2(t, \alpha)$ (see Figure 12.91). It can be seen in Figure 12.92c that the peaks and troughs on the contour map cannot be reached at the same circumferential point.

In this section, we introduced all the cases for superposing two surface waves. Superposing is used to reduce the size of the matrix of the harmonics and accelerate the matrix computation in acoustic noise analysis. Superposing of the harmonics will be utilized in the next chapter.

12.7 Excitation of Mode Shapes

Fundamentals of surface waves, mode shapes, and decomposition of forcing harmonics have been presented in the previous sections. Those sections help readers develop a solid base for the discussion and analysis of noise and vibration of SRMs in the next chapter. In this section, the relationship between mode shapes and forcing harmonics will be discussed. One of the most important topics in this section is that only a match between the mode shape (circumferential mode, and axial mode) and the shape of the forcing harmonic (circumferential order and axial order) can lead to excitation of that mode shape. Please note that there are two categories of displacements: static and dynamic. Static displacement means that there exists deformation but no vibration. Dynamic displacement means that there is vibration associated with a certain mode shape. Only the natural frequency of the mode shape, which is close to the forcing frequency of the forcing harmonic, can lead to resonance.

We would like to remind that in this chapter and the next, a few indices are used, such as u, v, $circ$, ax, and q. These indices represent the indexes in different domains. For example, u represents the index for the temporal domain. u can be either positive or negative, or even zero. v is used to denote the index in the circumferential domain. Similar to u, v can be any integer. $circ$ is also the index in the circumferential domain; however, it cannot be negative. $circ$ is the absolute value of v. Similarly, q, also a temporal index, is the absolute value of u. ax is the index along the axial direction and can only be a positive integer. The indices, as summarized in Table 12.6, can be used to express the forcing or natural frequency, angular frequency, mode shape, harmonic contents of radial force, forcing shape, displacement, and sound power. We will use the indices in Table 12.6 frequently in the rest of this chapter, and also in the next chapter.

12.7.1 Revisiting Spatial and Temporal Waves

Previously, the concepts of harmonic contents, rotation of harmonics, standing waves, shapes of the harmonics, and forcing frequencies have been explained in detail. In this section, four cases of surface waves shown in Figure 12.93 will be used to summarize what has been discussed. The circumferential modes ($circ$) for these four cases are all 2. For Case 1 in Figure 12.93a, the forcing frequency is 1 Hz. Case 2 in Figure 12.93b has a forcing frequency of 2 Hz. Case 3 and Case 4 in Figure 12.93c and d are both standing waves. The forcing frequency for Case 3 is 1 Hz and Case 4 is 2 Hz.

The harmonics, if plotted on a u-v plane, are shown in Figure 12.94. The harmonic of Case 1 rotates in the negative direction (clockwise) since $sgn(v)sgn(u)$ is greater than zero, as shown in Figure 12.94a. Similarly, the harmonic for Case 2, in Figure 12.94b, rotates in the negative direction, as well. As discussed earlier, a standing wave is the result of the superposition of harmonics. Figure 12.94c shows that the harmonics in Case 3, a standing wave, appear in four quadrants. Likewise, the harmonics of Case 4 are also in four quadrants.

TABLE 12.6

Summary of Indices Used in NVH Analysis

Symbol of Index	Description	Range
u	Index along temporal direction	Any integer
$q\ (q = |u|)$	Index along temporal direction	Non-negative integer
v	Index along circumferential direction	Any integer
$circ\ (circ = |v|)$	Index along circumferential direction	Non-negative integer
ax	Index along axial direction	Positive integer

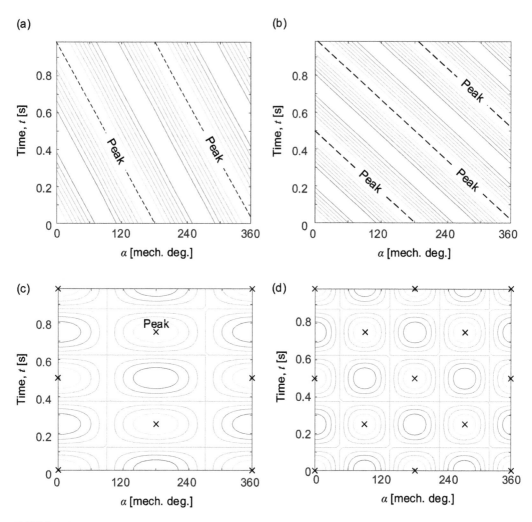

FIGURE 12.93
Four cases: (a) Case 1: 1 Hz, *circ* = 2, rotating wave, (b) Case 2: 2 Hz, *circ* = 2, rotating wave, (c) Case 3: 1 Hz, *circ* = 2, standing wave, and (d) Case 4: 2 Hz, *circ* = 2, standing wave.

Figure 12.95 shows different representations for Case 1. The temporal wave of $d(t, \alpha = 0°)$, shown on the left-hand side of Figure 12.95, describes the varying displacement at one circumferential position ($\alpha = 0°$) over the time domain from 0 to 1 s. The middle set of plots in Figure 12.95 is the spatial waves at 9 different time points. The set of plots on the right-hand side in Figure 12.95 is the same as the middle one, but plotted in polar coordinates. From this set of plots, shape of the harmonic can be better seen.

For the surface waves we have studied, they contain information in two domains: temporal and spatial. The wave for $d(t, \alpha = 0°)$ shows the forcing frequency of the harmonic, which is 1 Hz. However, this plot shows no information regarding the shape of the

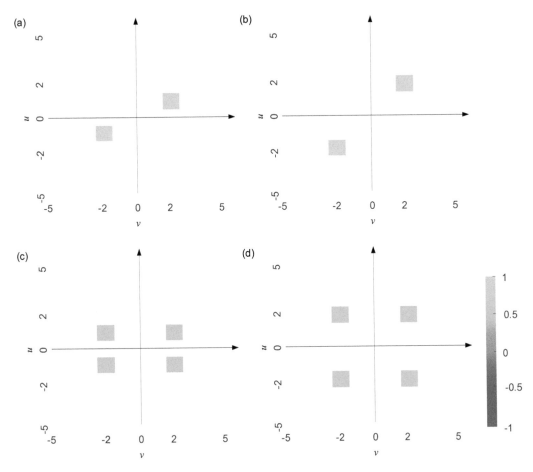

FIGURE 12.94
FFT decomposition: (a) 1 Hz, *circ* = 2, rotating wave, (b) 2 Hz, *circ* = 2, rotating wave, (c) 1 Hz, *circ* = 2, standing wave, and (d) 2 Hz, *circ* = 2, standing wave.

harmonic because the shape of the harmonic can be seen on the spatial or circumferential domain. Then $d(t = 0 \text{ s}, \alpha)$ shows the spatial wave at $t = 0$ s. However, $d(t = 0 \text{ s}, \alpha)$ shows no information about the value of forcing frequency—how frequently the magnitude of the displacement changes at a certain spatial position.

It can also be seen from Figure 12.96 that the "oval shape" plotted in the polar system in Case 2 rotates faster than Case 1 due to a higher forcing frequency. The plots of $d(t, \alpha = 0°)$ in Case 3 (Figure 12.97) and Case 4 (Figure 12.98) have similar trends as in Case 1 and Case 2, respectively. However, they are standing waves; hence, they are not traveling in the spatial domain. The magnitude of the peaks for the standing waves in Figures 12.97 and 12.98, also changes with time, as shown in Figure 12.99. For Case 1 and Case 2, the locations of the peaks change with time.

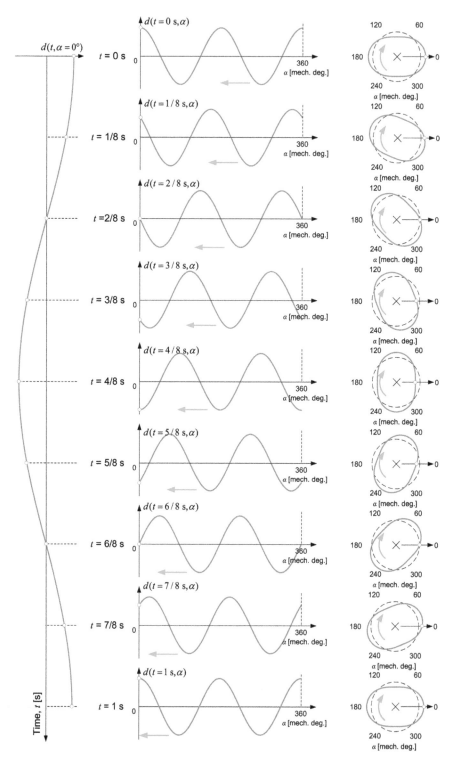

FIGURE 12.95
Description of temporal and spatial waves, Case 1.

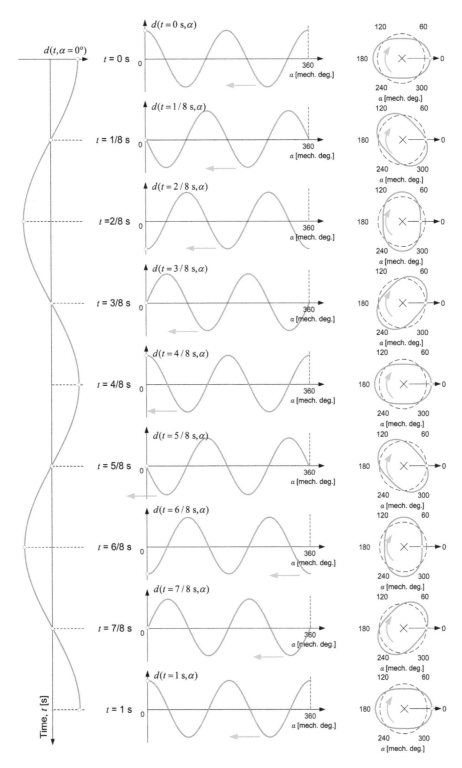

FIGURE 12.96
Description of temporal and spatial waves, Case 2.

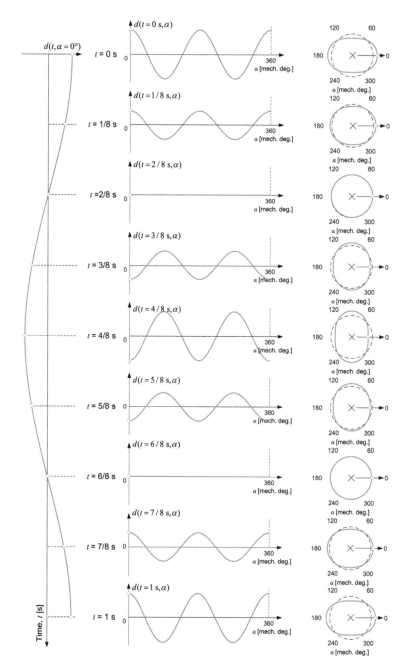

FIGURE 12.97
Description of temporal and spatial waves, Case 3.

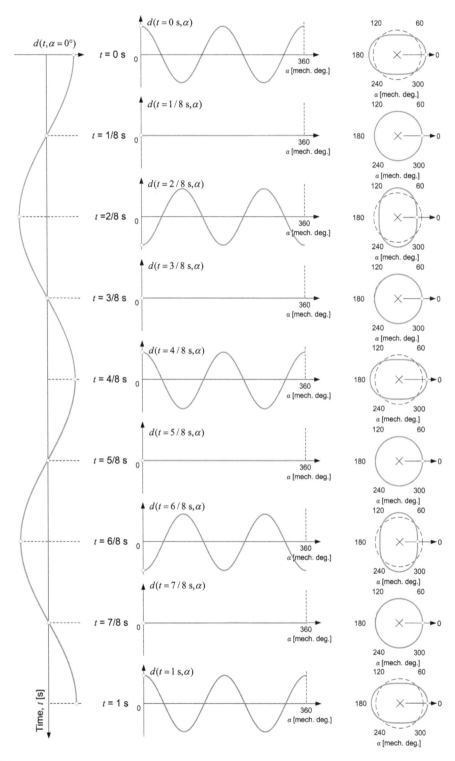

FIGURE 12.98
Description of temporal and spatial waves, Case 4.

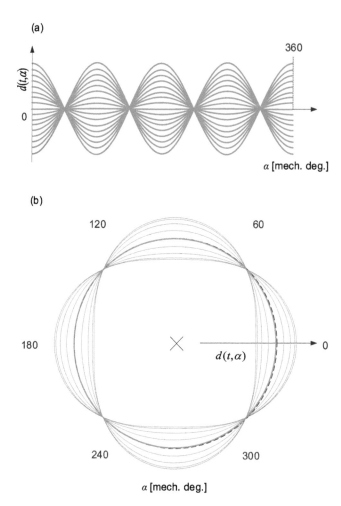

FIGURE 12.99
Description of standing waves, Case 3, Case 4: (a) Cartesian system and (b) polar system.

These analyses reveal that the forcing frequency is a quantity related to the temporal domain. The shape, or circumferential order of a harmonic is related to the spatial or circumferential domain. The harmonic can rotate in the circumferential domain as in Case 1 and Case 2. The superposition of harmonics can also lead to a standing wave, which does not travel over the spatial domain.

12.7.2 Temporal Order

Temporal order is related to the frequency of a certain harmonic. The product of the absolute value of the temporal order, u and the mechanical frequency is the frequency of that harmonic:

$$f_{f(u)} = |u| f_{mech} \tag{12.90}$$

where f_{mech} is the mechanical frequency. As it was shown in Figure 12.56, after the FFT analysis, temporal orders can be positive or negative. This is the reason why the absolute

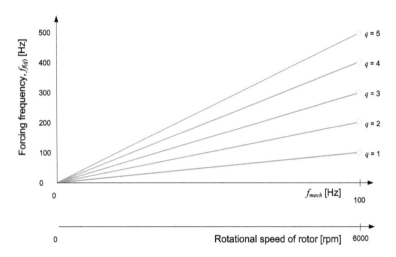

FIGURE 12.100
Relationship between f_{mech}, q and forcing frequency.

value of the temporal order is used in (12.90). For a rotating system, f_{mech} is the mechanical frequency of the rotor. The relationship between f_{mech} with the rotor speed, rpm is:

$$f_{mech} = \frac{rpm}{60} \ [\text{Hz}]$$

(12.91)

Thus, the equation relating the frequency of a certain harmonic with f_{mech}, is:

$$f_{f(u)} = |u| f_{mech} = |u| \frac{rpm}{60}$$

(12.92)

q is used to represent the absolute value of temporal order. Then, the equation for the forcing frequency is rewritten as:

$$f_{f(q)} = q f_{mech} = q \frac{rpm}{60}$$

(12.93)

Forcing frequency $f_{f(q)}$ is q-times the mechanical frequency, f_{mech}. As q increases and the rotor speed increases, $f_{f(q)}$ increases linearly, as shown in Figure 12.100. Whenever temporal order, u or q, equals zero, the forcing frequency is zero.

12.7.3 Spatial Order

If the surface wave is plotted in a polar system, spatial or circumferential order, v is related to the shape, as shown in Figure 12.101. When $v = 0$, the shape is a concentric circle with respect to the reference circle. When $v = \pm 1$, the shape is an eccentric circle with respect to the reference circle. When $v = \pm 2$, it has an oval shape. When $v = \pm 3$, it has a triangular shape. It can be seen that as the absolute value of v increases, the number of the shape's bulges increases.

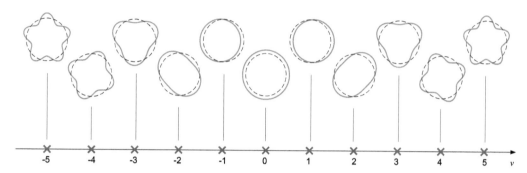

FIGURE 12.101
Relationship between circumferential order v and the shape.

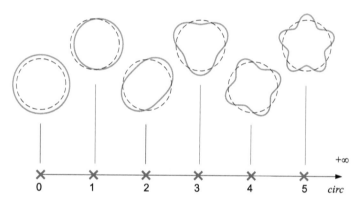

FIGURE 12.102
Relationship between *circ* and the shape of the order.

As discussed earlier, $\operatorname{sgn}(u)\operatorname{sgn}(v)$ determines the rotational direction of the shape. The positive or negative sign of either u or v itself does not solely determine the rotational direction of the shape. Thus, *circ* is used to represent the absolute value of v, which determines the shape of the order, as shown in Figure 12.102.

12.7.4 Harmonics in Four Quadrants

Figure 12.103 shows that the notation for the forcing frequency of harmonic content (u, v). $f_f(u, v)$ represents the forcing frequency of a certain harmonic, in which u is the temporal order and v is the spatial order. Forcing frequency is a function of u only because the frequency value in Hz can be solely determined by the product of u and f_{mech}. However, the information contained in circumferential order, v, is equally important. First, v determines the forcing shape of the harmonic, and the rotational direction of this harmonic is determined by $\operatorname{sgn}(u)\operatorname{sgn}(v)$. The forcing frequency can be calculated with the equation in Figure 12.103 by multiplying the mechanical frequency f_{mech} with the absolute value of u.

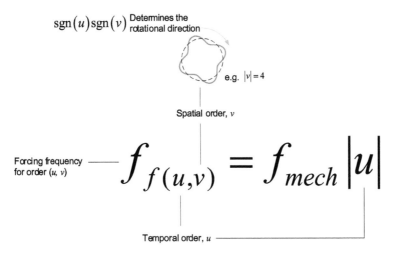

FIGURE 12.103
Notation for forcing frequency of Order (u, v).

As it was shown in Figure 12.21, if the order (u, v) of a certain forcing harmonic is located in the first quadrant, the forcing harmonic rotates in the negative direction (clockwise direction). If the forcing harmonic is in the second quadrant, the forcing harmonic rotates in the positive direction (counterclockwise direction), as shown in Figure 12.104. When v equals zero, the shape of the forcing harmonics changes uniformly in the circumference and has no rotation. As long as u does not equal to zero, the forcing harmonic has an associated forcing frequency. The forcing frequency can excite

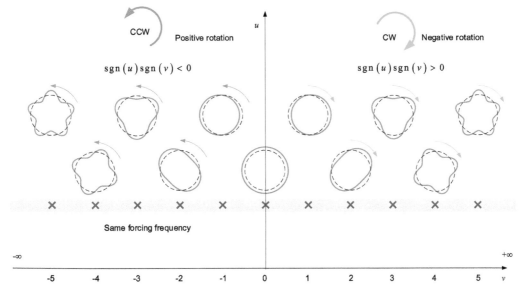

FIGURE 12.104
Selected forcing harmonics in the first and second quadrants.

a certain natural frequency when the circumferential order of the forcing frequency is the same as the circumferential mode of the structure.

When u equals zero, as shown in Figure 12.105, the forcing frequencies for all these orders are zero. This means that there exists no oscillation for any of these forcing harmonics. The harmonic with u equal to 0 only acts as a static force. The static force can be seen as the gravity in the forced damped spring-mass system (Section 12.3.3), which does not vary with time. The contour maps for circumferential modes, $circ = 1 \sim 6$, can be seen in Figure 12.106 for u equal to 0. For example, the contour in Figure 12.106b shows the magnitude of the forcing harmonic component in different circumferential positions. At a certain circumferential position, the magnitude of the harmonic component does not change with time. The contour of Figure 12.106b also shows that there are two peaks over the circumferential domain.

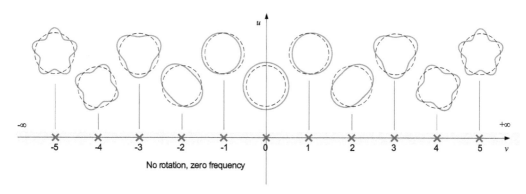

FIGURE 12.105
Temporal order u equals to 0.

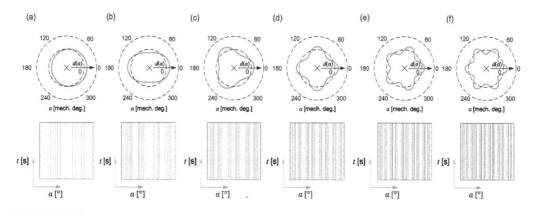

FIGURE 12.106
Relationship between waves plotted in polar system and their contour maps, $u = 0$: (a) $circ = 1$, (b) $circ = 2$, (c) $circ = 3$, (d) $circ = 4$, (e) $circ = 5$, and (f) $circ = 6$.

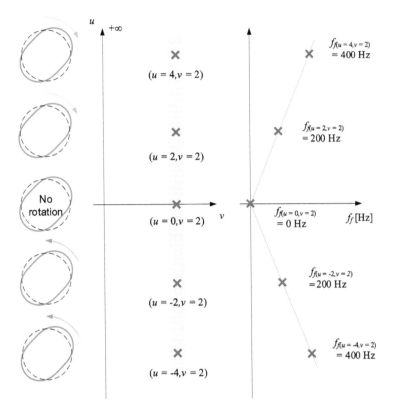

FIGURE 12.107
Orders with v equal to 2.

When comparing two harmonics, if their absolute values of v (*circ*), are the same, it means that the shapes of the forcing harmonics are the same. As shown in Figure 12.107, when v or *circ* equals 2, the forcing harmonic components all have the same oval-like shape. One major difference between the forcing harmonic components is the absolute value of u, i.e., q. As q increases, the forcing frequency of the forcing harmonic component increases if the rotor speed is fixed. Another difference is the rotational directions for the harmonics. As discussed earlier, the harmonics in the first quadrant rotate in the negative direction. The harmonics in the fourth quadrant rotate in the positive direction. The harmonic, located on the v-axis, does not rotate at all; hence, it has no forcing frequency.

Figure 12.108 shows a few harmonics with v equal to 0. It can be seen that, when v equals 0, the shapes of those harmonics are concentric. When u is non-zero, even though the harmonics do not rotate with time, they can still cause vibration. When u and v are both zero, the harmonic is called a DC component and it has no rotation.

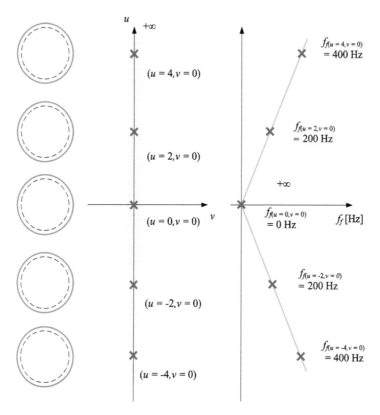

FIGURE 12.108
Orders with v equal to 0, no rotation for the harmonics.

Figure 12.109 shows the shapes for the forcing harmonic components with different temporal and circumferential orders in all four quadrants. The shapes in the first and the third quadrants rotate clockwise, or in the negative rotational direction. The shapes in the second and the fourth quadrants rotate counterclockwise, or in the positive rotational direction. When $v = 0$, the circumferential orders sit on the u axis and the forcing harmonic components all have a circular shape. When the temporal and circumferential order of the forcing harmonic component is at the origin ($u = 0$, and $v = 0$), the harmonic is the DC component. For SRMs, the DC component serves as a circumferentially-uniform radial force density acting on the air gap, and it does not change with time. For different forcing waves, when the absolute value of the spatial or circumferential order of the forcing harmonic component v is the same as the circumferential order of a certain mode, *circ*, they have the same shape. When the absolute value of u, or q, of forcing harmonics are the same, they have the same forcing frequency.

It can be seen in Figure 12.109 that, the forcing harmonics in the first quadrant is identical to the ones in the third when the absolute values of their corresponding temporal order and spatial order are the same. They have the same rotational directions. Similarly, the forcing harmonic in the second quadrant are identical to the ones in the fourth quadrant when the absolute values of their corresponding temporal order and spatial order are the same. Thus, the orders in the first and the fourth quadrants can be used to represent the forcing harmonic in all four quadrants.

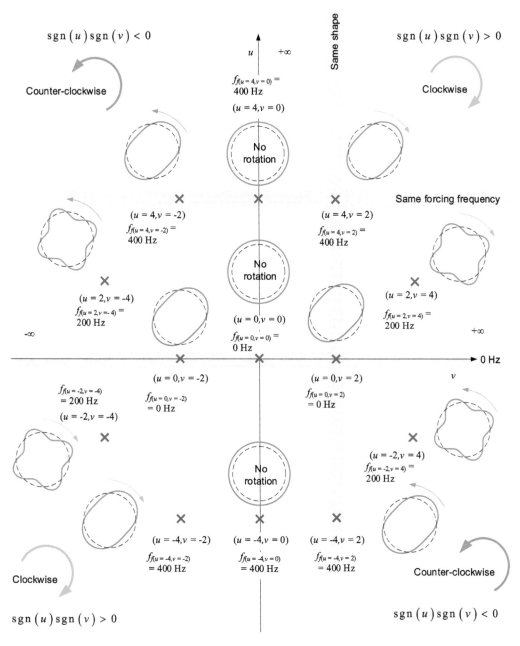

FIGURE 12.109
Polar plots of surface waves for selected orders in four quadrants.

Figure 12.110 shows the contour maps for a few typical harmonics in the first and the fourth quadrants. Figure 12.111 shows the surface waves for those harmonics. These two figures can be used to summarize the major discussions we have had in this chapter. Any harmonic, obtained from 2D FFT decomposition of a forcing waveform, has a few major features. These include the amplitude, temporal order (u), circumferential order (v), and phase angle. The displacement for a certain harmonic is proportional to its amplitude.

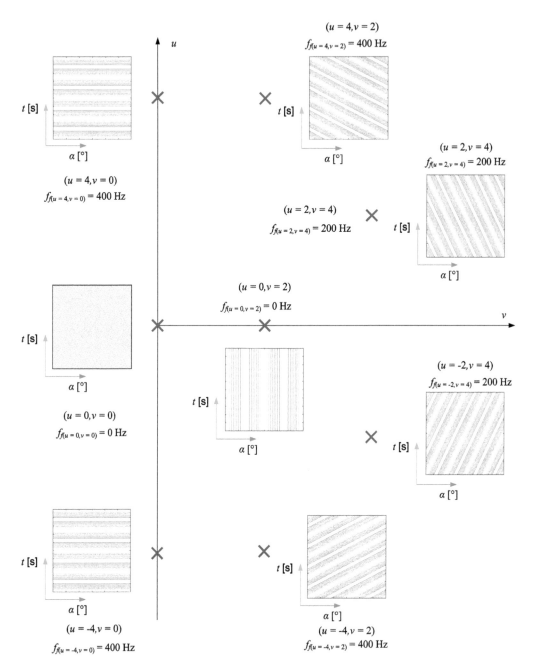

FIGURE 12.110
Contour plots of surface waves for selected orders in the first and fourth quadrants.

The phase angle doesn't have a direct impact on the shape, frequency, and direction of the rotation of the harmonic. But, as discussed in section 12.6, it is used for the superposition of harmonic contents. Temporal order, u, determines the forcing frequency. When the absolute value of u increases, at a fixed f_{mech}, the forcing frequency increases linearly. When u equals zero, the harmonic is located on the v-axis on the u-v plane, and the harmonic is static (has no vibration and no rotation). Here, we refer to vibration as the change in the

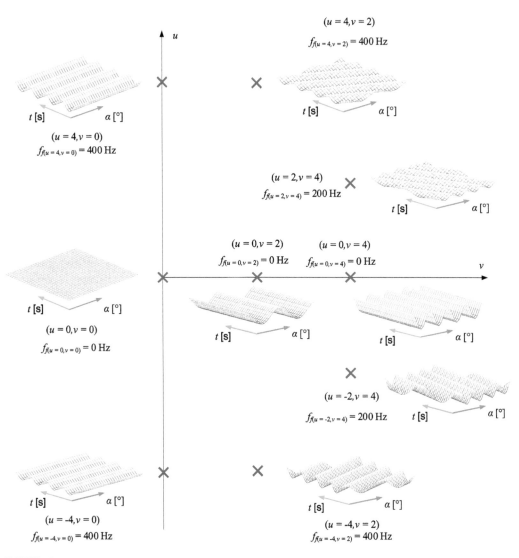

FIGURE 12.111
Mesh plots of surface waves for selected orders in the first and fourth quadrants.

amplitude of the shape of the forcing wave at a certain spatial position. Therefore, when a forcing harmonic rotates, it also vibrates if it is observed at a certain spatial position. Static harmonics affect the equilibrium positions only. The object is in its equilibrium position when the net force acting on it is zero.

As shown in Figures 12.110 and 12.111, the sgn(u)sgn(v) determines the rotational direction of the harmonic component. When sgn(u)sgn(v) is greater than zero, the harmonic rotates in the negative direction (clockwise). For instance, harmonics (4, 2) and (2, 4), in Figures 12.110 and 12.111, rotate in the negative direction. Conversely, if sgn(u)sgn(v) is negative, the harmonic rotates in the positive direction (counterclockwise). For instance, harmonics (−4, 2) and (−2, 4), in Figures 12.110 and 12.111, rotate in the positive direction. The harmonics located on the u-axis, for instance, (4, 0), (0, 0), and (−4, 0), do not rotate.

However, the harmonics (4, 0) and (−4, 0) still vibrate when referred to a certain spatial position. The DC component (0, 0) is static, which means it does not vibrate or rotate. The two harmonics, (0, 2) and (0, 4), sitting on the *v*-axis have their own distinctive shapes; however, they are static over time, thus they don't contribute to vibration and noise because the forcing frequency associated with them is zero.

In this section, the relationship between mode shapes and forcing harmonics were discussed. Temporal order, *q* (the absolute value of *u*), and spatial or circumferential order, *circ* (the absolute value of *v*), are discussed in detail, which will be heavily used in the next chapter. The temporal order *q* is directly linked to the forcing frequency and the circumferential order, *circ*, determines the shape of the forcing harmonic.

12.7.5 Calculation of Displacement

Earlier, the harmonic content of a forcing wave has been obtained using 2D FFT analysis. The natural frequency for a certain mode shape of different geometries has also been discussed earlier in this chapter. For the cylindrical model we used to represent the stator, if we consider the circumferential mode (*circ*) only, the force component with spatial order *v* only excites the circumferential mode *circ* with the same order ($|v| = circ$). The cylindrical shell can be excited in a similar way over the axial direction. As shown in Figure 12.112, when the harmonics have a certain pattern in the circumferential (*circ*) and axial (*ax*) directions, the cylindrical shell is excited.

For a cylindrical shell, its natural frequencies can be identified based on the circumferential mode (*circ*) and the axial mode (*ax*). For the harmonics of radial force density acting on the cylindrical shell, their forcing frequencies are solely determined by the absolute value of *u*, i.e., *q* ($q = |u|$). The circumferential mode, *circ*, doesn't affect the forcing frequency. To better illustrate the natural frequencies for the cylindrical shell and the forcing frequencies of harmonics, three indices (*q*, *circ*, *ax*) are used. For natural frequencies of the cylindrical shell, the index *q* is irrelevant and used only to facilitate matrix calculation:

$$f_{n(q,circ,ax)} = f_{n(circ,ax)} \tag{12.94}$$

For forcing frequencies, the index, *ax*, is added so that higher axial orders can be considered for a more accurate estimation of noise and vibration. The forcing frequencies of $f_{f(q,circ)}$ can be calculated by the equation in Figure 12.103. As previously mentioned, a certain forcing harmonic is determined by its shape in the circumferential and axial directions and by its forcing frequency. The circumferential and axial shapes are determined by the circumferential and axial orders. The forcing frequency is determined by the absolute value of the temporal order:

$$f_{f(q,circ,ax)} = f_{f(q)} = q f_{mech} \tag{12.95}$$

The angular frequencies for natural frequency and forcing frequency can be found as:

$$\begin{cases} \omega_{n(q,circ,ax)} = 2\pi f_{n(q,circ,ax)} = 2\pi f_{n(circ,ax)} \\ \\ \omega_{f(q,circ,ax)} = 2\pi f_{f(q,circ,ax)} = 2\pi q f_{mech} \end{cases} \tag{12.96}$$

It can be observed from (12.96) that *q* is eliminated from $f_{n(q,circ,ax)}$, yielding $f_{n(circ,ax)}$ because, for a cylindrical shell, its natural frequencies are solely determined by *circ* and *ax*. Again,

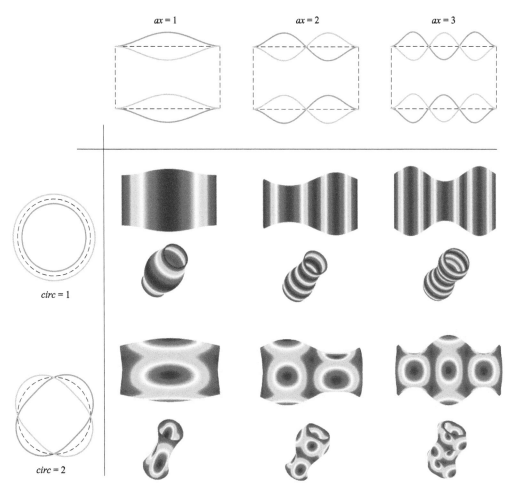

FIGURE 12.112
Selected deformations for the cylindrical shell.

adding another index, q, to the analytical expression is to facilitate matrix calculation. We will use this format (q, *circ*, *ax*) in the next chapter.

As discussed earlier, when the spatial order *circ* of a certain forcing harmonic is the same as the circumferential order *circ* of a mode shape, displacement occurs. The displacement of a forced damped spring-mass system has been shown in (12.41). The displacement $D_{(q, circ, ax)}$ caused by the harmonics (*circ*, *ax*) has a similar expression and it can be calculated as [7]:

$$D_{(q,circ,ax)} = \frac{F_{r(q,circ,ax)} / M}{\sqrt{\left(\omega_{n(q,circ,ax)}^2 - \omega_{f(q,circ,ax)}^2\right)^2 + 4\zeta_{(q,circ,ax)}^2 \omega_{f(q,circ,ax)}^2 \omega_{n(q,circ,ax)}^2}} \quad (12.97)$$

where M is the mass of the object, and $\zeta_{(q,circ,ax)}$ is the damping ratio associated with the mode (*circ*, *ax*). $F_{r(q, circ, ax)}$ is the amplitude of the corresponding harmonics of the radial force. The calculation of $F_{r(q, circ, ax)}$ and the damping ratio for SRMs will be discussed in the next chapter.

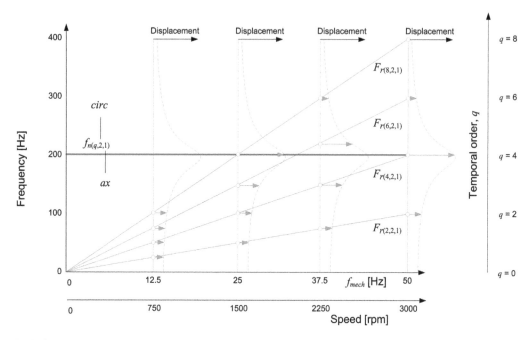

FIGURE 12.113
Displacements for four harmonics, $F_{r(2,2,1)}$, $F_{r(4,2,1)}$, $F_{r(6,2,1)}$, and $F_{r(8,2,1)}$, as the speed or f_{mech} increases.

Figure 12.113 shows an example of displacements for four harmonics of forcing wave $F_{r(q,\ circ,\ ax)}$, $F_{r(2,2,1)}$, $F_{r(4,2,1)}$, $F_{r(6,2,1)}$, and $F_{r(8,2,1)}$. It is assumed that these four harmonics have the same amplitude. It is assumed that the natural frequency $f_{n(q,2,1)}$ for Mode (2,1) is 200 Hz.

At 750 rpm, the mechanical frequency, f_{mech}, is 12.5 Hz. For $F_{r(2,2,1)}$, q equals 2 and, thus, the forcing frequency, $f_{f(2,2,1)}$ is 25 Hz. At 750 rpm, forcing frequency $f_{f(4,2,1)}$ is two times that of $f_{f(2,2,1)}$ and reaches 50 Hz. At 750 rpm, $f_{f(6,2,1)}$ equals 75 Hz and $f_{f(8,2,1)}$ equals 100 Hz. At 750 rpm, $f_{f(8,2,1)}$ is closer to the natural frequency of $f_{n(q,2,1)}$ for Mode (2,1) (200 Hz), thus it has higher displacement than the other three harmonics. At 1500 rpm, $f_{f(8,2,1)}$ reaches 200 Hz, which has the highest displacement among all four harmonics. When the rotor speed increases to 2250 rpm and f_{mech} reaches 37.5 Hz., then, at this speed, $f_{f(6,2,1)}$ is closer to the natural frequency of $f_{n(q,2,1)}$ for mode (2,1) and, hence, $F_{r(6,2,1)}$ generates the highest displacement. At 3000 rpm, the mechanical frequency f_{mech} is 50 Hz, and $f_{f(4,2,1)}$ has a forcing frequency of 200 Hz, which allows $F_{r(4,2,1)}$ to excite the highest displacement among the four harmonics at this speed. It can be seen that as the rotor speed increases from 0 to 3000 rpm, different harmonics can excite the corresponding natural frequency at different speeds.

This chapter has covered the fundamentals of surface waves and mechanical vibrations. In this chapter, mode shapes of straight beam, square plate, cylindrical shell, and circular plate are explained. Since SRMs can be modelled as layers of cylindrical shells, the mode shapes of cylindrical shells are discussed in detail. Fast Fourier Transform (FFT) and how it can be used in decomposing a complex surface wave, such as radial force density wave, have also been discussed. These fundamentals and tools lay the groundwork for further study of the acoustic noise behavior of SRMs found in the next chapter.

Questions and Problems

1. Undamped Spring-Mass System

   Given an undamped spring-mass system, as shown in Figure Q.12.1:

   $$\begin{cases} x(t) = X\cos(\omega_n t - \phi) \\ x(0) = 0.2 \ [\text{m}] \\ \dot{x}(0) = 0 \ [\text{m/s}] \end{cases}$$

   a. The mass of the weight is 1 kg and the stiffness of the spring is 10 N/mm. Please first calculate the natural frequency of the spring-mass system. Then calculate the amplitude and the phase angle of the motion with given initial conditions.

   b. Use MATLAB to repeat all the calculations above and plot the motion of the system over 0 ~ 0.2 s.

   c. For $\dot{x}(0) = 20$ [m/s], use the MATLAB code and plot the motion of the system over 0 ~ 0.2 s.

2. Viscous Damping

   We will add damping to the system discussed in Q1, as shown in Figure Q.12.2. The mass of the weight is 1 kg and the stiffness of the spring is 10 N/mm. The initial condition of the system is given as:

   $$\begin{cases} x(0) = 0.2 \ [\text{m}] \\ \dot{x}(0) = 0 \ [\text{m/s}] \end{cases}$$

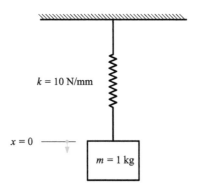

FIGURE Q.12.1
Undamped spring-mass system.

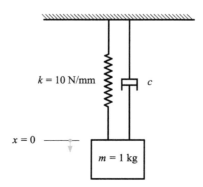

FIGURE Q.12.2
Damped spring-mass system.

The damping ratio has been defined in (12.28) as:

$$\varsigma = \frac{c}{2m\omega_n} = \frac{c}{2\sqrt{mk}}$$

a. When the damping ratio is greater than zero and less than one ($0 < \varsigma < 1$), the response of the system is expressed as given in (12.29):

$$x(t) = Ce^{-\varsigma\omega_n t}\cos(\omega_d t - \phi).$$

Given that $\varsigma = 0.1$, please calculate the amplitude C, the damped-circular frequency ω_d, the phase angle ϕ.

b. Use MATLAB to repeat the calculations in (a) and plot the response of the system over 0 ~ 0.4 s.

c. Use MATLAB to plot the response of the system for seven cases where $\varsigma = 0$, 0.01, 0.05, 0.1, 0.2, 0.4, 0.8 in one figure over 0 ~ 0.8 s.

d. When the damping ratio equals to one ($\varsigma = 1$), the response of the system is expressed as given in (12.32):

$$x(t) = e^{-\omega_n t}\left[x_0 + (v_0 + \omega_n x_0)t\right]$$

Given the same initial conditions, plot the response of the system using MATLAB.

e. Figure 12.27 shows four different responses for the system with the damping ratio equal to one. Please create MATLAB code to reproduce the plot.

f. When the damping ratio is greater than one ($\varsigma > 1$), the response of the system is expressed as given in (12.33) and (12.34):

$$
\begin{cases}
x(t) = C_1 e^{\left(-\varsigma+\sqrt{\varsigma^2-1}\right)\omega_n t} + C_2 e^{\left(-\varsigma-\sqrt{\varsigma^2-1}\right)\omega_n t} \\[2mm]
C_1 = \dfrac{x_0\omega_n\left(\varsigma+\sqrt{\varsigma^2-1}\right)+v_0}{2\omega_n\sqrt{\varsigma^2-1}} \\[4mm]
C_2 = \dfrac{-x_0\omega_n\left(\varsigma-\sqrt{\varsigma^2-1}\right)-v_0}{2\omega_n\sqrt{\varsigma^2-1}}
\end{cases}
$$

Plot the response of the system in one plot using MATLAB when $\varsigma = 2, 4, 6, 8, 10$.

g. Integrate the programs written in (b), (d) and (e) in MATLAB and plot the surface of response for the system as ς varies from 0.01 to 1.2 over 0 ~ 1.2 s.

3. Forced Damped Spring-Mass System

If a periodic force $F_0 \sin(\omega_f t)$ is added to the system discussed in question 2 (see Figure Q.12.3), the motion of the system can be described as given in (12.37):

$$
\ddot{x}(t) + 2\varsigma\omega_n\dot{x}(t) + \omega_n^2 x(t) = \frac{F_0}{m}\sin(\omega_f t)
$$

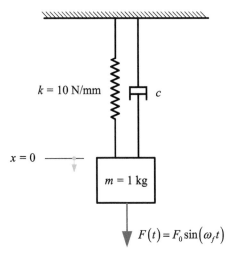

FIGURE Q.12.3
Forced damped spring-mass system.

The phase angle of the steady state of the system was quantified in (12.39):

$$\phi_s = \arctan\left[\frac{2\varsigma\dfrac{\omega_f}{\omega_n}}{1-\left(\dfrac{\omega_f}{\omega_n}\right)^2}\right]$$

The magnification factor was given in (12.43) as:

$$MF = \frac{1}{\sqrt{\left[1-\left(\dfrac{\omega_f}{\omega_n}\right)^2\right]^2 + \left(2\varsigma\dfrac{\omega_f}{\omega_n}\right)^2}}$$

a. Use MATLAB and plot the phase angle as ω_f/ω_n varies from 0.01 to 10 and ς varies from 0.01 to 1.2.

b. Similarly, use MATLAB and plot the magnification factor as ω_f/ω_n varies from 0.01 to 10 and ς varies from 0.01 to 1.2.

4. Mode Shapes of a Beam

Table 12.1 shows the functions describing the mode shapes for the constraints of supported-supported (s-s) and clamped-clamped (c-c). Use MATLAB to plot the mode shapes of Mode ax ($ax = 1, 2, 3,$ and 4) for s-s and c-c. The length of the beam, l, can be set to 1 meter. Please normalize the displacements of the mode shapes. The plots can be seen in Figure 12.30.

5. 1D FFT Analysis of Temporal Wave

A time-varying wave of radial force [N] at a fixed location has been given as follows:

$$f(t) = C_0 + f_1(t) + f_2(t) + f_3(t)$$

where the component functions are given as:

$$\begin{cases} f_1(t) = C_1\cos\left(2\pi f_{mech}tu_1 + \phi_1\right) \\ f_2(t) = C_2\cos\left(2\pi f_{mech}tu_2 + \phi_2\right) \\ f_3(t) = C_3\cos\left(2\pi f_{mech}tu_3 + \phi_3\right) \end{cases}$$

where f_{mech} equals 1 Hz. Please note that the subscripts of the component functions are just indices and do not indicate harmonic orders. The harmonic orders are given as:

$$\begin{cases} u_1 = 2 \\ u_2 = 5 \\ u_3 = 17 \end{cases}$$

The harmonics amplitudes are given as:

$$\begin{cases} C_0 = 5 \\ C_1 = 20 \\ C_2 = 8 \\ C_3 = 3 \end{cases}$$

The phase angles are given as:

$$\begin{cases} \phi_1 = -\dfrac{\pi}{6} \\ \phi_2 = \dfrac{\pi}{4} \\ \phi_3 = \dfrac{\pi}{2} \end{cases}$$

First use the following MATLAB code to formulate the wave $f(t)$ over 0 ~ 1 s. The initialization of the program is given:

```
%% Initialization

close all; clear all; clc

f1_time_phase_shift = -pi/6;
f2_time_phase_shift = pi/4;
f3_time_phase_shift = pi/2;

f1_time_order = 2;
f2_time_order = 5;
f3_time_order = 17;

altitude_shift = 5; % C0
f1_time_Ap = 20;
f2_time_Ap = 8;
f3_time_Ap = 3;

time_num = 1024; % 2^10 = 1024, total number of points in the FFT analysis
time_order_limit = 20;
time_step = 1/time_num; % unit: s
freq_mech = time_num/2^10; % mechanical frequency: 1 Hz
time_vector = [0:time_step:time_step*(time_num-1)];
```

Then the wave function can be formulated:

```
%% Formulation of waves

f1_time_wave  =  f1_time_Ap*cos  (2*pi*freq_mech*time_vector*f1_time_order  +
f1_time_phase_shift);

f2_time_wave  =  f2_time_Ap*cos(2*pi*freq_mech*time_vector*f2_time_order  +
f2_time_phase_shift);

f3_time_wave  =  f3_time_Ap*cos(2*pi*freq_mech*time_vector*f3_time_order  +
f3_time_phase_shift);

f_time_sum = f1_time_wave + f2_time_wave + f3_time_wave + altitude_shift;
```

Then use the following MATLAB code to perform FFT analysis over the wave function. The line, F_series = 2*abs(fft(f_time_sum)/time_num) is used to get the FFT results. 1/time_num is a scaling factor and is also the sampling interval, T. Assuming the analog signal is $x(t)$ with corresponding Fourier transform $X(\omega)$, sampling $x(t)$ with some sampling frequency results in a sequence of numbers: the sampled sequence $x[nT]$, where T is the sampling period and n is the sequence number. Let the sequence of complex numbers $Y(k)$ be the Fourier transform of the sampled sequence, i.e. the Discrete Fourier Transform (DFT) of $x[nT]$. If the sampling frequency is larger than twice the largest frequency in the signal $x(t)$, then the magnitude of $X(\omega)$ will be proportional to the magnitude of $Y(k)$. The proportionality factor turns out to be the sampling interval T. We can write: $X(\omega) = T*Y(k)$.

By multiplying the results with 2, normalization can be achieved. This way, the absolute value of the amplitude is calculated since the fft result is in complex form. The angle function returns the phase angles in radians. By taking F_series > 0.01, the major FFT components are utilized in the calculations.

```
%% FFT analysis, one-sided spectrum
F_series=2*abs(fft(f_time_sum)/time_num); % DC amplitude is doubled
Ang_series = angle(fft(f_time_sum)).*(F_series > 0.01)*180/pi; % mech. deg.
```

Please note that the first element in the array of F_series, is double the value of DC component

a. Use MATLAB and plot four subplots for $f(t)$, $f_1(t)$, $f_2(t)$ and $f_3(t)$ over time.

b. Use MATLAB and plot two subplots for one-sided spectrum of harmonic amplitudes and one-sided spectrum of phase angles respectively.

c. Rerun the FFT analysis in MATLAB using the command "fft(f_time_sum)/time_num." After FFT is done, use fftshift to shift the DC component to the center of the spectrum and plot two-sided spectrums for both the harmonic amplitudes and phase angles.

6. 1D FFT Analysis of Circumferential Wave

A wave of radial force [N] varying around a circumference at a certain time point has been given as follows:

$$f(\alpha) = C_0 + f_1(\alpha) + f_2(\alpha) + f_3(\alpha)$$

As discussed earlier, $\alpha \in [0, 2\pi)$ is the circumferential position. The base wavelength is 2π. The component functions are given as:

$$\begin{cases} f_1(\alpha) = C_1 \cos(\alpha v_1 + \phi_1) \\ f_2(\alpha) = C_2 \cos(\alpha v_2 + \phi_2) \\ f_3(\alpha) = C_3 \cos(\alpha v_3 + \phi_3) \end{cases}$$

Please note that the subscripts of the component functions are just indices and do not indicate spatial orders. The harmonic orders are given as:

$$\begin{cases} v_1 = 3 \\ v_2 = 6 \\ v_3 = 18 \end{cases}$$

The harmonics amplitudes are given as:

$$\begin{cases} C_0 = 4 \\ C_1 = 20 \\ C_2 = 4 \\ C_3 = 8 \end{cases}$$

The phase angles [radian] are given as:

$$\begin{cases} \phi_1 = -\dfrac{\pi}{3} \\ \phi_2 = 0 \\ \phi_3 = \dfrac{\pi}{3} \end{cases}$$

a. Similar to (a) and (b) in Q5, use MATLAB to formulate the waveform of $f(\alpha)$.
b. Use MATLAB to perform one-sided FFT analysis on $f(\alpha)$.
c. Use MATLAB to plot four subplots in a figure for $f(\alpha)$, $f_1(\alpha)$, $f_2(\alpha)$ and $f_3(\alpha)$ in a polar system.
d. Use MATLAB and plot two subplots in a figure for amplitudes and phase angles for the one-sided harmonic content of $f(\alpha)$, respectively.

7. 2D FFT of Surface Wave

This question is prepared to practice 2D FFT analysis. `fft2` in MATLAB can be used to implement 2D FFT analysis.

a. Figure 12.59 shows the component functions for the surface in Figure 12.58. The surface function is discussed in Section 12.5.2:

$$f(t,\alpha) = C_{0,0} + f_{3,1}(t,\alpha) + f_{3,6}(t,\alpha) + f_{7,1}(t,\alpha) + f_{7,6}(t,\alpha)$$

As discussed earlier, the component functions are:

$$\begin{cases} C_{0,0} = 10 \\ f_{3,1}(t,\alpha) = C_{3,1}\cos(2\pi f_{mech} \times 3 \times t + \alpha \times 1) \\ f_{3,6}(t,\alpha) = C_{3,6}\cos(2\pi f_{mech} \times 3 \times t + \alpha \times 6) \\ f_{7,1}(t,\alpha) = C_{7,1}\cos(2\pi f_{mech} \times 7 \times t + \alpha \times 1) \\ f_{7,6}(t,\alpha) = C_{7,6}\cos(2\pi f_{mech} \times 7 \times t + \alpha \times 6) \end{cases}$$

Please note that the phase angles are all zero. The amplitudes for the harmonic contents are given in the following matrix:

$$C_{(u,v)} = \begin{bmatrix} C_{3,1} & C_{3,6} \\ C_{7,1} & C_{7,6} \end{bmatrix} = \begin{bmatrix} 20 & 2 \\ 5 & 1 \end{bmatrix}$$

Please use MATLAB to formulate these four surfaces and combine them to generate the surface in Figure 12.58.

b. Use MATLAB to plot the four surfaces as shown in Figure 12.59 and the sum surface.

c. The following MATLAB code is given to perform two-sided FFT analysis over $f(t,\alpha)$ (variable: pl _ wave _ sum). The command fftshift and flipud are used to rearrange the results from fft2 to make sure that the lower harmonics are centralized and the harmonics with $u > 0$ and $v > 0$ are in the first quadrant, as shown in Figure 12.66.

```
%% FFT analysis

FFT2_raw = fft2(pl_wave_sum)/space_num/time_num;
FFT2_shifted = fftshift (FFT2_raw);
FFT2_abs_shifted = abs(FFT2_shifted);
FFT2_abs_shifted_cut = FFT2_abs_shifted (513-10:513+10,513-10:513+10);
FFT2_abs_shifted_cut_flipud = flipud (FFT2_abs_shifted_cut);
FFT2_abs_shifted_cut_flipud(FFT2_abs_shifted_cut_flipud<0.01) = NaN;

%% Phase angle

phase_angle_matrix = angle(FFT2_shifted).*(FFT2_abs_shifted > 0.01)*180/pi; %
mech. deg.
phase_angle_matrix(phase_angle_matrix==0) = NaN;
phase_angle_matrix(abs(phase_angle_matrix)<0.05) = 0;
phase_angle_matrix_cut = phase_angle_matrix (513-10:513+10,513-10:513+10);
phase_angle_matrix_cut_flipud = flipud (phase_angle_matrix_cut);
```

Please state the reason why the value 513 is used. Please use some other method to accomplish the function of this sentence

d. Figure 12.60 shows the heat map for the amplitudes for harmonic functions. `imagesc` in MATLAB can be used to generate the heat map. The following code is used to recreate Figure 12.61. Please make comments on the rotation directions and the phase angles of these harmonics.

```
figure (6)

vv = [-10 10];
uu = [10 -10];

d = imagesc (vv, uu, FFT2_abs_shifted_cut_flipud);
set(gca,'YDir','normal')
caxis([min(min(FFT2_abs_shifted_cut_flipud))
max(max(FFT2_abs_shifted_cut_flipud))])
colorbar;
set(d,'AlphaData',~isnan(FFT2_abs_shifted_cut_flipud))
xlabel('Spatial order, v')
ylabel('Temporal order, u')


figure (7)

vv = [-10 10];
uu = [10 -10];

d = imagesc (vv, uu, phase_angle_matrix_cut_flipud);
set(gca,'YDir','normal')
caxis([min(min(phase_angle_matrix_cut_flipud))
max(max(phase_angle_matrix_cut_flipud))])
colorbar;
set(d,'AlphaData',~isnan(phase_angle_matrix_cut_flipud))
xlabel('Spatial order, v')
ylabel('Temporal order, u')
```

8. Superposition of Surface Waves

Figures 12.82, 12.85, 12.86, and 12.92 show four cases of superposition of two surface waves. Please note that each figure contains three subplots: two for the components and one for the sum. The analytical expressions for these four cases are as follows:

$$d_{case\_1}(t,\alpha) = \cos(2\omega_{mech}t + 2\alpha) + \cos\left(-2\omega_{mech}t + 2\alpha + \frac{\pi}{3}\right)$$

$$d_{case\_2}(t,\alpha) = \cos(4\omega_{mech}t + 2\alpha) + \cos\left(-4\omega_{mech}t + 2\alpha + \frac{\pi}{3}\right)$$

$$d_{case\_3}(t,\alpha) = \cos(2\omega_{mech}t + 2\alpha) + 2\cos\left(-2\omega_{mech}t + 2\alpha + \frac{\pi}{3}\right)$$

$$d_{case\_4}(t,\alpha) = 2\cos(4\omega_{mech}t + 2\alpha) + \cos(2\alpha)$$

a. Please use the following MATLAB code to reproduce the four figures of Figures 12.82, 12.85, 12.86, and 12.92. In these figures, only contour plots are used. However, in the MATLAB program given below, both contour and mesh plots are generated. Please reproduce the program and update the values for u, v, amplitude and phase angles, as summarized in Table Q.12.1.

Please update the parameters of the two surfaces according to Table Q.12.2. Please make comments on how the phase angles of the two surfaces affect the sum surface.

b. Please modify the code in Question 7-3c and 7-4d to verify the amplitudes and phase angles for the harmonic contents.

TABLE Q.12.1

Parameters for Superposing the Waves

Parameter	Figure 12.82	Figure 12.85	Figure 12.86	Figure 12.92
time_order_1	2	4	2	4
sp_order_1	2	2	2	2
phase_angle_1	0	0	0	0
amplitude_1	1	1	1	2
time_order_2	−2	−4	−2	0
sp_order_2	2	2	2	2
phase_angle_2	$\pi/3$	$\pi/3$	$\pi/3$	0
amplitude_2	1	1	2	1

TABLE Q.12.2

Parameters for Updating

Parameter	Case 1	Case 2	Case 3	Case 4
time_order_1	2	2	2	2
sp_order_1	2	2	2	2
phase_angle_1	0	0	$-\pi$	$-\pi$
amplitude_1	1	1	1	1
time_order_2	−2	−2	−2	−2
sp_order_2	2	2	2	2
phase_angle_2	0	π	0	π
amplitude_2	1	1	1	1

```
%% Initialization

close all; clear all; clc

% Spatial dimension

space_num = 1024;
sp_step = 2*pi/space_num; % unit: s
sp_vector = [0:sp_step:sp_step*(space_num-1)];

% Temporal dimension

time_num = 1024;
time_step = 1/time_num; % unit: s
freq_mech = time_num/2^10; % 1 Hz
time_vector = [0:time_step:time_step*(time_num-1)];

% Generate temporal and spatial planes

[sp_plane, time_plane] = meshgrid(sp_vector,time_vector);

%% Prepare (progressive) surface waves

time_order_1 = 4; % u, to be updated
sp_order_1 = 2;   % v, to be updated
phase_angle_1 = 0; % to be updated
amplitude_1 = 2; % to be updated

time_order_2 = 0;  % u, to be updated
sp_order_2 = 2;    % v, to be updated
phase_angle_2 = 0; % to be updated
amplitude_2 = 1; % to be updated

pl_wave_1 = amplitude_1*cos(2*pi*time_plane*freq_mech*time_order_1 +
sp_plane*sp_order_1 + phase_angle_1);

pl_wave_2 = amplitude_2*cos(2*pi*time_plane*freq_mech*time_order_2 +
sp_plane*sp_order_2 + phase_angle_2);

pl_wave_sum = pl_wave_1 + pl_wave_2;

%% Plot surfaces

[xq,yq] = meshgrid(sp_vector (1:16:end), time_vector (1:16:end)); % reduce
mesh density for plotting purposes

figure (1)

sparse_pl_wave_1 = griddata(sp_vector,time_vector,pl_wave_1,xq,yq);
mesh(xq/2/pi*360, yq, sparse_pl_wave_1)
colorbar
caxis([min(min(pl_wave_sum)) max(max(pl_wave_sum))])
zlim([min(min(pl_wave_sum)) max(max(pl_wave_sum))]);
xlabel('Circumferential position, \alpha [mech. deg.]')
ylabel('Time, t [s]')
zlabel('f_1 (t, \alpha)')
set(gca,'XTick',[0:60:360])
title('Mesh plot of component surface wave');



figure (2)

sparse_pl_wave_2 = griddata(sp_vector,time_vector,pl_wave_2,xq,yq);
mesh(xq/2/pi*360, yq, sparse_pl_wave_2)
colorbar
caxis([min(min(pl_wave_sum)) max(max(pl_wave_sum))])
zlim([min(min(pl_wave_sum)) max(max(pl_wave_sum))]);
xlabel('Circumferential position, \alpha [mech. deg.]')
ylabel('Time, t [s]')
zlabel('f_2 (t, \alpha)')
set(gca,'XTick',[0:60:360])
title('Mesh plot of component surface wave');
```

References

1. Benaroya, H., and M. L. Nagurka. *Mechanical Vibration: Analysis, Uncertainties, and Control*. CRC Press, Boca Raton, FL, 2009.
2. Daniel, J. I. *Engineering Vibration*. Fourth ed. Pearson Education, Upper Saddle River, NJ, 2014.
3. Giancarlo, G. *Vibration Dynamics and Control*. Springer Science & Business Media, New York, 2008.
4. Wang, C. Y., and C. M. Wang. *Structural Vibration: Exact Solutions for Strings, Membranes, Beams, and Plates*. CRC Press, London, UK, 2013.
5. Arthur, W. L., and M. S. Qatu. *Vibration of Continuous Systems*. McGraw Hill Professional, New York, 2011.
6. Dunn, F., W. M. Hartmann, D. M. Campbell, and H. N. Fletcher. *Springer Handbook of Acoustics*. Springer, New York, 2015.
7. Jacek, F. G., C. Wang, and J. C. Lai. *Noise of Polyphase Electric Motors*. CRC Press, Boca Raton, FL, 2005.

13

<hr/>

Noise and Vibration in Switched Reluctance Machines

James Weisheng Jiang, Jianbin Liang, Jianning Dong,
Brock Howey, and Alan Dorneles Callegaro

CONTENTS

Acoustic noise and vibration is the most well-known issue in Switched Reluctance Machine (SRM). Due to the salient pole construction of SRMs, when a phase is excited with current, the flux penetrates into the rotor, mostly in the radial direction, and generates radial forces. These radial forces deform the stator core and frame, which results in vibration and acoustic noise. Acoustic noise and vibration in SRMs can be reduced through proper analysis and optimization. Noise and vibration in SRMs might not be a problem for some applications, for example a mining operation where the noise generated by the motor is less than the ambient noise. However, there are many occasions where the noise levels might need to be reduced to certain acceptable limits. For example, for an electric machine that is used as a traction motor for an electric vehicle or a hybrid electric passenger vehicle, if the forcing frequencies approaches the natural frequencies of the mechanical system, resonance can occur. The building noise and vibration energy will then be transmitted to the driver and passengers. In the design and control of SRMs used in noise-sensitive applications, reducing noise, vibration, and harshness (NVH) must be given a priority in the design process. In Chapter 12, the fundamentals and the tools to study noise and vibration were discussed. This chapter will first investigate the calculation of the natural frequencies for SRMs. Then, the features of radial force density will be discussed. Harmonic content of the radial force density waveform will be studied using a few SRM topologies. Towards the end of the chapter, a complete process used for acoustic analysis will be introduced. The procedures of acoustic study using numerical methods will be discussed as well. The chapter also includes a discussion of the current methods that are used for noise reduction in SRMs.

13.1 Sources of Noise

In the case of vehicles, sources of noise can be classified into a few categories: body, chassis, combustion engine, tires, electric drive, wind noise, and diesel knocking, as shown in Figure 13.1. The body and chassis of a car contribute mostly to low-frequency noise because of their relatively larger masses. The frequency range of acoustic noise from an engine depends not only on the engine type, structure, exhaust, and so forth, but also on the operating speed and load. The electric drive typically contributes to a higher frequency of noise than the combustion engine because of the high rotational speed of the electric machine and the switching frequency of the power converter. Noise and vibration generated by electric machines can be classified into four categories, as shown in Figure 13.2. When magnetic flux passes through the air gap between the stator and rotor in an SRM, radial force is produced. This radial force deforms the stator core and the frame, which results in vibration and acoustic noise.

Mechanical components and connections also contribute to the noise production in an SRM. For example, in a motor testing setup, shaft misalignment between the SRM under test and the load will contribute to noise production. Bearing problems, rotor unbalance, eccentricity between the rotor and stator, manufacturing and assembly faults, such as loose components, and improper motor installation all lead to worse noise and vibration output in a motor. While the rotor of an SRM rotates, it moves the air that it comes in contact with,

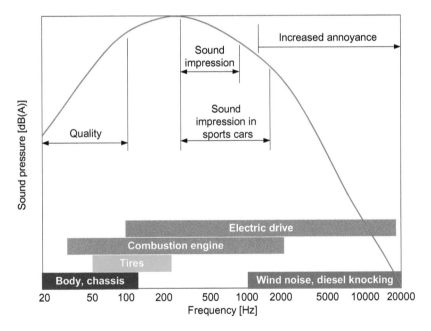

FIGURE 13.1
Noise sources and sound perception in automobiles. (From Bosing, M., Acoustic modeling of electrical drives: Noise and vibration synthesis based on force response superposition, Lehrstuhl und Institut für Stromrichtertechnik und Elektrische Antriebe, 2014.)

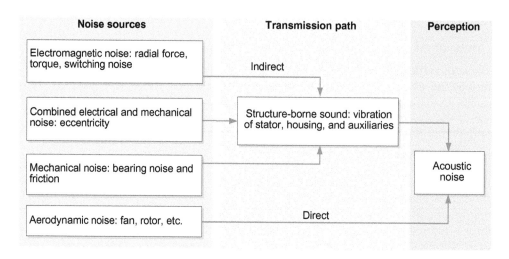

FIGURE 13.2
Sources of noise and vibration for motors. (From Gieras, J.F. et al., *Noise of Poly-Phase Electric Motors*, CRC Press, New York, 2005.)

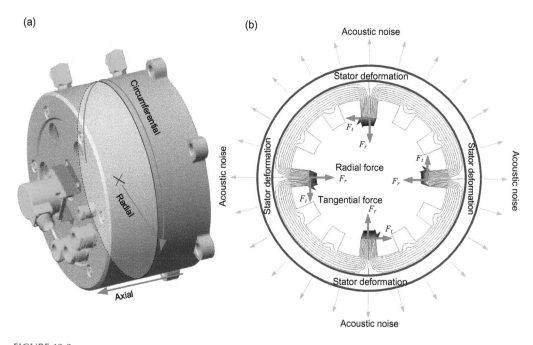

FIGURE 13.3
Generation of vibration and noise in switched reluctance motors: (a) definition of directions and (b) forces, deformation, and acoustic noise.

which is an aerodynamic source of noise that will increase mechanical losses (known specifically as windage loss). The motor controller switching action is also a source of noise. In a motor, vibration can be defined in three dimensions as shown in Figure 13.3a. Excitations from the harmonics of the radial force shown in Figure 13.3b are the major source of vibration and acoustic noise in the stator-frame system, which will be discussed thoroughly in this chapter. In the case of SRMs, the deformation of stator core due to radial forces, contributes significantly to noise and vibration.

As discussed in Chapter 12, vibrational resonance, in the form of increasing amplitude of oscillation for a system, happens when the natural frequencies of a system are close to forcing frequencies acting on the structure. The natural frequency is defined as the frequency at which a system's main mode of vibration oscillates without any external frequential forces. In a motor system, the stator-frame structure acts as the major source of noise. If the forcing frequencies of the radial force harmonics approach the natural frequencies of the stator-frame structure for a certain mode shape, resonance occurs in the stator-frame structure, causing a buildup of noise and vibration.

13.2 Procedures of Investigating Noise and Vibration

Figure 13.4 shows a flow chart for a process investigating noise and vibration of SRMs. This approach calculates the generated frequential forces acting on the motor throughout its operating range and completes a structural analysis to generate the natural frequencies. Then the information is combined to calculate the acoustic response of the motor. The process starts with the geometry parameters for the SRM. The radial force density waveform is generated based on the operating condition of the motor (speed and torque). The radial force density waveform can be obtained using the Maxwell stress tensor or from electromagnetic Finite Element Analysis (FEA). Once calculated, the radial force density waveform is decomposed into its harmonics and their corresponding circumferential orders and forcing frequencies. Then the geometries of a stator-frame system with material properties, such as mass, Young's modulus, and Poisson's ratio, are used to define the different vibrational modes and calculate their respective natural frequencies and damping ratios. Vibration and modal analysis can be conducted using either

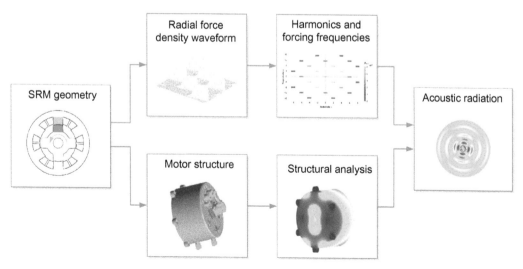

FIGURE 13.4
Process of investigating noise and vibration.

analytical methods or structural FEA (i.e. ACTRAN™). Finally, once the radial force harmonics and natural frequencies are determined, the overall noise radiation can be studied based on the vibration results by using analytical methods, experimental methods, or acoustic numerical methods.

13.3 Calculation of Natural Frequencies

In Chapter 12, the vibrational mode shapes of cylindrical shells were discussed. Conventionally, the natural frequencies of the stator of an electric motor can be estimated by treating the stator core as layers of cylindrical shells loaded with teeth and concentrated windings. First, a cylindrical shell model based on the Donnell-Mushtari theory [2] will be discussed. The frame is taken into account by summing up the lumped stiffness and mass with those of the stator core (assuming that the stator and the frame are rigidly connected).

13.3.1 Cylindrical Shell

Figure 13.5 shows an example of a circular cylindrical shell and its coordinate system. If only the circumferential mode (*circ*) is considered, the natural frequency of the stator system, $f_{n(circ)}$, for the circumferential mode *circ* is expressed as [2]:

$$f_{n(circ)} = \frac{1}{2\pi} \sqrt{\frac{K_{(circ)}}{M}} \tag{13.1}$$

where $K_{(circ)}$ (N/m) is the lumped stiffness of the circumferential mode *circ* and M is the lumped mass for the cylindrical shell. Since the stator yoke can be modeled as a cylindrical shell, (13.1) can be used to estimate the natural frequencies for the stator yoke:

$$f_{n(circ)} = \frac{1}{2\pi} \sqrt{\frac{K_{yoke(circ)}}{M_{yoke}}} \tag{13.2}$$

As the primary stiffness and mass source for the stator system, the stator yoke contributes the most to the natural frequencies. When the stator yoke is treated as a cylindrical shell, the lumped mass of the stator yoke in Figure 13.5 can be estimated as:

$$M_{yoke} = \pi \rho_s D_c L_s y_s \tag{13.3}$$

where ρ_s is the density of the stator core, D_c is the mean diameter of the stator yoke (as shown in Figure 13.5), L_s is the axial length of the stator core, and y_s is the stator yoke thickness. Please note that the stacking factor of the stator core can also be considered in estimating the mass of the stator yoke. If the stator yoke is regarded as an infinitely long cylindrical shell, according to the Donnell-Mushari theory, the lumped stiffness is solved as:

$$k_{yoke(circ)} = \frac{4\Omega_{(circ)}^2}{D_c} \frac{\pi L_s y_s E_s}{1 - v_s^2} \tag{13.4}$$

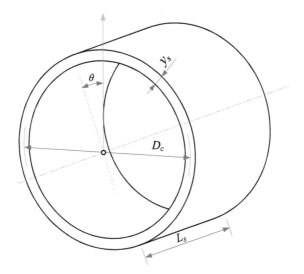

FIGURE 13.5
Circular cylindrical shell and its coordinate system.

where E_s and v_s are the elastic (or Young's) modulus and Poisson's ratio of the stator material; $\Omega_{(circ)}$ is the root of the characteristic equation of motion of the cylindrical shells [3]:

$$\begin{cases} \Omega_{(0)}{}^2 = 1, \quad \text{for } circ = 0 \\[2ex] \Omega_{(1)}{}^2 = \frac{1}{2}\left[1 + circ^2 + \kappa \cdot circ^4\right] + \\[2ex] \qquad \frac{1}{2}\sqrt{\left[1 + circ^2 + \kappa \cdot circ^4\right] - 4\kappa \cdot circ^6} \quad \text{for } circ = 1 \\[2ex] \Omega_{(circ)}{}^2 = \frac{1}{2}\left[1 + circ^2 + \kappa \cdot circ^4\right] - \\[2ex] \qquad \frac{1}{2}\sqrt{\left[1 + circ^2 + \kappa \cdot circ^4\right]^2 - 4\kappa.circ^6} \quad \text{for } circ > 1 \end{cases} \qquad (13.5)$$

where κ is a non-dimensional parameter and defined as:

$$\kappa = \frac{y_s}{\sqrt{3}D_c} \qquad (13.6)$$

The analysis of noise and vibration commonly uses three types of materials: fluid, isotropic solid, and porous materials. Some examples of properties that are determined for a fluid (such as air) are speed of sound and density. Examples of properties that are determined for an isotropic solid are the Young's modulus, density, and Poisson's ratio. These parameters are listed for a few materials in Table 13.1.

TABLE 13.1

Properties of Some Materials

Material	Young's Modulus [GPa]	Poisson's Ratio	Solid Density [g/cm³]
Aluminum (6160)	71	0.33	2.77
42CrMo4, steel	210	0.288	7.83
20C15, steel	210	0.288	7.85
280–520 N, steel	210	0.288	7.85
Electric steel stack	180	0.3	7.42
10JNEX900, electric steel lamination	210	0.33	7.49

Source: Michael, D.K., and Norton, P., *Fundamentals of Noise and Vibration Analysis for Engineers*, Cambridge University Press, Cambridge, UK, 2003.

13.3.2 Consideration of Teeth and Coils

Figure 13.6a shows a 3D model for an SRM stator. The stator poles are assumed to add additional mass without adding additional stiffness to the stator core. Figure 13.6b shows the stator-winding system for an SRM. The coils of the SRM are concentrated around the stator teeth, meaning that the teeth and coils do not form a stiff ring at the motor endings unlike motors with distributed windings. Therefore, it is assumed that they only provide additional mass instead of additional stiffness to the stator core. The natural frequency of the stator core with the teeth and the coils as pure mass is calculated as [2]:

$$f_{n(circ)} = \frac{1}{2\pi} \sqrt{\frac{K_{yoke(circ)}}{M_{yoke} + M_{teeth}}} \tag{13.7}$$

where $K_{yoke(circ)}$ and M_{yoke} are the lumped stiffness and mass of the stator yoke calculated from (13.3) and (13.4), respectively; M_{teeth} is the lumped mass of the teeth and coils, including the laminations, conductors, and insulation. Figure 13.7 shows the mode shapes for circumferential modes $circ = \{2, 3\}$, for the stator of a 6/4 SRM.

Encapsulated stators are extensively used in SRMs. The stator core and windings can be encapsulated together by an insulation material, which provides additional protection and enhanced heat dissipation. From structural and noise perspectives, the encapsulation also provides additional stiffness to the stator since it bonds the teeth and coils together to form a closed ring. If the encapsulated teeth with coils are simplified as a ring with uniform properties, which we call the teeth-coil region, its stiffness $K_{teeth(circ)}$ can be calculated using the same approach as the stator yoke:

$$K_{teech(circ)} = \frac{\Omega^2_{(circ)}}{R_t^2} \frac{E_t V_t}{1 - v_t^2} \tag{13.8}$$

where Ω_{circ} is solved from (13.5) with geometric parameters of the teeth-coil region; E_t is the equivalent elasticity modulus and v_t is the Poisson ratio. R_t and V_t are the mean radius and volume of the teeth-coil region, respectively, and they are calculated by:

$$R_t = D_c / 2 - y_s / 2 - h_s / 2 \tag{13.9}$$

$$V_t = 2\pi R_t h_s \left(L_s + 2h_{ov} \right) \tag{13.10}$$

where h_s is the tooth height, and h_{ov} is the one-sided axial overhang length of the winding ends.

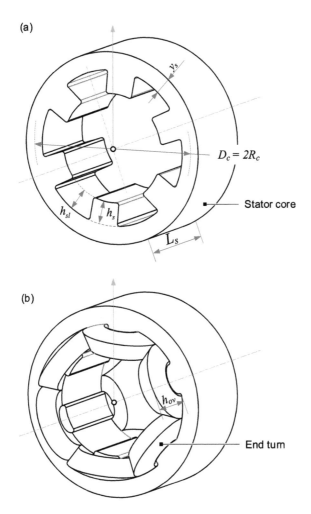

FIGURE 13.6
Examples of an SRM: (a) stator system and (b) stator-winding system.

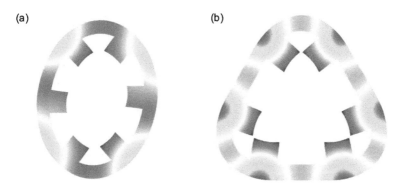

FIGURE 13.7
Mode shapes of SRM stator: (a) *circ* = 2 and (b) *circ* = 3.

13.3.3 Stator-Frame System

Figure 13.8 shows the stator-winding-frame system. The stator frames are often thin cylindrical structures with both ends constrained, which can be described as supported or clamped. The constraints of supported-supported, clamped-free, clamped-clamped structures have been discussed in Chapter 12. Under these conditions, the natural frequencies of the stator frame are related to both circumferential modes $circ = \{0, 1, 2...\}$ and axial modes $ax = \{1, 2, 3...\}$. Please note that axial mode starts from $ax = 1$, which was discussed in Chapter 12. If the frame is considered as a shell, based on the shell theory [2], for circumferential mode $circ$ and axial mode ax, the characteristic equation is written as:

$$\Omega_{(circ,ax)}^6 - \left(C_2 + \kappa\Delta C_2\right)\Omega_{(circ,ax)}^4 + \left(C_1 + \kappa\Delta C_1\right)\Omega_{(circ,ax)}^2 - \left(C_0 + \kappa\Delta C_0\right) = 0 \qquad (13.11)$$

According to the Donnell-Mushtari theory, the coefficients of (13.11) are listed as [2]:

$$\begin{cases} C_0 = \frac{1}{2}\left(1-v_f\right)\left[\left(1-v_f^2\right)\lambda^4 + \kappa^2\left(circ^2 + \lambda^2\right)^4\right] \\[2mm] C_1 = \frac{1}{2}\left(1-v_f\right)\left[\left(3+2v_f\right)\lambda^2 + circ^2 + \left(circ^2 + \lambda^2\right)^2 + \frac{3-v_f}{1-v_f}\kappa^2\left(circ^2 + \lambda^2\right)^3\right] \\[2mm] C_2 = 1 + \frac{1}{2}\left(3-v_f\right)\left(circ^2 + \lambda^2\right) + \kappa^2\left(circ^2 + \lambda^2\right)^2 \\[2mm] \Delta C_0 = 0 \\[1mm] \Delta C_1 = 0 \\[1mm] \Delta C_2 = 0 \end{cases} \qquad (13.12)$$

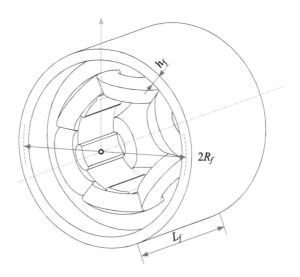

FIGURE 13.8
Stator-winding-frame system.

where v_f is the Poisson ratio of the frame. κ can be calculated using the geometric parameters of the frame:

$$\kappa = \frac{h_f}{2\sqrt{3}R_f} \tag{13.13}$$

where R_f and h_f are the mean radius and thickness of the frame, respectively, as shown in Figure 13.8. If the constraint of the frame is supported-supported, the non-dimensional parameter λ depends on the axial mode, ax and can be calculated as:

$$\lambda_{(ax)} = ax \cdot \pi \frac{R_f}{L_f} \tag{13.14}$$

where L_f is the frame length. If the frame is clamped on both ends, λ is modified as given in [4]:

$$\begin{cases} \lambda_{(ax)} = \dfrac{ax \cdot \pi R_f}{L_f - L_0} \\[3mm] L_0 = L_f \dfrac{0.3}{n+0.3} \end{cases} \tag{13.15}$$

For fixed values of $circ$ and ax, there are three roots for $\Omega_{(circ,ax)}^2$ in (13.11). The lowest one of them is associated with the natural frequency of the flexural vibration of the frame. The lumped stiffness $K_{frame(circ,ax)}$ and mass M_{frame} of the frame can be calculated as:

$$\begin{cases} K_{frame(circ,ax)} = \dfrac{\Omega_{(circ,ax)}^2}{R_f^2} \dfrac{E_f V_f}{1-v_f^2} \\[3mm] V_f = 2\pi R_f L_f h_f \\[2mm] M_{frame} = \rho_f V_f \end{cases} \tag{13.16}$$

where E_f and ρ_f are the elasticity modulus and mass density of the frame, respectively.

The natural frequency of the frame can be calculated as:

$$f_{n(circ,ax)} = \frac{1}{2} \frac{\Omega_{(circ,ax)}}{R_f} \sqrt{\frac{E_f}{\rho_f \left(1-v_f^2\right)}} \tag{13.17}$$

The frames in the electric motors usually have end bells (or end caps) at both ends constrained in the in-plane directions. Therefore, the frame can be modeled as a cylindrical shell with a clamped-clamped constraint. Figure 13.9 shows the two mode shapes of a typical stator-frame system.

Most electrical machines have the stator cores press-fit into the structural frames. This creates a mechanical coupling between the stator core and the frame. This is especially true for small machines, where each stator lamination is stamped for the entire stator cross section rather than for a segment of the stator as in large machines. For the stiffness and mass calculations, the stator core and the frame are both considered as cylindrical shells.

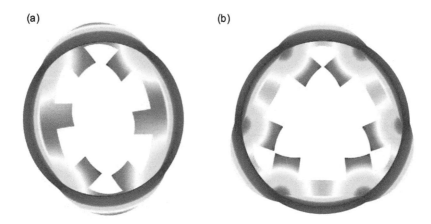

FIGURE 13.9
Mode shapes of stator-frame system: (a) *circ* = 2 and (b) *circ* = 3.

Therefore, the stator-frame system can be considered as a two-layer coaxial cylindrical shell. The resultant lumped stiffness and mass are calculated from each layer:

$$
\begin{cases}
K_{sum(circ,ax)} = K_{yoke(circ)} + K_{teeth(circ)} + \lambda_f K_{frame(circ,ax)} \\
M_{sum} = \lambda_{(circ)}\left(M_{yoke} + M_{teeth} + M_{frame} \right)
\end{cases}
\tag{13.18}
$$

As discussed earlier in this chapter, $K_{yoke(circ)}$, $K_{teeth(circ)}$, and $K_{frame(circ)}$ are the lumped stiffness for the stator yoke, teeth-coil region, and the frame, and M_{yoke}, M_{teeth}, and M_{frame} are the masses for the stator yoke, teeth-coil region, and the frame. The λ_f and λ_{circ} are factors related to the thickness of the frame, which can be calculated as:

$$
\begin{cases}
\lambda_f = \dfrac{R_f h_{ef}}{R_{ef} h_f} \\[2ex]
R_{ef} = \dfrac{R_f + R_c}{2} \\[2ex]
h_{ef} = h_f + y_s \\[2ex]
\lambda_{circ} = \begin{cases}
1, & circ = 0 \\[1ex]
\dfrac{circ^2 + 1}{circ^2}, & circ \geq 1
\end{cases}
\end{cases}
\tag{13.19}
$$

where R_f is the mean radius of the frame, h_f is the thickness of the frame, y_s is the stator yoke thickness, R_c is the mean radius of the stator yoke (as shown in Figure 13.8), and *circ* is the circumferential mode. Then, the natural frequency of the stator-frame system is calculated by:

$$
f_{n(circ,ax)} = \frac{1}{2\pi} \sqrt{\frac{K_{sum(circ,ax)}}{M_{sum}}}
\tag{13.20}
$$

The analytical method for calculating natural frequencies is discussed in this section. Later in this chapter, calculations for the natural frequencies using a numerical method will also be discussed.

13.4 Nodal Force Density

When the stator of an SRM is subjected to electromagnetic forces, the structure vibrates. The pressure and shear stresses caused by electromagnetic forces is called the force density, and for rotating machines, they are expressed with radial and tangential force densities. Both radial and tangential forces acting on the stator core can cause deformation of the stator structure, which results in structural vibration. However, the radial vibrations are a major source of the radiated noise. In this section, the characteristics of the radial force density and tangential force density will be discussed.

13.4.1 Radial Force Density

The air gap radial force density is defined as the magnetic force per unit area [N/m²]. It is often calculated from electromagnetic FEA models using either the virtual work method or the Maxwell stress tensor method as discussed in Chapter 2. The radial magnetic pressure or radial force density at a certain rotor position at time t and in the air gap spatial position of α can be calculated using the Maxwell stress tensor:

$$p_r(t,\alpha) = \frac{1}{2\mu_0}\left[B_r^2(t,\alpha) - B_t^2(t,\alpha)\right] \tag{13.21}$$

where B_r is the flux density in the radial direction, and B_t is the flux density in the tangential direction.

Figure 13.10 shows the surface wave of radial force density for a 6/4 SRM over spatial or circumferential position (mech. deg.) and time (or rotational angle in elec. deg. or mech. deg.). The phase excitation sequence is A-B-C. Thus, the phase excitation rotates in counterclockwise direction. When phase A is excited, two spikes in the radial force density waveform can be seen lasting for about 120 electrical degrees, which equals approximately 30 mechanical degrees. When the radial force density of phase A dies, phase B is excited, and two peaks at the locations of the two stator poles of phase B emerge. This pattern continues to Phase C, and one electrical cycle finishes with the radial force density of the phase C diminishing. By integrating the radial force density over the air gap surface, the total radial force, which causes radial deformation on the stator structure, can be calculated.

The radial force density $p_r(t,\alpha)$ has similar characteristics to that of a travelling wave. The magnitude of the waveform changes with time (at different rotor positions as the rotor rotates), as shown in Figure 13.11a, and with space (at different positions in the airgap), as shown in Figure 13.11b. Figure 13.11a shows that there is only one major pulsation on the radial force density waveform with time, because in every electrical cycle each phase is excited once. Figure 13.11b shows that there are two major

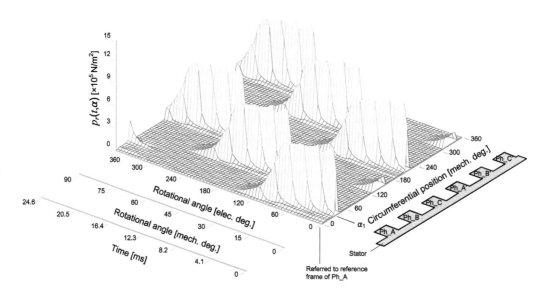

FIGURE 13.10
Air-gap radial force density of a 6/4 SRM in one electrical cycle.

peaks with the circumferential position at a certain time step. This happens because in a three-phase 6/4 SRM there are two stator poles per phase. As for a three-phase 24/16 SRM, eight poles in one phase conduct at the same time, so eight major peaks would appear at one time instant. One electrical cycle (360 elec. deg.) of a 24/16 SRM is 22.5 mech. deg.

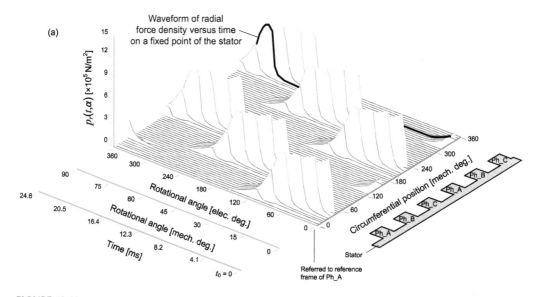

FIGURE 13.11
Decomposition of air-gap radial force density of a 6/4 SRM in one electrical cycle: (a) radial force density versus time at a fixed point on the stator.   (*Continued*)

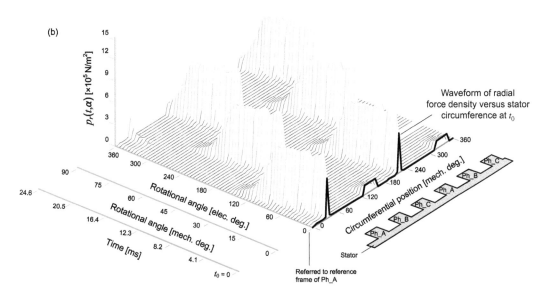

FIGURE 13.11 (Continued)
Decomposition of air-gap radial force density of a 6/4 SRM in one electrical cycle: (b) radial force density versus stator circumferential position at t_0.

13.4.2 Characteristics of Radial Force Density

For one mechanical revolution, a 6/4 SRM experiences 4 electrical cycles per phase, because the number of rotor poles, N_r is 4. As shown in Figure 13.12, for each electrical cycle (90 mech. deg.), there is 6 major surges on the radial force density waveform. For a mechanical cycle, there are 24 surges on the surface in total and 12 pairs of radial force surges acting on the stator core, as shown in Table 13.2.

As shown in Figure 13.13, this 2D map shows the radial force density in a 2D plane, whose axes are rotational angle and spatial angle. If we trace the radial force density only along the axis of rotation, it can be seen that the number of divisions of the radial force density over a mechanical cycle is $N_r N_{ph}$, which is the number of strokes. For a 6/4 SRM, $N_r N_{ph}$ equals 12.

Similarly, if we trace the radial force density only along the axis of spatial angle, it can be seen that the total number of divisions along this axis is N_s. For a 6/4 SRM, this value is 6. The rotational direction of phase excitation can also be interpreted from the map. The next electrical cycle appears at a higher rotational angle, which is due to the fact that the rotational directions of phase excitation and radial force waveform are counter-clockwise.

13.4.3 Tangential Force Density

Similarly, the tangential force density waveform for a 6/4 SRM over the spatial position and time can also be obtained, as shown in Figure 13.14. One obvious feature of the waveform is that when one stator pole is energized, a group of spikes can be observed.

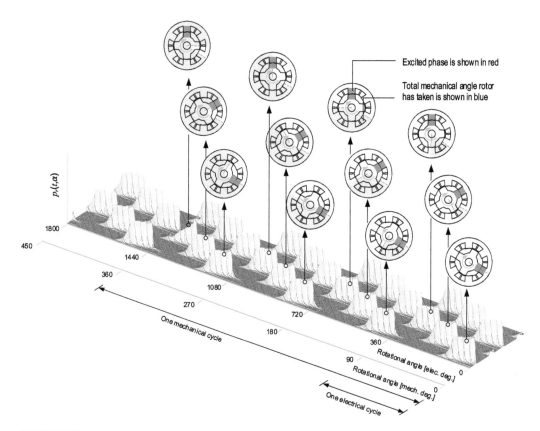

FIGURE 13.12
Radial force density waveform over one mechanical cycle in a 6/4 SRM.

TABLE 13.2

Map of Pulsations on the Radial Force Density Waveform for 6/4 SRM

			\multicolumn{12}{c}{Rotational Angle [mech. deg.]}											
			345	315	285	255	225	195	165	135	105	75	45	15
			4th Elec. Cycle			3rd Elec. Cycle			2nd Elec. Cycle			1st Elec. Cycle		
\multirow{6}{*}{Circumferential position [mech. deg.]}	\multirow{3}{*}{2nd pole}	330	Ph_C'			Ph_C'			Ph_C'			Ph_C'		
		270		Ph_B'			Ph_B'			Ph_B'			Ph_B'	
		210			Ph_A'			Ph_A'			Ph_A'			Ph_A'
	\multirow{3}{*}{1st pole}	150	Ph_C			Ph_C			Ph_C			Ph_C		
		90		Ph_B			Ph_B			Ph_B			Ph_B	
		30			Ph_A			Ph_A			Ph_A			Ph_A

As discussed in Chapter 2, by integrating the tangential force density over the air gap circumference and multiplying the result by the distance between the air gap and the center of the rotor, electromagnetic torque can be calculated. The spikes on the surface wave for the tangential force density in Figure 13.14 are ripples which cause torque ripple.

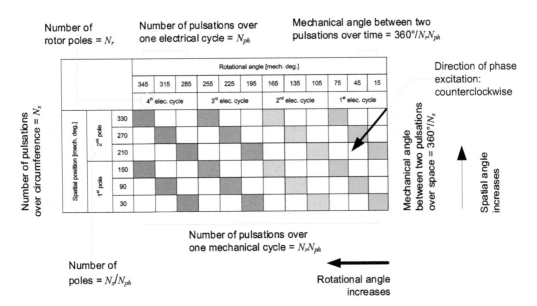

FIGURE 13.13
The map for pulsations of radial force density, 6/4 SRM.

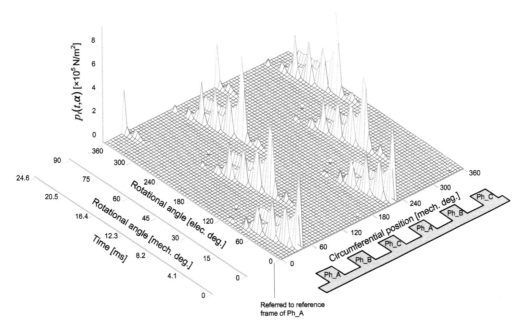

FIGURE 13.14
Air-gap tangential force density of a 6/4 SRM in one electrical cycle.

Figure 13.15a shows the variation of tangential force density over time and it can be seen that one stator pole is energized over an electrical cycle at a certain circumferential position. Since the surface wave of the tangential force density is obtained from a three-phase 6/4 SRM, it can be seen that, at each time point, there are two stator poles energized in Figure 13.15b. Even though the tangential force does not contribute to the vibration of

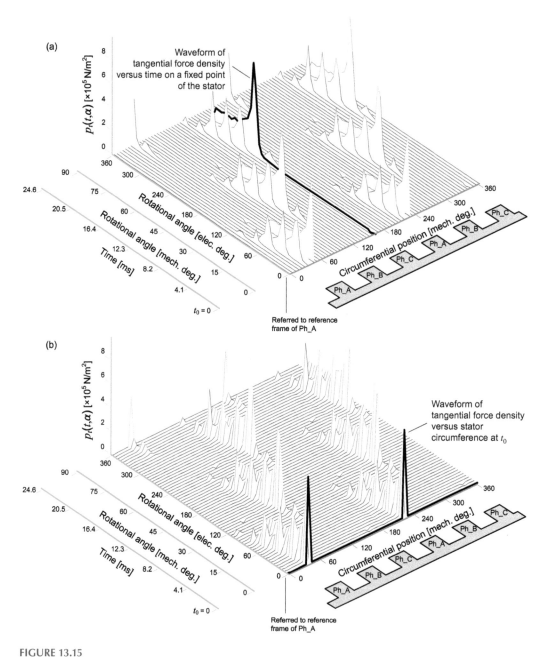

FIGURE 13.15
Decomposition of air-gap tangential force density of a 6/4 SRM in one electrical cycle: (a) tangential force density versus time at a fixed point on the stator and (b) tangential force density versus stator circumferential position at t_0.

the whole motor significantly, it can still cause deformation of the stator poles. So far, the features for radial and tangential force density have been discussed. As mentioned earlier, radial force contributes significantly to the noise and vibration of the motor. In the next section, a 2D FFT is used to identify the harmonics and forcing frequencies for the radial force density in SRMs.

13.5 Decomposition of Radial Force Density

As discussed in Chapter 12, a surface wave can be decomposed into a series of cosine functions using 2D FFT. In this section, 2D FFT is used to identify the harmonics of the surface wave for radial force density and it will help develop an understanding on the construction of the radial force density. This is important in the estimation of vibration and radiated noise, as well as the reduction of the noise and vibration. Some fundamental features of the harmonics will also be discussed.

13.5.1 2D Decomposition

As shown in Figure 13.16, if 2D FFT is performed on the radial force density wave, $p_r(t,\alpha)$, over one mechanical cycle, the harmonic content can be expressed by a series of cosine functions $p_{r(u,v)}$:

$$
\begin{aligned}
p_r(t,\alpha) &= \sum_{v=-\infty}^{\infty}\sum_{u=-\infty}^{\infty}\left(p_{r(u,v)}\right)\\
&= \sum_{v=-\infty}^{\infty}\sum_{u=-\infty}^{\infty}\left[P_{r(u,v)}\cos\left(\omega_{mech}ut+v\alpha+\phi_{(u,v)}\right)\right]\\
&= \sum_{v=-\infty}^{\infty}\sum_{u=-\infty}^{\infty}\left[P_{r(u,v)}\cos\left(2\pi f_{mech}ut+v\alpha+\phi_{(u,v)}\right)\right]
\end{aligned}
\tag{13.22}
$$

where u is the temporal order, v is the circumferential order, $P_{r(u,v)}$ is the amplitude, which is greater than zero for the harmonic order (u, v); ω_{mech} is the mechanical angular frequency, which can be calculated based on the motor speed. The $\phi_{(u,v)}$ is the phase angle for harmonic order (u, v), and f_{mech} is the mechanical frequency. Thus, the analytical expression of the surface wave for the harmonic order (u, v) is:

$$
p_{r(u,v)} = P_{r(u,v)}\cos\left(\omega_{mech}ut+v\alpha+\phi_{(u,v)}\right)
\tag{13.23}
$$

If the axial domain $(ax = 1, 2, \ldots)$ is considered, equation (13.22) can be expressed as:

$$
p_{r(t,\alpha)} = \sum_{u=-\infty}^{+\infty}\sum_{v=-\infty}^{+\infty}\sum_{ax=1}^{+\infty}p_{r(u,v,ax)}
\tag{13.24}
$$

As discussed in Chapter 12, the positive or negative sign before $P_{r(u,v)}$ is used to express the rotational directions. A positive sign means that the harmonic content rotates in a positive direction (counterclockwise). A negative sign indicates that the harmonic content rotates in a negative direction (clockwise). If there is no rotation for a certain harmonic content, no sign is added prior to the amplitude. For instance, the DC component, $p_{r(0,0)}$, has no positive or negative sign because the DC component does not vibrate or rotate.

As discussed in Chapter 12, the forcing frequency of a certain harmonic content can be expressed as:

$$
f_{f(u,v)} = |u|f_{mech} = qf_{mech}
\tag{13.25}
$$

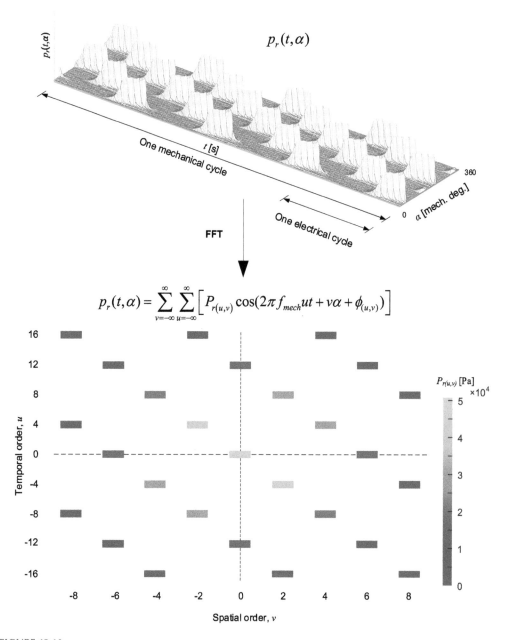

FIGURE 13.16
Decomposition of radial force density wave over one mechanical cycle using FFT, 6/4 SRM.

where q is the absolute value of temporal order u. As shown in Figure 13.16, 2D FFT is performed on the surface wave of radial force density over one mechanical cycle. If 2D FFT is performed on the wave over one electrical cycle, the temporal orders will be changed, as shown in Figure 13.17. The temporal order in this case is u_e, which is related to the electrical frequency. The relationship between u_e and u can be expressed using:

$$u = u_e N_r \tag{13.26}$$

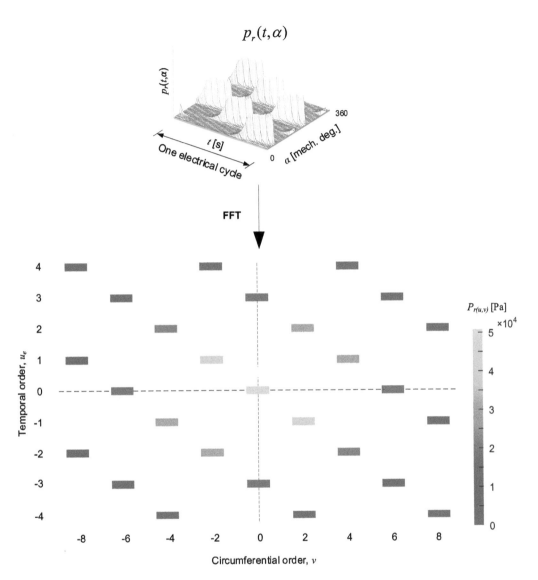

FIGURE 13.17
Decomposition of radial force density wave over one electrical cycle using FFT, $P_{r(u,v)}$, 6/4 SRM.

where N_r is the number of rotor poles. The relationship between the mechanical frequency and the electrical frequency is:

$$f_{elec} = N_r f_{mech}$$

Thus, the forcing frequency can also be calculated as:

$$f_{f(u,v)} = |u_e| N_r f_{mech} = |u_e| f_{elec} = |u| f_{mech} = q f_{mech} \tag{13.27}$$

To accelerate computation, 2D FFT can be applied over the radial force density for one electrical cycle rather than for one mechanical cycle. In this chapter, $f_{f(u,v)} = q f_{mech}$ will be used.

13.5.2 Harmonic Contents

Figure 13.18 shows the contour maps of the selected harmonics for the radial force density surface wave of a 6/4 SRM. The harmonic (12, 0) has a temporal order of 12, which means that its related forcing frequency is $12f_{mech}$. The circumferential order is 0, which can excite

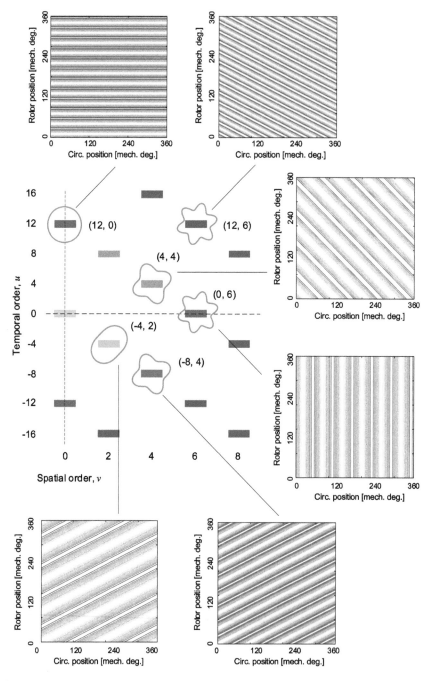

FIGURE 13.18
Contours for selected dominant harmonic contents, 6/4 SRM.

the vibration mode with a circumferential mode, $circ = 0$. The harmonic $(12, 6)$ has a temporal order 12 and a circumferential order, 6. Since the multiplication of the sign of the temporal order and the sign of the circumferential order is positive, the harmonic rotates in a negative direction as described in Section 12.2.4. It can excite circumferential mode, $circ = 6$. Its forcing frequency is also $12f_{mech}$. Please note that the summation of the harmonic $(12,6)$ and the harmonic $(-12,6)$ renders a standing wave. As described in Section 12.6.4, a standing wave is a wave in which its peaks or any other points on the wave do not move spatially.

The harmonic $(4, 4)$, rotating in a negative direction, can excite a vibration mode with a circumferential mode, $circ = 4$. The harmonic $(0, 6)$, sitting on the v axis, can cause deformation but not vibration, since the forcing frequency is zero and will not change with time. The harmonics $(-8, 4)$ and $(-4, 2)$ rotate in a positive direction. $(-8, 4)$ can excite the circumferential mode, $circ = 4$ and its forcing frequency is $8f_{mech}$. $(-4, 2)$ can excite the vibration mode with a circumferential mode, $circ = 2$ and its forcing frequency is $4f_{mech}$.

13.5.3 Rotation of the Forcing Harmonics

In order to explain the rotation of the force harmonics, we will use Figure 13.19 as an example. Figure 13.19a shows a motor with 6 stator poles but only two rotor poles. If the radial force is exerted on two of the stator poles as shown in Figure 13.20, the stator core is expected to deform in a manner as shown in Figure 13.19b. If the natural frequency for mode 2 of the stator is core is 2000 Hz, when the forcing frequency gets closer to 2000 Hz, the stator deformation increases leading to an increase in noise. Now, note that in Figure 13.19b, the stator structure only deforms in one direction. However, for a three-phase 6/4 SRM, when three phases are excited in a certain sequence over one electrical cycle, the stator structure will be distorted by a rotating radial force density, which results in a deformation of stator core in different directions, as show in Figure 13.21.

Figure 13.22 shows one mechanical cycle of a 6/4 SRM. At $\theta = 0$ mech. deg., the stator compresses vertically, the deformation of which is defined as $\varphi = 0°$ and used as a reference. When the rotor rotates in the clockwise direction for $\theta = 30$ mech. deg., using the reference just defined, it can be seen that the deformation occurs at $\varphi = 60°$. It appears as if the deformation rotates counterclockwise by 60 mech. deg. When the rotor rotates for another 30 mech. degrees from $\theta = 30°$ to $\theta = 60°$, the third phase is excited, and the deformation of the stator core happens at $\varphi = 120°$. When one electrical cycle finishes ($\theta = 90°$), the most significant deformation pattern of the rotor appears to have rotated 180 mech. deg. in the

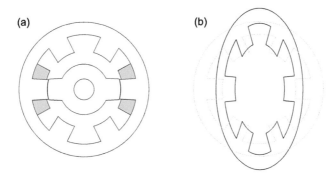

FIGURE 13.19
Radial force exerting on two stator poles: (a) original shape and (b) deformed stator core.

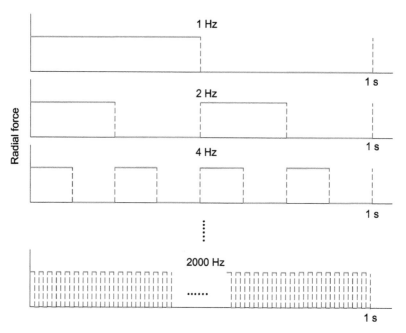

FIGURE 13.20
On and off periodic radial force with varying frequencies.

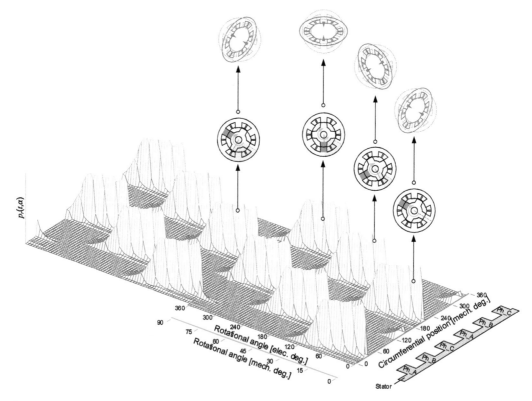

FIGURE 13.21
The deformation of stator core over one electrical cycle for a 6/4 SRM.

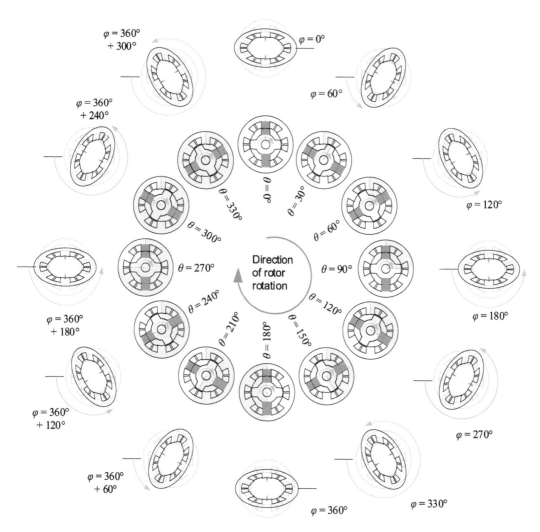

FIGURE 13.22
The deformation of stator core over one mechanical revolution for a 6/4 SRM.

opposite direction. Compared with a model that has two rotor poles (Figure 13.19), an operating SRM has a rotating deformation.

13.5.4 Spatial and Temporal Decomposition

For acoustic noise analysis, the radial force density waveform should be decomposed into harmonic components with different temporal and spatial orders. Figure 13.23 shows the case where the spatial FFT analysis is performed first. At each time step, the radial force density wave over the spatial position is extracted and a 1D FFT is performed on it. The histogram for the amplitudes of the harmonics can be seen in Figure 13.23b. Please note that the DC component is shifted to the center by using the techniques described in section 12.5.3.

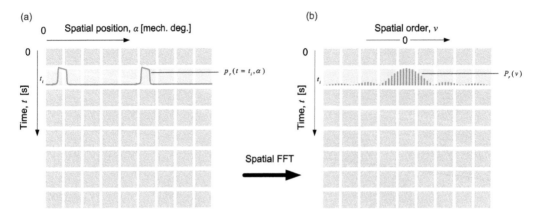

FIGURE 13.23
Spatial FFT over the surface wave: (a) surface wave and (b) results of spatial decomposition.

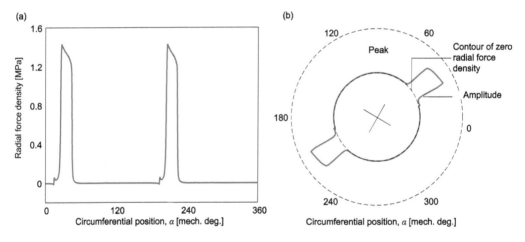

FIGURE 13.24
Radial force density versus circumferential position: (a) rectangular coordinate system and (b) polar system.

Figure 13.24 shows the waveform of radial force density versus circumferential position at one time step in both rectangular coordinate (see Figure 13.24a) and polar systems (see Figure 13.24b). Figure 13.25 shows the reproduction of the wave using its harmonic contents up to the 30th order. The red curve is the original data and the blue curve is reproduced using the first thirty harmonics. The curves of all the harmonics are shown in lighter blue. It can be seen that there is a good match between the original curve and the reproduced curve.

If 1D FFT is first performed over the time domain, as shown in Figure 13.26, the harmonics for the curve can be plotted against the temporal order. The red curve in Figure 13.27 shows the original data of radial force density versus rotor position at a certain circumferential position. The blue curve shows the reproduced curve using the first thirty

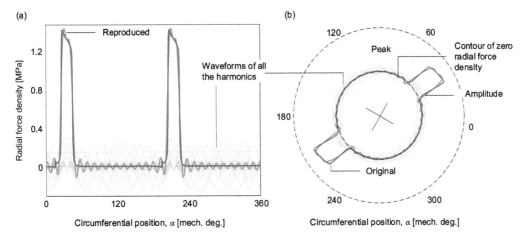

FIGURE 13.25
Decomposition and reproduction, radial force density versus circumferential position: (a) rectangular coordinate system and (b) polar system.

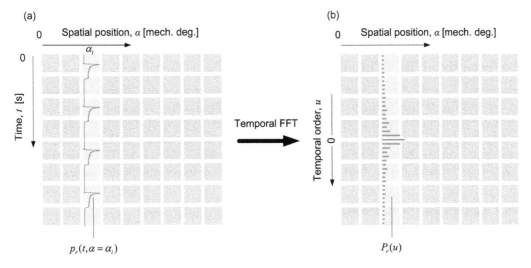

FIGURE 13.26
Temporal FFT over the surface wave: (a) surface wave and (b) results of temporal decomposition.

harmonics. The harmonics are shown in lighter blue. It can be seen that there is a good match between the original curve and the reproduced curve. As discussed in Chapter 12, the 2D FFT analysis can be done in a two-step manner. In fact, either spatial or temporal 1D FFT can be performed at first and the other afterwards.

In this section, the 2D decomposition of radial force density wave is studied. Some fundamentals for the harmonics are discussed. In the next section, we will study more of the properties of harmonics for both three-phase and four-phase SRMs.

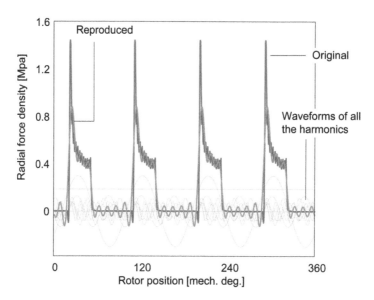

FIGURE 13.27
Decomposition and reproduction, radial force density versus rotor position.

13.6 Features of Harmonics for Radial Force Density Wave

Fundamentals of the harmonics for radial force density surface wave were discussed in the last section. In this section, the properties of harmonics for radial force density surface wave for three-phase and four-phase SRMs will be discussed. This section presents a method to locate the dominant harmonic content for conventional SRMs. It is assumed in this section that the rotor of the SRM rotates in clockwise direction. The directions of phase excitation and, hence, the radial force waveform are in counterclockwise.

13.6.1 Pattern for Dominant Harmonics

Figure 13.28 shows the dominant harmonic content for the surface wave of radial force density on a u-v plane over one mechanical cycle in a three-phase 6/4 SRM. The DC component is located at the origin where both temporal and spatial orders are zero. It does not rotate and has no forcing frequency. As discussed in Chapter 12, the rotational direction of the harmonics can be determined by $sgn(u)sgn(v)$. If $sgn(u)sgn(v)$ is positive, the rotational direction is negative (clockwise). If $sgn(u)sgn(v)$ is negative, the rotational direction is positive (counterclockwise). Please recall that counter clockwise is the positive rotational direction. The harmonic contents located in the first and the third quadrants rotate clockwise, or in the negative rotational direction. The harmonic contents located in the second and the fourth quadrants rotate counterclockwise, or in the positive rotational direction. The harmonic contents located on the u-axis (except for the DC component) have a concentric pattern of mode shape ($circ = 0$) with no rotation, but they have related forcing frequencies. The harmonic contents on the v-axis (except for DC component) have their own mode shapes, which are determined by the absolute value of v or $circ$, but they have no forcing frequencies.

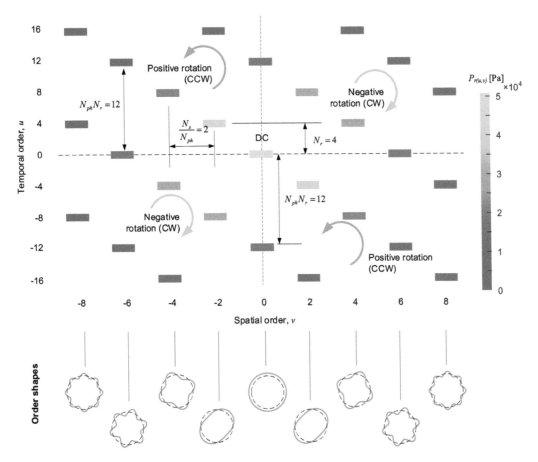

FIGURE 13.28
Amplitudes for dominant harmonic contents of radial force density, three-phase 6/4 SRM.

The layout of the harmonics in Figure 13.28 follow some specific pattern. Generally, the order (u, v) of the dominant harmonic contents for the radial force density can be expressed as:

$$\begin{cases} u = N_r N_{ph} j - k N_r \\ v = \dfrac{N_s}{N_{ph}} k \end{cases} \tag{13.28}$$

where N_r is the number of rotor poles, N_{ph} is the number of phases, N_s is the number of stator poles, j and k are integers ($k = \ldots, -2, -1, 0, 1, 2, \ldots$ and $j = \ldots, -2, -1, 0, 1, 2, \ldots$). The term N_s/N_{ph} is the number of magnetic poles in an SRM, as shown in Figure 13.13. The term $N_r N_{ph}$ in (13.28) is the number of torque pulsations over one mechanical cycle, as shown in Figure 13.13.

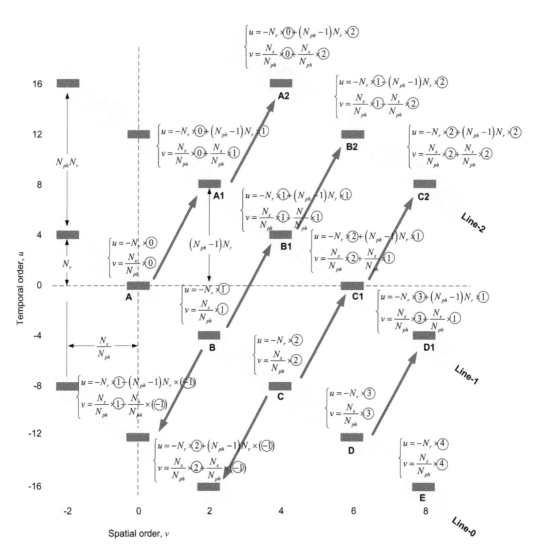

FIGURE 13.29
Study of the pattern for dominant harmonics in the diagonal direction.

The derivation of (13.28) can be explained using Figure 13.29, which shows the pattern of the dominant harmonics of the radial force density waveform of a three-phase 6/4 SRM. The orders for the harmonics in the diagonal Line-0 can be expressed using the following formula:

$$\begin{cases} u = -N_r \times m \\ v = \dfrac{N_s}{N_{ph}} \times m \end{cases} \qquad (13.29)$$

where m is an integer.

Moving from Line-0 to Line-1, the orders for the harmonics of Line-1 can be expressed based on the orders of Line 0 as:

$$\begin{cases} u = -N_r \times m + (N_{ph} - 1)N_r \times 1 \\ v = \dfrac{N_s}{N_{ph}} \times m + \dfrac{N_s}{N_{ph}} \times 1 \end{cases} \tag{13.30}$$

Similarly, the orders for the harmonics of Line-2 can be expressed as:

$$\begin{cases} u = -N_r \times m + (N_{ph} - 1)N_r \times 2 \\ v = \dfrac{N_s}{N_{ph}} \times m + \dfrac{N_s}{N_{ph}} \times 2 \end{cases} \tag{13.31}$$

Thus, the orders for the harmonics of Line-*j* can be generalized as:

$$\begin{cases} u = -N_r \times m + (N_{ph} - 1)N_r \times j \\ v = \dfrac{N_s}{N_{ph}} \times m + \dfrac{N_s}{N_{ph}} \times j \end{cases} \tag{13.32}$$

where *j* is also an integer. Equation (13.32) can be further simplified as:

$$\begin{cases} u = N_{ph}N_r \times j - N_r \times (m + j) \\ v = \dfrac{N_s}{N_{ph}} \times (m + j) \end{cases} \tag{13.33}$$

where *m* and *j* are the indices.

Figure 13.30 shows the layout from another perspective where the lines are defined in the vertical direction. The orders of the dominant harmonics for Line-0 can be expressed using:

$$\begin{cases} u = N_{ph}N_r \times j - N_r \times 0 \\ v = \dfrac{N_s}{N_{ph}} \times 0 \end{cases} \tag{13.34}$$

Based on the vertical Line-0, the orders of the dominant harmonics for Line-1 are:

$$\begin{cases} u = N_{ph}N_r \times j - N_r \times 1 \\ v = \dfrac{N_s}{N_{ph}} \times 1 \end{cases} \tag{13.35}$$

Similarly, on the vertical Line-2, the orders can be expressed using:

$$\begin{cases} u = N_{ph}N_r \times j - N_r \times 2 \\ v = \dfrac{N_s}{N_{ph}} \times 2 \end{cases} \tag{13.36}$$

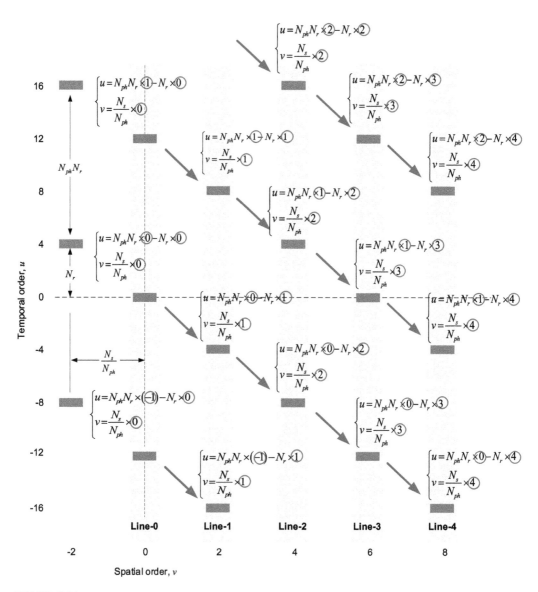

FIGURE 13.30
Study of the pattern for dominant harmonics in the vertical direction.

Thus, the orders for Line-k can be expressed using:

$$\begin{cases} u = N_{ph}N_r \times j - N_r \times k \\ \\ v = \dfrac{N_s}{N_{ph}} \times k \end{cases} \tag{13.37}$$

where k is an integer. Using (13.33) and (13.37), it can be concluded that $k = m + j$, where m and j are arbitrary integers. Equation (13.37) will be used in this chapter to express the orders of dominant harmonics for the radial force density because only one index, k, is needed for expressing the spatial order.

It can be seen from (13.37) that the series of circumferential orders $circ = (N_s/N_{ph})|k|$ determines the circumferential orders for harmonic contents of radial force density for an SRM. For instance, the harmonic contents of radial force density for 6/4 SRMs have a series of circumferential orders, $circ = \{0, 2, 4, 6, 8,...\}$ (multiples of 2).

The three-phase 12/8 SRM has four magnetic poles per phase, thus it has a series of circumferential orders, $circ = \{0, 4, 8, 12,...\}$ (multiples of 4) in its harmonic contents of radial force density wave. It can be seen that as the numbers of the stator and the rotor poles increase, it is less likely to excite smaller circumferential modes. For instance, the series for circumferential orders of 24/16 SRM is $circ = \{0, 8, 16, 24,...\}$ in its harmonic content. It is also interesting to note that circumferential order $circ = 0$ exists for all SRMs, which indicates that circumferential mode $circ = 0$ can always be excited.

13.6.2 Conversion of Harmonics

For a three-phase 6/4 SRM, the orders (u, v) for the dominant harmonics of the radial force density wave can be calculated as:

$$\begin{cases} u = 12j - 4k \\ v = \dfrac{6}{3}k = 2k \end{cases} \tag{13.38}$$

If the rotor rotates at 6000 rpm, the motor's rotational mechanical frequency, f_{mech} is 100 Hz. Thus, the forcing frequencies for dominant harmonics are summarized in Table 13.3. Again, the positive or negative sign of the amplitude represents the rotational direction of the harmonic. For a harmonic with a temporal order u, its forcing angular frequency $\omega_{f(u)}$ can be calculated as:

$$\omega_{f(u)} = 2\pi |u| f_{mech} = 2\pi q f_{mexh} = 2\pi q \frac{RPM}{60} \tag{13.39}$$

It can be observed that the harmonic $p_{r(12,0)}$ can excite the circumferential mode, $circ = 0$. Its forcing frequency, $f_{f(12,0)}$ is equal to $12 f_{mech}$. The harmonics $p_{r(8,2)}$ and $p_{r(-8,-2)}$ both have the same absolute value of u and can both excite the circumferential mode, $circ = 2$. They both have the same forcing frequency, $8 f_{mech}$.

The phase angles are important for the superposition of the harmonics. The principles of superposition of harmonic contents were discussed in Chapter 12. In the following analysis, the relationship between phase angles for different harmonics in an SRM will be discussed. Figure 13.31 shows the phase angles for the harmonic contents of a 6/4 SRM. It can be seen that the phase angles in the first and the third quadrants have opposite signs. Similarly, the phase angles in the second and the fourth are opposite. This is because the FFT results are two-sided. The harmonic content in the third quadrant can be directly added to the first quadrant. The harmonics in the second and the fourth are mirrored with respect to the origin (DC component).

Given the phase angle matrix:

$$\phi_{(u,v)} = -\phi_{(-u,-v)} \tag{13.40}$$

TABLE 13.3

Forcing Frequencies for Dominant Harmonics of Radial Force Density Wave of a 6/4 SRM, at 6000 rpm

			Circumferential Order, v						
			-6 ($k=-3$)	-4 ($k=-2$)	-2 ($k=-1$)	0 ($k=0$)	2 ($k=1$)	4 ($k=2$)	6 ($k=3$)
u	q	$f_{f(q)}$ [Hz]				$P_{r(u,v)}$			
16	16	16×100	1600			$+P_{r(16,-2)}$ ($j=1$)			$-P_{r(16,4)}$ ($j=2$)	
12	12	12×100	1200	$+P_{r(12,-6)}$ ($j=0$)			$P_{r(12,0)}$ ($j=1$)			$-P_{r(12,6)}$ ($j=2$)
8	8	8×100	800		$+P_{r(8,-4)}$ ($j=0$)			$-P_{r(8,2)}$ ($j=1$)		
4	4	4×100	400			$+P_{r(4,-2)}$ ($j=0$)			$-P_{r(4,4)}$ ($j=1$)	
0	0	0	0	$P_{r(0,-6)}$ ($j=-1$)			$P_{r(0,0)}$ (DC) ($j=0$)			$P_{r(0,6)}$ ($j=1$)
-4	4	4×100	400		$-P_{r(-4,-4)}$ ($j=-1$)			$+P_{r(-4,2)}$ ($j=0$)		
-8	8	8×100	800			$-P_{r(-8,-2)}$ ($j=-1$)			$+P_{r(-8,4)}$ ($j=0$)	
-12	12	12×100	1200	$-P_{r(-12,-6)}$ ($j=-2$)			$P_{r(-12,0)}$ ($j=-1$)			$+P_{r(-12,6)}$ ($j=0$)
-16	16	16×100	1600		$-P_{r(-16,-4)}$ ($j=-2$)			$+P_{r(-16,2)}$ ($j=-1$)		

(left-axis label: Temporal order, u)

and that:

$$\cos\left(2\pi f_{mech}ut + v\alpha + \phi_{(u,v)}\right) = \cos\left(2\pi f_{mech}(-u)t + (-v)\alpha - \phi_{(-u,-v)}\right) \qquad (13.41)$$

Thus, the radial force density wave for the 6/4 SRM can be approximated by:

$$
\begin{aligned}
p_r(t,\alpha) &= \sum_{u=-\infty}^{\infty}\sum_{v=-\infty}^{\infty} p_{r(u,v)} \\
&\approx p_{r(0,0)} + \left(p_{r(4,-2)} + p_{r(-4,2)}\right) + \left(p_{r(4,4)} + p_{r(-4,-4)}\right) + \left(p_{r(0,6)} + p_{r(0,-6)}\right) + \\
&\quad \left(p_{r(8,2)} + p_{r(-8,-2)}\right) + \left(p_{r(12,0)} + p_{r(-12,0)}\right) + \left(p_{r(8,-4)} + p_{r(-8,4)}\right) + \\
&\quad \left(p_{r(-12,6)} + p_{r(12,-6)}\right) + \left(p_{r(12,6)} + p_{r(-12,-6)}\right) + \left(p_{r(16,-2)} + p_{r(-16,2)}\right) + \left(p_{r(16,4)} + p_{r(-16,-4)}\right)
\end{aligned}
\qquad (13.42)
$$

For illustration purposes, we will investigate how to simplify (13.42) using one harmonic content. The same calculations can be applied to other harmonics as well. In (13.23), the analytical expression of the surface wave for the harmonic order (u, v) was defined as:

$$p_{r(u,v)} = P_{r(u,v)}\cos\left(\omega_{mech}ut + v\alpha + \phi_{(u,v)}\right) \qquad (13.43)$$

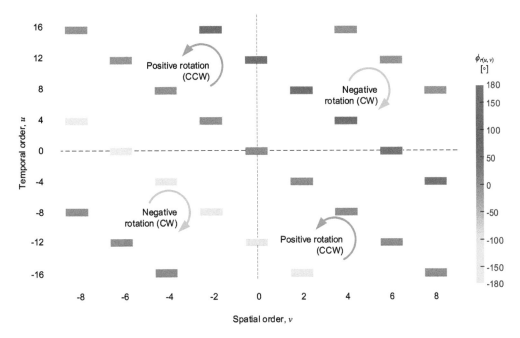

FIGURE 13.31
Phase angles for dominant harmonic contents of radial force density, three-phase 6/4 SRM.

For $(u, v) = (4, -2)$, the surface wave can be expressed as:

$$p_{r(4,-2)} = + P_{r(4,-2)} \cos\left(\omega_{mech}(4)t - 2\alpha + \phi_{(4,-2)}\right) \tag{13.44}$$

Since $(u, v) = (4, -2)$ is in the second quadrant, the harmonic rotates in the counterclockwise direction and, hence, plus sign is added to the magnitude. For $(u, v) = (-4, 2)$, the surface wave can be expressed as:

$$p_{r(-4,2)} = + P_{r(-4,2)} \cos\left(\omega_{mech}(-4)t + 2\alpha + \phi_{(-4,2)}\right) \tag{13.45}$$

Since $(u, v) = (-4, 2)$ is in the fourth quadrant, the harmonic rotates in the counterclockwise direction and, hence, a positive sign is added to the magnitude. According to Figure 13.31 and Equation (13.40), the relationship between the phase angles in (13.44) and (13.45) is:

$$\phi_{(4,-2)} = -\phi_{(-4,2)} \tag{13.46}$$

Therefore, (13.45) can be modified as:

$$\begin{aligned} p_{r(-4,2)} &= + P_{r(-4,2)} \cos\left(\omega_{mech}\left(-4\right)t + 2\alpha - \phi_{(4,-2)}\right) \\ &= + P_{r(-4,2)} \cos\left[-\left(\omega_{mech}\left(4\right)t - 2\alpha + \phi_{(4,-2)}\right)\right] \end{aligned} \tag{13.47}$$

Using (13.41), (13.47) can be simplified as:

$$p_{r(-4,2)} = + P_{r(-4,2)} \cos\left(\omega_{mech}\left(4\right)t - 2\alpha + \phi_{(4,-2)}\right) \tag{13.48}$$

Since $p_{r(-4,2)}$ and $p_{r(4,-2)}$ are symmetric, their magnitudes $P_{r(-4,2)}$ and $P_{r(4,-2)}$ are the same. Therefore,

$$p_{r(-4,2)} = +P_{r(4,-2)} \cos\left(\omega_{mech}(4)t - 2\alpha + \phi_{(4,-2)}\right) \tag{13.49}$$

For simplification, it is assumed that the cosine function is part of the magnitude when representing the harmonic. Hence, (13.49) can be expressed as:

$$p_{r(-4,2)} = +P_{r(4,-2)} \tag{13.50}$$

As a result,

$$p_{r(4,-2)} + p_{r(-4,2)} = +2P_{r(4,-2)} \tag{13.51}$$

If the same calculation is applied to the other harmonics, (13.42) can be expressed as:

$$p_r(t,\alpha) = \sum_{u=-\infty}^{\infty}\sum_{v=-\infty}^{\infty} P_{r(u,v)} \approx P_{r(0,0)} + 2P_{r(4,-2)} - 2P_{r(4,4)} + 2P_{r(0,6)} - 2P_{r(8,2)} + 2P_{r(12,0)} + 2P_{r(8,-4)}$$
$$+ 2P_{r(12,-6)} - 2P_{r(12,6)} + 2P_{r(16,-2)} - 2P_{r(16,4)} \tag{13.52}$$

The amplitudes for the dominant harmonics in the first and the second quadrants (axes included) are summarized in Table 13.4.

Similarly, the radial force density wave can be approximated by the dominant harmonic contents in the first and the fourth quadrants (axes included) ($v \geq 0$), as shown in Table 13.5:

$$p_r(t,\alpha) \approx P_{r(0,0)} + 2P_{r(0,6)} - 2P_{r(4,4)} - 2P_{r(8,2)} + 2P_{r(12,0)} - 2P_{r(12,6)} - 2P_{r(16,4)}$$
$$+ 2P_{r(-4,2)} + 2P_{r(-8,4)} + 2P_{r(-12,6)} + 2P_{r(-16,2)} \tag{13.53}$$

When the harmonic contents are converted to the first quadrant (axes included), the arrangement of the harmonics can be seen in Table 13.6. The $+2P_{r(-4,2)}$ rotates in the positive direction (counterclockwise). The circumferential order of $+2P_{r(-4,2)}$ is *circ* = 2. The forcing frequency of $+2P_{r(-4,2)}$ is $4 \times f_{mech}$ because its temporal order is $q = 4$. As interpreted from

TABLE 13.4

Expression of $P_{r(q,v)}$, the Amplitudes for the Harmonic Contents in the First and Second Quadrants, Axes Included, 6/4 SRM

				Spatial Order, v				
		−6	−4	−2	0	2	4	6
$P_{r(q,v)}$								
Temporal order $q = \|u\|$	16			$+2P_{r(16,-2)}$			$-2P_{r(16,4)}$	
	12	$+2P_{r(12,-6)}$			$2P_{r(12,0)}$			$-2P_{r(12,6)}$
	8		$+2P_{r(8,-4)}$			$-2P_{r(8,2)}$		
	4			$+2P_{r(4,-2)}$			$-2P_{r(4,4)}$	
	0				$2P_{r(0,0)}$			$2P_{r(0,6)}$

TABLE 13.5

Expression of $P_{r(u,circ)}$, the Amplitudes for the Harmonic Contents in the First and Fourth Quadrants, Axes Included, 6/4 SRM

		Circumferential Order, $\lvert v \rvert = circ$			
		0	2	4	6
$P_{r(u,\lvert v \rvert\,=\,circ)}$					
Temporal order u	16			$-2P_{r(16,4)}$	
	12	$2P_{r(12,0)}$			$-2P_{r(12,6)}$
	8		$-2P_{r(8,2)}$		
	4			$-2P_{r(4,4)}$	
	0	$P_{r(0,0)}$			$2P_{r(0,6)}$
	−4		$+2P_{r(-4,2)}$		
	−8			$+2P_{r(-8,4)}$	
	−12				$+2P_{r(-12,6)}$
	−16		$+2P_{r(-16,2)}$		

TABLE 13.6

Expression of $P_{r(q,circ)}$, the Amplitudes for the Harmonic Contents in the First Quadrant, Axes Included, 6/4 SRM

		Circumferential Order, $\lvert v \rvert = circ$			
		0	2	4	6
$P_{r(q,\,circ)}$					
Temporal order $\lvert u \rvert = q$	16		$+2P_{r(-16,2)}$	$-2P_{r(16,4)}$	
	12	$2P_{r(12,0)}$			$-2P_{r(12,6)}$ $\quad$ $+2P_{r(-12,6)}$ Standing wave
	8		$-2P_{r(8,2)}$	$+2P_{r(-8,4)}$	
	4		$+2P_{r(-4,2)}$	$-2P_{r(4,4)}$	
	0	$P_{r(0,0)}$			$2P_{r(0,6)}$

Table 13.6, $-2P_{r(4,4)}$ has the same forcing frequency as $+2P_{r(-4,2)}$, since $-2P_{r(4,4)}$ and $+2P_{r(-4,2)}$ are on the same row. However, $-2P_{r(4,4)}$ has a higher circumferential order of $circ = 4$. Highlighted in Table 13.6, $2P_{r(12,0)}$, does not rotate, but still vibrates. Highlighted in grey, $P_{r(0,0)}$ and $2P_{r(0,6)}$ do not rotate and do not vibrate because the forcing frequencies of both are zero. The sum of $-2P_{r(12,6)}$ and $+2P_{r(-12,6)}$ is a standing wave.

13.6.3 Three-Phase 6/4 SRM

Two major harmonics, $P_{r(12,0)}$ and $P_{r(-12,0)}$, of a 6/4 SRM can both excite the circumferential mode, $circ = 0$ as shown in Figure 13.32. For instance, if the rotor rotates at 6000 rpm, the

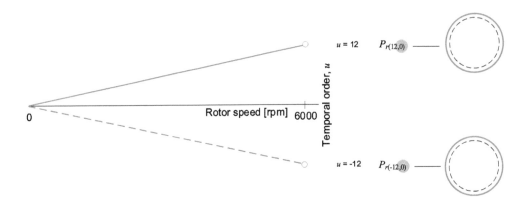

FIGURE 13.32
Excitation of circumferential mode, *circ* = 0, 6/4 SRM.

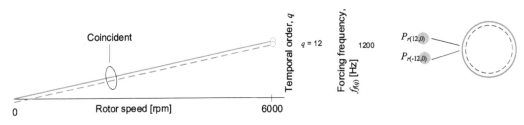

FIGURE 13.33
Forcing frequency of $P_{r(12,0)}$ and $P_{r(-12,0)}$, *circ* = 0, 6/4 SRM.

mechanical frequency is 100 Hz. Thus, the forcing frequency for $P_{r(12,0)}$ and $P_{r(-12,0)}$ are both 1200 Hz (=100 Hz × 12), as shown in Figure 13.33. Please note again that mode *circ* = 0 does not rotate. For a 6/4 SRM, the temporal order for the harmonics with *circ* = 0 can be generalized as:

$$u = 12j \hspace{4cm} (13.54)$$

where *j* is an integer. These harmonics can all excite circumferential mode 0.

Table 13.6 shows only a few selected harmonics. The series of temporal orders that excite the mode with *circ* = 0 can be *q* = {0, 12, 24, 36, 48...}, as discussed earlier. When the forcing frequency of the harmonic is close to the natural frequency (*circ* = 0), resonance occurs, leading to a larger displacement. For instance, assume that the natural frequency of the mode with *circ* = 0 is 3000 Hz. For the harmonic $P_{r(12,0)}$, its forcing frequency can be calculated as $12 \times f_{mech}$. When the motor rotates at around 15,000 rpm (f_{mech} = 250 Hz), the forcing frequency is 3000 Hz. This would cause a resonance.

As shown in Table 13.6, $+2P_{r(-16,2)}$, $-2P_{r(8,2)}$, and $+2P_{r(-4,2)}$ are the three major harmonics of a three-phase 6/4 SRM that can excite the circumferential mode, *circ* = 2, as shown in Figure 13.34. $-2P_{r(8,2)}$ rotates clockwise while $+2P_{r(-16,2)}$ and $+2P_{r(-4,2)}$ rotate counterclockwise. The forcing frequencies of the harmonics $+2P_{r(-16,2)}$, $-2P_{r(8,2)}$, and $+2P_{r(-4,2)}$ are shown in Figure 13.35 when the motor speed is 6000 rpm. The series of temporal orders that excite the mode with *circ* = 2 can be *q* = {4, 8, 16, 20, 28, 32, ...}. The series of temporal orders that excite the mode with *circ* = 4 is the same as the mode with *circ* = 2.

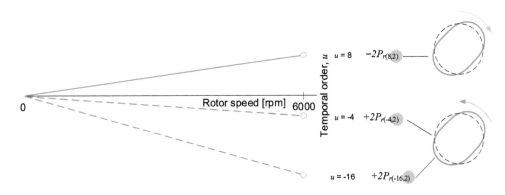

FIGURE 13.34
Excitation of circumferential mode, *circ* = 2, 6/4 SRM.

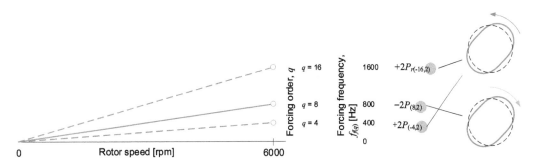

FIGURE 13.35
Forcing frequencies of $+2P_{r(-16,2)}$, $-2P_{r(8,2)}$ and $+2P_{r(-4,2)}$, *circ* = 2, 6/4 SRM.

Three major harmonics, $-2P_{r(16,4)}$, $+2P_{r(-8,4)}$ and $-2P_{r(4,4)}$, have the same circumferential order, *circ* = 4. These harmonic contents can excite the circumferential mode, *circ* = 4, as shown in Figure 13.36. Similarly, the forcing frequencies are shown in Figure 13.37 when the motor speed is 6000 rpm.

Figure 13.38 shows another pair of dominant harmonic contents, $-2P_{r(12,6)}$, and $+2P_{r(-12,6)}$. Even though $-2P_{r(12,6)}$ and $+2P_{r(-12,6)}$ have their own distinctive rotational directions,

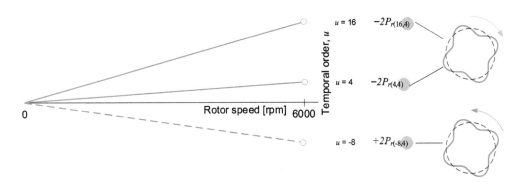

FIGURE 13.36
Excitation of circumferential mode, *circ* = 4, 6/4 SRM.

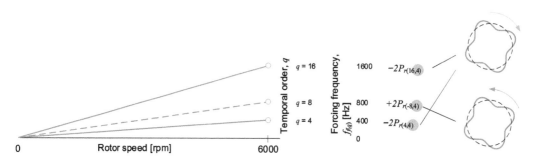

FIGURE 13.37
Forcing frequencies of $-2P_{r(16,4)}$, $+2P_{r(-8,4)}$ and $-2P_{r(4,4)}$, *circ* = 4, 6/4 SRM.

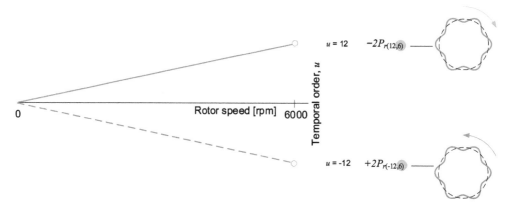

FIGURE 13.38
Excitation of circumferential mode, *circ* = 6, 6/4 SRM.

the superposition of the two harmonic contents yields a standing wave, as shown in Figure 13.39. The sum of $-2P_{r(12,6)}$, and $+2P_{r(-12,6)}$ does not rotate, but has the forcing frequency $12 \times f_{mech}$. If the motor speed is 6000 rpm, the standing surface wave has a forcing frequency of 1200 Hz.

As explained earlier, the harmonic content of radial force density for a three phase 6/4 SRM have a series of circumferential orders, *circ* = {0, 2, 4, 6,....} Low circumferential modes are more likely to appear in the audible range from 0 Hz to 20,000 Hz. Low circumferential modes (*circ* = {0, 2, 4, 6,...}) can be excited by the corresponding harmonic content with the same circumferential orders.

13.6.4 Three-Phase 12/8 SRM

For a three-phase 12/8 SRM, the harmonic orders (u, v) for the dominant harmonics of the radial force density wave can be calculated using (13.37):

$$\begin{cases} u = 24j - 8k \\ v = 4k \end{cases} \tag{13.55}$$

The heat map for dominant harmonic contents of radial force density for 12/8 SRM is shown in Figure 13.40. The series for circumferential orders of the harmonic content

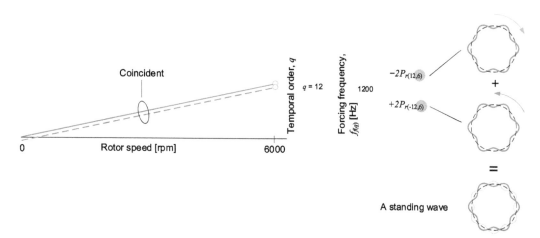

FIGURE 13.39
Forcing frequencies of $-2P_{r(12,6)}$, and $+2P_{r(-12,6)}$ *circ* = 6, 6/4 SRM.

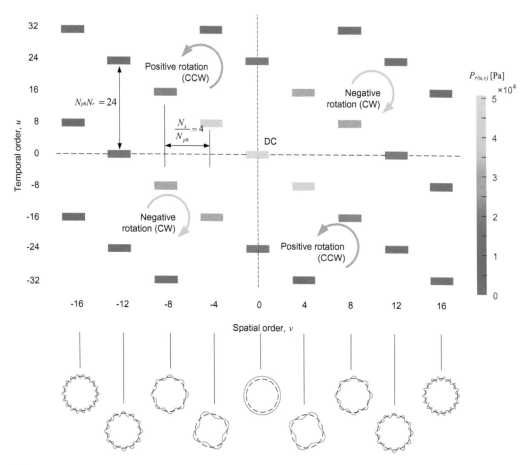

FIGURE 13.40
Amplitudes for dominant harmonic contents of radial force density, three-phase 12/8 SRM.

is $circ = \{0, 4, 8, 12,...\}$ (multiples of 4), because the 12/8 SRM has 4 magnetic poles. This means that a number of circumferential modes, such as $circ = \{2, 6, 10,...\}$ cannot be excited in a 12/8 SRM. Compared with Figure 13.28 for 6/4 SRM, the harmonic orders for a 12/8 SRM are more sparse. The distance along u-axis between two adjacent harmonic orders at a given spatial order is 24 ($N_{ph}N_r = 24$), compared to 12 for the 6/4 SRM. If converted to the first quadrant (axes included), the harmonic content can be organized as shown in Table 13.7. Figure 13.41 shows that the forcing frequency of the harmonics $P_{r(24,0)}$ and $P_{r(-24,0)}$ ($circ = 0$) for the 12/8 SRM is 2400 Hz at 6000 rpm. The harmonics of radial force density wave of the 12/8 SRM do not excite the circumferential mode, $circ = 2$. The harmonics, $+2P_{r(-32,4)}$, $-2P_{r(16,4)}$, and $+2P_{r(-8,4)}$, excite the circumferential mode, $circ = 4$, as shown in Figure 13.42. Figure 13.43 shows that the next immediate dominant harmonic contents are $-2P_{r(32,8)}$, $+2P_{r(-16,8)}$, and $-2P_{r(8,8)}$. As illustrated in Figure 13.44, $-2P_{r(24,12)}$ and $+2P_{r(-24,12)}$ in 12/8 SRM generate a standing wave.

TABLE 13.7

Expression of $P_{r(q, circ)}$, the Amplitudes for the Harmonic Contents in the First Quadrant, Axes Included, 12/8 SRM

		Circumferential Order, $|v| = circ$			
		0	4	8	12
$P_{r(q, circ)}$					
Temporal order $|u| = q$	32		$+2P_{r(-32,4)}$	$-2P_{r(32,8)}$	
	24	$2P_{r(24,0)}$			$-2P_{r(24,12)}$ $+2P_{r(-24,12)}$ Standing wave
	16		$-2P_{r(16,4)}$	$+2P_{r(-16,8)}$	
	8		$+2P_{r(-8,4)}$	$-2P_{r(8,8)}$	
	0	$P_{r(0,0)}$			$2P_{r(0,12)}$

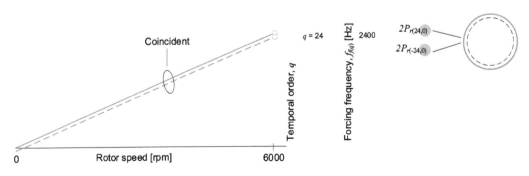

FIGURE 13.41
Forcing frequency of $P_{r(24,0)}$ and $P_{r(-24,0)}$, $circ = 0$, 12/8 SRM.

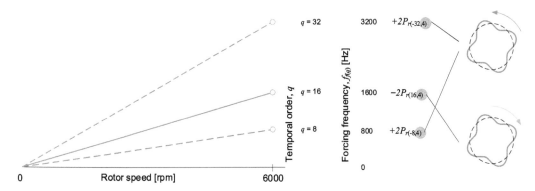

FIGURE 13.42
Forcing frequencies of $+2P_{r(-32,4)}$, $-2P_{r(16,4)}$ and $+2P_{r(-8,4)}$, $circ = 4$, 12/8 SRM.

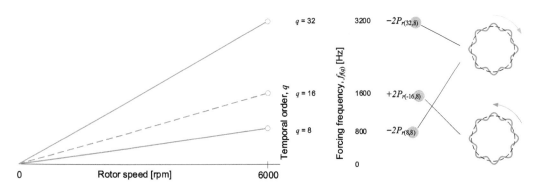

FIGURE 13.43
Forcing frequencies of $-2P_{r(32,8)}$, $+2P_{r(-16,8)}$ and $-2P_{r(8,8)}$, $circ = 8$, 12/8 SRM.

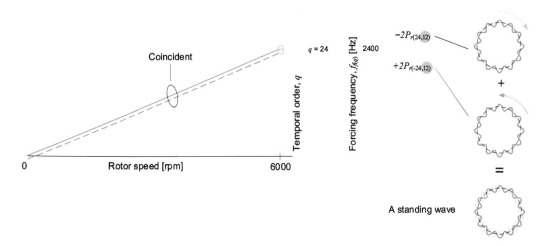

FIGURE 13.44
Forcing frequencies of $-2P_{r(24,12)}$, and $+2P_{r(-24,12)}$ $circ = 12$, 12/8 SRM.

13.6.5 Three-Phase 18/12 SRM

Figure 13.45 shows the heat map for the dominant harmonic contents of radial force density for the 18/12 SRM. It can be seen from the heat map that the series for the circumferential order of the harmonic content is $circ = \{0, 6, 12, 18,...\}$ (multiples of 6). The circumferential modes, $circ = \{2, 4, 8, 10,...\}$ cannot be excited for a 18/12 SRM. Compared with Figure 13.28 (6/4 SRM) and Figure 13.40 (12/8 SRM), the 18/12 SRM harmonic orders are even more sparse. The distance along the u-axis between two adjacent harmonic contents for a certain spatial harmonic is 36 ($N_{ph}N_r = 36$). For a three-phase 18/12 SRM, the harmonic orders (u, v) for the dominant harmonics of the radial force density wave can be calculated using (13.37):

$$\begin{cases} u = 36j - 12k \\ v = 6k \end{cases} \tag{13.56}$$

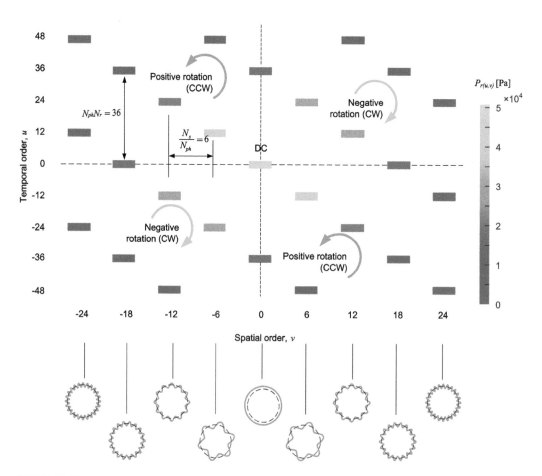

FIGURE 13.45
Amplitudes for dominant harmonic contents of radial force density, three-phase 18/12 SRM.

TABLE 13.8

Expression of $P_{r(q, circ)}$, the Amplitudes for the Harmonic Contents in the First Quadrant, Axes Included, 18/12 SRM

		Circumferential Order, $\lvert v \rvert = circ$			
		0	6	12	18
$P_{r(q, circ)}$					
Temporal order $\lvert u \rvert = q$	48		$+2P_{r(-48,6)}$	$-2P_{r(48,12)}$	
	36	$2P_{r(36,0)}$			$-2P_{r(36,18)}$ $\quad$ $+2P_{r(-36,18)}$ Standing wave
	24		$-2P_{r(24,6)}$	$+2P_{r(-24,12)}$	
	12		$+2P_{r(-12,6)}$	$-2P_{r(12,12)}$	
	0	$P_{r(0,0)}$			$2P_{r(0,18)}$

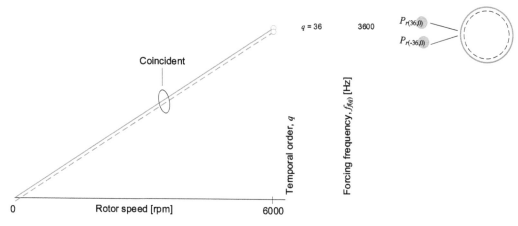

FIGURE 13.46
Forcing frequency of $P_{r(36,0)}$ and $P_{r(-36,0)}$, $circ = 0$, 18/12 SRM.

The harmonic content of 18/12 SRM can be summarized in the first quadrant, as shown in Table 13.8. Figure 13.46 shows the forcing frequency of $P_{r(36,0)}$ and $P_{r(-36,0)}$. As mentioned earlier, $P_{r(36,0)}$ and $P_{r(-36,0)}$ have no rotation, but have a forcing frequency ($36 \times f_{mech}$). Figure 13.47 shows that the forcing frequencies for the harmonic contents $+2P_{r(-48,6)}$, $-2P_{r(24,6)}$, and $+2P_{r(-12,6)}$, can excite the circumferential mode, $circ = 6$. At 6000 rpm, the forcing frequencies for $+2P_{r(-48,6)}$, $-2P_{r(24,6)}$, and $+2P_{r(-12,6)}$ are 4800 Hz, 2400 Hz, and 1200 Hz, respectively. The circumferential mode, $circ = 12$ in an 18/12 SRM, can be excited by $-2P_{r(48,12)}$, $+2P_{r(-24,12)}$, and $-2P_{r(12,12)}$, as shown in Figure 13.48. Similar to the previous three-phase topologies, $-2P_{r(36,18)}$, and $+2P_{r(-36,18)}$ can also generate a standing wave as shown in Figure 13.49. Again, the standing wave has no rotation but has a certain forcing frequency ($36 \times f_{mech}$).

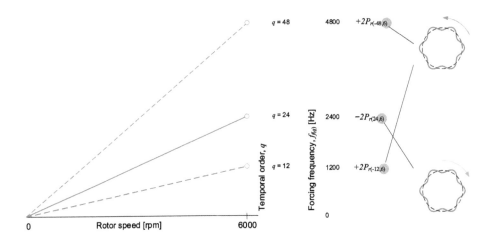

FIGURE 13.47
Forcing frequencies of $+2P_{r(-48,6)}$, $-2P_{r(24,6)}$ and $+2P_{r(-12,6)}$, *circ* = 6, 18/12 SRM.

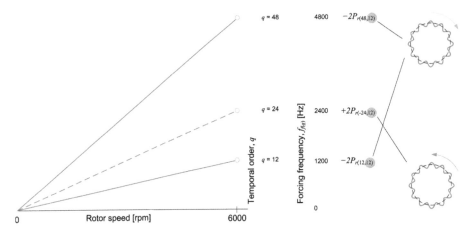

FIGURE 13.48
Forcing frequencies of $-2P_{r(48,12)}$, $+2P_{r(-24,12)}$ and $-2P_{r(12,12)}$, *circ* = 12, 18/12 SRM.

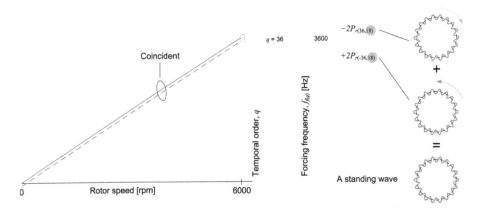

FIGURE 13.49
Forcing frequencies of $-2P_{r(36,18)}$, and $+2P_{r(-36,18)}$ *circ* = 18, 18/12 SRM.

13.6.6 Three-Phase 24/16 SRM

The dominant harmonic orders of the radial force density waveform of a three-phase 24/16 SRM can be calculated using (13.37):

$$\begin{cases} u = 48j - 16k \\ v = 8k \end{cases} \tag{13.57}$$

Figure 13.50 is the heat map for a few dominant harmonic orders for 24/16 SRM. A 24/16 SRM has the most sparse distribution of harmonics among the four three-phase SRM geometries discussed so far. The series of the circumferential orders for the harmonic content is $circ = \{0, 8, 16, 24,..\}$. (multiples of 8). The distance along the u-axis between two adjacent harmonic orders for a given spatial harmonic is 48 ($N_{ph}N_r = 48$) for a 24/16 SRM.

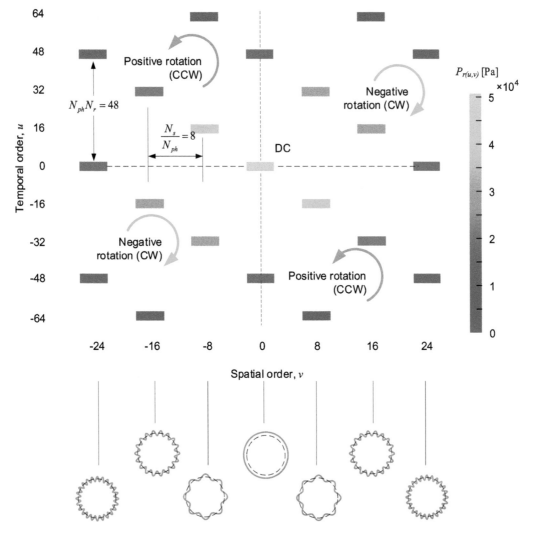

FIGURE 13.50
Amplitudes for dominant harmonic contents of radial force density, three-phase 24/16 SRM.

The dominant harmonic content of 24/16 SRMs can be grouped in the first quadrant, as shown in Table 13.9. $P_{r(48,0)}$ and $P_{r(-48,0)}$ with no rotational direction excite the circumferential mode, *circ* = 0. When the motor speed is 6000 rpm, the forcing frequency for $P_{r(48,0)}$ and $P_{r(-48,0)}$ is 4800 Hz, as shown in Figure 13.51. Figure 13.52 shows that $+2P_{r(-64,8)}$, $-2P_{r(32,8)}$, and $+2P_{r(-16,8)}$ with distinctive forcing frequencies can excite the circumferential mode, *circ* = 8. Figure 13.53 shows that $-2P_{r(64,16)}$, $+2P_{r(-32,16)}$, and $-2P_{r(16,16)}$, can excite the circumferential mode, *circ* = 16. The sum of the surface wave $-2P_{r(48,24)}$ and $+2P_{r(-48,24)}$ is a standing wave, which can excite the circumferential mode *circ* = 24, as shown in Figure 13.54.

TABLE 13.9

Expression of $P_{r(q,\ circ)}$, the Amplitudes for the Harmonic Contents in the First Quadrant, Axes Included, 24/16 SRM

		Circumferential Order, $|v|$ = circ			
		0	8	16	24
$P_{r(q,\ circ)}$					
Temporal order $|u|$ = q	64		$-2P_{r(-64,8)}$	$-2P_{r(64,16)}$	
	48	$2P_{r(48,0)}$			$-2P_{r(48,24)}$ $+2P_{r(-48,24)}$ Standing wave
	32		$-2P_{r(32,8)}$	$+2P_{r(-32,16)}$	
	16		$+2P_{r(-16,8)}$	$-2P_{r(16,16)}$	
	0	$P_{r(0,0)}$			$2P_{r(0,24)}$

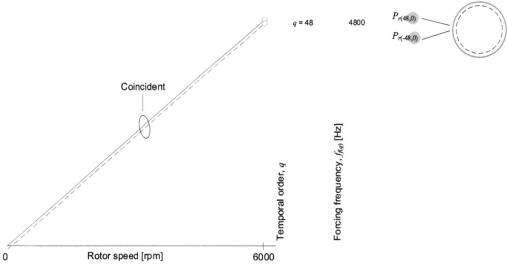

FIGURE 13.51
Forcing frequency of $P_{r(48,0)}$ and $P_{r(-48,0)}$, *circ* = 0, 24/16 SRM.

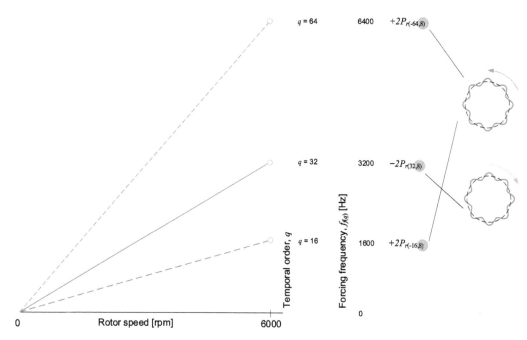

FIGURE 13.52
Forcing frequencies of $+2P_{r(-64,8)}$, $-2P_{r(32,8)}$ and $+2P_{r(-16,8)}$, $circ = 8$, 24/16 SRM.

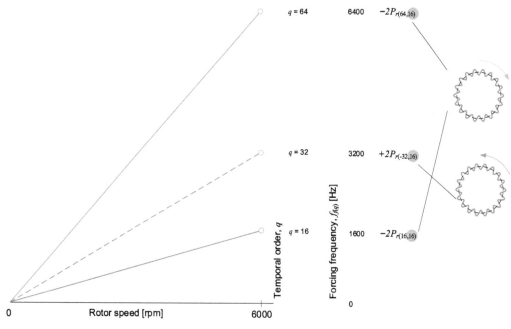

FIGURE 13.53
Forcing frequencies of $-2P_{r(64,16)}$, $+2P_{r(-32,16)}$ and $-2P_{r(16,16)}$, $circ = 16$, 24/16 SRM.

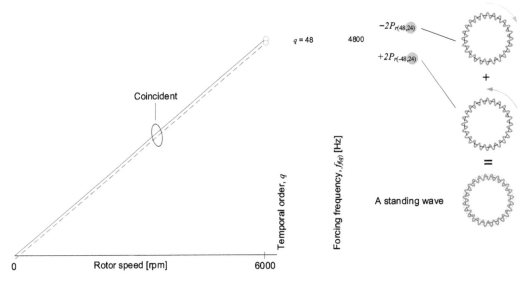

FIGURE 13.54

Forcing frequencies of $-2P_{r(48,24)}$ and $+2P_{r(-48,24)}$ *circ* = 24, 24/16 SRM.

So far, in this section, the harmonic content of four three-phase SRM topologies 6/4, 12/8, 18/12, and 24/16 SRM have been compared. Based on the discussions, it can be concluded that as the number of rotor poles and stator poles (or the number of magnetic poles) increase, the distribution of harmonic content on the *u-v* plane becomes more sparse. Table 13.10 compares the first four circumferential orders for the harmonic content of the radial force density wave. Within the frequency range from 0 Hz to 20,000 Hz, a number of low circumferential modes in a 6/4 SRM can be excited. In a 24/16 SRM, the first non-zero circumferential order is *circ* = 8.

Table 13.11 compares the four pairs of temporal orders for the harmonic content with *circ* = 0 for the four different three-phase SRM topologies. As the number of stator and rotor poles increase, the distance between two adjacent harmonics along the *u*-axis increases. As the number of phases is fixed, the distance increases linearly with the increase in the number of rotor poles. This affects the forcing frequency, since it is proportional to the mechanical frequency (calculated by motor speed divided by 60) and the absolute value of the temporal order ($|u|$).

TABLE 13.10

Comparison of Circumferential Orders for 6/4, 12/8, 18/12, and 24/16 SRMs

Geometry	Circumferential Order (*circ*)	Circumferential Orders, *circ* from 0 ~ 24												
		0	2	4	6	8	10	12	14	16	18	20	22	24
6/4 SRM	*circ* = 2*k*	√	√	√	√	√	√	√	√	√	√	√	√	√
12/8 SRM	*circ* = 4*k*	√		√		√		√		√		√		√
18/12 SRM	*circ* = 6*k*	√			√			√			√			√
24/16 SRM	*circ* = 8*k*	√				√				√				√

k is a positive integer.

TABLE 13.11

Comparison of Temporal Orders for 6/4, 12/8, 18/12, and 24/16 SRMs

Geometry	Distance between Two Adjacent Harmonics along *u*-Axis	Temporal Orders for First Four Pairs of Harmonic Contents for $circ = 0$ ($u = N_{ph}N_r \times j$, $k = 0$)			
6/4 SRM	$N_{ph}N_r = 12$	±12	±24	±36	±48
12/8 SRM	$N_{ph}N_r = 24$	±24	±48	±72	±96
18/12 SRM	$N_{ph}N_r = 36$	±36	±72	±108	±144
24/16 SRM	$N_{ph}N_r = 48$	±48	±96	±144	±192

13.6.7 Four-Phase SRMs

Figure 13.55 shows the radial force density of a four-phase 8/6 SRM over one electrical cycle. The 8/6 SRM has two magnetic poles similar to the three-phase 6/4 SRM. Thus, at each time step or at each rotor position, two projections (or peaks) can be seen in the radial force density along the circumferential position. Since the number of phases is four, four pulsations are observed in the circumferential position from 0 ~ 180 degrees and over one electrical cycle. It can be observed that there are 8 peaks in total over one electrical cycle for the four-phase 8/6 SRM.

Figure 13.56 shows the projections of radial force density wave over one mechanical cycle for the four-phase 8/6 SRM. Since the number of rotor poles is 6, there are six electrical cycles for each phase in one mechanical cycle. Figure 13.57 shows the dominant harmonic content of the radial force density wave of a 8/6 SRM.

Since a four-phase 8/6 SRM has 2 magnetic poles, the series of circumferential orders for radial force density harmonics is $circ = \{0, 2, 4, 6,...\}$ (multiples of 2). For a given circumferential order, for instance $circ = 0$, the distance between two adjacent harmonic orders along the *u*-axis is 24 ($N_{ph}N_r = 24$). The same distance for 6/4 SRM was 12 ($N_{ph}N_r = 12$). If the maximum speeds of the 6/4 and 8/6 SRMs are the same, the same forcing frequency range would contain more harmonics in a 6/4 SRM than a 8/6 SRM. The harmonic order

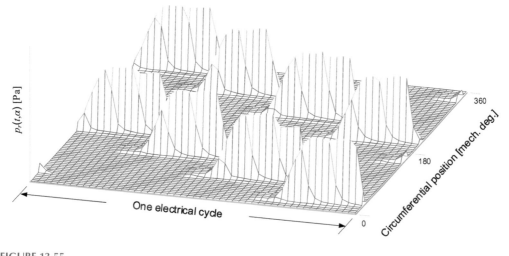

FIGURE 13.55
Radial force density over one electrical cycle for a four-phase 8/6 SRM.

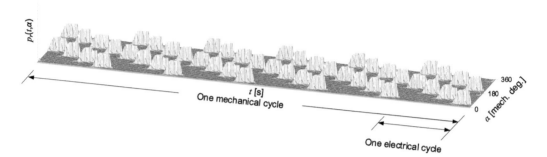

FIGURE 13.56
Radial force density over one mechanical cycle for a four-phase 8/6 SRM.

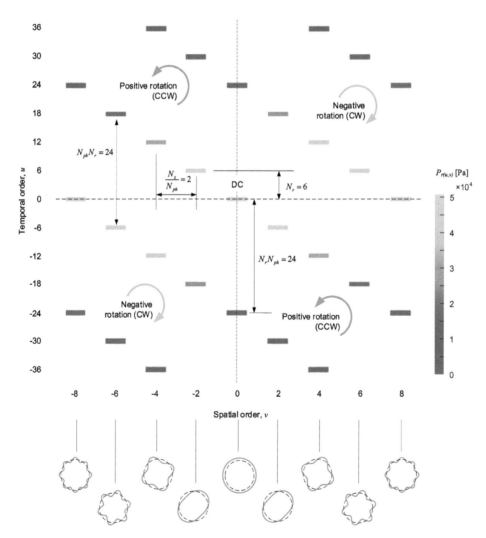

FIGURE 13.57
Amplitudes for dominant harmonic contents of radial force density, four-phase 8/6 SRM.

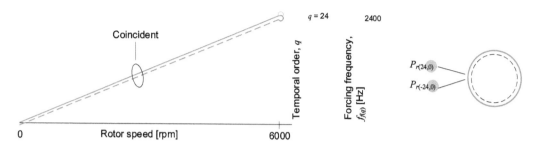

FIGURE 13.58
Forcing frequency of $P_{r(24,0)}$ and $P_{r(-24,0)}$, 8/6 SRM.

of the four-phase 8/6 SRM is more sparse than the three-phase 6/4 SRM. In other words, most of the harmonic content of the radial force density wave of a 8/6 SRM reach higher frequencies than 6/4 SRM for the same speed. Compared with the mode zero forcing frequencies in Figure 13.33 for a 6/4 SRM, the two dominant harmonics at $circ = 0$, $P_{r(24,0)}$ and $P_{r(-24,0)}$ in 8/6 SRM (see Figure 13.58) have much higher forcing frequencies than 6/4 SRM. At 6000 rpm, the forcing frequencies of $P_{r(24,0)}$ and $P_{r(-24,0)}$ in 8/6 SRM are 2400 Hz, double of $P_{r(12,0)}$ and $P_{r(-12,0)}$ in 6/4 SRM. In a four-phase 8/6 SRM, the orders (u, v) for the dominant radial force density harmonics can be obtained as:

$$\begin{cases} u = 24j - 6k \\ v = 2k \end{cases} \tag{13.58}$$

If all the harmonics in Figure 13.57 are converted to the first quadrant (axes included), the result will be as shown in Table 13.12.

Compared to Table 13.6 for the 6/4 SRM, a standing wave is generated in the 8/6 SRM from the sum of $-2P_{r(24,8)}$ and $+2P_{r(-24,8)}$ (see Table 13.12). In Table 13.12, there is another

TABLE 13.12

Expression of $P_{r(q,\,circ)}$, the Amplitudes for the Harmonic Contents in the First Quadrant, Axes Included, Four-Phase 8/6 SRM

$P_{r(q,\,circ)}$		Circumferential Order, $|v| = circ$				
		0	2	4	6	8
Temporal order $|u| = q$	36			$-2P_{r(36,4)}$ $+2P_{r(-36,4)}$		
	30		$+2P_{r(-30,2)}$		$-2P_{r(30,6)}$	
	24	$2P_{r(24,0)}$				$-2P_{r(24,8)}$ $+2P_{r(-24,8)}$
	18		$-2P_{r(18,2)}$		$+2P_{r(-18,6)}$	
	12			$-2P_{r(12,\,4)}$ $+2P_{r(-12,4)}$		
	6		$+2P_{r(-6,2)}$		$-2P_{r(6,\,6)}$	
	0	$P_{r(0,0)}$				$2P_{r(0,8)}$

standing wave that is generated from the sum of $-2P_{r(12,4)}$ and $+2P_{r(-12,4)}$. This happens because 8/6 SRM has an even number of phases. In Figure 13.57, the harmonics, $-P_{r(12,4)}$ and $+P_{r(-12,4)}$, are mirrored with respect to u axis. When all the harmonics are converted to the first quadrant, $-P_{r(12,4)}$ and $+P_{r(-12,4)}$ generate a standing wave.

Tables 13.13 and 13.14 show the harmonic content in the first quadrant for four-phase 16/12 and 24/18 SRMs. Similar to the three-phase topologies, for the four-phase SRMs, the higher the number of stator and rotor poles, the lower the excitation of low circumferential modes.

TABLE 13.13

Expression of $P_{r(q,\ circ)}$, the Amplitudes for the Harmonic Contents in the First Quadrant, Axes Included, Four-Phase 16/12 SRM

$P_{r(q, circ)}$		Circumferential Order, $|v| = circ$				
		0	4	8	12	16
Temporal order $|u| = q$	72			$-2P_{r(72,8)}$ $+2P_{r(-72,8)}$		
	60		$+2P_{r(-60,4)}$		$-2P_{r(60,12)}$	
	48	$2P_{r(48,0)}$				$-2P_{r(48,16)}$ $+2P_{r(-48,16)}$
	36		$-2P_{r(36,4)}$		$+2P_{r(-36,12)}$	
	24			$-2P_{r(24,8)}$ $+2P_{r(-24,8)}$		
	12		$+2P_{r(-12,4)}$		$-2P_{r12,12)}$	
	0	$P_{r(0,0)}$				$2P_{r(0,16)}$

TABLE 13.14

Expression of $P_{r(q,circ)}$, the Amplitudes for the Harmonic Contents in the First Quadrant, Axes Included, Four-Phase 24/18 SRM

$P_{r(u, circ)}$		Circumferential Order, $|v| = circ$				
		0	6	12	18	24
Temporal order $|u| = q$	108			$-2P_{r(108,12)}$ $+2P_{r(-108,12)}$		
	90		$+2P_{r(-90,6)}$		$-2P_{r(90,18)}$	
	72	$2P_{r(72,0)}$				$-2P_{r(72,24)}$ $+2P_{r(-72,24)}$
	54		$-2P_{r(54,6)}$		$+2P_{r(-54,18)}$	
	36			$-2P_{r(36,12)}$ $+2P_{r(-36,12)}$		
	18		$+2P_{r-18,6}$		$-2P_{r(18,18)}$	
	0	$P_{r(0,0)}$				$2P_{r(0,24)}$

This section has discussed the properties of harmonics for radial density surface wave for three-phase and four-phase SRMs. The harmonics of the radial force density surface wave excite different modes of the motor structure. The details of this phenomenon as well as an analysis of how the dominant harmonics affect the vibration and noise in SRMs are discussed in the following section.

13.7 Acoustic Analysis

Electromagnetic noise in internal-rotor SRMs is caused by the vibration of the stator structure due to the radial magnetic forces. In order to determine the radiated noise, it is important to know the intensity of the surface vibration (displacement, velocity) and how efficiently this vibration is converted into noise (sound radiation ratio).

13.7.1 Surface Displacement

In Chapter 12, the equations regarding the surface displacement have been discussed. Surface displacement is a function of the angular natural frequency and angular forcing frequency, which can be calculated as:

$$
\begin{cases}
\omega_{n(q,circ,ax)} = 2\pi f_{n(q,circ,ax)} = 2\pi f_{n(circ,ax)} \\
\omega_{f(q,circ,ax)} = 2\pi f_{f(q,circ,ax)} = 2\pi q f_{mech}
\end{cases}
\tag{13.59}
$$

where $f_{n(circ,ax)}$ is the natural frequency for vibration mode (*circ, ax*), which can be estimated analytically. q is the absolute value of the temporal order, u for a certain harmonic content of the radial force density, and f_{mech} is the mechanical frequency, which can be calculated based on the rotor speed. As discussed in Chapter 12, the surface displacement $D_{(q,circ,ax)}$ for vibration mode (*circ, ax*) caused by the harmonic content (q, *circ, ax*) is calculated as [2]:

$$
D_{(q,circ,ax)} = \frac{F_{r(q,circ,ax)} / M}{\sqrt{(\omega_{n(q,circ,ax)}^2 - \omega_{f(q,circ,ax)}^2)^2 + 4\zeta_{(q,circ,ax)}^2 \omega_{f(q,circ,ax)}^2 \omega_{n(q,circ,ax)}^2}}
\tag{13.60}
$$

where M is the lumped mass of the object, $\zeta_{(q,circ,ax)}$ is the damping ratio associated with the mode (*circ, ax*), which will be discussed later in this chapter, and $F_{r(q,circ,ax)}$ is the amplitude of the force component (q, *circ, ax*), which is calculated as:

$$
F_{r(q,circ,ax)} = \pi D L_s P_{r(q,circ,ax)}
\tag{13.61}
$$

where $P_{r(q,circ,ax)}$ is the amplitude of the harmonic content of the radial force density, which has been discussed in Section 13.6. Force density is in units of N/m². Therefore, to calculate the force, $P_{r(q,circ,ax)}$ is multiplied with the area of the surface where the radial force densities have been calculated (usually in the airgap). Hence, the term $\pi D L_s$ is the surface area where D is the inner diameter of the cylindrical shell (for an SRM, it is the inner diameter of the stator) and L_s is the axial length.

Equation (13.23) shows the function for any harmonic of the radial force density wave. If the term representing the surface area, πDL_s is multiplied with both sides of (13.23), the expression for the radial force wave can be obtained:

$$\mathcal{F}_{r(u,v)} = F_{r(u,v)} \cos(\omega_{mech}ut + v\alpha + \phi_{(u,v)}) \tag{13.62}$$

where $\mathcal{F}_{r(u,v)}$ represents the harmonics of 2D radial force wave at a certain speed and $F_{r(u,v)}$ is the amplitude of the harmonics. Similarly, (13.24) can be rewritten as:

$$\mathcal{F}_{r(t,\alpha)} = \sum_{u=-\infty}^{+\infty} \sum_{v=-\infty}^{+\infty} \sum_{ax=1}^{+\infty} \mathcal{F}_{r(u,v,ax)} \tag{13.63}$$

where $\mathcal{F}_{r(t,\alpha)}$ represents the radial force wave over one mechanical cycle.

13.7.2 Media Field and Radiation Ratio

The vibrations on the motor surface cause small fluctuations in the air pressure leading to sound pressure and noise. Generally speaking, the difference in the air pressure at a certain spatial point depends on the distance of the vibrating motor surface from that spatial point. The noise level can also be measured by sound power. Sound power is defined as the rate at which sound energy is emitted, reflected, transmitted, or received, per unit time. The measured sound power is not related to the distance between the vibrating motor surface and the measured point. Therefore, it is more reasonable to use the sound power rather than the sound pressure to evaluate the entire radiated noise from the surface of the vibrating structure. The relationship between the radiated sound power Π and the surface vibrations is characterized as the sound radiation ratio [2]:

$$\sigma = \frac{\Pi}{\rho_0 c_0 A_s \left\langle v^2 \right\rangle} \tag{13.64}$$

where ρ_0 and c_0 are the mass density of the air and the speed of sound in the air, respectively. A_s is the area of the sound radiation surface, and $\langle v^2 \rangle$ is the spatial averaged mean square vibration velocity on the surface. The vibration and sound radiation can be analyzed in the frequency domain by superimposing the sound power from each vibration mode.

The modal sound radiation ratio depends on the shape of surfaces and the vibration frequency. For its estimation, the stator can be simplified as a cylindrical shell, as shown in Figure 13.59. In Figure 13.59, a and l are the radius and the length of the shell respectively, and k_0 is the acoustic waveform number, which is related to the forcing frequency of the vibration ω_f as given in (13.65) [2].

$$k_0 = \frac{\omega_{f(q,circ,ax)}}{c_0} = \frac{2\pi f_{f(q,circ,ax)}}{c_0} = \frac{2\pi q f_{mech}}{c_0} \tag{13.65}$$

It can be seen from (13.65) that k_0 is proportional to the value of the forcing frequency and a function of temporal order, q. Again, c_0 is the speed of sound in the air, and k_r and k_z are the radial and axial components of k_0, respectively:

$$k_0 = \sqrt{k_z^2 + k_r^2} \tag{13.66}$$

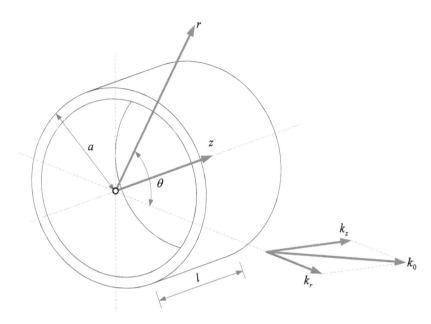

FIGURE 13.59
Cylindrical shell model for radiation ratio analysis.

In [5], the analytical expression for the modal radiation ratio of the finite length cylindrical shell was derived with respect to vibration frequency ω:

$$\sigma_{(circ,ax)} = \int_{-k_0}^{k_0} \frac{2k_0 l}{\pi^2 a\, k_r^2 \left| \dfrac{\mathrm{d}H_{circ}^{(2)}(k_r a)}{\mathrm{d}(k_r a)} \right|^2} \left[\frac{ax \cdot \pi / l}{k_z + ax \cdot \pi / l} \right]^2 \frac{\sin^2[(k_z - ax \cdot \pi / l)l / 2]}{[(k_z - ax \cdot \pi / l)l / 2]^2}\, dk_z \quad (13.67)$$

As shown in Figure 13.59, a is the radius of the shell, and l is the length of the shell, while *circ* and *ax* are the circumferential and the axial modes, respectively. Since k_0 is a function of temporal order, q, $\sigma_{(circ,ax)}$ also depends on q. Thus, hereafter, $\sigma_{(circ,ax,q)}$ is used to replace $\sigma_{(circ,ax)}$ to facilitate matrix computation. $H_{circ}^{(2)}(x)$, is the Hankel functions of the second kind [6]. The derivative of $H_{circ}^{(2)}(x)$ is given by (13.68):

$$\frac{\mathrm{d}H_{circ}^{(2)}(x)}{\mathrm{d}x} = \frac{H_{circ-1}^{(2)}(x) - H_{circ+1}^{(2)}(x)}{2} \quad (13.68)$$

Figure 13.60 shows the radiation ratio versus temporal order (q) and circumferential mode (*circ*) when the motor speed is 6000 rpm and for the first axial mode, $ax = 1$ only. Since the motor speed is fixed, the mechanical frequency, f_{mech} is fixed. As the circumferential mode, *circ* increases, in order to reach the radiation ratio of 1, temporal order q should increase as well. This implies that it is more difficult to radiate vibrations caused by a higher circumferential mode. A higher *circ* requires a higher q and, thus, a higher forcing frequency at a fixed motor speed to achieve a higher radiation ratio.

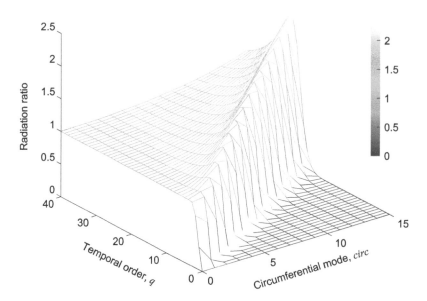

FIGURE 13.60
Radiation ratio versus temporal order and circumferential mode, $ax = 1$, motor speed = 6000 rpm ($f_{mech} = 100$ Hz).

13.7.3 Sound Radiation

When an elastic body or medium oscillates around its equilibrium position, vibration is generated. As vibration is transmitted through an elastic medium, such as solid, liquid or gas, it will create a variation in pressure. In air, frequencies in the approximate range of 20–20,000 Hz create changes in the air pressure that can be detected by human ears. Sound pressure level (SPL), sound intensity level (SIL), sound power level (SWL), and sound energy density (SED) can all be used to measure sound amplitude. Sound pressure is expressed in units of Pascal (Pa) and can be converted to SPL in units of decibel (dB).

Sound can be classified into four ranges, as summarized in Table 13.15. Figure 13.61 shows sound intensity and audibility zones as functions of frequency. The loudness contour is a measure of sound pressure over the frequency spectrum. A microphone is used to measure the loudness levels. Table 13.16 summarizes typical sound power levels for different scenarios. Noise is a type of sound perceived to be loud, unpleasant, unexpected, and/or undesired. There are two types of acoustic noise: pitched and unpitched noise.

TABLE 13.15

Sound Frequency Ranges

Type of Sound	Frequency	Activities
Infrasound	$0 < f < 20$ Hz	Seismic activities, communication between whales, elephants
Audible sound	20 Hz $< f < 20$ kHz	Audible range for human ears
Ultrasound	16 kHz $< f < 1.6$ GHz	Medical ultrasonic devices, plastic ultrasonic welding, sonar
Hyper sound	1 GHz $< f$	Aeronautics and aerospace

Source: Bertolini, T., and Fuchs, T., *Vibrations and Noises in Small Electric Motors Measurement, Analysis, Interpretation*, Ulm, Germany: Verlag onpact GmbH, 2012.

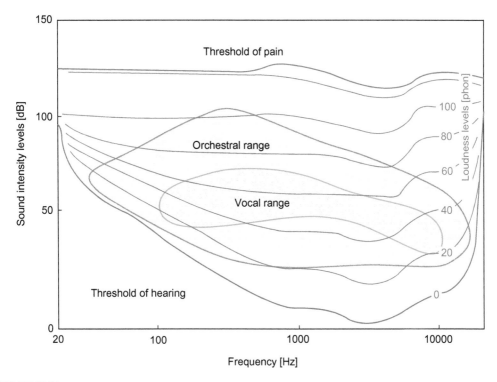

FIGURE 13.61
Sound intensity and audibility zone as a function of frequency. (From Gieras, J.F. et al., *Noise of Poly-Phase Electric Motors*, CRC Press, New York, 2005; Tong, W., *Mechanical Design of Electric Motors*, CRC Press, Boca Raton, FL, 2014.)

Pitched noise, also known as tonal noise, is a type of noise that occurs around some specific frequencies. Unpitched noise, also known as broadband noise, occupies a wide range in the frequency spectrum [3].

Once the modal radiation ratio $\sigma_{(q,circ,ax)}$ is calculated, the sound power for mode *circ* with forcing angular frequency ω_f can be calculated using [2]:

$$\Pi_{(q,circ,ax)} = \rho_0 c_0 \left(\frac{\omega_f D_{(q,circ,ax)}}{\sqrt{2}} \right)^2 \sigma_{(q,circ,ax)} A_s [\text{W}] \tag{13.69}$$

where A_s is the area of the sound radiation surface. If the axial mode is not considered, (13.69) can be expressed as:

$$\Pi_{(q,circ)} = \rho_0 c_0 \left(\frac{\omega_f D_{(q,circ)}}{\sqrt{2}} \right)^2 \sigma_{(q,circ)} A_s [\text{W}] \tag{13.70}$$

The sound power level resulting from different modes with the same forcing frequency can be superimposed together [2]. The sound power is a function of q, which can be calculated using:

$$\hat{\Pi}_{(q)} = \sum_{ax=1}^{ax\_\max} \sum_{circ=0}^{circ\_\max} \Pi_{(q,circ,ax)} [\text{W}] \tag{13.71}$$

TABLE 13.16

Typical Sound Power Level

Source of Noise	Sound Power Level [dB ref 10^{-12} W]	Sound Power [W]
Quietest audible sound for persons under normal conditions	10	10^{-11}
Whisper of one person	20	10^{-10}
Soft whisper, room in a quiet dwelling at midnight	30	10^{-9}
Refrigerator	40	10^{-8}
Modern elevator propulsion motor	50	10^{-7}
Bird singing	60	10^{-6}
Voice, conversation	70	10^{-5}
Pneumatic tools	80	10^{-4}
Lawn mower	90	10^{-3}
Car on highway	100	0.01
Chainsaw	110	0.1
Jackhammer	120	1
Machine gun	130	10
Jet plane taking off	150	1000

Source: Bertolini, T., and Fuchs, T., *Vibrations and Noises in Small Electric Motors Measurement, Analysis, Interpretation*, Ulm, Germany: Verlag onpact GmbH, 2012.

Since the forcing frequency is related to f_{mech}, $\hat{\Pi}_{(q)}$ is a function of mechanical frequency and, hence, speed. Similarly, if the axial mode is not considered, $\hat{\Pi}_{(q)}$ can be calculated using:

$$\hat{\Pi}_{(q)} = \sum_{circ=0}^{circ\_max} \Pi_{(q,circ)}[\text{W}] \tag{13.72}$$

When used in the units of Watts, the value of $\hat{\Pi}_{(q)}$ can vary greatly. Thus, it is more practical to use sound power level in units of dB. This unit conversion is achieved by logarithmic scaling of the superimposed sound power:

$$\text{SWL}_{(q)} = 10\lg\frac{\hat{\Pi}_{(q)}}{\Pi_{ref}}[\text{dB}] \tag{13.73}$$

where Π_{ref} is 10^{-12} W [2]. The sound pressure is dependent on the spatial location where the sound was emitted, and its average value can be calculated from the sound power according to the NEMA MG 1-2009 standard [9]:

$$\text{SPL}_{(q)} = \text{SWL}_{(q)} - 10\log_{10}\left\{2\pi\left[1 + \frac{\max(l_{shell}, 2r_{shell})}{2}\right]^2\right\}[\text{dB}] \tag{13.74}$$

In (13.74), SPL is the average sound pressure level in a free-field over a reflective plane on a hemispherical surface at 1-meter distance from the motor, plus half of the maximum linear dimension of the machine. In an electric motor, linear dimensions are l_{shell} and $2r_{shell}$, which are the axial length of the shell and the diameter of the shell, respectively. Since $f_{(q)} = qf_{mech}$, $SPL_{(q)}$ is a function of frequency $f_{(q)}$:

$$SPL\left(f_{(q)}\right) = SPL_{(q)} \tag{13.75}$$

It can be seen from (13.75), at one certain motor speed, SPL can be plotted over the frequency spectrum.

13.7.4 Perceived Sound Level

Up to now we have discussed how to calculate the physical noise caused by the magnetic forces. However, human ears have different levels of sensitivity depending on the frequency. To match our physiological perception of sound, weighting curves are commonly applied to the sound levels. In Figure 13.62, the A-weighting curve is approximately inverted from the 40 phon loudness contour. Phon is a unit of loudness level and it represents the SPL at a certain frequency that has an equal perceived loudness. The shape of the A-weighting curve is similar to the response of human ear at the lower noise levels. A-weighting curve is the predominant one in the electric machine industry out of the 4 weighting curves shown in Figure 13.62 [2]. The B-weighting curve was initially developed to cover the area between A and C curves. It was used heavily by the motor industry, but it is rarely used nowadays. The C-weighting curve is better at representing the human response to high noise levels. The D-weighting curve was developed to measure high level noise for aircraft. If the A-weighting curve is used to modify the SPL, the following equation can be used.

$$SPL_A\left(f_{(q)}\right) = SPL_A\left(f_{(q)}\right) + L_A\left(f\right) \tag{13.76}$$

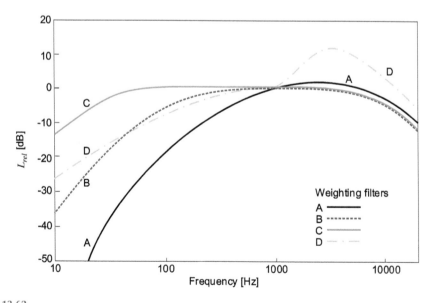

FIGURE 13.62

Weighting curves of A, B, C and D corrections. (From Bertolini, T., and Fuchs, T., *Vibrations and Noises in Small Electric Motors Measurement, Analysis, Interpretation*, Ulm, Germany: Verlag onpact GmbH, 2012; Fastl, H., and Zwicker, E., *Psychoacoustics: Facts and Models*, Springer Science & Business Media, Berlin, Germany, 2013.)

TABLE 13.17

Factors of A-Weighting Correction for Selected Frequencies

Frequency [Hz]	A-Weighting Correction [dB]	Frequency [Hz]	A-Weighting Correction [dB]
20	−50.5	800	−0.8
25	−44.7	1000	0.0
31.5	−39.4	1250	0.6
40	−34.6	1600	1.0
50	−30.2	2000	1.2
63	−26.2	2500	1.3
80	−22.5	3150	1.2
100	−19.1	4000	1.0
125	−16.1	5000	0.5
160	−13.4	6300	−0.1
200	−10.9	8000	−1.1
250	−8.6	10000	−2.5
315	−6.6	12500	−4.3
400	−4.8	16000	−6.6
500	−3.2	20000	−9.3

where $L_A(f)$ is the corresponding factor function for A-weighting correction, which can be calculated by using [2]:

$$\begin{cases} L_A(f) = 20\log_{10}\big(R_A(f)\big) + 2 \\ \\ R_A(f) = \dfrac{12194^2 \cdot f^4}{(f^2 + 20.6^2)\sqrt{(f^2 + 107.7^2)(f^2 + 737.9^2)}\,(f^2 + 12194^2)} \end{cases} \tag{13.77}$$

The factors of A-weighting correction for a number of selected frequencies can be found in Table 13.17. It can be seen from the table that humans are more sensitive to the noise with frequencies between 1000 Hz and 6300 Hz. Throughout the rest of this chapter, unless otherwise specified, SPL stands for SPL_A.

13.8 Estimation of the Damping Ratio

Damping dissipates mechanical energy when vibration occurs. As discussed in Chapter 12, the effect of the damping ratio on a damped spring-mass system is significant. When the damping ratio decreases, the amplitude of the resonance and, hence, vibration increases dramatically. Therefore, an accurate estimation of the damping ratio is important for the analysis of vibration and radiated noise. However, the damping ratio varies with vibration mode, which makes it even harder for the accurate estimation of modal damping ratio. In this section, the damping loss in SRMs as well as the effects and estimation of damping ratio will be discussed.

13.8.1 Damping Loss in SRMs

Damping ratio is commonly used to express the damping loss. When the damping ratio of a certain vibration mode increases, the damping loss due to that vibration mode reduces. The damping ratio is incorporated with several kinds of damping losses. As shown in Figure 13.63, there are three kinds of damping losses in an SRM, including structural damping loss, acoustic damping loss, and joint damping loss. The structural damping loss in SRMs is a kind of hysteresis damping. It changes with material properties and SRM

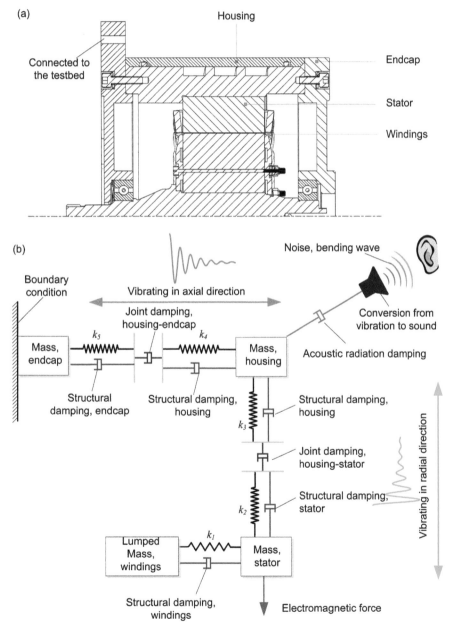

FIGURE 13.63
Damping loss in an SRM: (a) cross-section view of an SRM and (b) lumped damping loss model.

dimensions and geometries. The acoustic radiation damping is a kind of radiation loss from the SRM surface to the surrounding air. The joint damping loss is generated in the joints or interfaces between different components. When different components are rigidly joined together, the joint damping loss can be neglected.

As shown in Figure 13.63a, for the acoustic analysis for interior-rotor SRMs, the stator, windings, housing, and two end covers are considered. The shaft and rotor are not considered because the radial vibration of the stator is the main source of radiated noise. Figure 13.63b shows the lumped damping loss model. When the electromagnetic force is applied on the stator, the stator and the housing will vibrate in the radial direction. Hence, structural damping losses in the stator, windings, and housing are generated.

Small joint damping between the interface of the housing and stator can be neglected because the housing and the stator are typically connected via rigid press-fitting. The two end caps are connected to the housing by bolts and, thus, the end caps will have axial vibration. The joint damping loss in the interface between the end cap and housing can also be neglected because the bolted connection between these components is assumed to be rigid.

13.8.1.1 Structural Damping Loss

Ideally, engineering materials used in electric machines can be considered as elastic materials. The characteristics of the stress-strain curve of elastic materials are shown in Figure 13.64a. When the structures of electric machines are vibrating, the magnitude of the displacement is small. Therefore, the relationship between the strain ε and the stress σ can be described by a linear expression $\sigma = E\varepsilon$, in which E is the Young's modulus of the material. When subjected to cyclic loading, actual engineering materials also exhibit a type of structural internal damping causing energy losses. The structural damping is a kind of hysteresis loss in the hysteresis cycle, which is shown in Figure 13.64b.

The stress and strain histories in the time domain for different material are shown in Figure 13.65. Figure 13.65a shows the stress and strain of purely elastic materials, in which

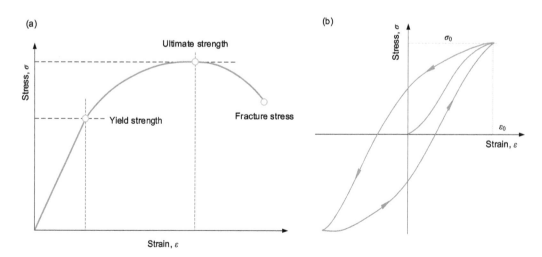

FIGURE 13.64
Relationship between stress and strain for elastic materials: (a) stress-strain curve and (b) hysteresis cycle in (σ, ε) plane.

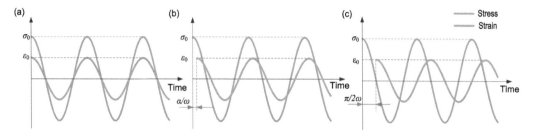

FIGURE 13.65
Stress and strain curves in time domain for different material: (a) purely elastic material, (b) actual engineering material, and (c) pure viscous material.

the stress and strain change in phase. For this kind of material, the strain appears simultaneously with the stress. For purely viscous materials, as shown in Figure 13.65c, the stain lags the stress by a 90° phase angle. For engineering materials used in electric machines, the hysteresis loss makes the strain lag the stress by a certain phase angle, as shown in Figure 13.65b. The phase angle depends on different materials used in the motor.

When the cyclic excitation load is applied on an SRM structure, the strain and stress can be expressed as:

$$\begin{cases} \varepsilon = \varepsilon_0 \cos(wt) \\ \sigma = \sigma_0 \cos(wt + \alpha) \end{cases} \tag{13.78}$$

It can also be expressed in the complex notation as follows:

$$\begin{cases} \varepsilon = \varepsilon_0 e^{iwt} \\ \sigma = \sigma_0 e^{i(wt+\alpha)} \end{cases} \tag{13.79}$$

The Young's modulus can be calculated using the ratio between the stress and the strain:

$$E = \frac{\sigma}{\varepsilon} = \frac{\sigma_0}{\varepsilon_0} \times e^{i\alpha} = E \times [\cos(\alpha) + i\sin(\alpha)] \tag{13.80}$$

Compared with an ideal engineering material, the Young's modulus for actual engineering material is in complex form with a real part and imaginary part. The real part E' and the imaginary part E'' can be expressed as:

$$\begin{cases} E' = E \times \cos(\alpha) \\ E'' = E \times \sin(\alpha) \end{cases} \tag{13.81}$$

The real part is a measure of the stiffness of the material while the imaginary part is a measure of the internal damping. The loss factor η, of the engineering material, is defined as the ratio between the imaginary part and the real part of the complex Young's modulus. The structural loss factors of different engineering materials used in SRMs are shown in Table 13.18 [12]. The structural loss factor is twice the structural damping ratio [11]. Thus, the structural damping ratio is calculated as:

TABLE 13.18

Structural Loss Factor of Some Engineering Materials

Material	Loss Factor η
Aluminum alloy	$0.0001 \sim 0.001$
Copper	$0.001 \sim 0.005$
Steel	$0.01 \sim 0.06$
Rubber	$0.01 \sim 3$
Cast iron	$0.001 \sim 0.08$

Source: Genta, G., *Vibration Dynamics and Control*, Springer, New York, 2009.

$$\zeta_s = \frac{E''}{2E'} = 0.5 \times \tan(\alpha) \tag{13.82}$$

13.8.1.2 Acoustic Damping Loss

The acoustic radiation damping is generated when fluid flows around the vibrating SRM structure. Figure 13.66a shows that air molecules are distributed uniformly around an SRM when it is static and does not vibrate. When an SRM is vibrating with a circumferential mode 2, as shown in Figure 13.66b, air rarefaction and compression appear. The region where air compression appears is close to the expanding part of the SRM structure. Similarly, the region where air rarefaction appears is close to the shrinking part of the SRM structure. Figure 13.66c shows this effect at different time instants during mode 2 vibration. The displacement of the air molecules around the SRM surface consumes some of the motor output power. The acoustic damping ratio can be calculated using:

$$\zeta_a = \frac{\rho_0 c_0 \sigma}{2\omega_f \rho_s} \tag{13.83}$$

where ρ_0 is the fluid density, c_0 is the speed of sound, σ is the radiation ratio ω_f is the angular forcing frequency, and ρ_s is the surface mass (mass per unit surface area) [3]. Calculation of the radiation ratio was presented in section 13.7.2.

As mentioned previously, if joint damping loss can be neglected, two kinds of damping ratios dominate when an SRM is vibrating: structural and acoustic damping. Because there is no interaction between these two kinds of damping losses, their corresponding damping ratio—structural damping ratio and acoustic damping ratio—are not related to each other. Therefore, the total modal damping ratio can be calculated by summing up the structural damping ratio and the acoustic damping ratio [3]:

$$\zeta = \zeta_a + \zeta_s \tag{13.84}$$

where ζ is the total modal damping ratio, ζ_a is the acoustic damping ratio, and ζ_s is the structural damping ratio.

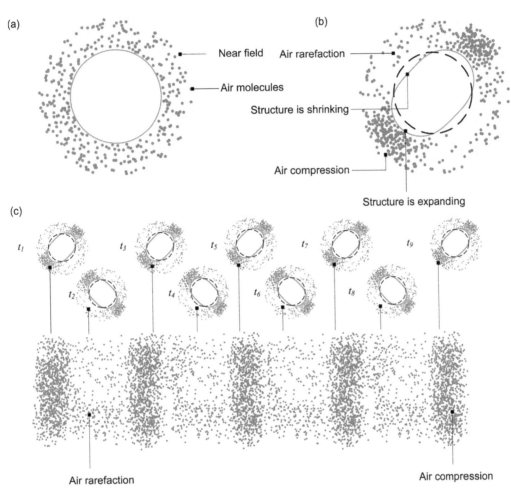

FIGURE 13.66
Displacement of air molecules when the SRM is vibrating: (a) before vibration, (b) SRM is vibrating with mode
circ = 2, and (c) the SRM is vibrating at different instant.

13.8.2 Effect of Damping on Vibration and Noise

Figure 13.67 shows the effect of damping ratio on surface displacement and sound pressure
of a 24/16 SRM at 2000 RPM. The simulations were conducted in ACTRAN®. All the mode
shapes are assumed to have the same damping ratio. The surface displacement is calculated based on the surface acceleration captured by the same accelerometer. The sound
pressure is also captured by the same virtual microphones. There is a noticeable difference
in the surface displacement and the sound pressure when considering different damping
ratios, as shown in Figure 13.67.

Taking the captured sound pressure in Figure 13.67b and applying 2D FFT analysis, the
SPL at different frequencies can be obtained. Although in some frequency components the
difference of the two damping ratios is small, for most frequency components the difference cannot be neglected.

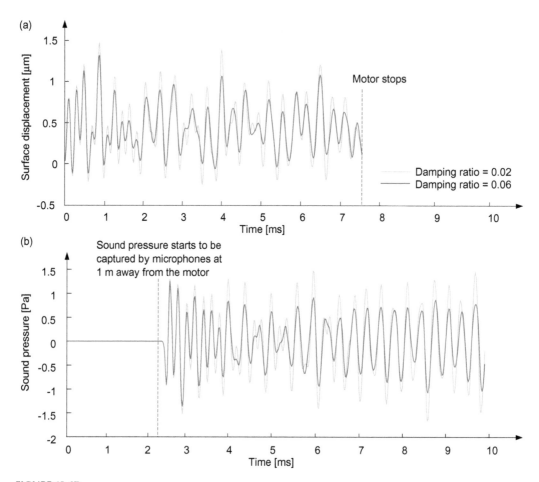

FIGURE 13.67
Effect of damping ratio on a 24/16 SRM at 2000 rpm: (a) surface displacement and (b) sound pressure.

13.8.3 Estimation of the Total Damping Ratio

A well-known expression for the estimation of total damping ratio in electric machines has been defined as:

$$\zeta_{(circ,1)} = \frac{1}{2\pi}(2.76 \times 10^{-5} f_{(circ,1)} + 0.062) \tag{13.85}$$

which is a fitting equation based on experimental results]. $f_{(circ,1)}$ represents the natural frequencies of a mode shape with a circumferential order *circ* and with the axial order of $ax = 1$. It is only applicable for medium- and small-size electric motors, and can only be used for mode shapes with an axial order 1.

An accurate estimation of the damping ratio can be obtained by conducting a modal test and generating the frequency response function (FRF). Two kinds of modal tests are commonly used: the hammer impulse test and the shaker test.

As shown in Figure 13.68a, the hammer test needs a hammer with a force gauge and accelerometers. When the hammer hits the surface of the stator structure, the force gauge captures the force while the accelerometer captures the surface acceleration. Although

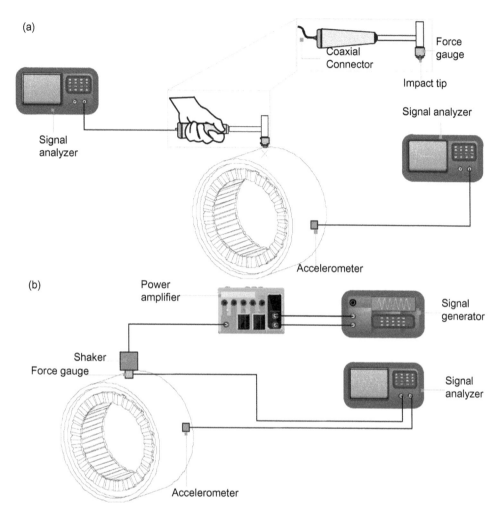

FIGURE 13.68
Modal test: (a) hammer impulse test and (b) shaker test.

this method is simple to perform, the biggest disadvantage is that the hammer can only generate an impulse signal (see Figure 13.69a). This impulse signal might have limited frequency range due to properties of the hammer.

On the other hand, the shaker test shown in Figure 13.68b can generate several kinds of force signals]. These force signals can cover a broader frequency range. The required equipment for the shaker test is similar to that of a hammer test. The shaker test also needs a force gauge to capture the force signal, an accelerometer to capture the surface acceleration, and a signal generator to generate different types of force signals. Four types of force signals are shown in Figure 13.69. Figure 13.69b1 and b2 shows the sinusoidal signal and its response, respectively. This sinusoidal signal is at a fixed frequency and, thus, many experiments must be conducted to analyze the system at different frequencies. Therefore, it might take a long time to obtain the response in a wide frequency range. In practice, the frequency steps should be selected carefully. The frequency steps should be small when the excitation frequency approaches the resonance region. The frequency steps can be larger when the excitation frequency is far from the resonant frequency. Figure 13.69c1

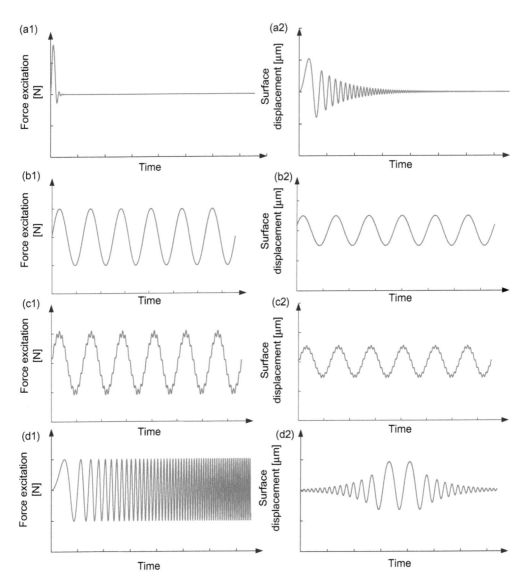

FIGURE 13.69
Types of excitation force signals and their responses: (a1) Impulse signal, (a2) response of impulse signal, (b1) sinusoidal signal, (b2) response of sinusoidal signal, (c1) Periodic signal, (c2) response of periodic signal, (d1) transient-chirp, and (d2) response of transient-chirp. (From Crocker, J., ed., *Handbook of Noise and Vibration Control*, John Wiley & Sons, Hoboken, NJ, 2007.)

and (c2) shows the excitation force of periodic signal and its response. Compared with sinusoidal signal, which only contains one frequency component, the periodic signal contains several harmonics. This can be an advantage since the excitation force signal can have several frequency components. The transient-chirp signal in Figure 13.69d1 is a kind of transient signal and its response in Figure 13.69d2 gives an approximate indication of the structure's response [14].

(a)   (b)   (c)

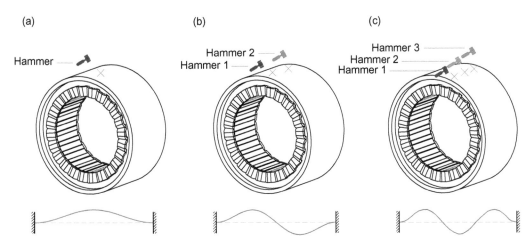

FIGURE 13.70
Number of hammers or shakers for the modal tests: (a) 1 hammer or shaker, (b) 2 hammers or shakers, and (c) 3 hammers or shakers.

Ideally, the number of hammers and shakers should be as many as possible to excite all the mode shapes of the structure. The number of hammers and shakers would be defined by the axial modes that need to be analyzed. As shown in Figure 13.70a, in order to excite the mode shapes with axial order 1, one hammer is needed. For mode shapes with axial order 2 and 3, at least 2 and 3 hammers or shakers are required, as shown in Figure 13.70b and c, respectively.

After the modal test, the force signal and the corresponding response can be captured. The FRF can be calculated by:

$$H(f) = \frac{v(f)}{F(f)} \tag{13.86}$$

where $H(f)$ is the FRF, $v(f)$ is the surface velocity in frequency domain, and $F(f)$ is the force applied on the SRM in frequency domain.

Figure 13.71a shows the FRF for a 24/16 SRM as simulated in ACTRAN®. The structural damping ratio of different modes can be obtained via FRF using the half-power method]. As shown in Figure 13.71b, the damped frequency of a certain mode shape is f_r with a peak value of Q, and $\sqrt{Q}$ is the half power of the peak value. The difference between Q and $\sqrt{Q}$ is approximately 3 dB. The damping ratio can be calculated as follows:

$$\varsigma = \frac{f_2 - f_1}{f_r} = \frac{\Delta f}{f_r} \tag{13.87}$$

where f_1 and f_2 are the frequencies whose amplitudes are $\sqrt{Q}$ [11].

In this section, the estimation of damping ratios was discussed. Structural damping losses and acoustic damping losses are generated when vibration occurs on an SRM. Generally, the structural losses dominate in the total damping loss. Taking a 24/16 SRM as an example, the effect of the damping ratio has been discussed. The effect of the damping ratio cannot be neglected because it affects the accuracy of the estimated noise and vibration.

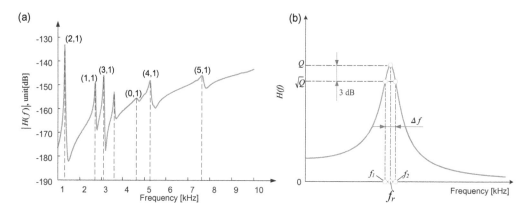

FIGURE 13.71

Frequency response function: (a) SRM and (b) magnification method for the estimation of damping ratio.

13.9 Selective Excitation

As discussed in Section 13.7, not every forcing frequency causes displacement for every circumferential mode. If only the first axial mode is considered, the circumferential order (*circ*) of the time-varying harmonic must be the same with the circumferential order of the vibration mode so that vibration is generated. Provided that the circumferential orders of the radial force density harmonic (simply referred to as "harmonic") and the vibration mode match, resonance can occur when the forcing frequency of the harmonic approaches the natural frequency of the vibration mode. In other words, a certain harmonic will not excite a certain mode shape if its circumferential order does not match the mode shape. For example, a certain harmonic with a circumferential order, 2, will only excite the mode shape with the same circumferential order, 2. Hence, this harmonic content will not excite the mode shapes with the circumferential order, 4 or 8. Therefore, the excitation of natural frequencies is selective. In this section, we will discuss the selective excitation of radial force harmonics in SRMs.

13.9.1 Densities of Forcing Frequencies

Figure 13.72 shows the distributions of forcing frequencies as the motor speed varies from 0 to 12,000 rpm (f_{mech} ranges from 0 Hz to 200 Hz), and as the temporal order q ranges from 0 to 100 for a 6/4 SRM. For a fixed value of q, the forcing frequency increases with the motor speed. The forcing frequency can be calculated as $q \times f_{mech}$. This means that as q increases, the forcing frequency increases for a constant motor speed. The oblique line, describing the variation of the forcing frequency with motor speed, will be referred to as a "ray" in this chapter.

If N_q (1, 2, 3, …) is used to represent the number of rays in Figure 13.72, q can be calculated as:

$$q = N_q N_r \tag{13.88}$$

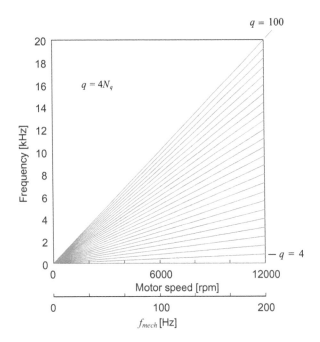

FIGURE 13.72
Forcing frequency versus motor speed (0 ~ 12,000 rpm) for the first 25 harmonic contents (Nq = 1, 2, ..., 25), 6/4 SRM, $q = 4 \times Nq$.

where N_r is the number of rotor poles. Equation (13.88) is based on the dominant harmonic contents of radial force density. As shown in Figure 13.28 for a 6/4 SRM, the difference between each temporal order, u is N_r since $q = |u|$, N_q in (13.88) presents the temporal orders, which can be excited by a certain SRM configuration.

For the first ray (N_q = 1) q equals 4. Its forcing frequency is $4f_{mech}$. The q for the second ray (N_q = 2) is 8. So as q ranges from 0 to 100 (included), the 6/4 SRM has 25 rays of forcing frequencies in total.

13.9.2 Densities of Natural Frequencies

The audible frequency range (0 Hz ~ 20 kHz) can be used as the relevant range of natural frequencies. Please note that higher natural frequencies (even higher than 20 kHz) can also be excited by much lower forcing frequencies. However, if the forcing frequency of a certain radial force harmonic is far away from the natural frequency, the displacement caused by that harmonic is fairly small; almost negligible. In this analysis, only the circumferential mode is considered, and the axial mode is assumed to be one. Figure 13.73 shows an example of the distribution of natural frequencies depicted using horizontal dashed lines.

As discussed earlier in this chapter, as the forcing frequency approaches the natural frequency, resonance can occur. However, since the excitation is selective, which means that the condition of matching mode shapes has to be satisfied, not every intersection generates resonance. This will be discussed in the next section.

The maximum motor speed also affects the excitation of vibration modes. For instance, if the maximum motor speed of the 6/4 SRM is 2000 RPM, as shown in Figure 13.74a, resonances can only happen to $circ$ = 4 and $circ$ = 2. When the maximum motor speed

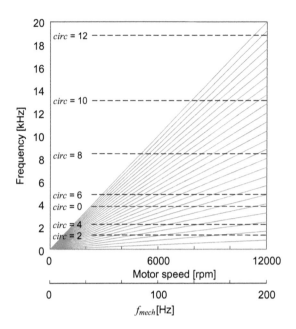

FIGURE 13.73
Forcing frequency versus motor speed (0 ~ 12,000 rpm) for the first 25 harmonic contents and the natural frequencies for $ax = 1$, 6/4 SRM, $q = 4 \times N_q$ ($N_q = 1, 2, \ldots, 25$).

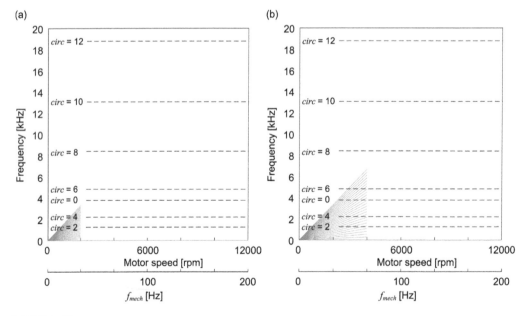

FIGURE 13.74
Forcing frequency versus motor speed as max. motor speed varies, 6/4 SRM: (a) 2000 rpm, (b) 4000 rpm.
(Continued)

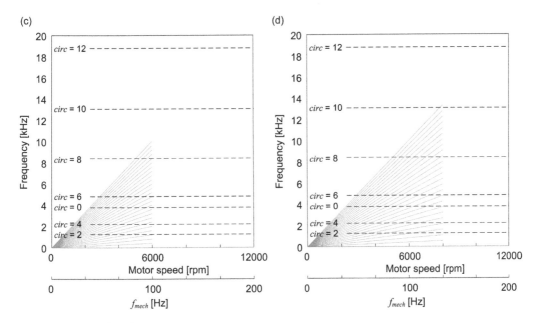

FIGURE 13.74 (Continued)
Forcing frequency versus motor speed as max. motor speed varies, 6/4 SRM: (c) 6000 rpm, and (d) 8000 rpm.

is 4000 RPM, four circumferential modes $circ = \{2, 4, 0, 6\}$ can experience resonances as shown in Figure 13.74b. If the maximum motor speed is further increased to 6000 RPM, resonances can also happen for $circ = 8$.

13.9.3 Density of Resonances

Figure 13.75 shows the excitations of four modes ($circ = \{0, 2, 4, 6\}$) with nine rays ($q = \{4, 8, 12, ..., 36\}$) for the three-phase 6/4 SRM. The rays represent the variations of forcing frequencies with the motor speed as temporal order, q increases. The horizontal lines represent the natural frequencies. The intersections, depicted by a sun icon, between the horizontal lines and the rays, are where resonance happens. Resonance occurs when the forcing frequency approaches the natural frequencies. However, since the excitation is selective, which means that the condition of matching mode shapes has to be satisfied, not every intersection generates resonance. Figure 13.33 shows that the ray of forcing frequency with $q = 12$ can excite the mode with $circ = 0$. This ray can also excite the mode with $circ = 6$, as shown in Figure 13.39. Figure 13.35 shows that the rays of forcing frequencies with $q = \{4, 8, 16\}$ can excite the mode with $circ = 2$. These three rays can also excite the mode with $circ = 4$, as shown in Figure 13.37. As shown in Figure 13.28, the harmonic content is repetitive. For a certain spatial harmonic, since the distance between the temporal harmonics is $N_{ph}N_{r}$, the rays of forcing frequencies with $q = \{12, 24, 36, ...\}$ can excite mode 6 in a 6/4 SRM. Similarly, the rays of forcing frequencies with $q = \{4, 8, 16, 20, 28, 32, ...\}$ can excite mode 4, since $circ = |v|$ and $q = |u|$.

It can be seen that $circ = \{0, 6\}$ can be excited by rays of forcing frequencies with $q = \{12, 24, 36\}$, as shown in red in Figure 13.75a. The resonances, represented by the sun icons, can be seen in Figure 13.75a. The mode shapes of $circ = 2$ can be excited by harmonics with temporal order $q = \{4, 8, 16, 20, 28, 32\}$, as shown in blue as in Figure 13.75b. The two colors

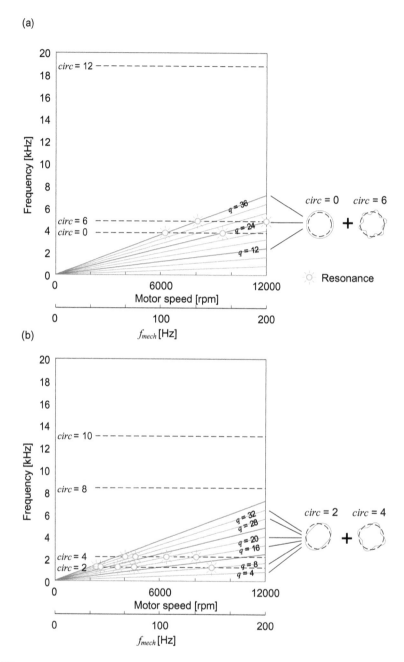

FIGURE 13.75
Excitations of circumferential modes ($circ < 10$), 6/4 SRM, $q = \{4, 8, ..., 36\}$ (first 9 rays): (a) $circ = \{0, 6\}$ and (b) $circ = \{2,4\}$.

of red and blue, which are used to differentiate the two groups of rays, will be used in the following discussion and in the questions at the end of the chapter. Similarly, the mode shapes of $circ = 4$ can be excited by harmonics with temporal order $q = \{16, 20, 28, 32\}$.

Figure 13.75a shows that there are four "suns" in total for the modes with $circ = \{0, 6\}$. The ray of forcing frequency with $q = 12$ can also excite the modes with $circ = \{0, 6\}$. However, given the speed range, the forcing frequency for $q = 12$ is not high enough to approach either the natural frequency of $circ = 0$ or $circ = 6$. Again, the absence of resonance does not mean that vibration does not happen. The ray of forcing frequency for $q = 12$ still causes first deformation and then vibration, but not resonance.

It is also noteworthy that in Figure 13.75a, the mode with $circ = 12$ can also be excited by the rays of forcing frequencies with $q = \{12, 24, 36\}$. Since the distance between the three rays and the natural frequency for $circ = 12$ is large, given the speed range of 0 rpm to 12000 rpm, this vibration mode has a relatively low contribution to the noise and vibration compared with the modes $circ = \{0, 6\}$. Other rays of forcing frequencies with higher q, for instance $q = \{48, 60, 72, 84, 96, \ldots\}$ can excite the mode with $circ = 12$ more and can even generate resonances. However, the amplitudes of these harmonics would be low and thus, their contribution to noise and vibration can be negligible.

As shown in Table 13.7, the circumferential modes $circ = \{0, 4, 8\}$ of a three-phase 12/8 SRM can be excited by the dominant radial force harmonics. As shown in Figure 13.76a, $circ = 0$ is excited by rays of forcing frequencies with $q = \{24, 48, 72\}$, as shown in red. The mode shapes of $circ = \{4, 8\}$ can be excited by harmonics with temporal order $q = \{8, 16, 32, 40, 56, 64\}$ as shown in blue in Figure 13.76b.

In this section, the concept of selective excitation was discussed using both three-phase 6/4 and 12/8 SRMs. The rays for forcing frequencies are selective when considering which circumferential modes they can excite. Each ray of forcing frequency can excite certain circumferential modes. Resonance is an extreme form of vibration. It happens when the forcing frequency approaches the natural frequency, given that the shape of the forcing harmonic matches the mode shape of that natural frequency. If the harmonic content of the radial force contributes greatly to noise and vibration, it could be due to resonance or because the amplitude of the harmonic content is high, or both, provided that the damping ratio is within a reasonable range.

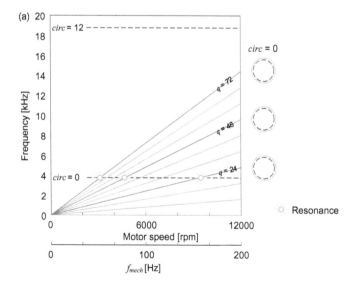

FIGURE 13.76
Excitations of circumferential modes ($circ < 12$) 12/8 SRM, $q = \{8, 16, \ldots, 72\}$ (first 9): (a) $circ = \{0, 12\}$ and (b) $circ = \{4, 8\}$.
(*Continued*)

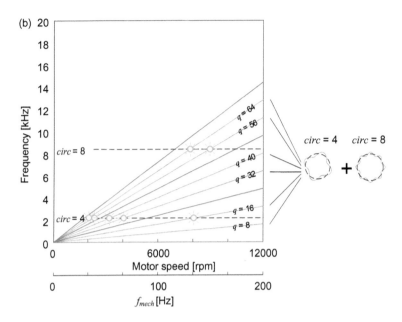

FIGURE 13.76 (Continued)
Excitations of circumferential modes (*circ* < 12), 12/8 SRM, $q = \{8, 16, ..., 72\}$ (first 9): (a) *circ* = $\{0, 12\}$ and (b) *circ* = $\{4, 8\}$.

13.10 Interpretation of Acoustic Results

The results of acoustic analysis can be presented in a number of ways. In this section, we will discuss a few methods of analyzing and interpreting acoustic noise and vibration results. This will help readers develop a comprehensive way to extract the characteristics of the acoustic behavior for a specific machine, which is important for the discussions on acoustic noise reduction.

13.10.1 SPL versus Rotor Speed

Figure 13.77 shows the format of SPL results for a 12/8 SRM operating at 7200 rpm. When the temporal order q varies from 0 to 160, the correlated forcing frequency increases linearly from 0 Hz to 20 kHz. In a 6/4 SRM, the series of q is $\{0, 4, 8, 12, ...\}$. and for the 12/8 SRM, the sequence of q is $\{0, 8, 16, 24, ...\}$. Since the sequence of q is discrete and, hence, SPL varies at discrete values of temporal order, stem plots are used for showing SPL versus frequency. The stem plot is preferred over continuous line plots to study the components of SPL in the frequency domain at a certain speed.

Figure 13.78 shows an example of SPL for $q = 0 \sim 36$ versus motor speed, ranging from 0 to 12,000 rpm which translates into a maximum mechanical frequency of 200 Hz (12,000 rpm/60 sec/min). There are three major dimensions in Figure 13.78: the motor

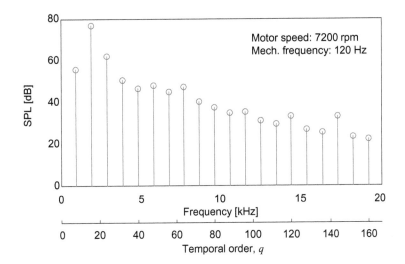

FIGURE 13.77
SPL versus frequency and temporal order ($q = 0 \sim 20$), 12/8 SRM, motor speed = 7200 rpm, mechanical frequency = 120 Hz.

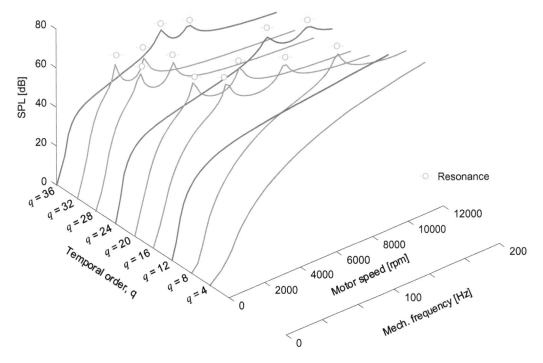

FIGURE 13.78
SPL ($q = 0 \sim 36$) versus motor speed, 6/4 SRM.

speed (mech. frequency), temporal order, q, and SPL. The forcing frequency can be added to the SPL plot as another dimension. Forcing frequency is the product of q and the mechanical frequency.

The red curves with $q = \{12, 24, 36\}$ in Figure 13.78 correspond to the red lines in Figure 13.75. The SPL curves in red in Figure 13.78 are caused by excitation of the modes with $circ = \{0, 6\}$, as shown in Figure 13.75a. Similarly, the blue curves with $q = \{4, 8, 16, 20, 28, 32\}$ in Figure 13.78 correspond to the blue lines in Figure 13.75b. The "sun" icons are also used in Figure 13.78 to describe the locations of resonances.

Figure 13.79 shows another representation of SPL values. Compared to Figure 13.78, the motor speed axis is replaced with the forcing frequency. Since forcing frequency is a function of the temporal order, it can be noticed that the SPL curves for lower values of the temporal orders, for instance, $q = \{4, 8, 12, 16\}$, only cover short spans over the forcing frequency. The merit of Figure 13.79 is that the excitation of circumferential modes can be easily seen from the peaks. The natural frequency of each mode in Figure 13.79 is at the same location as shown in Figure 13.73. In Figure 13.79, the effects of resonance and selective excitations can be seen more clearly.

For instance, the vibration mode with the circumferential order $circ = 2$ (the natural frequency is 1087.06 Hz as shown in Figure 13.73) is excited selectively by harmonics

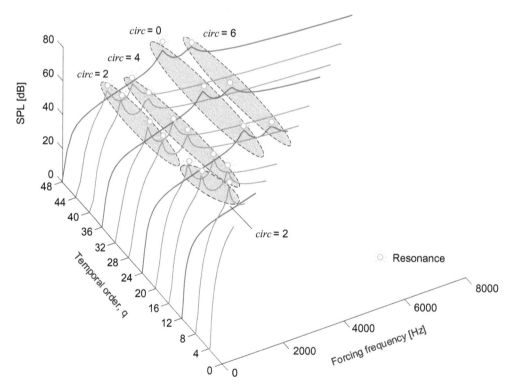

FIGURE 13.79
SPL versus temporal order ($q = 0 \sim 48$) and frequency, 6/4 SRM.

with temporal order, $q = \{8, 16, 20, 28, 32, 40, 44\}$, as discussed in Figure 13.75b. When $q = 8$, the forcing frequency range for the ray is 0–1600 Hz ($q \times 200$ Hz—highest mechanical frequency as shown in Figure 13.78). The natural frequency for the mode with $circ = 2$, 1087.06 Hz, is in between this range. Resonance happens when the forcing frequency for $q = 8$ approaches the natural frequency (1087.06 Hz), as illustrated by the spike on the graph. This is the location where resonance occurs (depicted by a "sun" icon).

The ray of forcing frequency with $q = 16$ can excite the mode with $circ = \{2, 4\}$, as shown in Figure 13.75b. For $q = 16$, the related forcing frequency range can be calculated as 0 rpm ~ 3200 rpm. The natural frequencies of the mode with $circ = \{2, 4\}$, 1087.06 Hz and 1943.83 Hz, both sit in this forcing frequency range. This indicates that resonances can happen at two different forcing frequencies. As shown in Figure 13.79, two spikes on the curve of SPL versus forcing frequency can be seen for $q = 16$.

For $q = 24$, the ray of the forcing frequency can excite the modes with $circ = \{0, 6\}$, as discussed in Figure 13.75a. The forcing frequency range for $q = 24$ is 0 Hz to 4800 Hz. The natural frequencies for the mode with $circ = \{0, 6\}$ are 3352 Hz and 4261 Hz. Both are within the forcing frequency range for $q = 24$. The two spikes on the SPL curve for $q = 24$ in Figure 13.79 are the two locations where resonances happen for the mode with $circ = \{0, 6\}$. Similarly, two spikes can be seen on the SPL curve for $q = 36$.

When motor speed ranges from 0 rpm to 12,000 rpm, the frequency ranges for $q = \{24, 36\}$ cover the natural frequencies for the modes with $circ = \{2, 4, 0, 6\}$. However, for each red SPL curve, only two spikes at the natural frequencies for $circ = \{0, 6\}$ can be seen, which happens because of the selective excitation. The radial force harmonics with $q = \{24, 36\}$ do not have the modes with $circ = \{2, 4\}$. Again, not every intersection between the natural frequency and the forcing frequency results in resonance.

13.10.2 Waterfall Diagram

If the values of SPL are presented using color scale on top of the rays of forcing frequencies in Figure 13.72, a waterfall diagram is generated, as shown in Figure 13.80. The horizontal axis is the motor speed and the vertical axis is the forcing frequency. The rays represent the forcing frequencies with different temporal orders. The color scale represents the SPL value at a certain speed and different forcing frequencies.

The rays of the harmonics with $q = \{4, 8, 16, 20, \ldots\}$ excite the circumferential modes with $circ = \{2, 4\}$ due to the selective excitation. This group of rays can be referred to as the first layer, as shown in Figure 13.80a. In Figure 13.80a, every intersection between the horizontal lines (natural frequencies) and the rays (forcing frequencies) represents resonance.

The rays of the harmonics with $q = \{12, 24, 36, \ldots\}$ excite the circumferential modes with $circ = \{0, 6\}$. This group of rays is defined as the second layer, as shown in Figure 13.80b. It can be seen that the rays in Figure 13.80b do not cause any excitation for $circ = \{2, 4\}$. In Figure 13.80a and b, every intersection between the horizontal lines (natural frequencies) and the rays (forcing frequencies) represents resonance. This is not the case for Figure 13.80c, because the excitation is selective.

Waterfall diagrams contain comprehensive data on the acoustic noise behavior of an SRM. As shown in Figure 13.81, a waterfall diagram can be represented from another perspective

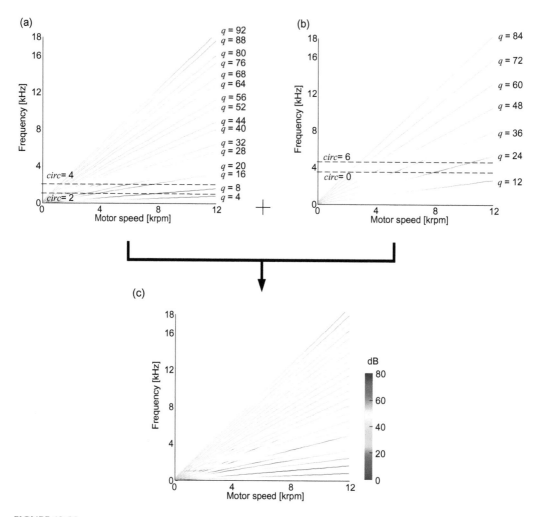

FIGURE 13.80
6/4 SRM waterfall diagram: (a) first layer, *circ* = {2, 4}, (b) second layer, *circ* = {0, 6}, and (c) combined.

to better locate the speed at which the motor has the highest sound pressure level. In addition, it might be challenging to identify the source (*circ* and *q*) for the peak SPL.

It can be seen from Figure 13.81 that the waterfall diagram is the projection of the SPL values on to motor speed (*x*) and frequency (*y*) plane. If the resolution of the waterfall diagram is low, some important information, which can be useful to interpret acoustic behavior, might be hard to identify. Therefore, looking at the SPL (*z*) versus motor speed (*x*), and SPL (*z*) versus frequency (*y*) curves shown in Figure 13.81 might be beneficial. These two diagrams can shed light on the acoustic noise behavior of the motor from two different perspectives.

Figure 13.82 shows the SPL versus motor speed. Similar to the waterfall diagram in Figure 13.80, Figure 13.82 can be decomposed into two layers. The layer in Figure 13.82b is for *circ* = {2, 4}. The layer in Figure 13.82c is for *circ* = {0, 6}. It can be observed from Figure 13.82a that almost every SPL curve versus motor speed has two peaks. This happens

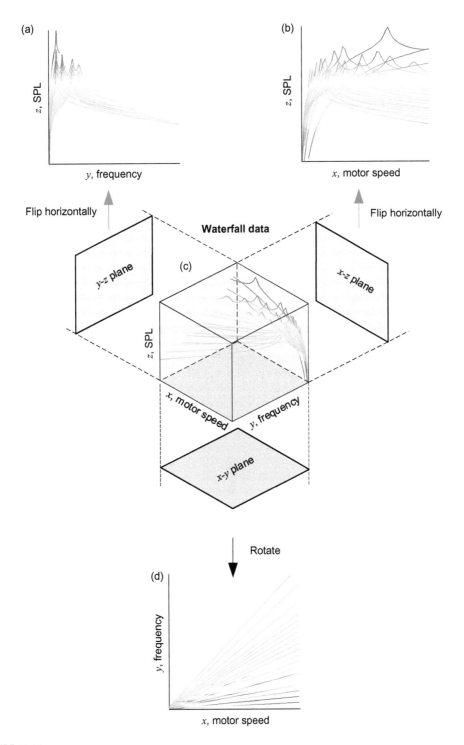

FIGURE 13.81
Projection of waterfall data onto difference planes: (a) SPL versus frequency, (b) SPL versus motor speed, (c) waterfall data, and (d) waterfall diagram, frequency versus motor speed.

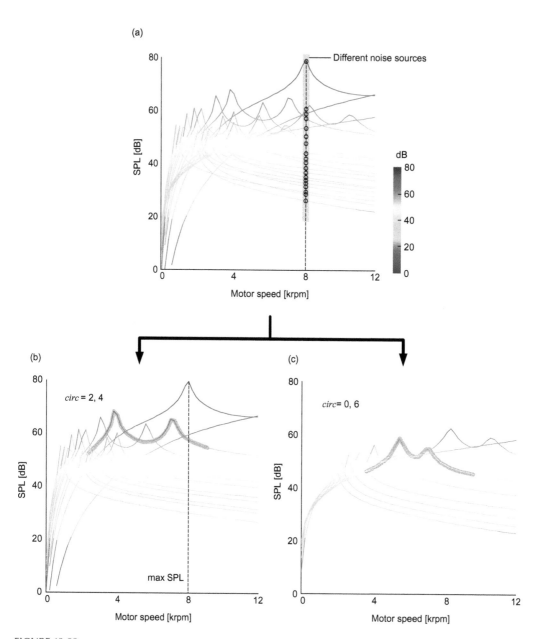

FIGURE 13.82
Decomposition of SPL versus motor speed, 6/4 SRM: (a) combined, (b) first layer, *circ* = {2, 4}, and (c) second layer, *circ* = {0, 6}.

because as the motor speed increases from 0 to the maximum speed, and the forcing frequency increases linearly. The peaks in Figure 13.82 make it easier to identify at which speed the motor has the highest acoustic noise.

As shown in Figure 13.80a, the natural frequency of $circ = 2$ is lower than $circ = 4$. As the forcing frequency increases, resonance happens for the circumferential mode $circ = 2$ first, then $circ = 4$. A similar trend can be observed for the modes with $circ = \{0, 6\}$ in Figure 13.82b. As shown in Figure 13.80b, the natural frequency of $circ = 6$ is higher than that of $circ = 0$. As the forcing frequency increases with the motor speed, the circumferential mode with $circ = 0$ experiences resonance earlier than $circ = 6$.

If the horizontal axis is switched from motor speed to frequency as shown in Figure 13.81, the resonance for one specific circumferential mode would have the same horizontal location. As it will be discussed in the next section, this makes noise source identification ($circ$ and q) simpler.

13.10.3 Calculation of Total SPL

Figure 13.82a shows that at a certain speed, there exists a number of SPL values with different temporal orders, and thus with different forcing frequencies, which can be treated as different sources for noise. For instance, at around 8000 rpm, the motor experiences the highest noise (SPL) over the speed range. Under the peak value, at the same speed, a number of other SPL values can be seen. These SPL values are also different sources of noise. Though lower than the peak SPL, they still contribute to the total noise at that speed.

The total sound pressure level is often a mixture of a few discrete sources. The sources are called incoherent sources [3] if they have different frequencies and random phase relations. The total sound pressure level (dB) can be seen as an overall impression of the noise. The total SPL is an addition of several sources of sounds which occur at the same time. The following formula can be used to add sounds on the decibel scale [3]:

$$SPL_{A\_total} = 10\log_{10}\left[\sum_{q=0}^{q\_max} 10^{\frac{SPL_{(q)}}{10}}\right] \tag{13.89}$$

Due to the logarithmic dB scale, it is important to note that if 60 dB is added to another 60 dB, the result is 63 dB instead of 120 dB. The dotted line in Figure 13.83 shows the total SPL versus the motor speed. The trend of total SPL is important because it shows the overall impression of the noise for the motor while the motor speed varies. In Figure 13.83, the dominant spike of total SPL happens at around 8000 rpm. To locate the source of this spike, the information in Figure 13.82a can be useful. It can be seen that either $circ = 2$ or $circ = 4$ is responsible for the noise.

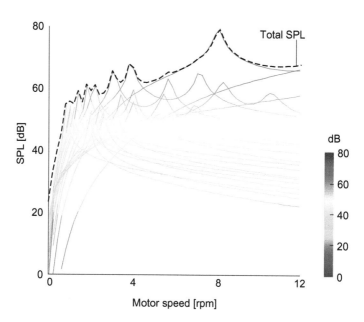

FIGURE 13.83
Total SPL versus motor speed, 6/4 SRM.

13.11 Source Identification and Vibration Reduction

In the previous section, we have discussed methods for interpreting acoustic noise results. From the SPL results, the sources of the noise can be determined. In this section, we will focus our discussion on the identification of sources for noise reduction.

13.11.1 Decomposition of Peaks in SPL Curves

As discussed earlier, if the motor operates mainly on a few operating points (speed points), such as in a washing machine, the plots of SPL versus frequency at a certain speed can be used (as in Figure 13.77). However, the plot of total SPL versus motor speed should be used if the motor operates over the full speed range, for example a traction motor.

The total SPL gives the overall impression of the noise. We need to determine the individual sources of the total SPL by studying the noise over the frequency spectrum. Figure 13.83 shows that the peak value of SPL occurs at 8200 rpm. As shown in Figure 13.84, the stem plot for SPL at 8200 rpm can be used to observe the SPL values over the frequency spectrum to observe which forcing frequency contributes the most to the SPL at 8200 rpm. The temporal order, q can be located accordingly by dividing the forcing frequency at the peak SPL by the mechanical frequency.

13.11.2 Source Identification

SPL versus motor speed diagrams (as in Figure 13.84) are helpful in identifying the motor speed points at which the motor experiences SPL spikes. In order to identify the source of

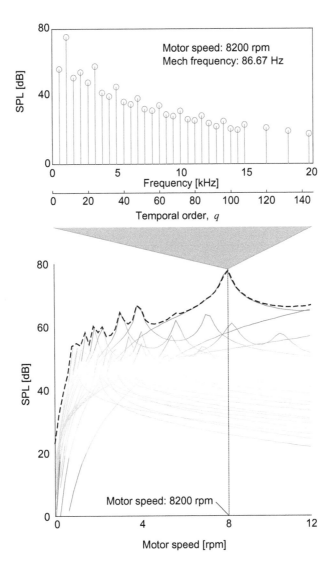

FIGURE 13.84
Total SPL at 8200 rpm, 6/4 SRM.

the noise and the exact locations of forcing frequencies, SPL versus frequency diagrams can be used (as in Figure 13.85). In Figure 13.85, the horizontal axis is the forcing frequency, and the vertical axis is the SPL value. Figure 13.85a shows that there are four SPL peaks. The locations of these four spikes coincide the natural frequencies of the modes with *circ* = {2, 4, 0, 6}. Among these four peaks, *circ* = 2 has the highest SPL value. As discussed previously in Figures 13.80 and 13.82, two layers can be identified due to selective excitation, as shown in Figure 13.85b and c.

In Figure 13.85b, two SPL peaks, caused by the excitations of the modes with *circ* = {2, 4}, can be seen. Please recall our discussion in Figure 13.75a where the rays of forcing frequencies with q = {4, 8, 16, 20, 28, 32, ...} excite the modes with *circ* = {2, 4} because of the selective excitation. The same phenomena can be observed in Figure 13.85b. The highest SPL peak is caused by the forcing frequency with q = 8. It can be concluded that highest SPL

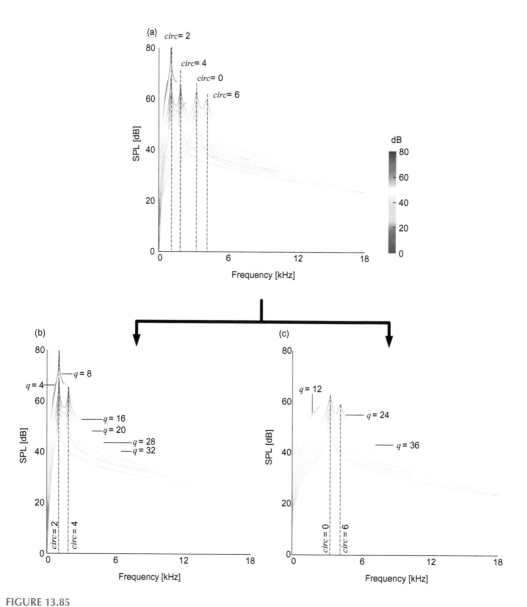

FIGURE 13.85
Decomposition of SPL versus frequency at 8200 rpm, 6/4 SRM: (a) combined, (b) the first layer, *circ* = {2, 4}, and (c) the second layer, *circ* = {0, 6}.

peak at 8200 rpm can be reduced by reducing the amplitude of the harmonic content with *circ* = 2 and *q* = 8. In Figure 13.85c, the modes with *circ* = {0, 6} cause the two SPL peaks. The forcing frequencies with *q* = {12, 24, 36, …} are responsible for the excitation of these modes, as discussed in Figure 13.75b.

13.11.3 Noise Reduction

There are a few different ways to reduce the noise level of an SRM. In this section, we group the methods of acoustic noise reduction into five major categories, as shown in Table 13.19.

TABLE 13.19

Groups of Factors for Reducing Acoustic Noise of an SRM

Group	Classification	Description
First	Configuration	N_r (number of rotor poles)
		N_s (number of stator poles)
		N_{ph} (number of phases)
Second	Major dimensions	D (bore diameter)
		D_s (stator outer diameter)
		h_s (stator pole height),
		β_r (rotor pole arc angle)
		β_s (stator pole arc angle)
		y_s (stator back iron thickness)
		τ_r (rotor taper angle)
		τ_s (stator taper angle)
		Lengths of stator and rotor cores
		Length of end turn
Third	Mechanical design	Housing (frame) design
		Bearings, shaft
		Cooling jacket (for liquid cooled motors)
Fourth	Geometric modification	Slits on the stator or rotor poles
		Cut outs, asymmetrical poles
		Skewed rotor or stator core
Fifth	Control	Optimization of the current waveform

The configuration of the SRM, for example, N_r and N_s, affects the noise and vibration level of the motor greatly. This is the first group of parameters for acoustic noise reduction, as shown in Table 13.19. As discussed earlier in this chapter, the SRM configuration determines which circumferential modes can be excited by the dominant harmonics of the radial force. Generally, as N_s and N_r increase, less circumferential modes can be excited, and the density of resonances decreases, as shown in Table 13.10.

The second group of parameters is the major dimensions of a motor. This includes the stack length, stator outer diameter, thickness of stator back iron, taper angles, and so on. This group of motor parameters affects the natural frequencies and damping ratios, and more importantly, the distribution of radial force harmonics. Stator and rotor geometries are the most crucial parameters of an SRM, as they define the motor performance that includes the output electromagnetic torque, efficiency, power density, and acoustic noise. The circumferential mode $circ = 2$ for a four-phase 8/6 SRM is highly dependent on the stator thickness y_s, while the stator taper angle τ_s has an influence on some certain circumferential modes [15,16]. Also, geometry parameters, such as the stator yoke thickness y_s, stator taper angle τ_s and stator pole arc angle β_s, and the stator outer diameter D_s are key parameters in terms of natural frequencies [16].

The third group of parameters focuses on mechanical design. The housing, shaft, bearings, and others affect the natural frequencies and the damping ratios of the vibrating structure. The stator frame has a direct effect on the stator's natural frequencies. Due to the additional material, the mass is higher when a frame is present. However, this is not always the case with the resultant stiffness. For example, radial ribs increase the natural frequencies while axial ribs can have an opposite effect. This means that an axial-rib frame contributes more to the mass than it does to the stiffness. Higher modal natural frequencies are generally desired since it reduces the resonance effect with lower forcing frequencies.

Additionally, smooth multilayer frames and screw-type frame can help improving the acoustic noise at high speeds and high-power applications [17].

The fourth group includes minor geometric modifications. For instance, by putting slits or cut outs on either the rotor or stator core, the distribution of the radial force harmonics can be modified. A skewed teeth stator can reduce the noise and vibration by reducing the radial force density [18]. However, skewing in SRM might reduce the average torque and increase the control complexity.

Last but not the least, by shaping the current waveform, the distribution of radial force harmonics can be manipulated. In terms of motor control, in [19] a two-stage turn-off methodology is proposed to cancel out the vibrations caused by the sudden change in radial force. The first stage is the freewheeling stage where zero volts is applied to the phase terminals, and in the second stage both switches are turned off and negative DC-link voltage is applied. Synchronizing the two stages can be challenging for the optimal vibration cancellation. Alternatively, to reduce the harmonic content of the phase voltage during the turn-off, the applied voltage can be shaped [20]. As a sudden decrease in the radial force causes the excitation of mode 0, techniques such as Direct Instantaneous Force Control (DIFC) can be implemented to overcome this problem [21]. In this approach, a force reference is generated for each phase individually to keep the sum of the total air gap force constant. Moreover, for a three-phase machine, the third harmonic and its multiples do not cancel out in the sum of total radial forces; therefore, in [22] a third-harmonic cancelation method is proposed.

13.12 Numerical Method for Acoustic Analysis

In the previous section, the analytical method for the prediction of noise and vibration was introduced. One of the major advantages of the analytical method is that it can provide a fast and computationally efficient method for acoustic prediction at both low and high frequencies. The use of the analytical method is therefore preferred in the design, optimization, and control of electric machines. However, this method usually incorporates a simplified geometry and is based on several assumptions. It can also be challenging to use the analytical method to simulate different boundary conditions, such as clamped-free and clamped-clamped. On the other hand, the numerical model is more precise because more structural details can be considered in the modeling. A complete assembly of vibrating structures, including the housing, stator, end caps, end plates, and shaft can be considered in the acoustic noise model, which can be hard to accomplish with an analytical method. A numerical model can also simulate various boundary conditions, including clamped-clamped, clamped-free, and others. Although the numerical method is more time-consuming and computationally inefficient in the high frequency range, its use may be necessary to verify the acoustic results obtained from the analytical method.

13.12.1 Procedures for Acoustic Numerical Modeling

Acoustic FEA modelling requires the 3D geometry of the motor. The mechanical system can be designed based on the geometries of an SRM's stator and rotor. The 3D geometry of the mechanical assembly can be created using CAD software, such as CREO® or SolidWorks®. The mechanical assembly can then be simplified as a vibrating structure, as shown in Figure 13.86. Many tools can be used for meshing the 3D geometry. For instance, ACTRAN®

(a) (b)

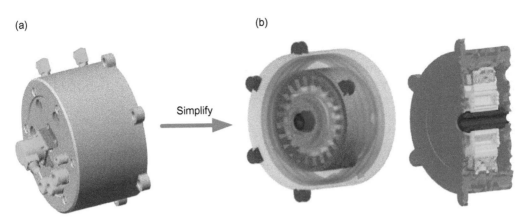

Simplify

FIGURE 13.86
Simplifying the complete mechanical construction into a vibrating structure: (a) mechanical structure and (b) vibrating structure.

can make use of meshed models generated with MSC NASTRAN®, ANSYS® Workbench, Mechanical APDL, ICEM CFD™, ANSA, PATRAN®, HyperMesh®, IDEAS®, and others.

As shown in Figure 13.87a, a complete vibrating structure is composed of the shaft, rotor, end plates, stator, end covers, windings, and the housing. The shaft is made from carbon steel, the end plates from brass, and the housing from aluminum. In this structure, the rotor is assumed to be static and the two end plates are fixed to either end of the rotor.

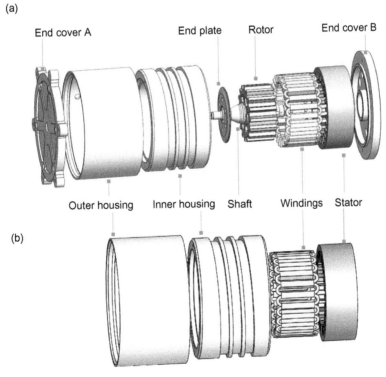

FIGURE 13.87
Exploded view of SRM structure: (a) a complete structure and (b) housing-stator-windings structure.

The nodal forces on the rotor are not taken into account in the model. Vibration of the housing and stator is the main source of the noise in SRMs [1,17,19,22,23]. Therefore, the vibrating structure in the acoustic noise model can be typically simplified by taking only the stator, housing, and windings into account, as shown in Figure 13.87. The simplified model in Figure 13.86 shows an exploded view of the vibrating structure. The SRM is fixed to a mounting surface at six locations.

The process of creating a coupled model and studying a motor's noise and vibration behavior using ACTRAN® is depicted in Figure 13.88. The time-domain electromagnetic analysis is carried out in JMAG®. The nodal force can be calculated by electromagnetic FEA based on the optimized geometries of the stator and rotor. Structural analysis is performed

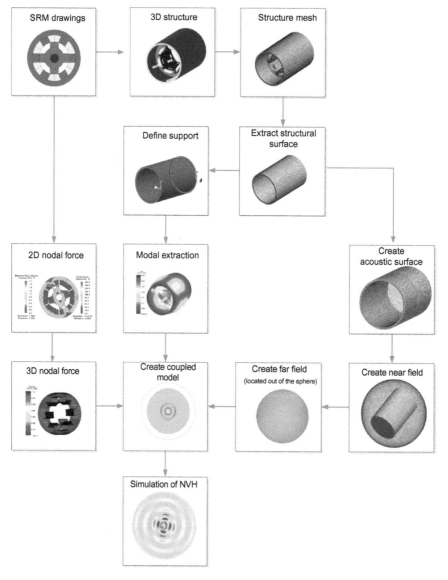

FIGURE 13.88
Process of creating a coupled model and studying acoustic noise behavior using JMAG® and ACTRAN®.

to obtain the vibration of the SRM. The calculation of the natural frequencies and the simulation of the vibration are based on a meshed structure, which is generated in a meshing tool using the 3D geometry of the SRM. The electromagnetic and structural numerical analyses are performed in JMAG and ACTRAN, respectively. The resulting natural frequencies and the nodal forces will be used as the input in the acoustic simulation.

As shown in Figure 13.89a, the near field is the field in which the nature of the sound wave depends on the vibration of SRM. The far field is the field in which the nature of the sound wave depends on the propagation medium, typically air. The near field will be modeled with finite elements while the far field is modeled with infinite elements [24]. Infinite elements are different from the finite elements, in that infinite elements have an exponential term in their shape functions, which better describe the unbounded property of the far field. The infinite element method is used to simulate the acoustic behavior in the far field and it has several advantages. It models the non-reflecting boundary conditions of the far field

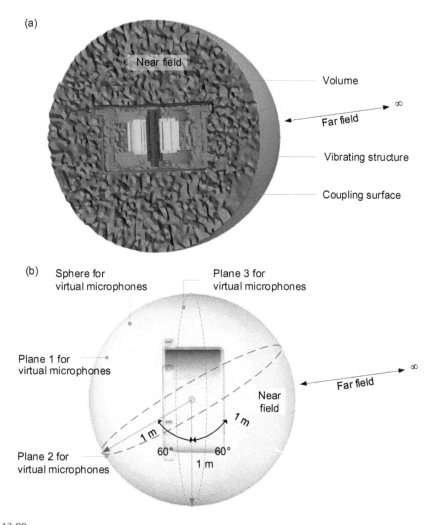

FIGURE 13.89
Coupled acoustic noise modeling in ACTRAN: (a) cross section of the acoustic noise model and (b) locations of the virtual microphones.

and provides a direct numerical estimate of the solution at all points in the far field. It has also been proven to be efficient in solving acoustic scattering problems [25–27].

Virtual microphones, shown in Figure 13.89b, can be positioned anywhere in the near field and far field. Ideally, as many virtual microphones as possible should be positioned in the model. In order to capture the sound pressure efficiently, a finite number of virtual microphones are located on the surface of a sphere surrounding the motor. Three planes passing through the motor and the vertical axis are positioned at 60-degree intervals. The arcs created by the intersection of each plane with the sphere defining the measurement surface help identify the position of the virtual microphones. The virtual microphones on these arcs are used to capture the sound pressure around the motor.

Because the measurement surface is defined by a sphere surrounding the center of the motor, the distance from each virtual microphone to the center of the motor is the same, as shown in Figure 13.90a. These three virtual microphones are located at 0°, 45°

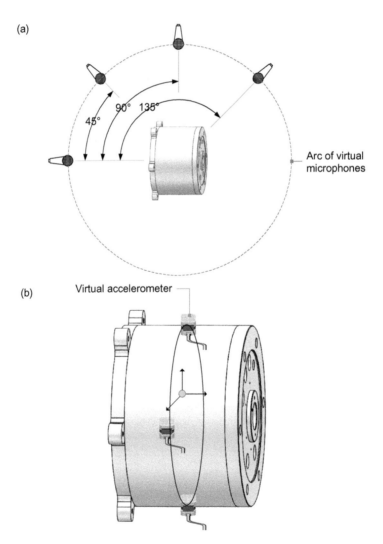

FIGURE 13.90
(a) Locations of the virtual microphones and (b) locations of the virtual accelerometers.

and 90°. Since there are differences in the amplitudes of the sound pressure for the three microphones, the positions of the microphones should be selected wisely in order to show the characteristics of the acoustic noise of the motor. Virtual accelerometers can also be mounted on the surface of the housing. Ideally, as many virtual accelerometers as possible should be included. For a cylindrical housing, the accelerometers can be placed in the same axial position on the motor (see Figure 13.90b). Circumferentially, accelerometers should be placed at two critical positions on the motor housing. The first is a position on the back iron aligned with a stator pole as most of the force on the stator is applied at the poles. The second is on the back iron between consecutive stator poles as these positions are the thinnest and therefore the weakest sections of the stator.

Different placement of the virtual microphones will affect the captured sound pressure. Figure 13.91 shows the exact positions of each microphone on a hemisphere as outlined in ISO 3744. In total, 20 virtual microphones are placed on the surface of a hemisphere with a radius of 1 m. Figure 13.91a and b illustrate the front view and top view of the hemisphere, respectively. For spherical measurement, the same arrangement of microphones can be mirrored on the other hemisphere.

13.12.2 Considerations in 3D Modeling

Compared to the analytical method, prediction of acoustic noise using the numerical method can be computationally expensive and time consuming. The numerical method solves equations at different nodes and mesh elements and, thus, it is necessary to

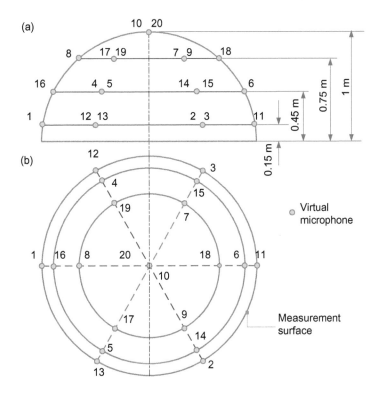

FIGURE 13.91
Positions of virtual microphones for a hemispherical measurement surface based on ISO 3744 standard: (a) front view and (b) top view.

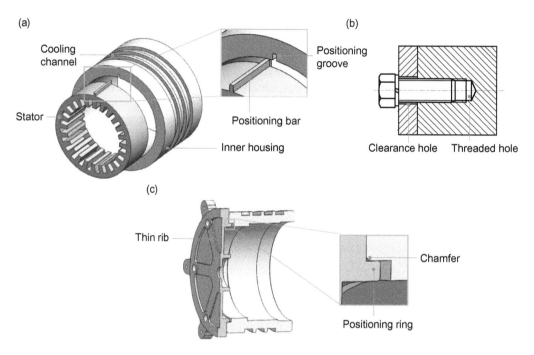

FIGURE 13.92
Small features for structural simplification: (a) positioning bar and groove, (b) clearance hole and threaded hole, and (c) thin rib, positioning ring and chamfer.

reduce the number of nodes and elements in order to reduce the complexity of the simulation. Since the geometry of the stator has a great effect on the performance of SRMs, it should contain minimal structural simplification in the model. The small features of the housing, on the other hand, can be simplified to reduce the computational cost and time. As shown in Figure 13.92, there are many small features on the housing, e.g. fillets, screw holes, welding and positioning grooves, and chamfers. These details require finer meshing, which will increase the number of nodes and mesh elements. These small features can be removed (see Figure 13.93). However, the coolant channel should not be simplified as it significantly affects the stiffness and the mass of the housing.

As mentioned previously, the geometry of the stator should be modelled as accurately as possible, because the match between the stator and the excitation force model is important for the accuracy of the acoustic noise prediction. As shown in Figure 13.94, the nodal forces in the electromagnetic model are applied on the meshed structure of the stator. The nodal forces appear as force vectors, which are the integration of the air gap magnetic pressure contributions over a local area. Any mismatch between the force model and the meshed structure will lead to inaccuracy in the simulation of the deformation and vibration of the stator structure and, thus, poor prediction of radiated noise. For this reason, a detailed model of the stator should be used. It should be noted that in Figure 13.94, a very fine mesh is used with numerous nodes and elements in the air gap. Therefore, the magnitude of the nodal force vectors is small. Once all the force vectors are integrated, the total electromagnetic force will be much higher.

Windings should also be modeled accurately. For concentrated windings in SRMs, the coils have lumped mass effect and weaker stiffness effect on the stator poles. The geometry of the coil should be modelled accurately to obtain an accurate volume and mass,

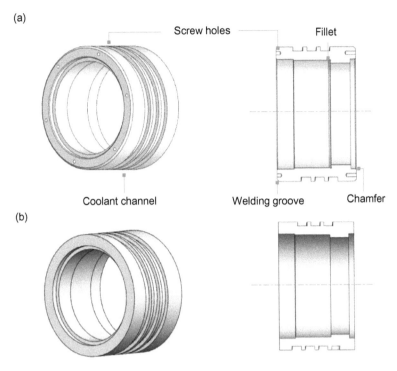

FIGURE 13.93
Structural simplification of inner housing: (a) original and (b) simplified.

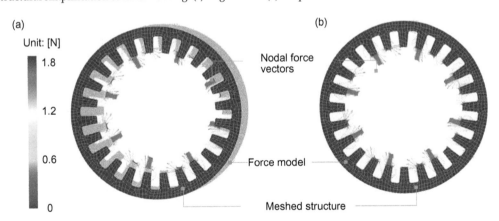

FIGURE 13.94
Match between vibrating structure and nodal force model: (a) mismatch and (b) match.

which will affect the calculation of the natural frequencies of the vibrating structure. As shown in Figure 13.95, the wires in each slot are bonded by two layers of insulation paper and epoxy resin. If a coil has no connection with other coils, it will vibrate separately, as shown in Figure 13.96. This will generate extra and unrealistic vibration modes of a single coil and will increase the computation of the acoustic simulation. In practice, the coils belonging the same phase are connected together. Therefore, in numerical analysis, all the coils are assumed to be integrated into a whole and, thus, the winding is supposed to vibrate as a whole as shown in Figure 13.96b.

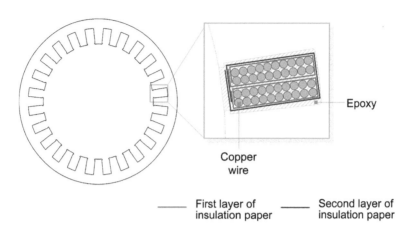

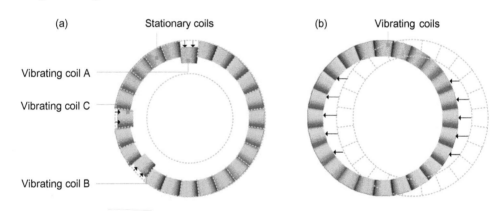

FIGURE 13.95
Modeling of windings.

FIGURE 13.96
Vibration mode of windings and coils: (a) a single coil and (b) windings.

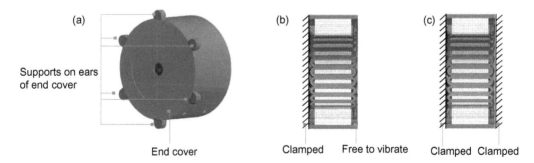

FIGURE 13.97
Boundary conditions of SRM: (a) actual support from the ears on the end cover, (b) clamped-free, and (c) clamped-clamped.

The boundary conditions affect the calculation of natural frequencies. In acoustic modeling, the boundary conditions should represent the actual configuration of the support. As shown in Figure 13.97, the SRM can be supported with clamped-free or clamped-clamped boundary conditions.

13.12.3 Contacts between Vibrating Components

The contacts between different parts affect the stiffness of the vibrating structure and, thus, the natural frequencies. As an example, Figure 13.98 shows the structure of a 24/16 SRM designed for a traction application. The stator is press fit into the inner housing. For small amplitudes of vibration, the contact between the stator and the inner housing can be considered as a bonded contact. The contact between the inner housing and the outer housing is defined as a bonded contact since the two housings are welded together. However, the end cover and the inner housing are connected by screws, which means that their contact cannot be considered a bonded contact.

Considering a different motor, Figure 13.99 shows the rotor-endcap subassembly in a 12/16 external rotor SRM designed for an e-bike application. In an external rotor design,

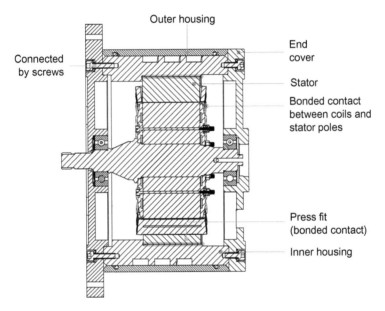

FIGURE 13.98
Contacts between parts in the 24/16 SRM.

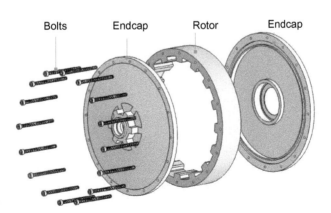

FIGURE 13.99
Rotor-endcaps subassembly in 12/16 external-rotor SRM.

the rotor is not supported rigidly by the shaft, which is different from that of internal-rotor SRM. As shown in Figure 13.99, the endcaps and the rotor of the 12/16 SRM are connected by bolts. The contact should not be defined as a bonded contact in which case all the nodes in the interfaces would be merged together. If all the nodes in the interfaces between the endcaps and the rotor are merged together, it would mean that all the nodes are connected with each other in the numerical acoustic simulation. This will lead to unrealistic stiffness and high natural frequencies for the external-rotor SRM.

As shown in Figure 13.100a, if a bonded contact is defined between the rotor and the endcaps, all the nodes in the interface between rotor and the endcaps will be merged. To avoid adding this artificial stiffness, only nodes in the bolt-area should be merged, as shown in Figure 13.100b, which more closely represents the actual contact and boundary conditions of a bolted connection.

In order to simulate the noise radiated from the SRM, the required element size (see Figure 13.101) of the SRM structure can be expressed as:

$$\varepsilon = \frac{1}{n} \times \lambda_B \tag{13.90}$$

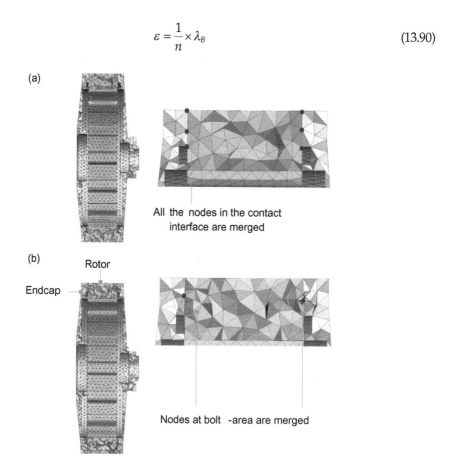

FIGURE 13.100
Contact and meshing in SRM structure with endcaps (shown on an external-rotor SRM): (a) all nodes in the contact interface are merged and (b) nodes at bolt-area are merged.

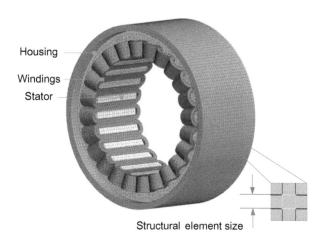

FIGURE 13.101
Element size of meshing of the vibrating structure.

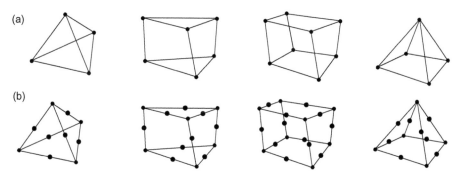

FIGURE 13.102
(a) Linear elements and (b) quadratic elements.

where λ_B is the wavelength of the bending wave and n is the number of elements per wavelength. Bending wave is a kind of structural acoustics and it is the main cause for the radiation of acoustic noise from vibrating structures.

The number of elements per wavelength, n is determined by the type of elements in the mesh of the SRM structure. For linear elements, n is typically 8 to 10. For quadratic elements, n is typically 4 to 6. Figure 13.102 shows linear elements and quadratic elements. For linear tetrahedron element, shown in Figure 13.102a, four interpolation nodes are required, and the shape function is a first-order function. For a quadratic tetrahedron element (see Figure 13.102b), 10 nodes are required along with a second-order shape function.

The wavelength λ_B can be calculated as:

$$\lambda_B = \frac{C_B}{f}$$

(13.91)

where C_B is the speed of the bending sound wave, and f is the maximum frequency in the acoustic simulation.

The speed of bending wave, C_B, can be calculated by:

$$C_B = \left(1.8 \times C_L \times t \times f\right)^{0.5} \tag{13.92}$$

where C_L is longitudinal sound wave, and t is the thickness of the housing.

The speed of the longitudinal wave, C_L, is calculated as:

$$C_L = \sqrt{\frac{E}{\rho \pi \left(1 - v^2\right)}} \tag{13.93}$$

where E, ρ and v are the Young's modulus, density and Poisson's ratio of the housing material, respectively.

13.12.4 Calculation of Nodal Force

The nodal forces are obtained through 2D electromagnetic FEA analysis, using electromagnetic software (e.g., JMAG®. ANSYS Maxwell®, Flux® etc.). Figure 13.103 shows the 2D simulation of nodal force in JMAG on a 6/4 SRM. The nodal force is calculated by JMAG® using the spatial and time distribution of air-gap flux density. The nodal force changes with time and spatial position. The meshing of the rotor and stator in the electromagnetic analysis is generated in JMAG®. Fine meshing is applied in the air gap,

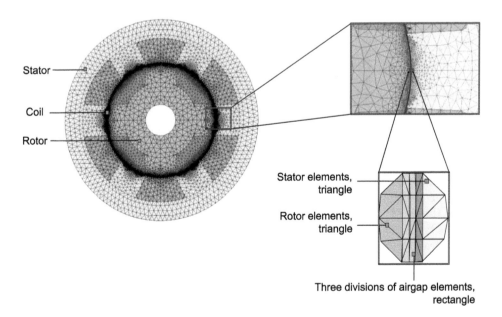

FIGURE 13.103
2D electromagnetic analysis of an SRM.

rotor, and stator pole tips, while a coarse mesh is generated in the areas that are further away from the air gap.

Mesh quality is important for the accuracy of the electromagnetic FEA simulation. The air gap has the finest meshing and uses rectangular elements. In order to determine the element size in the air gap, the circumferential length and the radial length of the rectangular elements should be determined. The circumferential length of the rectangle element, $E_{Airgap, Circ}$, can be determined by:

$$E_{Airgap,Circ} = \frac{L_{Airgap,Circ}}{N_{CircDivisions}} \tag{13.94}$$

where $L_{Airgap,Circ}$ is the circumference length of the airgap, and $N_{CircDivisions}$ is the number of circumferential divisions.

Similarly, the radial length of the rectangle element, $E_{Airgap,Radial}$, can be determined by:

$$E_{Airgap,Radial} = \frac{L_{Airgap,Length}}{N_{RadDivisions}} \tag{13.95}$$

where $L_{Airgap,Radial}$ is the radial length of the air gap and $N_{RadDivisions}$ is the number of radial divisions. As shown in Figure 13.103, the number of radial divisions is 3.

The circumferential length of the air gap, $L_{Airgap,Circ}$, can be calculated by:

$$L_{Airgap,Circ} = 2\pi R_{Airgap,Radius} \tag{13.96}$$

where $R_{Airgap,Radius}$ is the radius of the air gap. The number of the circumferential divisions, $N_{CircDivisions}$, is calculated by:

$$N_{CircDivisions} = N_{RotorPoles} N_{CalcsPerCycle} N_{Scaling} \tag{13.97}$$

where $N_{RotorPoles}$ is the number of rotor poles, $N_{CalcsPerCycle}$ is the number of points per electric cycle, and $N_{Scaling}$ is the scaling factor.

The scaling factor $N_{Scaling}$ can be determined by optimizing the aspect ratio, α, as follows:

$$\alpha = \frac{E_{Airgap,Circ}}{E_{Airgap,Radial}} \tag{13.98}$$

where $E_{Airgap,Circ}$, is the circumferential length of the rectangular element, $E_{Airgap,Radial}$ is the radial length of the rectangular element.

In radial flux machines, the radial length of the air gap element usually has a small value. Theoretically, the ideal rectangular element is a square element and, hence, the ideal aspect ratio is 1. However, if the aspect ratio is 1, the circumferential length of the element in the air gap will be as small as its radial length. This would increase the total element size leading to a long computation time. Therefore, a proper aspect ratio should be defined by the designers by tuning the scaling factor, $N_{scailing}$ based on the meshing quality and computation cost.

The rotor tip should also contain a fine mesh whose element size can be calculated by:

$$E_{Edge,Rotor} = \frac{2\pi R_{Rotor}}{N_{CircDivisions}}$$
(13.99)

where R_{rotor} is the outer radius of the rotor.

The element size of the stator tip can be calculated by

$$E_{Edge,Stator} = \frac{2\pi R_{Stator}}{N_{CircDivisions}}$$
(13.100)

where R_{stator} is the stator inner radius.

Then the 2D nodal force (see Figure 13.104a) is extended in the axial direction of the stator core. The 3D nodal force vector shown in Figure 13.104b has two directions of force component: the radial force component and tangential force component. In the numerical modeling presented, both radial and tangential force components applied on the stator are considered in the acoustic simulation. However, compared to tangential force, the radial force dominates the acoustic noise in SRMs [28,18].

13.12.5 Results and Discussion

In this section, the numerical solution of the acoustic noise model of a 6/4 SRM will be discussed. In addition, the acoustic noise results of a 24/16 traction SRM will be presented. In both cases, the techniques discussed in this section applied to develop the acoustic noise models of the motors in ACTRAN®.

The dimensional and the winding parameters for the electromagnetic and acoustic simulation of the 6/4 SRM are shown in Table 13.20 and 13.21, respectively. The stator, housing, and windings are included in the simulation. The geometry of the stator is created in JMAG®. The housing of the 6/4 SRM is simplified as a thin cylindrical shell. The calculation of the natural frequencies and the simulation of their modal shapes are simulated in ACTRAN® based on different boundary conditions. Clamped-clamped (C-C) boundary condition has been applied to model the support of the structure. The material properties of the SRM are summarized in Table 13.22.

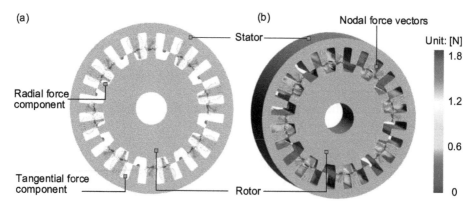

FIGURE 13.104
3D nodal force vectors generated by extending 2D FEA results into 3D, 3000 rpm: (a) 2D FEA results and (b) 3D nodal force.

TABLE 13.20

Dimensional Parameters of the 6/4 SRM

Parameters	Value
Number of stator poles	6
Air gap length	0.3 mm
Stator pole height	19.7 mm
Stator back iron thickness	15 mm
Stator pole arc angle	30°
Stator tapper angle	3°
Radius of the fillet at the bottom of the stator pole	2 mm
Stator stack length	75 mm
Number of rotor poles	4
Shaft diameter	25 mm
Rotor back iron thickness	17.5
Rotor pole height	15 mm
Rotor pole arc angle	32°
Rotor taper angle	2.6°
Radius of the fillet at the tip of the rotor pole	2 mm
Frame thickness	7 mm
Frame length	253 mm

TABLE 13.21

Winding Parameters of 6/4 SRM

Parameter	Value
Number of turns per pole	150
Number of strands in hand in one conductor	4
Bare copper diameter	0.5105 mm (AWG 24)
Slot fill factor	0.3788
Conductor area of one wire	0.2047 mm^2
Slot area	648.46 mm^2
Mean length of turn	256.05 mm
Axial length of one-sided end connection	20 mm

TABLE 13.22

Material Properties of the SRM

Part	Material	Density (kg/m^3)	Young's Modulus (GPa)	Poisson Ratio
Frame	Aluminum	2770	71	0.33
Stator core	Silicon iron stack	7420	180	0.3
Winding	Copper/insulation	4264	12	0.3

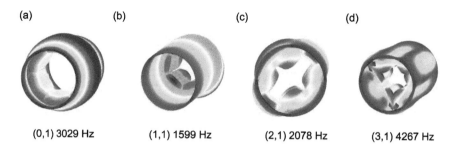

(a)   (b)   (c)   (d)

(0,1) 3029 Hz   (1,1) 1599 Hz   (2,1) 2078 Hz   (3,1) 4267 Hz

FIGURE 13.105
Examples of mode shapes for a stator-frame system, 6/4 SRM: (a) (*circ, ax*) = (0,1), (b) (1,1), (c) (2,1), and (d) (3,1).

The natural frequencies and the modal shapes are calculated and simulated in ACTRAN®. The modal shapes of modes (*circ, ax*) = (0,1)–(3,1) are shown in Figure 13.105. Mode (0,1) represents the circumferential mode *circ* = 0 and axial mode *ax* = 1. For 6/4 SRM, the circumferential modes *circ* = {0, 2, 4} have significant role on the vibration of the structure (see Table 13.4), which leads to radiated noise.

Following the presented procedures, the acoustic noise simulation can be performed in ACTRAN®. The vibrating structure is excited by transient nodal force; generated using JMAG® electromagnetic FEA for a certain operating point. A snap shot of the simulation result of the 6/4 SRM at a certain time is shown in Figure 13.106. The displacement of the

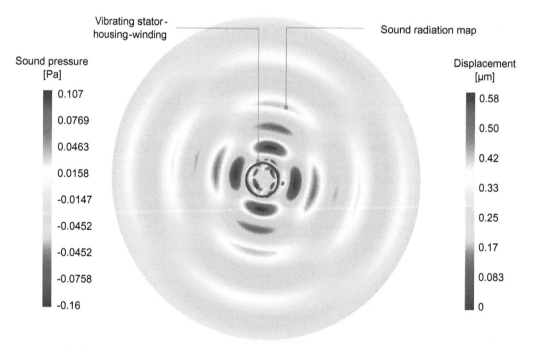

FIGURE 13.106
Vibration of the 6/4 SRM and the sound radiation map.

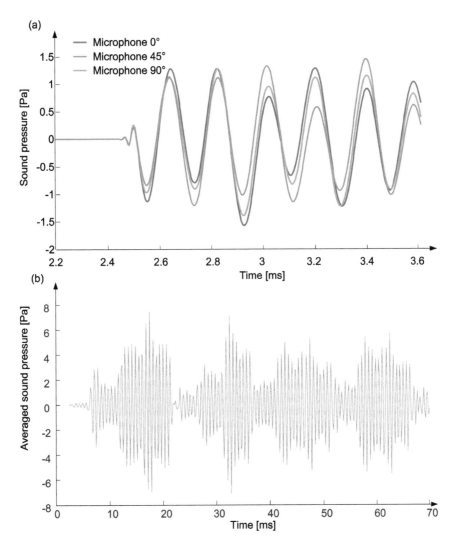

FIGURE 13.107
Sound pressure measured for a 24/16 traction SRM: (a) sound pressure for virtual microphones positioned at different locations and (b) averaged sound pressure on a sphere surface.

vibrating structure and the radiated sound pressure is shown. The sound radiation map shows that the noise is radiated from the vibrating structure to the far field through waves. The maximum sound pressure at this time interval is 0.107 Pa.

Numerical acoustic noise analysis has also been conducted on a 24/16 traction SRM. Figure 13.107a shows the sound wave, which is radiated from the motor as captured by three virtual microphones. The virtual microphones are located in the same arc on a sphere. The radius of the sphere is 1 m (see Figure 13.90a). The captured sound pressure is 0 at the first several milliseconds. Starting from around 2.5 ms, the captured sound wave starts to change due to the vibrating structure. Forty microphones in total are defined on the surface

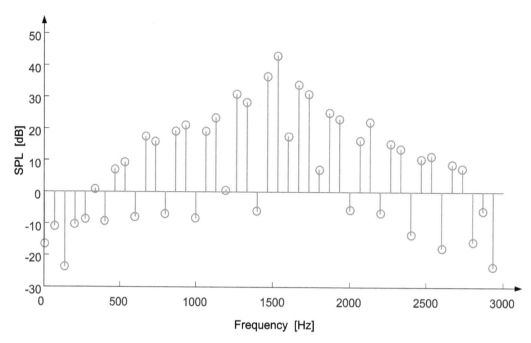

FIGURE 13.108
SPL of 6/4 SRM at 1000 rpm.

of a sphere to capture the sound pressure. The averaged captured sound pressure is shown in Figure 13.107b. Similar to the single sound pressure wave, the averaged sound pressure is 0 at the first several milliseconds. After about 2.5 ms, the averaged sound pressure changes over time. The maximum value of the averaged sound pressure is 8 Pa. The sound pressure radiated by the 24/16 traction SRM is much larger than that of the 6/4 SRM in Figure 13.106 because of its higher rated power.

The sound pressure shown in Figure 13.107 is in time domain. In order to see the frequency component of the sound pressure, FFT is used to transform the time-domain sound pressure into frequency domain. Figure 13.108 shows the SPL results of the 6/4 SRM at 1000 rpm.

The numerical method discussed in this section is used to verify the analytical method presented earlier for the prediction of acoustic noise in SRMs. The comparison of SPL between numerical method and analytical method for a 24/16 SRM at 2000 rpm is shown in Figure 13.109. Observing the peak and trend of the SPL of each method illustrates a good match.

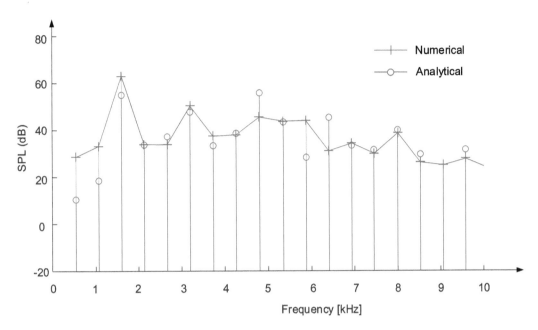

FIGURE 13.109
SPL comparison of analytical method and numerical method for a 24/16 SRM, 2000 rpm.

13.13 Computational Process of Acoustic Noise

This section aims to help establish a better understanding of the analysis of noise and vibration in SRM. The entire process for noise and vibration analysis of SRMs will be summarized and presented. Figure 13.110 shows the synthesis of acoustic noise analysis. The FFT results of the radial force surface wave provide the information on the forcing shapes, the related forcing frequencies, and the forcing amplitudes. The stator-frame system can be modelled as layers of cylindrical shells. Mode shapes related natural frequencies and damping ratios can be obtained either using the analytical method or FEA method. The forcing harmonics cause vibration when the forcing shape matches the mode shape. When the forcing frequency approaches the natural frequency, resonance occurs.

The formats of some commonly used matrices in the noise and vibration analysis are shown in Figure 13.111. If 2D FFT is applied to the surface wave of the radial force density, the results for the amplitudes of the harmonic contents can be expressed in the format shown in Figure 13.111a. This type of matrix has two dimensions: u and v. Similarly, the

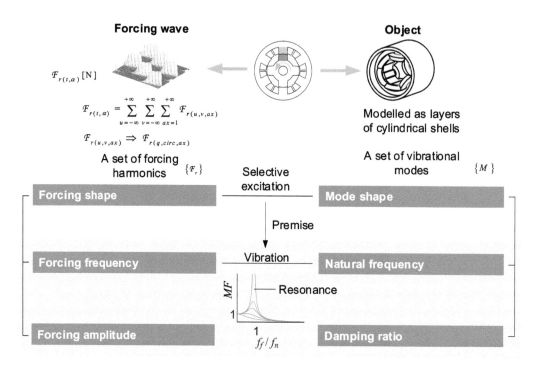

FIGURE 13.110
Synthesis of acoustic analysis.

related forcing frequencies for the harmonics can be organized in the same format. When all the harmonics in the four quadrants of a *u-v* plane are converted to the first quadrant, the converted harmonics can have the format shown in Figure 13.111b. This converted matrix has two dimensions as well: *q* and *circ*. To facilitate the calculation, the 2D matrices can be populated into 3D matrices by adding the third dimension, the axial mode (*ax*), as shown in Figure 13.111c and d.

Figure 13.112 shows the entire process for noise and vibration analysis of an SRM, operating at a certain motor speed. It has five major blocks:

- Forcing frequencies and harmonics
- Natural frequencies and damping ratios
- Media field
- Displacement
- Sound radiation and human sense

In the block calculating forcing frequencies and harmonics, 2D FFT is first used to decompose the radial force density wave. The motor speed is used to calculate the forcing frequency. Then the matrices of $f_{f(u,v)}[\text{Hz}]$ and $P_{r(u,v)}[\text{Pa}]$ are converted to $f_{f(q,circ)}[\text{Hz}]$ and $P_{r(q,circ)}[\text{Pa}]$, which are both in the first quadrant on the *u-v* plane. The forcing frequency

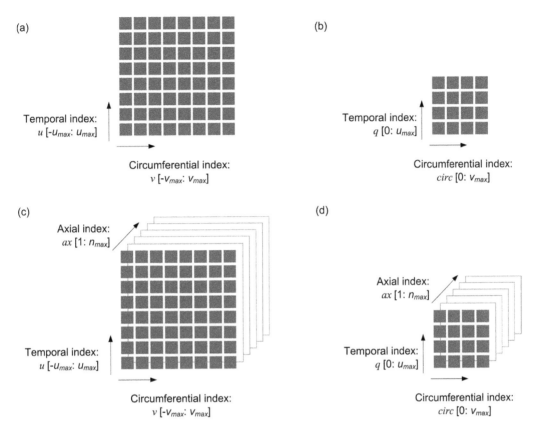

FIGURE 13.111
Two configurations of matrices used in acoustic noise computation: (a) u-v matrix, before conversion, (b) q-$circ$ matrix, before conversion, (c) u-v matrix, after conversion, and (d) q-$circ$ matrix, after conversion.

matrix $f_{f(q,circ)}$ [Hz] is then converted to the angular frequency $\omega_{f(q,circ)}$ [rad/s]. $F_{r(q,circ)}$ [N] can be obtained from the matrix of $P_{r(q,circ)}$ [Pa]. If the axial mode is considered, then the matrices have to be populated over the axial mode. If 3D FEA simulation is used to obtain the data for radial force, 3D FFT needs to be used to analyze the data because the data has four dimensions (time, circumferential, axial, radial pressure). It should be noted that 3D FEA simulation will be computationally more expensive.

In the block for natural frequencies and damping ratios, the matrices for $f_{n(circ,ax)}$ [Hz] and $\zeta_{(circ,ax)}$ are obtained first. They both have two dimensions: circumferential mode and axial mode. To facilitate a similar matrix format as the forcing frequencies, the third dimension, temporal order q is added. $f_{n(q,circ,ax)}$ [Hz] and $\zeta_{n(q,circ,ax)}$ are generated by populating $f_{n(circ,ax)}$ [Hz] and $\zeta_{(circ,ax)}$ over q. Then the matrix of natural frequencies is converted to the matrix of angular frequencies, $\omega_{n(q,circ,ax)}$ [rad/s].

In the media field block, the matrix for the radiation ratio, $\sigma_{(q,circ,ax)}$, is obtained. It can be used to calculate the acoustic damping and later to calculate the sound power in the sound radiation and human sense block. The motor speed is used to calculate the radiation ratio.

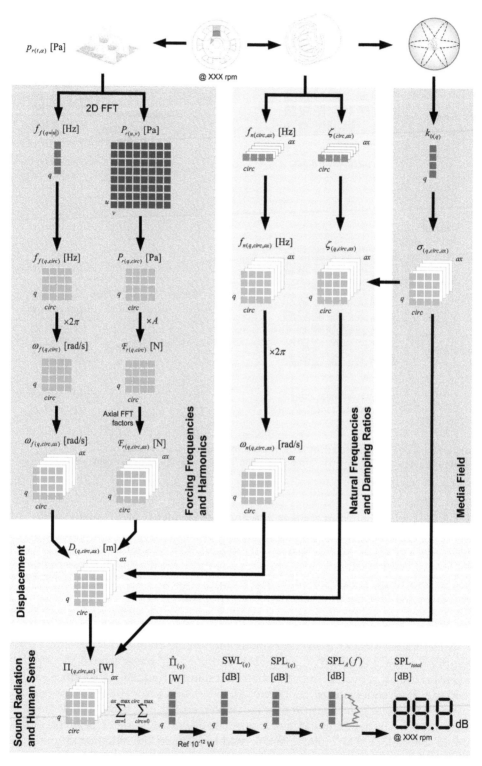

FIGURE 13.112
Computational process for acoustic analysis of an SRM.

In the block calculating displacements, the displacements caused by the all harmonics are calculated. The major inputs for calculating the displacements are $\omega_{f(q,circ,ax)}$ [rad/s], $F_{r(q,circ,ax)}$ [N], $\omega_{n(q,circ,ax)}$ [rad/s], and $\zeta_{(q,circ,ax)}$. The results are also populated in a matrix format as in $D_{(q,circ,ax)}$ [m].

In the sound radiation and human sense block, $D_{(q,circ,ax)}$ and $\sigma_{(q,circ,ax)}$ are used to calculate $\Pi_{(q,circ,ax)}$ [W]. $\Pi_{(q,circ,ax)}$ [W] is the sound power caused by each harmonic content in $F_{r(q,circ,ax)}$ [N]. By summing over circumferential mode and axial mode, $\hat{\Pi}_{(q)}$ [W] is obtained. $\hat{\Pi}_{(q)}$ is a function of the temporal order q. Then, the sound power matrix $\hat{\Pi}_{(q)}$ is converted to the sound power level vector, $SWL_{(q)}$ [dB]. $SWL_{(q)}$ is then converted to the sound pressure level vector $SPL_{(q)}$ [dB]. By applying the A-weighting correction, $SPL_{(q)}$ is converted to $SPL_A(f)$ [dB], since the forcing frequency, f is function of q. Then, $SPL_A(f)$ can be plotted over the forcing frequency spectrum. By summing the noise from different sources with varying frequencies, a single value of SPL_{total} [dB] can be obtained for the given operating point of the motor.

If this whole process is repeated for different motor speeds, the waterfall and SPL diagrams at different speed and forcing frequencies can be generated. At the end of this chapter, five questions are provided to practice the computational process for acoustic analysis of an SRM. The five questions are implemented using MATLAB® scripts.

Please note that the process shown in Figure 13.112 can be used for noise and vibration reduction as well. Methods to reduce the noise and vibration can be classified based on this map. For instance, starting from the motor configuration, by increasing the numbers of stator poles and rotor poles, and the number of phases, the noise can be decreased. By modifying the housing dimensions and materials, the natural frequencies and damping ratios of the stator-frame system can be adjusted. If the configuration of the motor is fixed, for example, Ns, Nr, and N_{ph}, by changing the major dimensions (stator taper angle, stator back iron thickness, and so on), the amplitudes of the radial force harmonics can be modified as well. Finally, the amplitudes of the harmonics can also be manipulated by optimizing the current control and the shape of the phase current.

Questions and Problems

1. Introduction to Problems of This Chapter

   The questions of this chapter cover the entire process of the acoustic analysis, as shown in Figure Q.13.1. The list of the topics addressed in the questions are:

2. Calculation of natural frequencies

3. Generation of radial force density waves and 2D FFT decomposition

4. Calculation of surface displacement

5. Calculation of radiation ratio

6. Calculation of SPL

   In 2, the natural frequencies are calculated analytically. Axial FFT decomposition is neglected, and the damping ratios are estimated. The outputs of Q1 are vectors of the natural frequencies and damping ratios.

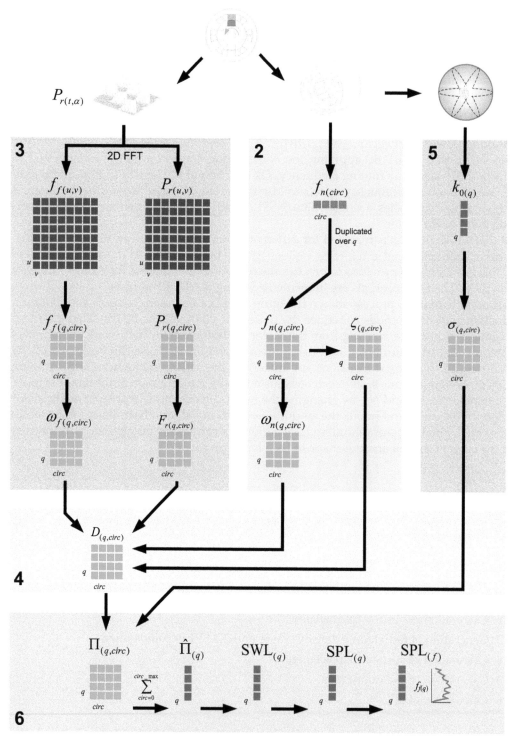

FIGURE Q.13.1
Process of acoustic analysis used in the questions of this chapter.

In 3, a pseudo-radial force waveform is used. 2D FFT is employed to obtain the harmonic contents and related forcing frequencies. The outputs of 3 are the matrices for the amplitudes of the dominant harmonic contents and the forcing frequencies.

The results of 2 and 3 are used to calculate the vector of surface displacements in 4. In 5, the radiation ratios at different modes are calculated analytically. The outputs of 4 and 5 are used to calculate the SPL in 6.

Please note that you need to execute the MATLAB scripts from 2 to 6 in a single file. When you copy the scripts to the MATLAB editor, please follow the order: A, B, C, ... L. It is also recommended that you convert the scripts from each question into a MATLAB function so that each may be called from a single, higher-order script.

2. Calculation of Natural Frequencies

It is assumed that the frame-stator system (called frame later) of the motor is a cylindrical shell. The properties of the frame with the clamped-clamped constraint are given in Figure Q.13.2 and Table Q.13.1.

First, we review the analytical method for calculating the natural frequencies of a cylindrical shell. The natural frequency of the cylindrical shell can be calculated from (13.17):

$$f_{n(circ,ax)} = \frac{1}{2} \frac{\Omega_{(circ,ax)}}{R_f} \sqrt{\frac{E_f}{\rho_f \left(1 - v^2_{\,f}\right)}}$$

The non-dimensional frequency parameter $\Omega_{(circ,ax)}$ can be solved using (13.11):

$$\Omega^6_{(circ,ax)} - \left(C_2 + \kappa \Delta C_2\right)\Omega^4_{(circ,ax)} + \left(C_1 + \kappa \Delta C_1\right)\Omega^2_{(circ,ax)} - \left(C_0 + \kappa \Delta C_0\right) = 0$$

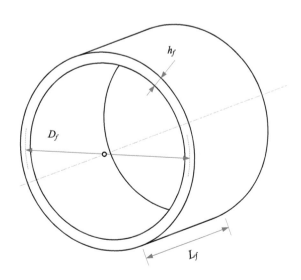

FIGURE Q.13.2
Circular cylindrical shell and its coordinate system.

TABLE Q.13.1

Values of the Cylindrical Shell Used for Calculating Natural Frequencies

Description	Symbol	Value
Medium diameter of frame	D_f	290 mm
Mean radius of the frame	R_f	145 mm
Shell thickness	h_f	10 mm
Axial length	L_f	600 mm
Material of the frame	N/A	Steel
Mass density	ρ_f	7.85 g/cm³
Modulus of elasticity of the frame	E_f	200 GPa
Poisson ratio	v_f	0.33

The above equation can then be seen as a third-order polynomial for $\Omega^2_{(circ,ax)}$. If x is used to represent $\Omega^2_{(circ,ax)}$, the equation can be rewritten as:

$$x^3 - \left(C_2 + \kappa\Delta C_2\right)x^2 + \left(C_1 + \kappa\Delta C_1\right)x - \left(C_0 + \kappa\Delta C_0\right) = 0$$

The smallest real root from the equation is the solution for $\Omega^2_{(circ,ax)}$. The roots in MATLAB can be used to solve for the equation. Since the boundary condition for the frame is clamped-clamped, the constants for the above equation are given by (13.12):

$$\begin{cases} C_0 = \dfrac{1}{2}\left(1-v_f\right)\left[\left(1-v_f^2\right)\lambda^2 + \kappa^2\left(circ^2 + \lambda^2\right)^4\right] \\[2mm] C_1 = \dfrac{1}{2}\left(1-v_f\right)\left[\left(3+2v_f\right)\lambda^2 + circ^2 + \left(circ^2 + \lambda^2\right)^2 + \dfrac{3-v_f}{1-v_f}\kappa^2\left(circ^2 + \lambda^2\right)^3\right] \\[2mm] C_2 = 1 + \dfrac{1}{2}\left(3-v_f\right)\left(circ^2 + \lambda^2\right) + \kappa^2\left(circ^2 + \lambda^2\right) \\[2mm] \Delta C_0 = 0 \\[2mm] \Delta C_1 = 0 \\[2mm] \Delta C_2 = 0 \end{cases}$$

$\lambda_{(ax)}$, which is axial mode dependent, can be solved by using (13.15):

$$\lambda_{(ax)} = ax.\pi\,\frac{R_f}{L_f - L_0}$$

L_0 is the correction to the length for clamped-clamped condition as given in (13.15):

$$L_0 = L_f\,\frac{0.3}{n+0.3}$$

a. Calculate the natural frequencies of the frame using MATLAB. The circumferential order, *circ*, ranges from 0 to 5, and the axial order, *ax*, is from 1 to 4.

b. Compare the analytical solutions with the FEA results shown in Figures 12.39 through 12.44 in Chapter 12. Calculate the difference between the analytical solutions and the FEA results.

c. Rank the analytical results from smallest to highest and determine which modes have the highest and lowest frequencies.

d. Update the parameters of Table Q.13.2 in the MATLAB code. Rerun the script and calculate the natural frequencies (*circ*, ranging from 0 to 12 and *ax* = 1.). Please note that the natural frequencies calculated here will be used in the following questions.

3. FFT Decomposition of Pseudo-Radial Force Density

A radial force density surface has three dimensions: (a) circumferential position, (b) time, and (c) radial force density. Figure Q.13.3 shows the pseudo-radial force waveforms for a 6/4 SRM when three phases, A, B, and C, are conducting separately. These three waves exist on the axis of circumferential domain. Since radial force density rises and falls over time, another time factor, as shown in Figure Q.13.4, is also applied whenever there is a radial

TABLE Q.13.2

Updated Values of the Cylindrical Shell Used for Calculating Natural Frequencies

Description	Symbol	Value
Medium diameter of frame	D_f	220 mm
Shell thickness	h_f	80 mm
Mass density	ρ_f	2.7 g/cm³

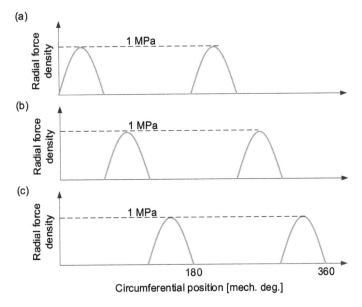

FIGURE Q.13.3
Pseudo-radial force density waves: (a) Phase A, (b) Phase B, and (c) Phase C.

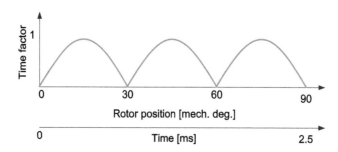

FIGURE Q.13.4
Using sinusoidal waveforms to represent radial force for 6/4 SRM.

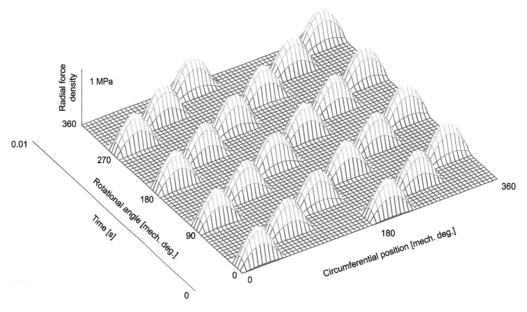

FIGURE Q.13.5
Pseudo radial force density surface over one mechanical cycle.

electromagnetic pulsation. The surface of radial force density can be seen over one mechanical cycle and 360 mechanical degrees of circumferential position in Figure Q.13.5. This surface waveform will be later used in the FFT analysis.

a. The following MATLAB script can be used to generate the surface waveform. Please reproduce the script below and use it to plot Figure Q.13.5.

```
%% B: Create radial force density surface

Ns = 6;              % Number of stator poles
Nr = 4;              % Number of rotor poles
Nph = 3;             % Number of phases
pp = Ns/Nph/2;       % Number of pole pairs
P_peak = 10^6;       % Unit: Pa. Peak value: 1 MPa
Pa_dominant = 200;   % Filter can be adjusted
rpm = 3000;          % Unit: rpm
f_base = rpm/60;
delta = round(2^10/Nph/Nr);
q_max = round(20000/f_base);

if q_max > 400
    q_max = 400;
else
end

% q_max = 30*pp;
circ_max = 12;
radial_pressure_elec_revo = [];
radial_pressure_mech_revo = [];
T_mech_rev = 0.01; % Unit: s

t_series_elec_revo = [0:T_mech_rev/delta/Nph/Nr:T_mech_rev/Nr-
T_mech_rev/delta/Nph/Nr]; % One electrical cycle

t_series_mech_revo = [0:T_mech_rev/delta/Nph/Nr:T_mech_rev-
T_mech_rev/delta/Nph/Nr];    % One mechanical cycle

sp_series = 0:2*pi/delta/Ns:2*pi-2*pi/delta/Ns; % Spatial position, v: 0 ~
360 mech. deg.

radial_wave = abs(sin(Ns/2*sp_series)); % Populate over space

for j = 1:Nph

    ph_A = (sawtooth(Ns/Nph*sp_series + 2*(-j+Nph)*pi/Nph)+1)/2;

    radial_A = P_peak * (ph_A > (Nph-1)/Nph).*radial_wave; % Unit: Pa

    if j < Nph

        for i = 1: round(length(t_series_elec_revo)/Nph)

            radial_pressure_elec_revo = [radial_pressure_elec_revo;radial_A];
% Populate over time

        end

    else

        for i = 1: length(t_series_elec_revo) -
round(length(t_series_elec_revo)/Nph)*(Nph - 1)
```

```
            radial_pressure_elec_revo = [radial_pressure_elec_revo;radial_A];
% Populate over time

        end
    end
end

for i = 1:Nr

    radial_pressure_mech_revo =
[radial_pressure_mech_revo;radial_pressure_elec_revo];

end

% Apply factor over time

factor_over_time = abs(sin(t_series_mech_revo/T_mech_rev*Nph*Nr*pi))';
matrix_over_time = [];

for i = 1:length(sp_series)

    matrix_over_time = [matrix_over_time,factor_over_time];

end

radial_pressure_mech_revo = matrix_over_time.*radial_pressure_mech_revo;
[sp_plane,time_plane] = meshgrid(sp_series (1:2^3:end), t_series_mech_revo
(1:2^3:end)); % reduce mesh density for plotting purpose
radial_surface_sparse =
griddata(sp_series,t_series_mech_revo,radial_pressure_mech_revo,sp_plane,time
_plane);

figure (index)
index = index + 1;

mesh(sp_plane/2/pi*360, time_plane, radial_surface_sparse)
zlim([0 P_peak*1.1]);
xlabel('Cirumferential position, \alpha [mech. deg.]')
ylabel('Time, t [s]')
zlabel('P (\alpha, t) [Pa]')
set(gca,'XTick',[0:60:360])
```

b. Using a 2D FFT function in MATLAB, perform the FFT analysis over the surface wave of radial_pressure_mech_revo. Plot the heat map of the center,
   (DC_row-q_max:DC_row+q_max,   DC_col-circ_max:DC_col+circ_
   max) on the u-v plane (hint: in the first quadrant, $u > 0$, $v > 0$) for the amplitudes and phase angles of the dominant harmonics (>200 Pa) respectively,
   using imagesc. DC_col is the column index of the DC component, and
   DC_row is the row index of the DC component. pp is the number of pole pairs,
   and q_max is the maximum number of temporal order q. circ_max is the
   maximum number of the circumferential order, *circ*.

Please note that MATLAB commands sflipud and fliplr are used to manip-
ulate the matrix of the surface to make sure the top-right is the first quadrant
($u > 0, v > 0$).

c. FFT _ shifted _ flipud _ abs _ cut and phase _ angle _ matrix _
cut store the amplitudes and the phase angles for the dominant harmonic
contents of the radial force density wave. Please use part of these dominant
harmonics ($u = -12, -11,..., 12; v = -6, -5,....,6$) to reproduce the surface for the
radial force density. In addition, plot the surface for each dominant harmonic
content ($u = -12, -11,..., 12; v = -6, -5,....,6$) using mesh and contour plots,
respectively.

d. Write a MATLAB script that will superimpose and plot any two dominant harmonic contents. Please plot the superposition for the following pairs of dominant harmonics (u, v):

- (12, 0) and (−12, 0)
- (8, 2) and (−8, −2)
- (12, 6) and (−12, 6)

e. Write a MATLAB script that will convert and organize the amplitudes of all the
dominant harmonics obtained in (b) in the first quadrant (u, v axes included).
A similar matrix can be found in Table 13.6. Please also plot the heat map for
the converted dominant harmonics.

4. Calculation of Surface Displacement

In this question, the displacement of each harmonic content is computed. The previous question determined the amplitude of $P_{r(q,circ)}$ for the harmonic content of a
pseudo-radial force density wave. The result is first used to calculate the amplitude of the force component by multiplying the amplitude for the harmonic content of the radial force density with the surface area:

$$F_{r(q,circ)} = \pi \left(D_f - h_f\right) L_f P_{r(q,circ)}\left[N\right]$$

a. It is assumed that the radial force acts on the inner diameter of the cylindrical
shell, as shown in Figure Q.13.2. In this question, only the first axial mode is
considered ($ax = 1$). The mode shapes used in this question are Mode ($circ$, 1)
with $circ$ ranges from 0 to 12. The natural frequencies calculated in 2-d are to be
used in the following questions. The following equation estimates the damping ratio for different modes using (13.85):

$$\zeta_{(circ,ax=1)} = \frac{1}{2\pi}\left(2.76 \times 10^{-5} f_{(circ,ax=1)} + 0.062\right)$$

Please note that the dimensions of the object are updated, as shown in Table Q.13.2. The MATLAB script for calculating the aforementioned details is provided below:

```
%% G: Recalculate natural frequencies circ = 0:12, ax = 1

vf = 0.33;
Df = 220/1000;          % Diameter
hf = 80/1000;           % Thickness
Lf = 300/1000;          % Axial length
Rf = 1/2*Df;  % Mean frame radius
rho = 2700;             % Unit: kg/m^3
Ec = 200*10^9;          % Unit: Pa, Modulus of elasticity
ax_max = 1;             % Highest axial order considered
omega_matrix = [];
freq_matrix = [];

% Loops for calculating natural frequencies for different orders

for i = 1:circ_max + 1

    circ = i-1; % Circumferential order starts from zero

    for j = 1:ax_max

        ax = j; % Axial order starts from one

        L0 = Lf*0.3/(ax + 0.3);
        lambda = ax*pi*Rf/(Lf - L0);
        kappa = sqrt(hf^2/12/Rf^2);
        circ_ind=circ;
        kappa_sqr=kappa^2;

        % Calculate the coefficients for the third order polynomial

        C0 = 0.5*(1-vf)*((1-
vf^2)*lambda^4+kappa_sqr*(circ_ind^2+lambda^2)^4);
        C1 = 0.5*(1-
vf)*((3+2*vf)*lambda^2+circ_ind^2+(circ_ind^2+lambda^2)^2+(3-vf)/(1-
vf)*kappa_sqr*(circ_ind^2+lambda^2)^3);
        C2 = 1+0.5*(3-
vf)*(circ_ind^2+lambda^2)+kappa_sqr*(circ_ind^2+lambda^2)^2;

        % Solve the third order polynomial

        omega_mn = sqrt(min(roots([1, -C2, C1, -C0])));

        % Calculate natural frequency

        fq_mn = 1/2/pi*omega_mn/Rf*sqrt(Ec/rho/(1-vf^2));

        % Save the solution in matrices

        omega_matrix (j,i) = omega_mn;
        freq_matrix (j,i) = fq_mn;

    end
end
```

```
figure (index)
index = index + 1;
stem ([0:circ_max],freq_matrix/1000)
xlabel ('Circumferential order, circ')
ylabel ('Natural frequency [kHz]')
% Prepare matrices for natural frequencies, damping ratios, and radial force
fn_circ = freq_matrix;
fn_q_circ = [];
for i = 1:q_max + 1

    fn_q_circ = [fn_q_circ ; fn_circ];

end

damping_q_circ = (2.76/10^5*fn_q_circ + 0.062)/2/pi;
Fr_q_circ = PPr_q_circ*(Df - hf)*pi*Lf;
```

Please reproduce the MATLAB script above.

b. The function for calculating the displacement was given by (13.60):

$$D_{(q,circ,ax=1)} = \frac{F_{r(q,circ,ax=1)} / M}{\sqrt{\left(\omega_{n(q,circ,ax=1)}^2 - \omega_{f(q,circ,ax=1)}^2\right)^2 + 4\zeta_{(q,circ,ax=1)}^2 \omega_{f(q,circ,ax=1)}^2 \omega_{n(q,circ,ax=1)}^2}} [m]$$

It is assumed that the motor speed is 3000 rpm. Thus the mechanical frequency can be calculated using:

$$f_{mech} = \frac{rpm}{60} = \frac{3000}{60} = 50[Hz]$$

The forcing angular frequency may be calculated as:

$$\omega_{f(q,cric,ax=1)} = 2\pi q f_{mech} [rads/s]$$

The natural frequencies, $\omega_{n(q,cric,ax=1)}$, and damping ratios, $\zeta_{n(q,cric,ax=1)}$, are calculated in the first part of this question. Please write a MATLAB script to calculate the displacements for the different harmonic content obtained from 3-e. Please also plot the heat map for the displacements for the different harmonic content (q, circ).

5. Calculation of Radiation ratio

In this question, the radiation ratios of the shell are calculated at different circumferential modes ($circ = 0 \sim 12$) with axial mode, $ax = 1$. k_0 is the acoustic waveform number, which is related to the forcing angular frequency of the vibration as given by (13.65):

$$K_0 = \frac{\omega_{f(q,cric,ax=1)}}{C_0} = \frac{2\pi f_{f(q,cric,ax=1)}}{C_0} = \frac{2\pi q f_{mech}}{C_0}$$

C_0 is the speed of sound in the air, which can be calculated using:

$$c_0 = 331.3\sqrt{1 + \frac{T(°C)}{273.15}}$$

It is assumed that the ambient temperature is 20°C.

k_r and k_z are the radial and axial components of the acoustic wave number respectively, as shown in (13.66):

$$k_0 = \sqrt{k_z^2 + k_r^2}$$

r_{shell} and l_{shell} are the radius and the length of the shell, respectively.

The radiation ratio of the finite length cylindrical shell can be calculated using (13.67):

$$\sigma_{(q,circ,ax=1)} = \int_{-k_0}^{k_0} \frac{2k_0 l_{shell}}{\pi^2 r_{shell} k_r^2 \left| \dfrac{dH_{circ}^{(2)}(k_r r_{shell})}{d(k_r r_{shell})} \right|^2} \left[\frac{ax \cdot \pi / l_{shell}}{k_z + ax \cdot \pi / l_{shell}} \right]^2 \frac{\sin^2[(k_z - ax \cdot \pi / l_{shell})l_{shell} / 2]}{[(k_z - ax \cdot \pi / l_{shell})l_{shell} / 2]^2} dk_z$$

where r_{shell} is the radius of the shell, l_{shell} is the length of the shell, and $circ$ and ax are the circumferential mode and the axial mode, respectively.

The MATLAB script used for calculating the radiation ratios is provided below:

```
%% I: Calculate radiation ratio

T_celcius = 20;
c0 = 331.3*sqrt((T_celcius+273.15)/273.15); % m/s
% q_max = 15;
l_shell = Lf;
a_shell = (Df + hf)/2;
ax = 1; % Only axial mode, ax = 1, is considered
rad_eff_q_circ = [];

for circ_i = 1: circ_max + 1

    circ = circ_i - 1;

    for q_i = 1:q_max + 1

        q = q_i - 1;

        k0 = 2*pi*q*f_mech/c0;
        N_kz = 200;                         % half of number of
integration steps
        kzi = -k0:k0/N_kz:k0;               % integration variables
        kri = sqrt(k0^2-kzi.^2);

        Hm2i = 0.5*(besselh(circ-1, 2, kri.*a_shell)-besselh(circ+1, 2,
kri.*a_shell)); % differential

        int_kz = 2*k0*l_shell/pi^2/a_shell./kri.^2./(abs(Hm2i)).^2.*...
            (ax*pi/l_shell./(kzi+ax*pi/l_shell)).^2.*...
            (sin((kzi-ax*pi/l_shell).*l_shell./2)).^2./((kzi-
ax*pi/l_shell).*l_shell./2).^2;

        int_kz(isnan(int_kz)) = 0;          % remove NaN from the array

        rad_eff_q_circ(q_i, circ_i) = trapz(kzi, int_kz); % trapezoidal
integration
    end
end

rad_eff_q_circ_flipud = flipud (rad_eff_q_circ);

%%%%%% heat map for radiation eff of dominant harmonic contents + mask

figure (index)
index = index + 1;
rad_eff_q_circ_flipud_dominant = rad_eff_q_circ_flipud.*dominant_mask;
c = imagesc (circcirc, qq, rad_eff_q_circ_flipud_dominant);
set(gca,'YDir','normal')
caxis([0 max(max(rad_eff_q_circ_flipud_dominant))])
colormap winter; colorbar;
set(c,'AlphaData',~isnan(rad_eff_q_circ_flipud_dominant))
xlabel('Circumferential order, circ')
ylabel('Temporal order, q')
title ('Heat map for radiation ratio of dominant harmonics')
```

a. Please reproduce the provided script in MATLAB.

b. For q _ max = 200, please plot the radiation ratios of different circumferential modes versus forcing frequency.

6. Calculation of SPL

Using the angular forcing frequencies, $\omega_{f(q,circ,ax=1)}$ and displacements, $D_{(q,circ,ax=1)}$ calculated in 4 and the corresponding radiation ratios, $\sigma_{(q,circ,ax=1)}$ obtained in 5, the sound power can be calculated using (13.69):

$$\Pi_{(q,circ,ax=1)} = \rho_0 c_0 \left(\frac{w_{f(q,circ,ax=1)} D_{(q,circ,ax=1)}}{\sqrt{2}} \right)^2 \sigma_{(q,circ,ax=1)} A_s$$

$$= \rho_0 c_0 \left(\frac{2\pi q f_{mech} D_{(q,circ,ax=1)}}{\sqrt{2}} \right)^2 \sigma_{(q,circ,ax=1)} A_s [\text{W}]$$

where A_s is the area of the sound radiation surface, which can be calculated as:

$$A_s = \pi r_{shell}^2 l_{shell} [\text{m}^2]$$

a. The sound power resulting from different circumferential modes with the same forcing frequency or the same temporal order, *q*, can be superimposed together. Please write a MATLAB script to calculate the sound power [W] and plot the heat map of the sound power versus *circ* (horizontal axis) and *q* (vertical axis). Please flip the matrix for the results of the sound power along its non-main diagonal and plot the heat map of the flipped matrix versus *q* (vertical axis) and *circ* (horizontal axis).

b. Superimpose the sound power with different circumferential modes with the same *q* using (13.72):

$$\hat{\Pi}(q) = \sum_{circ-0}^{circ\_max} \Pi(q, circ, ax = 1) [\text{W}]$$

The sound power level, SWL, is obtained by logarithm scaling the superimposed sound power using (13.73):

$$\text{SWL}_{(q)} = 10 \lg \frac{\hat{\Pi}_{(q)}}{\Pi_{ref}} [\text{dB}]$$

where Π_{ref} is 10^{-12} W.

SPL_A is the average sound pressure level in a free-field over a reflective plane on a hemispherical surface at 1-meter distance from the machine, which can be calculated using (13.74):

$$\text{SPL}_{(q)} = \text{SWL}_{(q)} - 10 \log_{10} \left\{ 2\pi \left[1 + \frac{\max(l_{shell}, 2r_{shell})}{2} \right]^2 \right\} [\text{dB}]$$

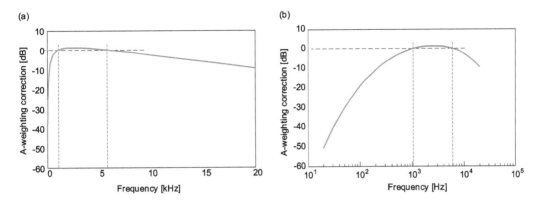

FIGURE Q.13.6
A-weighting correction: (a) linear scale and (b) logarithmic scale.

Calculate SPL_A and plot SPL_A versus frequency [Hz] and q (dual x-axis) using a stem plot in MATLAB. Figure Q.13.6 shows the trend of an A-weighting correction as frequency increases in both a linear scale and logarithmic scale.

c. Change the motor speed from 3000 rpm to 6000 rpm and 9000 rpm, and run the scripts from 3 to 6. Compare the SPL stem plots of the three speeds.

References

1. M. Bosing, Acoustic modeling of electrical drives: Noise and vibration synthesis based on force response superposition, Lehrstuhl und Institut für Stromrichtertechnik und Elektrische Antriebe, 2014.
2. J. F. Gieras, C. Wang, and J. C. Lai, Noise of Poly-Phase Electric Motors, New York: CRC Press, 2005.
3. D. K. Michael and P. Norton, *Fundamentals of Noise and Vibration Analysis for Engineers*, Cambridge, UK: Cambridge University Press, 2003.
4. R. N. Arnold and G. B. Warburton, The flexural vibrations of thin cylinders, *Proc. Inst. Mech. Eng.*, vol. 167, no. 1, pp. 62–80, 1953.
5. C. Wang and J. C. S. Lai, The sound radiation efficiency of finite length circular cylindrical shells under mechanical excitation II: Limitations of the infinite length model, *J. Sound Vib.*, vol. 241, no. 5, pp. 825–838, 2001.
6. E. W. Weisstein, Hankel function of the second kind [Online]. Available: http://mathworld. wolfram.com/HankelFunctionoftheSecondKind.html (Accessed December 2017).
7. W. Tong, *Mechanical Design of Electric Motors*, Boca Raton, FL: CRC Press, 2014.
8. T. Bertolini and T. Fuchs, *Vibrations and Noises in Small Electric Motors Measurement, Analysis, Interpretation*, Ulm, Germany: Verlag onpact GmbH, 2012.
9. National Electrical Manufacturers Association, Motors and Generators, 527 MG 1-2009, 2009.
10. H. Fastl and E. Zwicker, *Psychoacoustics: Facts and Models*, Berlin, Germany: Springer Science & Business Media, 2013.
11. C. W. De Silva, *Vibration: Fundamentals and Practice*, Boca Raton, FL: CRC Press, 2006.
12. G. Genta, *Vibration Dynamics and Control*, New York: Springer, 2009.
13. S. J. Yang, *Low-Noise Electrical Motors*, New York: Oxford University Press, 1981.

14. J. Crocker, ed., *Handbook of Noise and Vibration Control*, Hoboken, NJ: John Wiley & Sons, 2007.
15. R. S. Colby, F. M. Mottier, T. J. E. Miller, Vibration modes and acoustic noise in a four-phase switched reluctance motor, *IEEE Trans. Ind. Appl.*, vol. 32, no. 6, pp. 1357–1364, 1996.
16. M. Besbes, C. Picod, F. Camus, and M. Gabsi, Influence of stator geometry upon vibratory behavior and electromagnetic performances of switched reluctance motors, *IEE Proc.-Electr. Power Appl.*, vol. 145, no. 5, pp. 462–468, 1998.
17. S. M. Castano, B. Bilgin, E. Fairall, and A. Emadi, Acoustic noise analysis of a high-speed high power switched reluctance machine: Frame effects, in *IEEE Trans. Energy Convers.*, vol. 31, no. 1, pp. 69–77, 2016.
18. C. Gan, J. Wu, M. Shen, S. Yang, Y. Hu, and W. Cao, Investigation of skewing effects on the vibration reduction of three-phase switched reluctance motors, *IEEE Trans. Magn.*, vol. 51, no. 9, 2015.
19. Z. Q. Zhu, X. Liu, and Z. Pan, Analytical model for predicting maximum reduction levels of vibration and noise in switched reluctance machine by active vibration cancellation, *IEEE Trans. Energy Convers.*, vol. 26, no. 1, pp. 36–45, 2011.
20. A. Tanabe and K. Akatsu, Vibration reduction method in SRM with a smoothing voltage commutation by PWM, *9th Int. Conf. Power Electron.: ECCE Asia "Green World with Power Electron".* ICPE 2015-ECCE Asia, pp. 600–604, 2015.
21. A. Hofmann, A. Al-Dajani, M. Bosing, and R. W. De Doncker, Direct instantaneous force control: A method to eliminate mode-0-borne noise in switched reluctance machines, *Proc. 2013 IEEE Int. Electr. Mach. Drives Conf. IEMDC 2013*, pp. 1009–1016, 2013.
22. M. Takiguchi, H. Sugimoto, N. Kurihara, and A. Chiba, Acoustic noise and vibration reduction of SRM by elimination of third harmonic component in sum of radial forces, *IEEE Trans. Energy Convers.*, vol. 30, no. 3, pp. 883–891, 2015.
23. J. Dong, J. W. Jiang, B. Howey et al., Hybrid acoustic noise analysis approach of conventional and mutually coupled switched reluctance motors, *IEEE Trans. Energy Convers.*, vol. 32, no. 3, pp. 1042–1051, 2017.
24. D. d'Udekem, et al., Numerical prediction of the exhaust noise transmission to the interior of a trimmed vehicle by using the finite/infinite element method. No. 2011-01-1710. SAE Technical Paper, 2011.
25. S. Marburg and B. Nolte, *Computational Acoustics of Noise Propagation in Fluids: Finite and Boundary Element Methods*, Vol. 578. Berlin, Germany: Springer, 2008.
26. R. J. Astley and J. P. Coyette, The performance of spheroidal infinite elements, *Int. J. Numer. Meth. Eng.*, vol. 52, no. 12, pp. 1379–1396, 2001.
27. R. J. Astley and J. Coyette, Conditioning of infinite element schemes for wave problems, *Int. J. Numer. Meth. Bio. Eng.*, vol. 17, no. 1, pp. 31–41, 2000.
28. W. de S. Clarence, *Vibration Damping, Control, and Design*. Vancouver, Canada: CRC Press, 2007.

14

Thermal Management of Switched Reluctance Machines

Yinye Yang, Jianbin Liang, Elizabeth Rowan, and James Weisheng Jiang

CONTENTS

14.1 Introduction

Thermal management is a complex but crucial aspect in electric machine design. For example, high temperature results in higher resistance and, hence, higher copper losses in the windings. Thermal management is also a limiting factor for motor's torque-speed profile, as shown in Figure 14.1. A motor's transient torque-speed profile is greater than the continuous operation capability. The operating time for the area in between the continuous and transient limits is a function of the thermal constraints.

An ineffective cooling system can lead to faster insulation degradation and increased winding failure rates due to increased thermal cycles and peak temperatures. It is a rule of thumb that a 10° Celsius increase in temperature will reduce the life expectancy of the windings by half [1]. Materials with poor thermal conductivity such as adhesives and

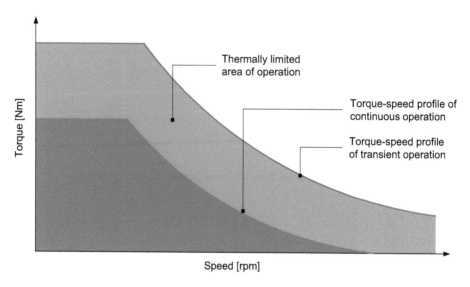

FIGURE 14.1
Idealized continuous and transient operating envelopes for electric machines.

winding insulations may also result in local heat concentration. Therefore, effective thermal management improves machine performance and ensures reliable operation, while poor thermal management results in performance degradation, accelerated machine wear, and may even cause machine breakdown.

SRMs, compared to other electric machine types, have the simplest construction because they incorporate no magnets or windings on the rotor. In general, rotors are more difficult to cool during the operation due to the lack of accessibility. From a thermal point of view, this provides an advantage to switched reluctance machines compared to other machine types, as there is no additional loss generated by the rotor magnets or windings. Concerns about demagnetization or reduction in the flux density of permanent magnets at high temperatures, which often limit the maximum operating temperature, do not exist in switched reluctance machines. SRMs are only limited by the stator winding and, therefore, have the highest capacity in terms of high temperature operation [2]. This chapter discusses various aspects of switched reluctance machine thermal management fundamentals, techniques, design approaches and considerations.

14.2 Loss Generation

Temperature rise in a switched reluctance machine is caused by power losses generated by different physical mechanisms. The loss generation mechanisms are discussed in Chapter 2 and will be briefly revisited here before focusing on the details of thermal management systems. Losses in a switched reluctance machine are typically classified as copper, iron (or core), mechanical, and excess losses. Figure 14.2 illustrates the loss distribution of a typical switched reluctance machine. The copper loss results from Joule heating due to the resistivity of the conductors. Iron losses have two components: eddy current and hysteresis losses. Eddy current losses are induced by the time varying magnetic field and are a result of the finite conductivity of the electrical steel. As discussed in Chapter 6, hysteresis

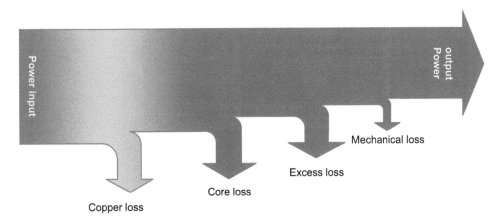

FIGURE 14.2
Distribution of losses for a typical switched reluctance machine.

losses related to the change in the alignment of the magnetic domains in the direction of the external magnetic field. Mechanical losses, in particular friction and windage losses, also contribute to the motor temperature rise. To fully capture the electromagnetic losses, additional mechanisms that effect the losses need to be considered, such as the skin effect, proximity effect, and excess loss.

14.2.1 Copper Loss

Copper loss is a major loss component in all electric machines, especially at higher torque operation. Electric machines with rotor windings (e.g. brush DC, induction, wound field synchronous, and claw-pole machines) feature copper losses in both the stator and rotor, while switched reluctance machines (SRMs) only have copper losses on the stator, as there are no conductors on the rotor.

Frequency dependent skin and proximity effects add to the overall copper losses by effectively reducing the cross-sectional area where the current flows through it. The skin effect is caused by eddy currents induced by a conductor's own flux linkage. The proximity effect is caused by the flux from neighboring conductors [3]. The skin and proximity effects cause a non-uniform current distribution in conductors, which becomes more extreme as frequency increases. In practice, the copper loss is calculated using the winding resistance. At low frequency, the copper losses are dominated by the Joule effect that is modelled with the DC resistance. Frequency dependent effects are captured by introducing equivalent AC resistances [4]:

$$P_{cu} = mI_p^2 \left(R_{dc} + R_{skin}(f) + R_{proximity}(f) \right) \tag{14.1}$$

where m is the number of phases, I_p is the rms value of the phase current, f is the frequency, R_{dc}, R_{skin}, and $R_{proximity}$ are DC, skin, and proximity components of the phase resistance, respectively.

The DC component of phase resistance is found through the resistance equation, which is a function of the conductor geometry. The AC components are either found analytically [3,4] or estimated with finite element analysis (FEA). The magnetic field strength is low in the end winding region as compared to the active region. Therefore, the proximity effect

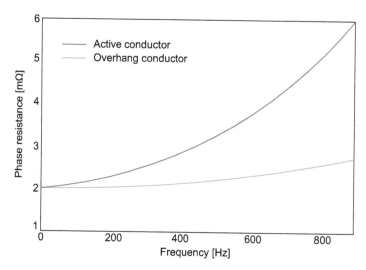

FIGURE 14.3
Frequency impact on the phase resistance in an electric machine with rectangular wires. (From Nalakath, S. et al., Modeling and analysis of AC resistance of a permanent magnet machine for online estimation purposes, in *Proceedings of the IEEE Energy Conversion Congress and Exposition*, Montreal, QC, September 2015, pp. 314–319.)

in the overhang region is typically negligible. The skin effect is approximated as constant throughout the overall length of the conductor. The impact of frequency on the phase resistance of the active and overhang regions are shown in Figure 14.3. It should be noted that the phase resistance is also marginally influenced by excitation current especially when the core is saturated [3].

The copper loss increases as the resistance increases with temperature. The temperature of the overhang conductors is typically higher than the active conductors. The active conductors in the slots are surrounded by electrical steel, which has higher thermal conductivity [5]. This provides better heat dissipation in the active region of the coils. Therefore, the heat generated by copper losses is modelled separately for the active and overhang regions. The effect of temperature on copper loss model is accounted by calculating resistivity as a function of temperature.

$$\rho_T = \rho_0 \left[1 + \alpha \left(T - T_0\right)\right] \tag{14.2}$$

where ρ_0 is the resistivity at the initial temperature T_0, α is the temperature coefficient, and T is the final temperature. The updated resistivity is used to compute the resistance components in (14.1).

Figure 14.4 plots the copper loss contour map of an example SRM with a peak power of 60 kW [6]. The motor is designed with stranded wires; therefore, the copper losses due to skin and proximity effect are neglected. The copper loss contour plot shows a range of 0 to 8820.8 W on the scale from blue to red. The peak copper losses occur at the peak torque region where the phase current is high. The root mean square (RMS) value of the phase current reduces at higher speeds due to the higher induced voltage. This results in lower copper losses.

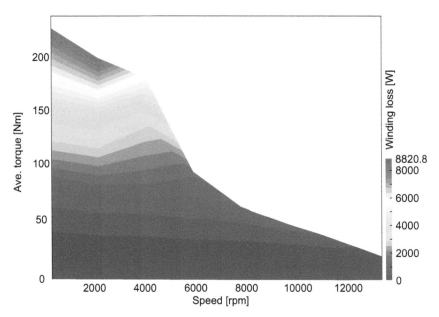

FIGURE 14.4
Torque-speed profile with contours of copper loss of a 60 kW SRM. (From Jiang, W., *Three-Phase 24/16 Switched Reluctance Machine for Hybrid Electric Powertrains: Design and Optimization*, PhD Dissertation, McMaster University, March 2016.)

14.2.2 Iron Losses

Iron loss is another major loss component in switched reluctance machines. It is typically dominant when a machine operates at higher frequency, for instance, at high speed. Losses are greater at high speed because the main iron loss components, hysteresis and eddy current losses, are a function of electrical frequency. The Steinmentz equation is primarily used to quantify the iron or core loss. The coefficients of the Steinmetz equation are generally treated as constants. However, variable coefficients are required for predicting core loss over a wide operating region. The variations of hysteresis and eddy current coefficients with respect to flux density for a specific electrical steel are shown in Figure 14.5. Modified equations are used to account for pulse width modulation (PWM) effects and arbitrary waveforms. The version of the Steinmentz equation that takes harmonic components into account is [7]:

$$P_{core} = \sum_{n} K_{hn} B_n^{1.6} nf + K_{en} B_n^2 (nf)^2 \tag{14.3}$$

where n is the harmonic index, K_h and K_e are hysteresis and eddy current loss coefficients, respectively, B is flux density, and f is frequency.

Iron loss is distributed throughout the core depending on the amplitude and frequency of the local flux density waveform. Iron losses occur primarily on the stator. The stator teeth and back iron are the major core loss regions. The loss in tooth tips or pole shoes is relatively low and it can be neglected when considering thermal effects [5]. In comparison, the rotor core loss is typically smaller since the frequency of the magnetic flux density

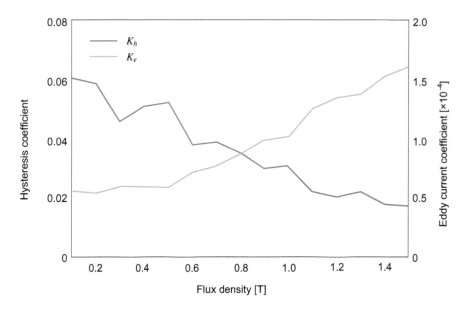

FIGURE 14.5
Estimated hysteresis and eddy current loss coefficients. (From Yang, Y. et al., *IET Electric. Syst. Transp.*, 7, 104–116, 2017.)

variation in the rotor core is lower compared to that of the stator. It should be noted that air gap and current harmonics yield alternating rotor flux harmonics that cause additional iron losses. These losses are more prevalent in wound field synchronous machines, for example, where the rotor laminations tend to be made from thicker steel to reduce manufacturing costs [10]. However, these effects can be ignored in some other machines. For example, the large effective air gap in surface permanent magnet machines shields the rotor back iron from flux harmonics.

A simple approach to include the core loss in the thermal analysis is to consider two heat sources; one is in the stator and other is in the rotor [5]. A detailed approach is to consider many heat sources corresponding to regions classified based on the same core loss density [7]. The values of core losses can be found from analytical equations. Many of the commercial software packages also provide core loss calculation as part of their post processing. Figures 14.6 and 14.7 illustrate the stator and rotor iron loss contour plot of an SRM. It can be observed that the losses increase as current and speed increase for both the stator and the rotor.

14.2.3 Mechanical Losses

Mechanical losses in an electric machine consist of friction and windage losses. Friction losses are mainly caused by the bearings. The resulting heat and the heat passing to ambient via the bearings increase the local temperature. This tends to degrade the lubricant and reduce the life of the bearing.

The aerodynamic drag experienced by the rotor periphery and the cooling fan causes windage loss. These losses can be minimized by using high quality bearings, lubricants and high-performance fan designs. A general expression for friction and windage losses for small machines is [11]:

$$P_{fw} = 2D^3Ln^3 \times 10^{-6} + K_{fb}G \, n \times 10^{-3} \tag{14.4}$$

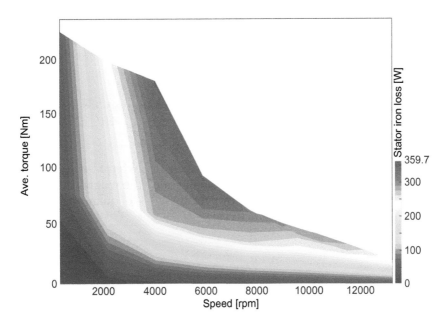

FIGURE 14.6
Torque-speed profile with contours of stator iron losses of a 60 kW SRM. (From Jiang, J.W. et al., *IEEE Trans. Transp. Electrif.*, 3, 76–85, 2017.)

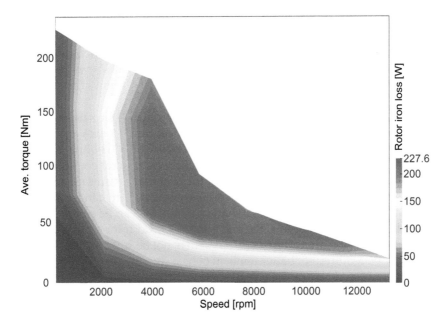

FIGURE 14.7
Torque-speed profile with contours of rotor iron losses for a 60 kW SRM. (From Jiang, J.W. et al., *IEEE Trans. Transp. Electrif.*, 3, 76–85, 2017.)

where D, L, and G are the outer diameter of the rotor and the length and weight of rotor, respectively, n is the rotational speed in RPM, and K_{fb} is the frictional loss coefficient.

14.2.4 Excess Losses

The core loss equation can be rewritten to include anomalous or excess losses:

$$P_{core} = P_{hysteresis} + P_{eddy} + P_{excess} \tag{14.5}$$

Excess losses arise from complex interaction between the external magnetic field and atomic scale interactions in the steel. On these small scales, the magnetic field still interacts with the steel atoms to produce losses; however, the fields of quantum mechanics and statistical dynamics are required to characterize the behavior.

14.3 Cooling Techniques

14.3.1 Types of Cooling Methods

Cooling technologies can be classified according to the mode of heat transfer: conduction, natural convection, forced convection, radiation, and evaporative cooling. They can also be classified according to cooling fluid: water, air, oil, and phase change materials. The selected cooling technology will depend on cooling requirements for the machine and on which parts are being targeted for cooling: stator core, stator winding, or end windings. Some of these technologies can be incorporated simultaneously to minimize the temperature of the various critical locations, such as hot spots. This section discusses various commonly-used cooling technologies for electric machines. Figure 14.8 shows the classification of cooling techniques in terms of the cooling targets. Figure 14.9 shows the classification of cooling methods based on cooling agents.

The transfer of heat generated within an SRM to an external heat sink depends on various factors, such as the mode of heat transfer, the effective heat transfer area and geometry, the working fluid used for cooling, the flow rate, and temperature of the cooling media. The simplest cooling technique is dissipating the heat to ambient by natural convection. Typically, the heat dissipation can be improved by increasing the heat transfer surface area with added fins on the housing. A more complicated forced air-cooling system can be used to further increase the heat dissipation. For example, a shaft mounted fan can be employed to enhance the heat transfer from the housing fins, the end windings, and rotor surfaces. However, for high current densities, using air as the cooling fluid may not be sufficient, and some form of liquid cooling may be required for better removal of heat. Typical rules of thumb for cooling techniques and associated heat transfer coefficients are listed in Table 14.1 [12]. Higher heat transfer coefficients enable higher current density and, hence, higher machine output power; however, at the expense of higher system complexity and energy cost.

In terms of the cooling location, stator core cooling is the most commonly used technology in electric machines. It can be divided into direct cooling where the heat is removed directly from the stator core, indirect cooling where the heat is transferred radially and/or axially to the external surfaces and dissipated by the cooling fluid, or it can be a combination of both.

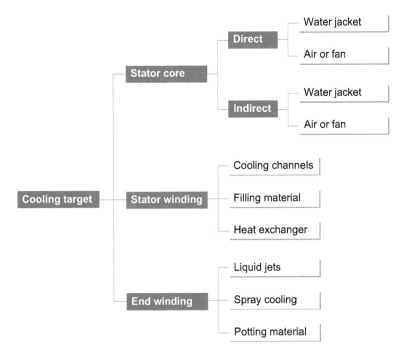

FIGURE 14.8
Cooling technologies depending on cooling target.

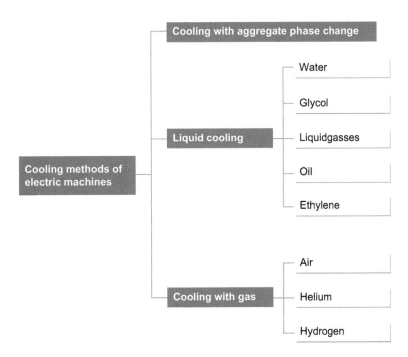

FIGURE 14.9
Types of cooling methods depending on cooling agents.

TABLE 14.1

Rules of Thumb for Cooling Type and Heat Transfer Coefficients

	Current Density Range [A/mm²]	Heat Transfer Coefficient [W/(m²·K)]	Cooling Efficiency	System Complexity	Energy Cost
Natural convection	1.5–5	5–10	Low	Simple	None
Forced convection	5–10	10–300	Medium	Medium	Low
Liquid cooling	10–30	50–20,000	High	Complex	High

Source: Staton, D., Thermal analysis of traction motors, presented in *IEEE Transportation Electrification Conference and Expo*, Dearborn, MI, 2014.

14.3.2 Air Cooling

For low power motors or motors for stationary industrial applications, the motor can be cooled without using a fan. For SRMs, the rotor's salient poles can work like a fan moving the air inside the motor when the rotor rotates. But for many applications, either an internal or an external fan is needed.

Air cooling of the motor is referred to as induced ventilation if either an internal or external fan creates reduced air pressure inside the motor leading to the suction of air into the motor. The internal or external fan is located near the air outlet. Figure 14.10 shows an example of induced ventilation with an internal fan driven directly by the motor. The fan can also be external, as shown in Figure 14.11. The hot air is pushed out of the motor by the external fan.

When either internal or external fan sucks the air from outside and forces it through the motor, the air cooling is called forced ventilation. The internal or external fan is located near the air inlet. As shown in Figure 14.12, the internal fan, driven by the motor via the main shaft, creates lower pressure at the air inlet and high pressure to push the air through the motor to cool it. Figure 14.13 shows an example of forced ventilation with external fan. The induced and forced ventilation are different because of the location of the fan. The fan

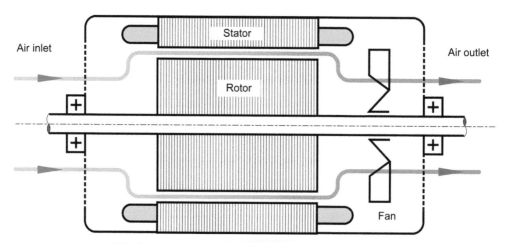

FIGURE 14.10
Induced ventilation with internal fan. (From Sawhney, A.K., *A Course in Electrical Machine Design*, Dhanpai Rai & Co., 2010.)

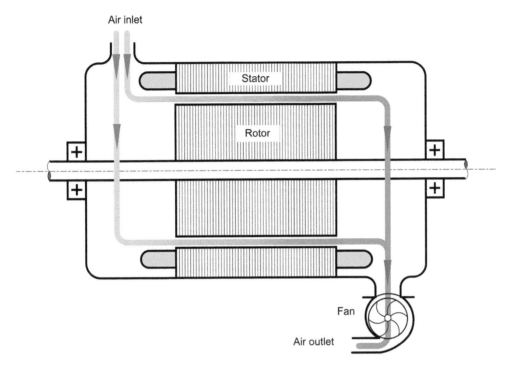

FIGURE 14.11
Induced ventilation with external fan. (From Sawhney, A.K., *A Course in Electrical Machine Design*, Dhanpai Rai & Co., 2010.)

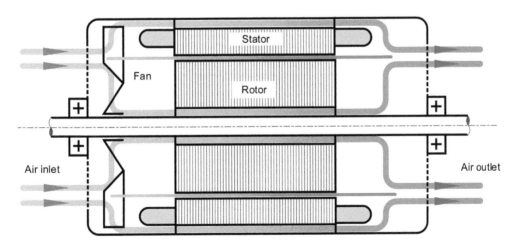

FIGURE 14.12
Forced ventilation with internal fan. (From Sawhney, A.K., *A Course in Electrical Machine Design*, Dhanpai Rai & Co., 2010.)

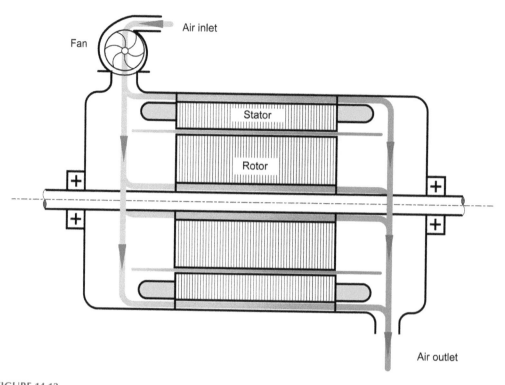

FIGURE 14.13
Forced ventilation with external fan. (From Sawhney, A.K., *A Course in Electrical Machine Design*, Dhanpai Rai & Co., 2010.)

is close to the air outlet in the induced ventilation while the fan is located near the air inlet in the forced ventilation.

The amount of ambient air used in induced ventilation is higher than forced ventilation. This is because forced ventilation takes in low-temperature ambient air that has higher density. Induced ventilation pushes out high-temperature air, which was heated up by the motor [13]. Compared with motors with an internal fan, motors using external fans can be quieter. Generally, the diameter of the internal fan is almost equal to the inner diameter of the housing, which can be a major source for noise during operation. The external fan can be a smaller size, which helps reduce the noise generated by the fan [13].

The fan is one of the sources of noise in electric machines. If the motor is not totally enclosed or air-tight, noise generated inside the motor will also have a path to radiate outward. To reduce noise emitted by the machine, air cooling may be achieved with an air tight motor by mounting an outer layer of air ducts around the circumference of the stator.

14.3.3 Water Jacket Cooling

Water jacket cooling, as shown in Figure 14.14, is a commonly-used indirect cooling technique. It enables effective heat transfer from the active part of stator to the coolant. Different fluids can be used as coolant other than water. The advantages of using a water jacket may include higher power-to-frame size ratio, lower noise level, higher efficiency, and a completely enclosed environment. Also, the removed heat is not directly dissipated into the environment. Properties of liquid coolants are summarized in Table 14.2. It should be

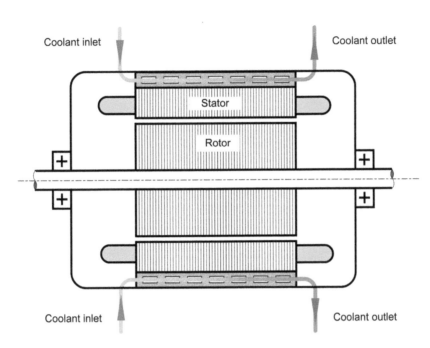

FIGURE 14.14
Water jacket cooling. (From Sawhney, A.K., *A Course in Electrical Machine Design*, Dhanpai Rai & Co., 2010.)

TABLE 14.2

Common Types of Fluids

Type of Fluid	Thermal Conductivity [W/(m·K)]	Specific heat [kJ/kg·K]	Density [kg/m³]	Kinematic Viscosity [m²/sec]
Air (sea level)	0.0264	1.0057	1.1174	1.57×10^{-5}
Brayco Micronic	0.1344	1.897	835	1.35×10^{-5}
Dynalene HF-LO	0.1126	2.019	778	3.2×10^{-6}
EGW 50/50	0.37	3.0	1088	7.81×10^{-6}
EGW 60/40	0.34	3.2	1100	1.36×10^{-5}
Engine Oil	0.147	1.796	899	4.28×10^{-3}
Mobil Jet Oil	0.149	1.926	1014	1.88×10^{-4}
Paratherm LR	0.1532	1.925	778	3.43×10^{-6}
PGW 50/50	0.35	3.5	1050	1.9×10^{-5}
PGW 60/40	0.28	3.25	1057	3.31×10^{-5}
RF 245 FA	0.014	0.9749	1051	1.027×10^{-5}
Silicone KF96	0.15	1.5	1000	8×10^{-5}
Skydrol 500-4	0.1317	1.75	1000	3.5×10^{-5}
Water	0.56	4.217	1000	1.78×10^{-6}

Source: Toliyat, H. and Kliman, B., *Handbook of Electric Motors*, Boca Raton, FL: CRC Press, 2004.

noted that the cooling performance of the water jacket depends on the coolant properties, such as thermal conductivity, specific heat, density and viscosity. Thermal conductivity is a measurement of the coolant's capability to conduct heat. The specific heat is defined as the amount of heat per unit mass required to raise the temperature by one degree Celsius. Kinematic viscosity is a measurement of the coolant's viscosity. Therefore, specific heat, density and kinematic viscosity are important in the calculation or simulation of the inlet and outlet pressure, and of the flow rate of the coolant in the cooling channel.

As shown in Figure 14.15, different configurations of water jacket channels and frame structures can be applied including helical ducts, circumferential channels, meander shape of the ducts, and axial serpentine channels. The configuration and the number of cooling paths determine the cooling efficiency as well as the pressure drop from the inlet to the outlet. Disadvantages of using a water jacket include higher manufacturing cost, requirement for an auxiliary system to provide the coolant, risk of corrosion inside the water circuit, risk of leaks, and additional maintenance.

Figure 14.16 shows an example of a motor housing with water jacket. The water jacket consists of a helical-shaped coolant channel that has one inlet and one outlet. The cooling sub-assembly includes the inner housing, coolant flow path, and outer housing. The helical water jacket has four pitches and has a rectangular cross section. The amount of thermal energy removed from the motor while it is operating depends on the coolant properties, such as the inlet temperature, coolant flow rate, and coolant pressure.

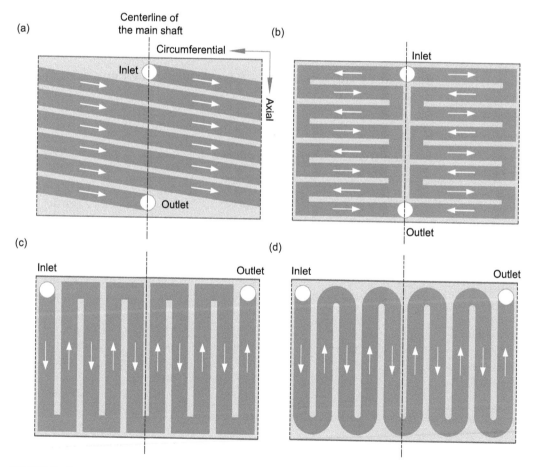

FIGURE 14.15
Various water jacket cooling configurations and paths: (a) helical channels, (b) circumferential channels, (c) meander axial channels, and (d) axial serpentine channels. (From Yang, Y. et al., *IET Electric. Syst. Transp.*, 7, 104–116, 2017.)

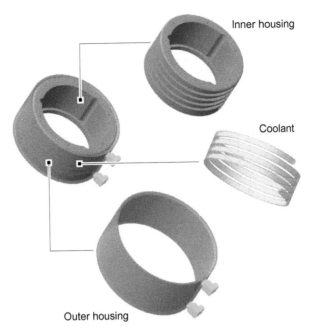

FIGURE 14.16
Cooling sub-assembly of an SRM. (From M. Kasprzak, J. W. Jiang, B. Bilgin, and A. Emadi, "Thermal analysis of a three-phase 24/16 switched reluctance machine used in HEVs," in *Proceedings of the IEEE Energy Conversion Congress and Exposition*, Milwaukee, WI, September 2016, pp. 1–7.)

14.3.4 Direct Liquid Cooling

For direct stator core cooling, the cooling liquid is either sprayed directly on to the stator and rotor, or pumped through the rotor shaft or stator core. Direct stator core cooling may be utilized when water jacket cooling is not sufficient, and a significant temperature gradient might exist between the stator outer surface and the inner stator core. For the same pressure drop in the channels, a machine with direct stator core cooling channels can remove significantly more heat from the stator back iron compared to ones with indirect cooling technologies [15]. For direct cooling, coolants should not be electrically conductive. Otherwise, they might cause short circuit. Thus, water should not be used for direct cooling. The selection of the coolant, such as mineral oil, is based on the performance and cost.

Placing the cooling channels in the stator slots with direct contact to the windings can also achieve a good cooling effect. Furthermore, filling materials with high thermal conductivity but low electromagnetic conductivity, such as resin, epoxy, polymers, and thermoplastic, can be applied to fill the air voids and allow better heat dissipation.

Another critical location that sometimes requires additional cooling efforts in electric machines is the end winding. End winding cooling techniques include spray cooling, liquid jets, and using thermally conductive material between the end windings and the frame. An example of spray cooling can be seen in Figure 14.17. The coolant is sprayed onto the rotor core, end winding, and stator core to cool them down.

14.3.5 Other Methods of Cooling

Evaporative cooling is another effective cooling technique that uses two-phase-flow cooling systems to enhance heat dissipation in electric machines. It has been successfully

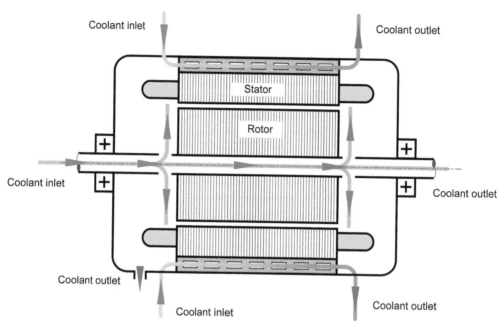

FIGURE 14.17
Spray and water jacket cooling. (From Sawhney, A.K., *A Course in Electrical Machine Design*, Dhanpai Rai & Co., 2010.)

applied in large electric machines. This method has significant cooling capability and high reliability, but has the inconvenience of requiring a high quantity of coolant, which makes it impractical for small size machines [16].

Among other cooling techniques, piezoelectric fans have been implemented to replace liquid cooling systems and to improve fault-tolerance and reliability [17]. Temperature sensitive ferrofluids (TSFF) can be used to cool the end windings. TSFF can avoid extra complexity and costs that exists in a conventional liquid cooling system, such as filters, sensors, and pumping devices. [18]. Totally immersed electric motors have been investigated such that the rotor and stator surfaces are directly flushed by hydraulic oil to increase the heat dissipation [19]. New motor geometry designs using stator flux barriers for cooling have been proposed to enhance the cooling area and, therefore, increase cooling efficiency [20].

14.4 Thermal Modeling

Prediction of temperature distribution in an electric motor using thermal modeling is key to designing thermal management systems. Common methods for thermal analysis include lumped parameter thermal networks (LPTN), FEA, and computational fluid dynamics (CFD). FEA is typically used to model conduction in solids along with empirical relations for the convection boundaries; whereas CFD predicts fluid flow in complicated motor geometries. LPTN offers a quick method for determining the temperature distribution within electric machines and allows the user to rapidly determine the changes resulting from variations in input parameters. LPTN models usually consist of different components, such as thermal resistances and thermal capacitances, being lumped into simplified areas to represent a more complex geometry. The geometric parameters, such as the number of poles, stator teeth, rotor teeth, and magnets, if applicable, all affect the thermal circuit and resistances between connected nodes in LPTN.

An LPTN is similar to an electrical circuit, where voltages represent the temperatures, currents represent the heat flow, and electrical resistances and capacitances represent the thermal resistance and capacitance, respectively [21]. Heat transfer coefficients can be used to calculate the thermal resistances due to convection. Empirical formulas to calculate the heat transfer coefficients have been developed. These formulas are applicable to different areas of the motor (e.g. across the air gap), and heat transfer through the internal end space air [22].

Unlike the other motor types, the rotor core of an SRM only consists of electrical steel and does not contain any magnet or coils. This allows for a simplified LPTN. Typically, the components in an SRM can be lumped in different partial models, such as a 2D model, where the heat flux flows radially through the cross section of the machine or a 3D model that takes the heat distribution in the axial direction into account. The 3Dnetworks contain many additional nodes to encompass components in the axial direction.

14.4.1 Types of Heat Transfer

As discussed earlier, there are three major types of heat transfer: conduction, convection, and radiation. These three heat transfer types all exist in an SRM, as shown in Figure 14.18. The conduction of heat occurs when there is direct physical contact between two solid objects with different temperatures. Among solids, liquids, and gases, solids are generally better conductors. Figure 14.18 shows that conduction dominates in the heat transfer between housing and stator back iron, shaft and rotor back iron, and so forth. The transfer of energy between a solid object and a fluid environment is called convection. For convection, there exists a continuous flow and circulation of fluid. Examples of convection can also be seen in Figure 14.18 in between ambient and the motor housing, the air gap and stator and rotor teeth. It can also be observed that radiation almost goes hand in hand with convection. That is because radiation is a form of electromagnetic wave that does not need any media for heat conduction. In reality, the energy flow within a motor is much more complicated than what is shown in Figure 14.18, but the energy flow paths shown in Figure 14.18 are the most significant ones.

14.4.2 Layers in SRM Thermal Model

Figure 14.19 shows the layers of an SRM when building the thermal model of the motor. The inner most layer is the shaft, which does not generate any heat. The shaft can be modeled as a cylinder. The layer next to the shaft is the rotor back iron, which is a heat source. On the rotor, the heat comes mostly from rotor iron loss. The third layer is the rotor teeth. The fourth layer is the air gap, in which radiation and convection are the major methods for heat transfer. The fifth layer is a combination of the winding and stator teeth. The sixth layer is the stator back iron. A large amount of the heat also comes from the copper loss and stator iron loss. The seventh layer is the frame, which may have a water jacket. The majority of the heat transfer between the sixth and seventh layers is via conduction.

The stator teeth can be modelled in a manner similar to the rotor teeth. The modeling of the winding can be complex, because it is composed of a number of coils and slot liners. Additionally, the varnish used for encapsulating the winding also plays an important

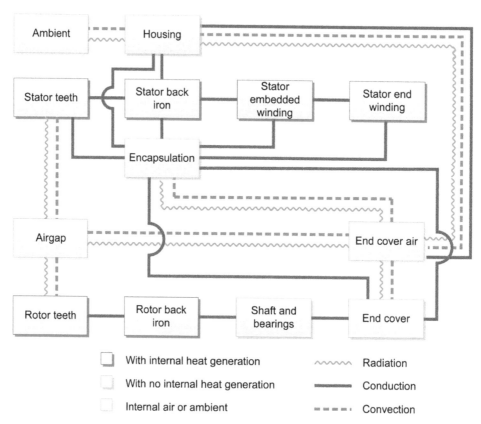

FIGURE 14.18
Major paths of heat flow in an SRM.

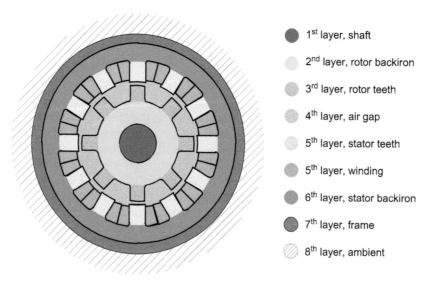

FIGURE 14.19
Layers of SRM for thermal modelling in radial direction.

role in the heat transfer. Modeling the stator back iron can be relatively easy. Unlike the rotor back iron, the stator back iron has physical contacts with the coils (via insulating slot liners) and the housing, provided that the motor has a housing. There are many motor designs with the stator core exposed to the ambient. If the housing has coolant channels or a water jacket, the thermal modeling of the housing can be complicated.

The last layer is the ambient, which is the external environment the motor is operating in. The heat transfer between the motor and the ambient is convection and radiation. The type of cooling method used affects the thermal network greatly. For each of the cooling methods discussed earlier in this chapter, the thermal network is slightly different. For instance, if induced air cooling is used, convective heat transfer will occur between the internal and external air.

Heat transfer for an electric motor also occurs in the axial direction. Figure 14.20 shows the cross section of an SRM in the axial direction. The rotor core, end winding, and stator core have convective heat transfer with the air in the end cover. Since an SRM rotor works like a fan, the air circulation and heat transfer between the air gap and the end cover air must be accounted for in thermal modeling. The heat traveling from the shaft to the housing via bearings should also be modelled. The axial thermal distribution for the housing can be complex if a water jacket is used.

14.4.3 SRM Thermal Network in Motor-CAD

Motor-CAD is a commercially available thermal analysis software that uses a 3D lumped parameter circuit model. It is capable of modelling steady-state, simple transient, and duty-cycle transient thermal characteristics. Interior rotor SRMs are a standard motor type available with a pre-generated thermal network in Motor-CAD. The user is required to input geometric parameters in order to approximate the temperature variation in an SRM. The axial and radial cross section views of an SRM in Motor-CAD can be seen in Figure 14.21. Figure 14.22 shows the thermal network layout for an SRM in Motor-CAD. In the network,

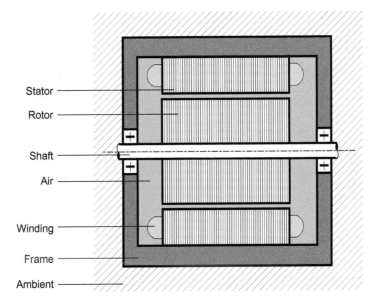

Stator

Rotor

Shaft

Air

Winding

Frame

Ambient

FIGURE 14.20
An example of SRM cross section in axial direction.

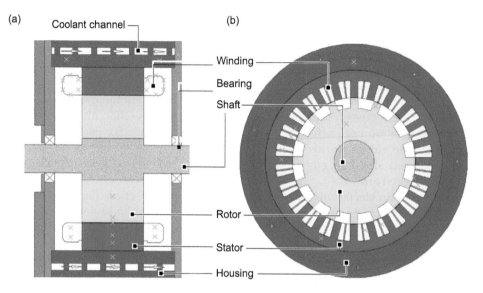

FIGURE 14.21
An SRM Geometry in Motor-CAD: (a) axial and (b) radial. (From Kasprzak, M. et al., Thermal analysis of a three-phase 24/16 switched reluctance machine used in HEVs, in *Proceedings of the IEEE Energy Conversion Congress and Exposition*, Milwaukee, WI, September 2016, pp. 1–7.)

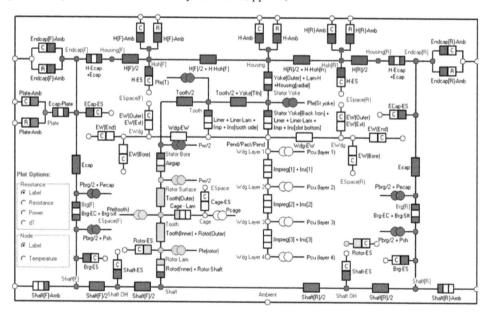

FIGURE 14.22
Example LPTN for an electric machine, in the schematic view in Motor-CAD. (From Boglietti, A. et al., *IEEE Trans. Ind. Electr.*, 56(3), 871–882, 2009.)

each node, which represents a specific spatial position in a motor's geometry, is connected to other nodes using a calculated thermal resistance value. Motor dimensions and material properties are used to calculate the thermal resistance.

The embedded LPTN model can predict the thermal variation for each component. For instance, Figure 14.23 shows a simulation result for an SRM operating at 3000 rpm using

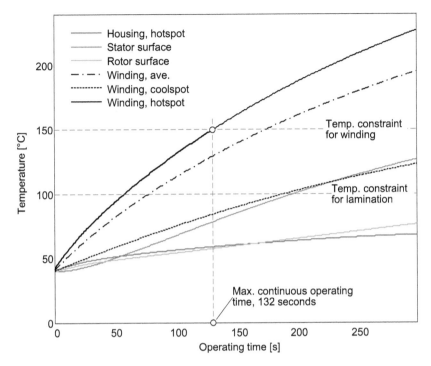

FIGURE 14.23

Temperature variation for major components of the SRM, at 3000 rpm, peak torque, with coolant inlet temperature set at 65°C, and ambient temperature set at 40°C. (From Kasprzak, M. et al., Thermal analysis of a three-phase 24/16 switched reluctance machine used in HEVs, in *Proceedings of the IEEE Energy Conversion Congress and Exposition*, Milwaukee, WI, September 2016, pp. 1–7.)

Motor-CAD. For this motor, the hotspot on the winding (maximum, 150°C) is the limiting factor. The maximum continuous operating time for the motor at the peak torque and 3000 rpm is 132 seconds. This happens because the copper loss for the peak torque operation at this speed is much higher than either the stator or rotor iron losses. Hence, the winding temperature rises faster than the core temperature. The specific temperature at each node within the axial and radial cross sections of the SRM at 132 seconds (the maximum continuous operating time for this operating point) can be seen in Figure 14.24.

A downside to the LPTN method of thermal analysis is that it involves substantial effort to create an accurate model [21]. The LPTN can calculate the heat transfer and temperature distribution much quicker than CFD or FEA methods, but it can be time consuming to create the thermal network [23]. Further efforts are required through validation from either simulation or experimental measurement. It can be challenging to modify the LPTN to be applicable to various types of machines. In addition, the heat transfer between nodes is calculated based on the thermal properties of the materials and fluids. Since some material properties are temperature dependent, calibration might be required to modify the thermal network to apply it to different operating conditions.

14.4.4 Numerical Modeling Methods

Numerical modeling methods are also proven tools for predicting the temperature within electric machines. The two most common examples of numerical modeling methods are finite element analysis and computational fluid dynamics. While these methods can be

(a)   (b)

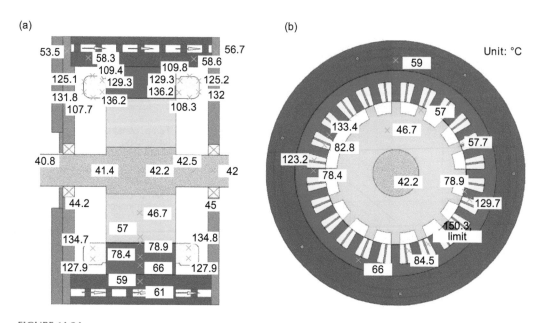

FIGURE 14.24

Temperature distribution for continuous operation at 132 seconds, 3000 rpm, and peak torque with coolant inlet temperature set at 65°C, ambient temperature set at 40°C: (a) axial cross section, and (b) radial cross section. (From Kasprzak, M. et al., Thermal analysis of a three-phase 24/16 switched reluctance machine used in HEVs, in *Proceedings of the IEEE Energy Conversion Congress and Exposition*, Milwaukee, WI, September 2016, pp. 1–7.)

more accurate in predicting temperatures within electric machines, they are generally more time consuming due to higher complexity and necessity for more detailed input. Each thermal analysis type has its own advantages and weaknesses.

Finite element analysis involves dividing the components within a model into many tiny meshed nodes and elements to analyze the changes across those dimensions. Many commercially available software packages are capable of performing 2D or 3D thermal FEA analyses through steady-state and transient simulations. Results from thermal FEA are similar to that of LPTN models since common inputs are applied. However, the processing time is a downside of FEA. Thermal analysis in FEA has an advantage when the geometry becomes too complex to model using a LPTN [21].

2D simulations can be used to represent the temperatures within a machine when a small temperature difference is assumed across the axial components. FEA models are usually simplified to be partial fractional models, representative of the full geometry by applying periodic patterns in order to speed up the processing time. Heat transfer coefficient values and thermal resistances are required as inputs to calculate the convective heat transfer between surfaces and air or fluids. Losses and heat transfer boundary values are also inputs and are determined through analytical calculations, numerical simulations, or experimental testing [10]. Material properties of components and the geometrical dimensions determine the thermal resistance due to conduction within the machine. The more detailed the mesh that is applied to the model, the more accurate the temperature prediction will be. However, the computation time will also increase with complexity. FEA thermal analysis can also be used to validate an LPTN model under development.

Computational fluid dynamics is usually a more accurate temperature prediction method when compared with the LPTN and FEA methods. It can be used to determine heat transfer boundaries for inputs into LPTN and FEA methods, and has the ability to perform simulations to predict the fluid flow characteristics and to optimize cooling methodologies (e.g. water jacket, air cooled, etc.) [24].

Conventional CFD simulations only solve for the fluid flow and do not cover the solid domains. The loss values evaluated from electromagnetic calculations are incorporated in the CFD simulations as surface heat flux or constant surface temperature boundary conditions. Several CFD thermal analyses with different levels of complexity can be implemented to simulate the liquid flow around different components of the machine such as in the air gap, end windings, or cooling channels. Steady state CFD simulations can be used to characterize the heat transfer performance of cooling channels with different shapes.

The simulation results can be improved by applying the conjugate heat transfer method, where the solid domains have to be modelled in addition to the fluid domains. Then, the heat losses can be defined in the solid domains as volumetric sources obtained from electromagnetic simulations [25]. Due to advancements in computational power in recent years, using CFD for thermal analysis of electric machines has become a trend.

CFD thermal analysis has an advantage when focusing on simultaneous air flows and mass transfer with a mix of laminar and turbulent states within a machine, and when designing channels and patterns for coolant flow [24]. Similar to how material properties are inputs into geometry components in FEA, fluid properties and specifications are inputs to CFD, including volume flow rate, inlet temperatures, and pressure. A drawback of CFD thermal analysis is that simulation takes much longer to process depending on the complexity of the geometry, mesh density, and the computational processing power available. These factors limit the users because changes in input parameters cannot be observed quickly during the design of the thermal management system of the motor.

14.5 Thermal Measurement

Both LPTN analysis and numerical modeling methods provide profound insight and guidance in temperature prediction and thermal management of electric machines. However, certain degrees of error or inaccuracy always exist due to model limitations, parameter deviation, various assumptions, simplified boundary conditions, and so on. Thus, it is often necessary to calibrate and verify the thermal analysis through experimental tests. In principle, two types of experiments are conducted:

1. Experiments that are designed to determine machine thermal parameters including thermal capacitance, thermal resistance, thermal conductivity, convection heat transfer coefficients, radiation heat transfer coefficients, and more. These can either be performed on a complete motor to draw conclusions on similar motors or on part of the motors such as segmented stator structures and end windings.

2. Experiments that are conducted to verify the analysis or simulation results, typically in the form of prototype motors.

In terms of temperature monitoring, direct thermal sensor measurements are widely used in industrial applications and laboratory experiments. Thermal sensors are applied at different locations, such as on the windings, end caps, housing, and cooling jacket. Commonly used thermal sensors include thermocouples, thermistors, resistance thermometers, and so on. Thermocouples are based on the principle that the junction of two dissimilar metals generates a voltage that increases with temperature. They have the advantage of wide measurement range, fast response time, and simple configuration. However, they only have medium accuracy and sensitivity, and are prone to electromagnetic interference (EMI) due to low signal strength. Thermistors are made of metal oxides whose resistance increases or decreases with increasing temperature. They also have either a positive temperature coefficient (PTC) or negative temperature coefficient (NTC). Compared to thermocouples, thermistors may provide a more accurate, sensitive, and robust signal output, and they have the advantages of simple construction, small size, low thermal mass, low cost, and greater immunity to electromagnetic noise. However, due to their narrow linear range, thermistors have limited useful temperature span [26]. Resistance thermometers, or resistance temperature detectors (RTDs), are similar to thermistors since they utilize resistance to measure temperature. The resistances of RTDs increases with temperature; hence, they are PTC sensors. RTDs are the most accurate, sensitive, and stable temperature sensors for industrial applications; however, they are higher priced and require external power sources. Table 14.3 compares three kinds of temperature sensors commonly used in industrial application.

Temperature sensors can be attached directly to the stationary housing, stator, or winding of an electric machine. Figure 14.25 shows a few locations for thermal measurements

TABLE 14.3

Comparison of Temperature Sensors

	Thermistors	Thermocouples	RTD
Material	Semiconductor	Junction between two different metals	Metal, e.g. platinum
Accuracy	Medium	Low	High
Stability	Medium	Low	Good
Linearity	Medium	Low	Good
Temperature range	≤200°C	Several thousand °C	−200°C to 800°C
Price	Cheapest	Moderate	Expensive

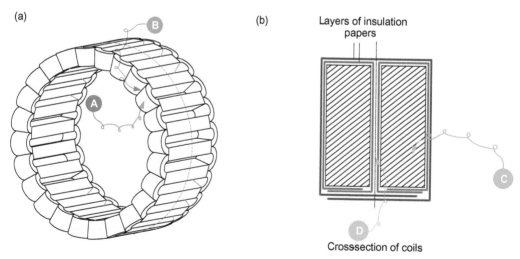

FIGURE 14.25

Locations for thermal readings: (a) end turns (A: on top of end turns; B: in between end turns) and (b) embedded (C: center of the coil; D: in between coils).

FIGURE 14.26
A stator-winding subassembly with thermal sensors for the winding (small green cables are for thermal measurements).

on motor windings. For instance, if a thermistor is used, it can be glued to the top of the end turn (A) or in between end turns (B), as shown in Figure 14.25a. Figure 14.25b shows that the thermistors can also be inserted in the coil (C) or in between coils (embedded in between the insulation papers as in D). This location can give accurate internal thermal measurements as the motor operates. Figure 14.26 shows a stator-winding subassembly with thermistors installed for the thermal readings from the windings.

Figure 14.27 shows the locations for thermal measurements on the stator core. Temperature sensors can be glued to either the stator teeth or the stator back iron, as shown in Figure 14.28. Figure 14.29 shows the locations for thermal measurements on the housing, and Figure 14.30 shows the thermal measurement points for the coolant inlet and outlet.

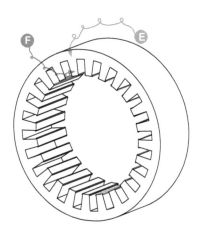

FIGURE 14.27
Locations for thermal reading on the stator (E: stator back iron; F: stator tooth).

(a)   (b)

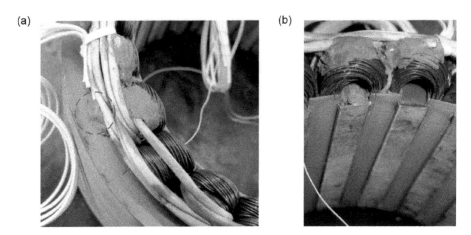

FIGURE 14.28
Thermistors glued to the stator surface: (a) back iron and (b) stator teeth.

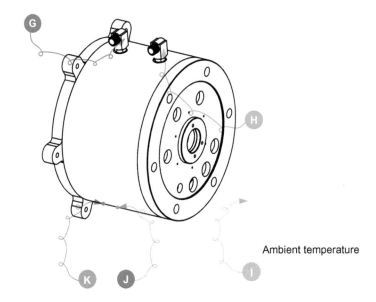

Ambient temperature

FIGURE 14.29
Locations for thermal readings on the housing (G: coolant inlet; H: coolant outlet; I: ambient temperature; J: frame surface, directly on top of the cooling channel; K: frame surface, directly on top of the cooling channel wall).

The thermal measurement on a spinning rotor is more difficult. Figure 14.31 shows four locations for thermistors employed on a rotor. Three sensors are located on the surface of the rotor while a sensor is embedded in the rotor for the temperature measurement on the inner core. A slip ring, as shown in Figure 14.32, is used to transfer the signals out from the spinning rotor. The structure of the slip ring consists of two parts: the stationary part and the spinning part. The rotating wires of temperature sensors can be transmitted through the holes of the main shaft. After that, the rotating wires are connected to the spinning part of slip ring. The stationary part of the slip ring is used to connect the rotating wires and the stationary data acquisition unit. Figure 14.33 shows the slip ring installed on the motor.

In cases where direct sensor installation on the rotor is difficult or for permanent magnet machines, where it is difficult to attach the sensors to the magnets, indirect

FIGURE 14.30
Thermal measurements for the coolant outlets.

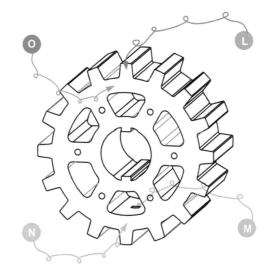

FIGURE 14.31
Locations for thermal measurements on the rotor (L: rotor tooth; M: rotor back iron, directly beneath rotor tooth, embedded in the rotor core; N: rotor back iron, directly beneath rotor tooth, surface of the rotor core; O: rotor back iron, in between rotor teeth, surface of the rotor core).

FIGURE 14.32
Slip ring.

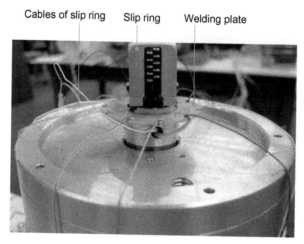

Cables of slip ring   Slip ring   Welding plate

FIGURE 14.33
Installation of a slip ring.

temperature measurement can be performed with infrared thermography and wireless communication [27]. Temperature estimation techniques based on flux observers, signal injection, and lumped-parameter thermal networks have also been developed to ensure design redundancy in case of sensor failures [28,29].

Questions and Problems

1. Table Q.14.1 summarizes some key operating parameters of an electric motor. Please answer the following questions.
   a. What cooling methods can be used for the motor?
   b. What will happen if the RMS current increases?
2. Air cooling and liquid cooling are commonly used in electric machines Please compare air and liquid cooling methods.
3. Thermocouples and thermistors can be used to measure the temperature of electric motors. Table Q.14.2 summarizes some key specs for a thermocouple and a thermistor. Please choose one to measure the temperature of the winding in the following locations: (a) embedded in between coils, (b) end turns, (c) inside of a coil and explain why.

TABLE Q.14.1

Key Operating Parameters of a Motor

Continuous Power Rating	10 kW
RMS current	20 A
Max. current density	6 A/mm²

TABLE Q.14.2

Comparisons of a Thermal Couple and a Thermistor

	Thermocouple	Thermistor
Temperature range	0°C–482°C	–55°C to 200°C
Accuracy	1.75%	1%
Probe diameter [mm]	6	2.41

4. Given a traction motor used in an HEV powertrain, the motor's winding end turns experience some overheating. Spray cooling should be applied to solve this problem and cool down the temperature of the end turn. Please answer the following questions.

   a. Choose a liquid either water or automatic transmission fluid (ATF) and justify your choice.

   b. If water must be used as the cooling agent, what is required for the frame?

   c. Why is water cooling not commonly used for vehicle traction motors?

References

1. H. Toliyat and B. Kliman, *Handbook of Electric Motors*. Boca Raton, FL: CRC Press, 2004.
2. M. Krishnamurthy, C. S. Edrington, A. Emadi, P. Asadi, M. Ehsani, and B. Fahimi, "Making the case for applications of switched reluctance motor technology in automotive products," *IEEE Trans. Power Electron.*, 21(3), 659–675, 2006.
3. S. Nalakath, M. Preindl, B. Bilgin, B. Cheng, and A. Emadi, "Modeling and analysis of AC resistance of a permanent magnet machine for online estimation purposes," in *Proceedings of the IEEE Energy Conversion Congress and Exposition*, Montreal, QC, September 2015, pp. 314–319.
4. S. Sudhoff, *AC conductor losses in Power Magnetic Devices: A Multi-Objective Design Approach*. Hoboken, NJ: John Wiley & Sons, 2014.
5. G. D. Demetriades and H. Z. De La Parra, "A real-time thermal model of a permanent-magnet synchronous motor," *IEEE Trans. Power Electron.*, 25(2), 463–474, 2010.
6. W. Jiang, *Three-Phase 24/16 Switched Reluctance Machine for Hybrid Electric Powertrains: Design and Optimization*, PhD Dissertation, McMaster University, March 2016.
7. A. Ridge, R. McMahon, and H. P. Kelly, "Detailed thermal modelling of a tubular linear machine for marine renewable generation," in *Proceedings of the IEEE International Conference on Industrial Technology*, Cape Town, February 2013, pp. 1886–1891.
8. Y. Yang, B. Bilgin, M. Kasprzak, S. Nalakath, H. Sadek, M. Preindl, J. Cotton, N. Schofield, and A. Emadi, "Thermal management of electric machines," *IET Electric. Syst. Transp.*, 7(2), 104–116, 2017.
9. J. W. Jiang, B. Bilgin, and A. Emadi, "Three-phase 24/16 switched reluctance machine for hybrid electric powertrains: design and optimization," *IEEE Trans. Transp. Electrif.*, vol. 3, no. 1, 76–85, 2017.
10. P. Rasilo, A. Belahcen, and A. Arkkio, "Importance of iron-loss modeling in modelling in simulation of wound-field synchronous machines," *IEEE Trans. Magn.*, 48(9), 2495–2504, 2012.

11. R. J. Wang and G. C. Heyns, "Thermal analysis of a water-cooled interior permanent magnet traction machine," in *Proceedings of the IEEE International Conference on Industrial Technology*, Cape Town, February 2013, pp. 416–421.

12. D. Staton, "Thermal Analysis of Traction Motors," presented in *IEEE Transportation Electrification Conference and Expo*, Dearborn, MI, 2014.

13. A. K. Sawhney, *A Course in Electrical Machine Design*, Dhanpai Rai & Co, New Delhi, India, 2010.

14. M. Kasprzak, J. W. Jiang, B. Bilgin, and A. Emadi, "Thermal analysis of a three-phase 24/16 switched reluctance machine used in HEVs," in *Proceedings of the IEEE Energy Conversion Congress and Exposition*, Milwaukee, WI, September 2016, pp. 1–7.

15. Z. Huang, S. Nategh, V. Lassila, M. Alaküla, and J. Yuan, "Direct oil cooling of traction motors in hybrid drives," in *Proceedings of the IEEE International Electric Vehicle Conference*, Greenville, SC, May 2012, pp. 1–8.

16. M. R. Guechi, P. Desevaux, P. Baucour, C. Espanet, R. Brunel, and M. Poirot, "On the improvement of the thermal behavior of electric motors," in *Proceedings of the IEEE Energy Conversion Congress and Exposition*, Denver, CO, September 2013, pp. 1512–1517.

17. G. M. Gilson, S. J. Pickering, D. B. Hann, and C. Gerada, "Piezoelectric fan cooling: a novel high reliability electric machine thermal management solution," *IEEE Trans. Ind. Electron.*, 60(11), 4841–4851, 2013.

18. G. Karimi-Moghaddam, R. D. Gould, S. Bhattacharya, and D. D. Tremelling, "Thermomagnetic liquid cooling: A novel electric machine thermal management solution," in *Proceedings of the IEEE Energy Conversion Congress and Exposition*, Pittsburgh, PA, September 2014, pp. 1482–1489.

19. P. Ponomarev, M. Polikarpova, and J. Pyrhönen, "Thermal modeling of directly-oil-cooled permanent magnet synchronous machine," in *Proceedings of the International Conference on Electrical Machines, Marseille*, September 2012, pp. 1882–1887.

20. A. Nollau and D. Gerling, "Novel cooling methods using flux-barriers," in *Proceeding of the International Conference on Electrical Machines*, Berlin, November 2014, pp. 1328–1333.

21. A. Boglietti, A. Cavagnino, D. Staton, M. Shanel, M. Mueller, and C. Mejuto, "Evolution and modern approaches for thermal analysis of electrical machines," *IEEE Trans. Ind. Electr.*, 56(3), 871–882, 2009.

22. J. Lindström, "Development of an Experimental Permanent-Magnet Motor Drive", Chalmers University of Technology, Göteborg, Sweden, Technical Report, April 1999.

23. D. Kim, J. Jung, S. Kwon, and J. Hong, "Thermal analysis using equivalent thermal network in IPMSM," in *Proceedings of the International Conference on Electrical Machines and Systems*, Hankou Wuhan, October 2008, pp. 3162–3165.

24. K. N. Srinivas and R. Arumugam, "Analysis and characterization of switched reluctance motors: Part II - Flow, thermal and vibration analyses," *IEEE Trans. Magn.*, 41(4), 1321–1332, 2005.

25. M. Schrittwieser, A. Marn, E. Farnleitner, and G. Kastner, "Numerical analysis of heat transfer and flow of stator duct models," *IEEE Trans. Ind. Appl.*, 50(1), 226–233, 2014.

26. L. Michalski, *Temperature Measurement*. New York: John Wiley & Sons, 2002.

27. M. Ganchev, B. Kubicek, and H. Kappeler, "Rotor temperature monitoring system," in *Proceedings of the International Conference on Electrical Machines*, Rome, Italy, October 2010, pp. 1–5.

28. A. Specht, O. Wallscheid, and J. Böcker, "Determination of rotor temperature for an interior permanent magnet synchronous machine using a precise flux observer," in *Proceedings of the International Power Electronics Conference*, Hiroshima, Japan, May 2014, pp. 1501–1507.

29. O. Wallscheid and J. Böcker, "Global identification of a low-order lumped-parameter thermal network for permanent magnet synchronous motors," *IEEE Trans. Energy Conv.*, 31(1), 354–365, 2016.

15

Axial Flux Switched Reluctance Machines

Jianing (Joanna) Lin

CONTENTS

15.1 Introduction

Switched reluctance machines (SRM) can fall into three classifications based on the direction of flux through the air gap: radial, transversal, and axial flux. Figure 15.1 presents a coordinate system to help understand the basic structural differences between radial and axial flux SRMs. The transversal motor is a configuration between the other two. Both radial and axial flux SRMs lie centralized around the z-axis, which is also called the axial direction. θ indicates the direction of rotation. It can be either positive or negative, depending on the operation. The r represents the radial direction. In Figure 15.1a, it can be observed that for the radial flux SRM (RFSRM), when stator and rotor poles are aligned, the flux generated by the coils travels mainly in the radial axis through the air gap

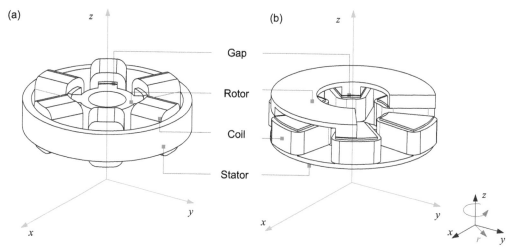

FIGURE 15.1
Comparison of RFSRM and AFSRM in the coordinate system: (a) RFSRM and (b) AFSRM.

between the stator and rotor poles. As for the axial flux SRM (AFSRM) in Figure 15.1b, the flux travels mainly along axial direction between the stator and rotor poles.

The first axial flux machine was developed in 1821 by British physicist Michael Faraday. He invented a primitive disc motor that was the early form of an electric machine. However, AFMs were quickly over shadowed by radial flux machines (RFM) mainly due to the large unbalanced attraction forces between the rotor and stator, as well as the difficulties in manufacturing. The problem of large unbalanced attraction has since been solved by utilizing a sandwich structure, such as the double-rotor external-rotor structure, or double-stator internal-rotor structure. With the improvement in manufacturing techniques and the appearance of new materials, axial flux machines, especially axial flux SRMs, have begun to attract industry and academia again.

Compared with RFSRM, AFSRM has the advantage of higher torque and power density [1,2] and an adjustable air gap. They can also be designed in special shapes to maintain larger diameter to length ratio, which is especially suitable for certain applications. For example, electric motors for in-wheel propulsion in electrified vehicles usually require a short planar motor. AFSRM's potential of high torque density is especially important when a powertrain has limited packaging space. In addition, some special structures of axial flux motor could benefit the machine with short iron flux paths. Minimizing the iron flux path length helps reducing the core losses. Some AFSRM configurations also provide flux path isolation, which significantly reduces mutual coupling between the coils and associated losses. This flux path isolation structure also allows a large degree of freedom in the choice of control strategy [3].

Because of its simple structure, high power density, and unique disc shape, AFSRMs are a strong candidate for electric vehicles, and various other types of applications, such as flywheels, wind generators, pumps, fans, valves, centrifuges, machine tools, robots, and industrial equipment [4]. However, similar to all types of SRMs, AFSRMs can suffer from high torque ripple, and vibration and acoustic noise. These detriments could be more obvious for AFSRMs when compared against conventional RFSRMs.

15.2 Topologies and Structures of AFSRMs

The classification of AFSRMs can follow the same rules as other types of axial flux machines, like an axial flux permanent magnet brushless machine. According to the number of air gaps in the axial direction, the classification can be divided to single-sided, double-sided, and multistage AFSRMs [4]. As shown in Figure 15.2, a double-sided sandwich structure is now the mainstream configuration for AFSRMs, because it can provide a simple structure and balanced attraction forces in the axial direction. The double-sided structure could be configured either with an external double rotor or internal single rotor. The different structures could lead to totally different electromagnetic, thermal, and mechanical performance.

Single-sided AFSRMs are more conventional. Motors of this configuration benefit from high diameter to length ratio. However, as it is asymmetric in the axial direction, this leads to high unbalanced attraction forces on the stator and the rotor. This configuration requires more mechanical rigidity, notably on the bearings to handle the attraction force and keep the air gap constant. Single-sided axial flux machines also have lower torque production capability. Figure 15.3a shows a typical structure of a single-sided AFSRM. In this structure, the rotor is locked on the shaft; therefore, the rotor and the shaft rotate together while the cage and the stator stay stationary.

Double-sided AFSRMs are more popular due to their relatively simple structure, as well as balanced axial forces. As stated earlier, there are two main configurations: double rotor (external rotor) or double stator (internal rotor). Figure 15.3b is a typical double stator (internal rotor) AFSRM, in which the shaft is locked together with the rotor, and the outer mount is stationary with the stator. Figure 15.3c shows the AFSRM with double rotor configuration where the stator is located between the two rotors. Figure 15.3d presents a multi-sided AFSRM.

In addition, it is also possible to classify the AFSRMs based on its winding configuration (conventional or toroidal windings), and based on different stator and rotor structures. In Figure 15.4, some other popular types of AFSRM are shown, which are well presented in the literature. The main components of these configurations and the flux paths can also be seen.

Figure 15.4a presents an AFSRM with C-shaped stator cores, which was first studied in [5]. A similar C-shaped stator core SRM is discussed in [8], which is actually a RFSRM. Unlike conventional SRMs, the windings could be wound on each stator segment in this configuration.

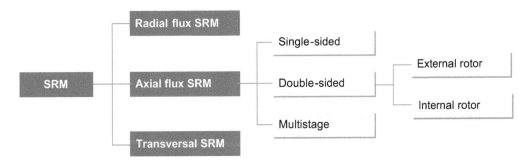

FIGURE 15.2
Classification of axial flux SRMs.

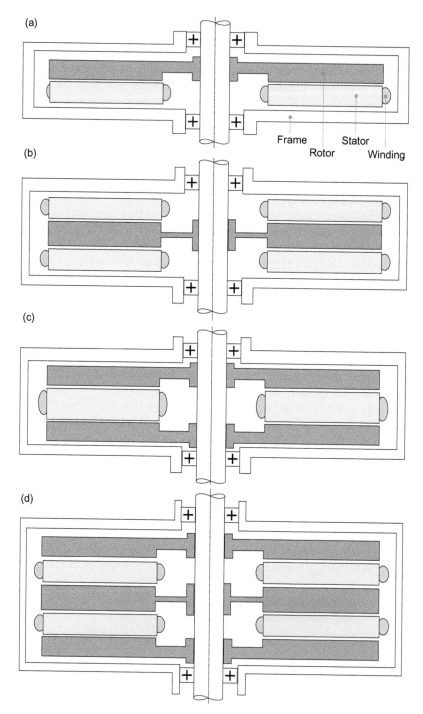

(a)

Frame   Stator
Rotor   Winding

(b)

(c)

(d)

FIGURE 15.3

Basic topologies of AFSRM machines: (a) single-sided machine, (b) double-stator double-sided machine, (c) double-rotor double-sided machine, and (d) multi-sided machine.

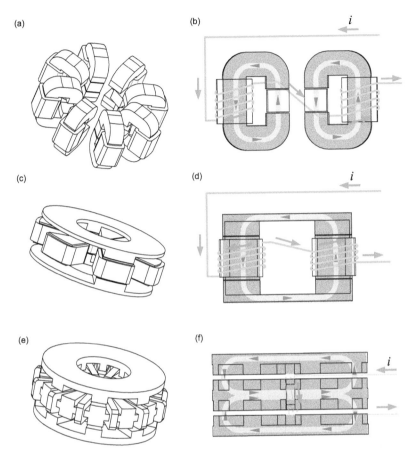

FIGURE 15.4
Examples of typical structures of AFSRM: (a) AFSRM with C-shaped stator core. (From Labak, A. and Kar, N.C., A novel five-phase pancake shaped switched reluctance motor for hybrid electric vehicles, in *Proceedings of the Vehicle Power and Propulsion Conference*, Dearborn, MI, September 2009, pp. 494–499.), (b) flux path of (a), (c) conventional wound AFSRM. (From Shibamoto, T. et al., A design of axial-gap switched reluctance motor for in-wheel direct drive EV, in *Proceedings of IEEE International Conference on Electrical Machines*, Marseille, France, September 2012, pp. 1158–1163.), (d) flux path of (c), (e) toroidal wound AFSRM. (From Madhavan, R. and Fernandes, B.G., A novel axial flux segmented SRM for electric vehicle application, in *Proceedings of IEEE International Conference on Electrical Machines*, Rome, Italy, September 2010, pp. 1–6.), and (f) flux path of (e).

This significantly reduces the winding complexity. Figure 15.4c and e have similar rotor structures, but different stator topologies. In Figure 15.4c, the winding is wound around the stator poles, but in Figure 15.4e toroidal winding is applied. These examples show the design flexibility that is possible with AFSRMs. With the developments in material and manufacturing technologies, more AFSRM types and geometries will come out in the near future.

15.3 Operating Principles and Analytical Modeling of AFSRMs

15.3.1 Principles of Operation

As a type of reluctance motor, AFSRMs still follow the "minimum reluctance principle." When the stator coils are excited, the flux generated has the tendency to move the rotor from a position with a larger magnetic reluctance to a position with the minimum.

For purpose of illustration and explanation, the following study concentrates on an AFSRM with the configuration as shown in Figure 15.5 with one stator and two rotors. This is a three-phase 6-stator/4-rotor pole machine. For this sandwich structure, the stator is in the middle of the two rotors. The rotor pole is a simple, single solid piece. Each stator pole has two parts. The first part is surrounded by the coils, and the second part acts like a back iron and provides a path for the magnetic flux.

In RFSRMs, flux distribution in the axial direction can be typically ignored and, hence, two-dimensional model is usually applied with good accuracy. However, due to the axially asymmetric flux distribution in AFSRMs, a three-dimensional model is usually required for an accurate estimation of performance.

For the AFSRM shown in Figure 15.5, the mutual inductance between phases is negligible and can be ignored. There is also enough distance between adjacent rotor poles, and as such they can be considered as isolated from one another. Put simply, there is only one rotor pair within the effective attraction region of an active phase. Therefore, a single pole model can be applied for this machine for 3D finite element analysis (FEA). By reducing the complexity of the model and running FEA on a single pole, computational time and memory utilization can be improved during AFSRM analysis.

The flux path at the aligned position is shown in Figure 15.6. Figure 15.6a shows the flux path from FEA. Figure 15.6b represents the simplified sectional cutting view in the *yz* plane with the coordinate system shown in Figure 15.6a. Figure 15.6c shows the simplified view from the *y*-axis.

In Figure 15.6b, it can be observed that the flux generated by stator coil travels axially through the right part of stator core, then axially across the right-upper air gap, and radially through the rotor core. After that, it goes across the left-upper air gap back to the other part of stator core. Following this, it goes across the left-lower air gap,

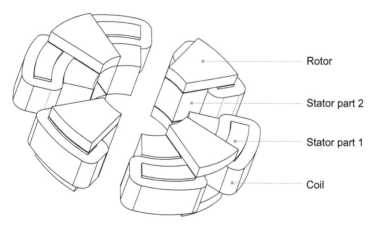

FIGURE 15.5
A typical 3-phase 6/4 axial flux SRM.

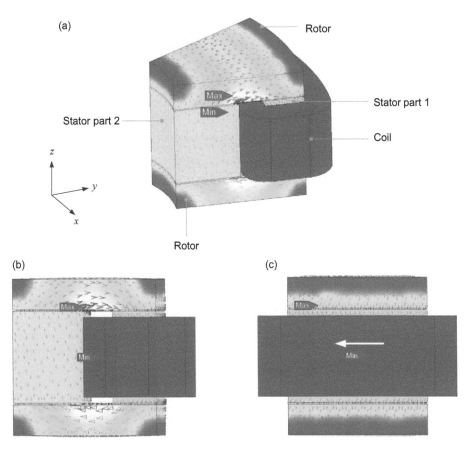

FIGURE 15.6
Different views of one aligned pole pair with the coordinate system shown in (a): (a) single stator and rotor pole pair, (b) simplified sectional cutting view in yz plane, and (c) cutting view in y direction.

then radially through the lower rotor, and finally crosses the fourth air gap before returning back to the right part of stator core. Figure 15.6c shows more clearly the relative position between stator and rotor at the aligned position. Flux goes axially along the z-axis, while the white arrow shows the direction of current in the coil. For this AFSRM, flux is in 3D, but Figure 15.6c only shows the y-axis view of the flux loop. In Figure 15.7, a single pole pair at unaligned position is presented. In this figure, the flux generated by coil has to go through a large air gap before penetrating into the rotor. After the flux enters the rotor, it travels radially inside the bottom rotor in the direction into the paper. Then after crossing the other air gap, flux goes to the other part of stator, which is not visible in Figure 15.7. After the third large air gap, flux goes to the upper rotor and travels in radial direction again in the direction out from the paper. Then the flux goes through the last air gap and enters back to the part of the stator, which is surrounded by the coil as shown in the Figure 15.7. As the actual flux path is in 3D, the described flux path is not in a loop in this 2D drawing. The large air gap results in high magnetic reluctance. As the flux penetrates into the rotor, force is applied on the rotor pole and, hence, the rotor tends to move towards the stator pole to reduce the reluctance of the magnetic circuit.

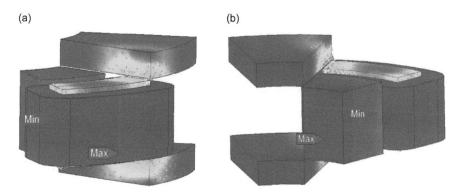

FIGURE 15.7
Flux distribution at the unaligned position: (a) interior view and (b) exterior view.

If the three-dimensional machine is expanded in the circumferential direction, the relative positions and corresponding inductance profile of the AFSRM would be as in Figure 15.8. It can be observed that the inductance profile is almost the same as a conventional radial flux SRM. Since the mutual inductance is negligible, the same control strategies as a conventional RFSRM and conventional asymmetric bridge converter can still be applied to AFSRMs.

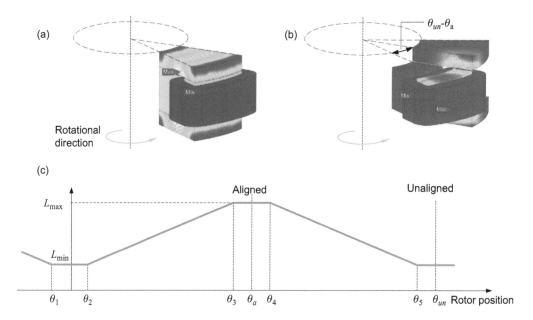

FIGURE 15.8
Relative position between stator pole and rotor pole, and change of phase inductance: (a) aligned position, (b) unaligned position, and (c) ideal phase inductance profile.

15.3.2 Derivation of Output Equation

The design equation is well developed for conventional radial flux SRMs. In [9], the output power is derived based on special magnetic loading and special electric loading. Similarly in [10], the output torque was developed as a function of motor geometry parameters. As for axial flux configuration, the output equation has been investigated in [1,11,12], but they are all for axial flux PM machines. For AFSRMs, an output power equation has been presented in [13]. The output equation is the fundamental expression of how the geometry parameters of an AFSRM is related to output power. It can also give us a better understanding about the difference between a RFSRM and AFSRM.

Figure 15.9 shows typical air gap structures of radial flux and axial flux switched reluctance machines. The effect of pole arcs has been neglected for simplicity. The arrows represent the direction of flux in the air gap.

The output power equation represents the relationship between the output power and various geometric and control parameters of the motor, such as inner and outer diameter, stack length, pole height, pole arc angles, and number of turns. The derivation will be based on the axial flux SRM configuration shown in Figure 15.5.

Figure 15.10 shows the geometric parameters of this machine. This AFSRM has one stator disc in the middle between two rotor discs. The two rotor discs rotate at the same speed and are always aligned with each other. Not always, but in many cases, the stator and rotor core share the same inner and outer diameters, the stator and rotor pole arc angles also share the same dimensions for most designs.

Output power can be derived from the input power as shown in (15.1).

$$P_{out} = k_e \cdot k_d \cdot V_{DC} \cdot i \cdot m \tag{15.1}$$

where k_e is the efficiency ratio, V_{DC} is the input DC voltage, i is the phase current, and m is the number of simultaneous active phases in an instance. The duty cycle ratio k_d is defined as:

$$k_d = \frac{\theta_c N_r}{2\pi} \cdot \frac{N_s}{2} \tag{15.2}$$

where θ_c is the conduction angle in mechanical degrees, N_s and N_r are stator and rotor pole numbers respectively.

(a)   (b)

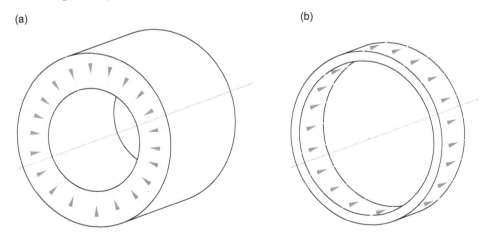

FIGURE 15.9
Direction of the flux in the air gap: (a) RFSRM and (b) AFSRM.

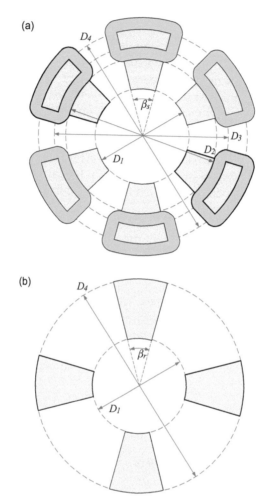

FIGURE 15.10
Cross section of the AFSRM stator and rotor: (a) stator and (b) rotor.

As shown in Figure 15.11, either in motoring or generating mode, the operation of the machine can be regarded as a repeating process where the rotor position changes from unaligned position to aligned position or the other way around. V_{DC} can be expressed by:

$$V_{DC} = R_s i + \frac{d\lambda}{dt} \approx \frac{d\lambda}{dt} \tag{15.3}$$

$$V_{DC} \cong \frac{d\lambda}{dt} = \frac{\lambda_a - \lambda_u}{\theta_c/\omega_m} \tag{15.4}$$

where θ_c is the angle when the flux linkage varies from λ_a to λ_u. λ_a and λ_u are the maximum and minimum phase flux linkages, which can also be regarded as the phase flux linkage at aligned and unaligned positions. The ω_m is the mechanical angular speed, which can be derived as a function of the rotor speed in *rpm*:

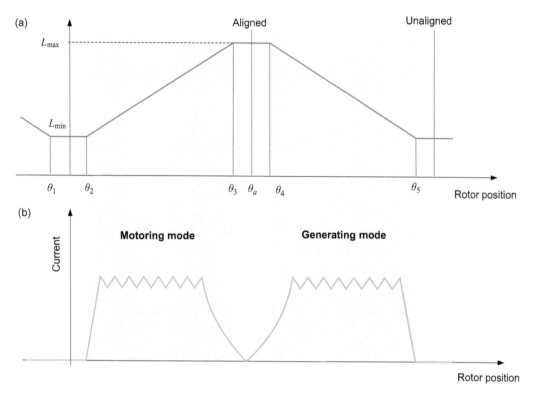

FIGURE 15.11
Variation of inductance as a function of rotor position in a typical AFSRM: (a) flux linkage and (b) current.

$$\omega_m = \frac{2\pi \times rpm}{60} \tag{15.5}$$

It can be noted from Figure 15.11 that the most significant change in the flux linkage or inductance profile happens only from position θ_2 to θ_3 or θ_4 to θ_5. Here, θ_2 and θ_5 corresponds to the position when rotor pole start to overlap with the stator pole, while θ_3 and θ_4 usually coincide with the aligned position θ_a. Therefore, the angle between θ_2 and θ_3 or between θ_4 and θ_5 can be regarded as the smaller of stator pole arc or rotor pole arc angles. Since the stator and rotor pole arcs usually share the same value for AFSRMs, stator pole arc angle β_s is used here to be consistent with the output equation of conventional RFSRMs, which usually has smaller stator pole arc angle. Therefore, $\theta_c = \beta_s$. Then, the voltage in (15.4) can be reorganized as:

$$V_{DC} \cong \frac{d\lambda}{dt} = \frac{\lambda_a - \lambda_u}{\beta_s/\omega_m} \tag{15.6}$$

At the aligned position, for the AFSRM considered here, the flux path has been discussed in Figure 15.6. Therefore, phase flux linkage can be written as:

$$\lambda_a = \phi \; T_{ph} = B \frac{\pi(D_4^2 - D_3^2)}{4} \beta_s T_{ph} \tag{15.7}$$

where the term $\pi(D_4{}^2 - D_3{}^2)\beta_s/4$ is the area surrounded by the phase coils. D_3 and D_4 have been shown in Figure 15.10. The flux is the product of the magnetic flux density B and the area calculated above. The flux linkage is the product of the flux and the number of turns per phase T_{ph}. Since saturation happens at aligned position, B is the saturation flux density. Combining equations from (15.1) to (15.7), the output power is then expressed as:

$$P_{out} = k_e\, k_d\, \frac{\pi^2}{120}\left(1 - \frac{\lambda_u}{\lambda_a}\right) B\ (D_4{}^2 - D_3{}^2)\ T_{ph}N\ i\ m \qquad (15.8)$$

This equation is similar to the classical power equation of RFSRMs. From (15.8), for a given magnetic loading at a certain speed, the dimensions, D_3 and D_4 of the outer part of stator have a direct relationship with the amount of output power. To fully utilize the machine for electrical and mechanical power conversion, the duty cycle is usually set as 1 for initial sizing. Efficiency ratio k_e needs to include all the losses; however, to simplify the sizing, it is also set at 1 at the initial stage. Therefore, initial sizing is for a primary sizing estimation. Obtaining accurate geometric details requires more analysis and optimization. Compared to the conventional RFSRM output power equation, the output power equation of this particular AFSRM in (15.8) mainly depends on the stator diameter, and, unlike RFSRMs, this value is independent of the stack length. However, the derivation of the power equation focuses more on magnetic loading, as B is the saturation flux density of the material. For the copper to provide significant electromotive force (EMF), the electrical load needs to be considered. For this AFSRM, the slot area for the copper depends not only on the space between D_2 and D_3, but also the stator stack length. From this point of view, the output power also relies on the stack length of the machine.

In order to determine the final machine geometry, D_1, D_2, D_3, and D_4, as well as stack length need to be decided. For this particular AFSRM, equal amount of flux will go through two parts of the stator core. Therefore, they should have equal area or volume:

$$\pi \frac{D_4^2 - D_3^2}{4}\beta_s = \pi \frac{D_2^2 - D_1^2}{4}\beta_s \qquad (15.9)$$

In practice, the distance between D_2 and D_3 should be limited otherwise it may lead to poor utilization of the core material, especially for the rotors. Therefore, this distance is usually small compared to other dimensions. For the initial estimation, it's assumed that $D_2 \approx D_3$. Hence,

$$D_2 = D_3 = \sqrt{\frac{D_4^2 + D_1^2}{2}} \qquad (15.10)$$

The diameter ratio $K = D_4/D_1$, which can also be written as $K = D_{out}/D_{in}$, is one of the most important parameters in the design of an axial flux machine. Reorganizing the term $(D_4^2 - D_3^2)$ in (15.8) with (15.9) and (15.10), and using the definition of the diameter ratio K results in:

$$D_4^2 - D_3^2 = D_4^2 - \frac{D_4^2 + D_1^2}{2} = \frac{D_4^2 - D_1^2}{2} = \frac{\left(\dfrac{D_4^2}{D_1^2} - 1\right)}{2\dfrac{D_4^2}{D_1^2}}D_4^2 = \frac{K^2 - 1}{2K^2}D_4^2 \qquad (15.11)$$

Therefore, (15.8) can be rewritten in terms of the diameter ratio and the core outer diameter as:

$$P_{out} = K_a \cdot \left(\frac{K^2 - 1}{2K^2} \right) D_4^2 \qquad (15.12)$$

where $K_a = k_e \cdot k_d \cdot \frac{\pi^2}{120}(1 - \frac{\lambda_u}{\lambda_a}) \cdot B \cdot T_{ph} \cdot N \cdot i \cdot m$.

The total stack length of the motor is the sum of:

- twice the air gap length;
- stator stack length;
- twice the rotor stack length.

If the boundary condition and requirements are given, the power equation in (15.12) can be used to initialize the design process. However, there are still too many unknown parameters. In order to obtain a detailed analysis, FEA is still necessary.

15.4 Challenges in AFSRMs

15.4.1 Torque Ripple Reduction in AFSRMs

Reduction of torque ripple has been a popular topic for SRMs. However, previous research and development focused primarily on RFSRMs, and how to improve their geometry and optimize control strategies. As for AFSRMs, work on torque ripple reduction is limited. In [7], an axial flux segmented SRM (AFSSRM) is first designed to replace permanent magnet (PM) machines in EVs. Then in [14], two rotors of this AFSSRM are circumferentially displaced with respect to each other to reduce the torque ripple. After that, in [15], it was stated that the segmented rotor AFSRM has higher average torque and lower torque ripple compared to toothed-rotor AFSRM. In [16], the influence of the stator pole and rotor pole arc angles on the average torque and the torque ripple is studied. Similarly in [17], the skewed stator segments are applied, and torque ripple reduction has been achieved during the phase commutation. In [18,19], the focus was on special winding configurations to improve the AFSRM inductance ratio and, therefore, reduce torque ripple.

15.4.2 Winding Configuration for Copper Loss Reduction

Reducing copper losses in electric machines is always an important issue. It not only improves motor efficiency, but also reduces temperature rise. Copper loss reduction in AFSRMs has been investigated in previous research studies. For example, for an external rotor structure, toroidal-winding has the potential to reduce the end winding length [20] and, hence, to decrease copper losses and provide more space for motor core material. This improves the torque production with the same space and thermal constraints. The work in [21–23] aimed to design AFSRMs with improved output torque while considering lower losses.

15.4.3 Soft Magnetic Composites for AFSRMs

For most RFSRMs, the typical core material is electrical steel laminations alloyed with silicon. However, electrical steel laminations are generally limited to a two-dimensional flux path. The flux path of an AFSRM is usually three-dimensional. Applying electrical steel lamination in an AFSRM will lead to high design complexity and iron losses.

When the electromagnetic field is optimized in AFSRM, the results usually yield a rather complicated geometry, for which the traditional machine manufacturing techniques could be challenging to apply. In the early days, challenges in manufacturing have been the biggest issue slowing down the development of AFSRMs. Today, soft magnetic composite (SMC) materials make it possible to explore new geometries and structures [24]. SMCs consist of metal particles coated in an insulated film [25]. SMC can be easily compacted to form complicated geometries. This unique property of SMC makes it attractive for the relatively complicated geometries of AFSRM.

Unlike electrical steel laminations, SMCs have three-dimensional isotropic ferromagnetic properties. This isotropic nature leads to relatively low core losses. This property is especially important in medium and high frequency applications, such as in high-speed electric motors. Lower core losses also lead to the possibility of improved thermal characteristics.

15.4.4 Other Considerations for AFSRMs

Since AFSRMs are exposed to electromagnetic forces in three dimensions, the axial component of electromagnetic force is relatively much larger than that of a RFSRM. Such an axial force will cause vibration and distortions. Since the air gap of AFSRM is small, the vibration in the axial direction may result in changes in the electromagnetic performance. Furthermore, if the distortion is too large, physical contact between stator and rotor is a possibility. Therefore, structural analysis for an AFSRM is crucial.

The construction and assembly of axial flux motors involves a larger number of parts than that of a comparable radial flux machine, but permits a simpler winding process [26]. Usually, in order to facilitate assembly, some modifications may be needed. These changes should not affect the electromagnetic performance significantly, but allows for an easier assembly process. In some axial flux machines, the motor is built from segmented components. These segments are placed on a plate, usually made of aluminum, which has very low magnetic permeability. The well-developed automatic winding process can be easily adopted for AFSRMs to achieve a relatively high copper fill factor.

Bearing selection for AFSRMs also need special consideration. Unlike conventional RFSRMs, there is a much higher axial force between the stator and rotor, and as such this force will be transferred to the bearing. Selecting the proper bearing to handle this axial loading is an important consideration to ensure the safe operation of the machine.

15.5 Suitable Applications for AFSRMs

Axial flux machines offer an alternative to conventional electric machines. AFSRMs have a distinct advantage of simple structure and high reliability. Besides, they also possess properties of traditional axial flux machines, such as higher power density, larger diameter to length ratio, adjustable air gap, and reduced iron losses, making it suitable for certain applications. Currently, axial flux permanent magnet brushless machines have already attracted attention in some commercial applications, such as renewable power generation, electric traction, and elevators. However, the cost of a PM machine depends significantly on rare-earth metal prices. The robustness and reliability of SRM, combined with the special properties associated with axial flux machines, makes AFSRMs a desirable candidate for these applications.

The special construction of an axial flux machine and a wind turbine makes the mechanical integration between these two components easier. A simplified integrated structure of a wind turbine and an AFSRM is shown in Figure 15.12. Research studies on axial flux wind turbines have been going on for some time [27–30]. Some organizations, such as Kestrel Wind Turbines Ltd. and Hurricane Wind Power provide commercial products or support on axial flux generators for wind turbines. Even though the PM axial flux machines dominate the products in the industry and research in the academia, there are active studies on AFSRM generators in wind turbines. In [31], an axial flux switched reluctance generator is simulated. It was concluded that the AFSRM is superior over other generators for a wide range of wind loading conditions. The same conclusion has been validated in [32] via testing. However, more research and development is still necessary before moving forward to commercializing axial flux switched reluctance generators.

The simple structure, ruggedness, fault-tolerant operation capability, wide-speed range, and low manufacturing cost of AFSRMs also make it attractive for electric vehicle (EV) and hybrid electric vehicle (HEV) applications. The large diameter to

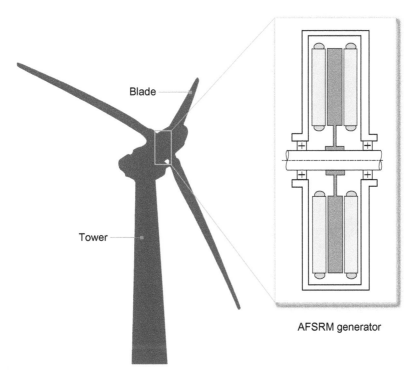

FIGURE 15.12
A simplified wind turbine power generator with double-stator AFSRM.

length ratio and adjustable air gap in AFSRMs makes it suitable for in-wheel direct-drive propulsion systems for EVs and HEVs. An in-wheel direct-drive system with an integrated AFSRM is shown in Figure 15.13. Compared to RFSRMs, AFSRMs have the advantage of higher torque/power density [19,20]. Since the space for EV and HEV propulsion is limited, AFSRM's potential for high torque density is even more relevant. However, for in wheel propulsion, there is a big concern with unsprung mass.

There are numerous publications discussing AFSRMs for EVs and HEVs. A novel pancake-shape AFSRM for HEVs has been presented in [5] where the advantages of high power density and reduced acoustic noise have been highlighted. The C-shaped stator also reduces the manufacturing complexity.

Another popular AFSRM structure is presented in [7], which is a double-rotor axial flux segmented SRM. The study of this three-phase double-rotor 12/8 AFSRM is based on the Indian Driving Cycle. A relatively high torque to weight ratio of 1.82 Nm/kg has been achieved in this application. Reduction in torque ripple on this type of AFSRM has been accomplished by circumferentially displacing the two rotors [14]. The torque density can be further improved by using higher pole numbers in a 12/16 topology as in [22]. Similar to [14,17], the two stators can be skewed with respect to each other in the circumferential direction in a two-stator one-rotor AFSRM topology to reduce torque ripple.

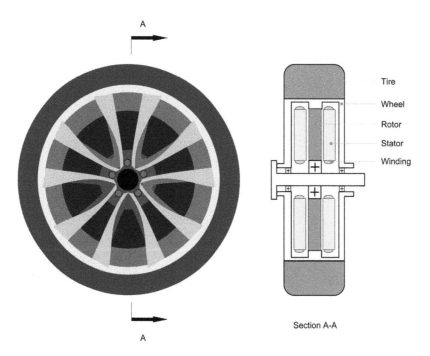

FIGURE 15.13
Direct driven wheel with integrated double-stator single-rotor AFSRM.

Another potential application for AFSRMs is elevators. Its high torque density and flat shape makes it feasible for direct-drive applications, by which the machine room can be eliminated leading to space and cost savings. For the Monospace™-lift designed by Kone Elevators, an axial flux motor solution is used [33]. In this design, the machine is directly integrated with the rail of the elevator. In [34], the design and construction of two twin prototypes of slotless axial flux permanent-magnet motor drives have been presented for elevator systems without the machine room.

Questions and Problems

1. Compare AFSRMs and RFSRMs in terms of structure, direction of magnetic flux, location of the coils, and components? For which applications AFSRM are more suitable? What are the advantages and drawbacks of AFSRMs?
2. A three-phase double-stator single-rotor AFSRM will be designed with a rated speed of 2000 rpm. The geometric parameters, D_1, D_2, D_3, D_4, stator and rotor pole arcs (β_s and β_r), stator and rotor stack lengths L_r, L_s, and air gap g, as well as number of turns per phase T_{ph} are listed in Table Q.15.1. Magnetization characteristics of the core material are also given in Figure Q.15.1. It is suggested that the motor operates at the knee point of the magnetization curve at the aligned position.

   a. Calculate the inductance at the aligned position.
   b. Find out the phase current to create required flux density at the aligned position. Also calculate the current density.

TABLE Q.15.1

Geometric Parameters

Parameter	Value
D_1	40 mm
D_2	70 mm
D_3	90 mm
D_4	110 mm
$\beta_s = \beta_r$	30°
L_s	60 mm
L_r	20 mm
g	0.5 mm
T_{ph}	32

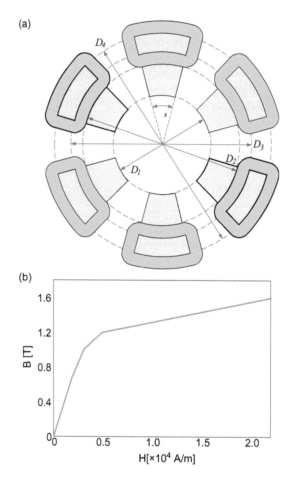

FIGURE Q.15.1
Cross section of the AFSRM and material characteristics: (a) geometry and (b) material B-H curve.

References

1. S. Huang, J. Luo, F. Leonardi, and T. A. Lipo, "A comparison of power density for axial flux machines based on general purpose sizing equations," *IEEE Trans. Energy Convers.*, vol. 14, no. 2, pp. 185–192, 1999.
2. K. Sitapati and R. Krishnan, "Performance comparisons of radial and axial field, permanent-magnet, brushless machines," *IEEE Trans. Ind. Appl.*, vol. 37, no. 5, pp. 1219–1226, 2001.
3. A. Nelson, and M. Chow, "Electric vehicles and axial flux permanent magnet motor propulsion systems," *IEEE Ind. Electron. Soc. Newslett.*, vol. 46, no. 4, pp. 3–6, 1999.
4. J. F. Gieras, R.-J. Wang, and M. J. Kamper, *Applications in Axial Flux Permanent Magnet Brushless Machines*, 2nd ed. Dordrecht, the Netherlands: Springer, 2008.
5. A. Labak and N. C. Kar, "A novel five-phase pancake shaped switched reluctance motor for hybrid electric vehicles," in *Proceedings of the IEEE Vehicle Power and Propulsion Conference*, Dearborn, MI, September 2009, pp. 494–499.
6. T. Shibamoto, K. Nakamura, H. Goto, and O. Ichinokura, "A design of axial-gap switched reluctance motor for in-wheel direct drive EV," in *Proceedings of IEEE International Conference on Electrical Machines*, Marseille, France, September 2012, pp. 1158–1163.

7. R. Madhavan and B. G. Fernandes, "A novel axial flux segmented SRM for electric vehicle application," in *Proceedings of IEEE International Conference on Electrical Machines*, Rome, Italy, September 2010, pp. 1–6.

8. S. H. Mao and M. C. Tsai, "A novel switched reluctance motor with C-core stators," *IEEE Trans. Magn.*, vol. 41, no. 12, pp. 4413–4420, 2005.

9. R. Krishnan, *Switched Reluctance Motor Drives: Modeling, Simulation, Analysis, Design, and Applications*, Boca Raton, FL: CRC Press, 2001.

10. T. J. E. Miller, *Switched Reluctance Motors and their Control*, Hillsboro, OH, Magna Physics; New York, Oxford University Press, 1993.

11. C. C. Chan, "Axial-field electrical machines—Design and applications," *IEEE Trans. Energy Convers.*, vol. EC-2, no. 2, pp. 294–300, 1987.

12. A. Parviainen, Design of axial-flux permanent-magnet low-speed machines and performance comparison between radial-flux and axial-flux machines, PhD dissertation, Lappeenranta University of Technology, Lappeenranta, Finland, 2005.

13. R. Krishnan, M. Abouzeid, and X. Mang, "A design procedure for axial field switched reluctance motors," in *Proceedings of IEEE Industry Applications Society Annual Meeting*, October 1990, pp. 241–246.

14. R. Madhavan and B. G. Fernandes, "A novel technique for minimizing torque ripple in axial flux segmented rotor SRM," in *Proceedings of IEEE Energy Conversion Congress and Expo*, Phoenix, AZ, September 2011, pp. 3383–3390.

15. R. Madhavan and B. G. Fernandes, "Comparative analysis of axial flux SRM topologies for electric vehicle application," in *Proceedings of IEEE International Conference on Power Electronics, Drives and Energy System*, Bengaluru, India, December 2012, pp. 1–6.

16. R. Madhavan and B. G. Fernandes, "Performance improvement in the axial flux-segmented rotor-switched reluctance motor," *IEEE Trans. Energy Convers.*, vol. 29, no. 3, pp. 641–651, 2014.

17. J. Ma, R. Qu, and J. Li, "Optimal design of axial flux switched reluctance motor for electric vehicle application," in *Proceedings of the IEEE International Conference on Electrical Machines and Systems*, Hangzhou, China, October 2014, pp. 1860–1865.

18. A. Labak and N. C. Kar, "Novel approaches towards leakage flux reduction in axial flux switched reluctance machines," *IEEE Trans. Magn.*, vol. 49, no. 8, pp. 4738–4741, 2013.

19. A. Labak, and N. C. Kar, "Design and prototyping a novel 5-phase pancake shaped axial flux SRM for electric vehicle application through dynamic FEA incorporating flux-tube modeling," *IEEE Trans. Ind. Appl.*, vol. 49, no. 3, pp. 1276–1288, 2013.

20. C. Du-Bar, "Design of an axial flux machine for an in-wheel motor application," MS thesis, Department of Energy and Environment, Chalmers University of Technology, Göteborg, Sweden, 2011.

21. H. Arihara and K. Akatsu, "Characteristics of axial type switched reluctance motor," in *Proceedings of IEEE Energy Conversion Congress and Expo*, Phoenix, AZ, September 2011, pp. 3582–3589.

22. R. Madhavan and B. G. Fernandes, "Axial flux segmented SRM with a higher number of rotor segments for electric vehicles," *IEEE Trans. Energy Convers.*, vol. 28, no. 1, pp. 203–213, 2013.

23. B. Wang, D. H. Lee, and J. W. Ahn, "Characteristic analysis of a novel segmental rotor axial field switched reluctance motor with single teeth winding," in *Proceedings of the IEEE International Conference on Industrial Technology*, Busan, Korea, February/March 2014, pp. 175–180.

24. G. S. Liew, N. Ertugrul, W. L. Soong, and D. B. Gehlert, "Analysis and performance evaluation of an axial-field brushless PM machine utilizing soft magnetic composites," in *Proceedings of IEEE International Electric Machines and Drives Conference*, Antalya, Turkey, May 2007, vol. 1, pp. 153–158.

25. H. Shokrollahi and K. Janghorban, "Soft magnetic composite materials (SMCs)," *J. Mater. Process. Technol.*, vol. 189, no. 1–3, pp. 1–12, 2007.

26. T. Lambert, M. Biglarbegian, and S. Mahmud, "A novel approach to the design of axial-flux switched-reluctance motors," *Machines*, vol. 3, no. 1, pp. 27–54, 2015.

27. E. Spooner and B. J. Chalmers, "'TORUS': A slotless, toroidal-stator, permanent-magnet generator," *Proc. Inst. Elect. Eng.*, vol. 139, pt. B, pp. 497–506, 1992.

28. B. J. Chalmers and E. Spooner, "An axial-flux permanent-magnet generator for a gearless wind energy system," *IEEE Trans. Energy Convers.*, vol. 14, no. 2, pp. 251–257, 1999.
29. L. Hansen, *Conceptual Survey of Generators and Power Electronics for Wind Turbines*, Roskilde, Denmark, Risø National Laboratory, 2001.
30. M. Dhifli, H. Bali, Y. Laoubi, G. Verez, Y. Amara, and G. Barakat, "Modeling and prototyping of axial flux permanent magnet machine for small wind turbine," in *Proceedings of the International Conference on Electrical Sciences and Technologies in Maghreb*, Tunis, Tunisia, November 2014, pp. 1–7.
31. M. Abouzeid, "The use of an axial field-switched reluctance generator driven by wind energy," *Renew. Energy*, vol. 6, no. 5–6, pp. 619–622, 1995.
32. M. Abou-Zaid, M. El-Attar, and M. Moussa, "Analysis and performance of axial field switched reluctance generator," in *Proceedings of IEEE International Electric Machines and Drives Conference*, Seattle, WA, May 1999, pp. 141–143.
33. H. Hakala, "Integration of motor and hoisting machine change the elevator business," in *Proceedings of the International Conference on Electrical Machines*, Helsinki, Finland, August 2000, Vol. 3, pp. 1242–1245.
34. R. L. Ficheux, F. Caricchi, F. Crescimbini, and O. Honorati, "Axial-flux permanent-magnet motor for direct-drive elevator systems without machine room," *IEEE Transactions on Industry Applications*, vol. 37, no. 6, pp. 1693–1701, 2001.

16

Switched Reluctance Motor and Drive Design Examples

Berker Bilgin, James Weisheng Jiang, and Alan Dorneles Callegaro

CONTENTS

Electric motor shipment is expected to increase significantly in the next decade with the increase in the demand for electrification. Due to their high torque density and higher efficiency especially in the low- to medium-speed range, permanent magnet machines are currently utilized heavily in residential, commercial, and transportation sectors. However, permanent magnet machines mostly employ rare-earth magnets. As explained in Chapter 1, China holds 84% of the world production of rare earth materials. This value is based on China's Ministry of Land and Resources production quotas. Around 20% of China's rare earth oxide was produced illegally and the total production was well beyond the 2015 quota. China also holds 76% of the permanent magnet production. Rare-earth magnets are usually the largest cost component of a permanent magnet motor. This is related to the high price of these materials. In addition, the price of rare-earth materials is highly volatile and, being the largest producer of rare-earth materials, China has significant control over the price and availability of these materials in the global market. With the increasing demand for high-efficiency motors in industrial, residential, commercial, and transportation sectors, and in energy generation, the price of rare-earth elements and supply chain issues are of significant concerns.

There are also production and environmental issues for rare-earth materials. They are difficult to mine, and it is usually hard to find them in high enough concentration so that the extraction process is economically viable. Mining and refining of rare-earth materials require significant amount of capital and expertise. Depending on the location and the production capacity, the extraction and processing of rare-earth materials might typically need capital between $100 million to $1 billion. Furthermore, the processing of rare-earth elements into high-purity rare-earth oxides is a highly specialized chemical process and it requires significant know-how in mineral processing. Producing high-quality permanent magnets from rare-earth materials require expertise and know-how, as well. It should also be noted that mining and extraction of rare-earth materials can have adverse environmental effects, which also increase the cost of production.

As a consequence, even though permanent magnet machines provide high torque density and high efficiency, the price volatility, supply chain issues, and environmental concerns for rare-earth materials are of significant concerns for high-energy rare-earth magnets. In the long run, these problems will force the electric motor industry to seek alternatives such as SRMs. SRMs are exceptionally attractive for the industry to respond to the increasing demand for high-efficiency, high-performance, and low-cost electric motors. Furthermore, SRMs will replace many permanent magnet motors in mass production for many applications due to their advantages.

In this chapter, two SRM designs will be presented which are proposed to replace the permanent magnet machines in a residential HVAC application and a hybrid-electric propulsion application. The SRMs are designed competitively based on performance requirements of the permanent magnet motors.

The converter topology that is used in SRM drives is typically different from that of permanent magnet and induction motor drives. Asymmetric bridge converters are commonly applied in SRM drives. In permanent magnet and induction machines, three-phase full bridge converters are mainly used. Utilizing a different converter is not a big challenge for the adoption and mass manufacturing of SRM drives. In fact, in this chapter, we will present the design of a high-power converter for switched reluctance motor drives which utilizes specialized modules. In these modules, the switches and diodes for one phase asymmetric bridge converter are packaged together. This enables a more compact converter design for switched reluctance motor drives.

16.1 Switched Reluctance Motor for a Residential HVAC System

HVAC applications account for 63% of the electric motor energy consumption in the residential sector. As it was shown in Figure 1.6 in Chapter 1, residential HVAC applications include central air conditioners (CAC), heat pumps, furnace blower fans, room air conditioners (RAC), and dehumidifiers. In the residential sector, furnace blower fans are the fourth-largest electrical energy consumer after the central air conditioner compressors, refrigerator/freezer compressors, and the heat pump compressors. The average operation hours of furnace fans is 1870 hours and average unit energy consumption is 678 kWh per year [1]. The majority of these motors are less efficient single-phase induction motors. However, as of 2012, 34% of these motors have been switched to more efficient, variable speed motors, mostly using rare-earth permanent magnets.

SRMs are expected to replace many furnace blower motors since they are cheaper to produce, and do not depend on the supply of rare-earth magnets. SRMs are robust, which is a beneficial characteristic for long-term continuous use. In permanent magnet machines, the magnets can lose magnetization at elevated temperatures. Due to lack of rotor excitation, SRM can provide robust operation in high-temperature conditions.

16.1.1 Household Furnace HVAC Application

Household forced-air heating ventilation and air conditioning (HVAC) systems are used to control the heating and cooling temperatures in the majority of residential houses. The furnace typically uses natural gas, heating oil, or electricity to heat the air circulating in the household and blow it through a ventilation system that outputs to different areas of the home. As shown in Figure 16.1, in a forced-air furnace heating system, air is circulated around the house through a network of vents installed in the walls and floors of the house during its construction. The

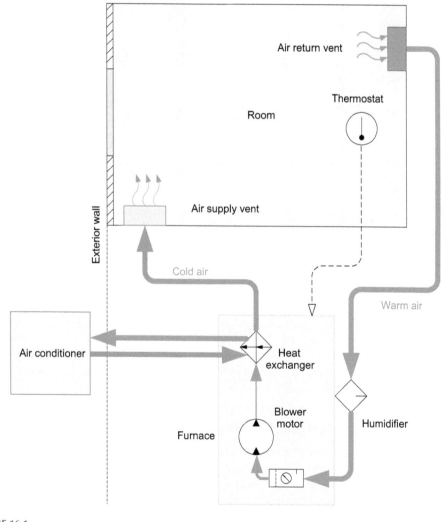

FIGURE 16.1

Typical HVAC household ventilation diagram with central air conditioning (CAC). (From Kasprzak, M., 6/14 switched reluctance machine design for household HVAC system applications, MASc thesis, Department of Mechanical Engineering, McMaster University, Hamilton, Canada, 2016.)

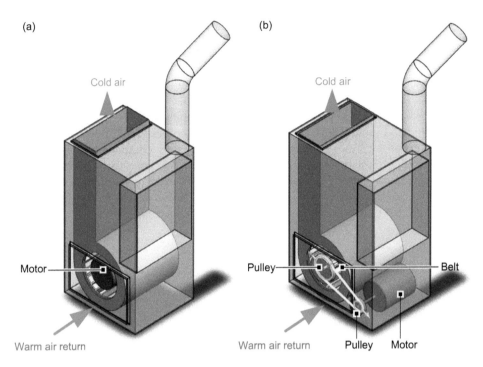

FIGURE 16.2
Furnace blower motor mounting styles: (a) direct drive and (b) belt drive. (From Kasprzak, M., 6/14 switched reluctance machine design for household HVAC system applications, MASc thesis, Department of Mechanical Engineering, McMaster University, Hamilton, Canada, 2016.)

furnace blower motor and its fan draw air from cold-air return vents throughout the house [3]. The cold air being returned to the furnace system passes through a filter to remove dust particles from the air, before heating and pumping it back to the rest of the house.

Furnace blower motors are typically mounted in two styles as shown in Figure 16.2. The first method is direct drive, where the output shaft of the electric motor is coupled directly to the blower fan. The second method uses a pulley system and a belt drive to connect the offset electric motor to the fan in the center of the blower.

The furnace is controlled by a thermostat, which is centrally located in the house, so that it can take the best average temperature of the house. When the room temperature deviates from the desired set temperature, it switches on the furnace to cool or heat the re-circulating air to change the room temperature back to the desired value. For this purpose, either on/off control or variable speed control is applied.

16.1.2 Commercial Permanent-Magnet Furnace Blower Motor

In this section, a commercial permanent-magnet furnace blower motor will be presented, which will be used to define the design requirement of the switched reluctance furnace blower motor. GENTEQ Evergreen 6110E from Regal Beloit is a 1HP surface mounted permanent magnet synchronous furnace blower motor and it is widely used in the residential HVAC aftermarket. The 6110E incorporates a National Electrical Manufacturers Association (NEMA) 48 Standard Frame for use with a bellyband clamp mount. This standardization allows the motor to be applied in any application with the same power rating and mounting style. Figure 16.3 shows the main dimensions of the GENTEQ Evergreen 6110E. The motor

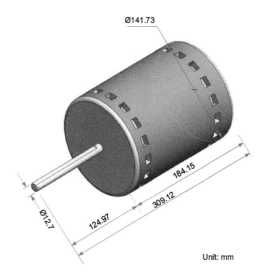

Ø141.73

Ø12.7

124.97

309.12

184.15

Unit: mm

FIGURE 16.3
3D CAD representation of the GENTEQ Evergreen 6110E furnace blower motor. (From Kasprzak, M., 6/14 switched reluctance machine design for household HVAC system applications, MASc thesis, Department of Mechanical Engineering, McMaster University, Hamilton, Canada, 2016.)

is cooled by air-over convection from the ambient temperature. It does not employ forced cooling. The product information provided by the motor manufacturer is shown in Table 16.1.

The motor housing and the name plate can be observed in Figure 16.4a. The housing is thin steel material with aluminum end caps at the top and bottom. The motor controller can be seen mounted in the bottom of the motor below the stack in Figure 16.4b.

TABLE 16.1

GENTEQ Product information

Motor type	Brushless DC
Available horsepower	1 HP
Speed range	600–1200 rpm
Peak motor efficiency	80%
Average power (cooling CFM)	413 W
Average power (continuous fan CFM)	95 W
Application	Retrofit for indoor standard X13 ECM motors
Stock No.	6110E
HP	1
Volts	115 AC
Max current	10.9 A
Rotation	CCW/CW
Motor length (including shaft)	12.17"
Motor housing length	7.25"
Approximate weight	19.3 lbs

Source: Genteq, "Evergreen EM spec sheet," November 5, 2014, Available: https://www.genteqmotors.com/Products/Aftermarket/Evergreen_EM/; Genteq, "Evergreen multi product brochure," September 28, 2015, Available: https://www.genteqmotors.com/Products/Aftermarket/Evergreen_EM/; Genteq, "Evergreen EM brochure," February 19, 2013, Available: http://www.swimming-pool-pump.com/.

(a)   (b)

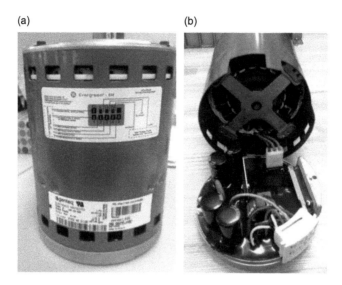

FIGURE 16.4
6110E Housing: (a) motor housing and (b) motor controller mounted in the bottom below the stack. (From Kasprzak, M., 6/14 switched reluctance machine design for household HVAC system applications, MASc thesis, Department of Mechanical Engineering, McMaster University, Hamilton, Canada, 2016.)

The 6110E furnace blower motor has 18 stator poles. The stator has plastic end caps on each side that cover the back iron and stator teeth. The coil windings are wrapped around the plastic end cap at each tooth. In the slots, an insulating liner is used that separates the coil from the sides of the stator tooth. Hence, the stator has two plastic end rings and 18 insulating slot liners.

The motor has an interior rotor with surface mounted permanent magnets. The total number of magnet poles is 12. Therefore, 6110E is a surface permanent magnet motor with 18 stator slots and 12 rotor poles.

The motor drive is a six-switch inverter with an integrated rectifier module. The circuit has a bridge diode rectifier which can be modeled to find the DC-link voltage based on the AC-line voltage and peak power of the motor, which is specified as 745 W. The AC voltage from the wall is 115 V, 60 Hz. The drive circuit has four parallel-connected capacitors at the output of the rectifier rated at 200 V, 1000 μF with a 130 mΩ resistance. A simulation model of the rectifier has been created using the drive parameters and the DC-link voltage was calculated as 163 V at the peak power. Various experiments have been conducted on the 6110E furnace blower motor to define the design requirements of the SRM for the furnace blower application.

16.1.3 Switched Reluctance Motor Design for the Furnace Blower Application

The design constraints of the SRM are limited to the same outer diameter and axial stack length of the commercial furnace blower motor and the same input power, while still being capable of meeting the performance requirements of the target motor. This would make the SRM sizing suitable for installing in the household HVAC application. Table 16.2 summarizes the design constraints for the SRM based on the experiments and analysis conducted on the commercial furnace blower motor.

TABLE 16.2

Design Constraints for the Switched Reluctance Furnace Blower Motor

Geometry	
Stator outer diameter	139.21 mm
Axial stack length	74 mm
Shaft diameter	12.7 mm
Wire size	AWG 19
Power	
DC link voltage	163 V
RMS phase current	4.29 A
Performance	
Operating speed range	600–1200 rpm
Peak torque	5.5 Nm at 1200 rpm
Power rating	745.7 W

Source: Kasprzak, M., 6/14 switched reluctance machine design for household HVAC system applications, MASc thesis, Department of Mechanical Engineering, McMaster University, Hamilton, Canada, 2016.

The SRM pole configuration selected for the design has 6 stator poles and 14 rotor poles with an interior rotor. A configuration with a higher number of rotor poles than the stator poles is selected to achieve lower torque ripples.

The design optimization process is performed through a number of optimization steps. Throughout the optimization, various parameters including stator and rotor pole heights, stator and rotor back iron thicknesses, pole arc angles, air gap length, stator and rotor pole taper angles, and number of turns per coil are adjusted to satisfy the design requirements and dimensional constraints. Table 16.3 summarizes the overall dimensions and design parameters of the 6/14 furnace blower SRM.

TABLE 16.3

6/14 Furnace Blower Motor Parameters

Number of rotor poles	14
Shaft diameter	12.7 mm
Stack length	74 mm
Number of stator poles	6
Airgap length	0.4 mm
Stator outer diameter	139.21 mm
Wire gauge	AWG 19
Insulation type	Heavy
Number of turns per coil	115
Number of strands per coil	1
Wire fill factor	37%
Number of phases	3
Winding style	Conventional series
Lamination material	Cogent NO30-1600
Shaft material	S45C

Source: Kasprzak, M., 6/14 switched reluctance machine design for household HVAC system applications, MASc thesis, Department of Mechanical Engineering, McMaster University, Hamilton, Canada, 2016.

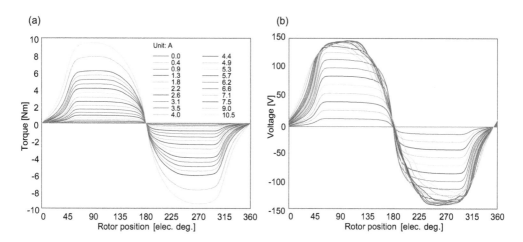

FIGURE 16.5
Simulated static characteristic waveforms of 6/14 HVAC SRM at different excitation currents: (a) electromagnetic torque and (b) phase voltage at 1200 rpm.

Figure 16.5 shows the static characteristic waveforms of the HVAC SRM at 1200 rpm and varying phase current levels from 0 to 10.5 A. It can be observed that the torque increases as the phase current increases; however, phase voltage stays constant after a certain phase current. As discussed in Chapter 2, this is because the co-energy increases, and the magnetic stored energy decreases as the level of saturation increases in SRM.

Once the 6/14 furnace blower SRM was fully characterized, the dynamic current at various operating points have been calculated to define the torque-speed profile of the motor. For this purpose, the conduction angles have been optimized at each operating point to maximize the average torque and minimize the torque ripple. For each operating point, copper loss, stator iron loss, and rotor iron loss are calculated from finite element analysis (FEA) simulations using the optimized current waveforms. Figure 16.6 shows the torque speed profile and performance contour plots of the 6/14 furnace blower SRM. As shown in Figure 16.6a, the efficiency of the SRM is around 85% at 1200 rpm. The peak efficiency is above 90%, which is achieved at higher speeds. As shown in Figure 16.6b, losses are higher at the low speed region. This is because the copper losses dominate at high-torque conditions.

The furnace blower SRM satisfies the design requirements as shown in Figure 16.6c. It delivers the required torque at 1200 rpm and it satisfies the root mean square (RMS) current requirement. Figure 16.6d shows the RMS torque ripple contour. It can be observed that the highest torque ripple occurs at the medium speed region. In the 600–1200 rpm operating range of the motor, the torque ripples are lower. This can be attributed to the novel 6/14 configuration and the optimization of the control parameters.

A thermal analysis of the 6/14 SRM was performed using a lumped parameter thermal network in MotorCAD software to further validate its suitability for the application. Similar to the commercial permanent magnet HVAC motor, the 6/14 furnace blower SRM does not have a forced cooling method. It employs only air-over cooling in ambient room temperatures. The temperature constraint on the coil windings was set at 100°C, which is quite conservative. Typical coil winding insulation is rated around 180°C, but as this motor will be used in a household application, and to reduce the risk of damage to the motor, a safety factor is applied. The maximum temperature constraint was used to test the maximum operating time of the furnace blower SRM under its peak torque performance at rated speed.

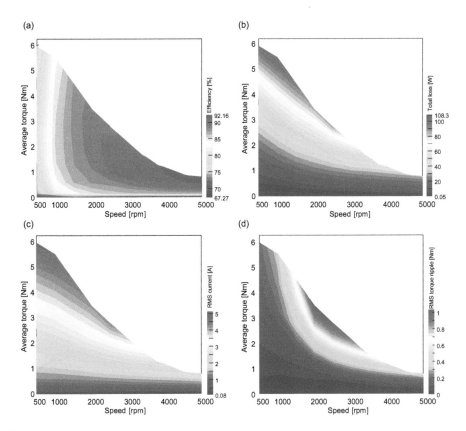

FIGURE 16.6
Simulated torque-speed characteristics of 6/14 HVAC SRM: (a) efficiency contour, (b) total loss contour, (c) RMS current contour, and (d) RMS torque ripple contour.

The temperature rise of the major components in the 6/14 furnace blower SRM while operating at the rated speed and peak torque is shown in Figure 16.7. The maximum temperature in the windings is observed to reach 100°C at 2550 seconds when the motor operates at the peak torque condition continuously. This maximum operating time is more than suitable for the operating point tested as it was the stall torque of the commercial permanent magnet furnace blower motor.

16.1.4 Mechanical Design of the Furnace Blower SRM

After electromagnetic and thermal analysis, through mechanical analysis, it was validated that the cut outs, which were applied on the rotor to reduce the rotating mass and inertia, and the overall weight of the motor, satisfy the mechanical stress requirements. With the furnace blower SRM capable of meeting the performance requirements, and thermal and mechanical stress analysis, the full construction of the motor assembly was completed. Figure 16.8 shows the exploded CAD view of the 6/14 furnace blower SRM components.

The 6/14 SRM has the same housing and shaft as the GENTEQ 6110E furnace blower motor. The SRM stator outer diameter was designed to fit the housing of the original motor. By using the original frame, furnace blower SRM can still be mounted using the NEMA 48 standard belly band mount. The end caps and the x-mount from the GENTEQ motor were used in the furnace blower SRM. The GENTEQ motor stator was found to have a plastic end

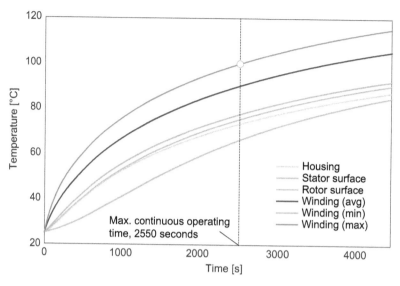

FIGURE 16.7
Simulated temperature variation for major components of the HVAC SRM at rated speed (1200 rpm) and peak torque (5.5 Nm).

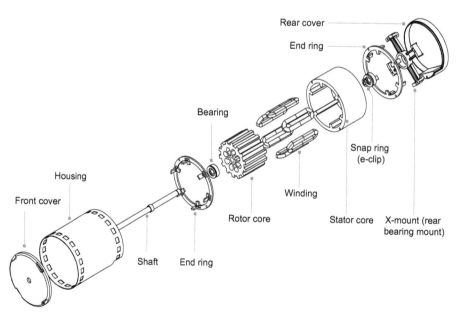

FIGURE 16.8
Exploded CAD view of 6/14 furnace blower SRM components.

ring on either side of the stator stack which covered the entire front surface. These end rings were redesigned in CAD to have the same axial dimensions. The concentrated windings wrap around the plastic end rings making each turns around the tooth. As per the design specifications, the novel SRM has the same housing and the shaft as the original GENTEQ motor. The x-mount and the plastic end ring can also be observed in the prototyped furnace blower SRM. Figure 16.9 shows the prototyped 6/14 furnace blower SRM.

HVAC SRM
prototype

GENTEQ
Evergreen 6110E

FIGURE 16.9
Prototype 6/14 furnace blower SRM.

16.2 Switched Reluctance Motor Design for a Hybrid Electric Propulsion Drive

Hybrid electric vehicles (HEVs) combine the propulsion from internal combustion engines and electric motors. HEVs have been proven to be an excellent alternative to the conventional gasoline-powered vehicles, because they incorporate engines with smaller sizes, produce less emissions, and improve fuel economy [7–10]. The world's first mass-produced HEV is the Toyota Prius, which has been manufactured since 1997. The Prius family has been evolving for more than 10 years, which now includes a plug-in hybrid electric vehicle option as well. The Prius hybrid is ranked as the best-sold HEV in the United States [11]. Figure 16.10 shows that the Prius family has been outselling its competitors since 2000 in the US HEV market.

An electric machine designed for an HEV should have the following features:

1. high torque and high-power density,
2. high torque at low-speed for starting and hill climbing,
3. high power at high-speed for cruising,
4. wide speed range including low-speed for urban and high-speed for highway driving,
5. wide constant power operation capability,
6. high efficiency over wide speed and torque ranges,
7. high reliability and robustness for specific vehicular working environment, and
8. reasonable cost [7,12].

Interior permanent magnet synchronous motors (IPMSMs) have been widely used as traction motors for HEV applications. Currently, the Prius powertrain uses two IPMSMs (one traction motor and one generator). However, the permanent magnets used in IPMSMs contain rare-earth materials. The prices of rare-earth materials have been suffering fierce fluctuations, and

(a)

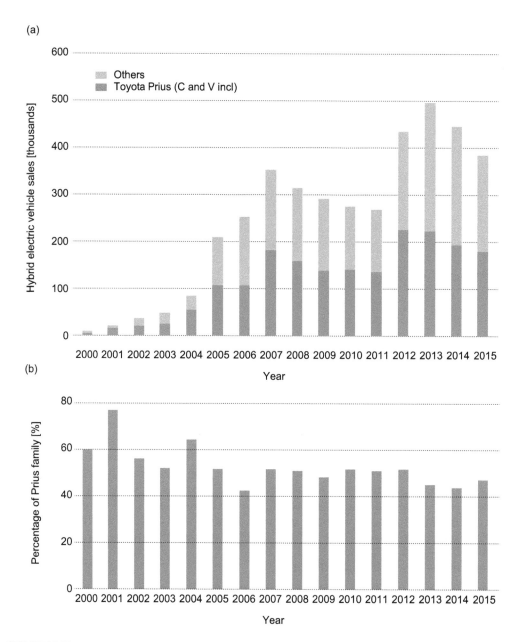

FIGURE 16.10
Hybrid electric vehicle sales in the U.S. from 2000 to 2015: (a) sales and (b) market share of Prius family. (From afdc.energy.gov, Alternative fuels data center: Maps and data, 2015, Available: https://www.afdc.energy.gov/data/10301.)

their supply has not been stable. SRM is an excellent candidate for HEV applications. SRM has several advantages: (i) no rare-earth materials, (ii) better operation at high temperatures, (iii) high rotational speeds, and (iv) robustness [9–12]. In this section, the design of a 60 kW SRM will be presented that replaces the traction motor used in the 2010 Toyota Prius. The SRM is competitive in terms of output torque, power density, and efficiency.

16.2.1 Hybrid Electric Vehicle Application

Toyota Hybrid System (THS) II is the hybrid electric powertrain used in Toyota HEVs. THS has the following energy-saving characteristics: (a) when the engine is idling, it stops automatically, (b) kinetic energy during deceleration and braking can be converted into electrical energy by the electric machines, (c) engine operates at high-efficiency zone more frequently, and (d) fuel economy is improved by employing electric motors. The electric motors can supplement the engine or even run the vehicle in conditions when engine efficiency is low. They can generate electricity when the engine efficiency is high.

Figure 16.11 shows the diagrams of HEV powertrain systems employed by Toyota Hybrid System. The 2004 Toyota Prius uses one power-split device, also known as the planetary-gear set, as shown in Figure 16.11a. The power-split device consists of a sun gear, planet gears, and a carrier. Planet gears are mounted on the carrier. The gasoline engine, connected to the carrier of the power-split device, is the primary source of traction power, and serves to both power the vehicle as well as charge the battery. MG1, connected to the sun gear of the power split device, is a synchronous AC permanent magnet motor, which functions as the generator to charge the battery. It acts as the engine starter as well. MG2, also an IPMSM, connected to the ring gear of the power split device, functions as the traction motor and powers the vehicle directly [7,11]. The 2010 Toyota Prius employs two sets of reduction planetary gears, as shown in Figure 16.11b.

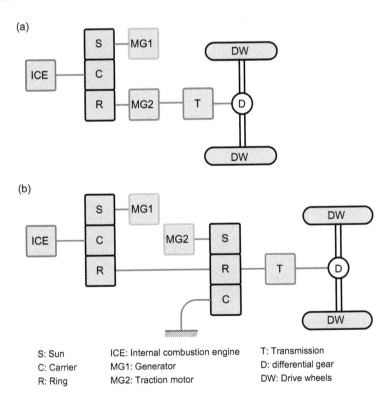

S: Sun	ICE: Internal combustion engine	T: Transmission
C: Carrier	MG1: Generator	D: differential gear
R: Ring	MG2: Traction motor	DW: Drive wheels

FIGURE 16.11
Diagrams of HEV powertrain systems: (a) 2004 Toyota Prius transmission system, and (b) 2010 Toyota Prius and Camry hybrid transmission system. (From Bilgin, B. et al., *IEEE Trans. Transp. Electrif.*, 1, 4–17, 2015.)

16.2.2 Commercial Permanent-Magnet Traction Motor

Figure 16.12a shows the torque-speed profile for the IPMSM traction motor used in the 2010 Prius. The maximum speed of the motor is 13,500 rpm. The peak torque rating is 207 Nm. The motor's peak power rating is 60 kW. Figure 16.12b shows that the highest efficiency of the motor is in the speed range from 5000 to 7000 rpm. The characteristics of the traction motor is summarized in Table 16.4. The one-eighth geometry of the Prius traction motor can be seen in Figure 16.13. It can be seen that the IPMSM has 48 slots and 8 poles. V-shape magnets are used in the design.

Three key geometric parameters for the 24/16 SRM to satisfy are, stator outer diameter (264 mm), motor axial length (108 mm), and slot fill factor (0.54). The 24/16 SRM is proposed to achieve the same torque-speed envelope of the IPMSM used as the traction motor in the Prius powertrain.

16.2.3 Switched Reluctance Motor Design for HEV Traction Application

The comparison between the design constraints of the SRM and the IPMSM is shown in Table 16.5. The number of the stator poles is 24 and number of the rotor poles is 16. High numbers of the stator and rotor poles are selected to reduce the torque ripple, and noise and vibration. Series winding connection is used for the coils of each phase. Genetic algorithm is used in the optimization of the motor design. Two objectives of the optimization are maximizing the torque and minimizing the torque ripple. Geometric parameters, such as stator and rotor pole heights, back iron thicknesses, and pole arc angles are optimized based on the optimization results of various motor designs. The lamination used for both the stator and rotor cores is Cogent NO10, which has a thickness of 0.1 mm. SURALAC 9000, a type of bonding coating, is used on the surface of the lamination. It bonds the lamination sheet together.

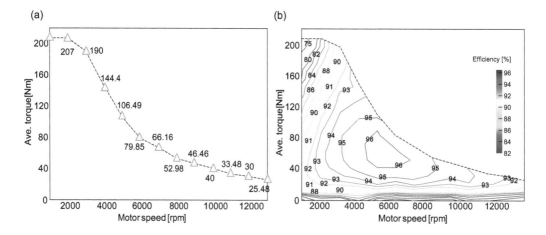

FIGURE 16.12

Motor performance of Prius 2010 traction motor: (a) torque-speed profile and (b) efficiency map. (From Jiang, J.W. et al., Design optimization of switched reluctance machine using genetic algorithm, in *Proceedings of the IEEE International Electric Machines and Drives Conference (IEMDC)*, Coeur d'Alene, ID, May 2015, pp. 1671–1677; Burress, T. et al., *Evaluation of the 2010 Toyota Prius Hybrid Synergy Drive System*, 2011. [Online]. (Accessed: October 24, 2017.))

TABLE 16.4

Characteristics of 2010 Prius Traction Motor

Parameter	Value
Stator outer diameter [mm]	264
Motor axial length [mm]	108
Stator interior diameter [mm]	161.9
Rotor outer diameter [mm]	160.4
Rotor lamination interior diameter [mm]	51
Air gap length [mm]	0.73
Lamination thickness [mm]	0.305
Motor casing diameter [mm]	275
Motor casing axial length [mm]	161
Number of stator slots	48
Slot fill factor	0.54
Turns per coil	11
Coils in series per phase	8
Number of wires in parallel	12
Wire size [AWG]	20
Motor peak power [kW]	60
Motor peak torque [Nm]	207
Maximum rotational speed [rpm]	13,500

Source: Jiang, J.W. et al., Design optimization of switched reluctance machine using genetic algorithm, in *Proceedings of the IEEE International Electric Machines and Drives Conference (IEMDC)*, Coeur d'Alene, ID, May 2015, pp. 1671–1677; Burress, T. et al., *Evaluation of the 2010 Toyota Prius Hybrid Synergy Drive System*, 2011 [Online]. (Accessed: October 24, 2017); Kiyota, K. and Chiba, A., *IEEE Trans. Ind. Appl.*, 48, 2303–2309, 2012.

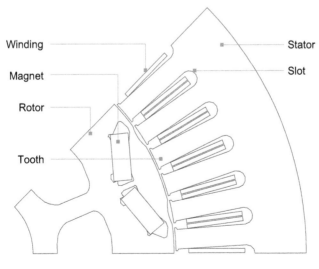

FIGURE 16.13

One-eighth geometry of 2010 Prius traction motor. (From Burress, T. et al., *Evaluation of the 2010 Toyota Prius Hybrid Synergy Drive System*, 2011. [Online]. (Accessed: October 24, 2017.))

TABLE 16.5

Design Constraints for 24/16 SRM used for HEV

Parameter	2010 Prius	24/16 SRM
Stator outer diameter [mm]	264	264
Stator core mass [kg]	10.36	13.10
Total mass of magnets [kg]	0.768	N/A
Motor axial length [mm]	108	108
Air gap length [mm]	0.73	0.5
Lamination thickness [mm]	0.305	0.1
Slot fill factor	0.54	0.54
Wire size [AWG]	20	20
DC side voltage [V]	650	650
Maximum peak current [A]	240	240
RMS current constraint [A]	140	140
Motor peak power rating [kW]	60	60

Source: Jiang, J.W. et al., "Three-phase 24/16 switched reluctance machine for a hybrid electric powertrain," *IEEE Trans. Transp. Electrif.*, 3, 76–85, 2017.

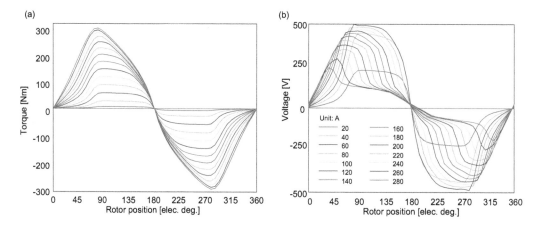

FIGURE 16.14

Simulated static characteristic waveforms of the 24/16 HEV SRM at different excitation currents: (a) electromagnetic torque and (b) phase voltage at 3000 rpm.

Figure 16.14 shows the static characteristic waveforms of the 24/16 HEV SRM at 3000 rpm and varying phase current levels from 0 to 240 A. The static characteristic waveforms of the 24/16 SRM are used to define the torque-speed profile of the motor. Genetic Algorithm (GA)-based multi-objective optimization method is used to maximize the output torque and minimize the torque ripple for each selected operating point. The copper loss and iron loss are calculated using the JMAG® with optimized current waveforms as inputs. Figure 16.15 shows the motor performance for the 24/16 SRM.

Figure 16.15a shows the torque-speed profile with efficiency contours. It can be seen that the highest efficiency of the motor reaches over 97% in the speed range of 6000 rpm to the maximum speed. Please note that mechanical loss is not included in the calculation of the motor efficiency. Figure 16.15b shows the contours of total loss for the motor.

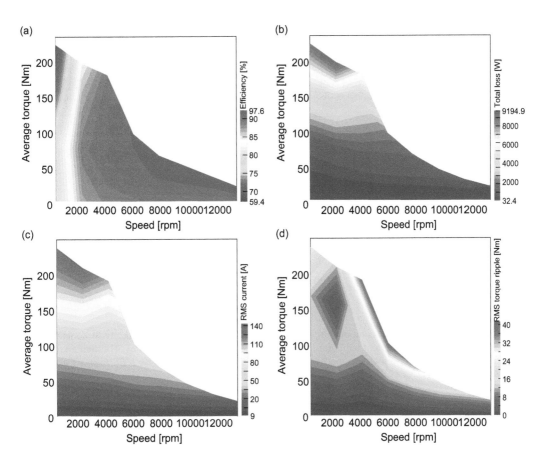

FIGURE 16.15
Simulated torque-speed characteristics of 6/14 HVAC SRM: (a) efficiency contour, (b) total loss contour, (c) RMS current contour, and (d) RMS torque ripple contour.

Similarly, the loss in the high-torque region is much higher than the rest because of the high copper loss in that region. Figure 16.15c shows the contour of RMS phase current. The value of RMS phase current is used as a constraint for the control. This constraint basically defines the envelop for the torque-speed curve of an SRM. Figure 16.15d shows the overall torque quality for the 24/16 SRM. The torque quality is measured using RMS torque ripple.

MotorCAD® is used to perform the thermal analysis for the 24/16 SRM. The loss data are extracted from JMAG® simulations and used in MotorCAD®. Water jacket can also be defined in the software as well. The temperature constraint used for the winding is 150°C. 100°C is the temperature constraint for the lamination. At 3000 rpm, which is the base speed for the motor, the winding first reaches its temperature constraint of 150°C. As shown in Figure 16.16, the maximum continuous operating time for the motor at 3000 rpm and peak torque is 132 seconds [17].

16.2.4 Mechanical Design of the HEV Traction SRM

The exploded view of the mechanical design for the 24/16 SRM is shown in Figure 16.17. It has four major subassemblies: housing subassembly, stator-winding subassembly,

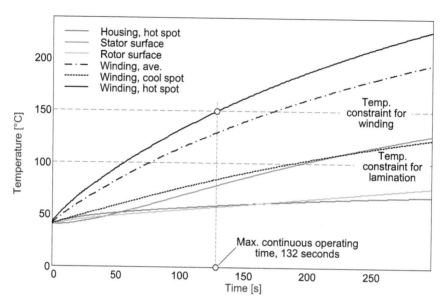

FIGURE 16.16
Temperature variation for major components of the SRM, at 3,000 rpm, peak torque, with coolant inlet temperature set at 65°C, and ambient temperature set at 40°C. (From Kasprzak, M. et al., Thermal analysis of a three-phase 24/16 switched reluctance machine used in HEVs, in *Proceedings of the IEEE Energy Conversion Congress and Exposition (ECCE)*, Milwaukee, WI, 2016, pp. 1–7.)

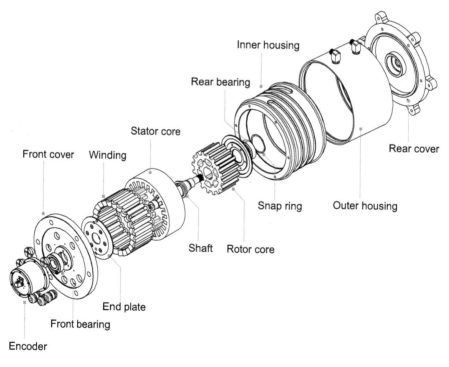

FIGURE 16.17
Exploded CAD view of 6/14 furnace blower SRM components.

rotor-shaft subassembly, and extended-shaft subassembly. The housing subassembly is comprised of inner housing, outer housing, and two end covers. The water jacket is formed in between the inner and the outer housings. The stator-winding subassembly consists of the stator core and the winding. As mentioned earlier, 0.1 mm thick lamination steel is used for both the stator and the rotor cores to reduce the eddy current loss and improve the efficiency. The rotor core, end plates and the main shaft are assembled using the keys and keyways and bolts. The extended-shaft subassembly owns the majority of the data-reading components: the encoder, the resolver, and the slip ring. The slip ring is used to channel out the wires for the thermistors, which are embedded in the rotor core and are used to obtain the thermal measurements from the rotor.

Figure 16.18 shows the prototyped 24/16 HEV SRM. Figure 16.18a and c show the rotor core and the shaft. The wires coming out from the shaft, as shown in Figure 16.18a, are connected to the thermistors, which are used to measure the rotor temperature at different

(a)

(b)

(c)

(d)

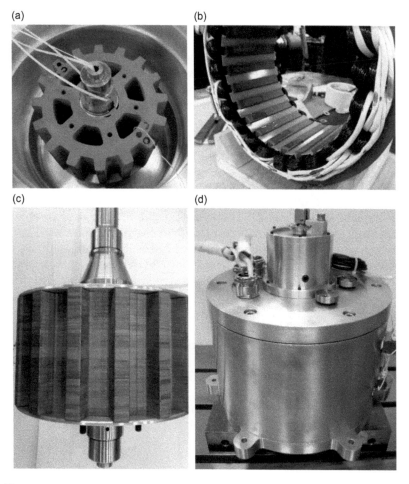

FIGURE 16.18
Prototype 24/16 SRM used as traction motor for HEV application: (a) rotor core, (b) stator-winding subassembly, (c) rotor-shaft subassembly, and (d) SRM assembly.

locations. The cut-outs are added to the rotor core to reduce the mass of the assembly. Figure 16.18b shows the prototyped stator-winding subassembly. As mentioned earlier, the winding has three phases and for each phase, the coils are connected in series. Figure 16.18d shows the assembly of the whole 24/16 SRM.

16.3 A High-Power Converter Design for Switched Reluctance Motor Drives

This section presents the fundamental aspects of a high-power inverter design for a four-phase SRM. The section also includes a discussion on how the main system components, such as power semiconductor, DC-link capacitors, heat sink, bus bar, and the inverter enclosure, can work together to improve the overall inverter performance.

16.3.1 Power Semiconductor Devices

High-speed and high-power switched reluctance motors usually have low inductance values. To illustrate this effect Figure 16.19 shows the flux linkage plot for an SRM rated at 22,000 rpm and 150 kW. The aligned and unaligned inductances are 542 and 85 μH respectively. At high-speed, the inverter is operating in single-pulse mode, which should not be a problem from the switching losses point of view. However, at low speeds, the phase current is controlled to its reference value by either hysteresis or PWM control. In this case, the switching frequency is defined by the hysteresis band or the PWM carrier frequency.

In that sense, the power semiconductor has to be selected according to the voltage and current ratings and, more importantly, to the maximum power dissipation capability. Metal–Oxide–Semiconductor Field-Effect Transistor (MOSFET) devices are able to operate at higher switching frequencies when compared to Insulated-Gate Bipolar Transistor (IGBT) technology, which presents higher switching losses. The latter is suitable when low-power is involved, while the former is intended for a high-power application. Generally speaking, torque ripple and current traceability would benefit from a high switching frequency at low-speed conditions. However, if hysteresis control is implemented, the sampling frequency would also have to be high to keep the phase current within the lower and upper hysteresis band limits. In that case, the microcontroller will have less computation time available, which limits the number of operations that can be performed by the interrupt routine.

For the design example presented here, the IGBT module illustrated in Figure 16.20 is chosen. It consists of an EconoDUAL™ package, made by Infineon Technologies, with an asymmetric bridge, instead of the conventional half-bridge. This module is rated at 1200 V and 400 A and it was specially designed for SRMs applications.

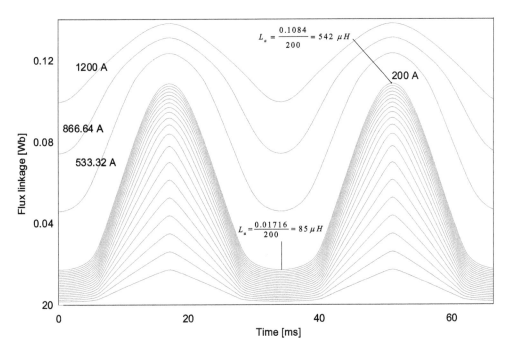

FIGURE 16.19
Example of a flux linkage profile for a high-power SRM.

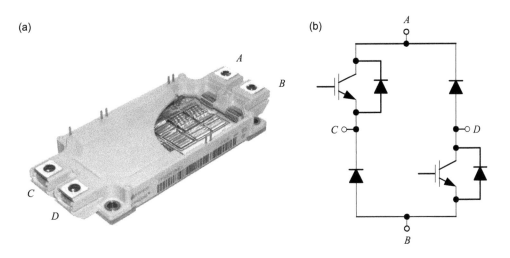

FIGURE 16.20
Infineon EconoDUAL™ asymmetric bridge IGBT module for high-power SRM applications: (a) module and (b) circuit.

16.3.2 Heat Sink for Thermal Management

Once the power semiconductor is selected, the next step is the heat sink design. This is an important component of the inverter that can dictate its overall performance, power density, and volume. A good knowledge of conduction and switching losses is necessary to build an accurate model to estimate the losses at the nominal operating condition. From the device's datasheet, the V/I characteristics for conduction losses and the turn-on and turn-off curves can be obtained and incorporated into a simulation model in software such as MATLAB®/Simulink or PSIM™.

With the maximum losses dissipated by each switch and diode, the heat sink design takes place. There are several options for a heat sink such as natural convection or forced air cooling, or water cooling. A much higher power density can be achieved with water cooling topologies. The simulation analysis of a water cooling plate is done by Computational Fluid Dynamics (CFD). As shown in Figure 16.21, the location of the heat sources inside the IGBT module must be identified for an optimal cooling design. Location and size of the heat sources are important since the cooling lines can be placed exactly under those hot spots. The simulation results for the maximum junction temperature of both diode and IGBT are shown in Figure 16.22.

16.3.3 DC-Link Filter Capacitor

Another crucial component of a high-power inverter is the DC-link capacitor. Ideally, a power semiconductor bridge requires an ideal voltage source. The DC-link capacitor is placed as close as possible to the asymmetric bridge input terminal in order to decouple the effects of the inductance from the connector, cabling and the DC voltage source. In the presence of stray inductance, the current ripple generated by the inverter, which is the result of the inductive characteristic of the load, can cause voltage spikes that can damage the

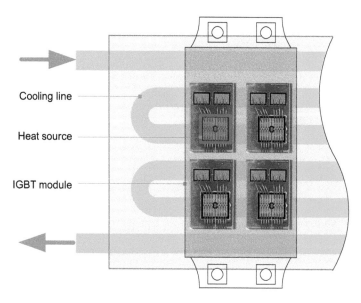

FIGURE 16.21
Example of a water cooling heat sink design for high-power Inverters.

Heatsink max temperature: 69.7 °C
Diode junction temperature: 95.5 °C
IGBT junction temperature: 93.1 °C

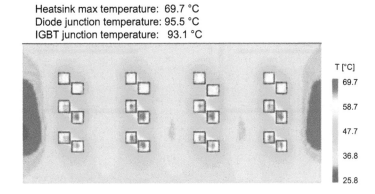

FIGURE 16.22
Maximum junction temperature for diode and IGBT at nominal power operating condition (R-TOOLS™ Heat sink Simulator from MERSEN).

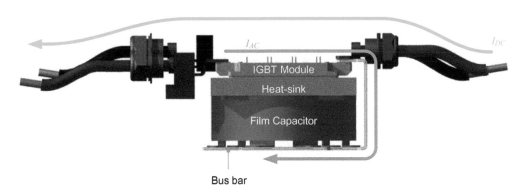

FIGURE 16.23
DC-link capacitor current flow.

power semiconductor device. Therefore, as illustrated in Figure 16.23, the filter capacitors provide a low impedance path for the high-frequency currents. This also eliminates most of the current ripple at the DC voltage source which can be a problem for batteries. Thus, the major aspect of capacitor sizing becomes the RMS current value. As a rule of thumb, for SRM drives, the DC-link capacitor RMS current can reach approximately 90% of the output current [18].

16.3.4 Bus Bar for Component Interconnections

Along with a water cooling solution, a bus bar plays a role in the power density of the inverter. It should be optimized to find the most flexible solution for the connection between the input voltage source, IGBT modules, and DC-link capacitors according to the location and the geometry of each individual component. Moreover, the bus bar also improves the voltage source characteristic by minimizing the stray inductance between the module terminals and DC-link capacitors. Some examples of bus bar designs are illustrated in Figure 16.24. The busbar type B in Figure 16.24b is applied to this design example.

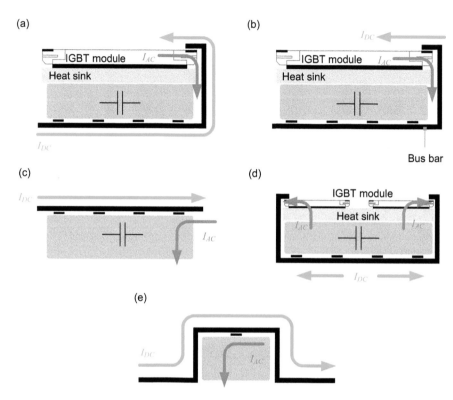

FIGURE 16.24
DC-link capacitor current flow: (a) type A, (b) type B, (c) type C, (d) type D, and (e) type E. (From Callegaro, A.D. et al., *IEEE Trans. Power Electron.*, 33(3), 2354–2367, 2018.)

Important aspects of a bus bar are the thickness of the conductor and the shape of the corners. These two points are shown in Figure 16.25. As previously mentioned, the major part of the AC current flows through the bus bar conductor; therefore, skin effect must be taken into consideration to avoid overheating and possible damage to the insulation material. Moreover, sharp corners and bends can result in eddy currents and voltage drops and, as a result, losses and heat generation can be produced. A comprehensive bus bar design methodology can be found in [19].

The final design of the bus bar considering the maximum ratings of the IGBT is shown in Figure 16.26a while a 3D view with the main components of the inverter is illustrated in Figure 16.26b. The two DC input connectors were symmetrically arranged for a better current distribution. The IGBT module corresponds to one phase of the SRM. This inverter was designed with a four-phase machine in mind. The heat sink is shared with the DC-link capacitors to remove potential heat generation.

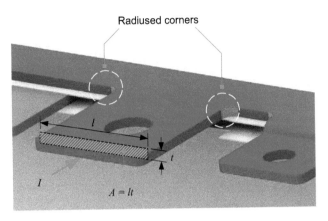

FIGURE 16.25
Bus bar thickness and corners considerations for eddy current minimization.

(a)   (b)

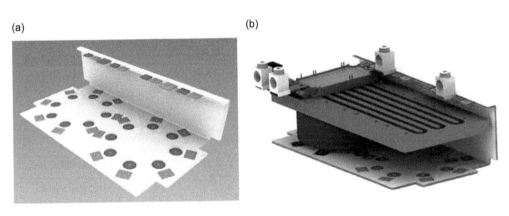

FIGURE 16.26
3D models of components: (a) bus bar final design and (b) high power inverter design.

16.3.5 Inverter Enclosure

The inverter enclosure design can vary depending on the application requirements, such as pollution degree, vibration levels, water resistance capability. For the application presented here, a water-tight solution was considered since water cooling plate does not require air exchange with the external ambient. The exploded view in Figure 16.27 illustrates how the inverter components are put together. The heat sink also has structural function since it is attached to the enclosure frame through four bolts which provide a solid structure where all the remaining components are attached. The final prototype is shown in Figure 16.28.

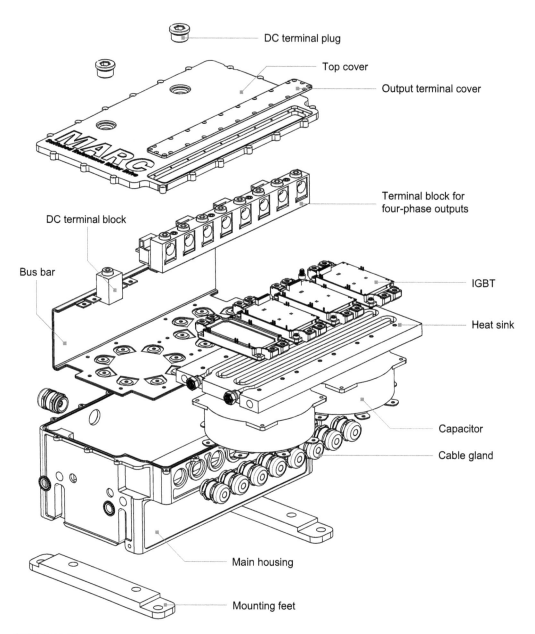

FIGURE 16.27
Exploded view of the high-power inverter design for SRM applications.

(a) (b)

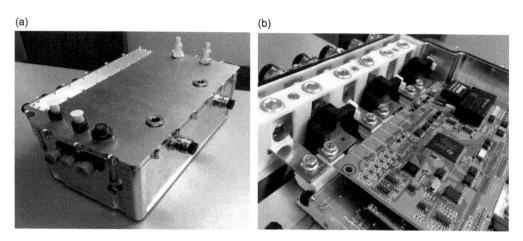

FIGURE 16.28
Final prototype of a four-phase SRM inverter: (a) housing and (b) high-power inverter design.

References

1. Navigant Consulting, Inc., "Energy savings potential and opportunities for high-efficiency electric motors in residential and commercial equipment," U.S. Department of Energy, Washington, DC, 2013.
2. M. Kasprzak, "6/14 switched reluctance machine design for household HVAC system applications," MASc thesis, Department of Mechanical Engineering, McMaster University, Hamilton, Canada, 2016.
3. Direct Energy, "Air conditioner and furnaces: How do they work together?" Direct Energy, 2016. Available: https://www.directenergy.com/learning-center/home-improvement/how-air-conditioner-furnace-work-together (Accessed October 10, 2017).
4. Genteq, "Evergreen EM spec sheet," November 5, 2014. Available: https://www.genteqmotors.com/Products/Aftermarket/Evergreen_EM/ (Accessed March 23, 2016).
5. Genteq, "Evergreen multi product brochure," September 28, 2015. Available: https://www.genteqmotors.com/Products/Aftermarket/Evergreen_EM/ (Accessed March 23, 2016).
6. Genteq, "Evergreen EM brochure," February 19, 2013. Available: http://www.swimming-pool-pump.com/ (Accessed March 23, 2016).
7. A. Emadi, *Advanced Electric Drive Vehicles*, Boca Raton, FL: CRC Press, 2014.
8. A. Emadi, "Transportation 2.0," *IEEE Power & Energy Magazine*, vol. 9, no. 4, pp. 18–29, 2011.
9. B. Bilgin and A. Emadi, "Electric motors in electrified transportation: A step toward achieving a sustainable and highly efficient transportation system," *IEEE Power Electronics Magazine*, vol. 1, no. 2, pp. 10–17, 2014.
10. A. Emadi, *Energy-Efficient Electric Motors: Selection and Applications*, New York: Marcel Dekker, 2004.
11. afdc.energy.gov, "Alternative fuels data center: Maps and data," 2015. Available: https://www.afdc.energy.gov/data/10301 (Accessed February 1, 2018).
12. B. Bilgin, P. Magne, P. Malysz, Y. Yang, V. Pantelic, M. Preindl, A. Korobkine, W. Jiang, M. Lawford, and A. Emadi, "Making the case for electrified transportation," *IEEE Transactions on Transportation Electrification*, vol. 1, no. 1, pp. 4–17, 2015.
13. J. W. Jiang, B. Bilgin, B. Howey and A. Emadi, "Design optimization of switched reluctance machine using genetic algorithm," in *Proceedings of the IEEE International Electric Machines and Drives Conference (IEMDC), Coeur d'Alene, ID, May 2015, pp. 1671–1677*.

14. T. Burress, S. Campbell, C. Coomer, C. Ayers, A. Wereszczak, and J. Cunningham, Oak Ridge National Laboratory, *Evaluation of the 2010 Toyota Prius Hybrid Synergy Drive System*, 2011. [Online]. (Accessed: October 24, 2017).

15. K. Kiyota and A. Chiba, "Design of switched reluctance motor competitive to 60-kW IPMSM in third-generation hybrid electric vehicle," *IEEE Transactions on Industry Applications*, vol. 48, no. 6, pp. 2303–2309, 2012.

16. J. W. Jiang, B. Bilgin, and A. Emadi, "Three-phase 24/16 switched reluctance machine for a hybrid electric powertrain," *IEEE Transactions on Transportation Electrification*, vol. 3, no. 1, pp. 76–85, 2017.

17. M. Kasprzak, J. W. Jiang, B. Bilgin, and A. Emadi, Thermal analysis of a three-phase 24/16 switched reluctance machine used in HEVs, in *Proceedings of the IEEE Energy Conversion Congress and Exposition (ECCE)*, Milwaukee, WI, 2016, pp. 1–7.

18. H. Wu, D. Winterborne, M. Ma, V. Pickert and J. Widmer, "DC link capacitors for traction SRM drives in high-temperature automotive environments: A review of current issues and solutions," *IET Hybrid and Electric Vehicles Conference 2013 (HEVC 2013)*, London, UK, 2013, pp. 1–6.

19. A. D. Callegaro, J. Guo, M. Eull, B. Danen, J. Gibson, M. Preindl, B. Bilgin, and A. Emadi, "Bus bar design for high-power inverters," *IEEE Transactions on Power Electronics*, vol. 33, no. 3, pp. 2354–2367, 2018.

Index

Note: Page numbers in italic and bold refer to figures and tables respectively.